Number Theory Revealed: A Masterclass

Number Theory Revealed: A Masterclass

Andrew Granville

American Mathematical Society
Providence, Rhode Island

Cover design by Marci Babineau.

Front cover image of Srinivasa Ramanujan in the playing card: Oberwolfach Photo Collection, https://opc.mfo.de/; licensed under Creative Commons Attribution Share Alike 2.0 Germany, https://creativecommons.org/licenses/by-sa/2.0/de/deed.en.

Front cover image of Andrew Wiles in playing card, credit: Alain Goriely.

2010 *Mathematics Subject Classification.* Primary 11-01, 11A55, 11B30, 11B39, 11D09, 11D25, 11N05, 11N25.

For additional information and updates on this book, visit
www.ams.org/bookpages/mbk-127

Library of Congress Cataloging-in-Publication Data
Names: Granville, Andrew, author.
Title: Number theory revealed : a masterclass / Andrew Granville.
Description: Providence, Rhode Island : American Mathematical Society, [2019] | Includes bibliographical references and index.
Identifiers: LCCN 2019029046 | ISBN 9781470441586 (hardcover) | ISBN 9781470454241 (ebook)
Subjects: LCSH: Number theory. | AMS: Number theory – Instructional exposition (textbooks, tutorial papers, etc.). | Number theory – Elementary number theory {For analogues in number fields, see 11R04} – Continued fractions {For approximation results, see 11J70} [See also 11K50, 30B70, 40A15]. | Number theory – Sequences and sets – Arithmetic combinatorics; higher degree uniformity. | Number theory – Sequences and sets – Fibonacci and Lucas numbers and polynomials and generalizations. | Number theory – Diophantine equations [See also 11Gxx, 14Gxx] – Linear equations. | Number theory – Diophantine equations [See also 11Gxx, 14Gxx] – Quadratic and bilinear equations. | Number theory – Diophantine equations [See also 11Gxx, 14Gxx] – Cubic and quartic equations. | Number theory – Multiplicative number theory – Distribution of primes. | Number theory – Multiplicative number theory – Distribution of integers with specified multiplicative constraints.
Classification: LCC QA241 .G647 2019 | DDC 512.7–dc23
LC record available at https://lccn.loc.gov/2019029046

Softcover ISBN: 978-1-4704-6370-0

Copying and reprinting. Individual readers of this publication, and nonprofit libraries acting for them, are permitted to make fair use of the material, such as to copy select pages for use in teaching or research. Permission is granted to quote brief passages from this publication in reviews, provided the customary acknowledgment of the source is given.

Republication, systematic copying, or multiple reproduction of any material in this publication is permitted only under license from the American Mathematical Society. Requests for permission to reuse portions of AMS publication content are handled by the Copyright Clearance Center. For more information, please visit www.ams.org/publications/pubpermissions.

Send requests for translation rights and licensed reprints to reprint-permission@ams.org.

© 2019 by the American Mathematical Society. All rights reserved.
The American Mathematical Society retains all rights
except those granted to the United States Government.
Printed in the United States of America.

∞ The paper used in this book is acid-free and falls within the guidelines
established to ensure permanence and durability.
Visit the AMS home page at https://www.ams.org/

10 9 8 7 6 5 4 3 2   24 23 22 21 20 19

Dedicated to my beloved wife, Marci. Writing this book has had its challenges. Being the spouse of the author, while writing this book, has also had its challenges.

The enchanting charms of this sublime science
reveal themselves only to those who have the
courage to go deeply into it.

CARL FRIEDRICH GAUSS, 1807

Contents

Preface	xvii
Gauss's *Disquisitiones Arithmeticae*	xxiii
Notation	xxv
The language of mathematics	xxvi
Prerequisites	xxvii
Preliminary Chapter on Induction	**1**
0.1. Fibonacci numbers and other recurrence sequences	1
0.2. Formulas for sums of powers of integers	3
0.3. The binomial theorem, Pascal's triangle, and the binomial coefficients	4
Appendices for Preliminary Chapter on Induction	
0A. A closed formula for sums of powers	9
0B. Generating functions	11
0C. Finding roots of polynomials	15
0D. What is a group?	19
0E. Rings and fields	22
0F. Symmetric polynomials	25
0G. Constructibility	30
Chapter 1. The Euclidean algorithm	**33**
1.1. Finding the gcd	33
1.2. Linear combinations	35
1.3. The set of linear combinations of two integers	37
1.4. The least common multiple	39

1.5.	Continued fractions	39
1.6.	Tiling a rectangle with squares	41

Appendices for Chapter 1:

1A.	Reformulating the Euclidean algorithm	45
1B.	Computational aspects of the Euclidean algorithm	51
1C.	Magic squares	54
1D.	The Frobenius postage stamp problem	57
1E.	Egyptian fractions	59

Chapter 2. Congruences   61

2.1.	Basic congruences	61
2.2.	The trouble with division	64
2.3.	Congruences for polynomials	66
2.4.	Tests for divisibility	66

Appendices for Chapter 2:

2A.	Congruences in the language of groups	71
2B.	The Euclidean algorithm for polynomials	75

Chapter 3. The basic algebra of number theory   81

3.1.	The Fundamental Theorem of Arithmetic	81
3.2.	Abstractions	83
3.3.	Divisors using factorizations	85
3.4.	Irrationality	87
3.5.	Dividing in congruences	88
3.6.	Linear equations in two unknowns	90
3.7.	Congruences to several moduli	92
3.8.	Square roots of 1 $(\bmod n)$	94

Appendices for Chapter 3:

3A.	Factoring binomial coefficients and Pascal's triangle modulo p	99
3B.	Solving linear congruences	104
3C.	Groups and rings	109
3D.	Unique factorization revisited	112
3E.	Gauss's approach	116
3F.	Fundamental theorems and factoring polynomials	117
3G.	Open problems	123

Chapter 4. Multiplicative functions   127

4.1.	Euler's ϕ-function	128
4.2.	Perfect numbers. "The whole is equal to the sum of its parts."	129

Contents   xi

Appendices for Chapter 4:
    4A. More multiplicative functions   134
    4B. Dirichlet series and multiplicative functions   140
    4C. Irreducible polynomials modulo p 144
    4D. The harmonic sum and the divisor function   147
    4E. Cyclotomic polynomials   153

Chapter 5. The distribution of prime numbers   155
    5.1. Proofs that there are infinitely many primes   155
    5.2. Distinguishing primes   157
    5.3. Primes in certain arithmetic progressions   159
    5.4. How many primes are there up to x?   160
    5.5. Bounds on the number of primes   163
    5.6. Gaps between primes   165
    5.7. Formulas for primes   167

Appendices for Chapter 5:
    5A. Bertrand's postulate and beyond   171
    Bonus read: A review of prime problems   175
        Prime values of polynomials in one variable   175
        Prime values of polynomials in several variables   177
        Goldbach's conjecture and variants   179
    5B. An important proof of infinitely many primes   182
    5C. What should be true about primes?   187
    5D. Working with Riemann's zeta-function   192
    5E. Prime patterns: Consequences of the Green-Tao Theorem   198
    5F. A panoply of prime proofs   202
    5G. Searching for primes and prime formulas   204
    5H. Dynamical systems and infinitely many primes   208

Chapter 6. Diophantine problems   215
    6.1. The Pythagorean equation   215
    6.2. No solutions to a Diophantine equation through descent   218
    6.3. Fermat's "infinite descent"   220
    6.4. Fermat's Last Theorem   221

Appendices for Chapter 6:
    6A. Polynomial solutions of Diophantine equations   225
    6B. No Pythagorean triangle of square area via Euclidean
          geometry   229
    6C. Can a binomial coefficient be a square?   233

Chapter 7. Power residues ... 235
   7.1. Generating the multiplicative group of residues ... 236
   7.2. Fermat's Little Theorem ... 237
   7.3. Special primes and orders ... 240
   7.4. Further observations ... 240
   7.5. The number of elements of a given order, and primitive roots ... 241
   7.6. Testing for composites, pseudoprimes, and Carmichael numbers ... 245
   7.7. Divisibility tests, again ... 246
   7.8. The decimal expansion of fractions ... 246
   7.9. Primes in arithmetic progressions, revisited ... 248

Appendices for Chapter 7:
   7A. Card shuffling and Fermat's Little Theorem ... 252
   7B. Orders and primitive roots ... 258
   7C. Finding nth roots modulo prime powers ... 265
   7D. Orders for finite groups ... 269
   7E. Constructing finite fields ... 273
   7F. Sophie Germain and Fermat's Last Theorem ... 278
   7G. Primes of the form $2^n + k$ ... 280
   7H. Further congruences ... 284
   7I. Primitive prime factors of recurrence sequences ... 290

Chapter 8. Quadratic residues ... 295
   8.1. Squares modulo prime p ... 295
   8.2. The quadratic character of a residue ... 297
   8.3. The residue -1 ... 300
   8.4. The residue 2 ... 301
   8.5. The law of quadratic reciprocity ... 303
   8.6. Proof of the law of quadratic reciprocity ... 305
   8.7. The Jacobi symbol ... 307
   8.8. The squares modulo m ... 309

Appendices for Chapter 8:
   8A. Eisenstein's proof of quadratic reciprocity ... 315
   8B. Small quadratic non-residues ... 319
   8C. The first proof of quadratic reciprocity ... 323
   8D. Dirichlet characters and primes in arithmetic progressions ... 326
   8E. Quadratic reciprocity and recurrence sequences ... 333

Chapter 9. Quadratic equations		337
9.1.	Sums of two squares	337
9.2.	The values of $x^2 + dy^2$	340
9.3.	Is there a solution to a given quadratic equation?	341
9.4.	Representation of integers by $ax^2 + by^2$ with x, y rational, and beyond	344
9.5.	The failure of the local-global principle for quadratic equations in integers	345
9.6.	Primes represented by $x^2 + 5y^2$	345

Appendices for Chapter 9:

9A.	Proof of the local-global principle for quadratic equations	348
9B.	Reformulation of the local-global principle	353
9C.	The number of representations	356
9D.	Descent and the quadratics	360

Chapter 10. Square roots and factoring		365
10.1.	Square roots modulo n	365
10.2.	Cryptosystems	366
10.3.	RSA	368
10.4.	Certificates and the complexity classes P and NP	370
10.5.	Polynomial time primality testing	372
10.6.	Factoring methods	373

Appendices for Chapter 10:

10A.	Pseudoprime tests using square roots of 1	376
10B.	Factoring with squares	380
10C.	Identifying primes of a given size	383
10D.	Carmichael numbers	387
10E.	Cryptosystems based on discrete logarithms	391
10F.	Running times of algorithms	393
10G.	The AKS test	395
10H.	Factoring algorithms for polynomials	399

Chapter 11. Rational approximations to real numbers		403
11.1.	The pigeonhole principle	403
11.2.	Pell's equation	406
11.3.	Descent on solutions of $x^2 - dy^2 = n$, $d > 0$	410
11.4.	Transcendental numbers	411
11.5.	The *abc*-conjecture	414

Appendices for Chapter 11:

11A.	Uniform distribution	418
11B.	Continued fractions	423
11C.	Two-variable quadratic equations	438
11D.	Transcendental numbers	439

Chapter 12. Binary quadratic forms — 443

12.1.	Representation of integers by binary quadratic forms	444
12.2.	Equivalence classes of binary quadratic forms	446
12.3.	Congruence restrictions on the values of a binary quadratic form	447
12.4.	Class numbers	448
12.5.	Class number one	449

Appendices for Chapter 12:

12A.	Composition rules: Gauss, Dirichlet, and Bhargava	456
12B.	The class group	465
12C.	Binary quadratic forms of positive discriminant	468
12D.	Sums of three squares	471
12E.	Sums of four squares	475
12F.	Universality	479
12G.	Integers represented in Apollonian circle packings	482

Chapter 13. The anatomy of integers — 487

13.1.	Rough estimates for the number of integers with a fixed number of prime factors	487
13.2.	The number of prime factors of a typical integer	488
13.3.	The multiplication table problem	491
13.4.	Hardy and Ramanujan's inequality	492

Appendices for Chapter 13:

13A.	Other anatomies	493
13B.	Dirichlet L-functions	497

Chapter 14. Counting integral and rational points on curves, modulo p — 501

14.1.	Diagonal quadratics	501
14.2.	Counting solutions to a quadratic equation and another proof of quadratic reciprocity	503
14.3.	Cubic equations modulo p	504
14.4.	The equation $E_b : y^2 = x^3 + b$	505
14.5.	The equation $y^2 = x^3 + ax$	507
14.6.	A more general viewpoint on counting solutions modulo p	509

Appendices for Chapter 14:
  14A. Gauss sums — 511

Chapter 15. Combinatorial number theory — 515
  15.1. Partitions — 515
  15.2. Jacobi's triple product identity — 517
  15.3. The Freiman-Ruzsa Theorem — 519
  15.4. Expansion and the Plünnecke-Ruzsa inequality — 522
  15.5. Schnirel'man's Theorem — 523
  15.6. Classical additive number theory — 525
  15.7. Challenging problems — 528

Appendices for Chapter 15:
  15A. Summing sets modulo p — 530
  15B. Summing sets of integers — 532

Chapter 16. The p-adic numbers — 535
  16.1. The p-adic norm — 535
  16.2. p-adic expansions — 536
  16.3. p-adic roots of polynomials — 537
  16.4. p-adic factors of a polynomial — 539
  16.5. Possible norms on the rationals — 541
  16.6. Power series convergence and the p-adic logarithm — 542
  16.7. The p-adic dilogarithm — 545

Chapter 17. Rational points on elliptic curves — 547
  17.1. The group of rational points on an elliptic curve — 548
  17.2. Congruent number curves — 551
  17.3. No non-trivial rational points by descent — 553
  17.4. The group of rational points of $y^2 = x^3 - x$ — 553
  17.5. Mordell's Theorem: $E_A(\mathbb{Q})$ is finitely generated — 554
  17.6. Some nice examples — 558

Appendices for Chapter 17:
  17A. General Mordell's Theorem — 561
  17B. Pythagorean triangles of area 6 — 563
  17C. 2-parts of abelian groups — 565
  17D. Waring's problem — 566

Hints for exercises — 569

Recommended further reading — 583

Index — 585

Preface

This is a modern introduction to number theory, aimed at several different audiences: students who have little experience of university level mathematics, students who are completing an undergraduate degree in mathematics, as well as students who are completing a mathematics teaching qualification. Like most introductions to number theory, our contents are largely inspired by Gauss's *Disquisitiones Arithmeticae* (1801), though we also include many modern developments. We have gone back to Gauss to borrow several excellent examples to highlight the theory.

There are many different topics that might be included in an introductory course in number theory, and others, like the law of quadratic reciprocity, that surely must appear in any such course. The first dozen chapters of the book therefore present a "standard" course. In the *masterclass* version of this book we flesh out these topics, in copious appendices, as well as adding five additional chapters on more advanced themes. In the *introductory* version we select an appendix for each chapter that might be most useful as supplementary material.[1] A "minimal" course might focus on the first eight chapters and at least one of chapters 9 and 10.[2]

Much of modern mathematics germinated from number-theoretic seed and one of our goals is to help the student appreciate the connection between the relatively simply defined concepts in number theory and their more abstract generalizations in other courses. For example, our appendices allow us to highlight how modern algebra stems from investigations into number theory and therefore serve as an introduction to algebra (including rings, modules, ideals, Galois theory, p-adic numbers,...). These appendices can be given as additional reading, perhaps as student projects, and we point the reader to further references.

Following Gauss, we often develop examples *before* giving a formal definition and a theorem, firstly to see how the concept arises naturally, secondly to conjecture a theorem that describes an evident pattern, and thirdly to see how a proof of the theorem emerges from understanding some non-trivial examples.

---

[1] In the main text we occasionally refer to appendices that only appear in the *masterclass* version.
[2] Several sections might be discarded; their headings are in ***bold italics***.

Why study number theory? Questions arise when studying any subject, sometimes fascinating questions that may be difficult to answer precisely. Number theory is the study of the most basic properties of the integers, literally taking integers apart to see how they are built, and there we find an internal beauty and coherence that encourages many of us to seek to understand more. Facts are often revealed by calculations, and then researchers seek proofs. Sometimes the proofs themselves, even more than the theorems they prove, have an elegance that is beguiling and reveal that there is so much more to understand. With good reason, Gauss called number theory the *"Queen of Mathematics"*, ever mysterious, but nonetheless graciously sharing with those that find themselves interested. In this first course there is much that is accessible, while at the same time natural, easily framed, questions arise which remain open, stumping the brightest minds.

Once celebrated as one of the more abstract subjects in mathematics, today there are scores of applications of number theory in the real world, particularly to the theory and practice of computer algorithms. Best known is the use of number theory in designing cryptographic protocols (as discussed in chapter 10), hiding our secrets behind the seeming difficulty of factoring large numbers which only have large prime factors.

For some students, studying number theory is a life-changing experience: They find themselves excited to go on to penetrate more deeply, or perhaps to pursue some of the fascinating applications of the subject.

Why give proofs? We give proofs to convince ourselves and others that our reasoning is correct. Starting from agreed upon truths, we try to derive a further truth, being explicit and precise about each step of our reasoning. A proof must be readable by people besides the author. It is a way of communicating ideas and needs to be persuasive, not just to the writer but also to a mathematically literate person who cannot obtain further clarification from the writer on any point that is unclear. It is not enough that the writer believes it; it must be clear to others. The burden of proof lies with the author.

The word "proof" can mean different things in different disciplines. In some disciplines a "proof" can be several different examples that justify a stated hypothesis, but this is inadequate in mathematics: One can have a thousand examples that work as predicted by the hypothesis, but the thousand and first might contradict it. Therefore to "prove" a theorem, one must build an incontrovertible argument up from first principles, so that the statement must be true in every case, assuming that those first principles are true.

Occasionally we give more than one proof of an important theorem, to highlight how inevitably the subject develops, as well as to give the instructor different options for how to present the material. (Few students will benefit from seeing *all* of the proofs on their first time encountering this material.)

Motivation. Challenging mathematics courses, such as point-set topology, algebraic topology, measure theory, differential geometry, and so on, tend to be dominated at first by formal language and requirements. Little is given by way of motivation. Sometimes these courses are presented as a prerequisite for topics that will come later. There is little or no attempt to explain what all this theory is good for or why it was developed in the first place. Students are expected to subject themselves to the course, motivated primarily by trust.

How boring! Mathematics surely should not be developed only for those few who already know that they wish to specialize and have a high tolerance for boredom. We should help our students to appreciate and cherish the beauty of mathematics. Surely courses should be motivated by a series of interesting questions. The right questions will highlight the benefits of an abstract framework, so that the student will wish to explore even the most rarified paths herself, as the benefits become obvious. Number theory does not require much in the way of formal prerequisites, and there are easy ways to justify most of its abstraction.

In this book, we hope to capture the attention and enthusiasm of the reader with the right questions, guiding her as she embarks for the first time on this fascinating journey.

Student expectations. For some students, number theory is their first course that formulates abstract statements of theorems, which can take them outside of their "comfort zone". This can be quite a challenge, especially as high school pedagogy moves increasingly to training students to learn and use sophisticated techniques, rather than appreciate how those techniques arose. We believe that one can best use (and adapt) methods if one fully appreciates their genesis, so we make no apologies for this feature of the elementary number theory course. However this means that some students will be forced to adjust their personal expectations. Future teachers sometimes ask why they need to learn material, and take a perspective, so far beyond what they will be expected to teach in high school. There are many answers to this question; one is that, in the long term, the material in high school will be more fulfilling if one can see its long-term purpose. A second response is that every teacher will be confronted by students who are bored with their high school course and desperately seeking harder intellectual challenges (whether they realize it themselves or not); the first few chapters of this book should provide the kind of intellectual stimulation those students need.

Exercises. Throughout the book, there are a lot of problems to be solved. Easy questions, moderate questions, hard questions, exceptionally difficult questions. No one should do them all. The idea of having so many problems is to give the teacher options that are suitable for the students' backgrounds:

An unusual feature of the book is that exercises appear embedded in the text.[3] This is done to enable the student to complete the proofs of theorems as one goes along.[4] This does not require the students to come up with new ideas but rather to follow the arguments given so as to fill in the gaps. For less experienced students it helps to write out the solutions to these exercises; more experienced students might just satisfy themselves that they can provide an appropriate proof.

---

[3]Though they can be downloaded, as a separate list, from www.ams.org/granville-number-theory.
[4]Often students have little experience with proofs and struggle with the level of sophistication required, at least without adequate guidance.

Other questions work through examples. There are more challenging exercises throughout, indicated by the symbol † next to the question numbers, in which the student will need to independently bring together several of the ideas that have been discussed. Then there are some really tough questions, indicated by the symbol ‡, in which the student will need to be creative, perhaps even providing ideas not given, or hinted at, in the text.

A few questions in this book are open-ended, some even phrased a little misleadingly. The student who tries to develop those themes her- or himself, might embark upon a rewarding voyage of discovery. Once, after I had set the exercises in section 9.2 for homework, some students complained how unfair they felt these questions were but were silenced by another student who announced that it was so much fun for him to work out the answers that he now knew what he wanted to do with his life!

At the end of the book we give hints for many of the exercises, especially those that form part of a proof.

Special features of our syllabus. Number theory sometimes serves as an introduction to "proof techniques". We give many exercises to practice those techniques, but to make it less boring, we do so while developing certain themes as the book progresses, for examples, the theory of recurrence sequences, and properties of binomial coefficients. We dedicate a preliminary chapter to induction and use it to develop the theory of sums of powers. Here is a list of the main supplementary themes which appear in the book:

Special numbers: Bernoulli numbers; binomial coefficients and Pascal's triangle; Fermat and Mersenne numbers; and the Fibonacci sequence and general second-order linear recurrences.

Subjects in their own right: Algebraic numbers, integers, and units; computation and running times: Continued fractions; dynamics; groups, especially of matrices; factoring methods and primality testing; ideals; irrationals and transcendentals; and rings and fields.

Formulas for cyclotomic polynomials, Dirichlet L-functions, the Riemann zeta-function, and sums of powers of integers.

Interesting issues: Lifting solutions; polynomial properties; resultants and discriminants; roots of polynomials, constructibility and pre-Galois theory; square roots (mod n); and tests for divisibility.

Fun and famous problems like the abc-conjecture, Catalan's conjecture, Egyptian fractions, Fermat's Last Theorem, the Frobenius postage stamp problem, magic squares, primes in arithmetic progressions, tiling with rectangles and with circles.

Our most unconventional choice is to give a version of Rousseau's proof of the law of quadratic reciprocity, which is directly motivated by Gauss's proof of Wilson's Theorem. This proof avoids Gauss's Lemma so is a lot easier for a beginning student than Eisenstein's elegant proof (which we give in section 8.10 of appendix 8A). Gauss's original proof of quadratic reciprocity is more motivated by the introductory material, although a bit more complicated than these other two proofs.

We include Gauss's original proof in section 8.14 of appendix 8C, and we also understand $(2/n)$ in his way, in the basic course, to interest the reader. We present several other proofs, including a particularly elegant proof using Gauss sums in section 14.7.

Further exploration of number theory. There is a tremendous leap in the level of mathematical knowledge required to take graduate courses in number theory, because curricula expect the student to have taken (and appreciated) several other relevant courses. This is a shame since there is so much beautiful advanced material that is easily accessible after finishing an introductory course. Moreover, it can be easier to study other courses, if one already understands their importance, rather than taking it on trust. Thus this book, *Number Theory Revealed*, is designed to lead to two subsequent books, which develop the two main thrusts of number theory research:

In *The distribution of primes: Analytic number theory revealed*, we will discuss how number theorists have sought to develop the themes of chapter 5 (as well as chapters 4 and 13). In particular we prove the prime number theorem, based on the extraordinary ideas of Riemann. This proof rests heavily on certain ideas from complex analysis, which we will outline in a way that is relevant for a good understanding of the proofs.

In *Rational points on curves: Arithmetic geometry revealed*, we look at solutions to Diophantine equations, especially those of degree two and three, extending the ideas of chapter 12 (as well as chapters 14 and 17). In particular we will prove Mordell's Theorem (developed here in special cases in chapter 17) and gain a basic understanding of modular forms, outlining some of the main steps in Wiles's proof of Fermat's Last Theorem. We avoid a deep understanding of algebraic geometry, instead proceeding by more elementary techniques and a little complex analysis (which we explain).

References. There is a list of great number theory books at the end of our book and references that are recommended for further reading at the end of many chapters and appendices. Unlike most textbooks, I have chosen to not include a reference to every result stated, nor necessarily to most relevant articles, but rather focus on a smaller number that might be accessible to the reader. Moreover, many readers are used to searching online for keywords; this works well for many themes in mathematics.[5] However the student researching online should be warned that Wikipedia articles are often out of date, sometimes misleading, and too often poorly written. It is best to try to find relevant articles published in expository research journals, such as the *American Mathematical Monthly*,[6] or posted at arxiv.org which is "open access", to supplement the course material.

The cover (designed by Marci Babineau and the author).

In 1675, Isaac Newton explained his extraordinary breakthroughs in physics and mathematics by claiming, "*If I have seen further it is by standing on the shoulders*

---

[5]Though getting just the phrasing to find the right level of article can be challenging.

[6]Although this is behind a paywall, it can be accessed, like many journals, by logging on from most universities, which have paid subscriptions for their students and faculty.

of Giants." Science has always developed this way, no more so than in the theory of numbers. Our cover represents five giants of number theory, in a fan of cards, each of whose work built upon the previous luminaries.

Modern number theory was born from PIERRE DE FERMAT's readings of the ancient Greek texts (as discussed in section 6.1) in the mid-17th century, and his enunciation of various results including his tantalizingly difficult to prove "Last Theorem." His "Little Theorem" (chapter 7) and his understanding of sums of two squares (chapter 9) are part of the basis of the subject.

The first modern number theory book, Gauss's *Disquisitiones Arithmeticae*, on which this book is based, was written by CARL FRIEDRICH GAUSS at the beginning of the 19th century. As a teenager, Gauss rethought many of the key ideas in number theory, especially the law of quadratic reciprocity (chapter 8) and the theory of binary quadratic forms (chapter 12), as well as inspiring our understanding of the distribution of primes (chapter 5).

Gauss's contemporary SOPHIE GERMAIN made perhaps the first great effort to attack Fermat's Last Theorem (her effort is discussed in appendix 7F). Developing her work inspired my own first research efforts.

SRINIVASA RAMANUJAN, born in poverty in India at the end of the 19th century, was the most talented untrained mathematician in history, producing some extraordinary results before dying at the age of 32. He was unable to satisfactorily explain many of his extraordinary insights which penetrated difficult subjects far beyond the more conventional approaches. (See appendix 12F and chapters 13, 15, and 17.) Some of his identities are still inspiring major developments today in both mathematics and physics.

ANDREW WILES sits atop our deck. His 1994 proof of Fermat's Last Theorem built on the ideas of the previous four mentioned mathematicians and very many other "giants" besides. His great achievement is a testament to the success of science building on solid grounds.

Thanks. I would like to thank the many inspiring mathematicians who have helped me shape my view of elementary number theory, most particularly Bela Bollobas, Paul Erdős, D. H. Lehmer, James Maynard, Ken Ono, Paulo Ribenboim, Carl Pomerance, John Selfridge, Dan Shanks, and Hugh C. Williams as well as those people who have participated in developing the relatively new subject of "additive combinatorics" (see sections 15.3, 15.4, 15.5, and 15.6). Several people have shared insights or new works that have made their way into this book: Stephanie Chan, Leo Goldmakher, Richard Hill, Alex Kontorovich, Jennifer Park, and Richard Pinch. The six anonymous reviewers added some missing perspectives and Olga Balkanova, Stephanie Chan, Patrick Da Silva, Tristan Freiberg, Ben Green, Mariah Hamel, Jorge Jimenez, Nikoleta Kalaydzhieva, Dimitris Koukoulopoulos, Youness Lamzouri, Jennifer Park, Sam Porritt, Ethan Smith, Anitha Srinivasan, Paul Voutier, and Max Wenqiang Xu kindly read subsections of the near-final draft, making valuable comments.

Gauss's *Disquisitiones Arithmeticae*

In July 1801, Carl Friedrich Gauss published *Disquisitiones Arithmeticae*, a book on number theory, written in Latin. It had taken five years to write but was immediately recognized as a great work, both for the new ideas and its accessible presentation. Gauss was then widely considered to be the world's leading mathematician, and today we rate him as one of the three greatest in history, alongside Archimedes and Sir Isaac Newton.

The first four chapters of *Disquisitiones Arithmeticae* consist of essentially the same topics as our course today (with suitable modifications for advances made in the last two hundred years). His presentation of ideas is largely the model upon which modern mathematical writing is based. There follow several chapters on quadratic forms and then on the rudiments of what we would call Galois theory today, most importantly the constructibility of regular polygons. Finally, the publisher felt that the book was long enough, and several further chapters did not appear in the book (though Dedekind published Gauss's disorganized notes, in German, after Gauss's death).

One cannot overestimate the importance of *Disquisitiones* to the development of 19th-century mathematics. It led, besides many other things, to Dirichlet's formulation of ideals (see sections 3.19, 3.20 of appendix 3D, 12.8 of appendix 12A, and 12.10 of appendix 12B), and the exploration of the geometry of the upper half-plane (see Theorem 1.2 and the subsequent discussion).

As a young man, Dirichlet took his copy of *Disquisitiones* with him wherever he went. He even slept with it under his pillow. As an old man, it was his most prized possession even though it was in tatters. It was translated into French in 1807, German in 1889, Russian in 1959, English only in 1965, Spanish and Japanese in 1995, and Catalan in 1996!

Disquisitiones is no longer read by many people. The notation is difficult. The assumptions about what the reader knows do not fit today's reader (for example, neither linear algebra nor group theory had been formulated by the time Gauss wrote his book, although *Disquisitiones* would provide some of the motivation for developing those subjects). Yet, many of Gauss's proofs are inspiring, and some have been lost to today's literature. Moreover, although the more advanced two-thirds of *Disquisitiones* focus on binary quadratic forms and have led to many of today's developments, there are several themes there that are not central to today's research. In the fourth book in our trilogy (!), *Gauss's* Disquisitiones Arithmeticae *revealed*, we present a reworking of Gauss's classic, rewriting it in modern notation, in a style more accessible to the modern reader. We also give the first English version of the missing chapters, which include several surprises.

Notation

$\mathbb{N}$ – The *natural numbers*, $1, 2, 3, \ldots$.

$\mathbb{Z}$ – The *integers*, $\ldots, -3, -2, -1, 0, 1, 2, 3, \ldots$.

Throughout, all variables are taken to be integers, unless otherwise specified.

Usually p, and sometimes q, will denote prime numbers.

$\mathbb{Q}$ – The *rational numbers*, that is, the fractions a/b with $a \in \mathbb{Z}$ and $b \in \mathbb{N}$.

$\mathbb{R}$ – The *real numbers*.

$\mathbb{C}$ – The *complex numbers*.

mean that we sum, or product, the summand over the integer values of some variable, satisfying certain conditions.

Brackets and parentheses: There are all sorts of brackets and parentheses in mathematics. It is helpful to have protocols with them that take on meaning, so we do not have to repeat ourselves too often, as we will see in the notation below. But we also use them in equations; usually we surround an expression with "(" and ")" to be clear where the expression begins and ends. If too many of these are used in one line, then we might use different sizes or even "{" and "}" instead. If the brackets have a particular meaning, then the reader will be expected to discern that from the context.

$A[x]$ — The set of *polynomials* with coefficients from the set A, that is, $f(x) = \sum_{i=0}^{d} a_i x^i$ where each $a_i \in A$. Mostly we work with $A = \mathbb{Z}$.

$A(x)$ —The set of *rational functions* with coefficients from the set A, in other words, functions $f(x)/g(x)$ where $f(x), g(x) \in A[x]$ and $g(x) \neq 0$.

$[t]$ — The *integer part* of t, that is, the largest integer $\leq t$.

$\{t\}$ — The *fractional part of* (real number) t, that is, $\{t\} = t - [t]$. Notice that $0 \leq \{t\} < 1$.

(a, b) — The greatest common divisor of a and b.

$[a, b]$ — The least common multiple of a and b.

$b|a$ — Means b divides a.

$p^k \| a$ — Means p^k divides a, but not p^{k+1} (where p is prime). In other words, k is the "exact power" of p dividing a.

$I(a, b)$ — The set $\{am + bn : m, n \in \mathbb{Z}\}$, which is called the *ideal* generated by a and b over $\mathbb{Z}$.

log — The logarithm in base e, the natural logarithm, which is often denoted by "ln" in earlier courses.

Parity – The *parity* of an integer is either even (if it is divisible by 2) or odd (if it is not divisible by 2).

The language of mathematics

"By a *conjecture* we mean a proposition that has not yet been proven but which is favored by some serious evidence. It may be a significant amount of computational evidence, or a body of theory and technique that has arisen in the attempt to settle the conjecture.

An *open question* is a problem where the evidence is not very convincing one way or the other.

A *theorem*, of course, is something that has been proved. There are important theorems, and there are unimportant (but perhaps curious) theorems.

The distinction between open question and conjecture is, it is true, somewhat subjective, and different mathematicians may form different judgements concerning a particular problem. We trust that there will be no similar ambiguity concerning the theorems."

—— Dan Shanks [**Sha85**, p. 2]

Today we might add to this a *heuristic* argument, in which we explore an open question with techniques that help give us a good idea of what to conjecture, even if those techniques are unlikely to lead to a formal proof.

Prerequisites

The reader should be familiar with the commonly used sets of numbers $\mathbb{N}$, $\mathbb{Z}$, and $\mathbb{Q}$, as well as polynomials with integer coefficients, denoted by $\mathbb{Z}[x]$. Proofs will often use the *principle of induction*; that is, if $S(n)$ is a given mathematical assertion, dependent on the integer n, then to prove that it is true for all $n \in \mathbb{N}$, we need only prove the following:

- $S(1)$ is true.
- $S(k)$ is true implies that $S(k+1)$ is true, for all integers $k \geq 1$.

The example that is usually given to highlight the principle of induction is the statement "$1 + 2 + 3 + \cdots + n = \frac{n(n+1)}{2}$" which we denote by $S(n)$.[1] For $n = 1$ we check that $1 = \frac{1 \cdot 2}{2}$ and so $S(1)$ is true. For any $k \geq 1$, we assume that $S(k)$ is true and then deduce that

$$
\begin{aligned}
1 + 2 + 3 + \cdots + (k+1) &= \underbrace{(1 + 2 + 3 + \cdots + k)}_{} + (k+1) \\
&= \frac{k(k+1)}{2} + (k+1) \quad \text{as } S(k) \text{ is true} \\
&= \frac{(k+1)(k+2)}{2} ;
\end{aligned}
$$

that is, $S(k+1)$ is true. Hence, by the principle of induction, we deduce that $S(n)$ is true for all integers $n \geq 1$.

To highlight the technique of induction with more examples, we develop the theory of sums of powers of integers (for example, we prove a statement which gives a formula for $1^2 + 2^2 + \cdots + n^2$ for each integer $n \geq 1$) in section 0.1 and give formulas for the values of the terms of recurrence sequences (like the Fibonacci numbers) in section 0.2.

---

[1] There are other, easier, proofs of this assertion, but induction will be the only viable technique to prove some of the more difficult theorems in the course, which is why we highlight the *technique* here.

Induction and the least counterexample: Induction can be slightly disguised. For example, sometimes one proves that a statement $T(n)$ is true for all $n \geq 1$, by supposing that it is false for some n and looking for a contradiction. If $T(n)$ is false for some n, then there must be a least integer m for which $T(m)$ is false. The trick is to use the assumption that $T(m)$ is false to prove that there exists some smaller integer k, $1 \leq k < m$, for which $T(k)$ is also false. This contradicts the minimality of m, and therefore $T(n)$ must be true for all $n \geq 1$. Such proofs are easily reformulated into an induction proof:

Let $S(n)$ be the statement that $T(1), T(2), \ldots, T(n)$ all hold. The induction proof then works for if $S(m-1)$ is true, but $S(m)$ is false, then $T(m)$ is false and so, by the previous paragraph, $T(k)$ is false for some integer k, $1 \leq k \leq m-1$, which contradicts the assumption that $S(m-1)$ is true.

A beautiful example is given by the statement, "*Every integer > 1 has a prime divisor.*" (A *prime* number is an integer > 1, such that the only positive integers that divide it are 1 and itself.) Let $T(n)$ be the statement that n has a prime divisor, and let $S(n)$ be the statement that $T(2), T(3), \ldots, T(n)$ all hold. Evidently $S(2) = T(2)$ is true since 2 is prime. We suppose that $S(k)$ is true (so that $T(2), T(3), \ldots, T(k)$ all hold). Now:

Either $k+1$ is itself a prime number, in which case $T(k+1)$ holds and therefore $S(k+1)$ holds.

Or $k+1$ is not prime, in which case it has a divisor d which is not equal to either 1 or $k+1$, and so $2 \leq d \leq k$. But then $S(d)$ holds by the induction hypothesis, and so there is some prime p, which divides d, and therefore divides $k+1$. Hence $T(k+1)$ holds and therefore $S(k+1)$ holds.

(The astute reader might ask whether certain "facts" that we have used here deserve a proof. For example, if a prime p divides d, and d divides $k+1$, then p divides $k+1$. We have also assumed the reader understands that when we write "d divides $k+1$" we mean that when we divide $k+1$ by d, the remainder is zero. One of our goals at the beginning of the course is to make sure that everyone interprets these simple facts in the same way, by giving as clear definitions as possible and outlining useful, simple deductions from these definitions.)

Chapter 0

Preliminary Chapter on Induction

Induction is an important proof technique in number theory. This preliminary chapter gives the reader the opportunity to practice its use, while learning about some intriguing number-theoretic concepts.

0.1. Fibonacci numbers and other recurrence sequences

The *Fibonacci numbers*, perhaps the most famous sequence of integers, begin with

$$F_0 = 0, \ F_1 = 1, \ F_2 = 1, \ F_3 = 2, \ 3, \ 5, \ 8, \ 13, \ 21, \ 34, \ 55, \ 89, \ 144, \ 233, \ldots.$$

The Fibonacci numbers appear in many places in mathematics and its applications.[1] They obey a rule giving each term of the Fibonacci sequence in terms of the recent history of the sequence:

$$F_n = F_{n-1} + F_{n-2} \text{ for all integers } n \geq 2.$$

We call this a *recurrence relation*. It is not difficult to find a formula for F_n:

$$(0.1.1) \qquad F_n = \frac{1}{\sqrt{5}} \left(\left(\frac{1+\sqrt{5}}{2} \right)^n - \left(\frac{1-\sqrt{5}}{2} \right)^n \right) \text{ for all integers } n \geq 0,$$

where $\frac{1+\sqrt{5}}{2}$ and $\frac{1-\sqrt{5}}{2}$ each satisfy the equation $x+1 = x^2$. Having such an explicit formula for the Fibonacci numbers makes them easy to work with, but there is a problem. It is not obvious from this formula that every Fibonacci number is an integer; however that does follow easily from the original recurrence relation.[2]

---

[1] Typically when considering a biological process whose current state depends on its past, such as evolution, and brain development.

[2] It requires quite sophisticated ideas to decide whether a given complicated formula like (0.1.1) is an integer or not. Learn more about this in appendix 0F on symmetric polynomials.

Exercise 0.1.1. (a) Use the recurrence relation for the Fibonacci numbers, and induction to prove that every Fibonacci number is an integer.
(b) Prove that (0.1.1) is correct by verifying that it holds for $n = 0, 1$ and then, for all larger integers n, by induction.

Exercise 0.1.2. Use induction on $n \geq 1$ to prove that
(a) $F_1 + F_3 + \cdots + F_{2n-1} = F_{2n}$ and
(b) $1 + F_2 + F_4 + \cdots + F_{2n} = F_{2n+1}$.

The number $\phi = \frac{1+\sqrt{5}}{2}$ is called the *golden ratio*; one can show that F_n is the nearest integer to $\phi^n/\sqrt{5}$.

Exercise 0.1.3. (a) Prove that ϕ satisfies $\phi^2 = \phi + 1$.
(b) Prove that $\phi^n = F_n \phi + F_{n-1}$ for all integers $n \geq 1$, by induction.

Any sequence $x_0, x_1, x_2, \ldots$, for which the terms x_n, with $n \geq 2$, are defined by the equation

(0.1.2) $$x_n = ax_{n-1} + bx_{n-2} \text{ for all } n \geq 2,$$

where a, b, x_0, x_1 are given, is called a *second-order linear recurrence sequence*. Although this is a vast generalization of the Fibonacci numbers one can still prove a formula for the general term, x_n, analogous to (0.1.1): We begin by factoring the polynomial

$$x^2 - ax - b = (x - \alpha)(x - \beta)$$

for the appropriate $\alpha, \beta \in \mathbb{C}$ (we had $x^2 - x - 1 = (x - \frac{1+\sqrt{5}}{2})(x - \frac{1-\sqrt{5}}{2})$ for the Fibonacci numbers). If $\alpha \neq \beta$, then there exist coefficients c_α, c_β for which

(0.1.3) $$x_n = c_\alpha \alpha^n + c_\beta \beta^n \text{ for all } n \geq 0.$$

(In the case of the Fibonacci numbers, we have $c_\alpha = 1/\sqrt{5}$ and $c_\beta = -1/\sqrt{5}$.) Moreover one can determine the values of c_α and c_β by solving the simultaneous equations obtained by evaluating the formula (0.1.3) at $n = 0$ and $n = 1$, that is,

$$c_\alpha + c_\beta = x_0 \quad \text{and} \quad c_\alpha \alpha + c_\beta \beta = x_1.$$

Exercise 0.1.4. (a) Prove (0.1.3) is correct by verifying that it holds for $n = 0, 1$ (with x_0 and x_1 as in the last displayed equation) and then by induction for $n \geq 2$.
(b) Show that c_α and c_β are uniquely determined by x_0 and x_1, provided $\alpha \neq \beta$.
(c) Show that if $\alpha \neq \beta$ with $x_0 = 0$ and $x_1 = 1$, then $x_n = \frac{\alpha^n - \beta^n}{\alpha - \beta}$ for all integers $n \geq 0$.
(d) Show that if $\alpha \neq \beta$ with $y_0 = 2, y_1 = a$ with $y_n = ay_{n-1} + by_{n-2}$ for all $n \geq 2$, then $y_n = \alpha^n + \beta^n$ for all integers $n \geq 0$.

The $\{x_n\}_{n \geq 0}$ in (c) is a *Lucas sequence*, and the $\{y_n\}_{n \geq 0}$ in (d) its *companion sequence*

Exercise 0.1.5.[3] (a) Prove that $\alpha = \beta$ if and only if $a^2 + 4b = 0$.
(b)† Show that if $a^2 + 4b = 0$, then $\alpha = a/2$ and $x_n = (cn + d)\alpha^n$ for all integers $n \geq 0$, for some constants c and d.
(c) Deduce that if $\alpha = \beta$ with $x_0 = 0$ and $x_1 = 1$, then $x_n = n\alpha^{n-1}$ for all $n \geq 0$.

Exercise 0.1.6. Prove that if $x_0 = 0$ and $x_1 = 1$, if (0.1.2) holds, and if α is a root of $x^2 - ax - b$, then $\alpha^n = \alpha x_n + bx_{n-1}$ for all $n \geq 1$.

---

[3]In this question, and from here on, induction should be used at the reader's discretion.

0.2. Formulas for sums of powers of integers

When Gauss was ten years old, his mathematics teacher aimed to keep his class quiet by asking them to add together the integers from 1 to 100. Gauss did this in a few moments, by noting if one adds that list of numbers to itself, but with the second list in reverse order, then one has

$$1 + 100 = 2 + 99 = 3 + 98 = \cdots = 99 + 2 = 100 + 1 = 101.$$

That is, twice the asked-for sum equals 100 times 101, and so

$$1 + 2 + \cdots + 100 = \frac{1}{2} \times 100 \times 101.$$

This argument generalizes to adding up the natural numbers less than any given N, yielding the formula[4]

$$(0.2.1) \qquad \sum_{n=1}^{N-1} n = \frac{(N-1)N}{2}.$$

The sum on the left-hand side of this equation varies in length with N, whereas the right-hand side does not. The right-hand side is a formula whose value varies but has a relatively simple structure, so we call it a *closed form* expression. (In the prerequisite section, we gave a less interesting proof of this formula, by induction.)

Exercise 0.2.1. (a) Prove that $1 + 3 + 5 + \cdots + (2N-1) = N^2$ for all $N \geq 1$ by induction.
(b) Prove the formula in part (a) by the young Gauss's method.
(c) Start with a single dot, thought of as a 1-by-1 array of dots, and extend it to a 2-by-2 array of dots by adding an appropriate row and column. You have added 3 dots to the original dot and so $1 + 3 = 2^2$.

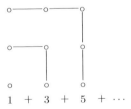

In general, draw an N-by-N array of dots, and add an additional row and column of dots to obtain an $(N+1)$-by-$(N+1)$ array of dots. By determining how many dots were added to the number of dots that were already in the array, deduce the formula in (a).

Let $S = \sum_{n=1}^{N-1} n^2$. Using exercise 0.2.1 we can write each square, n^2, as the sum of the odd positive integers $\leq 2n$. Therefore $2m-1$ appears $N-m$ times in the sum for S, and so

$$S = \sum_{m=1}^{N-1} (2m-1)(N-m) = -N \sum_{m=1}^{N-1} 1 + (2N+1) \sum_{m=1}^{N-1} m - 2S.$$

---

[4]This same idea appears in the work of Archimedes, from the third century B.C. in ancient Greece.

Using our closed formula for $\sum_m m$, we deduce, after some rearrangement, that

$$\sum_{n=1}^{N-1} n^2 = \frac{(N-1)N(2N-1)}{6},$$

a closed formula for the sum of the squares up to a given point. There is also a closed formula for the sum of the cubes:

(0.2.2) $$\sum_{n=1}^{N-1} n^3 = \left(\frac{(N-1)N}{2}\right)^2.$$

This is the square of the closed formula (0.2.1) that we obtained for $\sum_{n=0}^{N-1} n$. Is this a coincidence or the first hint of some surprising connection?

Exercise 0.2.2. Prove these last two formulas by induction.

These three examples suggest that there are closed formulas for the sums of the kth powers of the integers, for every $k \geq 1$, but it is difficult to guess exactly what those formulas might look like. Moreover, to hope to prove a formula by induction, we need to have the formula at hand.

We will next find a closed formula in a simpler but related question and use this to find a closed formula for the sums of the kth powers of the integers in appendix 0A. We will go on to investigate, in section 7.34 of appendix 7I, whether there are other amazing identities for sums of different powers, like

$$\sum_{n=1}^{N-1} n^3 = \left(\sum_{n=1}^{N-1} n\right)^2.$$

0.3. The binomial theorem, Pascal's triangle, and the binomial coefficients

The *binomial coefficient* $\binom{n}{m}$ is defined to be the number of different ways of choosing m objects from n. (Therefore $\binom{n}{m} = 0$ whenever $m < 0$ or $m > n$.) From this definition we see that the binomial coefficients are all integers. To determine $\binom{5}{2}$ we note that there are 5 choices for the first object and 4 for the second, but then we have counted each pair of objects twice (since we can select them in either order), and so $\binom{5}{2} = \frac{5 \times 4}{2}$. It is arguably nicer to write 5×4 as $\frac{5 \times 4 \times 3 \times 2 \times 1}{3 \times 2 \times 1} = \frac{5!}{3!}$ so that $\binom{5}{2} = \frac{5!}{3!2!}$. One can develop this proof to show that, for any integers $0 \leq m \leq n$, one has the very neat formula[5]

(0.3.1) $$\binom{n}{m} = \frac{n!}{m!(n-m)!}, \text{ where } r! = r \cdot (r-1) \cdots 2 \cdot 1.$$

From this formula alone it is not obvious that the binomial coefficients are integers.

Exercise 0.3.1. (a) Prove that $\binom{n+1}{m} = \binom{n}{m} + \binom{n}{m-1}$ for all integers m, and all integers $n \geq 0$.
(b) Deduce from (a) that each $\binom{n}{m}$ is an integer.

---

[5] We prefer to work with the closed formula $27!/(15!12!)$ rather than to evaluate it as 17383860, since the three factorials are easier to appreciate and to manipulate in subsequent calculations, particularly when looking for patterns.

0.3. The binomial theorem, Pascal's triangle, and the binomial coefficients

Pascal's triangle is a triangular array in which the $(n+1)$st row contains the binomial coefficients $\binom{n}{m}$, with m increasing from 0 to n, as one goes from left to right:

$$\begin{array}{c}1\\ 1\ 1\\ 1\ 2\ 1\\ 1\ 3\ 3\ 1\\ 1\ 4\ 6\ 4\ 1\\ 1\ 5\ 10\ 10\ 5\ 1\\ 1\ 6\ 15\ 20\ 15\ 6\ 1\\ \ldots \text{etc.}\end{array}$$

The addition formula in exercise 0.3.1(a) yields a rule for obtaining a row from the previous one, by adding any two neighboring entries to give the entry immediately below. For example the third entry in the bottom row is immediately below 5 and 10 (to either side) and so equals $5 + 10 = 15$. The next entry is $10 + 10 = 20$, etc.

The *binomial theorem* states that if n is an integer ≥ 1, then

$$\boxed{(x+y)^n = \sum_{m=0}^{n} \binom{n}{m} x^{n-m} y^m.}$$

Exercise 0.3.2.† Using exercise 0.3.1(a) and induction on $n \geq 1$, prove the binomial theorem.

Notice that one can read off the coefficients of $(x+y)^n$ from the $(n+1)$st row of Pascal's triangle; for example, reading off the bottom row above (which is the 7th row down of Pascal's triangle), we obtain

$$(x+y)^6 = x^6 + 6x^5y + 15x^4y^2 + 20x^3y^3 + 15x^2y^4 + 6xy^5 + y^6.$$

In the previous section we raised the question of finding a closed formula for the sum of n^k, over all positive integers $n < N$. We can make headway in a related question in which we replace n^k with a different polynomial in n of degree k, namely the binomial coefficient

$$\binom{n}{k} = \frac{n(n-1)\cdots(n-k+1)}{k!}.$$

This is a polynomial of degree k in n. For example, we have $\binom{n}{3} = \frac{n^3}{6} - \frac{n^2}{2} + \frac{n}{3}$, a polynomial in n of degree 3. We can identify a closed formula for the sum of these binomial coefficients, over all positive integers $n < N$, namely:

(0.3.2) $$\sum_{n=0}^{N-1} \binom{n}{k} = \binom{N}{k+1}$$

for all N and $k \geq 0$. For $k=2, N=6$, this can be seen in the following diagram:

$$
\begin{array}{ccccccc}
& & & 1 & & & \\
& & 1 & & 1 & & \\
& \textcircled{1} & & 2 & & 1 & \\
& 1 & \textcircled{3} & & 3 & & 1 \\
1 & & 4 & \textcircled{6} & & 4 & 1 \\
1 & 5 & & 10 & \textcircled{10} & 5 & 1 \\
1 & 6 & 15 & & \textcircled{\mathbf{20}} & 15 & 6 & 1
\end{array}
$$

so that $1 + 3 + 6 + 10$ equals 20.

Exercise 0.3.3. Prove (0.3.2) for each fixed $k \geq 1$, for each $N \geq k+1$, using induction and exercise 0.3.1. You might also try to prove it by induction using the idea behind the illustration in the last diagram.

If we instead display Pascal's triangle by lining up the initial 1's and then summing the diagonals,

$$
\begin{array}{cccccc}
1 & & & & & \\
1 & 1 & & & & \\
1 & 2 & 1 & & & \\
1 & 3 & 3 & \textcircled{1} & & \\
1 & 4 & \textcircled{6} & 4 & 1 & \\
1 & \textcircled{5} & 10 & 10 & \ldots & \\
\textcircled{1} & 6 & 15 & \ldots & &
\end{array}
$$

the sums are $1, 1, 1+1, 1+2, 1+3+1, 1+4+3, 1+5+6+1, \ldots$ which equal $1, 1, 2, 3, 5, 8, 13, \ldots$, the Fibonacci numbers. It therefore seems likely that

$$(0.3.3) \qquad F_n = \sum_{k=0}^{n-1} \binom{n-1-k}{k} \text{ for all } n \geq 1.$$

Exercise 0.3.4. Prove (0.3.3) for each integer $n \geq 1$, by induction using exercise 0.3.1(a).

Articles with further thoughts on factorials and binomial coefficients

[1] Manjul Bhargava, *The factorial function and generalizations*, Amer. Math. Monthly **107** (2000), 783–799 (preprint).
[2] John J. Watkins, chapter 5 of *Number theory. A historical approach*, Princeton University Press, 2014.

Additional exercises

Exercise 0.4.1. (a) Prove that for all $n \geq 1$ we have
$$\begin{pmatrix} 1 & 1 \\ 1 & 0 \end{pmatrix}^n = \begin{pmatrix} F_{n+1} & F_n \\ F_n & F_{n-1} \end{pmatrix}.$$
(b) Deduce that $F_{n+1}F_{n-1} - F_n^2 = (-1)^n$ for all $n \geq 1$.
(c) Deduce that $F_{n+1}^2 - F_{n+1}F_n - F_n^2 = (-1)^n$ for all $n \geq 0$.

Exercise 0.4.2.† Deduce from (0.1.1) that the Fibonacci number F_n is the nearest integer to $\phi^n/\sqrt{5}$, for each integer $n \geq 0$, where the constant $\phi := \frac{1+\sqrt{5}}{2}$. This *golden ratio* appears in art and architecture when attempting to describe "perfect proportions".

Exercise 0.4.3. Prove that $F_n^2 + F_{n+3}^2 = 2(F_{n+1}^2 + F_{n+2}^2)$ for all $n \geq 0$.

Exercise 0.4.4. Prove that for all $n \geq 1$ we have
$$F_{2n-1} = F_{n-1}^2 + F_n^2 \quad \text{and} \quad F_{2n} = F_{n+1}^2 - F_{n-1}^2.$$

Exercise 0.4.5. Use (0.1.1) to prove the following:
(a) For every r we have $F_n^2 - F_{n+r}F_{n-r} = (-1)^{n-r}F_r^2$ for all $n \geq r$.
(b) For all $m \geq n \geq 0$ we have $F_m F_{n+1} - F_{m+1}F_n = (-1)^n F_{m-n}$.

Exercise 0.4.6. Let $u_0 = b$ and $u_{n+1} = a u_n$ for all $n \geq 0$. Give a formula for all u_n with $n \geq 0$.

The expression 011010 is a *string of 0's and 1's*. There are 2^n strings of 0's and 1's of length n as there are two possibilities for each entry. Let A_n be the set of strings of 0's and 1's of length n which contain no two consecutive 1's. Our example 011010 does not belong to A_6 as the second and third characters are consecutive 1's, whereas 01001010 is in A_8. Calculations reveal that $|A_1| = 2$, $|A_2| = 3$, and $|A_3| = 5$, data which suggests that perhaps $|A_n| = F_{n+2}$, the Fibonacci number.

Exercise 0.4.7.† (a) If $0w$ is a string of 0's and 1's of length n, prove that $0w \in A_n$ if and only if $w \in A_{n-1}$.
(b) If $10w$ is a string of 0's and 1's of length n, prove that $10w \in A_n$ if and only if $w \in A_{n-2}$.
(c) Prove that $|A_n| = F_{n+2}$ for all $n \geq 1$, by induction on n.

Exercise 0.4.8.† Prove that every positive integer other than the powers of 2 can be written as the sum of two or more consecutive integers.

Exercise 0.4.9. Prove that $\binom{n}{m}\binom{n-m}{a-m} = \binom{a}{m}\binom{n}{a}$ for any integers $n \geq a \geq m \geq 0$.

Exercise 0.4.10.† Suppose that a and b are integers and $\{x_n : n \geq 0\}$ is the second-order linear recurrence sequence given by (0.1.2) with $x_0 = 0$ and $x_1 = 1$.
(a) Prove that for all non-negative integers m we have
$$x_{m+k} = x_{m+1}x_k + bx_m x_{k-1} \text{ for all integers } k \geq 1.$$
(b) Deduce that
$$x_{2n+1} = x_{n+1}^2 + bx_n^2 \quad \text{and} \quad x_{2n} = x_{n+1}x_n + bx_n x_{n-1} \quad \text{for all natural numbers } n.$$

Exercise 0.4.11. Suppose that the sequences $\{x_n : n \geq 0\}$ and $\{y_n : n \geq 0\}$ both satisfy (0.1.2) and that $x_0 = 0$ and $x_1 = 1$, whereas y_0 and y_1 might be anything. Prove that
$$y_n = y_1 x_n + by_0 x_{n-1} \text{ for all } n \geq 1.$$

Exercise 0.4.12. Suppose that $x_0 = 0$, $x_1 = 1$, and $x_{n+2} = ax_{n+1} + bx_n$. Prove that for all $n \geq 1$ we have
(a) $(a+b-1)\sum_{j=1}^{n} x_j = x_{n+1} + bx_n - 1$;
(b) $a(b^n x_0^2 + b^{n-1}x_1^2 + \cdots + bx_{n-1}^2 + x_n^2) = x_n x_{n+1}$;
(c) $x_n^2 - x_{n-1}x_{n+1} = (-b)^{n-1}$.

Exercise 0.4.13. Suppose that $x_{n+2} = ax_{n+1} + bx_n$ for all $n \geq 0$.
(a) Show that
$$\begin{pmatrix} x_{n+2} & x_{n+1} \\ x_{n+1} & x_n \end{pmatrix} = \begin{pmatrix} a & b \\ 1 & 0 \end{pmatrix}^n \begin{pmatrix} x_2 & x_1 \\ x_1 & x_0 \end{pmatrix} \text{ for all } n \geq 0.$$
(b) Deduce that $x_{n+2}x_n - x_{n+1}^2 = c(-b)^n$ for all $n \geq 0$ where $c := x_2 x_0 - x_1^2$.
(c) Deduce that $x_{n+1}^2 - ax_{n+1}x_n - bx_n^2 = -c(-b)^n$.

Other number-theoretic sequences can be obtained from linear recurrences or other types of recurrences. Besides the Fibonacci numbers, there is another sequence of integers that is traditionally denoted by $(F_n)_{n\geq 0}$: These are the *Fermat numbers*, $F_n = 2^{2^n} + 1$ for all $n \geq 0$ (see sections 3.11 of appendix 3A, 5.1, 5.25 of appendix 5H, etc.).

Exercise 0.4.14. Show that if $F_0 = 3$ and $F_{n+1} = F_n^2 - 2F_n + 2$, then $F_n = 2^{2^n} + 1$ for all $n \geq 0$.

Exercise 0.4.15. (a) Show that if $M_0 = 0$, $M_1 = 1$, and $M_{n+2} = 3M_{n+1} - 2M_n$ for all integers $n \geq 0$, then $M_n = 2^n - 1$ for all integers $n \geq 0$. The integer M_n is the nth *Mersenne number* (see exercise 2.5.16 and sections 4.2, 5.1, etc.).
(b) Show that if $M_0 = 0$ with $M_{n+1} = 2M_n + 1$ for all $n \geq 0$, then $M_n = 2^n - 1$.

Exercise 0.4.16.[‡] We can reinterpret exercise 0.4.3 as giving a recurrence relation for the sequence $\{F_n^2\}_{n\geq 0}$, where F_n is the nth Fibonacci number; that is,
$$F_{n+3}^2 = 2F_{n+2}^2 + 2F_{n+1}^2 - F_n^2 \text{ for all } n \geq 0.$$
Here F_{n+3}^2 is described in terms of the last three terms of the sequence; this is called a *linear recurrence of order* 3. Prove that for any integer $k \geq 1$, the sequence $\{F_n^k\}_{n\geq 0}$ satisfies a linear recurrence of order $k + 1$.

How to proceed through this book. It can be challenging to decide what proof technique to try on a given question. There is no simple guide—practice is what best helps decide how to proceed. Some students find Zeitz's book [**Zei17**] helpful as it exhibits all of the important techniques in context. I like Conway and Guy's [**CG96**] since it has lots of great questions, beautifully discussed with great illustrations, and introduces quite a few of the topics from this book.

A paper that questions one's assumptions is

[1] Richard K. Guy *The strong law of small numbers*, Amer. Math. Monthly, **95** (1988), 697–712.

Appendix 0A. A closed formula for sums of powers

In chapter 0, we discussed closed form expressions for sums of powers. We will prove here that there is such a formula for the sum of the kth power of the integers up to a given point, developing themes from earlier in this chapter.

0.5. Formulas for sums of powers of integers, II

Our goal in this section is to use our formula (0.3.2) for summing binomial coefficients, to find a formula for summing powers of integers. For example, since

$$n^3 = 6\binom{n}{3} + 6\binom{n}{2} + \binom{n}{1},$$

we can use (0.3.2) with $k = 3$, 2, and 1, respectively, to obtain

$$\sum_{n=0}^{N-1} n^3 = 6\sum_{n=0}^{N-1} \binom{n}{3} + 6\sum_{n=0}^{N-1} \binom{n}{2} + \sum_{n=0}^{N-1} \binom{n}{1}$$
$$= 6\binom{N}{4} + 6\binom{N}{3} + \binom{N}{2}.$$

Summing these three multiples of binomial coefficients gives the formula for the sum of the cubes of the integers up to $N - 1$, which we encountered in section 0.2. To make this same technique work to sum n^k, for arbitrary integer $k \geq 1$, we need to show that x^k can be expressed as a sum of fixed multiples of the binomial coefficients $\binom{x}{k}, \ldots, \binom{x}{1}$, where by $\binom{x}{k}$ we mean the polynomial

$$\binom{x}{k} = \frac{x(x-1)\cdots(x-(k-1))}{k!}.$$

Notice that if we substitute $x = n$ into this expression, we obtain the binomial coefficient $\binom{n}{k}$.

9

Proposition 0.5.1. *Any polynomial $f(x) \in \mathbb{Z}[x]$ of degree $k \geq 0$ can be written as a sum of integer multiples of the binomial coefficients $\binom{x}{k}, \ldots, \binom{x}{1}, \binom{x}{0}$.*

Proof. By induction on k. The result is immediate for $k = 0$. Otherwise, suppose that $f(x)$ has leading coefficient ax^k; then subtract $a \cdot k! \cdot \binom{x}{k}$, which also has leading coefficient ax^k. The resulting polynomial, $g(x) = f(x) - a \cdot k! \cdot \binom{x}{k}$, has degree $k-1$ so can be written as $c_0 \binom{x}{0} + \cdots + c_{k-1} \binom{x}{k-1}$ by the induction hypothesis. But then $f(x) = c_0 \binom{x}{0} + \cdots + c_k \binom{x}{k}$, with $c_k = a \cdot k!$, as desired. $\square$

In particular, there are integers $c_0, c_1, \ldots, c_k$ for which

$$(0.5.1) \qquad x^k = c_k \binom{x}{k} + \cdots + c_1 \binom{x}{1} + c_0 \binom{x}{0}.$$

One can then immediately deduce, from (0.3.2), that

$$\sum_{n=0}^{N-1} n^k = c_k \sum_{n=0}^{N-1} \binom{n}{k} + \cdots + c_1 \sum_{n=0}^{N-1} \binom{n}{1} + c_0 \sum_{n=0}^{N-1} \binom{n}{0}$$

$$= c_k \binom{N}{k+1} + \cdots + c_1 \binom{N}{2} + c_0 \binom{N}{1}.$$

Expanding out the binomial coefficients, this gives the desired closed form expression for $\sum_{n=0}^{N-1} n^k$, a polynomial in N of degree $k+1$.

There is a difficulty. We proved that the c_j exist but did not show how to determine them. We can do this by successively substituting in $x = 0$, then $x = 1$, then $\ldots, x = k - 1$ into (0.5.1), since one obtains

$$0^k = c_k \cdot 0 + \cdots + c_1 \cdot 0 + c_0,$$

and so $c_0 = 0$; then

$$1^k = c_k \cdot 0 + \cdots + c_2 \cdot 0 + c_1 + c_0,$$

and so $c_1 = 1$; and then $c_2 = 2^k - 2$, $c_3 = 3^k - 3 \cdot 2^k + 3$, etc. We end this appendix with a particularly challenging exercise.

Exercise 0.5.1.[‡] (a) Establish that (0.5.1) holds with

$$c_m = m^k - \binom{m}{1}(m-1)^k + \binom{m}{2}(m-2)^k - \cdots + (-1)^{m-2}\binom{m}{m-2}2^k + (-1)^{m-1}m,$$

for all $m \geq 1$ and for all $k \geq 1$. The integers $c_m/m!$ are the *Stirling numbers of the second kind*, usually denoted by $S_2(k,m)$. They arise in several interesting combinatorial settings; for example, $S_2(k,m)$ is the number of ways to partition a set of k objects into m non-empty subsets.

(b) Deduce that, for any given integer $k \geq 0$, there exist rational numbers $a_0, a_1, \ldots, a_{k+1}$ for which $\sum_{n=0}^{N-1} n^k = a_0 + a_1 N + \cdots + a_{k+1} N^{k+1}$ for all integers $N \geq 1$.

Exercise 0.5.2. Prove that $c_j/j!$ is an integer for all $j \geq 0$ in (0.5.1).

Exercise 0.5.3.[†] Let $f(x) \in \mathbb{C}[x]$. Prove that $f(n)$ is an integer for all integers n if and only if $f(x) = \sum_m a_m \binom{x}{m}$ where the a_m are all integers.

We will return to this topic, finding an elegant description of the rational numbers a_j by introducing the Bernoulli numbers in the next appendix, appendix 0B.

Appendix 0B. Generating functions

The *generating function* (or *generating series*) of a given sequence of numbers $a_0, a_1, \ldots$ is the *power series*

$$a_0 + a_1 x + a_2 x^2 + \cdots$$

involving a variable x, where the nth term is $a_n x^n$. We now see how generating functions allow us to provide alternative, elegant proofs of the results of this chapter. We begin with an alternative proof of (0.3.2) that exhibits the power of constructing generating functions.

Exercise 0.6.1. (a) Prove that for every integer $k \geq 0$ one has

$$\frac{1}{(1-t)^{k+1}} = \binom{k}{k} + \binom{k+1}{k}t + \binom{k+2}{k}t^2 + \cdots + \binom{k+m}{k}t^m + \cdots.$$

(b) Prove that (0.3.1) follows by equating the coefficient of t^{N-k-1} on either side of

$$\frac{1}{(1-t)^{k+1}} \cdot \frac{1}{(1-t)} = \frac{1}{(1-t)^{k+2}}.$$

(c) Multiply this identity through by $1 - t$ and reprove the formula in exercise 0.3.1.(a) by equating the coefficients on each side.

0.6. Formulas for sums of powers of integers, III

The *Bernoulli numbers*, B_n, are the coefficients in the power series:

$$\frac{X}{e^X - 1} = \sum_{n \geq 0} B_n \frac{X^n}{n!}.$$

They are a sequence of numbers that occur in all sorts of interesting contexts in number theory. The first few Bernoulli numbers are $B_0 = 1$, $B_1 = -\frac{1}{2}$, $B_2 = \frac{1}{6}$, $B_3 = 0$, $B_4 = -\frac{1}{30}$, $B_5 = 0$, $B_6 = \frac{1}{42}$, $B_7 = 0$, $B_8 = -\frac{1}{30}$, $B_9 = 0$, $B_{10} = \frac{5}{66}, \ldots$.

From this data we can make a few guesses as to what they look like in general:

- If n is odd and > 1, then $B_n = 0$. This is easily proved since

$$\sum_{\substack{n \geq 0 \\ n \text{ odd}}} 2B_n \frac{X^n}{n!} = \sum_{\substack{n \geq 0 \\ n \text{ odd}}} B_n \frac{X^n}{n!} - \sum_{\substack{n \geq 0 \\ n \text{ odd}}} B_n \frac{(-X)^n}{n!}$$

$$= \frac{X}{e^X - 1} - \frac{(-X)}{e^{-X} - 1} = \frac{X}{e^X - 1} - \frac{(Xe^X)}{e^X - 1} = -X.$$

Comparing the coefficients of X^n on either side of this equation, we conclude that $B_1 = -\frac{1}{2}$ and $B_n = 0$ for all odd $n \geq 3$.

- The B_n are rational. We expand the power series

$$1 = \frac{e^X - 1}{X} \cdot \frac{X}{e^X - 1} = \sum_{r \geq 0} \frac{X^r}{(r+1)!} \cdot \sum_{s \geq 0} B_s \frac{X^s}{s!}$$

$$= \sum_{n \geq 0} \left(\sum_{\substack{r,s \geq 0 \\ r+s=n}} \frac{(n+1)!}{(r+1)!s!} B_s \right) \frac{X^n}{(n+1)!}$$

and compare the coefficients of X^n on either side to obtain that $B_0 = 1$ and $\sum_{s=0}^{n} \binom{n+1}{s} B_s = 0$ for each $n \geq 1$. This can be rewritten as

$$B_n = -\frac{1}{n+1} \sum_{s=0}^{n-1} \binom{n+1}{s} B_s \text{ for each } n \geq 1.$$

We can then deduce by induction on $n \geq 1$ that the B_n are rational, since we have given B_n as a finite sum of rational numbers times Bernoullli numbers B_s with $s < n$.

Next we define the *Bernoulli polynomials*, $B_n(t)$, as the coefficients in the power series:

$$\frac{Xe^{tX}}{e^X - 1} = \sum_{n \geq 0} B_n(t) \frac{X^n}{n!},$$

and therefore $B_n(0) = B_n$. To verify that these are really polynomials, note that

$$\sum_{k \geq 0} B_k(t) \frac{X^k}{k!} = e^{tX} \cdot \frac{X}{e^X - 1} = \sum_{m \geq 0} \frac{(tX)^m}{m!} \cdot \sum_{n \geq 0} B_n \frac{X^n}{n!} = \sum_{m \geq 0} \sum_{n \geq 0} B_n t^m \frac{X^{m+n}}{m!n!}.$$

Here we change variable, writing $k = m + n$, and then the coefficient of $X^k/k!$ is

(0.6.1) $$B_k(t) = \sum_{\substack{m,n \geq 0 \\ m+n=k}} \frac{k!}{m!n!} B_n t^m = \sum_{n=0}^{k} \binom{k}{n} B_n t^{k-n}.$$

We have done all this preliminary work so as to prove the following extraordinary formula for the sum of the mth powers of the positive integers $< N$.

Theorem 0.1. *For any integers $k \geq 1$ and $N \geq 1$ we have*

$$\sum_{n=0}^{N-1} n^{k-1} = \frac{1}{k}(B_k(N) - B_k).$$

Proof. If N is an integer ≥ 1, then

$$\sum_{k\geq 0}(B_k(N) - B_k)\frac{X^k}{k!} = \frac{X(e^{NX} - 1)}{e^X - 1} = X\sum_{n=0}^{N-1} e^{nX}$$

$$= X\sum_{n=0}^{N-1}\sum_{j\geq 0}\frac{(nX)^j}{j!} = \sum_{j\geq 0}\left(\sum_{n=0}^{N-1} n^j\right)\frac{X^{j+1}}{j!}.$$

Therefore for any integer $N \geq 1$ we obtain

$$\sum_{k\geq 0}(B_k(N) - B_k)\frac{X^k}{k!} = \sum_{k\geq 1}\left(k\sum_{n=0}^{N-1} n^{k-1}\right)\frac{X^k}{k!}$$

by letting $k = j + 1$. The result follows by comparing the coefficients on both sides. $\square$

Negative powers. A key quantity in number theory is the infinite sum

$$\zeta(k) = 1 + \frac{1}{2^k} + \frac{1}{3^k} + \frac{1}{4^k} + \cdots$$

which we define for each integer $k \geq 2$. This is called the *Riemann zeta-function* even though it was first explored by Euler more than a hundred years earlier. Each of these sums is convergent, as each $1/m^k \leq 1/m^2$ for all $m \geq 1, k \geq 2$, and $1/m^2 < \int_{m-1}^m dt/t^2$, so that

$$1 + \frac{1}{2^k} + \frac{1}{3^k} + \cdots \leq 1 + \frac{1}{2^2} + \frac{1}{3^2} + \cdots < 1 + \int_1^\infty \frac{dt}{t^2} = 2.$$

We will make a few observations about the values of these sums:

Exercise 0.6.2. (a) Prove that $\sum_{m\geq 2}\frac{1}{m(m-1)} = 1$.
(b) Prove that $\sum_{k\geq 2}\frac{1}{m^k} = \frac{1}{m(m-1)}$.
(c) Deduce that $\sum_{k\geq 2}(\zeta(k) - 1) = 1$.

Exercise 0.6.3. Let $\mathcal{P}$ be the set of perfect powers > 1. Let $\mathcal{N}$ be the set of integers > 1 that are not perfect powers (so that $\mathcal{P} \cup \mathcal{N}$ is a partition of the integers > 1).
(a) Prove that $\mathcal{P} = \{n^k : n \in \mathcal{N}$ and $k \geq 2\}$ and $\{n^k : n \in \mathcal{N}$ and $k \geq 1\} = \{m \geq 2\}$.
(b) Prove that $\sum_{P \in \mathcal{P}}\frac{1}{P-1} = \sum_{k\geq 2}\sum_{n\in\mathcal{N}}\sum_{j\geq 1}\frac{1}{n^{jk}}$.
(c) Deduce that $\sum_{P\in\mathcal{P}}\frac{1}{P-1} = 1$.
This result was communicated by Goldbach to Euler in 1744.

0.7. The power series view on the Fibonacci numbers

An alternate view on Fibonacci numbers, and indeed all second-order linear recurrence sequences, is via their generating functions. For Fibonacci numbers we study the generating function

$$\sum_{n\geq 0} F_n x^n,$$

which is a power series in x. Remembering that $F_n - F_{n-1} - F_{n-2} = 0$ for all $n \geq 2$ we then have

$$(1 - x - x^2) \sum_{n \geq 0} F_n x^n = F_0 + (F_1 - F_0)x + (F_2 - F_1 - F_0)x^2$$

$$+ \cdots + (F_n - F_{n-1} - F_{n-2})x^n + \cdots$$

$$= 0 + (1 - 0)x + 0 \cdot x^2 + \cdots + 0 \cdot x^n + \cdots = x.$$

Hence if $\alpha = \frac{1+\sqrt{5}}{2}$ and $\beta = \frac{1-\sqrt{5}}{2}$, then

$$\sum_{n \geq 0} F_n x^n = \frac{x}{1 - x - x^2} = \frac{1}{\alpha - \beta} \left(\frac{\alpha x}{1 - \alpha x} - \frac{\beta x}{1 - \beta x} \right)$$

$$= \frac{1}{\alpha - \beta} \left(\sum_{m \geq 1} \alpha^m x^m - \sum_{m \geq 1} \beta^m x^m \right) = \sum_{m \geq 1} \frac{\alpha^m - \beta^m}{\alpha - \beta} x^m,$$

and the result (0.1.1) follows, again. Note that if $x_n = ax_{n-1} + bx_{n-2}$, then

$$(1 - at - bt^2) \sum_{n \geq 0} x_n t^n = x_0 + (x_1 - ax_0)t.$$

The sequence $\{x_n\}_{n \geq 0}$ is again determined by the values of a, b, x_0, and x_1.

Exercise 0.7.1.[†] Use this to deduce (0.1.3) when $a^2 + 4b \neq 0$, and exercise 0.1.5(c) when $a^2 + 4b = 0$.

Both of these methods generalize to arbitrary linear recurrences of degree n, as follows.

Theorem 0.2. *Suppose that $a_1, a_2, \ldots, a_d$ and $x_0, x_1, \ldots, x_{d-1}$ are given and that*

$$x_n = a_1 x_{n-1} + a_2 x_{n-2} + \cdots + a_d x_{n-d} \quad \text{for all } n \geq d.$$

Factor the following polynomial into linear factors as

$$X^d - a_1 X^{d-1} - a_2 X^{d-2} + \cdots - a_{d-1} X - a_d = \prod_{j=1}^{k} (X - \alpha_j)^{e_j}.$$

Then there exist polynomials $P_1, \ldots, P_k$, each P_j of degree $\leq e_j - 1$, such that

(0.7.1) $$x_n = \sum_{j=1}^{k} P_j(n) \alpha_j^n \quad \text{for all } n \geq 0.$$

The coefficients of the P_j (and the polynomials P_j themselves) can be determined by solving the linear equations obtained by taking this for $n = 0, 1, 2, \ldots, d-1$.

Exercise 0.7.2.[‡] Prove that (0.7.1) holds.

Exercise 0.7.3.[‡] Let $(x_n)_{n \geq 0}$ be the sequence which begins $x_0 = 0, x_1 = 1$ and then $x_n = ax_{n-1} + bx_{n-2}$ for all $n \geq 2$. Its companion sequence, $(y_n)_{n \geq 0}$, begins $y_0 = 2, y_1 = x_2$ and then $y_n = ay_{n-1} + by_{n-2}$ for all $n \geq 2$. For example, $x_n = 2^n - 1$ has companion sequence $y_n = 2^n + 1$.
(a) Prove that $y_n = \alpha^n + \beta^n$ for all $n \geq 0$ and also that $y_n = x_{2n}/x_n$.
(b) Let $z_0 = -1$ and $z_n = -bz_{n-1}$ for all $n \geq 1$. Give an explicit formula for z_n.
(c) Prove that $x_{m+2n} = y_n x_{m+n} + z_n x_m$ for all $m, n \geq 0$.
(d) Deduce that $F_{n+6} = 4F_{n+3} + F_n$ for all $n \geq 0$.

Appendix 0C. Finding roots of polynomials

In the remaining appendices of this preliminary chapter (chapter 0) we introduce several important themes in number theory that do not often appear in a first course but will be of interest to some readers. We also take some time to introduce some basic notions of algebra that appear (sometimes in disguise) throughout this and subsequent number theory courses. To begin with we discuss the famous question of techniques for factoring polynomials into their linear factors.

The reader knows that the roots of a quadratic polynomial $ax^2 + bx + c = 0$ are

$$\frac{-b \pm \sqrt{\Delta}}{2a}, \quad \text{where} \quad \Delta := b^2 - 4ac,$$

is called the *discriminant* of our polynomial, $ax^2 + bx + c$. The easy way to prove this is to put the equation into a form that is easy to solve: Divide through by a, to get $x^2 + (b/a)x + c/a = 0$, so that the leading coefficient is 1. Next make the change of variable, $y = x + b/2a$, to obtain

$$y^2 - \Delta/4a^2 = 0.$$

Having removed the y^1 term, we can simply take square roots to obtain the possibilities, $\pm\sqrt{\Delta}/2a$, for y, and hence we obtain the possible values for x (since $x = y - b/2a$). Can one similarly find the roots of a *cubic*?

0.8. Solving the general cubic

We can certainly begin solving cubics in the same way as we approached quadratics.

Exercise 0.8.1. Show that the roots of any given cubic polynomial, $Ax^3 + Bx^2 + Cx + D$ with $A \neq 0$, can be obtained from the roots of some cubic polynomial of the form $x^3 + ax + b$, by adding $B/3A$ to each root. Moreover write a and b explicitly as functions of A, B, and C.

We wish to find the roots of $x^3 + ax + b = 0$ for arbitrary a and b (which then allows us to determine the roots of an arbitrary cubic polynomial, by exercise 0.8.1). This does not look so easy since we cannot simply take cube roots unless $a = 0$. Cardano's trick (1545) is a little surprising: Write $x = u + v$ so that

$$x^3 + ax + b = (u+v)^3 + a(u+v) + b = (u^3 + v^3 + b) + (u+v)(3uv + a).$$

This equals 0 when

$$u^3 + v^3 = -b \quad \text{and} \quad 3uv = -a.$$

These conditions imply the simultaneous equations

$$u^3 + v^3 = -b \quad \text{and} \quad u^3 v^3 = -a^3/27,$$

so that, as a polynomial in X we have

$$(X - u^3)(X - v^3) = X^2 + bX - a^3/27.$$

Using the formula for the roots of a quadratic polynomial yields

(0.8.1) $$u^3, v^3 = \frac{-b \pm \sqrt{b^2 + 4a^3/27}}{2} = \frac{-b \pm \sqrt{\Delta/(-27)}}{2},$$

where $\Delta := -4a^3 - 27b^2$ is the discriminant of our polynomial, $x^3 + ax + b$. (The definition and some uses of discriminants are discussed in detail in section 2.11 of appendix 2B.)

All real numbers have a unique real cube root, call it t, and then the other cube roots are ωt and $\omega^2 t$, where ω is a cube root of 1; for instance we may take $\omega = e^{2i\pi/3} = \frac{-1+\sqrt{-3}}{2}$. Therefore if U and V are the real cube roots in (0.8.1), so that $-3UV$ is real and therefore equal to a, then the possible solutions to $u^3 + v^3 = -b$ together with $3uv = -a$ are

$$(u, v) = (U, V), \ (\omega U, \omega^2 V), \ \text{and} \ (\omega^2 U, \omega V).$$

This implies that the roots of $x^3 + ax + b$ are given by

$$U + V, \ \omega U + \omega^2 V, \ \text{and} \ \omega^2 U + \omega V.$$

The roots of a quadratic polynomial were obtained in terms of integers and square roots of integers. We have just seen that the roots of a cubic polynomial can be obtained in terms of integers, square roots, and finally cube roots. How about the roots of a quartic polynomial? Can these be found in terms of integers, square roots, cube roots, and fourth roots? And are there analogous expressions for the roots of quintics and higher degree polynomials?

0.9. Solving the general quartic

This is bound to be technically complicated, so much so that it is arguably more interesting to know that it can be done rather than actually doing it, so we just sketch the proofs:

We begin, as above, by rewriting the equation in the form $x^4 + ax^2 + bx + c = 0$. Following Ferrari (1550s) we add an extra variable y to obtain the equation

(0.9.1) $$(x^2 + a + y)^2 = (a + 2y)x^2 - bx + ((a+y)^2 - c)$$

and then select y so as to make the right-hand side the square of a linear polynomial $rx + s \in \mathbb{C}[x]$, in which case $(x^2 + a + y)^2 = (rx+s)^2$, so that x is a root of one of the quadratic polynomials

$$(x^2 + a + y) \pm (rx+s).$$

A quadratic polynomial is the square of a linear polynomial in x if and only if its discriminant equals 0. The right side of (0.9.1) has discriminant

$$b^2 - 4(a+2y)((a+y)^2 - c),$$

a cubic polynomial in y. We can find the roots, y, of this cubic polynomial by the method explained in the previous section. Given these roots, we can determine the possible values of r and s, and then we can solve for x to find the roots of the original equation.

Example. The roots of $X^4 + 4X^3 - 37X^2 - 100X + 300$. Letting $x = X+1$ yields $x^4 - 43x^2 - 18x + 360$. Proceeding as above leads to the cubic equation $2y^3 - 215y^2 + 6676y - 64108 = 0$. Dividing through by 2 and then changing variable $y = t + 215/6$ gives the cubic $t^3 - (6169/12)t - (482147/108) = 0$. This has discriminant $-4(6169/12)^3 + 27(482147/108)^2 = -(2310)^2$. Hence $u^3, v^3 = (482147 \pm 27720\sqrt{-3})/216$. Unusually this has an exact cube root in terms of $\sqrt{-3}$; that is, $u, v = \omega^*(-37 \pm 40\sqrt{-3})/6$, where ω^* denotes ω to some power. Now $-3(-37+40\sqrt{-3})/6 \cdot (-37-40\sqrt{-3})/6 = -6169/12 = a$. Therefore we can take $u, v = (-37 \pm 40\sqrt{-3})/6$, and the roots of our cubic are $t = u + v = -37/3$, $\omega u + \omega^2 v = 157/6$, $\omega^2 u + \omega v = -83/6$ so that $y = 47/2, 62, 22$. From these, Ferrari's equation becomes $(x^2 - 39/2) = \pm(2x + 9/2)$ for $y = 47/2$ and so the possible roots $-5, 3; -4, 6$; or $(x^2 + 19) = \pm(9x+1)$ for $y = 62$ and so the possible roots $-5, -4; 3, 6$; or $(x^2 - 21) = \pm(x+9)$ for $y = 22$ and so the possible roots $-5, 6; 3, -4$. For each such y we get the same roots $x = 3, -4, -5, 6$, yielding the roots $X = 2, -5, -6, 5$ of the original quartic.

Example. A fun example is to find the fifth roots of unity, other than 1. That is those x satisfying $\frac{x^5-1}{x-1} = x^4 + x^3 + x^2 + x + 1 = 0$. Proceeding as above we find the four roots

$$\frac{\sqrt{5} - 1 \pm \sqrt{-2\sqrt{5} - 10}}{4}, \quad \frac{-\sqrt{5} - 1 \pm \sqrt{2\sqrt{5} - 10}}{4}.$$

0.10. Surds

A *surd* is a square root or a cube root or a higher root, that is, an nth root for some number $n \geq 2$. We have shown above that the roots of degree 2, 3, and 4 polynomials can be determined by taking a combination of surds. We would like to show something similar for polynomials of degree 5 and higher, which is the focus of a course on Galois theory.

Gauss's favorite example of surds was the expression for $\cos \frac{2\pi}{2^k}$, which we denote by $c(k)$. A double angle formula states that $\cos 2\theta = 2\cos^2 \theta - 1$, and so taking $\theta = 2\pi/2^k$ we have

$$c(k-1) = 2c(k)^2 - 1,$$

which may be rewritten as $c(k) = \frac{1}{2}\sqrt{2 + 2c(k-1)}$. Note that $c(k) \geq 0$ for $k \geq 2$ and $c(2) = 0$. Hence

$$c(3) = \frac{1}{2}\sqrt{2}, \quad c(4) = \frac{1}{2}\sqrt{2 + \sqrt{2}}, \quad c(5) = \frac{1}{2}\sqrt{2 + \sqrt{2 + \sqrt{2}}}$$

and so we deduce by induction that

$$\cos\left(\frac{2\pi}{2^k}\right) = \frac{1}{2}\underbrace{\sqrt{2 + \sqrt{2 + \sqrt{2 + \sqrt{2 + \cdots + \sqrt{2}}}}}}_{k-2 \text{ times}} \quad \text{for each } k \geq 3.$$

Why does expressing the roots of polynomials in terms of surds seem like a good idea? Are the roots, given explicitly as in the second example above, any more enlightening than simply saying that one has a root of the original equation? We can give arbitrarily good approximations to the value of any given irrational (and rather rapidly using the right software), so what is really the advantage of expressing the roots of polynomials in terms of surds? The answer is more to do with our comfort with certain concepts, and aesthetics, than any intrinsic notion. In the rather sophisticated *Galois theory* there are identifiable differences between these different types of expressions, but such concepts are best left to a more advanced course.

One can learn much more about these beautiful classical themes by studying the first six chapters of [**Tig16**].

References discussing solvability of polynomials

[1] Raymond G. Ayoub, *On the nonsolvability of the general polynomial*, Amer. Math. Monthly **89** (1982), 397–401.

[2] Harold M. Edwards, *The construction of solvable polynomials*, Bull. Amer. Math. Soc. **46** (2009), no. 3, 397–411.

[3] Blair K. Spearman and Kenneth S. Williams, *Characterization of solvable quintics $x^5 + ax + b$*, Amer. Math. Monthly 101 (1994), 986–992.

Appendix 0D. What is a group?

Mathematical objects are often structured into *groups*. It is easiest to prove results for arbitrary groups, so that these results apply for all examples of groups that arise.[6] Many of the main theorems about groups were first proved in a number theory context and then found to apply elsewhere.

0.11. Examples and definitions

The main examples of groups that you have encountered so far are *additive groups* such as the integers, the rationals, the complex numbers, polynomials of a given degree, and matrices of given dimensions; and *multiplicative groups* such as the non-zero rationals, the non-zero complex numbers, and invertible square matrices of given dimension. (The integers mod p, a notion we will introduce in chapter 2, also give rise to both an additive and a multiplicative group.)

We will now give the definition of a group—keep in mind the objects named in the last paragraph and the usual operations of addition and multiplication:

A *group* is defined to be a set of objects G and an operation, call it $*$, such that:

(i) If $a, b \in G$, then $a * b \in G$. We say that G is *closed* under $*$.

(ii) If $a, b, c \in G$, then $(a * b) * c = a * (b * c)$; that is, when multiplying three elements of G together it does not matter which pair we multiply first. We say that G is *associative*.

(iii) There exists an element $e \in G$ such that for every $a \in G$ we have $a * e = e * a = a$. We call e the *identity element* of G for $*$. (For a group in which "$*$" is

---

[6]One can waste a lot of energy giving the same proof, with minor variations, in each situation where a group arises. Gauss wrote *Disquisitiones* before the abstract notion of a group was formulated and therefore *does* give very similar proofs in different places when dealing with different examples of groups.

much like addition we typically denote the identity by 0; for a group in which "$*$" is much like multiplication we typically denote the identity by 1.)

(iv) For every $a \in G$ there exists $b \in G$ such that $a*b = b*a = e$. We call b the *inverse* of a. (For a group in which "$*$" is much like addition we write $-a$ for the inverse of a; for a group in which "$*$" is much like multiplication we write a^{-1} for the inverse of a.)

One can check that the examples of groups given above satisfy these criteria. We have given examples of both finite and infinite groups. Notice that neither the integers nor the polynomials form multiplicative groups, because there is no inverse to 2 in the integers and no inverse to x amongst the polynomials.

There is one familiar property of numbers and polynomials that is not used in the definition of a group, and that is that $a*b = b*a$, that a and b *commute*. Although this often holds, there are some simple counterexamples, for instance most pairs of 2-by-2 matrices *do not commute*: For example,

$$\begin{pmatrix} 1 & 1 \\ 0 & 1 \end{pmatrix} \begin{pmatrix} 1 & 0 \\ -1 & 2 \end{pmatrix} = \begin{pmatrix} 0 & 2 \\ -1 & 2 \end{pmatrix} \quad \text{whereas} \quad \begin{pmatrix} 1 & 0 \\ -1 & 2 \end{pmatrix} \begin{pmatrix} 1 & 1 \\ 0 & 1 \end{pmatrix} = \begin{pmatrix} 1 & 1 \\ -1 & 1 \end{pmatrix}.$$

We develop the full theory for 2-by-2 matrices in the next section. If all pairs of elements of a group commute, then we call the group *commutative* or *abelian*. Typically we use multiplicative notation for groups that are non-commutative. It will be useful to develop a theory that works for non-commutative, as well as commutative, groups

A given group G can contain other, usually smaller, groups H, which are called *subgroups*. Every group G contains the subgroup given by the identity element, $\{0\}$ (the *trivial subgroup*), and also the subgroup G itself. It can also contain others; any subgroup other than G itself is a *proper subgroup*. For example the additive group of integers mod 6 with elements $\{0, 1, 2, 3, 4, 5\}$ contains the four subgroups

$$\{0\}, \ \{0, 3\}, \ \{0, 2, 4\}, \ \{0, 1, 2, 3, 4, 5\}.$$

The middle two are non-trivial, proper subgroups. Note that every group, and so subgroup, contains the identity element. Infinite groups can also contain subgroups; indeed

$$\mathbb{C} \supset \mathbb{R} \supset \mathbb{Q} \supset \mathbb{Z}.$$

Exercise 0.11.1. Prove that if G a subgroup of $\mathbb{Z}$ under addition, then either $G = \{0\}$ or $G = m\mathbb{Z} := \{mn : n \in \mathbb{Z}\}$ for some integer $m \geq 1$.

0.12. Matrices usually don't commute

Let $\mathcal{M}_2(\mathbb{C})$ be the set of 2-by-2 matrices with entries in $\mathbb{C}$, and then define

$$\text{Comm}(M) := \{A \in \mathcal{M}_2(\mathbb{C}) : AM = MA\}$$

for each $M \in \mathcal{M}_2(\mathbb{C})$. It is evident that if M is a multiple of the identity matrix I, then M commutes with all of $\mathcal{M}_2(\mathbb{C})$. Otherwise $\text{Comm}(M)$ forms a 2-dimensional subspace of (the 4-dimensional) $\mathcal{M}_2(\mathbb{C})$, as we now prove:

Proposition 0.12.1. *If M is not a multiple of the identity matrix, then*

$$\text{Comm}(M) = \{rI + sM : r, s \in \mathbb{C}\}.$$

0.12. Matrices usually don't commute

Exercise 0.12.1. Let M be an n-by-n matrix.
(a) Prove that if A and B commute with M, then so does $rA + sB$ for any complex numbers r and s. (We call $rA + sB$ a *linear combination* of A and B.)
(b) Prove that M^k commutes with M, for all k.
(c) Deduce that all linear combinations of $I, M, \ldots, M^{n-1}$ belong to $\text{Comm}(M)$.

Exercise 0.12.2. Let $M = \begin{pmatrix} a & b \\ c & d \end{pmatrix}$.
(a) Prove that M is not a multiple of I if and only if at least one of $a \neq d$, $b \neq 0$, $c \neq 0$ holds.
(b) Prove that if $a \neq d$, then for any matrix A there exists $r, s \in \mathbb{C}$ such that $A - rI - sM$ has zeros down the diagonal.
(c) Prove that if $b \neq 0$, then for any matrix A there exists $r, s \in \mathbb{C}$ such that $A - rI - sM$ has zeros throughout the top row.
(d) Prove that if $c \neq 0$, then for any matrix A there exists $r, s \in \mathbb{C}$ such that $A - rI - sM$ has zeros throughout the first column

Proof of Proposition 0.12.1. It is evident that I and M commute with M, and hence M commutes with any linear combination of I and M by exercise 0.12.1. We now show that these are the only matrices that commute with M.

Let $M = \begin{pmatrix} a & b \\ c & d \end{pmatrix}$. If $A \in \text{Comm}(M)$, then $B = A - rI - sM \in \text{Comm}(M)$ for any r and $s \in \mathbb{C}$, by exercise 0.12.1.

If $a \neq d$, then we select r and s as in exercise 0.12.2(a) so that $B = \begin{pmatrix} 0 & x \\ y & 0 \end{pmatrix}$ for some $x, y \in \mathbb{C}$. As $B \in \text{Comm}(M)$, we have

$$\begin{pmatrix} cx & dx \\ ay & by \end{pmatrix} = \begin{pmatrix} 0 & x \\ y & 0 \end{pmatrix}\begin{pmatrix} a & b \\ c & d \end{pmatrix} = BM = MB = \begin{pmatrix} a & b \\ c & d \end{pmatrix}\begin{pmatrix} 0 & x \\ y & 0 \end{pmatrix} = \begin{pmatrix} by & ax \\ dy & cx \end{pmatrix}.$$

Comparing the off-diagonal terms on the left- and right-hand ends of the equation forces $x = y = 0$ (as $a \neq d$), so that $B = 0$, and therefore $A = rI + sM$.

If $a = d$, $b \neq 0$, and $M \neq aI$, then B may be written in the form $B = \begin{pmatrix} 0 & 0 \\ x & y \end{pmatrix}$ for some $x, y \in \mathbb{C}$, by exercise 0.12.2(b), so that

$$\begin{pmatrix} 0 & 0 \\ ax + cy & bx + dy \end{pmatrix} = \begin{pmatrix} 0 & 0 \\ x & y \end{pmatrix}\begin{pmatrix} a & b \\ c & d \end{pmatrix} = BM = MB$$
$$= \begin{pmatrix} a & b \\ c & d \end{pmatrix}\begin{pmatrix} 0 & 0 \\ x & y \end{pmatrix} = \begin{pmatrix} bx & by \\ dx & dy \end{pmatrix}.$$

Comparing the terms in the top row on the left- and right-hand ends of the equation forces $x = y = 0$ (as $b \neq 0$), so that $B = 0$, and therefore $A = rI + sM$.

If $a = d$, $b = 0$, $c \neq 0$, and $M \neq aI$, then we may proceed analogously to the previous paragraph. Alternatively we may note that the result for M^T (the transpose of M) is given by the previous paragraph and then follows for M since $BM = MB$ if and only if $B^T M^T = M^T B^T$.

Finally if $a = d$ and $b = c = 0$, then M is a multiple of I. □

Appendix 0E. Rings and fields

In section 0.11 of appendix 0D we introduced the notion of a group and gave various examples. The real numbers are a set of objects that remarkably support two different groups: There is both the additive group and the multiplicative group acting on the non-zero real numbers, and this partly explains why they play such a fundamental role in mathematics. In this appendix, we formalize these notions and the key differences between the structure of the real numbers and of the integers. This will allow us to better identify the properties of many important types of numbers that arise in number theory.

0.13. Mixing addition and multiplication together: Rings and fields

A set of numbers A which, like the reals, has an additive group on A with identity element 0 and a commutative multiplicative group on $A \setminus \{0\}$ and for which the two groups interact according the *distributive* properties,

$$a \times (b+c) = (a \times b) + (a \times c) \text{ and } (a+b) \times c = (a \times c) + (b \times c),$$

is called a *field*. The reals $\mathbb{R}$ are an example, as are $\mathbb{C}$ and $\mathbb{Q}$. A field provides the most convenient situation in which to do arithmetic.

Exercise 0.13.1. Prove that $a \times 0 = 0$ for all $a \in A$, when A is a field.

However the integers, $\mathbb{Z}$, which are also vital to arithmetic, do not form a field: Although they form a group under addition, they do not form a group under multiplication, since not every integer has a multiplicative inverse within the integers (for example, the multiplicative inverse of 2 is 1/2 which is not an integer). But you can multiply integers together, and the integers possess a multiplicative identity, 1, so they have some of the properties of a multiplicative group, but not all. The integers are an example of a *ring*, which is a set of objects that form an additive group, are closed under multiplication, and have a multiplicative identity, 1, as well as satisfying the above distributive properties.[7] Thus $\mathbb{Z}$ is a commutative ring.

The set of even integers, $2\mathbb{Z}$, narrowly fails being a ring; it simply lacks the multiplicative identity. The polynomials with integer coefficients, $\mathbb{Z}[x]$, form a ring. Indeed if A is a commutative ring, then $A[x]$ is also a commutative ring.

For a given ring or field A and object α that is not in A, we are often interested in what type of mathematical object is created by adjoining α to A. This may be done in more than one way:

- $A[\alpha]$, which denotes polynomials in α with coefficients in A, that is, expressions of the form $a_0 + a_1\alpha + \cdots + a_d\alpha^d$ for any $d \geq 0$ where each $a_i \in A$.
- $A(\alpha)$, which denotes rational functions in α with coefficients in A, or, more simply, quotients u/v with $u, v \in A[\alpha]$ and $v \neq 0$.

For example we will prove in section 3.4 that $\sqrt{2}$ is irrational (that is, $\sqrt{2} \notin \mathbb{Q}$, and so $\sqrt{2} \notin \mathbb{Z}$), so we would like to understand the sets $\mathbb{Q}(\sqrt{2})$ and $\mathbb{Z}[\sqrt{2}]$.

Exercise 0.13.2. (a) Prove that $\mathbb{Z}[\sqrt{2}] = \{a + b\sqrt{2} : a, b \in \mathbb{Z}\}$ and that $\mathbb{Z}[\sqrt{2}]$ is a ring.
(b) Prove that $\mathbb{Q}(\sqrt{2}) = \mathbb{Q}[\sqrt{2}] = \{a + b\sqrt{2} : a, b \in \mathbb{Q}\}$ and that $\mathbb{Q}(\sqrt{2})$ is a field.

0.14. Algebraic numbers, integers, and units, I

If $f(x) = \sum_{j=0}^{d} f_j x^j$ where $f_d \neq 0$, then $f(x)$ has *degree* d and *leading coefficient* f_d. We say that $f(x)$ is *monic* if $f_d = 1$. If all of the coefficients, f_j, of $f(x)$ are integers, then we write $f(x) \in \mathbb{Z}[x]$. The *content* of $f(x) \in \mathbb{Z}[x]$ is the largest integer that divides all of its coefficients. If m is the content of $f(x)$, then $f(x) = mg(x)$ for some $g(x) \in \mathbb{Z}[x]$ of content 1. Obviously m divides every value $f(n)$ with $n \in \mathbb{Z}$ but there could be other integers that also have this property. For example, $f(x) = x^2 + x + 2$ has content 1, but 2 divides $f(n)$ for every integer n.

We call $\alpha \in \mathbb{C}$ an *algebraic number* if it is a root of a polynomial $f(x) \in \mathbb{Z}[x]$, with integer coefficients. If f is monic, then α is an *algebraic integer*. We call $f(x)$ the *minimal polynomial* for α if f is the polynomial with integer coefficients, of smallest degree, with positive leading coefficient and of content 1, for which $f(\alpha) = 0$. Minimal polynomials are irreducible in $\mathbb{Z}[x]$, for if $f(x) = g(x)h(x)$ with $g(x), h(x) \in \mathbb{Z}[x]$, then $g(\alpha)h(\alpha) = f(\alpha) = 0$ so that either $g(\alpha) = 0$ or $h(\alpha) = 0$; and therefore α is a root of a polynomial of lower degree than f, contradicting minimality.

Exercise 0.14.1. Let $f(x)$ be the minimal polynomial of an algebraic number α.
(a) Prove that if $g(x)$ is a polynomial with integer coefficients for which $g(\alpha) = 0$, then $f(x)$ divides $g(x)$. (You may use Proposition 2.10.1 of appendix 2B in your proof.)
(b) Prove that if $f(x)$ divides $g(x) \in \mathbb{Z}[x]$ and g is monic, then f is monic. Deduce that if $g(\alpha) = 0$, then α is an algebraic integer.
(c) Prove that if $g(\alpha) = 0$ and g is irreducible, then $g = \kappa f$ for some constant $\kappa \neq 0$.
(d) Prove that $f(x)$ is the only minimal polynomial of α.
(e) Prove that $(x - \alpha)^2$ does not divide $f(x)$.

---

[7]For the sake of comparison, a ring does not necessarily have two of the properties of a field: The numbers in a ring do not necessarily have a multiplicative inverse, and they do not necessarily commute when multiplying them together.

Exercise 0.14.2. Prove that if α is an algebraic number and a root of $f(x) \in \mathbb{Z}[x]$ where f has leading coefficient a, then $a\alpha$ is an algebraic integer.

Exercise 0.14.3. What are the algebraic integers in $\mathbb{Q}$?

Exercise 0.14.4.   (a) Prove that $\mathbb{Z}[\sqrt{d}]$ is a subset of the algebraic integers.
(b) Prove that $\mathbb{Z}[\sqrt{2}]$ is the set of algebraic integers in $\mathbb{Q}(\sqrt{2})$.
(c) Prove that $\frac{1+\sqrt{5}}{2}$ is an algebraic integer.

If α is an algebraic integer, then so is $m\alpha + n$ for any integers m, n, for if $f(x)$ is the minimal polynomial of α and has degree d, then $F(x) := m^d f(\frac{x-n}{m})$ is a monic polynomial in $\mathbb{Z}[x]$ with root $m\alpha + n$.

If α is a non-zero algebraic number, with minimal polynomial $f(x)$ of degree d, then $1/\alpha$ is a root of $x^d f(1/x)$.

Exercise 0.14.5.   (a) Prove that $1/\alpha$ has minimal polynomial $x^d f(1/x)$.
(b) Prove that α and $1/\alpha$ are both algebraic integers if and only if f is monic and $f(0) = 1$ or -1. In this case α and $1/\alpha$ are called *units*.

Another way to view this is that α is a unit if and only if α divides 1, for if $\beta = 1/\alpha$, then $\alpha\beta = 1$ and α and β are both algebraic integers.

Exercise 0.14.6. Suppose that α and β are algebraic integers such that α divides β, and β divides α. Prove that there exists a unit u for which $\beta = u\alpha$

In the section 0.17 of appendix 0F we will prove that if α and β are algebraic numbers, then so are $\alpha + \beta$ and $\alpha\beta$. Moreover if α and β are algebraic integers, then so are $\alpha + \beta$ and $\alpha\beta$.

Exercise 0.14.7.   (a) Prove that if α is an algebraic number, then $\mathbb{Q}(\alpha)$ is a field.
(b) Prove that if α is an algebraic integer, then $\mathbb{Z}[\alpha]$ is a ring.

Do there exist numbers α that are *irrational* (that is, that are not rational), so that, for instance, $\mathbb{Q}(\alpha)$ is not the same thing as $\mathbb{Q}$? To determine what numbers are irrational we should first classify, in a useful way, the rational numbers. The minimal polynomial of a rational number p/q with $(p, q) = 1$ is $qx - p \in \mathbb{Z}[x]$. Therefore one way to show that an algebraic number is irrational is to show that its minimal polynomial has degree > 1. Therefore given a polynomial, say $x^2 - 2$, we have to decide whether it is the minimal polynomial for some number, or perhaps prove that it is irreducible (so that it cannot have a rational root). We will develop number-theoretic tools to do this. Another way to find irrational numbers is to perhaps show that there are numbers that are not the roots of *any* polynomial in $\mathbb{Z}[x]$. Such numbers are not only irrational but are not even algebraic numbers and so are called *transcendental*. It is not too difficult to prove that transcendental numbers exist (by the "diagonalization argument" given in section 11.16 of appendix 11D), but it is rather more subtle to determine an actual transcendental number (though we will do so, using number-theoretic ideas in chapter 11).

For much much more on university level algebra, much of which stems from number theory, the reader might care to look at the excellent textbook [**DF04**] by Dummit and Foote or the more advanced but number theoretic [**IR90**].

Appendix 0F. Symmetric polynomials

It is difficult to work with algebraic numbers since one cannot necessarily evaluate them precisely. For example the golden ratio, $\frac{1+\sqrt{5}}{2}$, can easily be well-approximated, but how can you determine its precise value (since it is irrational)?

We can often avoid working with the actual algebraic numbers themselves, but rather work with the set of all of the roots of the minimal polynomial. For example, the formula (0.1.1) for the nth Fibonacci number involves both the golden ratio and the other root of its minimum polynomial $x^2 - x - 1$. It was Sir Isaac Newton who recognized that a function that is symmetric in all of the roots of a given polynomial is a rational number.

0.15. The theory of symmetric polynomials

We say that $P(x_1, x_2, \ldots, x_n)$ is a *symmetric polynomial* if
$$P(x_k, x_2, \ldots, x_{k-1}, x_1, x_{k+1}, \ldots, x_n) = P(x_1, x_2, \ldots, x_n) \text{ for each } k.$$
Here we swapped x_1 and x_k and kept everything else the same.

Exercise 0.15.1. Show that for any permutation σ of $1, 2, \ldots, n$ and any symmetric polynomial P we have $P(x_{\sigma(1)}, x_{\sigma(2)}, \ldots, x_{\sigma(n)}) = P(x_1, x_2, \ldots, x_n)$.

Theorem 0.3 (The fundamental theorem of symmetric polynomials). *For a given monic polynomial $f(x) = \sum_{i=0}^{d} a_i x^i$ with integer coefficients, each symmetric polynomial in the roots of f, with integer coefficients, can be expressed as a polynomial in the a_i with rational coefficients.*

Proof. Let $f(x) = \prod_{i=1}^{d}(x - \alpha_i)$. We begin by proving the claim for the
$$s_k := \sum_{i=1}^{d} \alpha_i^k \quad \text{for each} \quad k \geq 0.$$

Multiplying out $f(x) = \prod_{i=1}^{d}(x-\alpha_i)$ we have

$$\sum_i \alpha_i = -a_1, \quad \sum_{i<j}\alpha_i\alpha_j = a_2, \quad \sum_{i<j<k}\alpha_i\alpha_j\alpha_k = -a_3, \quad \ldots, \quad \alpha_1\alpha_2\ldots\alpha_n = \pm a_n.$$

Then, since $\frac{f'(x)}{f(x)} = \sum_{i=1}^{d}\frac{1}{x-\alpha_i}$, we have

$$\frac{\sum_{j=0}^{d}ja_jx^{d-j}}{\sum_{i=0}^{d}a_ix^{d-i}} = \frac{x^{d-1}}{x^{d-1}}\cdot\frac{f'(1/x)}{xf(1/x)} = \sum_{i=1}^{d}\frac{1}{1-\alpha_ix} = \sum_{i=1}^{d}\sum_{k\geq 0}(\alpha_ix)^k = \sum_{k\geq 0}s_kx^k.$$

This implies that

$$\sum_{j=0}^{d}(d-j)a_{d-j}x^j = \sum_{i=0}^{d}a_{d-i}x^i \cdot \sum_{k\geq 0}s_kx^k = \sum_{N\geq 0}\left(\sum_{\substack{i+k=N\\0\leq i\leq d}}a_{d-i}s_k\right)x^N.$$

Comparing the coefficients of x^N, we obtain (as $a_d = 1$)

$$s_N = -\sum_{i=1}^{\min\{d,N\}}a_{d-i}s_{N-i} + \begin{cases}(d-N)a_{d-N} & \text{if } N < d,\\ 0 & \text{if } N \geq d.\end{cases}$$

Hence, by induction on N, we see that the s_N are polynomials in the a_j.

We now sketch a proof of Newton's result for arbitrary symmetric polynomials, by showing that every symmetric polynomial in the roots of f, with integer coefficients, can be written as a polynomial in the s_k with integer coefficients. We proceed by induction on the number r of variables in the monomials of the symmetric polynomial; that is, we select the monomials $c\alpha_{i_1}^{k_1}\alpha_{i_2}^{k_2}\ldots\alpha_{i_r}^{k_r}$ in f, with each $k_i \geq 1$, for which r is maximal. In the $r = 1$ case, our polynomial is simply a linear combination of the s_k. Suppose that $r > 1$. If the k_i are distinct in such a monomial, we subtract $cs_{k_1}s_{k_2}\ldots s_{k_r}$ and we are left with various cross terms but, in all of which, two or more of the variables α_j are equal. If the k_i are not all distinct, then we subtract $s_{k_1}s_{k_2}\ldots s_{k_r}/\prod_i m_i!$ where $m_i := \#\{j : k_j = i\}$, to obtain various cross terms, with the same property. Hence, in the remaining expression, each monomial contains fewer variables and the result follows by induction. □

Exercise 0.15.2. If f is not monic, develop analogous results by working with $g(x)$ defined by $g(a_dx) = a_d^{d-1}f(x)$.

Example. Look at $\sum_{i,j,k}\alpha_i\alpha_j^2\alpha_k^3$. Subtract $s_1s_2s_3$ and we have to account for the cases where $i = j$ or $i = k$ or $j = k$. Hence what remains is

$$-\sum_{i,k}\alpha_i^3\alpha_k^3 - \sum_{i,j}\alpha_i^4\alpha_j^2 - \sum_{i,j}\alpha_i\alpha_j^5 + 2s_6,$$

where in the first sum we have $i = j$, in the second $i = k$, in the third $j = k$, and in the last $i = j = k$ (the coefficients being chosen by inclusion-exclusion). Proceeding the same way again we have

$$\sum_{i,j} \alpha_i^4 \alpha_j^2 = s_4 s_2 - s_6, \quad \sum_{i,j} \alpha_i \alpha_j^5 = s_1 s_5 - s_6, \quad \text{and} \quad \sum_{i,k} \alpha_i^3 \alpha_k^3 = (s_3^2 - s_6)/2,$$

the last since in s_3^2 the cross term $\alpha_i^3 \alpha_k^3$ appears also as $\alpha_k^3 \alpha_i^3$. Collecting this all together yields

$$\sum_{i,j,k} \alpha_i \alpha_j^2 \alpha_k^3 = s_1 s_2 s_3 - s_1 s_5 - s_2 s_4 - s_3^2/2 + 9 s_6/2.$$

Throughout these calculations, the sum of the indices in each term is 6, the degree of the original polynomial.

0.16. Some special symmetric polynomials

If α and β are the two roots of a monic quadratic polynomial with integer coefficients, then $x_n = (\alpha^n - \beta^n)/(\alpha - \beta)$ is a symmetric function in α and β and hence must be an integer by the fundamental theorem of symmetric polynomials. (We saw in exercise 0.1.4(c) that this is the nth term of the general second-order linear recurrence sequence that starts $0, 1$.)

If α is a root of an irreducible polynomial $f(x) = a \prod_{i=1}^{d} (x - \alpha_i)$, then there are two symmetric polynomials of particular interest:

The *trace* of α is $\alpha_1 + \alpha_2 + \cdots + \alpha_d$, the sum of the roots of f.

The *norm* of α is $\alpha_1 \alpha_2 \ldots \alpha_d$, the product of the roots of f.

By the fundamental theorem of symmetric polynomials, the trace and the norm of an irreducible polynomial are rational numbers.

Exercise 0.16.1. Show that if $f(t) = \prod_{i=1}^{k} (t - \alpha_i) \in \mathbb{Z}[t]$, then $\prod_{j=1}^{d} f'(\alpha_j)$ is an integer, by using the theory of symmetric polynomials.

Using the product rule we see that

$$f'(t) = a \sum_{j=1}^{k} \prod_{\substack{1 \leq i \leq k \\ i \neq j}} (t - \alpha_i) \quad \text{and so} \quad f'(\alpha_j) = a \prod_{\substack{1 \leq i \leq k \\ i \neq j}} (\alpha_j - \alpha_i).$$

We deduce that

$$\prod_{j=1}^{d} f'(\alpha_j) = a^d \prod_{1 \leq i < j \leq d} (-(\alpha_i - \alpha_j)^2).$$

This is a symmetric polynomial in the roots α_i, and so by Newton's fundamental theorem of symmetric polynomials it must be a rational number.

Let's evaluate this product for the quadratic polynomial $ax^2 + bx + c$. If this has roots α and β, then $ax^2 + bx + c = a(x - \alpha)(x - \beta)$ and so

$$a(\alpha + \beta) = -b \quad \text{and} \quad a\alpha\beta = c.$$

Therefore

$$a^2 (\alpha - \beta)^2 = (a(\alpha + \beta))^2 - 4a(a\alpha\beta) = b^2 - 4ac,$$

the discriminant of the polynomial, $ax^2 + bx + c$.

For the cubic polynomial $x^3 + ax + b = (x-\alpha)(x-\beta)(x-\gamma)$ we have
$$\alpha + \beta + \gamma = 0, \quad \alpha\beta + \alpha\gamma + \beta\gamma = a, \quad \text{and} \quad \alpha\beta\gamma = -b.$$
But then $\gamma = -(\alpha + \beta)$ so that $\alpha^2 + \alpha\beta + \beta^2 = -a$ and $\alpha\beta(\alpha+\beta) = b$. Therefore
$$(-1)^3((\alpha-\beta)(\gamma-\beta)(\alpha-\gamma))^2 = -((\alpha-\beta)(\alpha+2\beta)(2\alpha+\beta))^2 = 4a^3 + 27b^2,$$
which is the discriminant of the polynomial $x^3 + ax + b$.

A beautifully symmetric function is given by the Vandermonde matrix. The 3-by-3 version is
$$\begin{pmatrix} 1 & 1 & 1 \\ x & y & z \\ x^2 & y^2 & z^2 \end{pmatrix},$$
which has determinant $(x-y)(y-z)(z-x)$. This is not quite a symmetric function since swapping any two variables multiplies the determinant by -1. (This is also apparent when swapping any two columns of the matrix.) One intuitive way to see that $(x-y)(y-z)(z-x)$ is the determinant is by showing that each factor separately and together divides the determinant. To begin with, if $x = y$, then the determinant equals 0 as the first two columns are now equal, which implies that $x - y$ must be a factor of the determinant. Similarly $x - z$ and $z - y$ also divide the determinant. If x, y, and z are variables, then these expressions do not have any common factors, and so their product divides the determinant. This product has degree three (adding the degrees of the variables), as does the determinant, so they can differ by at most a constant factor. The constant factor can be determined by checking the coefficient of a particular monomial on both sides. For example $x^0 y^1 z^2$ only arises in the determinant from multiplying out the terms of the main diagonal and therefore has coefficient 1, and one can equally look for how this monomial arises in the product.[8]

Exercise 0.16.2. Use the same argument to explain that the determinant of *Vandermonde matrix* V, where $V_{i,j} = \alpha_i^{j-1}$, $1 \leq i, j \leq d$, is $\prod_{1 \leq i < j \leq d}(\alpha_i - \alpha_j)$.

Exercise 0.16.3. Prove Theorem 0.2 when each $e_j = 1$ (assuming exercise 0.16.2).

Now
$$(V^T V)_{i,k} = \sum_{j=1}^{d}(V^T)_{i,j} V_{j,k} = \sum_{j=1}^{d} V_{j,i} V_{j,k} = \sum_{j=1}^{d} \alpha_j^{i-1} \alpha_j^{k-1} = s_{i+k-2}$$
for $1 \leq i, k \leq d$. Hence $(\det V)^2$ is the determinant of the matrix with (i,k)th entry s_{i+k-2}.

Exercise 0.16.4.[‡] (This question requires some knowledge of linear algebra.) Suppose that M is an n-by-n matrix.
  (a) Prove that if M is a diagonal matrix in which all the diagonal entries are distinct, then Comm(M) equals the set of diagonal matrices.
  (b) Use exercise 0.16.2 to show that the set of diagonal matrices is then given by $\{a_0 I + a_1 M + \cdots + a_{n-1} M^{n-1} : \text{each } a_j \in \mathbb{C}\}$.

---

[8]Diehard algebraists might be uncomfortable with this discussion since we ignore ideals that arise from the gcds of the polynomial factors, but these details can all be justified.

(c) Now let M, N, and T be n-by-n matrices with T invertible. Prove that M and N commute if and only if $T^{-1}MT$ and $T^{-1}NT$ commute.

(d) Prove that if M is an n-by-n matrix with n distinct eigenvalues, then $\operatorname{Comm}(M) = \{a_0 I + a_1 M + \cdots + a_{n-1} M^{n-1} : \text{each } a_j \in \mathbb{C}\}$.

0.17. Algebraic numbers, integers, and units, II

We are now in a position to prove some of the claims of section 0.14 of appendix 0E. Suppose that α and β are algebraic integers with minimal polynomials f and g. Then
$$\prod_{\substack{u:\ f(u)=0 \\ v:\ g(v)=0}} (x - (u+v)) = \prod_{u:\ f(u)=0} g(x - u),$$
by exercise 0.14.1(d). This is a symmetric polynomial in the roots u of f and so, by the fundamental theorem of symmetric polynomials, this is a monic polynomial with integer coefficients having root $\alpha + \beta$, and so $\alpha + \beta$ is an algebraic integer.

Exercise 0.17.1. (a)† Prove that if $\alpha \neq 0$ and β are algebraic integers, then $\alpha\beta$ is also an algebraic integer.
(b) Prove that if $\alpha \neq 0$ and β are algebraic numbers, then $\alpha + \beta$ and $\alpha\beta$ are algebraic numbers.

Exercise 0.17.2. Prove that if $\alpha_1, \ldots, \alpha_k$ are algebraic numbers, then $\mathbb{Q}(\alpha_1, \ldots, \alpha_k)$ is a field. These are the *number fields*.

Let $\overline{\mathbb{Q}}$ denote the set of all algebraic numbers. Evidently if K is any number field, then $K \subset \overline{\mathbb{Q}}$. It is not difficult to prove that $\overline{\mathbb{Q}}$ is itself a field. Similarly if A is the set of all algebraic integers, then A is a ring and the algebraic integers inside a given number field K form a subring, which is precisely $K \cap A$. However identifying the elements of $K \cap A$ explicitly can be rather more challenging, as we saw in exercise 0.14.4.

Rather more interestingly, the roots of any polynomial with coefficients in $\overline{\mathbb{Q}}$ all belong to $\overline{\mathbb{Q}}$.

Proposition 0.17.1. *Suppose that $f(x) \in \overline{\mathbb{Q}}[x]$ and that $f(\rho) = 0$. Then $\rho \in \overline{\mathbb{Q}}$. We say that $\overline{\mathbb{Q}}$ is* algebraically closed.

Proof. Suppose that $f(x) = a_0 + a_1 x + \cdots + a_d x^d$ so that each a_j is an algebraic number. Suppose that a_j has minimal polynomial $g_j(x)$; and let A_j be the set of roots of $g_j(x)$. Then $f(x)$ divides the polynomial
$$F(x) := \prod_{\alpha_0 \in A_0} \prod_{\alpha_1 \in A_1} \cdots \prod_{\alpha_d \in A_d} (\alpha_0 + \alpha_1 x + \cdots + \alpha_d x^d)$$
which is a symmetric polynomial in the elements of each A_j with $0 \leq j \leq d$ and therefore belongs to $\mathbb{Q}[x]$ by the law of symmetric polynomials. Any root of $f(x)$ is a root of $F(x)$ and therefore must be an algebraic number. □

For further development of these ideas see chapter 8 of [**Tig16**].

Appendix 0G. Constructibility

0.18. Constructible using only compass and ruler

The ancient Greeks were interested in what could be constructed using only a straight edge (sometimes called an "unmarked ruler", or just plain "ruler") and a compass. Three questions stumped them:

Quadrature of the circle:

Draw a square that has area equal to that of a given circle.

To draw a square whose area is π (the same area as a circle of radius 1), we need to be able to draw a square with sides of length x, where
$$x \text{ is a root of the equation } x^2 - \pi.$$

Duplication of the cube:

Construct a cube that has twice the volume of a given cube.

If the original cube has side length 1 (and so volume 1), we would need to be able to construct a cube with sides of length x, where
$$x \text{ is a root of the equation } x^3 - 2.$$

Trisection of the angle:

Construct an angle which is one third the size of a given angle.

Constructing an angle θ is as difficult as constructing a right-angled triangle containing that angle, that is, the triangle with side lengths $\sin\theta$, $\cos\theta$, 1. Therefore if we start with angle 3θ and wish to determine the angle θ, then we will need to be able to determine $\cos\theta$ from $\cos 3\theta$ and $\sin 3\theta$. But these are linked by the formula $\cos 3\theta = 4\cos^3\theta - 3\cos\theta$; that is, we need to find the root $x = 2\cos\theta$ of $x^3 - 3x - A$ where $A = 2\cos 3\theta$. For example, if $\theta = \pi/9$, we will need to be able to construct a right-angled triangle with a side of length $x/2$ where
$$x \text{ is a root of the equation } x^3 - 3x - 1.$$

0.18. Constructible using only compass and ruler

We need to understand the algebra of points that are constructed from given points and lengths by "ruler and compass". Our tools are:

- *An unmarked ruler*, which allows us to draw the line between any two given points and to extend that line as far as we like.
- *A compass*, which allows us to draw the circle centered on one given point, of radius a given length, or the distance between two given points.

Proposition 0.18.1. *Given a set of points on lines and a set of lengths, any new points that can be constructed from these, using only ruler and compass, will have coordinates that can be determined as the roots of degree-one or degree-two polynomials, whose coefficients are rational functions of the already known coordinates.*

Proof. The lines are defined by pairs of points: Given the points $A = (a_1, a_2)$ and $B = (b_1, b_2)$ the line between them is $(b_1 - a_1)(y - a_2) = (b_2 - a_2)(x - a_1)$.

Exercise 0.18.1. Show that the coefficients of the equation of this line can be determined by a degree-one equation in already known coordinates.

Exercise 0.18.2. Prove that any two (non-parallel) lines intersect in a point that can be determined by a degree-one equation in the coefficients of the equations of the lines.

Given a length r and a point $C = (c_1, c_2)$, we can draw the circle $(x - c_1)^2 + (y - c_2)^2 = r^2$ centered at C of radius r.

Exercise 0.18.3. Prove that the points of intersection of this circle with a given line can be given by a degree-two equation in already known coordinates.

Exercise 0.18.4. Prove that the points of intersection of two circles can be given by a degree-two equation in already known coordinates.

Combining all these exercises implies Proposition 0.18.1. □

We sketch here how one uses Proposition 0.18.1 to show that the Greeks were stumped by their three questions for good reason—none of the three were possible. Proposition 0.18.1 implies that we can draw a square that has area equal to that of a given circle if and only if π can be obtained in terms of a (finite) succession of roots of linear or quadratic polynomials whose coefficients are already constructed. If this can be done, then π would be the root of some polynomial (perhaps of high degree); in other words π would be an algebraic number. However Lindemann proved, in 1882, that π is transcendental (as we will discuss in more detail in section 11.17 of appendix 11D).

If α is obtained from a (finite) succession of roots of linear or quadratic polynomials whose coefficients are already constructed, then α is not only an algebraic number but one can show that its minimal polynomial has degree which is a power of 2. Both $x^3 - 2$ and $x^3 - 3x - 1$ are irreducible (which can be shown using Theorem 3.4; see exercise 3.4.4), and so these are the minimum polynomials for their roots (by exercise 0.14.1(c)). Therefore one cannot duplicate the cube, nor trisect the angle $\pi/3$, since the roots of these irreducible polynomials of degree three do not have minimum polynomials that have degrees that are a power of 2.

For further development of these ideas see section 13.3 of [**DF04**] or section 9.11 of [**IR90**].

Chapter 1

The Euclidean algorithm

1.1. Finding the gcd

Most readers will know the Euclidean algorithm, used to find the greatest common divisor (gcd) of two given integers. For example, to determine the greatest common divisor of 85 and 48, we begin by subtracting the smaller from the larger, 48 from 85, to obtain $85 - 48 = 37$. Now $\gcd(85, 48) = \gcd(48, 37)$, because the common divisors of 48 and 37 are precisely the same as those of 85 and 48, and so we apply the algorithm again to the pair 48 and 37. So we subtract the smaller from the larger to obtain $48 - 37 = 11$, so that $\gcd(48, 37) = \gcd(37, 11)$. Next we should subtract 11 from 37, but then we would only do so again, and a third time, so let's do all that in one go and take $37 - 3 \times 11 = 4$, to obtain $\gcd(37, 11) = \gcd(11, 4)$. Similarly we take $11 - 2 \times 4 = 3$, and then $4 - 3 = 1$, so that the gcd of 85 and 48 is 1. This is the Euclidean algorithm that you might already have seen,[1] but did you ever prove that it really works?

To do so, we will first carefully define terms that we have implicitly used in the above paragraph, perhaps mathematical terms that you have used for years (such as "divides", "quotient", and "remainder") without a formal definition. This may seem pedantic but the goal is to make sure that the rules of basic arithmetic are really established on a sound footing.

Let a and b be given integers. We say that *a is divisible by b*, or that *b divides a*,[2] if there exists an integer q such that $a = qb$. For convenience we write "$b \mid a$".[3,4] We now set an exercise for the reader to check that the definition allows one to manipulate the notion of division in several familiar ways.

Exercise 1.1.1. In this question, and throughout, we assume that a, b, and c are integers.
  (a) Prove that if b divides a, then either $a = 0$ or $|a| \geq |b|$.

---

[1] There will be a formal discussion of the Euclidean algorithm in appendix 1A.
[2] One can also say *a is a multiple of b* or *b is a divisor of a* or *b is a factor of a*.
[3] And if b does not divide a, we write "$b \nmid a$".
[4] One reason for giving a precise mathematical definition for division is that it allows us to better decide how to interpret questions like, "What is 1 divided by 0?" or "What is 0 divided by 0?"

(b) Deduce that if $a|b$ and $b|a$, then $b = a$ or $b = -a$ (which, in future, we will write as "$b = \pm a$").
(c) Prove that if a divides b and c, then a divides $bx + cy$ for all integers x, y.
(d) Prove that a divides b if and only if a divides $-b$ if and only if $-a$ divides b.
(e) Prove that if a divides b, and b divides c, then a divides c.
(f) Prove that if $a \neq 0$ and ac divides ab, then c divides b.

Next we formalize the notion of "dividing with remainder".

Lemma 1.1.1. *If a and b are integers, with $b \geq 1$, then there exist unique integers q and r, with $0 \leq r \leq b - 1$, such that $a = qb + r$. We call q the "quotient", and r the "remainder".*

Proof by induction. We begin by proving the existence of q and r. For each $b \geq 1$, we proceed by induction on $a \geq 0$. If $0 \leq a \leq b - 1$, then the result follows with $q = 0$ and $r = a$. Otherwise assume that the result holds for $0, 1, 2, \ldots, a - 1$, where $a \geq b$. Then $a - 1 \geq a - b \geq 0$ so, by the induction hypothesis, there exist integers Q and r, with $0 \leq r \leq b - 1$, for which $a - b = Qb + r$. Therefore $a = qb + r$ with $q = Q + 1$.

If $a < 0$, then $-a > 0$ so we have $-a = Qb + R$, for some integers Q and R, with $0 \leq R \leq b - 1$, by the previous paragraph. If $R = 0$, then $a = qb$ where $q = -Q$ (and $r = 0$). Otherwise $1 \leq R \leq b - 1$ and so $a = qb + r$ with $q = -Q - 1$ and $1 \leq r = b - R \leq b - 1$, as required.

Now we show that q and r are unique. If $a = qb + r = Qb + R$, then b divides $(q - Q)b = R - r$. However $0 \leq r, R \leq b - 1$ so that $|R - r| \leq b - 1$, and $b \mid R - r$. Therefore $R - r = 0$ by exercise 1.1.1(a), and so $Q - q = 0$. In other words $q = Q$ and $r = R$; that is, the pair q, r is unique. □

An easier, but less intuitive, proof. We can add a multiple of b to a to get a positive integer. That is, there exists an integer n such that $a + nb \geq 0$; any integer $n \geq -a/b$ will do. We now subtract multiples of b from this number, as long as it remains positive, until subtracting b once more would make it negative. In other words we now have an integer $a - qb \geq 0$, which we denote by r, such that $r - b < 0$; in other words $0 \leq r \leq b - 1$. □

Exercise 1.1.2. Suppose that $a \geq 1$ and $b \geq 2$ are integers. Show that we can write a in base b; that is, show that there exist integers $a_0, a_1, \ldots \in [0, b-1]$ for which $a = a_d b^d + a_{d-1} b^{d-1} + a_1 b + a_0$.

We say that d is a *common divisor* of integers a and b if d divides both a and b. We are mostly interested in the *greatest common divisor* of a and b, which we denote by $\gcd(a, b)$, or more simply as (a, b).[5,6]

We say that a is *coprime* with b, or that a and b are *coprime integers*, or that a and b are *relatively prime*, if $(a, b) = 1$.

---

[5] In the UK this is known as the *highest common factor* of a and b and is written $\text{hcf}(a, b)$.

[6] When $a = b = 0$, every integer is a divisor of 0, so there is no greatest divisor, and therefore $\gcd(0, 0)$ is undefined. There are often one or two cases in which a generally useful mathematical definition does not give a unique value. Another example is 0 divided by 0, which we explore in exercise 1.7.1. For aesthetic reasons, some authors choose to assign a value which is consistent with the theory in one situation but perhaps not in another. This can lead to artificial inconsistencies which is why we choose to leave such function-values undefined.

1.2. Linear combinations

Corollary 1.1.1. *If $a = qb + r$ where a, b, q, and r are integers, then*
$$\gcd(a, b) = \gcd(b, r).$$

Proof. Let $g = \gcd(a, b)$ and $h = \gcd(r, b)$. Now g divides both a and b, so g divides $a - qb = r$ (by exercise 1.1.1(c)). Therefore g is a common divisor of both r and b, and therefore $g \leq h$. Similarly h divides both b and r, so h divides $qb + r = a$ and hence h is a common divisor of both a and b, and therefore $h \leq g$. We have shown that $g \leq h$ and $h \leq g$, which together imply that $g = h$. □

Corollary 1.1.1 justifies the method used to determine the gcd of 85 and 48 in the first paragraph of section 1.1 and indeed in general:

Exercise 1.1.3. Use Corollary 1.1.1 to prove that the Euclidean algorithm indeed yields the greatest common divisor of two given integers. (You might prove this by induction on the smallest of the two integers.)

Exercise 1.1.4. Prove that $(F_n, F_{n+1}) = 1$ by induction on $n \geq 0$.

1.2. Linear combinations

The Euclidean algorithm can also be used to determine a linear combination[7] of a and b, over the integers, which equals $\gcd(a, b)$; that is, one can always use the Euclidean algorithm to find integers u and v such that

(1.2.1) $$au + bv = \gcd(a, b).$$

Let us see how to do this in an example, by finding integers u and v such that $85u + 48v = 1$; remember that we found the gcd of 85 and 48 at the beginning of section 1.1. We retrace the steps of the Euclidean algorithm, but in reverse: The final step was that $1 = 1 \cdot 4 - 1 \cdot 3$, a linear combination of 4 and 3. The second to last step used that $3 = 11 - 2 \cdot 4$, and so substituting $11 - 2 \cdot 4$ for 3 in $1 = 1 \cdot 4 - 1 \cdot 3$, we obtain
$$1 = 1 \cdot 4 - 1 \cdot 3 = 1 \cdot 4 - 1 \cdot (11 - 2 \cdot 4) = 3 \cdot 4 - 1 \cdot 11,$$
a linear combination of 11 and 4. This then implies, since we had $4 = 37 - 3 \cdot 11$, that
$$1 = 3 \cdot (37 - 3 \cdot 11) - 1 \cdot 11 = 3 \cdot 37 - 10 \cdot 11,$$
a linear combination of 37 and 11. Continuing in this way, we successively deduce, using that $11 = 48 - 37$ and then that $37 = 85 - 48$,
$$1 = 3 \cdot 37 - 10 \cdot (48 - 37) = 13 \cdot 37 - 10 \cdot 48$$
$$= 13 \cdot (85 - 48) - 10 \cdot 48 = 13 \cdot 85 - 23 \cdot 48;$$
that is, we have the desired linear combination of 85 and 48.

To prove that this method always works, we will use Lemma 1.1.1 again: Suppose that $a = qb + r$ so that $\gcd(a, b) = \gcd(b, r)$ by Corollary 1.1.1, and that we have $bu - rv = \gcd(b, r)$ for some integers u and v. Then

(1.2.2) $\gcd(a, b) = \gcd(b, r) = bu - rv = bu - (a - qb)v = b(u + qv) - av,$

---

[7] A *linear combination* of two given integers a and b, over the integers, is a number of the form $ax + by$ where x and y are integers. This can be generalized to yield a linear combination $a_1 x_1 + \cdots + a_n x_n$ of any finite set of integers, $a_1, \ldots, a_n$. Linear combinations are a key concept in linear algebra and appear (without necessarily being called that) in many courses.

the desired linear combination of a and b. This argument forms the basis of our proof of (1.2.1), but to give a complete proof we proceed by induction on the smaller of a and b:

> **Theorem 1.1.** *If a and b are positive integers, then there exist integers u and v such that*
> $$au + bv = \gcd(a, b).$$

Proof. Interchanging a and b if necessary we may assume that $a \geq b \geq 1$. We shall prove the result by induction on b. If $b = 1$, then b only has the divisor 1, so that
$$\gcd(a, 1) = 1 = 0 \cdot a + 1 \cdot 1.$$
We now prove the result for $b > 1$: If b divides a, then
$$\gcd(b, a) = b = 0 \cdot a + 1 \cdot b.$$
Otherwise b does not divide a and so Lemma 1.1.1 implies that there exist integers q and r such that $a = qb + r$ and $1 \leq r \leq b - 1$. Since $1 \leq r < b$ we know, by the induction hypothesis, that there exist integers u and v for which $bu - rv = \gcd(b, r)$. The result then follows by (1.2.2). □

We now establish various useful properties of the gcd:

Exercise 1.2.1. (a) Prove that if d divides both a and b, then d divides $\gcd(a, b)$.
(b) Deduce that d divides both a and b if and only if d divides $\gcd(a, b)$.
(c) Prove that $1 \leq \gcd(a, b) \leq |a|$ and $|b|$.
(d) Prove that $\gcd(a, b) = |a|$ if and only if a divides b.

Exercise 1.2.2. Suppose that a divides m, and b divides n.
(a) Deduce that $\gcd(a, b)$ divides $\gcd(m, n)$.
(b) Deduce that if $\gcd(m, n) = 1$, then $\gcd(a, b) = 1$.

Exercise 1.2.3. Show that Theorem 1.1 holds for any integers a and b that are not both 0. (It is currently stated and proved only for positive integers a and b.)

Corollary 1.2.1. *If a and b are integers for which $\gcd(a, b) = 1$, then there exist integers u and v such that*
$$au + bv = 1.$$

This is one of the most useful results in mathematics and has applications in many areas, including in safeguarding today's global communications. For example, we will see in section 10.3 that to implement RSA, a key cryptographic protocol that helps keep important messages safe in our electronic world, one uses Corollary 1.2.1 in an essential way. More on that later, after developing more basic number theory.

Exercise 1.2.4. (a) Use exercise 1.1.1(c) to show that if $au + bv = 1$, then $(a, b) = (u, v) = 1$.
(b) Prove that $\gcd(u, v) = 1$ in Theorem 1.1.

Corollary 1.2.2. *If $\gcd(a, m) = \gcd(b, m) = 1$, then $\gcd(ab, m) = 1$.*

Proof. By Theorem 1.1 there exist integers r, s, u, v such that
$$au + mv = br + ms = 1.$$

1.3. The set of linear combinations of two integers

Therefore
$$ab(ur) + m(bvr + aus + msv) = (au + mv)(br + ms) = 1,$$
and the result follows from exercise 1.2.4(a). □

Corollary 1.2.3. *We have* $\gcd(ma, mb) = m \cdot \gcd(a, b)$ *for all integers* $m \geq 1$.

Proof. By Theorem 1.1 there exist integers u, v such that $au + bv = \gcd(a, b)$. Now $\gcd(ma, mb)$ divides ma and mb so it divides $mau + mbv = m \cdot \gcd(a, b)$. Similarly $\gcd(a, b)$ divides a and b, so that $m \cdot \gcd(a, b)$ divides ma and mb, and therefore $\gcd(ma, mb)$ by exercise 1.2.1(a). The result follows from exercise 1.1.1(b), since the gcd is always positive. □

Exercise 1.2.5. (a) Show that if A and B are given integers, not both 0, with $g = \gcd(A, B)$, then $\gcd(A/g, B/g) = 1$.
(b) Prove that any rational number u/v where $u, v \in \mathbb{Z}$ with $v \neq 0$ may be written as r/s where r and s are coprime integers with $s > 0$. This is called a *reduced fraction*.

1.3. The set of linear combinations of two integers

Theorem 1.1 states that the greatest common divisor of two integers is a linear combination of those two integers. This suggests that it might be useful to study *the set of linear combinations*
$$I(a, b) := \{am + bn : m, n \in \mathbb{Z}\}$$
of two given integers a and b.[8] We see that $I(a, b)$ contains 0, a, b, $a + b$, $a + 2b$, $2b + a$, $a - b$, $b - a$, ... and any sum of integer multiples of a and b, so that $I(a, b)$ is closed under addition. Let $I(a) := I(a, 0) = \{am : m \in \mathbb{Z}\}$, the set of integer multiples of a. We now prove that $I(a, b)$ can be described as the set of integer multiples of $\gcd(a, b)$, a set which is easier to understand:

Corollary 1.3.1. *For any given non-zero integers a and b, we have*
$$\{am + bn : m, n \in \mathbb{Z}\} = \{gk : k \in \mathbb{Z}\}$$
where $g := \gcd(a, b)$; *that is,* $I(a, b) = I(g)$. *In other words, there exist integers m and n with* $am + bn = c$ *if and only if* $\gcd(a, b)$ *divides* c.

Proof. By Theorem 1.1 we know that there exist $u, v \in \mathbb{Z}$ for which $au + bv = g$. Therefore $a(uk) + b(vk) = gk$ so that $gk \in I(a, b)$ for all $k \in \mathbb{Z}$; that is, $I(g) \subset I(a, b)$. On the other hand, as g divides both a and b, there exist integers A, B such that $a = gA$, $b = gB$, and so any $am + bn = g(Am + Bn) \in I(g)$. That is, $I(a, b) \subset I(g)$. The result now follows from the two inclusions. □

It is instructive to see how this result follows directly from the Euclidean algorithm: In our example, we are interested in $\gcd(85, 48)$, so we will study $I(85, 48)$, that is, the set of integers of the form
$$85m + 48n.$$

---
[8]This is usually called the *ideal* generated by a and b in $\mathbb{Z}$ and denoted by $\langle a, b \rangle_\mathbb{Z}$. The notion of an ideal is one of the basic tools of modern algebra, as we will discuss in appendix 3D.

The first step in the Euclidean algorithm was to write $85 = 1 \cdot 48 + 37$. Substituting this in above yields
$$85m + 48n = (1 \cdot 48 + 37)m + 48n = 48(m+n) + 37m,$$
and so $I(85, 48) \subset I(48, 37)$. In the other direction, any integer in $I(48, 37)$ can be written as
$$48a + 37b = 48a + (85 - 48)b = 85b + 48(a - b),$$
and so belongs to $I(85, 48)$. Combining these last two statements yields that
$$I(85, 48) = I(48, 37).$$
Each step of the Euclidean algorithm leads to a similar equality, and so we get
$$I(85, 48) = I(48, 37) = I(37, 11) = I(11, 4) = I(4, 3) = I(3, 1) = I(1, 0) = I(1).$$
To truly justify this we need to establish an analogous result to Corollary 1.1.1:

Lemma 1.3.1. *If $a = qb + r$ where a, b, q, and r are integers, then $I(a, b) = I(b, r)$.*

Proof. We begin by noting that
$$am + bn = (qb + r)m + bn = b(qm + n) + rm$$
so that $I(a, b) \subset I(b, r)$. In the other direction
$$bu + rv = bu + (a - qb)v = av + b(u - qv)$$
so that $I(b, r) \subset I(a, b)$. The result follows by combining the two inclusions. $\square$

We have used the Euclidean algorithm to find the gcd of any two given integers a and b, as well as to determine integers u and v for which $au + bv = \gcd(a, b)$. The price for obtaining the actual values of u and v, rather than merely proving the existence of u and v (which is all that was claimed in Theorem 1.1), was our somewhat complicated analysis of the Euclidean algorithm. However, if we *only wish to prove* that such integers u and v exist, then we can do so with a somewhat easier proof: [9]

Non-constructive proof of Theorem 1.1. Let h be the smallest positive integer that belongs to $I(a, b)$, say $h = au + bv$. Then $g := \gcd(a, b)$ divides h, as g divides both a and b.

Now $a = a \cdot 1 + b \cdot 0$ so that $a \in I(a, b)$, and $1 \leq h \leq a$ by the definition of h. Therefore Lemma 1.1.1 implies that there exist integers q and r, with $0 \leq r \leq h-1$, for which $a = qh + r$. Therefore
$$r = a - qh = a - q(au + bv) = a(1 - qu) + b(-qv) \in I(a, b),$$
which contradicts the minimality of h, unless $r = 0$; that is, h divides a. An analogous argument reveals that h divides b, and so h divides g by exercise 1.2.1(a).

---

[9] We will now prove the *existence* of u and v by showing that their non-existence would lead to a contradiction. We will develop other instances, as we proceed, of both constructive and non-constructive proofs of important theorems.

Which type of proof is preferable? This is somewhat a matter of taste. The non-constructive proof is often shorter and more elegant. The constructive proof, on the other hand, is practical—that is, it *gives* solutions. It is also "richer" in that it develops more than is (immediately) needed, though some might say that these extras are irrelevant.

Which type of proof has the greatest clarity? That depends on the *algorithm* devised for the constructive proof. A compact algorithm will often cast light on the subject. But a cumbersome one may obscure it. In this case, the Euclidean algorithm is remarkably simple and efficient ([**Sha85**], p. 11).

Hence g divides h, and h divides g, and g and h are both positive, so that $g = h$ as desired. $\square$

We say that the integers a, b, and c are *relatively prime* if $\gcd(a,b,c) = 1$. We say that they are *pairwise coprime* if $\gcd(a,b) = \gcd(a,c) = \gcd(b,c) = 1$. For example, $6, 10, 15$ are relatively prime, but they are not pairwise coprime (since each pair of integers has a common factor > 1).

Exercise 1.3.1. Suppose that a, b, and c are non-zero integers for which $a + b = c$.
(a) Show that a, b, c are relatively prime if and only if they are pairwise coprime.
(b) Show that $(a,b) = (a,c) = (b,c)$.
(c) Show that the analogy to (a) is false for integer solutions a, b, c, d to $a + b = c + d$ (perhaps by constructing a counterexample).

1.4. The least common multiple

The *least common multiple*[10] of two given integers a and b is defined to be the smallest positive integer that is a multiple of both a and b. We denote this by $\mathrm{lcm}[a, b]$ (or simply $[a, b]$). We now prove the counterpart to exercise 1.2.1(a):

Lemma 1.4.1. $\mathrm{lcm}[a,b]$ *divides integer* m *if and only if* a *and* b *both divide* m.

Proof. Since a and b divide $\mathrm{lcm}[a,b]$, if $\mathrm{lcm}[a,b]$ divides m, then a and b both divide m, by exercise 1.1.1(e).

On the other hand suppose a and b both divide m, and write $m = q \, \mathrm{lcm}[a,b] + r$ where $0 \leq r < \mathrm{lcm}[a,b]$. Now a and b both divide m and $\mathrm{lcm}[a,b]$ so they both divide $m - q \, \mathrm{lcm}[a,b] = r$. However $\mathrm{lcm}[a,b]$ is defined to be the smallest positive integer that is divisible by both a and b, which implies that r must be 0. Therefore $\mathrm{lcm}[a,b]$ divides m. $\square$

The analogies to exercise 1.2.1(d) and Corollary 1.2.3 for lcms are given by the following two exercises:

Exercise 1.4.1. Prove that $\mathrm{lcm}[m,n] = n$ if and only if m divides n.

Exercise 1.4.2. Prove that $\mathrm{lcm}[ma, mb] = m \cdot \mathrm{lcm}[a,b]$ for any positive integer m.

1.5. *Continued fractions*

Another way to write Lemma 1.1.1 is that for any given integers $a \geq b \geq 1$ with $b \nmid a$, there exist integers q and r, with $b > r \geq 1$, for which

$$\frac{a}{b} = q + \frac{r}{b} = q + \frac{1}{\frac{b}{r}}.$$

This is admittedly a strange way to write things, but repeating this process with the pair of integers b and r, and then again, will eventually lead us to an interesting representation of the original fraction a/b. Working with our original example, in which we found the gcd of 85 and 48, we can represent $85 = 48 + 37$ as

$$\frac{85}{48} = 1 + \frac{1}{\frac{48}{37}},$$

---

[10]Sometimes called the *lowest common multiple*.

and the next step, $48 = 37 + 11$, as
$$\frac{48}{37} = 1 + \frac{1}{\frac{37}{11}}, \text{ so that } \frac{85}{48} = 1 + \frac{1}{\frac{48}{37}} = 1 + \frac{1}{1 + \frac{1}{\frac{37}{11}}}.$$

The remaining steps of the Euclidean algorithm may be rewritten as
$$\frac{37}{11} = 3 + \frac{1}{\frac{11}{4}}, \quad \frac{11}{4} = 2 + \frac{1}{\frac{4}{3}}, \quad \text{and} \quad \frac{4}{3} = 1 + \frac{1}{3},$$
so that
$$\frac{85}{48} = 1 + \cfrac{1}{1 + \cfrac{1}{3 + \cfrac{1}{2 + \cfrac{1}{1 + \frac{1}{3}}}}}.$$

This is the *continued fraction* for $\frac{85}{48}$ and is conveniently written as $[1, 1, 3, 2, 1, 3]$. Notice that this is the sequence of quotients a_i from the various divisions; that is,
$$\frac{a}{b} = [a_0, a_1, a_2, \ldots, a_k] := a_0 + \cfrac{1}{a_1 + \cfrac{1}{a_2 + \cfrac{1}{a_3 + \cdots + \frac{1}{a_k}}}}.$$

The a_i's are called the *partial quotients* of the continued fraction.

Exercise 1.5.1. (a) Show that if $a_k > 1$, then $[a_0, a_1, \ldots, a_k] = [a_0, a_1, \ldots, a_k - 1, 1]$.
(b) Prove that the set of positive rational numbers are in 1-1 correspondence with the finite length continued fractions that do not end in 1.

We now list the rationals that correspond to the first few entries in our continued fraction $[1, 1, 3, 2, 1, 3]$. We have $[1] = 1, [1, 1] = 2$, and
$$1 + \cfrac{1}{1 + \frac{1}{3}} = \frac{7}{4}, \quad 1 + \cfrac{1}{1 + \cfrac{1}{3 + \frac{1}{2}}} = \frac{16}{9}, \quad 1 + \cfrac{1}{1 + \cfrac{1}{3 + \cfrac{1}{2 + \frac{1}{1}}}} = \frac{23}{13}.$$

These yield increasingly good approximations to $85/48 = 1.770833\ldots$, that is, in decimal notation,
$$1, \ 2, \ 1.75, \ 1.777\ldots, \ 1.7692\ldots.$$

We call these p_j/q_j, $j \geq 1$, the *convergents* for the continued fraction, defined by
$$\frac{p_j}{q_j} = [a_0, a_1, a_2, \ldots, a_j],$$
since they converge to $a/b = p_k/q_k$ for some k. Do you notice anything surprising about the convergents for $85/48$? In particular the previous one, namely $23/13$? When we worked through the Euclidean algorithm we found that $13 \cdot 85 - 23 \cdot 48 = 1$ — could it be a coincidence that these same numbers show up again in this new context? In section 1.8 of appendix 1A we show that this is no coincidence; indeed we always have
$$p_j q_{j-1} - p_{j-1} q_j = (-1)^{j-1},$$
so, in general, if $u = (-1)^{k-1} q_{k-1}$ and $v = (-1)^k p_{k-1}$, then
$$au + bv = 1.$$

When one studies this in detail, one finds that the continued fraction is really just a convenient reworking of the Euclidean algorithm (as we explained it above) for finding u and v. Bachet de Meziriac[11] introduced this method to Renaissance mathematicians in the second edition of his brilliantly named book *Pleasant and delectable problems which are made from numbers* (1624). Such methods had been known from ancient times, certainly to the Indian scholar Āryabhata in 499 A.D., probably to Archimedes in Syracuse (Greece) in 250 B.C., and possibly to the Babylonians as far back as 1700 B.C.[12]

1.6. *Tiling a rectangle with squares*[13]

Given a 48-by-85 rectangle we will tile it, greedily, with squares. The largest square that we can place inside a 48-by-85 rectangle is a 48-by-48 square. This 48-by-48 square goes from top to bottom of the rectangle, and if we place it at the far right, then we are left with a 37-by-48 rectangle to tile, on the left.

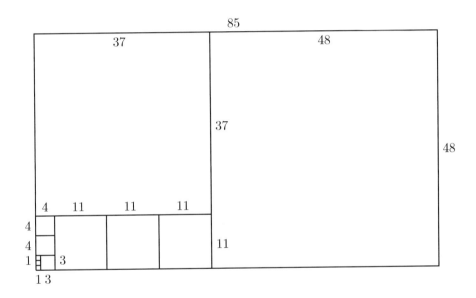

Figure 1.1. Partitioning a rectangle into squares, using the Euclidean algorithm.

If we place a 37-by-37 square at the top of this rectangle, then we are left with an 11-by-37 rectangle in the bottom left-hand corner. We can now place three 11-by-11 squares inside this, leaving a 4-by-11 rectangle. We finish this off with two 4-by-4 squares, one 3-by-3 square, and finally three 1-by-1 squares.

---

[11] The celebrated editor and commentator on Diophantus, whom we will meet again in chapter 6.

[12] There are Cuneiform clay tablets from this era that contain related calculations. It is known that after conquering Babylon in 331 B.C., Alexander the Great ordered his archivist Callisthenes and his tutor Aristotle to supervise the translation of the Babylonian astronomical records into Greek. It is therefore feasible that Archimedes was introduced to these ideas from this source. Indeed, Pythagoras's Theorem may be misnamed as the Babylonians knew of integer-sided right-angled triangles like 3, 4, 5 and 5, 12, 13 more than one thousand years before Pythagoras (570–495 B.C.) was born.

[13] Thanks to Dusa MacDuff and Dylan Thurston for bringing my attention to this beautiful application.

The area of the rectangle can be computed in terms of the areas of each of the squares; that is,
$$85 \cdot 48 = 1 \cdot 48^2 + 1 \cdot 37^2 + 3 \cdot 11^2 + 2 \cdot 4^2 + 1 \cdot 3^2 + 3 \cdot 1^2.$$
What has this to do with the Euclidean algorithm? Hopefully the reader has recognized the same sequence of numbers and quotients that appeared above, when we computed the gcd(85, 48). This is no coincidence. At a given step we have an a-by-b rectangle, with $a \geq b \geq 1$, and then we can remove q b-by-b squares, where $a = qb + r$ with $0 \leq r \leq a - 1$ leaving an r-by-b rectangle, and so proceed by induction.

Exercise 1.6.1. Given an a-by-b rectangle show how to write $a \cdot b$ as a sum of squares, as above, in terms of the partial quotients and convergents of the continued fraction for a/b.

Exercise 1.6.2.   (a) Use this to show that $F_{n+1}F_n = F_n^2 + F_{n-1}^2 + \cdots + F_0^2$, where F_n is the nth Fibonacci number (see section 0.1 for the definition and a discussion of Fibonacci numbers and exercise 0.4.12(b) for a generalization of this exercise).
(b)† Find the correct generalization to more general second-order linear recurrence sequences.

Additional exercises

Exercise 1.7.1.   (a) Does 0 divide 0? (Use the definition of "divides".)
(b) Show that there is no unique meaning to 0/0.
(c) Prove that if b divides a and $b \neq 0$, then there is a unique meaning to a/b.

Exercise 1.7.2. Prove that if a and b are not both 0, then $\gcd(a, b)$ is a positive integer.

Exercise 1.7.3.† Prove that if m and n are coprime positive integers, then $\frac{(m+n-1)!}{m!n!}$ is an integer.

Exercise 1.7.4. Suppose that $a = qb + r$ with $0 \leq r \leq b - 1$.
(a) Let $[t]$ be the *integer part* of t, that is, the largest integer $\leq t$. Prove that $q = [a/b]$.
(b) Let $\{t\}$ to be the *fractional part* of t, that is, $\{t\} = t - [t]$. Prove that $r = b\{r/b\} = b\{a/b\}$.
(Beware of these functions applied to negative numbers: e.g., $[-3.14] = -4$ *not* -3, and $\{-3.14\} =$ .86 *not* .14.)

Exercise 1.7.5.†   (a) Show that if n is an integer, then $\{n + \alpha\} = \{\alpha\}$ and $[n + \alpha] = n + [\alpha]$ for all $\alpha \in \mathbb{R}$.
(b) Prove that $[\alpha + \beta] - [\alpha] - [\beta] = 0$ or 1 for all $\alpha, \beta \in \mathbb{R}$, and explain when each case occurs.
(c) Deduce that $\{\alpha\} + \{\beta\} - \{\alpha + \beta\} = 0$ or 1 for all $\alpha, \beta \in \mathbb{R}$, and explain when each case occurs.
(d) Show that $\{\alpha\} + \{-\alpha\} = 1$ unless α is an integer in which case it equals 0.
(e) Show that if $a \in \mathbb{Z}$ and $r \in \mathbb{R} \setminus \mathbb{Z}$, then $[r] + [a - r] = a - 1$.

Exercise 1.7.6. Suppose that d is a positive integer and that $N, x > 0$.
(a) Show that there are exactly $[x]$ positive integers $\leq x$.
(b) Show that kd is the largest multiple of d that is $\leq N$, where $k = [N/d]$.
(c) Deduce that there are exactly $[N/d]$ positive integers $n \leq N$ which are divisible by d.

Exercise 1.7.7. Prove that $\sum_{k=0}^{n-1} [a + \frac{k}{n}] = [na]$ for any real number a and integer $n \geq 1$.

Exercise 1.7.8. Suppose that $a + b = c$ and let $g = \gcd(a, b)$. Prove that we can write $a = gA$, $b = gB$, and $c = gC$ where $A + B = C$, where A, B, and C are pairwise coprime integers.

Exercise 1.7.9. Prove that if $(a, b) = 1$, then $(a + b, a - b) = 1$ or 2.

Exercise 1.7.10.† Prove that for any given integers $b > a \geq 1$ there exists an integer solution u, w to $au - bw = \gcd(a, b)$ with $0 \leq u \leq b - 1$ and $0 \leq w \leq a - 1$.

Exercise 1.7.11.† Show that if $\gcd(a, b) = 1$, then $\gcd(a^k, b^\ell) = 1$ for all integers $k, \ell \geq 1$.

Exercise 1.7.12. Let m and n be positive integers. What fractions do the two lists $\frac{1}{m}, \ldots, \frac{m-1}{m}$ and $\frac{1}{n}, \ldots, \frac{n-1}{n}$ have in common (when the fractions are reduced)?

Exercise 1.7.13. Suppose m and n are coprime positive integers. When the fractions $\frac{1}{m}, \frac{2}{m}, \ldots, \frac{m-1}{m}, \frac{1}{n}, \ldots, \frac{n-1}{n}$ are put in increasing order, what is the shortest distance between two consecutive fractions?

Given a 7-liter jug and a 5-liter jug one can measure 1 liter of water as follows: Fill the 5-liter jug, and pour the contents into the 7-liter jug. Fill the 5-liter jug again, use this to fill the 7-liter jug, so we are left with 3 liters in the 5-liter jug and the 7-liter jug is full. Empty the 7-liter jug, pour the contents of the 5-liter jug into the 7-liter jug, and refill the 5-liter jug. We now have 3 liters in the 7-liter jug. Fill the 7-liter jug using the 5-liter jug; we have poured 4 liters from the 5-liter jug into the 7-liter jug, so that there is just 1 liter left in the 5-liter jug! Notice that we filled the 5-liter jug 3 times and emptied the 7-liter jug twice, and so we used here that $3 \times 5 - 2 \times 7 = 1$. We have wasted 2×7 liters of water in this process.

Exercise 1.7.14. (a) Since $3 \times 7 - 4 \times 5 = 1$ describe how we can proceed by filling the 7-liter jug each time rather than filling the 5-liter jug.
(b) Can you measure 1 liter of water using a 25-liter jug and a 17-liter jug?
(c)† Prove that if m and n are positive coprime integers then you can measure one liter of water using an m liter jug and an n liter jug?
(d) Prove that one can do this wasting less than mn liters of water.

Exercise 1.7.15. Can you weigh 1 lb of tea using scales with 25-lb and 17-lb weights?

The definition of a set of linear combinations can be extended to an arbitrary set of integers (in place of the set $\{a, b\}$); that is,

$$I(a_1, \ldots, a_k) := \{a_1 m_1 + a_2 m_2 + \cdots + a_k m_k : m_1, m_2, \ldots, m_k \in \mathbb{Z}\}.$$

Exercise 1.7.16. Show that $I(a_1, \ldots, a_k) = I(g)$ for any non-zero integers $a_1, \ldots, a_k$, where we have $g = \gcd(a_1, \ldots, a_k)$.

Exercise 1.7.17.† Deduce that if we are given integers $a_1, a_2, \ldots, a_k$, not all zero, then there exist integers $m_1, m_2, \ldots, m_k$ such that

$$m_1 a_1 + m_2 a_2 + \cdots + m_k a_k = \gcd(a_1, a_2, \ldots, a_k).$$

We say that the integers $a_1, a_2, \ldots, a_k$ are *relatively prime* if $\gcd(a_1, a_2, \ldots, a_k) = 1$. We say that they are *pairwise coprime* if $\gcd(a_i, a_j) = 1$ whenever $i \neq j$. Note that $6, 10, 15$ are relatively prime, but not pairwise coprime (since each pair of integers has a common factor > 1).

Exercise 1.7.18. Prove that if $g = \gcd(a_1, a_2, \ldots, a_k)$, then $\gcd(a_1/g, a_2/g, \ldots, a_k/g) = 1$.

Exercise 1.7.19.† (a) Prove that $abc = [a, b, c] \cdot \gcd(ab, bc, ca)$.
(b)‡ Prove that if $r + s = n$, then

$$a_1 \cdots a_n = \operatorname{lcm}\left[\prod_{i \in I} a_i : I \subset \{1, \ldots, n\}, |I| = r\right] \cdot \gcd\left(\prod_{j \in J} a_j : J \subset \{1, \ldots, n\}, |J| = s\right).$$

Throughout this book we will present more challenging exercises in the final part of each chapter. If some of the questions are part of a consistent subject, then they will be presented as a separate subsection:

Divisors in recurrence sequences

We begin by noting that for any integer $d \geq 1$ we have the polynomial identity

(1.7.1) $\qquad x^d - y^d = (x-y)(x^{d-1} + x^{d-2}y + \cdots + xy^{d-2} + y^{d-1})$.

Hence if r and s are integers, then $r - s$ divides $r^d - s^d$. (This also follows from Corollary 2.3.1 in the next chapter.)

Exercise 1.7.20. (a) Prove that if $m|n$, then $2^m - 1$ divides $2^n - 1$.
(b)† Prove that if $n = qm + r$ with $0 \leq r \leq m - 1$, then there exists an integer Q such that
$$2^n - 1 = Q(2^m - 1) + (2^r - 1) \qquad \text{(and note that } 0 \leq 2^r - 1 < 2^m - 1\text{)}.$$
(c)† Use the Euclidean algorithm to show that $\gcd(2^n - 1, 2^m - 1) = 2^{\gcd(n,m)} - 1$.
(d) What is the value of $\gcd(N^a - 1, N^b - 1)$ for arbitrary integer $N \neq -1, 0$, or 1?

In exercise 0.4.15(a) we saw that the Mersenne numbers $M_n = 2^n - 1$ (of the previous exercise) are an example of a second-order linear recurrence sequence. We will show that an analogous result holds for *any* second-order linear recurrence sequence that begins $0, 1, \ldots$. For the rest of this section we assume that a and b are coprime integers with $x_0 = 0$, $x_1 = 1$ and that $x_n = ax_{n-1} + bx_{n-2}$ for all $n \geq 2$.

Exercise 1.7.21. Use exercise 0.4.10(a) to show that $\gcd(x_m, x_n) = \gcd(x_m, x_{m+1}x_{n-m})$ whenever $n \geq m$.

Exercise 1.7.22.† Prove that if $m|n$, then $x_m|x_n$; that is, $\{x_n : n \geq 0\}$ is a *division sequence*.

Exercise 1.7.23.† Assume that $(a,b) = 1$.
(a) Prove that $\gcd(x_n, b) = 1$ for all $n \geq 1$.
(b) Prove that $\gcd(x_n, x_{n-1}) = 1$ for all $n \geq 1$.
(c) Prove that if $n > m$, then $(x_n, x_m) = (x_{n-m}, x_m)$.
(d) Deduce that $(x_n, x_m) = x_{(n,m)}$.

Exercise 1.7.24.† For any given integer $d \geq 2$, let $m = m_d$ be the smallest positive integer for which d divides x_m. Prove that d divides x_n if and only if m_d divides n.

It is sometimes possible to reverse the direction in the defining recurrence relation for a recurrence sequence; that is, if $b = 1$, then (0.1.2) can be rewritten as $x_{n-2} = -ax_{n-1} + x_n$. So if $x_0 = 0$ and $x_1 = 1$, then $x_{-1} = 1, x_{-2} = -a, \ldots$. We now try to understand the terms x_{-n}.

Exercise 1.7.25. Let us suppose that $x_n = ax_{n-1} + x_{n-2}$ for all integers n, both positive and negative, with $x_0 = 0$ and $x_1 = 1$. Prove, by induction on $n \geq 1$, that $x_{-n} = (-1)^{n-1}x_n$ for all $n \geq 2$.

Appendix 1A. Reformulating the Euclidean algorithm

In section 1.5 we saw that the Euclidean algorithm may be usefully reformulated in terms of continued fractions. In this appendix we reformulate the Euclidean algorithm in two further ways: firstly, in terms of matrix multiplication, which makes many of the calculations easier; and secondly, in terms of a dynamical system, which will be useful later when we develop similar ideas in a more general context.

1.8. Euclid matrices and Euclid's algorithm

In discussing the Euclidean algorithm we showed that $\gcd(85, 48) = \gcd(48, 37)$ from noting that $85 - 1 \cdot 48 = 37$. In this we changed our attention from the pair $85, 48$ to the pair $48, 37$. Writing this down using matrices, we performed this change via the map

$$\binom{85}{48} \to \binom{48}{37} = \begin{pmatrix} 0 & 1 \\ 1 & -1 \end{pmatrix} \binom{85}{48}.$$

Next we went from the pair $48, 37$ to the pair $37, 11$ via the map

$$\binom{48}{37} \to \binom{37}{11} = \begin{pmatrix} 0 & 1 \\ 1 & -1 \end{pmatrix} \binom{48}{37}$$

and then from the pair $37, 11$ to the pair $11, 4$ via the map

$$\binom{37}{11} \to \binom{11}{4} = \begin{pmatrix} 0 & 1 \\ 1 & -3 \end{pmatrix} \binom{37}{11}.$$

We can compose these maps so that

$$\binom{85}{48} \to \binom{48}{37} \to \binom{37}{11} = \begin{pmatrix} 0 & 1 \\ 1 & -1 \end{pmatrix} \binom{48}{37} = \begin{pmatrix} 0 & 1 \\ 1 & -1 \end{pmatrix} \cdot \begin{pmatrix} 0 & 1 \\ 1 & -1 \end{pmatrix} \binom{85}{48}$$

Appendix 1A. Reformulating the Euclidean algorithm

and then
$$\begin{pmatrix} 85 \\ 48 \end{pmatrix} \to \begin{pmatrix} 11 \\ 4 \end{pmatrix} = \begin{pmatrix} 0 & 1 \\ 1 & -3 \end{pmatrix} \begin{pmatrix} 37 \\ 11 \end{pmatrix} = \begin{pmatrix} 0 & 1 \\ 1 & -3 \end{pmatrix} \cdot \begin{pmatrix} 0 & 1 \\ 1 & -1 \end{pmatrix} \begin{pmatrix} 0 & 1 \\ 1 & -1 \end{pmatrix} \begin{pmatrix} 85 \\ 48 \end{pmatrix}.$$

Continuing on to the end of the Euclidean algorithm, via $11 = 2 \cdot 4 + 3$, $4 = 1 \cdot 3 + 1$, and $3 = 3 \cdot 1 + 0$, we have

$$\begin{pmatrix} 1 \\ 0 \end{pmatrix} = \begin{pmatrix} 0 & 1 \\ 1 & -3 \end{pmatrix} \begin{pmatrix} 0 & 1 \\ 1 & -1 \end{pmatrix} \begin{pmatrix} 0 & 1 \\ 1 & -2 \end{pmatrix} \begin{pmatrix} 0 & 1 \\ 1 & -3 \end{pmatrix} \begin{pmatrix} 0 & 1 \\ 1 & -1 \end{pmatrix} \begin{pmatrix} 0 & 1 \\ 1 & -1 \end{pmatrix} \begin{pmatrix} 85 \\ 48 \end{pmatrix}.$$

Since $\begin{pmatrix} 0 & 1 \\ 1 & -x \end{pmatrix} \begin{pmatrix} x & 1 \\ 1 & 0 \end{pmatrix} = I$ for any x, we can invert to obtain

$$\begin{pmatrix} 85 \\ 48 \end{pmatrix} = M \begin{pmatrix} 1 \\ 0 \end{pmatrix}$$

where

$$M = \begin{pmatrix} 1 & 1 \\ 1 & 0 \end{pmatrix} \begin{pmatrix} 1 & 1 \\ 1 & 0 \end{pmatrix} \begin{pmatrix} 3 & 1 \\ 1 & 0 \end{pmatrix} \begin{pmatrix} 2 & 1 \\ 1 & 0 \end{pmatrix} \begin{pmatrix} 1 & 1 \\ 1 & 0 \end{pmatrix} \begin{pmatrix} 3 & 1 \\ 1 & 0 \end{pmatrix}.$$

Here we used that the inverse of a product of matrices is the product of the inverses of those matrices, in reverse order. If we write

$$M := \begin{pmatrix} \alpha & \beta \\ \gamma & \delta \end{pmatrix}$$

where $\alpha, \beta, \gamma, \delta$ are integers (since the set of integer matrices are closed under multiplication), then

$$\alpha\delta - \beta\gamma = \det M = (-1)^6 = 1,$$

since M is the product of six matrices, each of determinant -1, and the determinant of the product of matrices equals the product of the determinants. Now

$$\begin{pmatrix} 85 \\ 48 \end{pmatrix} = M \begin{pmatrix} 1 \\ 0 \end{pmatrix} = \begin{pmatrix} \alpha & \beta \\ \gamma & \delta \end{pmatrix} \begin{pmatrix} 1 \\ 0 \end{pmatrix} = \begin{pmatrix} \alpha \\ \gamma \end{pmatrix}$$

so that $\alpha = 85$ and $\gamma = 48$. This implies that

$$85\,\delta - 48\,\beta = 1;$$

that is, the matrix method gives us the solution to (1.2.1) without extra effort.

If we multiply the matrices defining M together in order, we obtain the sequence

$$\begin{pmatrix} 1 & 1 \\ 1 & 0 \end{pmatrix}, \begin{pmatrix} 1 & 1 \\ 1 & 0 \end{pmatrix} \begin{pmatrix} 1 & 1 \\ 1 & 0 \end{pmatrix} = \begin{pmatrix} 2 & 1 \\ 1 & 1 \end{pmatrix}, \begin{pmatrix} 2 & 1 \\ 1 & 1 \end{pmatrix} \begin{pmatrix} 3 & 1 \\ 1 & 0 \end{pmatrix} = \begin{pmatrix} 7 & 2 \\ 4 & 1 \end{pmatrix}$$

and then

$$\begin{pmatrix} 16 & 7 \\ 9 & 4 \end{pmatrix}, \begin{pmatrix} 23 & 16 \\ 13 & 9 \end{pmatrix}, \begin{pmatrix} 85 & 23 \\ 48 & 13 \end{pmatrix}.$$

We notice that the columns give us the numerators and denominators of the convergents of the continued fraction for $85/48$, as discussed in section 1.5.

We can generalize this discussion to formally explain the Euclidean algorithm:

> Let $u_0 := a \geq u_1 := b \geq 1$. Given $u_j \geq u_{j+1} \geq 1$:
> - Let $a_j = [u_j/u_{j+1}]$, an integer ≥ 1.
> - Let $u_{j+2} = u_j - a_j u_{j+1}$ so that $0 \leq u_{j+2} \leq u_{j+1} - 1$.
> - If $u_{j+2} = 0$, then $g := \gcd(a,b) = u_{j+1}$, and terminate the algorithm.
> - Otherwise, repeat these steps with the new pair u_{j+1}, u_{j+2}.

The first two steps work by Lemma 1.1.1, the third by exercise 1.1.3. We end up with the continued fraction

$$a/b = [a_0, a_1, \ldots, a_k]$$

for some $k \geq 0$. The convergents $p_j/q_j = [a_0, a_1, \ldots, a_j]$ are most easily calculated by matrix arithmetic as

$$(1.8.1) \qquad \begin{pmatrix} p_j & p_{j-1} \\ q_j & q_{j-1} \end{pmatrix} = \begin{pmatrix} a_0 & 1 \\ 1 & 0 \end{pmatrix} \begin{pmatrix} a_1 & 1 \\ 1 & 0 \end{pmatrix} \cdots \begin{pmatrix} a_j & 1 \\ 1 & 0 \end{pmatrix}$$

so that $a/g = p_k$ and $b/g = q_k$, where $g = \gcd(a,b)$.

Exercise 1.8.1. Prove that this description of the Euclidean algorithm really works.

Exercise 1.8.2. (a) Show that $p_j q_{j-1} - p_{j-1} q_j = (-1)^{j+1}$ for all $j \geq 0$.
(b) Explain how to use the Euclidean algorithm, along with (1.8.1), to determine, for given positive integers a and b, an integer solution u, v to the equation $au + bv = \gcd(a,b)$.

Exercise 1.8.3. With the notation as above, show that $[a_k, \ldots, a_0] = a/c$ for some integer c for which $0 < c < a$ and $bc \equiv (-1)^k \pmod{a}$.

Exercise 1.8.4. Prove that for every $n \geq 1$ we have

$$\begin{pmatrix} F_{n+1} & F_n \\ F_n & F_{n-1} \end{pmatrix} = \begin{pmatrix} 1 & 1 \\ 1 & 0 \end{pmatrix}^n,$$

where F_n is the nth Fibonacci number.

My favorite open question in this area is Zaremba's conjecture: He conjectured that there is an integer $B \geq 1$ such that for every integer $n \geq 2$ there exists a fraction m/n, where m is an integer, $1 \leq m \leq n-1$, coprime with n, for which the continued fraction $m/n = [a_0, a_1, \ldots, a_k]$ has each partial quotient $a_k \leq B$. Calculations suggest one can take $B = 5$.

1.9. Euclid matrices and ideal transformations

In section 1.3 we used Euclid's algorithm to transform the basis of the ideal $I(85, 48)$ to $I(48, 37)$, and so on, until we showed that it equals $I(1, 0) = I(1)$. The transformation rested on the identity

$$85m + 48n = 48m' + 37n', \text{ where } m' = m + n \text{ and } n' = n;$$

a transformation we can write as

$$(m, n) \to (m', n') = (m, n) \begin{pmatrix} 1 & 1 \\ 1 & 0 \end{pmatrix}.$$

The transformation of linear forms can then be seen by

$$48m' + 37n' = (m', n')\begin{pmatrix}48\\37\end{pmatrix} = (m, n)\begin{pmatrix}1 & 1\\1 & 0\end{pmatrix}\begin{pmatrix}48\\37\end{pmatrix} = (m, n)\begin{pmatrix}85\\48\end{pmatrix} = 85m + 48n.$$

The inverse map can be found simply by inverting the matrix:

$$\begin{pmatrix}m'\\n'\end{pmatrix} \to \begin{pmatrix}m\\n\end{pmatrix} = \begin{pmatrix}0 & 1\\1 & -1\end{pmatrix}\begin{pmatrix}m'\\n'\end{pmatrix}.$$

These linear transformations can be composed by multiplying the relevant matrices, which are the same matrices that arise in the previous section, section 1.8. For example, after three steps, the change is

$$(m, n) \to (m_3, n_3) = (m, n)\begin{pmatrix}7 & 2\\4 & 1\end{pmatrix},$$

so that $11m_3 + 4n_3 = 85m + 48n$.

Exercise 1.9.1. (a) With the notation of section 1.8, establish that $xu_j + yu_{j+1} = ma + nb$ where the variables x and y are obtained from the variables m and n by a linear transformation.
(b) Deduce that $I(u_j, u_{j+1}) = I(a, b)$ for $j = 0, \ldots, k$.

1.10. The dynamics of the Euclidean algorithm

We now explain a dynamical perspective on the Euclidean algorithm, by focusing on each individual transformation of the pair of numbers with which we work. In our example, we began with the pair of numbers $(85, 48)$, subtracted the smaller from the larger to get $(37, 48)$, and then swapped the order to obtain $(48, 37)$. Now we begin with the fraction $x := 85/48$; the first step transforms $x \to y := x-1 = 37/48$, and the second transforms $y \to 1/y = 48/37$. The Euclidean algorithm can easily be broken down into a series of steps of this form:

$$\frac{85}{48} \to \frac{37}{48} \to \frac{48}{37} \to \frac{11}{37} \to \frac{37}{11} \to \frac{26}{11} \to \frac{15}{11} \to \frac{4}{11}$$
$$\to \frac{11}{4} \to \frac{7}{4} \to \frac{3}{4} \to \frac{4}{3} \to \frac{1}{3} \to \frac{3}{1} \to \frac{2}{1} \to \frac{1}{1} \to \frac{0}{1}.$$

It is possible that the map $x \to x - 1$ is repeated several times consecutively (for example, as we went from $37/11$ to $4/11$), the number of times corresponding to the quotient, $[x]$. On the other hand, the map $y \to 1/y$ is not immediately repeated, since repeating this map sends y back to y, which corresponds to swapping the order of a pair of numbers twice, sending the pair back to their original order.

These two linear maps correspond to our matrix transformations:

$x \to x - 1$ corresponds to $\begin{pmatrix}1 & -1\\1 & 0\end{pmatrix}$, so that $\begin{pmatrix}37\\48\end{pmatrix} = \begin{pmatrix}1 & -1\\1 & 0\end{pmatrix}\begin{pmatrix}85\\48\end{pmatrix}$;

and $y \to 1/y$ corresponds to $\begin{pmatrix}0 & 1\\1 & 0\end{pmatrix}$, so that $\begin{pmatrix}48\\37\end{pmatrix} = \begin{pmatrix}0 & 1\\1 & 0\end{pmatrix}\begin{pmatrix}37\\48\end{pmatrix}$.

The Euclidean algorithm is therefore a series of transformations of the form $x \to x - 1$ and $y \to 1/y$ and defines a finite sequence of these transformations that begins with any given positive rational number and ends with 0. One can invert

1.10. The dynamics of the Euclidean algorithm

that sequence of transformations, to transformations of the form $x \to x+1$ and $y \to 1/y$, to begin with 0 and to end at any given rational number.

Determinant 1 transformations. Foreshadowing later results, it is more useful to develop a variant on the Euclidean algorithm in which the matrices of all of the transformations have determinant 1. To begin with, we break each transformation down into the two steps:

- Beginning with the pair $85, 48$ the first step is to subtract 1 times 48 from 85, and in general we subtract q times b from a. This transformation is therefore given by
$$\begin{pmatrix} a \\ b \end{pmatrix} \to \begin{pmatrix} 1 & -q \\ 0 & 1 \end{pmatrix} \begin{pmatrix} a \\ b \end{pmatrix}, \text{ and notice that } \begin{pmatrix} 1 & -q \\ 0 & 1 \end{pmatrix} = \begin{pmatrix} 1 & 1 \\ 0 & 1 \end{pmatrix}^{-q}.$$

- The second step swaps the roles of $37 (= 85 - 48)$ and 48, corresponding to a matrix of determinant -1. Here we do something unintuitive which is to change 48 to -48, so that the matrix has determinant 1:
$$\begin{pmatrix} 37 \\ 48 \end{pmatrix} \to \begin{pmatrix} 0 & -1 \\ 1 & 0 \end{pmatrix} \begin{pmatrix} 37 \\ 48 \end{pmatrix}, \text{ and more generally } \begin{pmatrix} a \\ b \end{pmatrix} \to \begin{pmatrix} 0 & -1 \\ 1 & 0 \end{pmatrix} \begin{pmatrix} a \\ b \end{pmatrix}.$$

One then sees that if $g = \gcd(a, b)$ and $a/b = [a_0, \ldots, a_k]$, then

$$\begin{pmatrix} 0 \\ g \end{pmatrix} = \begin{pmatrix} 1 & 1 \\ 0 & 1 \end{pmatrix}^{-a_k} \begin{pmatrix} 0 & -1 \\ 1 & 0 \end{pmatrix} \begin{pmatrix} 1 & 1 \\ 0 & 1 \end{pmatrix}^{-a_{k-1}} \cdots \begin{pmatrix} 0 & -1 \\ 1 & 0 \end{pmatrix} \begin{pmatrix} 1 & 1 \\ 0 & 1 \end{pmatrix}^{-a_0} \begin{pmatrix} a \\ b \end{pmatrix}.$$

We write $S := \begin{pmatrix} 1 & 1 \\ 0 & 1 \end{pmatrix}$ and $T := \begin{pmatrix} 0 & 1 \\ -1 & 0 \end{pmatrix}$. Taking inverses here we get

$$\begin{pmatrix} a \\ b \end{pmatrix} = S^{a_0} T S^{a_1} T \cdots S^{a_{k-1}} T S^{a_k} \begin{pmatrix} 0 \\ g \end{pmatrix}.$$

If a and b are coprime, then this implies that

$$(1.10.1) \qquad S^{a_0} T S^{a_1} T \cdots S^{a_{k-1}} T S^{a_k} = \begin{pmatrix} c & a \\ d & b \end{pmatrix}$$

for some integers c and d. The left-hand side is the product of determinant one matrices, and so the right-hand side also has determinant one; that is, $cb - ad = 1$. This is therefore an element of $\mathrm{SL}(2, \mathbb{Z})$, the subgroup (under multiplication) of 2-by-2 integer matrices of determinant one; more specifically

$$\mathrm{SL}(2, \mathbb{Z}) := \left\{ \begin{pmatrix} \alpha & \beta \\ \gamma & \delta \end{pmatrix} : \alpha, \beta, \gamma, \delta \in \mathbb{Z}, \; \alpha\delta - \beta\gamma = 1 \right\}.$$

Theorem 1.2. *Each matrix in* $\mathrm{SL}(2, \mathbb{Z})$ *can be represented as* $S^{e_1} T^{f_1} \cdots S^{e_r} T^{f_r}$ *for integers* $e_1, f_1, \ldots, e_r, f_r$.

Proof. Suppose that we are given $\begin{pmatrix} x & a \\ y & b \end{pmatrix} \in \mathrm{SL}(2, \mathbb{Z})$. Taking determinants we see that $bx - ay = 1$. Therefore $\gcd(a, b) = 1$, and so above we saw how to construct an element of $\mathrm{SL}(2, \mathbb{Z})$ with the same last column. In Theorem 3.5 we will show that every other integer solution to $bx - ay = 1$ is given by $x = c - ma, y = d - mb$ for some integer m. Therefore

$$\begin{pmatrix} x & a \\ y & b \end{pmatrix} = \begin{pmatrix} c & a \\ d & b \end{pmatrix} \begin{pmatrix} 1 & 0 \\ -m & 1 \end{pmatrix}.$$

One can easily verify that

$$T^{-1} = \begin{pmatrix} 0 & -1 \\ 1 & 0 \end{pmatrix}, \text{ so that } T^{-1}ST = \begin{pmatrix} 1 & 0 \\ -1 & 1 \end{pmatrix},$$

and therefore

$$\begin{pmatrix} 1 & 0 \\ -m & 1 \end{pmatrix} = (T^{-1}ST)^m = T^{-1}S^mT.$$

Combining these last two statements together with (1.10.1) completes the proof of the theorem. □

Appendix 1B. Computational aspects of the Euclidean algorithm

In this appendix we study the speed at which the Euclidean algorithm works. First we look for simple ways to speed the algorithm up. Secondly we establish how to formulate a practical way to measure how fast an algorithm works and identify and analyze the running time of the worst case scenario for the Euclidean algorithm.

1.11. Speeding up the Euclidean algorithm

There are various simple steps that can be used to speed up the Euclidean algorithm. For example if one of the two initial numbers is odd, then we know that the gcd is odd, so we can divide out any factor of 2 that we come across while implementing the algorithm.[14] Hence in our favorite example, since $48 = 2^4 \cdot 3$ and 85 is odd, we have

$$(85, 48) \to (85, 3) \to (1, 3) = 1$$

which is much simpler. Another popular option is to allow minus signs: In each step of the Euclidean algorithm we replace a and b, by b and r, the remainder when a is divided by b. If r is "large", that is, r is close to b, then we can replace r by $r - b$ (which is negative) or, ignoring the minus sign, by $b - r$. Hence as $85 - 2 \cdot 48 = -11$, we have

$$(85, 48) \to (48, 11) \to (11, 4) \to (4, 1) = 1.$$

Again this is faster than the usual Euclidean algorithm. In practice one tries to combine these two ideas, and typically this speeds up Euclid's algorithm by a factor of 2 or more.

---

[14]This is particularly easy if the numbers are represented in base 2, as on a computer, since then one can simply remove any trailing 0's.

1.12. Euclid's algorithm works in "polynomial time"

In section 1.8 of appendix 1A we gave notation for Euclid's algorithm, and we now write $v_n = u_{j+2-n}$ so that we take the numbers in Euclid's algorithm in reverse order. Thus $v_0 = 0$ and $v_1 = g$. Now each $a_j \geq 1$, which implies that $u_j = a_j u_{j+1} + u_{j+2} \geq u_{j+1} + u_{j+2}$, and therefore

$$v_{i+2} \geq v_{i+1} + v_i$$

for all $i \geq 0$.

Exercise 1.12.1.[†] (a) Prove that if F_n is the nth Fibonacci number, then $v_n \geq F_n$ for all n.
(b)[‡] Show that this inequality cannot be improved, in general.
(c) Show that if we apply the usual Euclid algorithm to $a \geq b \geq 1$, then it terminates in $\leq \frac{\log a}{\log \phi} + 2$ steps, where $\phi = \frac{1+\sqrt{5}}{2}$.

It is important to determine whether a given algorithm is practical, and to do that we figure out how many steps that algorithm takes. In exercise 1.12.1, we have just seen that the Euclidean algorithm on two integers $a \geq b \geq 1$ takes $\leq c \log a$ steps.[15] Each step of the algorithm invokes Lemma 1.1.1, and finding q and r can be done in $\leq c \log a$ *bit operations*.[16] Hence in total the Euclidean algorithm works in $\leq c(\log a)^2$ bit operations.

We want to decide whether or not this means that the Euclidean algorithm is efficient. Any algorithm computes some function f, which maps a given input to some value. If the input is n, then to write it down, say in binary, takes $c \log n$ bits. Hence we cannot expect an algorithm to work any quicker than this. A *polynomial time algorithm* is one that, given an input of d binary digits, computes the output in time $\leq c d^A$ bit operations, for some contants $c > 0$ and $A > 0$. A polynomial time algorithm is considered to be an efficient algorithm. The Euclidean algorithm is a polynomial time algorithm (with $A = 2$) and therefore is efficient.

Exercise 1.12.2. Find two coprime four-digit integers a and b for which the Euclidean algorithm works in (a) as few steps as possible and (b) as many steps as possible.

In exercise 1.12.1, particularly part (b), we have focused on the "worst case", the slowest possible example in Euclid's algorithm. It is more instructive, but more difficult, to study what "typically happens". This is explained by the *Gauss-Kuzmin law*: For a typical fraction a/b, the proportion of partial quotients in the continued fraction of a/b which equal k is roughly[17]

$$\frac{1}{\log 2} \log\left(\frac{(k+1)^2}{k(k+2)}\right);$$

so a typical continued fraction has roughly 41.5% 1's, 17% 2's, 9.3% 3's, 5.9% 4's, 4% 5's, etc., and has length roughly $c_0 \log a$ where $c_0 = \frac{12}{\pi^2} \log 2 = 0.8427\ldots$. This

---

[15]Here, and henceforth, c stands for some positive constant. We do not try to specify its value since that would be complicated and is not really the point. The point is to understand how the running time of an algorithm varies with the size of the input. The value of c might even be different from one line to the next, not to confuse the reader but simply to reiterate that its exact value is irrelevant.

[16]A *bit operation* is the addition, subtraction, or shifting of two bits, each 0 or 1, in computer machine code.

[17]The words "typical" and "roughly" here are deliberately vague, as correctly formulating this result would require a substantial amount of notation. Our formulation gives the flavor of the correct result.

1.12. Euclid's algorithm works in "polynomial time"

is not much smaller than the upper bound that we obtained for the worst case in exercise 1.12.1(c). (The constant there is $1/\log \phi = 2.0780\ldots$.)

A careful analysis of the Gauss-Kuzmin law suggests that typically, the largest partial quotient in the continued fraction of a/N, with $1 \leq a \leq N-1$, is about $\frac{12}{\pi^2} \log N$. It also suggests Zaremba's 1971 conjecture that, for every integer $N > 1$, the partial quotient of the continued fraction of at least one of the a/N is ≤ 5. This is an open problem. The best result we have was given by Bourgain and Kontorovich in 2014: They showed that the partial quotients of the continued fraction of at least one of the a/N are all ≤ 50, for 100% of integers N.[18]

In this subsection we have not been too clear about what base we have used for our logarithms (nor does it matter since we have been working with ratios of logarithms, which take the same value, no matter what base). In advanced mathematics we typically work with the *natural logarithm*, which is the logarithm to the base e, where

$$e := 1 + \frac{1}{1} + \frac{1}{2!} + \frac{1}{3!} + \frac{1}{4!} + \frac{1}{5!} + \cdots.$$

All logarithms in this book are natural logarithms. There is good reason for this: Typically this is first justified by noting that e^x is the only function $f(x)$ for which $\frac{df}{dx} = f$, or perhaps because, when optimizing compound interest, one learns that $(1 + \frac{1}{N})^N \to e$ as $N \to \infty$. We shall see a number-theoretic justification when we study the distribution of primes in chapter 5.

---

[18]Beware, this does not mean all N, nor "all N from some point onwards". In fact, "100% of integers N" means that the proportion of integers for which this fails tends to 0 as $N \to \infty$.

Appendix 1C. Magic squares

1.13. Turtle power

The only 3-by-3 (normal) magic square up to isomorphism

2	9	4
7	5	3
6	1	8

was given to mankind by a turtle from the river Lo, in the days of the legendary Emperor Yu of China ($\approx$ 2200 B.C.). It is *magic* because it is a square array of distinct integers in which all rows, columns, and main diagonals have the same sum.

By the 12th century, the Jaina in India had found a 4-by-4 normal magic square:

7	12	1	14
2	13	8	11
16	3	10	5
9	6	15	4

Albrecht Dürer engraved *Melencolia I* in 1514 (part of which we display on the left), including a 4-by-4 magic square containing the numbers 15 and 14 (which therefore gave the date) in the center of the bottom row. Each row, column, and diagonal sums to 34, and also the inner square of integers and the four corners. Any two symmetrically placed integers, around the center, sum to 17.

1.14. Latin squares

A *magic square* is an n-by-n square array, in which the rows and columns all sum up to the same amount. All 3-by-3 magic squares may be parametrized in terms of 5 variables, x_1, x_2, x_3, a, b:

x_1	$x_2 + b$	$x_3 + a$
$x_3 + b$	$x_1 + a$	x_2
$x_2 + a$	x_3	$x_1 + b$

Figure 1.2. The general magic square of order 3.

A magic square is *normal* if it contains each of the integers between 1 and n^2. One can construct normal magic squares of every order $n \geq 3$. However, we do not know how many there are of each order.

Some seemingly new magic squares are obtained from old ones by a simple transformation: For example, we can rotate the turtle's magic square by $90°$, or we can reflect it though a horizontal line, to obtain

6	7	2
1	5	9
8	3	4

and

4	9	2
3	5	7
8	1	6

In order to make progress with the problem of classifying magic squares we will write the integers in the turtle's magic square, minus 1, in base 3. Similarly we write the Jaina magic square in base 4:

01	22	10
20	11	02
12	00	21

and

12	23	00	31
01	30	13	22
33	02	21	10
20	11	32	03

In both cases it becomes obvious that all of the row and column sums must be the same, since each possible digit appears exactly once in each row and column.

Exercise 1.13.1. Prove that if A is a magic square, then $mA + n$ is also a magic square for any integers m, n, where $(mA + n)_{i,j} = mA_{i,j} + n$ for all i, j.

1.14. Latin squares

A *Latin square* is an n-by-n square of integers, in which each row and each column contains each of the numbers $\{0, 1, \ldots, n-1\}$ exactly once. Thus we have the following 3-by-3 Latin squares:

0	2	1
2	1	0
1	0	2

and

1	2	0
0	1	2
2	0	1

Three times the entries of the first plus the entries of the second plus 1 gives our original 3-by-3 magic square. Replacing $0, 1, 2$ by x_1, x_2, x_3 in the first and by $b, 0, a$ in the second and adding, we obtain the general magic square in Figure 1.2.

In general if we are given two n-by-n Latin squares A and B, then we can construct an n-by-n magic square M by the formula

$$M_{i,j} = nA_{i,j} + B_{i,j} + 1.$$

Since $0 \le A_{i,j}, B_{i,j} \le n-1$, therefore $1 \le M_{i,j} \le n^2$. Moreover, for each fixed j,

$$\sum_i M_{i,j} = n\sum_i A_{i,j} + \sum_i B_{i,j} + \sum_i 1 = n\sum_{k=0}^{n-1} k + \sum_{k=0}^{n-1} k + n = \frac{1}{2}(n^3 + n),$$

since each column of A and of B contains each of $\{0, 1, \ldots, n-1\}$ exactly once. The analogous argument shows that each row sum of M takes the same value.

In order that M is normal, the values of $nA_{i,j} + B_{i,j}$ must all be different, which is the same as saying that the pairs $(A_{i,j}, B_{i,j})$ are all different. Any two n-by-n Latin squares A and B in which no two positions have the same A- and B-values are called *orthogonal*, as in our example above.

Exercise 1.14.1. For a given integer $n > 1$, let $(\ell)_n$ be the least non-negative residue of ℓ (mod n).
 (a) Show that if $(ab, n) = 1$ and $A_{i,j} = (ai + bj)_n$, then A is an n-by-n Latin square.
 (b) Show that if, also, $(cd, n) = (ad - bc, n) = 1$ and $B_{i,j} = (ci + dj)_n$, then A and B are orthogonal n-by-n Latin squares.
 (c) Prove that there exist integers a, b, c, d for which $(abcd, n) = (ad - bc, n) = 1$ if and only if n is odd.
 (d) Deduce that there are n-by-n normal magic squares whenever n is odd and > 1.

It is an open question to count the number of pairs of orthogonal n-by-n Latin squares. However it is known that if $n = 10$ or 14 or any number of the form $4k + 10$ with $k \ge 0$, then there is at least one pair of orthogonal n-by-n Latin squares and therefore the corresponding magic square.

1.15. Factoring magic squares

Let A be an n-by-n normal, magic square, and let $B(r, s)$ be m-by-m normal, magic squares for $0 \le r, s \le n - 1$ (it could be that each $B(r, s) = B$). We construct an mn-by-mn magic square C as follows:

$$C_{(i-1)m+I, (j-1)m+J} = m^2(A_{i,j} - 1) + B(i, j)_{I, J}$$

for $1 \le i, j \le n$ and $1 \le I, J \le m$.

Exercise 1.15.1. (a) Verify that C is a normal magic square.
 (b) Suppose that we have constructed a normal magic square of order 8. Deduce that there are normal magic squares of order n, whenever n is divisible by 4.

Exercise 1.15.2. Let $n = 4m$ and define $O = \{0, \ldots, m-1\} \cup \{3m, \ldots, 4m-1\}$ whereas $I = \{m, \ldots, 3m - 1\}$. We define an n-by-n magic square A with (i, j)th entry, for $0 \le i, j \le n - 1$,

$$A(i, j) = \begin{cases} in + j + 1 & \text{if } i, j \in I \text{ or } i, j \in O, \\ (n - 1 - i)n + (n - 1 - j) + 1 & \text{if } i \in I \text{ and } j \in O, \text{ or if } i \in O \text{ and } j \in I. \end{cases}$$

Prove that A is a normal n-by-n magic square in which any two symmetrically placed integers, around the center, sum to $n^2 + 1$.

Reference for this appendix

[1] Harold M. Stark, chapter 4 of *An introduction to number theory*, MIT Press, Cambridge, Mass.-London, 1978.

There are many websites with constructions of magic and Latin squares.

Appendix 1D. The Frobenius postage stamp problem

1.16. The Frobenius postage stamp problem, I

The Frobenius postage stamp problem asks, given an unlimited supply of postage stamps worth a cents and b cents, what (precise) amounts of postage can we stick on an envelope? That is, what is the set of values

$$\mathcal{P}(a,b) := \{am + bn : \ m, n \in \mathbb{Z}, \ m, n \geq 0\}?$$

(To obtain $am + bn$ cents we take m stamps worth a cents and n stamps worth b cents.) We only allow non-negative coefficients for a and b in our linear combinations, whereas in $I(a, b)$ there is no such restriction.

We will begin with the example in which we have stamps of value 3 and 5 cents: Evidently we can make up any multiple of 3 cents and any multiple of 5 cents. But how about with a mixture of different types of stamps? When we start trying all possible combinations we find

$3 = 1 \cdot 3;\ 5 = 1 \cdot 5;\ 6 = 2 \cdot 3;\ 8 = 1 \cdot 3 + 1 \cdot 5;\ 9 = 3 \cdot 3;\ 10 = 2 \cdot 5;\ 11 = 2 \cdot 3 + 1 \cdot 5;$

and it seems that from here on we get every integer value:

$12 = 4 \cdot 3;\ 13 = 1 \cdot 3 + 2 \cdot 5;\ 14 = 3 \cdot 3 + 1 \cdot 5;\ 15 = 5 \cdot 3;\ 16 = 2 \cdot 3 + 2 \cdot 5;\ 17 = 4 \cdot 3 + 1 \cdot 5.$

To prove that we indeed get every integer value we study these representations and look for a pattern. A quick inspection and we see that if we wish to pay n cents postage and we already know how to pay $n - 3$ cents, then we simply add a 3-cent stamp. Hence once we know how to pay exactly 8 cents, exactly 9 cents, and exactly 10 cents, then we can deduce that every integer $n \geq 8$ belongs to $\mathcal{P}(3, 5)$.

Exercise 1.16.1. (a) Rewrite this last proof as a formal proof by induction to establish that $\{n \in \mathbb{Z}: \ n \geq 8\} \subset \mathcal{P}(3, 5)$.
(b) Suppose that $1 \leq a < b$. Assume there exists an integer N (which may depend on a and b), for which $N, N+1, \ldots, N+a-1$ all belong to $\mathcal{P}(a, b)$. Deduce that $\{n \in \mathbb{Z}: \ n \geq N\} \subset \mathcal{P}(a, b)$.

In section 3.25 of appendix 3G, we will develop a technique for establishing that this assumption holds for some integer N whenever $\gcd(a,b) = 1$.

We have now proven that $\mathcal{P}(3,5) = \{n \geq 0 : n \neq 1, 2, 4, \text{ or } 7\}$.

What is the connection between $\mathcal{P}(a,b)$ and $I(a,b)$? First we note that if $r = am + bn$, then $-r = a(-m) + b(-n)$, so that $I(a,b) \supset \mathcal{P}(a,b) \cup -\mathcal{P}(a,b)$. In our example above this means that $I(3,5) \supset \{n \in \mathbb{Z} : n \neq \pm 1, \pm 2, \pm 4, \pm 7\}$. It is not difficult to establish that $\pm 1, \pm 2, \pm 4, \pm 7 \in I(3,5)$, by constructing linear combinations like $1 = 2 \cdot 3 - 1 \cdot 5$, and so $I(3,5) = \mathbb{Z}$.

There is an easier way to show that $I(3,5) = \mathbb{Z}$: We know that there are two consecutive integers in $\mathcal{P}(a,b)$, for example $13 = 1 \cdot 3 + 2 \cdot 5$ and $14 = 3 \cdot 3 + 1 \cdot 5$. Subtracting one from the other we have

$$1 = 14 - 13 = (3 \cdot 3 + 1 \cdot 5) - (1 \cdot 3 + 2 \cdot 5) = 2 \cdot 3 - 1 \cdot 5.$$

Multiplying through by n, we can write any integer n as $(2n) \cdot 3 - n \cdot 5$ and so $I(3,5) = \mathbb{Z}$.

We proved above that $\mathcal{P}(3,5) = \{n \geq 0\} \setminus \mathcal{E}(3,5)$ where $\mathcal{E}(3,5) := \{1, 2, 4, 7\}$ is the set of *exceptional numbers* (that is, those non-negative integers not of the form $3m + 5n$ with $m, n \geq 0$). We notice that $\mathcal{E}(3,5)$ is a finite set and that if $r \in \mathcal{E}(3,5)$, then $r < 3 \cdot 5$ and $\gcd(r, 3 \cdot 5) = 1$. These properties, and more, hold for any $\mathcal{E}(a,b)$, where a and b are coprime integers, as we will see in sections 3.25 of appendix 3G and 15.9 of appendix 15B.

Exercise 1.16.2. (a) Prove that $\mathcal{E}(2,3) = \{1\}$.
(b) An integer n is a *power* if $n = m^k$ for some integers m and $k \geq 2$. Prove that we can write any power n as $n = a^2 b^3$ where a and b are integers.

Exercise 1.16.3. (a) Construct $\mathcal{E}(a,b)$ for various pairs of coprime integers a and b.
(b) Guess at a formula (in terms of a and b) for the largest element of $\mathcal{E}(a,b)$, that is, the largest positive integer not representable in the form $am + bn$ with $m, n \geq 0$. (Use your data from part (a).)
(c)[‡] Prove that your conjectural formula in (b) is true.

Appendix 1E. Egyptian fractions

1.17. Simple fractions

The ancient Egyptians represented all fractions, a/b with $(a,b) = 1$, as a sum of distinct fractions of the form $1/n$. Such a representation is called an *Egyptian fraction*, and each $1/n$ is a *unit fraction*. For example if n is odd and $n+1 = 2m$, then
$$\frac{2}{n} = \frac{1}{m} + \frac{1}{mn}.$$

Exercise 1.17.1. Deduce that $1/b$ with $b > 1$ may always be written as $1/m + 1/n$ with m and n distinct positive integers.

We have shown how to write a/b with $(a,b) = 1$ as a sum of unit fractions when $a = 1$ and $a = 2$. How about when $b > a \geq 3$?

Exercise 1.17.2. Suppose that $b > a \geq 2$ are positive coprime integers. Let $q = [b/a]$.
(a) Prove that $a/b = 1/(q+1) + A/B$ where A and B are positive integers with $A < a, B$.
(b) Deduce that a/b can be written as a sum of no more than a distinct unit fractions.

For $a = 1$ and $a = 2$ we have shown that a/b can be written as the sum of at most a distinct unit fractions (and this is trivially best possible). In exercise 1.17.2 we have shown that a/b can always be written as at most a distinct unit fractions, but is this best possible for larger a? We will restrict our attention to prime values of b, for if prime p divides b and $a/p = \sum_{j=1}^{k} 1/n_j$, a sum of k distinct unit fractions, then $a/b = \sum_{j=1}^{k} 1/m_j$, where each $m_j = n_j(b/p)$, which is also a sum of k distinct unit fractions.

For $a = 3$ one can show that $3/7$ cannot be written as the sum of two distinct unit fractions, and indeed $3/p$ cannot be written as the sum of two distinct unit fractions whenever p is a prime of the form $3m + 1$. (We will prove this in section 3.26 of appendix 3G.) Hence $3/b$ can always be written as the sum of at most 3 distinct unit fractions, and this is best possible. One can do better if b is of the form $3m - 1$, or even of the form $am - 1$:

$$\frac{a}{am-1} = \frac{1}{m} + \frac{1}{m(am-1)}.$$

For $a = 4$ one can show that $4/13$ cannot be written as the sum of two distinct unit fractions, nor any $4/p$ whenever p is a prime of the form $4m + 1$ (see exercise 3.26.1). Therefore the minimum number of distinct unit fractions needed to represent all unit fractions of the form $4/b$ is either 3 or 4. But which is it? The Erdős-Strauss conjecture states that $4/n$ can always be written as the sum of three unit fractions or less. Although this is an open problem, it is known to be true for all $n < 10^{14}$.

The representation above shows how to represent $4/p$ whenever p is a prime of the form $4m - 1$, and also whenever p is a prime of the form $3n - 1$ by adding $1/p$ to each side of the representation there for $3/p$. This leaves us to represent $4/p$ as a sum of at most three unit fractions, only for the primes p of the form $12k + 1$.

Chapter 2

Congruences

The key step in understanding the Euclidean algorithm, Lemma 1.1.1, shows that $\gcd(a,b)$ equals $\gcd(r,b)$, *because* b divides $a - r$. Inspired by how useful this observation is, Gauss developed the theory of when two given integers, like a and r, differ by a multiple of b:

2.1. Basic congruences

If m, b, and c are integers for which m divides $b - c$, then we write

$$b \equiv c \pmod{m}$$

and say that b and c are *congruent modulo m*, where m is the *modulus*.[1] The numbers involved should be integers, not fractions, and the modulus can be taken in absolute value; that is, $b \equiv c \pmod{m}$ if and only if $b \equiv c \pmod{|m|}$, by definition.

For example, $-10 \equiv 15 \pmod{5}$, and $-7 \equiv 15 \pmod{11}$, but $-7 \not\equiv 15 \pmod{3}$. Note that $b \equiv b \pmod{m}$ for all integers m and b.

The integers $\equiv a \pmod{m}$ are precisely those of the form $a + km$ where k is an integer, that is, $a, a+m, a+2m, \ldots$ as well as $a-m, a-2m, a-3m, \ldots$. We call this set of integers a *congruence class* or *residue class* mod m, and any particular element of the congruence class is a *residue*.[2]

For any given integers a and $m > 0$, there exists a unique pair of integers q and r with $0 \leq r \leq m-1$, for which $a = qm + r$, by Lemma 1.1.1. Therefore there exists a unique integer $r \in \{0, 1, 2, \ldots, m-1\}$ for which $a \equiv r \pmod{m}$. Moreover, if two integers are congruent mod m, then they leave the same remainder, r, when

---

[1] Gauss proposed the symbol $\equiv$ because of the analogies between equality and congruence, which we will soon encounter. To avoid ambiguity he made a minor distinction by adding the extra bar.

[2] The sequence of numbers $a, a+m, a+2m, \ldots$, in which we add m to the last number in the sequence to obtain the next one, is an *arithmetic progression*.

divided by m. We now prove a generalization of these last remarks:

Theorem 2.1. *Suppose that m is a positive integer. Exactly one of any m consecutive integers is $\equiv a \pmod{m}$.*

Two proofs.[3] Suppose we have the m consecutive integers $x, x+1, \ldots, x+m-1$.

Analytic proof: An integer n in the range $x \leq n < x+m$ is of the form $a + km$, for some integer k, if and only if there exists an integer k for which
$$x \leq a + km < x + m.$$
Subtracting a from each term here and dividing through by m, we find that this holds if and only if
$$\frac{x-a}{m} \leq k < \frac{x-a}{m} + 1.$$
Hence k must be an integer from an interval of length one which has just one endpoint included in the interval. Such an integer k exists and is unique; it is the smallest integer that is $\geq \frac{x-a}{m}$.

Exercise 2.1.1. Prove that for any real number t there is a unique integer in the interval $[t, t+1)$.

Number-theoretic proof: By Lemma 1.1.1 there exist integers q and r with $0 \leq r \leq m-1$, for which $a - x = qm + r$, with $0 \leq r \leq m-1$. Then $x \leq x + r \leq x + m - 1$ and $x + r = a - qm \equiv a \pmod{m}$, and so $x + r$ is the integer that we are looking for. We still need to prove that it is unique:

If $x + i \equiv a \pmod{m}$ and $x + j \equiv a \pmod{m}$, where $0 \leq i < j \leq m - 1$, then $i \equiv a - x \equiv j \pmod{m}$, so that m divides $j - i$, which is impossible as $1 \leq j - i \leq m - 1$. □

Exercise 2.1.2. Prove that m divides $(n-1)(n-2)\cdots(n-m)$ for every integer n and every integer $m \geq 1$.

Theorem 2.1 implies that any m consecutive integers yield a *complete set of residues* $\pmod{m}$; that is, every congruence class $\pmod{m}$ is represented by exactly one element of the given set of m integers. For example, every integer has a unique residue amongst

the least non-negative residues $\pmod{m}$: $\quad 0, 1, 2, \ldots, (m-1),$

as well as amongst

the least positive residues $\pmod{m}$: $\quad 1, 2, \ldots, m,$

and also amongst

the least negative residues $\pmod{m}$: $\quad -(m-1), -(m-2), \ldots, -2, -1, 0.$

For example, 2 is the least positive residue of $-13 \pmod 5$, whereas -3 is the least negative residue; and 5 is its own least positive residue mod 7, whereas -2 is the least negative residue. Notice that if the residue is not $\equiv 0 \pmod m$, then these residues occur in pairs, one positive and the other negative, and at least one of each

---

[3]Why give two proofs? Throughout this book we will frequently take the opportunity to give more than one proof of a key result. The idea is to highlight different aspects of the theory that are, or will become, of interest. Here we find both an analytic proof (meaning that we focus on the size or quantity of the objects involved) as well as a number-theoretic proof (in which we use their algebraic properties). Sometimes the interplay between these two perspectives can take us much further than either one alone.

2.1. Basic congruences

pair is $\leq m/2$ in absolute value. We call this the *absolutely least* residue (mod m) (and we select $m/2$, rather than $-m/2$, when m is even).[4] For example if $m = 5$, we can pair up the least positive residues and the least negative residues as

$$1 \equiv -4 \pmod 5, \ 2 \equiv -3 \pmod 5, \ 3 \equiv -2 \pmod 5, \ 4 \equiv -1 \pmod 5,$$

as well as the exceptional $5 \equiv 0 \pmod 5$. Hence the absolutely least residues (mod 5) are $-2, -1, 0, 1, 2$. Similarly the the absolutely least residues (mod 6) are $-2, -1, 0, 1, 2, 3$. More generally if $m = 2k + 1$ is odd, then the absolutely least residues (mod $2k+1$) are $-k, \ldots, -1, 0, 1 \ldots, k$; and if $m = 2k$ is even, then the absolutely least residues (mod $2k$) are $-(k-1), \ldots, -1, 0, 1 \ldots, k$.

We defined a *complete set of residues* to be any set of representatives for the residue classes mod m, one for each residue class. A *reduced set of residues* has representatives only for the residue classes that are coprime with m. For example $\{0, 1, 2, 3, 4, 5\}$ is a complete set of residues (mod 6), whereas $\{1, 5\}$ is a reduced set of residues, as 0, 2, and 4 are divisible by 2, and 0 and 3 are divisible by 3 and so are excluded.

Exercise 2.1.3. Suppose that $a_1, \ldots, a_m$ is a complete set of residues mod m. Prove that m divides $(n - a_1) \cdots (n - a_m)$ for every integer n.

Exercise 2.1.4. (a) Explain how "a number of the form $3n - 1$" means the same thing as "a number of the form $3n + 2$", using the language of congruences.
  (b) Prove that the set of integers in the congruence class $a \pmod d$ can be partitioned into the set of integers in the congruence classes $a \pmod{kd}$, $a + d \pmod{kd}, \ldots$ and $a + (k-1)d \pmod{kd}$.

Exercise 2.1.5. Show that if $a \equiv b \pmod m$, then $(a, m) = (b, m)$.

Exercise 2.1.6. Prove that if $a \equiv b \pmod m$, then $a \equiv b \pmod d$ for any divisor d of m.

Exercise 2.1.7. Satisfy yourself that addition and multiplication mod m are commutative.[5]

Exercise 2.1.8. Prove that the property of congruence modulo m is an *equivalence relation* on the integers. To prove this, one must establish

  (i) $a \equiv a \pmod m$;
  (ii) $a \equiv b \pmod m$ implies $b \equiv a \pmod m$;
  (iii) $a \equiv b \pmod m$ and $b \equiv c \pmod m$ imply $a \equiv c \pmod m$.

The equivalence classes are therefore the congruence classes mod m.

One consequence of this is that integers that are congruent modulo m have the same least residues modulo m, whereas integers that are not congruent modulo m have different least residues.

The main use of congruences is that it simplifies arithmetic when we are looking into questions about remainders. This is because the usual rules for addition, subtraction, and multiplication work for congruences. However, division is a little more complicated, as we will see in the next section.

---

[4] This is often called the *least residue in absolute value*.
[5] A mathematical operation is *commutative* if you get the same result no matter what order you take the input variables in. Thus, in $\mathbb{C}$, we have $x + y = y + x$ and $xy = yx$. There are common operations that are not commutative; for example $a - b \neq b - a$ in $\mathbb{C}$, unless $a = b$. Moreover multiplication in different settings might not be commutative, for example when we multiply 2-by-2 matrices, as we discovered, in detail, in section 0.12 of appendix 0D.

Lemma 2.1.1. *If* $a \equiv b \pmod{m}$ *and* $c \equiv d \pmod{m}$, *then*
$$a + c \equiv b + d \pmod{m},$$
$$a - c \equiv b - d \pmod{m},$$
$$\text{and} \quad ac \equiv bd \pmod{m}.$$

Proof. By hypothesis there exist integers u and v for which $a - b = um$ and $c - d = vm$. Therefore
$$(a + c) - (b + d) = (a - b) + (c - d) = um + vm = (u + v)m$$
so that $a + c \equiv b + d \pmod{m}$;
$$(a - c) - (b - d) = (a - b) - (c - d) = um - vm = (u - v)m$$
so that $a - c \equiv b - d \pmod{m}$; and
$$ac - bd = a(c - d) + d(a - b) = a \cdot vm + d \cdot um = (av + du)m$$
so that $ac \equiv bd \pmod{m}$. □

These are the rules of *modular arithmetic*.

Exercise 2.1.9. Under the hypothesis of Lemma 2.1.1, show that $ka + lc \equiv kb + ld \pmod{m}$ for any integers k and l.

Exercise 2.1.10. If $p|m$ and $m/p \equiv a \pmod{q}$, then prove that $m \equiv ap \pmod{q}$.

2.2. The trouble with division

Although the rules for addition, subtraction, and multiplication work for congruences as they do for the integers, reals, and most other mathematical objects we have encountered, the rule for division is more subtle. In the complex numbers, if we are given numbers a and $b \neq 0$, then there exists a unique value of c for which $a = bc$ (so that $c = a/b$), and therefore there is no ambiguity in the definition of division. We now look at the multiplication tables mod 5 and mod 6 to see whether this same property holds for modular arithmetic:

×	0	1	2	3	4
0	0	0	0	0	0
1	0	1	2	3	4
2	0	2	4	1	3
3	0	3	1	4	2
4	0	4	3	2	1

The multiplication table (mod 5).

Other than in the top row, we see that every congruence class mod 5 appears exactly once in each row of the table. For example, in the row corresponding to the multiples of 2, mod 5 we have 0, 2, 4, 1, 3, which implies that for each $a \pmod{5}$

2.2. The trouble with division

there exists a unique value of $c \pmod 5$ for which $a \equiv 2c \pmod 5$; that is, $c \equiv a/2 \pmod 5$. We read off

$$0/2 \equiv 0 \pmod 5,\ 1/2 \equiv 3 \pmod 5,\ 2/2 \equiv 1 \pmod 5,$$
$$3/2 \equiv 4 \pmod 5,\ \text{and}\ 4/2 \equiv 2 \pmod 5,$$

each division leading to a unique value. This is true in each row, so for every non-zero value of $b \pmod 5$ and every $a \pmod 5$, there exists a unique multiple of b, which equals a mod 5. Therefore division is well- (and uniquely) defined modulo 5.

However, the multiplication table mod 6 looks rather different.

×	0	1	2	3	4	5
0	0	0	0	0	0	0
1	0	1	2	3	4	5
2	0	2	4	0	2	4
3	0	3	0	3	0	3
4	0	4	2	0	4	2
5	0	5	4	3	2	1

The multiplication table (mod 6).

The row corresponding to the multiples of 5, mod 6, is 0, 5, 4, 3, 2, 1, so that each $b/5 \pmod 6$ is well-defined.

However, the row corresponding to the multiples of 2, mod 6, reads 0, 2, 4, 0, 2, 4. There is no solution to $1/2 \pmod 6$. On the other hand, for something as simple as $4/2 \pmod 6$, there are two different solutions: 5 (mod 6) as well as 2 (mod 6). Evidently it is more complicated to understand division mod 6 than mod 5.

We can obtain a hint of what is going on by applying exercise 2.1.4, which implies that the union of the sets of integers in the two arithmetic progressions 5 (mod 6) and 2 (mod 6) gives exactly the integers $\equiv 2 \pmod 3$. So we now have a unique solution to $4/2 \pmod 6$, albeit a congruence class belonging to a different modulus.

Exercise 2.2.1. Determine one congruence class which gives all solutions to 3 divided by 3 (mod 6). (In other words, find a congruence class $a \pmod m$ such that $3x \equiv 3 \pmod 6$ if and only if $x \equiv a \pmod m$.)

These issues with division arise when we try to solve equations by division: If we divide each side of $8 \equiv 2 \pmod 6$ by 2, we obtain the incorrect "$4 \equiv 1 \pmod 6$". We can correct this by dividing the modulus through by 2 also, so as to obtain $4 \equiv 1 \pmod 3$. Even this is not the whole story, for if we wish to divide both sides of $21 \equiv 6 \pmod 5$ through by 3, we cannot also divide the modulus, since 3 does not divide 5. However, in this case one does not need to divide the modulus through by 3, since $7 \equiv 2 \pmod 5$. So what is the general rule? We shall resolve all of these issues in Lemma 3.5.1, after we have developed a little more theory.

2.3. Congruences for polynomials

Let $\mathbb{Z}[x]$ denote the set of polynomials with integer coefficients. Using the above rules for congruences, one gets a very useful result for congruences involving polynomials:

Corollary 2.3.1. *If $f(x) \in \mathbb{Z}[x]$ and $a \equiv b \pmod{m}$, then $f(a) \equiv f(b) \pmod{m}$.*

Proof. Since $a \equiv b \pmod{m}$ we have $a^2 \equiv b^2 \pmod{m}$ by Lemma 2.1.1, and then

Exercise 2.3.1. Prove that $a^k \equiv b^k \pmod{m}$ for all integers $k \geq 1$, by induction.

Now, writing $f(x) = \sum_{i=0}^{d} f_i x^i$ where each f_i is an integer, we have

$$f(a) = \sum_{i=0}^{d} f_i a^i \equiv \sum_{i=0}^{d} f_i b^i = f(b) \pmod{m},$$

by Lemma 2.1.1. □

This result can be extended to polynomials in many variables.

Exercise 2.3.2. Deduce, from Corollary 2.3.1, that if $f(t) \in \mathbb{Z}[t]$ and $r, s \in \mathbb{Z}$, then $r - s$ divides $f(r) - f(s)$.

Therefore, for any polynomial $f(x) \in \mathbb{Z}[x]$, the sequence $f(0), f(1), f(2), \ldots$ modulo m is *periodic* of period m; that is, the values repeat every mth term in the sequence, repeating indefinitely. More precisely $f(n + m) \equiv f(n) \pmod{m}$ for all integers n.

Example. If $f(x) = x^3 - 8x + 6$ and $m = 5$, then we get the sequence

$$f(0), f(1), \ldots = 1, 4, 3, 4, 3, 1, 4, 3, 4, 3, 1 \ldots$$

and the first five terms $1, 4, 3, 4, 3$ repeat infinitely often. Moreover we get the same pattern if we run though the consecutive negative integer values for x.

Note that in this example $f(x)$ is never 0 or $2 \pmod{5}$. Thus none of the equations

$$x^3 - 8x + 6 = 0, \qquad y^3 - 8y + 1 = 0, \qquad \text{and} \qquad z^3 - 8z + 4 = 0$$

can have solutions in integers x, y, or z.

Exercise 2.3.3. Let $f(x) \in \mathbb{Z}[x]$. Suppose that $f(r) \not\equiv 0 \pmod{m}$ for all integers r in the range $0 \leq r \leq m - 1$. Deduce that there does not exist an integer n for which $f(n) = 0$.

2.4. Tests for divisibility

There are easy tests for divisibility based on ideas from this chapter. For example, writing an integer in decimal as[6]

$$a + 10b + 100c + \cdots,$$

---

[6] More precisely, $\sum_{i=0}^{d} a_i 10^i$ where each a_i is an integer in $\{0, 1, 2, \ldots, 9\}$ and $a_d \neq 0$. Why did we write the decimal expansion so informally in the text, when surely good mathematics is all about precision? While good mathematics is anchored by precision, mathematical writing also requires good communication—after all why shouldn't the reader understand with as little effort as possible?—and so we attempt to explain accurately with as little notation as possible.

2.4. Tests for divisibility

we employ Corollary 2.3.1 with $f(x) = a + bx + cx^2 + \cdots$ and $m = 9$, so that
$$a + 10b + 100c + \cdots = f(10) \equiv f(1) = a + b + c + \cdots \pmod{9}.$$
Therefore we can test whether the integer $a + 10b + 100c + \cdots$ is divisible by 9 by testing whether the much smaller integer $a + b + c + \cdots$ is divisible by 9. In other words, if an integer is written in decimal notation, then it is divisible by 9 if and only if the sum of its digits is divisible by 9. This same test works for divisibility by 3 (by exercise 2.1.6) since 3 divides 9. For example, to decide whether 7361842509 is divisible by 9, we need only decide whether $7+3+6+1+8+4+2+5+0+9 = 45$ is divisible by 9, and this holds if and only if $4 + 5 = 9$ is divisible by 9, which it obviously is.

One can test for divisibility by 11 in a similar way: Since $10 \equiv -1 \pmod{11}$, we deduce that $f(10) \equiv f(-1) \pmod{11}$ from Corollary 2.3.1, and so
$$a + 10b + 100c + \cdots \equiv a - b + c \cdots \pmod{11}.$$
Therefore 7361842509 is divisible by 11 if and only if $7-3+6-1+8-4+2-5+0-9 = 1$ is divisible by 11, which it is not.

One may determine similar (but slightly more complicated) rules to test for divisibility by *any* integer, though we will need to develop our theory of congruences. We return to this theme in section 7.7.

Exercise 2.4.1. (a) Invent tests for divisibility by 2 and 5 (easy).
(b) Invent tests for divisibility by 7 and 13 (similar to the above).
(c)† Create one test that tests for divisibility by 7, 11, and 13 simultaneously (assuming that one knows about the divisibility by 7, 11, and 13 of every non-negative integer up to 1000).

Additional exercises

Exercise 2.5.1. Prove that if a, b, and c are integers and $d = b^2 - 4ac$, then $d \equiv 0$ or $1 \pmod{4}$.

Exercise 2.5.2. Prove that if $N = a^2 - b^2$, then either N is odd or N is divisible by 4.

Exercise 2.5.3. (a) Prove that 2 divides $n(3n + 101)$ for every integer n.
(b) Prove that 3 divides $n(2n + 1)(n + 10)$ for every integer n.
(c) Prove that 5 divides $n(n + 1)(2n + 1)(3n + 1)(4n + 1)$ for every integer n.

Exercise 2.5.4. (a) Prove that, for any given integer $k \geq 1$, exactly k of any km consecutive integers is $\equiv a \pmod{m}$.
(b)† Let I be an interval of length N. Prove that the number of integers in I that are $\equiv a \pmod{m}$ is between $N/m - 1$ and $N/m + 1$.
(c) By considering the number of even integers in $(0, 2)$ and then in $[0, 2]$, show that (b) cannot be improved, in general.

Exercise 2.5.5. The *Universal Product Code* (that is, the bar code used to identify items in the supermarket) has 12 digits, each between 0 and 9, which we denote by $d_1, \ldots, d_{12}$. The first 11 digits identify the product. The 12th is chosen to be the least residue of
$$3d_1 - d_2 + 3d_3 - d_4 - \cdots - d_{10} + 3d_{11} \pmod{10}.$$
(a) Deduce that $d_1 + 3d_2 + d_3 + \cdots + d_{11} + 3d_{12}$ is divisible by 10.
(b) Deduce that if the scanner does not read all the digits correctly, then either the sum in (a) will not be divisible by 10 or the scanner has misread at least two digits.

Exercise 2.5.6. (a) Take $f(x) = x^2$ in Corollary 2.3.1 to determine the squares modulo m, for $m = 3, 4, 5, 6, 7, 8, 9$, and 10. ("The squares modulo m" are those congruence classes (mod m) that are equivalent to the square of at least one congruence class (mod m).)

(b) Show that there are no solutions in integers x, y, z to $x^2 + y^2 = z^2$ with x and y odd.
(c) Show that if $x^2 + y^2 = z^2$, then 3 divides xy.
(d) Show that there are no solutions in integers x, y, z to $x^2 + y^2 = 3z^2$ with $(x, y) = 1$.
(e) Show that there are no solutions in integers x, y, z to $x^2 + y^2 = 666z^2$ with $(x, y) = 1$.
(f) Prove that no integer $\equiv 7 \pmod{8}$ can be written as the sum of three squares of integers.

Exercise 2.5.7.[†] Show that if $x^3 + y^3 = z^3$, then 7 divides xyz.

Binomial coefficients modulo p

We will assume that p is prime for all of the next two sections.

Exercise 2.5.8. Use the formula for $\binom{p}{j}$ given in (0.3.1) to prove that p divides $\binom{p}{j}$ for all integers j in the range $1 \leq j \leq p-1$. This implies that $\frac{1}{p}\binom{p}{j}$ is an integer.

For $1 \leq j \leq p-1$ we can write $\binom{p-1}{j}$ as $\frac{p-1}{1}\frac{p-2}{2}\cdots\frac{p-j}{j}$. There is considerable cancelation when we reduce this latter expression mod p.

Exercise 2.5.9. (a) Prove that $\binom{p-1}{j} \equiv (-1)^j \pmod{p}$ for all j, $0 \leq j \leq p-1$.
(b) Prove that $\frac{1}{p}\binom{p}{j} \equiv (-1)^{j-1}/j \pmod{p}$ for all j, $1 \leq j \leq p-1$.

Exercise 2.5.10.[†] (a) Prove that $\binom{ap}{bp} \equiv \binom{a}{b} \pmod{p}$ whenever $a, b \geq 0$.
(b) Prove that $\binom{ap+c}{bp+d} \equiv \binom{a}{b} \cdot \binom{c}{d} \pmod{p}$ whenever $0 \leq c, d \leq p-1$. (Remember that $\binom{c}{d} = 0$ if $c < d$.)
(c) If $m = m_0 + m_1 p + m_2 p^2 + \cdots + m_k p^k$ and $n = n_0 + n_1 p + \cdots + n_k p^k$ are non-negative integers written in base p, deduce *Lucas's Theorem* (by induction on $k \geq 0$), that
$$\binom{n}{m} \equiv \binom{n_0}{m_0}\binom{n_1}{m_1}\binom{n_2}{m_2}\cdots\binom{n_k}{m_k} \pmod{p}.$$

One can extend the notion of congruences to polynomials with integer coefficients: For $f(x), g(x) \in \mathbb{Z}[x]$ we have $f(x) \equiv g(x) \pmod{m}$ if and only if there exists a polynomial $h(x) \in \mathbb{Z}[x]$ for which $f(x) - g(x) = mh(x)$. This notion can be extended even further to polynomials in several variables.

The binomial theorem for $n = 3$ gives
$$(x+y)^3 = x^3 + 3x^2y + 3xy^2 + y^3.$$
Notice that the two middle coefficients here are both 3, and so
$$(x+y)^3 \equiv x^3 + y^3 \pmod{3}.$$
Similarly
$$(x+y)^5 = x^5 + 5x^4y + 10x^3y^2 + 10x^2y^3 + 5xy^4 + y^5 \equiv x^5 + y^5 \pmod{5},$$
since all four of the middle coefficients are divisible by 5. This does not generalize to all exponents n, for example for $n = 4$ we have $(x+y)^4 \equiv x^4 + 2x^2y^2 + y^4 \pmod{4}$ which is not congruent to $x^4 + y^4 \pmod{4}$, but the above does generalize to all prime exponents, as we will see in the next exercise.

Exercise 2.5.11. Deduce from exercise 2.5.8 that $(x+y)^p \equiv x^p + y^p \pmod{p}$ for all primes p.[7]

Exercise 2.5.12. Prove that $(x+y)^{p-1} \equiv x^{p-1} - yx^{p-2} + \cdots - xy^{p-2} + y^{p-1} \pmod{p}$.

---
[7]This is sometimes called the *freshman's dream* or the *child's binomial theorem*, sarcastically referring to the unfortunately common mistaken belief that this works over $\mathbb{C}$, rather than the more complicated binomial theorem, as in section 0.3.

Exercise 2.5.13. Prove that $(x+y)^{p^k} \equiv x^{p^k} + y^{p^k} \pmod{p}$ for all primes p and integers $k \geq 1$.

Exercise 2.5.14. (a) Writing a positive integer $n = n_0 + n_1 p + n_2 p^2 + \cdots$ in base p, use exercise 2.5.13 to prove that
$$(x+y)^n \equiv (x+y)^{n_0}(x^p + y^p)^{n_1}(x^{p^2} + y^{p^2})^{n_2} \cdots \pmod{p}.$$
(b)† Reprove Lucas's Theorem (as in exercise 2.5.10(c)) by studying the coefficient of $x^m y^{n-m}$ in (a).

Exercise 2.5.15. (a) Prove that $(x+y+z)^p \equiv x^p + y^p + z^p \pmod{p}$.
(b) Deduce that $(x_1 + x_2 + \cdots + x_n)^p \equiv x_1^p + x_2^p + \cdots + x_n^p \pmod{p}$ for all $n \geq 2$.

The Fibonacci numbers modulo d

The Fibonacci numbers mod 2 are
$$0, 1, 1, 0, 1, 1, 0, 1, 1, \ldots.$$
We see that the Fibonacci numbers modulo 2 are periodic of period 3. The Fibonacci numbers mod 3 are
$$0, 1, 1, 2, 0, 2, 2, 1, 0, 1, 1, 2, 0, 2, 2, 1, \ldots$$
and so seem to be periodic of period 8. In exercise 1.7.24 we defined $m = m_d$ to be the smallest positive integer for which d divides F_m and showed that d divides F_n if and only if m_d divides n. In our two cases we therefore have $m_2 = 3$ which is the period and $m_3 = 4$ which is half the period.

In the next exercise we show that Fibonacci numbers (and other such sequences) are periodic mod d, for every integer $d > 1$, by using the *pigeonhole principle*. This states that if one puts $N+1$ letters into N pigeonholes, then, no matter how one does this, some pigeonhole will contain at least two letters.[8]

Exercise 2.5.16. (a) Prove that the pigeonhole principle is true.
We will now show that the Mersenne numbers $M_n := 2^n - 1$ are periodic mod d.
(b) Show that there exist two integers in the range $0 \leq r < s \leq d$ for which $M_r \equiv M_s \pmod{d}$.
(c) In exercise 0.4.15(b) we saw that the Mersenne numbers satisfy the recurrence $M_{n+1} = 2M_n + 1$. Use this to show that $M_{r+j} \equiv M_{s+j} \pmod{d}$ for all $j \geq 0$.
(d) Deduce that there exists a positive integer $p = p_d$, which is $\leq d$, such that $M_{n+p} \equiv M_n \pmod{d}$ for all $n \geq d$. That is, M_n is *eventually periodic* mod d with period $p_d \leq d$.

An analogous proof works for general second-order linear recurrence sequences, including Fibonacci numbers. For the rest of this section, we suppose a and b are integers and $\{x_n : n \geq 0\}$ is the second-order linear recurrence sequence given by
$$x_n = ax_{n-1} + bx_{n-2} \text{ for all } n \geq 2 \text{ with } x_0 = 0 \text{ and } x_1 = 1.$$

Exercise 2.5.17. (a) By using the pigeonhole principle creatively, prove that there exist two integers in the range $0 \leq r < s \leq d^2$ for which $x_r \equiv x_s \pmod{d}$ and $x_{r+1} \equiv x_{s+1} \pmod{d}$.
(b) Use the recurrence for the x_n to show that $x_{r+j} \equiv x_{s+j} \pmod{d}$ for all $j \geq 0$.
(c) Deduce that the x_n are eventually periodic mod d with period $\leq d^2$.
(d) Prove that m_d divides the period mod d.

---

[8] In French, this is the "principle of the drawers". What invocative metaphors are used to describe this principle in other languages?

We saw above that the Fibonacci numbers mod 3 have period 3^2-1, and further calculations reveal that the period mod d never seems to be larger than $d^2 - 1$, a small improvement over the bound that we obtained in exercise 2.5.17(c). In the next exercise we see how to obtain this bound, in general.

Exercise 2.5.18. (a) Show that if there exists a positive integer r for which $x_r \equiv x_{r+1} \equiv 0$ (mod d), then $x_n \equiv 0$ (mod d) for all $n \geq r$ so that the x_n are eventually periodic mod d with period 1.
(b) Now assume that there does not exist a positive integer r for which $x_r \equiv x_{r+1} \equiv 0$ (mod d). Modify the proof of exercise 2.5.17 to prove that the x_n are eventually periodic mod d with period $\leq d^2 - 1$.

It is possible to get a more precise understanding of the Fibonacci numbers and other second-order recurrences, mod d:

Exercise 2.5.19. In order to understand x_n (mod d), we take $m = m_d$ in the results of this exercise.
(a) Prove, by induction, that $x_{m+k} \equiv x_{m+1} x_k$ (mod x_m) for all $k \geq 0$.
(b) Deduce the same result from exercise 0.4.10.
(c) Deduce that if $n = qm + r$ with $0 \leq r \leq m - 1$, then $x_n \equiv (x_{m+1})^q x_r$ (mod x_m).

We will return to this result in chapter 7 where we study the powers mod n.

In exercise 0.1.5 we saw the importance of the discriminant[9] $\Delta := a^2 + 4b$ of the quadratic polynomial $x^2 - ax - b$. The rule for the x_n mod Δ is a little easier:

Exercise 2.5.20. Prove by induction that
(a) $x_{2k} \equiv ka(-b)^{k-1}$ (mod Δ) and $x_{2k+1} \equiv (2k+1)(-b)^k$ (mod Δ) for all $k \geq 0$ and
(b) $x_{2k} \equiv kab^{k-1}$ (mod a^2) and $x_{2k+1} \equiv b^k$ (mod a^2) for all $k \geq 0$.

Exercise 2.5.21. Suppose that the sequence $(u_n)_{n\geq 1}$ satisfies a dth-order linear recurrence (as defined in appendix 0B). Prove that for any integer $m > 1$, the u_n are eventually periodic mod m with period $\leq m^d - 1$. (We prove that this bound is best possible when m is prime in exercise 7.25.5.)

---

[9]The colon ":" plays many roles in the grammar of mathematics. Here it means that "Henceforth we define Δ to be ...."

Appendix 2A. Congruences in the language of groups

2.6. Further discussion of the basic notion of congruence

Congruences can be rephrased in the language of groups. The integers, $\mathbb{Z}$, form a group,[10] in which addition is the group operation. In exercise 0.11.1 of appendix 0D we proved that the non-trivial, proper subgroups of $\mathbb{Z}$ all take the form $m\mathbb{Z} := \{mn : n \in \mathbb{Z}\}$ for some integer $m > 1$, that is, the set of integers divisible by m. The congruence classes (mod m) are simply the *cosets* of $m\mathbb{Z}$ inside $\mathbb{Z}$:

$$0 + m\mathbb{Z},\ 1 + m\mathbb{Z},\ 2 + m\mathbb{Z}, \ldots,\ (m-1) + m\mathbb{Z},$$

where

$$j + m\mathbb{Z} := \{j + mn : n \in \mathbb{Z}\},$$

which is the set of integers belonging to the congruence class j (mod m). Notice that the m cosets of $m\mathbb{Z}$ are disjoint and their union gives all of $\mathbb{Z}$.

The group operation on $\mathbb{Z}$, namely addition, is inherited by the cosets of $m\mathbb{Z}$. For example, as $7 + 11 = 18$ in $\mathbb{Z}$, the same is true when we add together the relevant cosets of $m\mathbb{Z}$ in $\mathbb{Z}$; in other words,[11]

$$(7 + m\mathbb{Z}) + (11 + m\mathbb{Z}) = (18 + m\mathbb{Z}).$$

This new additive group is the *quotient group*

$$\mathbb{Z}/m\mathbb{Z}.$$

This is the beginning of the theory of quotient groups, which we develop in the next section.

---

[10] See appendix 0D for a discussion of the basic properties of groups.
[11] Throughout, we define the sum of two given sets A and B to be $A + B := \{a + b : a \in A, b \in B\}$, that is, the set of elements that can be represented as $a + b$ with $a \in A$ and $b \in B$. Note that an element may be represented more than once.

The reader should be aware that multiplication mod m (and, in particular, how its properties are inherited from $\mathbb{Z}$) does not fit into this discussion of additive quotient groups.

2.7. Cosets of an additive group

Suppose that H is a subgroup of an additive (and so abelian[12]) group G. A *coset* of H in G is given by the set
$$a + H := \{a + h : h \in H\}.$$
In Proposition 2.7.1 we will show, as in the example $m\mathbb{Z}$ of the previous section, that the cosets of H are all disjoint and their union gives G.

The *quotient group* G/H has as its elements the distinct cosets $a + H$ and inherits its group law from G, in this case addition, so that
$$(a + H) + (b + H) = (a + b) + H.$$

Proposition 2.7.1. *Let H be a subgroup of an additive group G. The cosets of H in G are disjoint, so that the elements of G/H are well-defined; and the addition law on G/H is also well-defined. If G is finite, then $|H|$ divides $|G|$ and $|G/H| = |G|/|H|$.*

Proof. If $a + H$ and $b + H$ have a common element c, then there exists $h_1, h_2 \in H$ such that $a + h_1 = c = b + h_2$. Therefore $b = a + h_1 - h_2 = a + h_0$ where $h_0 = h_1 - h_2 \in H$ since H is a group (and therefore closed under addition). Now if $h \in H$, then $b + h = a + (h_0 + h) \in a + H$, as $h_0 + h \in H$, so that $b + H \subset a + H$, and by the analogous argument $a + H \subset b + H$. We deduce that $a + H = b + H$. Hence the cosets of H are either identical or disjoint, which means that they partition G; therefore if G is finite, then $|H|$ divides $|G|$.

This also implies that if $c \in a + H$, then $c + H = a + H$. We wish to show that addition in G/H is well-defined. If $a + H$, $b + H$ are cosets of H, then we defined $(a + H) + (b + H) = (a + b) + H$, so we need to verify that the sum of the two cosets does not depend on the choice of representatives of the cosets. So, if $c \in a + H$ and $d \in b + H$, then there exists $h_1, h_2 \in H$ for which $c = a + h_1$ and $d = b + h_2$. Then $c + H = a + H$ and $d + H = b + H$. Moreover $c + d = a + b + (h_1 + h_2) \in a + b + H$, as H is closed under addition, and so $c + d + H = a + b + H$, as desired. Hence G/H is well-defined, and $|G/H| = |G|/|H|$ when G is finite. □

Example. $\mathbb{Z}$ is a subgroup of the additive group $\mathbb{R}$, and the cosets $a + \mathbb{Z}$ are given by all real numbers r that differ from a by an integer. Every coset $a + \mathbb{Z}$ has exactly one representative in any given interval of length 1, in particular the interval $[0, 1)$ where the coset representative is $\{a\}$, the fractional part of a. These cosets are well-defined under addition and yield the quotient group $\mathbb{R}/\mathbb{Z}$.

The *exponential map* $e : \mathbb{R} \to U := \{z \in \mathbb{C} : |z| = 1\}$, from the real numbers to the unit circle, is defined by $e(t) = e^{2i\pi t}$. Since $e(1) = 1$, therefore $e(n) = e(1)^n = 1$ for every integer n. Therefore if $b \in a + \mathbb{Z}$ so that $b = a + n$ for some integer n, then $e(b) = e(a + n) = e(a)e(n) = e(a)$, so the value of $e(t)$ depends only what

---

[12] A group G is called *abelian* or *commutative* if $ab = ba$ for all elements $a, b \in G$.

coset t belongs to in $\mathbb{R}/\mathbb{Z}$. Therefore we can think of the exponential map as the concatenation of two maps: firstly the natural quotient map from $\mathbb{R} \to \mathbb{R}/\mathbb{Z}$ (that is, $a \to a + \mathbb{Z}$) and then the map $e : \mathbb{R}/\mathbb{Z} \to U$. Picking the representatives $[0, 1)$ for $\mathbb{R}/\mathbb{Z}$, we see that the restricted map $e : [0, 1) \to U$ is 1-to-1.

By a slight abuse of terminology, we let $a \equiv b \pmod 1$, for real numbers a and b, if and only if a and b belong to the same coset of $\mathbb{R}/\mathbb{Z}$.

Exercise 2.7.1. Prove that $a \equiv b \pmod m$ if and only if a/m and b/m belong to the same coset of $\mathbb{R}/\mathbb{Z}$.

Exercise 2.7.2. (a) Prove that $t \equiv \{t\} \pmod 1$ for all real numbers t.
(b) Prove that the usual rules of addition, subtraction, and multiplication hold mod 1.
(c) Show that division is not always well-defined mod 1, by finding a counterexample.

2.8. A new family of rings and fields

We have seen, in Lemma 2.1.1, that the congruence classes mod m support both an additive and multiplicative structure.

Exercise 2.8.1. Prove that $\mathbb{Z}/m\mathbb{Z}$ is a ring for all integers $m \geq 2$.

To be a field, all the non-zero congruence classes of $\mathbb{Z}/m\mathbb{Z}$ would need to have a multiplicative inverse, but this is not the case for all m. For example we claim that 3 does not have a multiplicative inverse mod 15. If it did, say $3m \equiv 1 \pmod{15}$, then multiplying through by 5 we obtain $5 \equiv 5 \cdot 1 \equiv 5 \cdot 3m \equiv 0 \pmod{15}$, which is evidently untrue.

We call 3 and 5 *zero divisors* since they non-trivially divide 0 in $\mathbb{Z}/15\mathbb{Z}$.

Exercise 2.8.2. (a) Prove that if m is a composite integer > 1, then $\mathbb{Z}/m\mathbb{Z}$ has zero divisors.
(b) Prove that $\mathbb{Z}/m\mathbb{Z}$ is not a field whenever m is a composite integer > 1.
(c) Prove that if R is any ring with zero divisors, then R cannot be a field.

An *integral domain* is a ring with no zero divisors. Note that $\mathbb{Z}$ is an integral domain (hence the name) but is not a field.

If R is a commutative ring and $m \in R$, then mR is an additive subgroup of R, and the cosets of mR support a multiplicative structure. To see this, note that if $x \in a + mR$ and $y \in b + mR$, then $x = a + mr_1$ and $y = b + mr_2$ for some $r_1, r_2 \in R$, and so $xy = ab + mr$ where $r = ar_2 + br_1 + mr_1r_2$ which belongs to R, as R is closed under both addition and multiplication. That is, $xy \in ab + mR$. Hence R/mR inherits the multiplicative and distributive properties of R, as well as the identity element $1 + mR$; and so R/mR is itself a commutative ring.

2.9. The order of an element

If g is an element of a given group G, we define the *order* of g to be the smallest integer $n \geq 1$ for which $g^n = 1$, where 1 is the identity element of G. If n does not exist, then we say that g has infinite order (for example, 1 in the additive group $\mathbb{Z}$). We shall explore the multiplicative order of a reduced residue mod m, in detail, in chapter 7.

There is a beautiful observation of Lagrange which restricts the possible order of an element in any finite abelian group.

Theorem 2.2 (Lagrange). *If G is a finite abelian group, then the order of any element g of G divides $|G|$, the number of elements in G. Moreover, $g^{|G|} = 1$.*

Proof. Suppose that g has order n and let $H := \{1, g, g^2, \ldots, g^{n-1}\}$, a subgroup of G of order n. By Proposition 2.7.1 we deduce that $n = |H|$ divides $|G|$. Moreover if $|G| = mn$, then $g^{|G|} = g^{mn} = (g^n)^m = 1^m = 1$. □

Lagrange's Theorem actually holds for any finite group, non-abelian as well as abelian, as we will see in Corollary 7.23.1 of appendix 7D.

Appendix 2B. The Euclidean algorithm for polynomials

We use the Euclidean algorithm to find the greatest common divisor of two integers. There are sets of mathematical objects, other than the integers, in which an analogy to the Euclidean algorithm works, but to do so, there must be notions analogous to "divisor" and "greatest". Moreover we know that the Euclidean algorithm terminates in finitely many steps because there are only finitely many positive integers smaller than a given integer; the analogous statement is not true for the real numbers or the rationals. However, one can use the Euclidean algorithm to find the highest degree common polynomial factor of two given polynomials.

2.10. The Euclidean algorithm in $\mathbb{C}[x]$

Non-zero polynomials take the form

$$f(x) = a_0 + a_1 x + \cdots + a_d x^d \text{ where the leading coefficient } a_d \neq 0,$$

for some integer $d \geq 0$, where the a_j's belong to some set of mathematical objects, for example, $\mathbb{C}$, $\mathbb{Q}$, or $\mathbb{Z}$. We call $a_d x^d$ the *leading term*. It is often possible to assume, without loss of generality, that the polynomials involved are *monic* (that is, $a_d = 1$), since $f(x)$ has exactly the same roots as $cf(x)$, for any non-zero constant c (so we can divide $f(x)$ through by its leading coefficient).

Now suppose that $f(x), g(x) \in \mathbb{C}[x]$. We say that "$g(x)$ divides $f(x)$" if there exists a polynomial $h(x) \in \mathbb{C}[x]$, for which $f(x) = g(x)h(x)$. If $f(x) \neq 0$, then $\deg f = \deg g + \deg h$; in particular, $\deg f \geq \deg g$. The following is the polynomial analogy to Lemma 1.1.1:

Proposition 2.10.1. *Suppose that $f(x), g(x) \in \mathbb{C}[x]$, where $D := \deg(g) \geq 1$. There exist unique polynomials $q(x)$, $r(x) \in \mathbb{C}[x]$ for which*

$$f(x) = q(x)g(x) + r(x),$$

where $r(x)$ has degree $< D$.

Proof. We prove the existence of r by induction on the degree of f. Suppose that $g(x)$ has leading term $b_D x^D$ with $b_D \neq 0$. If $\deg(f) < D$, then let $q(x) = 0$ and $r(x) = f(x)$. Otherwise, suppose that $f(x)$ has leading term $a_d x^d$ with $d \geq D$ and let $F(x) := f(x) - (a_d/b_D)x^{d-D}g(x)$. The leading terms of $f(x)$ and $(a_d/b_D)x^{d-D}g(x)$ are both $a_d x^d$, and so F has lower degree than f. By induction we then know that there exists a polynomial $Q(x)$ for which $r(x) = F(x) - Q(x)g(x)$ has degree $< D$, and the result follows taking $q(x) = Q(x) + (a_d/b_D)x^{d-D}$.

To prove that r is unique, suppose that we also have $f(x) = Q(x)g(x) + R(x)$ where $R(x)$ has degree $< D$. Then

$$(r-R)(x) = (f(x) - q(x)g(x)) - (f(x) - Q(x)g(x)) = (Q-q)(x)g(x);$$

that is, $g(x)$ divides $(r-R)(x)$. Now $\deg(r-R) \leq \max\{\deg(r), \deg(R)\} < D = \deg g$, and so $r - R = 0$. Therefore $R = r$ and $Q = q$. □

Exercise 2.10.1. Prove that if $a \in \mathbb{C}$, then $x - a$ is a factor of $f(x)$ if and only if $f(a) = 0$.

The *greatest common divisor* of $f(x)$ and $g(x)$ is the monic polynomial $h(x)$ of highest degree which divides both $f(x)$ and $g(x)$. We denote this by

$$\gcd(f(x), g(x))_{\mathbb{C}[x]}.$$

One can develop a theory analogous to the integer case; for example, one can show that any common divisor of f and g must divide $h(x)$.

We now describe the Euclidean algorithm: Given $f(x), g(x) \in \mathbb{C}[x]$ for which $\deg(f) \geq \deg(g)$ we appeal to Proposition 2.10.1 so that there exists $q(x) \in \mathbb{C}[x]$ for which $r(x) = f(x) - q(x)g(x)$ has degree $< D$. We claim that

$$\gcd(f(x), g(x))_{\mathbb{C}[x]} = \gcd(g(x), r(x))_{\mathbb{C}[x]}.$$

To prove this, note that if $P(x)$ divides both $g(x)$ and $r(x)$, then it divides the linear combination $q(x)g(x) + r(x) = f(x)$. Alternatively if $P(x)$ divides both $f(x)$ and $g(x)$, then it divides the linear combination $f(x) - q(x)g(x) = r(x)$.

We deduce that the Euclidean algorithm in $\mathbb{C}[x]$ does indeed yield the greatest common divisor, since the sum of the degrees of the polynomials involved reduce by at least 1 at each step.

Exercise 2.10.2. Suppose the Euclidean algorithm gives the polynomials $F_0 = f, F_1 = g, F_2, F_3,$ ... . Prove that, for all $k \geq 0$, there exist $a_k(x), b_k(x) \in \mathbb{C}[x]$ for which $F_k(x) = a_k(x)f(x) + b_k(x)g(x)$.

In exercise 2.10.2 we proved there exist $a(x), b(x) \in \mathbb{C}[x]$ for which

$$a(x)f(x) + b(x)g(x) = \gcd(f(x), g(x))_{\mathbb{C}[x]},$$

which we will denote by $h(x)$. We would like to bound the degrees of the polynomials $a(x)$ and $b(x)$ (just as we controlled the sizes of the integers involved in exercise 1.7.10). By Proposition 2.10.1 there exists $q(x) \in \mathbb{C}[x]$ for which $A(x) := a(x) - q(x)g(x)$ has degree $< \deg(g)$. Let $B(x) := b(x) + q(x)f(x)$, so that

$$A(x)f(x) + B(x)g(x) = a(x)f(x) + b(x)g(x) = h(x).$$

Then $\deg(B(x)g(x)) \leq \max\{\deg(A(x)f(x)), \deg(h(x))\} < \deg(f) + \deg(g)$, and therefore $\deg(A) < \deg(g)$ and $\deg(B) < \deg(f)$.

Exercise 2.10.3. (a) Show that A and B (with these degree bounds) are unique, up to a scalar multiple.
(b) Write $f = hF$ and $g = hG$ where $h = (f, g)$. Prove that all solutions of $a(x)f(x) + b(x)g(x) = h(x)$ are given by $a = A + kG$ and $b = B - kF$ for any $k(x) \in \mathbb{C}[x]$.

All of the arguments in this section work for the polynomials whose coefficients come from any field, in place of $\mathbb{C}[x]$, for example $\mathbb{R}[x]$ or $\mathbb{Q}[x]$. In other words, we can state that for any two given $f(x), g(x) \in \mathbb{Q}[x]$ there exist $a(x), b(x) \in \mathbb{Q}[x]$ with $\deg(a) < \deg(g)$ and $\deg(b) < \deg(f)$, for which

$$\gcd(f(x), g(x))_{\mathbb{Q}[x]} = a(x)f(x) + b(x)g(x).$$

One interesting consequence of this construction is that if $f(x), g(x) \in \mathbb{Q}[x]$ have a common root in $\mathbb{C}$, then they have a common polynomial factor in $\mathbb{Q}[x]$.

Exercise 2.10.4. Suppose that $f(x), g(x) \in \mathbb{Z}[x]$. Prove that $\gcd(f(x), g(x))_{\mathbb{C}[x]} = 1$ if and only if $\gcd(f(x), g(x))_{\mathbb{Q}[x]} = 1$.

Exercise 2.10.5. (a) Explain why the proof of Proposition 2.10.1 works in any field in place of $\mathbb{C}$.
(b) Prove that the result holds with $f(x), g(x) \in \mathbb{Z}[x]$, whenever g is monic.
(c) When $f(x), g(x) \in \mathbb{Z}[x]$ and $g(x)$ has leading coefficient $c \neq 0$, show that the result follows with "$f(x) = q(x)g(x) + r(x)$" replaced by "$c^{1+\deg f - \deg g} f(x) = q(x)g(x) + r(x)$".

2.11. Common factors over rings: Resultants and discriminants

Now suppose that $f(x), g(x) \in \mathbb{Z}[x]$ have no common polynomial factors in $\mathbb{Z}[x]$. We just saw that there exist $A(x), B(x) \in \mathbb{Q}[x]$ with $\deg(A) < \deg(g)$ and $\deg(B) < \deg(f)$, for which

$$A(x)f(x) + B(x)g(x) = 1.$$

We multiply $A(x)$ and $B(x)$ through by the least common multiple of the denominators of their coefficients and then divide the resulting polynomials through by the greatest common divisor of their numerators. We therefore obtain $a(x), b(x) \in \mathbb{Z}[x]$ with $\deg(a) < \deg(g)$ and $\deg(b) < \deg(f)$, for which

$$a(x)f(x) + b(x)g(x) = R,$$

where $R \in \mathbb{Z}[x] \cap \mathbb{Q} = \mathbb{Z}$. We call R the *resultant* of f and g.[13]

A particularly interesting case is when $g(x) = f'(x)$. Then f and g have no common root if and only if f has no repeated factor; that is, $f(x)$ is the product of distinct linear polynomials over $\mathbb{C}[x]$. The resultant of f and f' is called the *discriminant* of f. For example, if $f(x) = ax^2 + bx + c$, then $f'(x) = 2ax + b$. Applying the Euclidean algorithm,

$$2f(x) - xf'(x) = bx + 2c, \quad \text{and then}$$
$$2a(bx + 2c) - bf'(x) = -\Delta, \quad \text{where } \Delta := b^2 - 4ac.$$

---
[13]This definition is only guaranteed to agree with the usual definition in algebra books when f and g are both monic. For the non-monic case, the algebra books tend to be misleading.

This yields that the discriminant
$$\Delta := b^2 - 4ac = -4a(ax^2 + bx + c) + (2ax + b)^2.$$

We have seen, in section 0.8, that the general cubic can be taken to be of the form $f(x) = x^3 + ax + b$, so that $f'(x) = 3x^2 + a$. Therefore the discriminant is
$$\Delta := -4a^3 - 27b^2 = 9(2ax - 3b)f(x) - (6ax^2 - 9bx + 4a^2)f'(x).$$

2.12. Euclidean domains

To have (the analogy to) the Euclidean algorithm in a given integral domain R we need something like Lemma 1.1.1 and Proposition 2.10.1; that is, when we divide a by non-zero b in R, then the remainder must be "smaller" than b. To be precise, R is a *Euclidean domain* if there exists a function $w : R \to \mathbb{Z}_{\geq 0}$ such that:

For any $a, b \in R$ with $b \neq 0$, there exists $q \in R$ such that if $r := a - qb$, then
$$w(r) < w(b).$$

Moreover we need that $w(r) = 0$ only if $r = 0$.

When $R = \mathbb{Z}$ we had $w(n) = |n|$; and when $R = \mathbb{C}[x]$ we had $w(f) = \deg f$.

Exercise 2.12.1. Prove that the Euclidean algorithm works in a Euclidean domain.

Let $R = \mathbb{Z}[i]$ and $w(a+bi) = |a+bi|$. To study divisibility by $\beta = a + bi \neq 0$ we move from the usual basis $1, i$ for the complex plane over $\mathbb{R}$, to the basis $a+bi, b-ai$ ($= -i(a+bi)$). Therefore if $\alpha = x + yi$, then

(2.12.1) $\qquad \alpha = X(a+bi) - Y(b-ai)$ where $x = aX - bY$ and $y = bX + aY$,

and so we can determine X and Y from given values of x and y by inverting this pair of simultaneous equations. (X and Y are real but not necessarily integers.)

To divide $\alpha = x + yi$ by $\beta = a + bi \neq 0$, we write α as in (2.12.1) and then select m to be the nearest integer to X and n to be the nearest integer to Y, so that $|m - X|, |n - Y| \leq \frac{1}{2}$. Let $q = m + ni$ so that
$$q\beta = (m+ni)(a+bi) = m(a+bi) - n(b-ai),$$
and therefore
$$r := \alpha - q\beta = (X(a+bi) - Y(b-ai)) - (m(a+bi) - n(b-ai))$$
$$= (X-m)(a+bi) + (n-Y)(b-ai),$$
and so
$$w(r) = |r| \leq |X-m||a+bi| + |n-Y||b-ai| \leq \frac{1}{2}|a+bi| + \frac{1}{2}|b-ai| = |a+bi| = w(\beta).$$

We cannot get equality here or else $a + bi$ and $b - ai = -i(a+bi)$ would be parallel, which is obviously nonsense. Hence $w(r) < w(\beta)$, as desired. $\square$

Exercise 2.12.2. Prove that $\mathbb{Z}[\omega]$ is a Euclidean domain, where $\omega = \frac{-1+\sqrt{-3}}{2}$.

2.12. Euclidean domains

Exercise 2.12.3. Suppose that R is a Euclidean domain (as defined above). Prove that for any $a, b \in R$ one can find $g \in R$ such that

- g divides both a and b using the Euclidean algorithm,
- there exists $u, v \in R$ for which $au + bv = g$, and
- if d also divides both a and b, then $w(d) \leq w(g)$.

We call g the greatest common divisor of a and b, measuring size using the function w.

Franz Lemmermeyer's survey *The Euclidean algorithm in algebraic number fields*, Exposition. Math. 13 (1995), no. 5, 38–416, gives more information on Euclidean domains.

Chapter 3

The basic algebra of number theory

A *prime number* is an integer $n > 1$ whose only positive divisors are 1 and n. Hence $2, 3, 5, 7, 11, \ldots$ are primes. An integer $n > 1$ is *composite* if it is not prime.[1]

Exercise 3.0.1. Suppose that p is a prime number. Prove that $\gcd(p, a) = 1$ if and only if p does not divide a.

3.1. The Fundamental Theorem of Arithmetic

Positive integers factor into primes, the basic building blocks out of which integers are made. Often, in school, one discovers this by factoring a given composite integer into two parts and then factoring each of those parts that are composite into two further parts, etc. For example $120 = 8 \times 15$, and then $8 = 2 \times 4$ and $15 = 3 \times 5$. Now 2, 3, and 5 are all primes, but $4 = 2 \times 2$ is not. Putting this altogether gives $120 = 2 \times 2 \times 2 \times 3 \times 5$. This can be factored no further since 2, 3, and 5 are all primes. It is not difficult to prove that this always works:

Exercise 3.1.1. Prove that any integer $n > 1$ can be factored into a product of primes.

We can factor 120 in other ways. For example $120 = 4 \times 30$, and then $4 = 2 \times 2$ and $30 = 5 \times 6$. Finally noting that $6 = 2 \times 3$, we eventually obtain the same factorization, $120 = 2 \times 2 \times 2 \times 3 \times 5$, of 120 into primes, even though we arrived at it in a different way. No matter how you go about splitting a positive integer up into its factors, you will always end up with the same factorization into primes.[2] If it is true that any two such factorizations are indeed the same and if we are given one factorization of n as $q_1 \cdots q_k$, then every prime factor p of n, found in any other way, must equal some q_i. This suggests that we will need to prove Theorem 3.1.

---

[1] Notice that 1 is neither prime nor composite, and the same is true of 0 and all negative integers.

[2] Recognizing that this claim needs a proof and then supplying a proof, is one of the great achievements of Greek mathematics. They developed an approach to mathematics which assures that theorems are established on a solid basis.

Theorem 3.1. *If prime p divides ab, then p must divide at least one of a and b.*

We will prove this in the next subsection. The necessity of such a result was appreciated by ancient Greek mathematicians, who went on to show that Theorem 3.1 is sufficient to establish that every integer has a unique factorization, as we will see. It is best to begin by making a simple deduction from Theorem 3.1:

Exercise 3.1.2. (a) Prove that if prime p divides $a_1 a_2 \cdots a_k$, then p divides a_j for some j, $1 \leq j \leq k$.
(b) Deduce that if prime p divides $q_1 \cdots q_k$ where each q_i is prime, then $p = q_j$ for some j, $1 \leq j \leq k$.

With this preparation we are ready to prove the first great theorem of number theory, which appears in Euclid's "*Elements*":[3]

Theorem 3.2 (The Fundamental Theorem of Arithmetic). *Every integer $n > 1$ can be written as a product of primes in a unique way (up to reordering).*

Proof. We first show that there is a factorization of n into primes and afterwards we will prove that it is unique. We prove this by induction on n: If n is prime, then we are done; since 2 and 3 are primes, this also starts our induction hypothesis. If n is composite, then it must have a divisor a for which $1 < a < n$, and so $b = n/a$ is also an integer for which $1 < b < n$. Then, by the induction hypothesis, both a and b can be factored into primes, and so $n = ab$ equals the product of these two factorizations. (For example, to prove the result for 1050, we note that $1050 = 15 \times 70$. We have already obtained the factorizations of 15 and 70, namely $15 = 3 \times 5$ and $70 = 2 \times 5 \times 7$, so that $1050 = 15 \times 70 = (3 \times 5) \times (2 \times 5 \times 7) = 2 \times 3 \times 5 \times 5 \times 7$.)

Now we prove that there is just one factorization for each $n \geq 2$. If this is not true, then let n be the smallest integer ≥ 2 that has two distinct factorizations,
$$p_1 p_2 \cdots p_r = q_1 q_2 \cdots q_s,$$
where the p_i and q_j are (not necessarily distinct) primes. Now prime p_r divides $q_1 q_2 \cdots q_s$, and so $p_r = q_j$ for some j, by exercise 3.1.2(b). Reordering the q_j if necessary we may assume that $j = s$, and if we divide through both factorizations by $p_r = q_s$, then we have two distinct factorizations of
$$n/p_r = p_1 p_2 \cdots p_{r-1} = q_1 q_2 \cdots q_{s-1}.$$
This contradicts the minimality of n unless $n/p_r = 1$. But then $n = p_r$ is prime, and by the definition (of primes) it can have no other factorization. □

The *Fundamental Theorem of Arithmetic* states that there is a unique way to break down an integer into its fundamental (i.e., irreducible) parts, and so every integer can be viewed simply in terms of these parts (i.e., its prime factors). On the other hand any finite product of primes equals an integer, so there is a 1-to-1 correspondence between positive integers and finite products of primes, allowing one to translate questions about integers into questions about primes and vice versa.

---

[3] When we write that a product of primes is "unique up to reordering" we mean that although one can write 12 as $2 \times 2 \times 3$ or $2 \times 3 \times 2$ or $3 \times 2 \times 2$, we think of all of these as the same product, since they involve the same primes, each the same number of times, differing only in the way that we order the prime factors.

It is useful to write the factorizations of natural numbers n in a standard form, like
$$n = 2^{n_2} 3^{n_3} 5^{n_5} 7^{n_7} \ldots,$$
where n_p denotes the exact number of times the prime p divides n. Since n is an integer, each $n_p \geq 0$, and only finitely many of the n_p are non-zero. Usually we write down only those prime powers where $n_p \geq 1$, for example $12 = 2^2 \cdot 3$ and $50 = 2 \cdot 5^2$. We will write $p^e \| n$ if p^e is the highest power of p that divides n; thus $5^2 \| 50$ and $11^1 \| 1001$.

Our proof of the *Fundamental Theorem of Arithmetic* is constructive but it does not provide an efficient way to find the prime factors of a given integer n. Indeed finding efficient techniques for factoring an integer is a difficult and important problem, which we discuss in chapter 10.[4]

In particular, the known difficulty of factoring large integers underlies the security of the RSA cryptosystem, which is discussed in section 10.3.

Exercise 3.1.3. (a) Prove that every natural number has a unique representation as $2^k m$ with $k \geq 0$ and m an odd natural number.
(b) Show that each integer $n \geq 3$ is either divisible by 4 or has at least one odd prime factor.
(c) An integer is *squarefree* if every prime in its factorization appears to the power 1. Prove that every non-zero integer can be written, uniquely, in the form mn^2 where m is a squarefree integer and n is a non-zero positive integer.
(d)[†] Deduce that every non-zero rational number can be written, uniquely, in the form mr^2 where m is a squarefree integer and r is a positive rational number.

Exercise 3.1.4. (a) Show that if all of the prime factors of an integer n are $\equiv 1 \pmod{m}$, then $n \equiv 1 \pmod{m}$. Deduce that if $n \not\equiv 1 \pmod{m}$ then n has a prime factor that is $\not\equiv 1 \pmod{m}$.
(b)[†] Show that if all of the prime factors of an integer n are $\equiv 1$ or $3 \pmod{8}$, then $n \equiv 1$ or $3 \pmod{8}$. Prove this with 3 replaced by 5 or 7.
(c)[†] Generalize this as much as you can to other moduli and other sets of congruence classes.

3.2. Abstractions

The ancient Greek mathematicians recognized that abstract lemmas allowed them to *prove* sophisticated theorems. For example, in the previous section we stated Theorem 3.1, a result whose formulation is not obviously relevant and yet was used to good effect. The archetypal lemma is known today as "Euclid's Lemma", an important result that first appeared in Euclid's "*Elements*" (Book VII, No. 32), and we will see that it is even more useful than Theorem 3.1:

> **Theorem 3.3** (Euclid's Lemma). *If c divides ab and $\gcd(c, a) = 1$, then c must divide b.*

---

[4]It is easy enough to multiply together two given integers. If the integers each have 50 digits, then one can obtain the product in about 3,000 steps (digit-by-digit multiplications) and this can be accomplished within a second on a computer. On the other hand, given the 100-digit product, how do we factor it to find the original two 50-digit integers? Trial division is too slow ... if every atom in the universe were a computer as powerful as any supercomputer, then most such products would not be factored before the end of the universe! This is why we need more sophisticated factoring methods, and although the best ones known today, implemented on the best computers, can factor a 100-digit number in reasonable time, they are currently incapable of factoring typical 200-digit numbers. (See sections 10.4 and 10.6 for further discussion on this theme.)

Proof of Euclid's Lemma. Since $\gcd(c, a) = 1$ there exist integers m and n such that $cm + an = 1$ by Theorem 1.1. Now c divides both c and ab, so that

$$c \text{ divides } c \cdot bm + ab \cdot n = b(cm + an) = b,$$

by exercise 1.1.1(c). □

This proof surprisingly uses, inexplicitly, the complicated construction from Euclid's algorithm. Now that we have proved Euclid's Lemma we proceed to

Deduction of Theorem 3.1. Suppose that prime p does not divide a (or else we are done), and so $\gcd(p, a) = 1$ (as seen in exercise 3.0.1). Taking $c = p$ in Euclid's Lemma, we deduce that p divides b. □

The hypothesis "$\gcd(c, a) = 1$" in Euclid's Lemma is necessary, as may be seen from the example in which 4 divides $2 \cdot 6$, but 4 does not divide either 2 or 6.

Now that we have completed the proof of the Fundamental Theorem of Arithmetic, we are ready to develop the basic number-theoretic properties of integers.[5] We begin by noting one further important consequence of Euclid's Lemma:

Corollary 3.2.1. *If $am = bn$, then $a/\gcd(a, b)$ divides n.*

Proof. Let $a/\gcd(a, b) = A$ and $b/\gcd(a, b) = B$ so that $(A, B) = 1$ by exercise 1.2.5(a) and $Am = Bn$. Therefore $A|Bn$ with $(A, B) = 1$, and so $A|n$ by Euclid's Lemma, as desired. We also observe that if we write $n = Ak$ for some integer k, then $m = Bn/A = Bk$. □

One consequence is a simple way to determine the least common multiple of two integers from knowing their greatest common divisor.

Corollary 3.2.2. *For any positive integers a and b we have $ab = \gcd(a, b) \cdot \text{lcm}(a, b)$.*

Proof. By definition, there exist integers m and n for which $am = bn = \text{lcm}[a, b]$. By Corollary 3.2.1 we know that $a/\gcd(a, b)$ divides n and so $L := b \cdot a/\gcd(a, b)$ divides $bn = \text{lcm}[a, b]$. Therefore $L \leq \text{lcm}[a, b]$. On the other hand L is a multiple of b, by definition, and of a, since $L = a \cdot b/\gcd(a, b)$. Therefore L is a common multiple of a and b, and so $L \geq \text{lcm}[a, b]$ by the definition of lcm. These two inequalities imply that $L = \text{lcm}[a, b]$, and the result follows by multiplying through by the denominator. □

We will see an easier proof of this elegant result in exercise 3.3.2.

Exercise 3.2.1. Suppose that $(a, b) = 1$. Prove that if a and b both divide m, then ab divides m.

---

[5] However if we wish to develop the analogy of this theory for more complicated sets of numbers, for example the numbers of the form $\{a + b\sqrt{d} : a, b \in \mathbb{Z}\}$ for some fixed large integer d, then Euclid's Lemma generalizes in a straightforward way, but the Fundamental Theorem of Arithmetic does not. We discuss this further in appendix 3F.

3.3. Divisors using factorizations

Suppose that[6]
$$n = \prod_{p \text{ prime}} p^{n_p}, \quad a = \prod_p p^{a_p}, \quad \text{and} \quad b = \prod_p p^{b_p}.$$

If $n = ab$, then
$$2^{n_2} 3^{n_3} 5^{n_5} \cdots = 2^{a_2} 3^{a_3} 5^{a_5} \cdots 2^{b_2} 3^{b_3} 5^{b_5} \cdots = 2^{a_2+b_2} 3^{a_3+b_3} 5^{a_5+b_5} \cdots.$$

As there is only one factorization into primes of a given positive integer, by the Fundamental Theorem of Arithmetic, we can equate the exact power of prime p dividing each side of the last equation, to deduce that
$$n_p = a_p + b_p \quad \text{for each prime } p.$$

As $a_p, b_p \geq 0$ for each prime p, therefore
$$0 \leq a_p, b_p \leq n_p \quad \text{for each prime } p.$$

On the other hand if $a = 2^{a_2} 3^{a_3} 5^{a_5} \cdots$ with each $0 \leq a_p \leq n_p$, then a divides n since we can construct the integer
$$b = 2^{n_2-a_2} 3^{n_3-a_3} 5^{n_5-a_5} \cdots$$

for which $n = ab$. We have therefore classified all of the possible (positive integer) divisors a of n.

This classification allows us to easily count the number of divisors a of n, since this is equal to the number of possibilities for the exponents a_p; and we have that each a_p is any integer in the range $0 \leq a_p \leq n_p$. There are, therefore, $n_p + 1$ possibilities for the exponent a_p, for each prime p, making
$$(n_2 + 1)(n_3 + 1)(n_5 + 1) \cdots$$

possible divisors in total. Hence if we write $\tau(n)$ for the number of divisors of n, then
$$\tau(n) = \prod_{\substack{p \text{ prime} \\ p^{n_p} \| n}} \tau(p^{n_p});$$

and $\tau(p^k) = k + 1$ for all integers $k \geq 0$. A function whose value at n equals the product of the values of the function at the exact prime powers that divide n is called a *multiplicative function* (which will be explored in detail in the next chapter).

As an example, we see that the divisors of $175 = 5^2 7^1$ are given by
$$5^0 7^0 = 1, \ 5^1 7^0 = 5, \ 5^2 7^0 = 25, \ 5^0 7^1 = 7, \ 5^1 7^1 = 35, \ 5^2 7^2 = 175;$$

in other words, they can all be factored as
$$5^0, 5^1, \text{ or } 5^2 \text{ times } 7^0 \text{ or } 7^1.$$

Therefore the number of divisors is $(2 + 1) \times (1 + 1) = 3 \times 2 = 6$.

Use the Fundamental Theorem of Arithmetic in all of the remaining exercises in this section.

---

[6]We suppress writing "prime" in the subscript of $\prod$, for convenience, at least when it should be obvious, from the context, that the parameter is only taking prime values.

Exercise 3.3.1. Use the description of the divisors of a given integer to prove the following: If $m = \prod_p p^{m_p}$ and $n = \prod_p p^{n_p}$ are positive integers, then (a) $\gcd(m,n) = \prod_p p^{\min\{m_p, n_p\}}$ and (b) $\text{lcm}[m,n] = \prod_p p^{\max\{m_p, n_p\}}$.

The method in exercise 3.3.1(a) for finding the gcd of two integers appears to be much simpler than the Euclidean algorithm. However, in order to make this method work, one needs to be able to factor the integers involved. We have not yet discussed techniques for factoring integers (though we will in chapter 10). Factoring is typically difficult for large integers. This difficulty limits when we can, in practice, use exercise 3.3.1 to determine gcds and lcms. On the other hand, the Euclidean algorithm is very efficient for finding the gcd of two given integers (as discussed in appendix 1B) without needing to know anything about those numbers.

Exercise 3.3.2. Deduce that $mn = \gcd(m,n) \cdot \text{lcm}[m,n]$ for all pairs of natural numbers m and n using exercise 3.3.1. (The proof in Corollary 3.2.2 is more difficult.)

In combination with the Euclidean algorithm, the result in exercise 3.3.2 allows us to quickly and easily calculate the lcm of any two given integers. For example, to determine $\text{lcm}[12, 30]$, we first use the Euclidean algorithm to show that $\gcd(12, 30) = 6$, and then $\text{lcm}[12, 30] = 12 \cdot 30 / \gcd(12, 30) = 360/6 = 60$.

Although we have already proved the results in the next exercise (exercise 1.2.1(a), Lemma 1.4.1, exercise 1.2.5(a), and Corollary 1.2.2), we can now reprove them more easily by using our description of the divisors of a given integer.

Exercise 3.3.3. (a) Prove that d divides $\gcd(a,b)$ if and only if d divides both a and b.
(b) Prove that $\text{lcm}[a,b]$ divides m if and only if a and b both divide m.
(c) Prove that if $(a,b) = g$, then $(a/g, b/g) = 1$.
(d) Prove that if $(a,m) = (b,m) = 1$, then $(ab, m) = 1$.
(e) Prove that if $(a,b) = 1$, then $(ab, m) = (a,m)(b,m)$.
(f)† Show that the hypothesis $(a,b) = 1$ is necessary in part (e), by constructing a counterexample to the conclusion when $(a,b) > 1$.

One can obtain the gcd and lcm for any number of integers by means similar to exercise 3.3.1:

Example. If $A = 504 = 2^3 \cdot 3^2 \cdot 7$, $B = 2880 = 2^6 \cdot 3^2 \cdot 5$, and $C = 864 = 2^5 \cdot 3^3$, then the greatest common divisor is $2^3 \cdot 3^2 = 72$ and the least common multiple is $2^6 \cdot 3^3 \cdot 5 \cdot 7 = 60480$. That is, if the powers of prime p that divide A, B, and C are a_p, b_p, and c_p, respectively, then the powers of p that divide the gcd and lcm are $\min\{a_p, b_p, c_p\}$ and $\max\{a_p, b_p, c_p\}$, respectively.

Exercise 3.3.4. Prove that $\gcd(a,b,c) = \gcd(a, \gcd(b,c))$ and $\text{lcm}[a,b,c] = \text{lcm}[a, \text{lcm}[b,c]]$.

Exercise 3.3.5. Prove that if each of $a, b, c, \ldots$ is coprime with m, then so is $abc \ldots$.

The representation of an integer in terms of its prime power factors can be useful when considering powers of integers:

Exercise 3.3.6. Prove that if prime p divides a^n, then p^n divides a^n.

Exercise 3.3.7. (a) Prove that a positive integer A is the square of an integer if and only if the exponent of each prime factor of A is even.
(b) Prove that if $a, b, c, \ldots$ are pairwise coprime, positive integers and their product is a square, then they are each a square.

(c) Prove that if ab is a square, then either $a = gA^2$ and $b = gB^2$, or $a = -gA^2$ and $b = -gB^2$, where $g = \gcd(a,b)$, for some coprime integers A and B.

Exercise 3.3.8. (a) Prove that a positive integer A is the nth power of an integer if and only if n divides the exponent of all of the prime power factors of A.
(b) Prove that if $a, b, c, \ldots$ are pairwise coprime, positive integers and their product is an nth power, then they are each an nth power.

3.4. Irrationality

One of the most beautiful applications of the Fundamental Theorem of Arithmetic is its use in showing that there are real irrational numbers,[7] the easiest example being $\sqrt{2}$:

Proposition 3.4.1. *The real number $\sqrt{2}$ is irrational. That is, there is no rational number a/b for which $\sqrt{2} = a/b$.*

Proof. We will assume that $\sqrt{2}$ is rational and find a contradiction. If $\sqrt{2}$ is rational, then we can write $\sqrt{2} = a/b$ where a and $b \geq 1$ are coprime integers by exercise 1.2.5(b). We have $a = b\sqrt{2} > 0$.

Now $a = b\sqrt{2}$ and so $a^2 = 2b^2$. If we factor

$$a = \prod_p p^{a_p} \text{ and } b = \prod_p p^{b_p}, \text{ then } \prod_p p^{2a_p} = a^2 = 2b^2 = 2\prod_p p^{2b_p},$$

where the a_p's and b_p's are all integers. The exponent of the prime 2 in the factorization of $a^2 = 2b^2$ is $2a_2 = 1 + 2b_2$ which is impossible (mod 2), giving a contradiction. Hence $\sqrt{2}$ cannot be rational. □

More generally we have, by a different proof,

Proposition 3.4.2. *If d is an integer for which $\sqrt{d}$ is rational, then $\sqrt{d}$ is an integer. Therefore if integer d is not the square of an integer, then $\sqrt{d}$ is irrational.*

Proof. We may write $\sqrt{d} = a/b$ where a and b are coprime positive integers, so that $a^2 = db^2$. Now $(a^2, b^2) = 1$ and a^2 divides db^2, which implies that a^2 divides d, by Euclid's Lemma. But then $d \leq db^2 = a^2 \leq d$, implying that $d = a^2$; that is, d is the square of an integer as claimed. □

Exercise 3.4.1. Give a proof of Proposition 3.4.2, which is analogous to the proof of Proposition 3.4.1 above.

Exercise 3.4.2.[†] Prove that $17^{1/3}$ is irrational (using the ideas of the proof of Proposition 3.4.1).

The proof of Proposition 3.4.2 generalizes to give a nice application of Euclid's Lemma to rational roots of arbitrary polynomials with integer coefficients:

Theorem 3.4 (The rational root criterion)**.** *Suppose $f(x)$ is a polynomial with integer coefficients, with leading coefficient a_d and last coefficient a_0. If $f(m/n) = 0$ where m and n are coprime integers, then m divides a_0 and n divides a_d.*

---

[7]That is, real numbers that are not rational.

Proof. Writing $f(x) = \sum_{j=0}^{d} a_j x^j$ where each $a_j \in \mathbb{Z}$ we have

$$a_d m^d + a_{d-1} m^{d-1} n + \cdots + a_1 m n^{d-1} + a_0 n^d = n^d f(m/n) = 0.$$

Reducing this equation mod n gives $a_d m^d \equiv 0 \pmod{n}$ as every other term on the left-hand side is divisible by n. This can be restated as n divides $a_d m^d$. By the hypothesis, we have $(n, m) = 1$ and so $(n, m^d) = 1$ by exercise 1.7.11. Therefore n divides $a_d m^d$ and $(n, m^d) = 1$, which implies that n divides a_d by Euclid's Lemma. We complete the proof by establishing

Exercise 3.4.3. Prove that m divides a_0 by reducing the above equation mod m. □

Corollary 3.4.1. *If a monic polynomial $f(x) \in \mathbb{Z}[x]$ has a rational root, then that root must be an integer.*

Proof. We have $a_d = 1$ as f is monic. Therefore $n = \pm 1$ in the rational root criterion, which implies that $m/n = \pm m$, an integer. □

We can apply Corollary 3.4.1 to the rational roots of the polynomial $x^n - d$, and so we deduce that if $d^{1/n}$ is rational, then $d^{1/n}$ is an integer (and therefore if $d^{1/n}$ is not an integer, then it is irrational), generalizing Proposition 3.4.2.

We have now proved that there exist infinitely many irrational numbers, the numbers $\sqrt{d}$ when d is not the square of an integer. This caused important philosophical conundrums for the early Greek mathematicians.[8]

Exercise 3.4.4. Prove that the polynomial $x^3 - 3x - 1$ is irreducible over $\mathbb{Q}$.

3.5. Dividing in congruences

We are now ready to return to the topic of dividing both sides of a congruence through by a given divisor, resolving the conundrums raised in section 2.2.

Lemma 3.5.1. *If d divides both a and b and $a \equiv b \pmod{m}$, then*

$$a/d \equiv b/d \pmod{m/g} \quad \text{where} \quad g = \gcd(d, m).$$

---

[8] Ancient Greek mathematicians did not think of numbers as an abstract concept, but rather as units of measurement. That is, one starts with fixed length measures and determines what lengths can be measured by a combination of those original lengths: A stick of length a can be used to measure any length that is a positive integer multiple of a (by measuring out k copies of length a, one after another). Theorem 1.1 can be interpreted as stating that if one has measuring sticks of length a and b, then one can measure length $\gcd(a, b)$ by measuring out u copies of length a and then v copies of length b, to get total length $au + bv = \gcd(a, b)$. One can then measure out any multiple of $\gcd(a, b)$ by copying the above construction that many times.

Pythagoras ($\approx$ 570–495 B.C.) traveled to Egypt and perhaps India in his youth on his quest for understanding. In 530 B.C. he founded a mystical sect in Croton, a Greek colony in southern Italy, which developed influential philosophical theories. Pythagoreans believed that numbers must be constructible in a finite number of steps from a finite given set of lengths and so erroneously concluded that no irrational number could be constructed in this way. However an isosceles right-angled triangle with two sides of length 1 has a hypotenuse of length $\sqrt{2}$, and so the Pythagoreans believed that $\sqrt{2}$ must be a rational number. When one of them proved Proposition 3.4.1 it contradicted their whole philosophy and so was suppressed, *"for the unspeakable should always be kept secret"*!

We looked at what types of lengths are "constructible" using only a compass and a straight edge in section 0.18 of appendix 0G. In fact, although the constructible lengths are quite restricted, they are, nonetheless, a far richer set of numbers than just the rational numbers.

The Pythagoreans similarly associated the four regular polygons that were then known (the *Platonic solids* after Plato) with the four "elements"—the tetrahedron with fire, the cube with earth, the octahedron with air, and the icosahedron with water—and so believed that there could be no others. They also suppressed their discovery of a fifth regular polygon, the dodecahedron.

3.5. Dividing in congruences

Proof. As d divides both a and b, we may write $a = dA$ and $b = dB$ for some integers A and B, so that $dA \equiv dB \pmod{m}$. Hence m divides $d(A - B)$ and therefore $\frac{m}{g}$ divides $\frac{d}{g}(A - B)$. Now $\gcd(\frac{m}{g}, \frac{d}{g}) = 1$ by exercise 1.2.5(a), and so $\frac{m}{g}$ divides $A - B$ by Euclid's Lemma. This is the result that was claimed. □

For example, $14 \equiv 91 \pmod{77}$. Now $14 = 7 \times 2$ and $91 = 7 \times 13$, and so we divide 7 out from 77 to obtain $2 \equiv 13 \pmod{11}$. More interestingly $12 \equiv 42 \pmod{15}$, and 6 divides both 12 and 42. However 6 does not divide 15, so we cannot divide this out from 15, but rather we divide out by $\gcd(15, 6) = 3$ to obtain $2 \equiv 7 \pmod{5}$.

Corollary 3.5.1. *Suppose that* $(a, m) = 1$.

(i) $u \equiv v \pmod{m}$ *if and only if* $au \equiv av \pmod{m}$.

(ii) *The residues*

(3.5.1) $$a.0, \ a.1, \ \ldots, \ a.(m-1)$$

form a complete set of residues $\pmod{m}$.

Proof. (i) The third congruence of Lemma 2.1.1 implies that if $u \equiv v \pmod{m}$, then $au \equiv av \pmod{m}$. In the other direction, we take a, b, d in Lemma 3.5.1 to equal au, av, a, respectively. Then $g = (a, m) = 1$, and so $au \equiv av \pmod{m}$ implies that $u \equiv v \pmod{m}$ by Lemma 3.5.1.

(ii) By part (i) we know that the residues in (3.5.1) are distinct mod m. Since there are m of them, they must form a complete set of residues $\pmod{m}$. □

Corollary 3.5.1(ii) states that the residues in (3.5.1) form a complete set of residues $\pmod{m}$. In particular one of them is congruent to 1 $\pmod{m}$; and so we deduce the following:

Corollary 3.5.2. *If* $(a, m) = 1$, *then there exists an integer* r *such that* $ar \equiv 1 \pmod{m}$. *We call* r *the* inverse of a $\pmod{m}$. *We denote this by* $1/a \pmod{m}$, *or* $a^{-1} \pmod{m}$; *some authors write* $\overline{a} \pmod{m}$.

Third proof of Theorem 1.1. [*For any positive integers* a, b, *there exist integers* u *and* v *such that* $au + bv = \gcd(a, b)$.] Let $g = \gcd(a, b)$ and write $a = gA, b = gB$ so that $(A, B) = 1$. By Corollary 3.5.2, there exists an integer r such that $Ar \equiv 1 \pmod{B}$, and so there exists an integer s such that $Ar - 1 = Bs$; that is, $Ar - Bs = 1$. Therefore $ar - bs = g(Ar - Bs) = g \cdot 1 = g = \gcd(a, b)$, as desired. □

This also goes in the other direction:

Second proof of Corollary 3.5.2. By Theorem 1.1 there exist integers u and v such that $au + mv = 1$, and so

$$au \equiv au + mv = 1 \pmod{m}.$$

Therefore u is the inverse of $a \pmod{m}$. □

Exercise 3.5.1. Assume that $(a, m) = 1$.
  (a) Prove that if b is an integer, then $a.0 + b, \; a.1 + b, \; \ldots, \; a(m-1) + b$ form a complete set of residues (mod m).
  (b) Deduce that for all given integers b and c, there is a unique value of x (mod m) for which $ax + b \equiv c$ (mod m).

If $(a, m) = 1$, then we can (unambiguously) express the root of $ax \equiv c$ (mod m) as ca^{-1} (mod m), or c/a (mod m); we take this to mean the residue class mod m which contains the unique value from exercise 3.5.1(b). For example $19/17 \equiv 11$ (mod 12). Such quotients share all the properties described in Lemma 2.1.1.

Exercise 3.5.2. Prove that if $\{r_1, \ldots, r_k\}$ is a reduced set of residues mod m and $(a, m) = 1$, then $\{ar_1, \ldots, ar_k\}$ is also a reduced set of residues mod m

Exercise 3.5.3. (a) Show that there exists r (mod b) for which $ar \equiv c$ (mod b) if and only if $\gcd(a, b)$ divides c.
  (b)† Prove that the solutions r are precisely the elements of a residue class mod $b/\gcd(a, b)$.

Exercise 3.5.4. Prove that if $(a, m) > 1$, then there does not exist an integer r such that $ar \equiv 1$ (mod m). (And so Corollary 3.5.2 could have been phrased as an "if and only if" condition.)

Exercise 3.5.5. Explain how the Euclidean algorithm may be used to efficiently determine the inverse of a (mod m) whenever $(a, m) = 1$. (Calculating the inverse of a (mod m) is an essential part of the RSA algorithm discussed in section 10.3.)

3.6. Linear equations in two unknowns

Given integers a, b, c, can we determine all of the integer solutions m, n to
$$am + bn = c \; ?$$

Example. To find all integer solutions to $4m + 6n = 10$, we begin by noting that we can divide through by 2 to get $2m + 3n = 5$. There is clearly a solution, $2 \cdot 1 + 3 \cdot 1 = 5$. Therefore
$$2m + 3n = 5 = 2 \cdot 1 + 3 \cdot 1,$$
so that $2(m-1) = 3(1-n)$. We therefore need to find all integer solutions u, v to
$$2u = 3v$$
and then the general solution to our original equation is given by $m = 1 + u$, $n = 1 - v$, as we run over the possible pairs u, v. Now $2|3v$ and $(2, 3) = 1$ so that $2|v$. Hence we may write $v = 2\ell$ for some integer ℓ and then deduce that $u = 3\ell$. Therefore all integer solutions to $4m + 6n = 10$ take the form
$$m = 1 + 3\ell, \; n = 1 - 2\ell, \text{ for some integer } \ell.$$

We can imitate this procedure to establish a general result:

Theorem 3.5. *Let a, b, c be given integers. There are solutions in integers m, n to $am + bn = c$ if and only if (a, b) divides c. Given a first solution, say r, s (which can be found using the Euclidean algorithm), all integer solutions to $am + bn = c$ are then given by the formula*
$$m = r + \frac{b}{(a,b)}\ell, \quad n = s - \frac{a}{(a,b)}\ell \quad \text{for some integer } \ell.$$

3.6. Linear equations in two unknowns

The full set of *real* solutions to $ax + by = c$ is given by

$$x = r + kb, \quad y = s - ka, \text{ where } k \text{ is an arbitrary real number.}$$

By Theorem 3.5 these are integer solutions exactly when $k = \ell/(a,b)$ for some $\ell \in \mathbb{Z}$.

In the discussion above we saw that it is best to "reduce" this to the case when $(a,b) = 1$.

Corollary 3.6.1. *Let a, b, c be given integers with $(a, b) = 1$. Given a first solution in integers r, s to $ar + bs = c$, all integer solutions to $am + bn = c$ are then given by the formula*

$$m = r + b\ell, \quad n = s - a\ell \quad \text{for some integer } \ell.$$

Deduction of Theorem 3.5 from Corollary 3.6.1. If there is a solution in integers m, n to $am + bn = c$, then $g := (a, b)$ divides a, b and $am + bn = c$, so we can write $a = Ag$, $b = Bg$, $c = Cg$ for some integers A, B, C with $(A, B) = 1$. We now determine the integer solutions to $Am + Bn = C$, where $(A, B) = 1$ by Corollary 3.6.1. □

Proof #1 of Corollary 3.6.1. If

$$am + bn = c = ar + bs,$$

then

$$a(m - r) = b(s - n).$$

We therefore need to find all integer solutions u, v to

$$au = bv.$$

In any given solution a divides v by Euclid's Lemma as $(a, b) = 1$, and so we may write $v = a\ell$ for some integer ℓ and deduce that $u = b\ell$. We then deduce the claimed parametrization of integer solutions to $am + bn = c$. □

Exercise 3.6.1. Show that if there exists a solution in integers m, n to $am + bn = c$ with $(a, b) = 1$, then there exists a solution with $0 \leq m < b$.

Proof #2 of Corollary 3.6.1. There is an inverse to $a \pmod{b}$, as $(a,b) = 1$; call it r. Let m be any integer $\equiv rc \pmod{b}$, so that $am \equiv arc \equiv c \pmod{b}$, and therefore there exists an integer n for which $am + bn = c$. The result follows. □

Exercise 3.6.2. (a) Find all solutions in integers m, n to $7m + 5n = 1$.
(b) Find all solutions in integers u, v to $7v - 5u = 3$.
(c) Find all solutions in integers j, k to $3j - 9k = 1$.
(d) Find all solutions in integers r, s to $5r - 10s = 15$.

Exercise 3.6.3. Show that a linear equation $am + bn = c$ where a, b, and c are given integers, cannot have exactly one solution in integers m, n.

An equation involving a congruence is said to be *solved* when integer values can be found for the variables so that the congruence is satisfied. For example $6x + 5 \equiv 13 \pmod{11}$ has the unique solution $x \equiv 5 \pmod{11}$, that is, all integers of the form $11k + 5$.

There is another way to interpret Theorem 3.5, which will prove to be the best reformulation to understand what happens with quadratic equations:

Exercise 3.6.4 (The local-global principle for linear equations). Let a, b, c be given non-zero integers. There are solutions in integers m, n to $am + bn = c$ if and only if there exist residue classes $u, v \pmod{b}$ such that $au + bv \equiv c \pmod{b}$.

"Global" refers to looking over the infinite number of possibilities for integer solutions, "local" to looking through the finite number of possibilities mod b. This exercise will be revisited in exercise 3.9.13.

3.7. Congruences to several moduli

What are the integers that satisfy given congruences to two different moduli?

Lemma 3.7.1. *Suppose that a, A, b, B are integers. There exists an integer x such that both $x \equiv a \pmod{A}$ and $x \equiv b \pmod{B}$ if and only if $b \equiv a \pmod{\gcd(A, B)}$. If there is such an integer x, then the two congruences hold simultaneously for all integers x belonging to a unique residue class $\pmod{\operatorname{lcm}[A, B]}$.*

Proof. The integers x for which $x \equiv a \pmod{A}$ may be written in the form $x = Ay + a$ for some integer y. We are therefore seeking solutions to $Ay + a = x \equiv b \pmod{B}$, which is the same as $Ay \equiv b - a \pmod{B}$. By exercise 3.5.3(a), this has solutions if and only $\gcd(A, B)$ divides $b - a$. Moreover exercise 3.5.3(b) implies that y is a solution if and only if it is of the form $u + n \cdot B/(A, B)$ for some initial solution u and any integer n. Therefore x must be of the form

$$x = Ay + a = A(u + n \cdot B/(A, B)) + a = v + n \cdot \operatorname{lcm}[A, B],$$

where $v = Au + a$ and since $A \cdot B/(A, B) = [A, B]$ by Corollary 3.2.2. □

The generalization of this last result is most elegant when we restrict to moduli that are pairwise coprime. We prepare with the following exercises:

Exercise 3.7.1. Determine all integers n for which $n \equiv 101 \pmod{7^{11}}$ and $n \equiv 101 \pmod{13^{17}}$, in terms of one congruence.

Exercise 3.7.2. Suppose that $a, b, c, \ldots$ are pairwise coprime integers.
  (a) Prove that if $a, b, c, \ldots$ each divide m, then $abc \ldots$ divides m.
  (b) Deduce that if $m \equiv n \pmod{a}$ and $m \equiv n \pmod{b}$ and $m \equiv n \pmod{c}$, $\ldots$, then $m \equiv n \pmod{abc \ldots}$.

> **Theorem 3.6** (The Chinese Remainder Theorem). *Suppose that $m_1, \ldots, m_k$ are a set of pairwise coprime positive integers. For any set of residue classes*
>
> $$a_1 \pmod{m_1}, \quad a_2 \pmod{m_2}, \ldots, a_k \pmod{m_k},$$
>
> *there exists a unique residue class $x \pmod{m}$ where $m = m_1 m_2 \ldots m_k$, for which*
>
> $$x \equiv a_j \pmod{m_j} \quad \text{for each } j.$$

Proof. We can map $x \pmod{m}$ to the vector $(x \pmod{m_1}, x \pmod{m_2}, \ldots, x \pmod{m_k})$. There are $m_1 m_2 \ldots m_k$ different such vectors and each different x mod m maps to a different one, for if $x \equiv y \pmod{m_j}$ for each j, then $x \equiv y \pmod{m}$ by exercise 3.7.2(b). Hence there is a suitable 1-to-1 correspondence between residue classes mod m and vectors, which implies the result. □

3.7. Congruences to several moduli

This is known as the *Chinese Remainder Theorem* because of the ancient Chinese practice (as discussed in Sun Tzu's 4th-century *Classic Calculations*) of counting the number of soldiers in a platoon by having them line up in three columns and seeing how many are left over, then in five columns and seeing how many are left over, and finally in seven columns and seeing how many are left over, etc. For instance, if there are a hundred soldiers, then there should be 1, 0, and 2 soldiers left over, respectively;[9] and the next smallest number of soldiers one would need for this to be true is 205 (since 205 is the next smallest positive integer $\equiv 100 \pmod{105}$). Presumably an experienced commander can eyeball the difference between 100 soldiers and 205! Primary school children in China learn a song that celebrates this contribution.

We can make the Chinese Remainder Theorem a practical tool by giving a formula to determine x, given $a_1, a_2, \ldots, a_k$: Since $(m/m_j, m_j) = 1$ there exists an integer b_j such that $b_j \cdot \frac{m}{m_j} \equiv 1 \pmod{m_j}$ for each j, by Corollary 3.5.2. Then

$$(3.7.1) \qquad \boxed{x \equiv a_1 b_1 \cdot \frac{m}{m_1} + a_2 b_2 \cdot \frac{m}{m_2} + \cdots + a_k b_k \cdot \frac{m}{m_k} \pmod{m}.}$$

This works because m_j divides m/m_i for each $i \neq j$ and so

$$x \equiv 0 + \cdots + 0 + a_j \cdot b_j \frac{m}{m_j} + 0 + \cdots + 0 \equiv a_j \cdot 1 \equiv a_j \pmod{m_j}$$

for each j. The b_j can all be determined using the Euclidean algorithm, so x can be determined rapidly in practice.

Exercise 3.7.3.[†] Use this method to give a general formula for $x \pmod{1001}$ when $x \equiv a \pmod 7$, $x \equiv b \pmod{11}$, and $x \equiv c \pmod{13}$.

Exercise 3.7.4.[†] Find the smallest positive integer n which can be written as $n = 2a^2 = 3b^3 = 5c^5$ for some integers a, b, c.

There is more discussion of the Chinese Remainder Theorem in section 3.14 of appendix 3B, in particular in the more difficult case in which the m_i's have common factors:

Exercise 3.7.5.[†] Given residue classes $a_1 \pmod{m_1}, \ldots, a_k \pmod{m_k}$ let $m = \text{lcm}[m_1, \ldots, m_k]$. Prove that there exists a residue class $b \pmod m$ for which $b \equiv a_j \pmod{m_j}$ for each j if and only if $a_i \equiv a_j \pmod{(m_i, m_j)}$ for all $i \neq j$.

Moreover in appendix 3C we explain how the Chinese Remainder Theorem can be extended to, and understood in, the more general and natural context of group theory.

Exercise 3.7.6. (a) Prove that each of $a, b, c, \ldots$ divides m if and only if $\text{lcm}[a, b, c, \ldots]$ divides m.
(b) Deduce that if $m \equiv n \pmod a$ and $m \equiv n \pmod b$ and $\ldots$, then $m \equiv n \pmod{\text{lcm}[a, b, \ldots]}$.
(c) Prove that if $b \pmod m$ in exercise 3.7.5 exists, then it is unique.

Exercise 3.7.7.[†] Let M, N, g be positive integers with $(M, N, g) = 1$. Prove that the set of residues $\{aN + bM \pmod g : 0 \leq a, b \leq g - 1\}$ is precisely g copies of the complete set of residues mod g.

---

[9]Since $100 \equiv 1 \pmod 3$, $\equiv 0 \pmod 5$, and $\equiv 2 \pmod 7$.

Exercise 3.7.8. (a) Prove that for any odd integer m there are infinitely many integers n for which $(n, m) = (n + 1, m) = 1$.
(b) Why is this false if m is even?
(c) Prove that for any integer m there are infinitely many integers n for which $(n, m) = (n + 2, m) = 1$.
(d)‡ Let $a_1 < a_2 < \cdots < a_k$ be given integers. Give an "if and only if" criterion in terms of the $a_i \pmod{p}$, for each prime p dividing m, to determine whether there are infinitely many integers n for which $(n + a_1, m) = (n + a_2, m) = \cdots = (n + a_k, m) = 1$.

Exercise 3.7.9. Prove that there exist one million consecutive integers, each of which is divisible by the cube of an integer > 1.

3.8. Square roots of $1 \pmod{n}$

We begin by noting

Lemma 3.8.1. *If p is an odd prime, then there are exactly two square roots of $1 \pmod{p}$, namely 1 and -1.*

Proof. If $x^2 \equiv 1 \pmod{p}$, then $p | (x^2 - 1) = (x-1)(x+1)$ and so p divides either $x - 1$ or $x + 1$ by Theorem 3.1. Hence $x \equiv 1$, or $-1 \pmod{p}$. □

There can be more than two square roots of 1 if the modulus is composite. For example, 1, 3, 5, and 7 are all roots of $x^2 \equiv 1 \pmod{8}$, while $1, 4, -4,$ and -1 are all roots of $x^2 \equiv 1 \pmod{15}$, and $\pm 1, \pm 29, \pm 34, \pm 41$ are all square roots of 1 $\pmod{105}$. How can we find all of these solutions?

By the Chinese Remainder Theorem, x is a root of $x^2 \equiv 1 \pmod{15}$ if and only if $x^2 \equiv 1 \pmod{3}$ and $x^2 \equiv 1 \pmod{5}$. But, by Lemma 3.8.1, this happens if and only if $x \equiv 1$ or $-1 \pmod{3}$ and $x \equiv 1$ or $-1 \pmod{5}$. There are therefore four possibilities for $x \pmod{15}$, given by making the choices

$x \equiv 1 \pmod{3}$ and $x \equiv 1 \pmod{5}$, which imply $x \equiv 1 \pmod{15}$;

$x \equiv -1 \pmod{3}$ and $x \equiv -1 \pmod{5}$, which imply $x \equiv -1 \pmod{15}$;

$x \equiv 1 \pmod{3}$ and $x \equiv -1 \pmod{5}$, which imply $x \equiv 4 \pmod{15}$;

$x \equiv -1 \pmod{3}$ and $x \equiv 1 \pmod{5}$, which imply $x \equiv -4 \pmod{15}$,

the last two giving the less obvious solutions. This proof generalizes in a straightforward way:

Proposition 3.8.1. *If m is an odd integer with k distinct prime factors, then there are exactly 2^k solutions $x \pmod{m}$ to the congruence $x^2 \equiv 1 \pmod{m}$.*

Proof. Lemma 3.8.1 proves the result for m prime. What if $m = p^e$ is a power of an odd prime p? If $x^2 \equiv 1 \pmod{p^e}$, then $p | (x^2 - 1) = (x-1)(x+1)$ and so p divides either $x - 1$ or $x + 1$ by Theorem 3.1. However p cannot divide both, or else p divides their difference, which is 2. Now suppose that p does not divide $x + 1$. Since $p^e | (x^2 - 1) = (x-1)(x+1)$ we deduce that $p^e | (x-1)$ by Euclid's Lemma. Similarly, if p does not divide $x - 1$, then $p^e | (x+1)$. Therefore $x \equiv -1$ or $1 \pmod{p^e}$.

Now, suppose that a is an integer for which

$$a^2 \equiv 1 \pmod{m},$$

3.8. Square roots of 1 (mod n)

where $m = p_1^{e_1} \ldots p_k^{e_k}$ where the p_j are distinct odd primes and the $e_j \geq 1$. By the Chinese Remainder Theorem, this is equivalent to a satisfying

$$a^2 \equiv 1 \pmod{p_j^{e_j}} \text{ for } j = 1, 2, \ldots, k.$$

By the first paragraph, this is, in turn, equivalent to

$$a \equiv 1 \text{ or } -1 \pmod{p_j^{e_j}} \text{ for } j = 1, 2, \ldots, k.$$

By the Chinese Remainder Theorem, each choice of $a \pmod{p_1^{e_1}}, \ldots, a \pmod{p_k^{e_k}}$ gives rise to a different value of $a \pmod m$ that will satisfy the congruence $a^2 \equiv 1 \pmod m$. Therefore there are exactly 2^k distinct solutions. $\square$

Proposition 3.8.1 is, in effect, an algorithm for finding all of the square roots of 1 (mod m), provided one knows the factorization of m. Conversely, in section 10.1, we will see that if we are able to find square roots mod m, then we are able to factor m.

Exercise 3.8.1. Prove that if $(x, 6) = 1$, then $x^2 \equiv 1 \pmod{24}$ *without working mod* 24. You are allowed to work mod 8 and mod 3.

Exercise 3.8.2. (a)† What are the roots of $x^2 \equiv 1 \pmod{2^e}$ for each integer $e \geq 1$? (This must be different from the odd prime case since $x^2 \equiv 1 \pmod 8$ has four solutions, $1, 3, 5, 7 \pmod 8$.)
(b)† Prove that if m has k distinct prime factors, there are exactly $2^{k+\delta}$ solutions $x \pmod m$ to the congruence $x^2 \equiv 1 \pmod m$, where, if $2^e \| m$, then $\delta = 0$ if $e = 0$ or 2, $\delta = -1$ if $e = 1$, and $\delta = 1$ if $e \geq 3$.
(c) Deduce that the product of the square roots of 1 (mod 2^e) equals 1 (mod 2^e) if $e \geq 3$.

Exercise 3.8.3.† Prove that the product of the square roots of 1 (mod m) equals 1 (mod m), unless $m = 4$ or $m = p^e$ or $m = 2p^e$ for some power p^e of an odd prime p, in which case it equals $-1 \pmod m$.

In Gauss's 1801 book he gives an explicit practical example of the Chinese Remainder Theorem. Before pocket watches and cheap printing, people were more aware of solar cycles and the moon's phases than what year it actually was. Moreover, from Roman times to Gauss's childhood, taxes were hard to collect since travel was difficult and expensive and so were not paid annually but rather on a multiyear cycle. Gauss explained how to use the Chinese Remainder Theorem to deduce the year in the Julian calendar from these three pieces of information:

• The *indiction* was used from 312 to 1806 to specify the position of the year in a 15-year taxation cycle. The indiction is $\equiv$ year $+ 3 \pmod{15}$.

• The moon's phases and the days of the year repeat themselves every 19 years.[10] The *golden number*, which is $\equiv$ year $+ 1 \pmod{19}$, indicates where one is in that cycle of 19 years (and is still used to calculate the correct date for Easter).

• The days of the week and the dates of the year repeat in cycles of 28 years in the Julian calender.[11] The *solar cycle*, which is $\equiv$ year $+ 9 \pmod{28}$, indicates where one is in this cycle of 28 years.

---

[10] Meton of Athens, in the 5th century BC, observed that 19 (solar) years is less than two hours out from being a whole number of lunar months.
[11] Since there are seven days in a week and leap years occur every four years.

Taking $m_1 = 15$, $m_2 = 19$, $m_3 = 28$, we observe that

$$b_1 \equiv \frac{1}{19 \cdot 28} \equiv \frac{1}{4 \cdot (-2)} \equiv -2 \pmod{15} \text{ and } b_1 \cdot \frac{m}{m_1} = -2 \cdot 19 \cdot 28 = -1064,$$

$$b_2 \equiv \frac{1}{15 \cdot 28} \equiv \frac{1}{(-4) \cdot 9} \equiv \frac{1}{2} \equiv 10 \pmod{19} \text{ and } b_2 \cdot \frac{m}{m_2} = 10 \cdot 15 \cdot 28 = 4200,$$

$$b_3 \equiv \frac{1}{15 \cdot 19} = \frac{1}{(14+1) \cdot 19} \equiv \frac{1}{14+19} \equiv \frac{1}{5} \equiv -11 \pmod{28} \text{ and } b_3 \cdot \frac{m}{m_3} = -3135.$$

Therefore if the indiction is a, the golden number is b, and the solar cycle is c, then the year is

$$\equiv -1064a + 4200b - 3135c \pmod{7980}.$$

Additional exercises

Exercise 3.9.1. Prove that if $2^n - 1$ is prime, then n must be prime.

Exercise 3.9.2. Suppose that $0 \leq x_0 \leq x_1 \leq \cdots$ is a division sequence (that is, $x_m | x_n$ whenever $m | n$; see exercise 1.7.22), with $x_{n+1} > x_n$ whenever $n \geq n_0$ (≥ 1). Prove that if x_n is prime for some integer $n > n_0^2$, then n is prime.

We can apply exercise 3.9.2 to the Mersenne numbers $M_n = 2^n - 1$, with $n_0 = 1$, so that if M_n is prime, then n is prime; and to the Fibonacci numbers with $n_0 = 2$, so that if F_n is prime, then n is prime or $n = 4$.

Exercise 3.9.3. We introduced the companion sequence $(y_n)_{n \geq 0}$ of the Lucas sequence $(x_n)_{n \geq 0}$ in exercise 0.1.4. Note that $y_1 = a$ does not necessarily divide $y_2 = a^2 + 2b$.
  (a)‡ Prove that y_m divides y_n whenever m divides n and n/m is odd.
  (b) Assume that $a > 1$ and $b > 0$. Deduce that if y_n is prime, then n must be a power of 2.
  (c) Deduce that if $2^n + 1$ is prime, then it must be a Fermat number.

Exercise 3.9.4.‡ Prove that the Fundamental Theorem of Arithmetic implies that for any finite set of primes $\mathcal{P}$, the numbers $\log p$, $p \in \mathcal{P}$, are linearly independent[12] over $\mathbb{Q}$.

Exercise 3.9.5.† Prove that $\gcd(a, b, c) \cdot \text{lcm}[a, b, c] = abc$ if and only if a, b, and c are pairwise coprime.

Exercise 3.9.6.† Prove that if a and b are positive integers whose product is a square and whose difference is a prime p, then $a + b = (p^2 + 1)/2$.

Exercise 3.9.7. Let p be an odd prime and x, y, and z pairwise coprime, positive integers.
  (a)† Prove that $\frac{z^p - y^p}{z - y} \equiv py^{p-1} \pmod{z - y}$.
  (b) Deduce that $\gcd(\frac{z^p - y^p}{z - y}, z - y) = 1$ or p.
(This problem is continued in exercise 7.10.6.)

Exercise 3.9.8. Suppose that $f(x) \in \mathbb{Z}[x]$ is monic and $f(0) = 1$. Prove that if $r \in \mathbb{Q}$ and $f(r) = 0$, then $r = 1$ or -1.

Exercise 3.9.9 (Another proof that $\sqrt{2}$ is irrational). Suppose that $\sqrt{2} = a/b$ where a and b are coprime integers, so that $a^2 = 2b^2$.
  (a) Prove that 3 cannot divide b, and so let $c \equiv a/b \pmod 3$.
  (b) Prove that $c^2 \equiv 2 \pmod 3$, and therefore obtain a contradiction.

---

[12] $x_1, \ldots, x_k$ are *linearly dependent* over $\mathbb{Q}$ if there exist rational numbers $a_1, \ldots, a_k$, which are not all zero, such that $a_1 x_1 + \cdots + a_k x_k = 0$. They are *linearly independent* over $\mathbb{Q}$ if they are not linearly dependent over $\mathbb{Q}$.

Exercise 3.9.10.‡ (a) Prove that $\sqrt{2}+\sqrt{3}$ is irrational.
(b) Prove that $\sqrt{a}+\sqrt{b}$ is irrational unless a and b are both squares of integers.

Exercise 3.9.11. Suppose that d is an integer and $\sqrt{d}$ is rational.
(a) Show that there exists an integer m such that $\sqrt{d}-m = p/q$ where $0 \leq p < q$ and $(p,q) = 1$.
(b) If $p \neq 0$, show that $\sqrt{d}+m = Q/p$ for some integer Q.
(c) Use (a) and (b) to establish a contradiction when $p \neq 0$.
(d) Deduce that $d = m^2$.

Reference on the many proofs that $\sqrt{2}$ is irrational

[1] John H. Conway and Joseph Shipman, *Extreme proofs I: The irrationality of $\sqrt{2}$*, Math. Intelligencer **35** (2013), 2–7.

We say that N can be represented by the linear form $ax + by$, if there exist integers m and n such that $am + bn = N$. The representation is *proper* if $(m,n) = 1$.

Exercise 3.9.12.† In this question we prove that if N can be represented by $ax + by$, then it can be represented properly. Let $A = a/(a,b)$ and $B = b/(a,b)$. Theorem 3.5 states that if $N = ar + bs$, then all solutions to $am + bn = N$ take the form $m = r + kB, n = s - kA$ for some integer k.
(a) Prove that $\gcd(m,n)$ divides N.
(b) Prove that at least one of A and B is not divisible by p, for each prime p.
(c) Prove that if $p \nmid A$, then there exists a residue class $k_p \pmod p$ such that $p | s - kA$ if and only if $k \equiv k_p \pmod p$. Therefore deduce that $p \nmid s - kA$ if $k \equiv k_p + 1 \pmod p$. Note an analogous result if $p|A$ (in which case $p \nmid B$).
(d) Deduce that there exists an integer k such that, for all primes p dividing N, either p does not divide $r + kB$ or p does not divide $s - kA$ (or both).
(e) Deduce that if $m = r + kB$ and $n = s - kA$, then N is properly represented by $am + bn$.

Exercise 3.9.13. Prove the following version of the local-global principle for linear equations (exercise 3.6.4): Let a,b,c be given integers. There are solutions in integers m, n to $am + bn = c$ if and only if for all prime powers p^e (where p is prime and e is an integer ≥ 1) there exist residue classes $u, v \pmod{p^e}$ for which $au + bv \equiv c \pmod{p^e}$.

Exercise 3.9.14. Find all solutions to $5a + 7b = 211$ where a and b are positive integers.

Exercise 3.9.15. Suppose that $f(x) \in \mathbb{Z}[x]$ and m and n are coprime integers.
(a) Prove that there exist integers a and b for which $f(a) \equiv 0 \pmod m$ and $f(b) \equiv 0 \pmod n$ if and only if there exists an integer c for which $f(c) \equiv 0 \pmod{mn}$, and show that we may take $c \equiv a \pmod m$ and $c \equiv b \pmod n$.
(b) Suppose that $p_1 < p_2 < \cdots < p_k$ are primes. Prove that there exist integers $a_1, \ldots, a_k$ such that $f(a_i) \equiv 0 \pmod{p_i}$ for $1 \leq i \leq k$ if and only if there exists an integer a such that $f(a) \equiv 0 \pmod{p_1 p_2 \ldots p_k}$.

Adding reduced fractions. A *reduced fraction* takes the form a/b where a and $b > 0$ are coprime integers. We wish to better understand adding reduced fractions.

Exercise 3.9.16.† Suppose that m and n are coprime integers.
(a) Prove that for any integer c there exist integers a and b for which $\frac{c}{mn} = \frac{a}{m} + \frac{b}{n}$.
(b) Prove that there are (unique) positive integers a and b for which $\frac{1}{mn} = \frac{a}{m} + \frac{b}{n} - 1$.

Exercise 3.9.17. Let m and n be given positive integers.
(a) Prove that for any integers a and b there exists an integer c for which $\frac{a}{m} + \frac{b}{n} = \frac{c}{L}$ where $L = \text{lcm}[m, n]$.

For the denominators 3 and 6, with $L = 6$, we have the example $\frac{1}{3} + \frac{1}{6} = \frac{1}{2}$, a case in which the sum has a denominator smaller than L when written as a reduced fraction. However $\frac{1}{3} + \frac{5}{6} = \frac{7}{6}$ so there are certainly examples with these denominators for which the sum has denominator L.

(b)† Show that $\text{lcm}[m, n]$ is the smallest positive integer L such that for all integers a and b we can write $\frac{a}{m} + \frac{b}{n}$ as a fraction with denominator L. (This is why $\text{lcm}[m, n]$ is sometimes called the *lowest* (or *least*) *common denominator* of the fractions $1/m$ and $1/n$.)

(c)† Show that if $\frac{a}{m}$ and $\frac{b}{n}$ are reduced fractions whose sum has denominator less than L, then there must exist a prime power p^e such that $p^e \| m$ and $p^e \| n$ for which p^{e+1} divides $an + bm$.

Appendix 3A. Factoring binomial coefficients and Pascal's triangle modulo p

3.10. The prime powers dividing a given binomial coefficient

Lemma 3.10.1. *The power of prime p that divides $n!$ is $\sum_{k\geq 1}[n/p^k]$. In other words*

$$n! = \prod_{p \ prime} p^{\left[\frac{n}{p}\right]+\left[\frac{n}{p^2}\right]+\left[\frac{n}{p^3}\right]+\cdots}.$$

Proof. We wish to determine the power of p dividing $n! = 1 \cdot 2 \cdot 3 \cdots (n-1) \cdot n$. If p^k is the power of p dividing m, then we will count 1 for p dividing m, then 1 for p^2 dividing $m, \ldots$, and finally 1 for p^k dividing m. Therefore the power of p dividing $n!$ equals the number of integers m, $1 \leq m \leq n$, that are divisible by p, plus the number of integers m, $1 \leq m \leq n$, that are divisible by p^2, plus $\ldots$. The result follows as there are $[n/p^j]$ integers m, $1 \leq m \leq n$, that are divisible by p^j for each $j \geq 1$, by exercise 1.7.6(c). $\square$

Exercise 3.10.1. Write $n = n_0 + n_1 p + \cdots + n_d p^d$ in base p so that each $n_j \in \{0, 1, \ldots, p-1\}$.
   (a) Prove that $[n/p^k] = (n - (n_0 + n_1 p + \cdots + n_{k-1} p^{k-1}))/p^k$.
   The *sum of the digits of n in base p* is defined to be $s_p(n) := n_0 + n_1 + \cdots + n_d$.
   (b) Prove that the exact power of prime p that divides $n!$ is $\frac{n-s_p(n)}{p-1}$.

Theorem 3.7 (Kummer's Theorem). *The largest power of prime p that divides the binomial coefficient $\binom{a+b}{a}$ is given by the number of carries when adding a and b in base p.*

99

Example. To recover the factorization of $\binom{14}{6}$ we add 6 and 8 in each prime base ≤ 14:

$$\begin{array}{cccccc} 0101 & 020 & 11 & 06 & 06 & 06 \\ \underline{1000_2} & \underline{022_3} & \underline{13_5} & \underline{11_7} & \underline{08_{11}} & \underline{08_{13}} \\ 1101 & 112 & 24 & 20 & 13 & 11 \end{array}$$

We see that there are no carries in base 2, 1 carry in base 3, no carries in base 5, 1 carry in base 7, 1 carry in base 11, and 1 carry in base 13, so we deduce that $\binom{14}{6} = 3^1 \cdot 7^1 \cdot 11^1 \cdot 13^1$.

Proof. For given integer $k \geq 1$, let $q = p^k$. Then let A and B be the least non-negative residue of a and b (mod q), respectively, so that $0 \leq A, B \leq q-1$. Note that A and B give the first k digits (from the right) of a and b in base p. If C is the first k digits of $a + b$ in base p, then C is the least non-negative residue of $a + b$ (mod q), that is, of $A + B$ (mod q). Now $0 \leq A + B < 2q$:

- If $A + B < q$, then $C = A + B$ and there is no carry in the kth digit when we add a and b in base p.

- If $A + B \geq q$, then $C = A + B - q$ and so there is a carry of 1 in the kth digit when we add a and b in base p.

We need to relate these observations to the formula in Lemma 3.10.1. The trick comes in noticing that $A = a - p^k \left[\frac{a}{p^k}\right]$, and similarly $B = b - p^k \left[\frac{b}{p^k}\right]$ and $C = a + b - p^k \left[\frac{a+b}{p^k}\right]$. Therefore

$$\left[\frac{a+b}{p^k}\right] - \left[\frac{a}{p^k}\right] - \left[\frac{b}{p^k}\right] = \frac{A + B - C}{p^k} = \begin{cases} 1 & \text{if there is a carry in the } k\text{th digit,} \\ 0 & \text{if not,} \end{cases}$$

and so

$$\sum_{k \geq 1} \left(\left[\frac{a+b}{p^k}\right] - \left[\frac{a}{p^k}\right] - \left[\frac{b}{p^k}\right] \right)$$

equals the number of carries when adding a and b in base p. However Lemma 3.10.1 implies that this also equals the exact power of p dividing $\frac{(a+b)!}{a!b!} = \binom{a+b}{a}$, and the result follows. □

Exercise 3.10.2. State, with proof, the analogy to Kummer's Theorem for trinomial coefficients $n!/(a!b!c!)$ where $a + b + c = n$.

Corollary 3.10.1. *If p^e divides the binomial coefficient $\binom{n}{m}$, then $p^e \leq n$.*

Proof. There are $k + 1$ digits in the base p expansion of n when $p^k \leq n < p^{k+1}$. When adding m and $n - m$ there can be carries in every digit except the $(k+1)$st (which corresponds to the number of multiples of p^k). Therefore there are no more than k carries when adding m to $n - m$ in base p, so that $p^e \leq p^k \leq n$ by Kummer's Theorem. □

Exercise 3.10.3. Prove that if $0 \leq k \leq n$, then $\binom{n}{k}$ divides $\text{lcm}[m : m \leq n]$.

3.11. Pascal's triangle modulo 2

In section 0.3 we explained the theory and practice of constructing Pascal's triangle. We are now interested in constructing Pascal's triangle modulo 2, mod 3, mod 4, etc. To do so one can either reduce the binomial coefficients mod m (for $m = 2, 3, 4, \ldots$) or one can rework Pascal's triangle, starting with a 1 in the top row and then obtaining a row from the previous one by adding the two entries immediately above the given entry, modulo m. For example, Pascal's triangle mod 2 starts with the rows

```
                    1
                  1   1
                1   0   1
              1   1   1   1
            1   0   0   0   1
          1   1   0   0   1   1
        1   0   1   0   1   0   1
```

It is perhaps easiest to visualize this by replacing 1 (mod 2) by a dark square and, otherwise, a white square, as in the following fascinating diagram:[13]

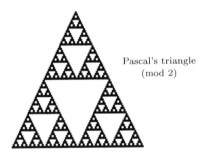

Pascal's triangle (mod 2)

One can see patterns emerging. For example the rows corresponding to $n = 1, 3, 7, 15, \ldots$ are all 1's, and the next rows, $n = 2, 4, 8, 16, \ldots$, start and end with a 1 and have all 0's in between. Even more: The two 1's at either end of row $n = 4$ seem to each be the first entry of a (four-line) triangle, which is an exact copy of the first four rows of Pascal's triangle mod 2, similarly the two 1's at either end of row $n = 8$ and the eight-line triangles beneath (and including) them. In general if T_k denotes the top 2^k rows of Pascal's triangle mod 2, then T_{k+1} is given by a triangle of copies of T_k, with an inverted triangle of zeros in the middle, as in the

---

[13]This and other images in this section reproduced with kind permission of Bill Cherowitzo.

following diagram:

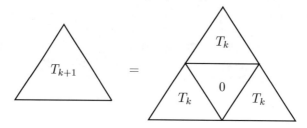

Figure 3.1. The top 2^{k+1} rows of Pascal's triangle mod 2, in terms of the top 2^k rows.

This is called *self-similarity*. One immediate consequence is that one can determine the number of 1's in a given row: If $2^k \leq n < 2^{k+1}$, then row n consists of two copies of row m ($:= n - 2^k$) with some 0's in between.

Exercise 3.11.1. Deduce that there are 2^k odd entries in the nth row of Pascal's triangle, where $k = s_2(n)$, the number of 1's in the binary expansion of n.

This self-similarity generalizes nicely for other primes p, where we again replace integers divisible by p by a white square, and those not divisible by p by a black square.

Pascal's triangle
(mod 3)

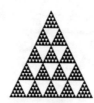
Pascal's triangle
(mod 5)

Pascal's triangle
(mod 7)

The top p rows are all black since the entries $\binom{n}{m}$ with $0 \leq m \leq n \leq p-1$ are never divisible by p. Let T_k denote the top p^k rows of Pascal's triangle. Then T_{k+1} is given by an array of p rows of triangles, in which the nth row contains n copies of T_k, with inverted triangles of 0's in between.

Pascal's triangle modulo primes p is a bit more complicated; we wish to color in the black squares with one of $p - 1$ colors, each representing a different reduced residue class mod p. Call the top row the 0th row, and the leftmost entry of each row its 0th entry. Therefore the mth entry of the nth row is $\binom{n}{m}$. By Lucas's Theorem (exercise 2.5.10) the value of $\binom{rp^k+s}{ap^k+b}$ (mod p), which is the bth entry of the sth row of the copy of T_k which is the ath entry of the rth row of the copies of T_k that make up T_{k+1}, is $\equiv \binom{r}{a}\binom{s}{b}$ (mod p). In other words, the values in the copy of T_k which is the ath entry of the rth row of the copies of T_k are $\binom{r}{a}$ times the values in T_k.

3.11. Pascal's triangle modulo 2

The odd entries in Pascal's triangle mod 4 make even more interesting patterns, but this will take us too far afield; see [1] for a detailed discussion.

Reading each row of Pascal's triangle mod 2 as the binary expansion of an integer, we obtain the numbers

$1, 11_2 = 3, 101_2 = 5, 1111_2 = 15, 10001_2 = 17, 110011_2 = 51, 10101ptestamps01_2 = 85,\ldots$

Do you recognize these numbers? If you factor them, you obtain

$$1, \quad F_0, \quad F_1, \quad F_0 F_1 \quad F_2, \quad F_0 F_2, \quad F_1 F_2, \quad F_0 F_1 F_2, \ldots$$

where $F_m = 2^{2^m} + 1$ are the Fermat numbers (introduced in exercise 0.4.14). It appears that all are products of Fermat numbers, and one can even guess at which Fermat numbers. For example the 6th row is $F_2 F_1$ and $6 = 2^2 + 2^1$ in base 2, whereas the 7th row is $F_2 F_1 F_0$ and $7 = 2^2 + 2^1 + 2^0$ in base 2, and our other examples follow this same pattern. This leads to the following challenging problem:

Exercise 3.11.2.[†] Show that the nth row of Pascal's triangle mod 2, considered as a binary number, is given by $\prod_{j=0}^{k} F_{n_j}$, where $n = 2^{n_0} + 2^{n_1} + \cdots + 2^{n_k}$, with $0 \leq n_0 < n_1 < \cdots < n_k$ (i.e., the binary expansion of n).[14]

References for this chapter

[1] Andrew Granville, *Zaphod Beeblebrox's brain and the fifty-ninth row of Pascal's triangle*, Amer. Math. Monthly **99** (1992), 318–331.

[2] Kathleen M. Shannon and Michael J. Bardzell, *Patterns in Pascal's Triangle - with a Twist - First Twist: What is It?*, Convergence (December 2004).

---

[14] An m-sided regular polygon with m odd is constructible with ruler and compass (see section 0.18 of appendix 0G) if and only if m is the product of distinct Fermat primes. Therefore the integers m created here include all of the odd m-sided, constructible, regular polygons.

Appendix 3B. Solving linear congruences

Gauss's approach to composite moduli in the Chinese Remainder Theorem uses methods that are different from those used today, but which are no less effective:

3.12. Composite moduli

If the modulus m is composite, then we can solve any linear congruence question, "one prime at a time", as in the following example: To solve

$$19x \equiv 1 \pmod{140}$$

we first do so (mod 2), as 2 divides 140, to get $x \equiv 1 \pmod 2$. Substituting $x = 1 + 2y$ into the original equation we get

$$38y \equiv -18 \pmod{140} \quad \text{or, equivalently,} \quad 19y \equiv -9 \pmod{70}.$$

Since 2 divides 70 we again view this (mod 2) to get $y \equiv 1 \pmod 2$. Substituting $y = 1 + 2z$ into this equation we get

$$38z \equiv -28 \pmod{70} \quad \text{and thus} \quad 19z \equiv -14 \pmod{35}.$$

Viewing this (mod 5) gives $-z \equiv 1 \pmod 5$, and so substitute $z = -1 + 5w$ to get

$$95w \equiv 5 \pmod{35} \quad \text{so that} \quad 19w \equiv 1 \pmod 7.$$

Therefore $5w \equiv 1 \pmod 7$ and so $w \equiv 3 \pmod 7$ which implies, successively, that

$$z \equiv -1 + 5 \cdot 3 \equiv 14 \pmod{35}, \quad y \equiv 1 + 2 \cdot 14 \equiv 29 \pmod{70},$$
$$\text{and } x \equiv 1 + 2 \cdot 29 \equiv 59 \pmod{140}.$$

3.13. Solving linear congruences with several unknowns

We restrict our attention to when there are as many congruences as there are unknowns, so that we aim to find all integer (vector) solutions $x \pmod{m}$ to $Ax \equiv b \pmod{m}$, where A is a given n-by-n matrix of integers and b is a given vector of n integers.

Let a_i be the ith column vector of A, and let
$$V_j = \{v \in \mathbb{R}^n : v \cdot a_i = 0 \text{ for all } i \neq j\}$$
be the set of vectors in $\mathbb{R}^n$ that are orthogonal to all the a_i other than a_j. Basic linear algebra gives us that V_j is itself a vector space of dimension $\geq n - (n-1) = 1$ and has a basis over $\mathbb{Q}$ made up of vectors with only integer entries (since we may multiply through any $\mathbb{Q}$-vector by some integer to make it into a $\mathbb{Z}$-vector). Hence we may take a non-zero vector in V_j with integer entries and divide through by the gcd of those entries to obtain a vector c_j whose entries are coprime. Therefore $c_j \cdot a_i = 0$ for all $i \neq j$. Let $d_j = c_j \cdot a_j \in \mathbb{Z}$. Let C be the matrix with ith row vector c_i, and let D be the diagonal matrix with (j,j)th entry d_j. Then
$$Dx = (CA)x = C(Ax) \equiv Cb \pmod{m}.$$

Let $y \equiv Cb \pmod{m}$, so that if $Ax \equiv b \pmod{m}$, then $Dx \equiv y \pmod{m}$. This has solutions if and only if there exists a solution x_j to $d_j x_j \equiv y_j \pmod{m}$ for each j. In exercise 3.5.3 we saw that there are solutions to this last congruence if and only if (d_j, m) divides y_j for each j, and we determined how to find all solutions.

If you have studied linear algebra, then your first impulse is to define $\Delta = |\det A|$, where $\det A$ is the determinant of A. If $\Delta \neq 0$, then we can solve the system of equations $Ay = b$ over the rationals by taking $C = \Delta \cdot A^{-1}$, so that $\Delta y = Cb$ (in which case, we have replaced each d_j by Δ). However, this might not be useful in trying to solve equations mod m for, if, say, m divides Δ, then our equation becomes $0 \equiv Cb \pmod{m}$. With Gauss's construction, the d_j all divide Δ but are not necessarily equal to it, which gives Gauss's method more flexibility. We now give two examples; in the first the $|d_j|$ all equal the determinant, but in the second they are smaller.

Example. We wish to solve $\begin{pmatrix} 1 & 3 & 1 \\ 4 & 1 & 5 \\ 2 & 2 & 1 \end{pmatrix} \begin{pmatrix} x_1 \\ x_2 \\ x_3 \end{pmatrix} \equiv \begin{pmatrix} 1 \\ 7 \\ 3 \end{pmatrix} \pmod{8}$. The matrix has determinant 15. We obtain, proceeding much like we would over the integers,

$$\begin{pmatrix} 15 & 0 & 0 \\ 0 & 15 & 0 \\ 0 & 0 & 15 \end{pmatrix} x = \begin{pmatrix} -9 & -1 & 14 \\ 6 & -1 & -1 \\ 6 & 4 & -11 \end{pmatrix} \begin{pmatrix} 1 & 3 & 1 \\ 4 & 1 & 5 \\ 2 & 2 & 1 \end{pmatrix} x$$

$$\equiv \begin{pmatrix} -9 & -1 & 14 \\ 6 & -1 & -1 \\ 6 & 4 & -11 \end{pmatrix} \begin{pmatrix} 1 \\ 7 \\ 3 \end{pmatrix} = \begin{pmatrix} 26 \\ -4 \\ 1 \end{pmatrix} \pmod{8},$$

so that $x \equiv \begin{pmatrix} -2 \\ 4 \\ -1 \end{pmatrix} \pmod{8}$. This gives all solutions mod 8.

Example. We wish to solve $\begin{pmatrix} 3 & 5 & 1 \\ 2 & 3 & 2 \\ 5 & 1 & 3 \end{pmatrix} \begin{pmatrix} x_1 \\ x_2 \\ x_3 \end{pmatrix} \equiv \begin{pmatrix} 4 \\ 7 \\ 6 \end{pmatrix}$ (mod 12). The matrix has determinant 28. We obtain, using Gauss's technique,

$$\begin{pmatrix} 4 & 0 & 0 \\ 0 & 7 & 0 \\ 0 & 0 & 28 \end{pmatrix} x = \begin{pmatrix} 1 & -2 & 1 \\ 1 & 1 & -1 \\ -13 & 22 & -1 \end{pmatrix} \begin{pmatrix} 3 & 5 & 1 \\ 2 & 3 & 2 \\ 5 & 1 & 3 \end{pmatrix} x$$

$$\equiv \begin{pmatrix} 1 & -2 & 1 \\ 1 & 1 & -1 \\ -13 & 22 & -1 \end{pmatrix} \begin{pmatrix} 4 \\ 7 \\ 6 \end{pmatrix} = \begin{pmatrix} -4 \\ 5 \\ 96 \end{pmatrix} \pmod{12},$$

and so, if we have a solution (x_1, x_2, x_3), then $x_1 \equiv -1 \pmod 3$, $x_2 \equiv -1 \pmod{12}$, and $x_3 \equiv 0 \pmod 3$. To obtain all solutions mod 12 we substitute $x_1 = 2 + 3t$, $x_2 = -1$, $x_3 = 3u$ into the original equations, so that $\begin{pmatrix} 3 & 1 \\ 2 & 2 \\ 5 & 3 \end{pmatrix} \begin{pmatrix} t \\ u \end{pmatrix} \equiv \begin{pmatrix} 1 \\ 2 \\ -1 \end{pmatrix}$ (mod 4) which is equivalent to $t \equiv u - 1 \pmod 4$. Therefore all solutions are given by

$$x_3 \equiv 0 \pmod 3 \quad \text{with} \quad x_1 \equiv x_3 - 1 \quad \text{and} \quad x_2 \equiv -1 \pmod{12}.$$

3.14. The Chinese Remainder Theorem in general

When the moduli are not coprime

We proved the Chinese Remainder Theorem for any two moduli in Lemma 3.7.1, but restricted to arbitrarily many pairwise coprime moduli in Theorem 3.6. We now give the full, more complicated, statement for arbitrary moduli:

> **Theorem 3.8** (The Chinese Remainder Theorem, revisited). *Suppose that we are given positive integers $m_1, m_2, \ldots, m_k$ and any residue classes*
>
> (3.14.1) $\qquad a_1 \pmod{m_1},\ a_2 \pmod{m_2}, \ldots, a_k \pmod{m_k}.$
>
> *There exists a unique residue class $x \bmod m = \mathrm{lcm}[m_1, m_2, \ldots, m_k]$ for which $x \equiv a_j \bmod m_j$ for each j if and only if $a_i \equiv a_j \bmod (m_i, m_j)$ for all $i \neq j$.*

Proof. If there is a solution, then $a_i \equiv x \equiv a_j \pmod{(m_i, m_j)}$ for all $i \neq j$, since (m_i, m_j) divides both m_i and m_j.

Now assume that $a_i \equiv a_j \pmod{(m_i, m_j)}$ for all $i \neq j$. We will prove the result by induction on $k \geq 1$. For $k = 1$ there is nothing to prove so now assume that $k \geq 2$. By the induction hypothesis there exists a unique residue class a_0 mod $m_0 := \mathrm{lcm}[m_2, \ldots, m_k]$, for which $a_0 \equiv a_j \pmod{m_j}$ for each $j \geq 2$.

Now $a_1 \equiv a_j \equiv a_0 \pmod{(m_1, m_j)}$ and so (m_1, m_j) divides $a_1 - a_0$, for each $j \geq 2$. Therefore $\mathrm{lcm}[(m_1, m_j) : 2 \leq j \leq k]$, which equals (m_1, m_0) (by exercise 3.14.1) divides $a_1 - a_0$; that is, $a_0 \equiv a_1 \pmod{(m_0, m_1)}$. By Lemma 3.7.1, there exists a unique residue class x mod $\text{lcm}[m_0, m_1] = m$, for which $x \equiv a_1 \pmod{m_1}$ and $x \equiv a_0 \pmod{m}$ which is $\equiv a_j \pmod{m_j}$ for each $j \geq 2$.

Moreover, exercise 3.7.6 also implies that if there is a residue class $x \pmod{m}$ which belongs to all of the residue classes in (3.14.1), then it is unique. □

Exercise 3.14.1. Prove that $\text{lcm}[\gcd(m, n_j) : 1 \leq j \leq k] = \gcd(m, \text{lcm}[n_j : 1 \leq j \leq k])$.

Finding the solution modulo m. The Chinese Remainder Theorem shows that there is a solution $x \pmod{m}$ but not how to find it efficiently. We discuss algorithms to do so in this subsection.

Example. Can one find integers z for which $z \equiv -4 \pmod{35}$, $z \equiv 17 \pmod{504}$, and $z \equiv 1 \pmod{16}$? The first two congruences combine to give $z \equiv 521 \pmod{2520}$. Combining this with the congruence $z \equiv 1 \pmod{16}$, we get $z \equiv 3041 \pmod{5040}$.

Given $a_1 \pmod{m_1}, \ldots, a_k \pmod{m_k}$ for which $a_i \equiv a_j \pmod{(m_i, m_j)}$ for all $i \neq j$ we now show how to find $x \bmod m = [m_1, \ldots, m_k]$ for which $x \equiv a_j \pmod{m_j}$ for each j: For each prime power $p^e \| m$ there exists an index $j(p)$ such that $p^e | m_{j(p)}$. We can then determine $x \pmod{m}$ from the original Chinese Remainder Theorem, using the congruences $x \equiv a_{j(p)} \pmod{p^e}$ for all prime powers $p^e \| m$

We now use this technique on the example above. The three congruences may be rewritten in terms of prime power divisors as

$z \equiv -4 \pmod{35} \Leftrightarrow z \equiv -4 \equiv 1 \pmod 5$ and $z \equiv -4 \equiv 3 \pmod 7$,

$z \equiv 17 \pmod{504} \Leftrightarrow z \equiv 1 \pmod 8$, $z \equiv -1 \pmod 9$, and $z \equiv 3 \pmod 7$,

$z \equiv 1 \pmod{16} \Leftrightarrow z \equiv 1 \pmod{16}$.

We see that $m = 2^4 \cdot 3^2 \cdot 5 \cdot 7 = 5040$, and the congruences with largest prime powers are

$z \equiv 1 \pmod{16}$, $z \equiv -1 \pmod 9$, $z \equiv 1 \pmod 5$, and $z \equiv 3 \pmod 7$.

These are consistent with all six congruences above and combine to give

$$z \equiv 3041 \pmod{5040}.$$

This algorithm requires one to factor the moduli which might be impractical with large moduli (see section 10.3).

We now proceed without factoring to combine the congruences $x \equiv a \pmod m$ and $x \equiv b \pmod n$, under the assumption that $a \equiv b \pmod g$ where $g = (m, n)$: We use the Euclidean algorithm to find integers r and s for which $mr + ns = g$. Then we can take $x \equiv c \pmod L$ where $L = [m, n]$ and

$$c = a + \frac{(b-a)}{g} \cdot mr = b + \frac{(a-b)}{g} \cdot ns,$$

since this construction immediately implies that $c \equiv a \pmod m$ and $c \equiv b \pmod n$. For a practical algorithm with more moduli simply combine the first two congruences, then the answer with the third congruence, etc.

A representation of the solution modulo m. Suppose that the moduli $m_1, \ldots, m_k$ are again pairwise coprime. Dividing through by m, the equation (3.7.1) is equivalent to
$$\frac{x}{m} \equiv \frac{a_1 b_1}{m_1} + \frac{a_2 b_2}{m_2} + \cdots + \frac{a_k b_k}{m_k} \pmod{1}.$$
The difference between the two sides is an integer, which we denote by n, so we let $x_1 = a_1 b_1 + nm_1$ and $x_j = a_j b_j$ for all $j \geq 2$, to obtain
$$\frac{x}{m} = \frac{x_1}{m_1} + \frac{x_2}{m_2} + \cdots + \frac{x_k}{m_k}.$$
If each m_j is a power of a different prime, then this proves that one can always decompose a fraction with a composite denominator $\prod_p p^{e_p}$ into a sum of fractions whose denominators are the prime powers p^{e_p} and whose numerators are fixed mod p^{e_p}.

Exercise 3.14.2. Find all integers n satisfying $13n \equiv 407 \pmod{175}$ and $55n \equiv 29 \pmod{63}$.

Exercise 3.14.3. Suppose that integer $m \geq 0$ is given. Prove that there exist infinitely many integers n such that $n + j$ is divisible by $m + j$ for $j = 1, 2, \ldots, 100$.

Appendix 3C. Groups and rings

For any ring A we define A^* to be the set of elements of A with a multiplicative inverse, so that A is an additive group and A^* forms a multiplicative group. Therefore, by Corollary 3.5.2 and exercise 3.5.4,

$$(\mathbb{Z}/m\mathbb{Z})^* = \{a \pmod{m} : (a,m) = 1\},$$

the reduced residues mod m, form a multiplicative group.

In order for A to be a field we need multiplication to be commutative and for A^* to equal $A \setminus \{0\}$. Multiplication is commutative in $\mathbb{Z}/m\mathbb{Z}$, and so we need only verify whether

$$\{a \pmod{m} : (a,m) = 1\} \text{ equals } \{a \pmod{m} : a \not\equiv 0 \pmod{m}\}.$$

If m has a non-trivial divisor d, then d belongs to the second set but not the first. Otherwise $m = p$ is a prime, and the two sets are obviously the same (since $1, 2, \ldots, p-1$ are all coprime to p). Hence $\mathbb{Z}/m\mathbb{Z}$ is a field if and only if $m = p$ is prime.

3.15. A direct sum

The points in $\mathbb{R}^2$ (also called the complex plane) are a 2-dimensional vector space over $\mathbb{R}$. These points, usually written (x, y), form an additive group, the operation of addition working separately on the x- and y-coordinates. One can view this as a "pasting together" of two copies of the additive group $\mathbb{R}$, technically called a "direct sum". More generally the *direct sum* of the groups $(G_1, *_1), (G_2, *_2), \ldots, (G_k, *_k)$ is denoted by

$$G_1 \oplus G_2 \oplus \cdots \oplus G_k$$

and has the group operation
$$(g_1, \ldots, g_k) + (h_1, \ldots, h_k) = (g_1 *_1 h_1, \ldots, g_k *_k h_k)$$
where $g_j, h_j \in G_j$ for each j.

Exercise 3.15.1. Verify that a direct sum of two groups indeed forms a group.

The *order* of an element g of a group G is the least positive integer m, if it exists, for which $g^m = 1$. If no such m exists, then g *has infinite order*. For example $3 \in (\mathbb{Z}/5\mathbb{Z})^*$ has order 4, whereas $-1 \in (\mathbb{Z}/n\mathbb{Z})^*$ has order 2 for all $n \geq 3$. In the additive group $(\mathbb{Z}, +)$, where 0 is the identity element, all the non-zero elements have infinite order.

Two groups G and H are *isomorphic*, and we write $G \cong H$ if there is a 1-to-1 correspondence $\phi : G \to H$ such that the group operation is conserved by the map. In other words, $\phi(a *_G b) = \phi(a) *_H \phi(b)$ for every $a, b \in G$, where $*_G$ is the group operation in G and $*_H$ is the group operation in H.

For example $(\mathbb{Z}/5\mathbb{Z})^* \cong \mathbb{Z}/4\mathbb{Z}$, as may be seen by mapping $\phi(2) = 1$ and then $\phi(4) = \phi(2^2) = 2 \cdot 1 = 2 \pmod 4$, $\phi(3) = \phi(2^3) = 3 \cdot 1 = 3 \pmod 4$, and $\phi(1) = \phi(2^4) = 4 \cdot 1 \equiv 0 \pmod 4$. To verify that the group operation is preserved by the map we have, for example $3 \cdot 4 \equiv 2 \pmod 5$ and $\phi(3) + \phi(4) \equiv 3 + 2 \equiv 1 \equiv \phi(2) \pmod 4$.

The Chinese Remainder Theorem states that there is a 1-to-1 correspondence between the residue classes $a \pmod m$ and the "vector" of residue classes $(a_1 \pmod{m_1}, a_2 \pmod{m_2}, \ldots, a_k \pmod{m_k})$, when the m_i's are pairwise coprime and their product equals m. Moreover the group operation of addition is conserved, so we can write

$$\mathbb{Z}/m\mathbb{Z} \cong \mathbb{Z}/m_1\mathbb{Z} \oplus \mathbb{Z}/m_2\mathbb{Z} \oplus \cdots \oplus \mathbb{Z}/m_k\mathbb{Z}$$
$$a \pmod m \leftrightarrow (a_1 \pmod{m_1}, a_2 \pmod{m_2}, \ldots, a_k \pmod{m_k}).$$

But this isomorphism goes beyond mere addition. It also works for multiplication performed componentwise; there is a 1-to-1 correspondence between the reduced residue classes modulo m and the reduced residue classes modulo the m_i which we write as

$$(\mathbb{Z}/m\mathbb{Z})^* \cong (\mathbb{Z}/m_1\mathbb{Z})^* \oplus (\mathbb{Z}/m_2\mathbb{Z})^* \oplus \cdots \oplus (\mathbb{Z}/m_k\mathbb{Z})^*.$$

When each m_i is a distinct prime power, p^{e_p}, so that $m = \prod_p p^{e_p}$, then we have

$$\mathbb{Z}/m\mathbb{Z} \cong \bigoplus_{p:\, p^{e_p} \| m} \mathbb{Z}/p^{e_p}\mathbb{Z} \quad \text{and} \quad (\mathbb{Z}/m\mathbb{Z})^* \cong \bigoplus_{p:\, p^{e_p} \| m} (\mathbb{Z}/p^{e_p}\mathbb{Z})^*.$$

Exercise 3.15.2.[†] Give an example of an additive group G and a subgroup H for which G is not isomorphic with $H \oplus G/H$.

3.16. The structure of finite abelian groups

An abelian group G is *generated* by $g_1, g_2, \ldots, g_k$ if every element takes the form

$$g_1^{a_1} g_2^{a_2} \cdots g_k^{a_k} \tag{3.16.1}$$

3.16. The structure of finite abelian groups

where each $a_j \in \mathbb{Z}$. The g_j are the *generators*, and we write $G = \langle g_1, g_2, \ldots, g_k \rangle$. If G is finite, then each element of G has an order, and if g_j has order m_j, then $g_j^{a_j} = g_j^{b_j}$ if and only if $a_j \equiv b_j \pmod{m_j}$. This implies that we may take each a_j in (3.16.1) to be between 0 and $m_j - 1$. A solution to $g_1^{e_1} g_2^{e_2} \cdots g_k^{e_k} = 1$ is a *multiplicative dependence* if not all of the $g_j^{e_j}$ equal 1. If the g_j have no multiplicative dependence, then they are *multiplicatively independent*, so that the exponents on the g_j work independently from each other, and therefore the group G is isomorphic to

$$\mathbb{Z}/m_1\mathbb{Z} \oplus \mathbb{Z}/m_2\mathbb{Z} \oplus \cdots \oplus \mathbb{Z}/m_k\mathbb{Z}.$$

We now show that every finite abelian group has this structure:

Theorem 3.9 (Fundamental Theorem of Abelian Groups). *Any finite abelian group G may be written as*

$$\mathbb{Z}/m_1\mathbb{Z} \oplus \mathbb{Z}/m_2\mathbb{Z} \oplus \cdots \oplus \mathbb{Z}/m_k\mathbb{Z}.$$

Proof. Every finite abelian group G contains a set of generators: One could simply take all of the elements of G, but we want a set of generators of minimal size. So suppose that $g_1, \ldots, g_k$ is a set of generators of G of minimal size. For each j, let m_j denote the order of g_j, so that $g_j^r = g_j^a$ where a is the least residue of $r \pmod{m_j}$. Therefore all of the elements of G can be expressed in the form

(3.16.2) $\qquad g_1^{a_1} g_2^{a_2} \cdots g_k^{a_k}$ with $0 \le a_j \le m_j - 1$ for each j.

If these are all distinct, then G is the direct sum of cyclic groups with generators $g_1, \ldots, g_k$, respectively. Otherwise two distinct elements of G in (3.16.2) must be equal. Dividing one by the other we determine an element $g_1^{a_1} \cdots g_k^{a_k}$ in (3.16.2) which equals 1, but where the a_i are not all 0 $\pmod{m_i}$. Moreover, as $g_i^{a_i} \ne 1$ if $a_i \not\equiv 0 \pmod{m_i}$, we deduce that at least two of the a_i's are not 0 $\pmod{m_i}$.

Select that set of k generators of G in which $a_1 + \cdots + a_k$ is minimal in this dependence. At least two of the a_i are non-zero, say $1 \le a_1 \le a_2$, so we replace our set of generators with another set of generators $h_1, \ldots, h_k$ where $h_1 = g_1 g_2$ and $h_j = g_j$ for all $j \ge 2$. The multiplicative dependence now becomes $h_1^{a_1} h_2^{a_2 - a_1} h_3^{a_3} \ldots h_k^{a_k} = 1$, contradicting the minimality of the exponents of the g_i-dependence. $\square$

As each $\mathbb{Z}/m_j\mathbb{Z}$ can be written as a direct sum of cyclic groups of prime power order (by the Chinese Remainder Theorem), the Fundamental Theorem of Abelian Groups implies we can write any finite abelian group G as a direct sum of cyclic groups of prime power order. Then the *p-part* of the group G can be written as

$$G_p := \mathbb{Z}/p^{e_1}\mathbb{Z} \oplus \mathbb{Z}/p^{e_2}\mathbb{Z} \oplus \cdots \oplus \mathbb{Z}/p^{e_\ell}\mathbb{Z} \text{ where } e_1 \ge e_2 \ge \cdots,$$

which is the subgroup of G of elements whose order is a power of p. If $n = |G|$, then

$$G = \bigoplus_{p | n} G_p.$$

Let n_1 be the product of the p^{e_1}s, then n_2 the product of the p^{e_2}s, etc., so, by the Chinese Remainder Theorem,

$$G \cong \mathbb{Z}/n_1\mathbb{Z} \oplus \mathbb{Z}/n_2\mathbb{Z} \oplus \cdots \oplus \mathbb{Z}/n_\ell\mathbb{Z} \text{ where } n_\ell | n_{\ell-1} | \ldots | n_2 | n_1.$$

We deduce that $g^{n_1} = 1$ for any $g \in G$.

Appendix 3D. Unique factorization revisited

Does an analogy to the Fundamental Theorem of Arithmetic hold for other sets of mathematical objects, other than the positive integers? For example, do polynomials factor in a unique way into irreducibles? Or numbers of the form $\{a + b\sqrt{d} : a, b \in \mathbb{Z}\}$? How about other simple arithmetic sets? How about group elements? Or specifically the permutation group?

3.17. The Fundamental Theorem of Arithmetic, clarified

Why did we only state the Fundamental Theorem of Arithmetic for positive integers? What about negative integers? To access the negative integers $-1, -2, \ldots$ we need to factor into primes and powers of -1, but the powers of -1 are not distinct, which seems like a big problem. However underlying this issue is that -1 is a *unit*, a number that divides 1. In the integers the only units are -1 and 1, and with this concept the Fundamental Theorem of Arithmetic can be neatly reformulated as follows:

Every non-zero integer can be written uniquely as a unit times a product of primes.

In other sets of numbers the units can be more complicated. For example in the set $\mathbb{Z}[i] := \{a + bi : a, b \in \mathbb{Z}\}$ we have the units i and $-i$ as well as 1 and -1. This is because i is a fourth root of 1, and similar remarks may be made about sets generated, in the analogous manner, by the nth roots of unity (that is, the roots of $x^n - 1$). More interestingly, there are domains in which there are **infinitely many** units and which are not roots of unity. For example in $\mathbb{Z}[\sqrt{2}] := \{a + b\sqrt{2} : a, b \in \mathbb{Z}\}$, we start with $3 + 2\sqrt{2}$ which divides $(3 + 2\sqrt{2})(3 - 2\sqrt{2}) = 9 - 8 = 1$ and then take $a + b\sqrt{2} = (3 + 2\sqrt{2})^k$ for any integer k.

Exercise 3.17.1. Prove that the numbers $(3 + 2\sqrt{2})^k$ with $k = 1, 2, 3, \ldots$ are distinct units in $\mathbb{Z}[\sqrt{2}]$.

The Fundamental Theorem of Algebra (Theorem 3.10) states that any polynomial $f(x) \in \mathbb{C}[x]$ has exactly d complex roots, counted with multiplicity. Therefore $f(x)$ factors into polynomials of the form $x - \alpha$ with $\alpha \in \mathbb{C}$, times a non-zero constant. Therefore polynomials can be uniquely factored, the role of the primes is played by the polynomials of the form $x - \alpha$, and the role of the units by the non-zero constants.

Exercise 3.17.2. Let R be a set of numbers containing 1 which is closed under multiplication. Prove that the units of R form a group.

Exercise 3.17.3. Let $f(x)$ be the minimum polynomial for $u \in R$.
(a) Prove that $x^d f(1/x)$ is the minimum polynomial for $1/u$.
(b) Deduce that u is a unit if and only if $f(0)$ equals 1 or -1.

3.18. When unique factorization fails

Let R be a set of numbers containing 1, which is closed under multiplication (for example, the positive integers). We call an element of R *irreducible* if it does not factor into two non-unit elements of R. We begin with a well-known example.

Exercise 3.18.1. The set of positive integers, $\mathcal{F}$, which are $\equiv 1 \pmod 4$, is closed under multiplication and contains 1. Note that 21 is irreducible in $\mathcal{F}$, despite not being prime in the positive integers. Show that factorization into irreducibles in not unique in $\mathcal{F}$.

This might seem to be an artificial example, so we give a more convincing one:

Proposition 3.18.1. *Let R be the ring $\mathbb{Z}[\sqrt{-5}] := \{a + b\sqrt{-5} : a, b \in \mathbb{Z}\}$. The elements $2, 3, 1+\sqrt{-5}$, and $1-\sqrt{-5}$ are all irreducible in R, and therefore 6 factors in two different ways,*

$$6 = 2 \cdot 3 = (1 + \sqrt{-5}) \cdot (1 - \sqrt{-5}),$$

into irreducibles of R. Hence there are elements of R that do not factor into irreducible elements of R in a unique way.

Exercise 3.18.2 (Proof of Proposition 3.18.1). Let's suppose that rational prime $p = (a + b\sqrt{-5})(c + d\sqrt{-5})$ with $(a, b) = (c, d) = 1$.
(a) By studying the coefficient of the imaginary part, prove that $c + d\sqrt{-5} = \pm(a - b\sqrt{-5})$.
(b) Deduce that $p = a^2 + 5b^2$ and that this is impossible for $p = 2$ and $p = 3$. Deduce that 2 and 3 are irreducible in R.
(c) Now assume that $1 + \sqrt{-5} = (a + b\sqrt{-5})(c + d\sqrt{-5})$. Multiply this with its complex conjugate to prove that $6 = (a^2 + 5b^2)(c^2 + 5d^2)$.
(d) Deduce that one of $a^2 + 5b^2$ and $c^2 + 5d^2$ equals 1, and therefore either $a+b\sqrt{-5}$ or $c+d\sqrt{-5}$ is a unit. Deduce that $1 + \sqrt{-5}$ and $1 - \sqrt{-5}$ are irreducible in R.

To fix what is wrong in these examples we will change our definition of a prime, using one of their properties over the integers which generalizes nicely into other rings. The idea is to work with the set of numbers that p divides, that is, the ideal $I_\mathbb{Z}(p)$, rather than the numbers that divide p.

3.19. Defining ideals and factoring

Let R be a set of numbers that is closed under addition and subtraction, for example $\mathbb{Z}, \mathbb{Q}, \mathbb{R}$, or $\mathbb{C}$, but not $\mathbb{N}$. We define the *ideal generated by* $a_1, \ldots, a_k$ over R to be the set of linear combinations of $a_1, \ldots, a_k$ with coefficients in R; that is,

$$I_R(a_1, \ldots, a_k) = \{r_1 a_1 + r_2 a_2 + \cdots + r_k a_k : r_1, \ldots, r_k \in R\}$$

($a_1, \ldots, a_k$ are not necessarily in R). In Corollary 1.3.1 and exercise 1.7.16 we saw that any ideal over $\mathbb{Z}$ can be generated by just one element, but this is not necessarily true when the a_i are taken from other domains. For example if $R = \mathbb{Z}[\sqrt{-5}]$, that is, the numbers of the form $u + v\sqrt{-5}$ where u and v are integers, then the ideal $I_R(2, 1 + \sqrt{-5})$ cannot be generated by just one element, as we will see below.

A *principal ideal* is an ideal that can be generated by just one element. As every ideal in $\mathbb{Z}$ is principal (exercise 1.7.16), $\mathbb{Z}$ is called a *principal ideal domain*.

Exercise 3.19.1. Prove that if $I_R(\alpha) = I_R(\beta)$, then $\beta = u\alpha$ for some unit $u \in R$.

Exercise 3.19.2. Prove that every Euclidean domain (as defined in section 2.12 of appendix 2B) is a principal ideal domain.

To prove that every principal ideal domain has unique factorization, we can show that an element is prime if and only if it is irreducible (see section 8.3 of [**DF04**]).

One can multiply ideals together by multiplying together pairs of elements of the ideals and establish that

$$I_R(\alpha, \beta) \cdot I_R(\gamma, \delta) = I_R(\alpha\gamma, \alpha\delta, \beta\gamma, \beta\delta).$$

Therefore if $n = ab$ in R, then $I_R(n) = I_R(a)I_R(b)$. All issues with units disappear for if I is an ideal and u a unit, then $I = uI$.

Prime ideals are ideals that cannot be factored into two other ideals. (We also call an element p of R *prime* if $I_R(p)$ is a prime ideal.) In some R, the notions of "irreducible" and "prime" are not in general the same.

In "number rings" R, all ideals can be factored into prime ideals in a unique way, and so we get unique factorization. Note though that prime ideals are no longer elements of the ring or even necessarily principal ideals of the ring.

In our example $6 = 2 \cdot 3 = (1 + \sqrt{-5}) \cdot (1 - \sqrt{-5})$ above, all of $2, 3, 1 + \sqrt{-5}, 1 - \sqrt{-5}$ are irreducibles of $\mathbb{Z}[\sqrt{-5}]$ but none generate prime ideals. In fact we can factor the ideals they generate into prime ideals as

$$I_R(2) = I_R(2, 1 + \sqrt{-5}) \cdot I_R(2, 1 - \sqrt{-5}),$$
$$I_R(3) = I_R(3, 1 + \sqrt{-5}) \cdot I_R(3, 1 - \sqrt{-5}),$$
$$I_R(1 + \sqrt{-5}) = I_R(2, 1 + \sqrt{-5}) \cdot I_R(3, 1 + \sqrt{-5}),$$
$$I_R(1 - \sqrt{-5}) = I_R(2, 1 - \sqrt{-5}) \cdot I_R(3, 1 - \sqrt{-5}).$$

None of these prime ideals are principal for if, say, $I_R(2, 1 + \sqrt{-5}) = I_R(\alpha)$, then 2 would factor into $\alpha\beta$ for some $\beta \in R$ which contradicts Proposition 3.18.1. These multiplications work out since, for example, the product

$$I_R(2, 1 + \sqrt{-5}) \cdot I_R(2, 1 - \sqrt{-5}) = I_R(4, 2(1 + \sqrt{-5}), 2(1 - \sqrt{-5}), 1^2 - \sqrt{-5}^2 = 6);$$

but then both 4 and $6 \in R$ and so $2 = 6 - 4 \in R$, which divides all these four basis elements and so the product equals $I_R(2)$.

Any prime of R is irreducible, but not vice versa. Davenport asked for the maximum number of prime ideal factors an irreducible integer can have in R.

Exercise 3.19.3. Prove that if $\alpha, \beta \in I_R$ and $r, s \in R$, then the linear combination $r\alpha + s\beta \in I_R$.

Exercise 3.19.4. Prove that if $a_1, \ldots, a_k \in R$, then $I_R(a_1, \ldots, a_k)$ is a ring.

Proposition 3.19.1. *If R is a principal ideal domain with $\alpha \in R$, then α is irreducible if and only if $I_R(\alpha)$ is a prime ideal. Moreover R has unique factorization.*

Proof. Suppose that $I_R(\alpha)$ factors into two non-trivial ideals. These must both be principal as R is a principal ideal domain; that is, there exists $\beta, \gamma \in R$ such that $I_R(\alpha) = I_R(\beta) I_R(\gamma) = I_R(\beta\gamma)$ and so $\alpha = u\beta\gamma$ for some unit u, by exercise 3.19.1. That is, α is reducible. On the other hand, if α is reducible, say $\alpha = \beta\gamma$, then $I_R(\alpha) = I_R(\beta) I_R(\gamma)$ and so $I_R(\alpha)$ is not a prime ideal.

$I_R(\alpha)$ can be uniquely factored into prime ideals, which must be principal since R is a principal ideal domain. We write $I_R(\alpha) = I_R(\pi_1) \cdots I_R(\pi_k) = I_R(\pi_1 \cdots \pi_k)$, where each π_j is irreducible and $\alpha = u\pi_1 \cdots \pi_k$ for some unit u, by exercise 3.19.1.

We need to prove that this factorization is unique. If $\alpha = \gamma_1 \ldots \gamma_\ell$, then $I_R(\pi_1) \cdots I_R(\pi_k) = I_R(\alpha) = I_R(\gamma_1) \cdots I_R(\gamma_\ell)$. By the unique factorization of prime ideals these can be paired off, say (perhaps after some rearrangement) $I_R(\pi_j) = I_R(\gamma_j)$ for each j with $k = \ell$. Then $\gamma_j = \pi_j u_j$ for some unit u_j, for each j, and so the two factorizations differ only by units. □

3.20. Bases for ideals in quadratic fields

One can determine a basis for a given ideal of $\mathbb{Z}[\sqrt{d}] := \{a + b\sqrt{d} : a, b \in \mathbb{Z}\}$ where d is a non-square integer, which takes a special and convenient form:

Every ideal I of $\mathbb{Z}[\sqrt{d}]$ is either principal or can be written in the form

$$I_\mathbb{Z}(s) \cdot I_\mathbb{Z}(b + \sqrt{d}, a),$$

for some integers s, a, b where a divides $b^2 - d$.

Exercise 3.20.1. Let I be an ideal of $\mathbb{Z}[\sqrt{d}]$, and let s be the smallest positive integer for which there is some $r + s\sqrt{d} \in I$.
 (a) Prove that if $u + v\sqrt{d} \in I$, then s divides v.
 (b)† Deduce that there exists an integer m for which $I = I_\mathbb{Z}(r + s\sqrt{d}, m)$.
 (c) Prove that s divides both r and m, and so deduce the claimed form of the ideal.
 (d) Prove that a divides $b^2 - d$.

Exercise 3.20.2. Let $R = \mathbb{Z}[\sqrt{d}]$ where d is a squarefree integer. Let $I = I(b + \sqrt{d}, a)$ where a divides $b^2 - d$.
 (a) Prove that I is principal if and only if $|a| = 1$ or $|b^2 - d|$.
 (b) Let $I^c := I(b - \sqrt{d}, a)$. Prove that $I \cdot I^c = (a)J$ where J is a principal ideal dividing (2).
 (c) Prove that I is a prime ideal if and only if I^c is a prime ideal.
 (d) Prove that if I is a non-principal prime ideal, then $I \cdot I^c = (p)$ where p is a prime number.

Suppose $p \nmid r + s\sqrt{d}$ but $p | r^2 - ds^2$. Then $I_R(p, r + s\sqrt{d})$ is principal if and only if there exist integers m, n with $p = m^2 - dn^2$ and $ms \equiv rn \pmod{p}$. In that case we have $I_R(p, r + s\sqrt{d}) = I_R(m + n\sqrt{d})$.

Therefore, for example $I_R(2, 1 + \sqrt{-5})$ and $I_R(3, 1 + \sqrt{-5})$ are non-principal since there do not exist integers m, n for which $m^2 + 5n^2 = 2$ or 3.

Appendix 3E. Gauss's approach

3.21. Gauss's approach to Euclid's Lemma

Gauss took a slightly different approach to the main technical part of chapter 3, developing his theory from a lemma that is equivalent to Euclid's Lemma (*if c divides ab and $(c,a) = 1$, then c divides b*):

Corollary 3.21.1. *Suppose that* $\gcd(a,c) = 1$. *Then a and c both divide n if and only if ac divides n.*

Deduction of Corollary 3.21.1 from Euclid's Lemma. Since a divides n we can write $n = ab$ for some integer b. Then c divides $n = ab$ with $(c,a) = 1$, and so c divides b by Euclid's Lemma. Hence we can write $b = cd$ for some integer d and so $n = acd$. This implies that ac must divide n. In the other direction note that a and c both divide ac, which divides n, and so they both divide n. □

Deduction of Euclid's Lemma from Corollary 3.21.1. Since a and c both divide ab and $(a,c) = 1$, therefore ac divides ab by Corollary 3.21.1. Therefore c divides b by exercise 1.1.1(e).

We also find another proof in exercise 3.2.1. □

Another proof of Corollary 3.2.2. [*We have $ab = (a,b)[a,b]$.*] Let $g = (a,b)$, and then $A = a/g$ and $B = b/g$, so that $(A,B) = 1$ by exercise 1.2.5(a). Now $(A,B) = 1$ and A and B both divide $[A,B]$, so AB divides $[A,B]$ by Corollary 3.21.1. However $[A,B] \leq AB$ and so we deduce that $[A,B] = AB$.

Therefore, by exercise 1.4.2, we obtain

$$ab = (gA)(gB) = g \cdot gAB = g \cdot g\operatorname{lcm}[A,B] = g \cdot \operatorname{lcm}[gA, gB] = g \cdot \operatorname{lcm}[a,b],$$

as claimed. □

Appendix 3F. Fundamental theorems and factoring polynomials

In this appendix we will prove the Unique Factorization Theorem for polynomials with complex coefficients, for polynomials with integer coefficients, and for polynomials mod p. In so doing we will introduce and discuss the Fundamental Theorem of Algebra. We then discuss resultants in more detail.

3.22. The number of distinct roots of polynomials

In the analogy to the Unique Factorization Theorem for polynomials with integer coefficients, the role of the primes is played by the irreducible polynomials f of content 1, and the role of the units, $\{1, -1\}$, is played by the integers. We prove this by induction on the degree of f. First, f can be written as a product of irreducibles, for either f is irreducible or f is the product of two polynomials in $\mathbb{Z}[x]$ of smaller degree ≥ 1, which can therefore be factored into irreducibles by the induction hypothesis, and so f is the concatenation of those two products. The proof that the factorization is unique is analogous to the proof over the integers:

Exercise 3.22.1 (Euclid's Lemma for polynomials). Suppose that $\gcd(f(x), g(x))_{\mathbb{C}[x]} = 1$ and $f(x)$ divides $g(x)h(x)$. Deduce that $f(x)$ divides $h(x)$.

The *Fundamental Theorem of Algebra* is the well-known result that a polynomial $f(x) = \sum_{j=0}^{d} f_j x^j$ with complex coefficients of degree d has exactly d roots (where the roots are counted according to their multiplicity; for example, $x(x-1)^2$ has the three roots 0, 1, and 1, the root 1 counted twice). This theorem, used without proof while a student learns basic mathematics, is rather more subtle to prove than one might guess. Four of the great mathematicians, Euler (1749), Lagrange (1772), Laplace (1795), and Gauss (1799) published purported proofs which had subtle errors, before the first correct proof appeared in 1814 due to Argand. The difficulty in proving the Fundamental Theorem of Algebra is the following first step, which appears to be innocuous but isn't.

Lemma 3.22.1. *If $f(x) \in \mathbb{C}[x]$ has degree $d \geq 1$, then $f(x)$ has at least one root in $\mathbb{C}$.*

We will not prove this but we will deduce the Fundamental Theorem from it:

Theorem 3.10 (The Fundamental Theorem of Algebra). *If $f(x) \in \mathbb{C}[x]$ has degree $d \geq 1$, then $f(x)$ has exactly d roots in $\mathbb{C}$, counted with multiplicity.*

Proof. By induction. For $d = 1$ we note that $ax + b$ has the unique root $-b/a$. For higher degree, f has a root α by Lemma 3.22.1. Let $g(x) = x - \alpha$ in Proposition 2.10.1 to obtain a polynomial $q(x)$ and a constant r for which $f(x) = (x-\alpha)q(x)+r$. Substituting in $x = \alpha$, we deduce that $r = f(\alpha) = 0$; that is, $f(x) = (x - \alpha)q(x)$. Now $q(x)$ has degree $d - 1$ so has $d - 1$ roots, counted with multiplicity, by the induction hypothesis. Therefore $f(x) = (x - \alpha)q(x)$ has $1 + (d - 1) = d$ roots, counted with multiplicity. □

We note that if $f(x) = \sum_{j=0}^{d} f_j x^j$, then

$$q(x) := \frac{f(x) - f(\alpha)}{x - \alpha} = \sum_{j=0}^{d} f_j \left(\frac{x^j - \alpha^j}{x - \alpha} \right) = \sum_{j=0}^{d} f_j (x^{j-1} + \alpha x^{j-2} + \cdots + \alpha^{j-1}).$$

If we do not assume Lemma 3.22.1, then our proof of Theorem 3.10 can be easily modified to prove the following unconditionally.

Theorem 3.11 (Weak Fundamental Theorem of Algebra, Lagrange (1772)). *If $f(x) \in \mathbb{C}[x]$ has degree $d \geq 1$, then $f(x)$ has no more than d distinct roots in $\mathbb{C}$.*

Proof. If f has no roots, then we are done. Otherwise if $f(x)$ has a root, call it α, then we write $f(x) = (x - \alpha)q(x)$ as in the proof of Theorem 3.10. Now $q(x)$ has degree $d - 1$ so, by the induction hypothesis, has $\leq d - 1$ distinct roots, which implies that $f(x) = (x - \alpha)q(x)$ has $\leq 1 + (d - 1)$ distinct roots. □

Exercise 3.22.2. (a) Deduce that every irreducible polynomial in $\mathbb{C}[x]$ has degree one.
(b) Prove that the set of d roots of $f(x)$ is unique.
(c) Prove that every irreducible polynomial in $\mathbb{R}[x]$ has degree one or two.

Following exercise 3.22.2, the Fundamental Theorem of Algebra implies that one can factor any polynomial of $\mathbb{C}[x]$ in a unique way into linear factors. This is much like the Fundamental Theorem of Arithmetic, the polynomial $x - \alpha$ taking the place of the primes and the elements of $\mathbb{C} \setminus \{0\}$ taking the place of $\{-1, 1\}$.

The Euclidean algorithm for polynomials

As discussed in section 2.10 of appendix 2B, the obvious analogy to Euclid's algorithm may be used to find $g(x)$, the polynomial of highest degree which divides two given polynomials $a(x)$ and $b(x)$ (in $\mathbb{C}[x]$). More precisely if

$$a(x) = \prod_{t \in \mathbb{C}} (x - t)^{\alpha_t} \quad \text{and} \quad b(x) = \prod_{t \in \mathbb{C}} (x - t)^{\beta_t}, \quad \text{then} \quad g(x) = \prod_{t \in \mathbb{C}} (x - t)^{\min\{\alpha_t, \beta_t\}}$$

where each α_t, β_t is an integer ≥ 0 and only finitely many are non-zero. The operations of the Euclidean algorithm respect the field, so if $a(x), b(x) \in \mathbb{Q}[x]$, then there exist $u(x), v(x) \in \mathbb{Q}[x]$, such that

(3.22.1) $$a(x)u(x) + b(x)v(x) = g(x).$$

As discussed in section 0.14 of appendix 0E, the minimum polynomial f for an algebraic number $\alpha \in \mathbb{C}$ is in $\mathbb{Z}[x]$, has content 1 and a positive leading coefficient, and is unique.

Proposition 3.22.1. *Let $\alpha \in \mathbb{C}$ and let $f(x) \in \mathbb{Z}[x]$ be the minimum polynomial for α. If $g(x) \in \mathbb{Z}[x]$ with $g(\alpha) = 0$, then $f(x)$ divides $g(x)$ in $\mathbb{Q}[x]$.*

Proof. By exercise 2.10.5(c) there exist $q(x), r(x) \in \mathbb{Z}[x]$ and $k \in \mathbb{Z}$ with $0 \leq \deg r \leq \deg f - 1$, such that $kg(x) = q(x)f(x) + r(x)$. Hence $r(\alpha) = kg(\alpha) - q(\alpha)f(\alpha) = 0$ and $r(x)$ has smaller degree than $f(x)$. As f is the minimum polynomial for α, the only possibility for $r(x)$ is 0. Hence $kg(x) = q(x)f(x)$ and the result follows. $\square$

Exercise 3.22.3. Show that if α has minimal polynomial $f(x)$ and β is another root of $f(x)$, then $f(x)$ is also the minimum polynomial for β.

Another result that will be useful is

Lemma 3.22.2. *If $f(x), g(x) \in \mathbb{Q}[x]$ are monic and $f(x)g(x) \in \mathbb{Z}[x]$, then $f(x)$ and $g(x) \in \mathbb{Z}[x]$.*

Proof. If the conclusion is false, then some coefficient of $f(x)$ is not an integer, so let p be a prime dividing the denominator of that coefficient of $f(x)$. Let p^a and p^b be the highest powers of p dividing the denominator of any coefficient of $f(x)$ and $g(x)$, respectively, so that $a \geq 1$ and $b \geq 0$. Therefore we may write $p^a f(x) \equiv f_d x^d + \cdots \pmod{p}$ where $f_d \not\equiv 0 \pmod{p}$, and similarly $p^b g(x) \equiv g_k x^k + \cdots \pmod{p}$ where $g_k \not\equiv 0 \pmod{p}$. Now $a + b \geq 1$ and $f(x)g(x) \in \mathbb{Z}[x]$ so that

$$0 \equiv p^{a+b}f(x)g(x) \equiv (f_d x^d + \cdots)(g_k x^k + \cdots) \equiv f_d g_k x^{d+k} + \cdots \pmod{p},$$

which implies that p divides $f_d g_k$, a contradiction. $\square$

Lemma 3.22.2 implies that we can conclude $f(x)$ divides $g(x)$ in $\mathbb{Z}[x]$ in Proposition 3.22.1.

Corollary 3.22.1 (Gauss's Lemma). *Suppose that $f(x) \in \mathbb{Z}[x]$ is monic. Then $f(x)$ is irreducible in $\mathbb{Z}[x]$ if and only if $f(x)$ is irreducible in $\mathbb{Q}[x]$.*

Proof. One direction is trivial. We need only prove that if $f(x)$ is irreducible in $\mathbb{Z}[x]$, then it is irreducible in $\mathbb{Q}[x]$. If not, then $f(x) = g(x)h(x)$ for some polynomials $g(x), h(x) \in \mathbb{Q}[x]$ which we may assume are monic by dividing them each by their leading coefficients. But then $g(x), h(x) \in \mathbb{Z}[x]$ by Lemma 3.22.2. $\square$

Unique factorization of polynomials modulo p

We can reduce polynomials in $\mathbb{Z}[x]$ to polynomials mod p by defining $f(x) \equiv g(x)$ (mod p) if and only if there exists $h(x) \in \mathbb{Z}[x]$ for which $f(x) - g(x) = ph(x)$.

If $g(x)$ is a polynomial whose coefficients belong to residue classes mod p, then we can write $g(x)$ as a polynomial with integer coefficients (by taking representatives of the residue classes), but with the understanding that we are studying $g(x)$ mod p. If $g(x)$ has leading term cx^d where $c \not\equiv 0$ (mod p), let $d \equiv 1/c$ (mod p), and so $G(x) := dg(x)$ (mod p) is monic.

Given polynomials $f(x), g(x)$ (mod p),[15] select $F(x), G(x) \in \mathbb{Z}[x]$ with G monic for which $F(x) \equiv f(x)$ (mod p) and $G(x) \equiv dg(x)$ (mod p). We then apply exercise 2.10.5(b) so there exists $Q(x), R(x) \in \mathbb{Z}[x]$ with $F = GQ + R$ and $\deg R < \deg G$. Hence if $q \equiv dQ$ (mod p) and $r \equiv R$ (mod p), we have $f \equiv gq + r$ (mod p) with $\deg r < \deg g$. This means that the Euclidean algorithm works for polynomials (mod p), that we can prove the analogy to Euclid's Lemma, and that polynomials mod p have unique factorization. (The readers should convince themselves of the details.)

3.23. Interpreting resultants and discriminants

In sections 2.10 and 2.11 of appendix 2B we noted that if f and g are polynomials with integer coefficients which have no common root, then there exist polynomials $a, b \in \mathbb{Z}[x]$ with $\deg a < \deg g$ and $\deg b < \deg f$ and a positive integer R for which

$$(3.23.1) \qquad a(x)f(x) + b(x)g(x) = R.$$

We call the smallest such R the resultant of f and g.

Now, let us suppose that there exists an integer m such that $f(m) \equiv g(m) \equiv 0$ (mod p). Substituting $x = m$ into (3.23.1) we see that p divides R. But what if f and g have no common root mod p but do have a common non-linear factor? For example, if $f(x) = x^2 - 2$ and $g(x) = 4x^2 + 1$, then they are both $\equiv x^2 + 1$ (mod 3), which has no roots mod 3, but then f and g do have a repeated factor mod 3, which is why 3 divides $R = g - 4f = 9$. This all follows from

Proposition 3.23.1. *Suppose that $f(x), g(x) \in \mathbb{Z}[x]$ have no common roots. Then prime p divides the resultant of f and g if and only if f and g have a common polynomial factor mod p.*

Proof. Suppose that f and g have a common factor $h(x)$ mod p (which is not necessarily a linear polynomial). Therefore $h(x)$ divides $af + bg = R$ mod p. But since $\deg R = 0$ this is impossible, unless $R \equiv 0$ (mod p).

Now suppose that prime p divides R so that $a(x)f(x) \equiv -b(x)g(x)$ (mod p). Hence $f(x)$ divides $b(x)g(x)$ (mod p), but f has higher degree than b and so it must have some factor in common with $g(x)$ (mod p), by the unique factorization of polynomials mod p. □

---

[15]More precisely, elements of $(\mathbb{Z}/p\mathbb{Z})[x]$.

This is an extraordinary result. It tells us that the prime factors of R are exactly the primes for which f and g have a common root mod p; and a study of the prime powers dividing R can tell us more about multiple factors.[16]

The discriminant of f is defined to be the resultant of f and f'. We immediately deduce the following:

Corollary 3.23.1. *Suppose that $f(x) \in \mathbb{Z}[x]$ has no repeated roots. Then prime p divides Δ, the discriminant of f if and only if f has a repeated polynomial factor mod p.*

A few examples: If $f(x) = ax^2 + bx + c$, then $f'(x) = 2ax + b$ and so
$$(2ax + b)(2ax + b) - 4a(ax^2 + bx + c) = b^2 - 4ac.$$
If $f(x) = x^3 + ax + b$, then $f'(x) = 3x^2 + a$ and so
$$9(3b - 2ax)(x^3 + ax + b) + (6ax^2 - 9bx + 4a^2)(3x^2 + a) = 4a^3 + 27b^2.$$

3.24. Other approaches to resultants and gcds

The polynomial common factor of highest degree of f and f' can be obtained by using the Euclidean algorithm but can also be described as follows: Suppose that $f(x) \in \mathbb{Z}[x]$ has an irreducible factor $p(x)$ and that $p(x)^e \| f(x)$. Therefore we can write $f(x) = p(x)^e g(x)$ where $p(x) \nmid g(x)$ and so
$$f'(x) = p(x)^e g'(x) + e p(x)^{e-1} g(x) = p(x)^{e-1}(p(x)g'(x) + eg(x))$$
Now $(p(x), p(x)g'(x) + eg(x)) = (p(x), eg(x)) = 1$ and so $p(x)^{e-1} \| f'(x)$. This implies that $p(x)^{e-1} \| (f'(x), f(x))$ and so
$$\gcd_{\mathbb{Z}[t]}(f(x), f'(x)) = \prod_{\substack{p(x)^e \| f(x) \\ p(x) \text{ irreducible}}} p(x)^{e-1}$$
and therefore is divisible by any repeated factor of $f(x)$.

We are more interested here in the case where we are given two polynomials that have no common factor. We begin with resultants. Suppose we are given polynomials $f(x) = \sum_{i=0}^d f_i x^i$ and $g(x) = \sum_{j=0}^D g_j x^j$ and we wish to know if there exist polynomials $a(x) = \sum_{i=0}^{D-1} a_i x^i$ and $b(x) = \sum_{j=0}^{d-1} b_j x^j$ for which $af + bg$ is a non-zero, integer. The polynomials a and b contain $d + D$ variables to be assigned values, and the polynomial $af + bg$ has degree $d + D - 1$ and so $d + D - 1$ coefficients that must be 0. Each of the coefficients is a linear polynomial in the a_i and b_j so we can determine whether or not there is a solution by linear algebra. This is all discussed in detail in exercises 29 to 32 of section 14.6 of [**DF04**]. They also prove some lovely formulas for R' (a small multiple of R):
$$R' = f_d^D g_D^d \prod_{i=1}^d \prod_{j=1}^D (\alpha_i - \beta_j) = f_d^D \prod_{i=1}^d g(\alpha_i) = (-1)^{dD} g_D^d \prod_{j=1}^D f(\beta_j),$$

---

[16] The usual textbook definition of resultant (see, e.g., exercise 29 of section 14.6 of [**DF04**]) allows one to find polynomials a and b for which $af + bg$ is a non-zero integer; call it R'. However R' can be a multiple of R. For example, if $f(x) = 3x$ and $g(x) = 3x + 1$, then we have $g - f = 1 = R$ but $R' = 3$. The reason for this is that the textbook definition also determines when there are common factors of the leading coefficients of f and g.

where the α_i are the roots of f and the β_j are the roots of g, both counted with multiplicity. As a consequence, if $f(x)$ has no repeated factors and $g = f'$, then

$$R' = f_d^D \prod_{i=1}^{d} f'(\alpha_i),$$

a multiple of the discriminant of f.

Additional exercises

Exercise 3.24.1. Suppose that $f(x) \in \mathbb{Z}[x]$.
(a) Show that if f has an integer root n, then $f(n) \equiv 0 \pmod{m}$ for any integer m.
(b) Suppose that f has a rational root r/s, where r and s are coprime integers. Show that if $(s, m) = 1$, then there exists an integer n such that $f(n) \equiv 0 \pmod{m}$.
(c) For each integer m give an example of a polynomial f which has a rational root r/s with $(r, s) = 1$, but for which there does not exist an integer n such that $f(n) \equiv 0 \pmod{m}$.

Exercise 3.24.2. Suppose that $f(x) \in \mathbb{Z}[x]$ has degree d, and let $\alpha, \beta, \gamma, \delta \in \mathbb{Z}$ with $\alpha\delta - \beta\gamma = 1$.
(a) Show that $f(x)$ is irreducible if and only if $x^d f(1/x)$ is irreducible.
(b) Show that $f(x)$ is irreducible if and only if $(\gamma x + \delta)^d f(\frac{\alpha x + \beta}{\gamma x + \delta})$ is irreducible. (Remark: The easy way to prove this is to know that all such transformations can be given as a composition of the transformations $z \to z \pm 1$ and $z \to -1/z$ as we saw in section 1.10.)

Exercise 3.24.3.
(a) Give examples of cubic polynomials in $\mathbb{Z}[x]$ with no roots in $\mathbb{Z}$.
(b) Give examples of cubic polynomials in $\mathbb{Z}[x]$ with all three roots in $\mathbb{Z}$.
(c) Give examples of cubic polynomials in $\mathbb{Z}[x]$ with exactly one root in $\mathbb{Z}$.
(d) Prove that there are no examples with precisely two roots in $\mathbb{Z}$.
(e) Answer these same questions for cubic polynomials when $\mathbb{Z}$ is replaced by $\mathbb{Z}/p\mathbb{Z}$ for some odd prime p.

Appendix 3G. Open problems

3.25. The Frobenius postage stamp problem, II

Suppose that a and b are positive coprime integers. We wish to better understand the sets

$$\mathcal{P}(a,b) := \{am + bn : m, n \in \mathbb{Z}, \ m, n \geq 0\} = \{N \geq 0\} - \mathcal{E}(a,b).$$

Theorem 3.12. *If $a > b \geq 1$ are coprime integers, then $ab - a - b$ is the largest integer in $\mathcal{E}(a,b)$.*

Proof. For a given integer N, let j be the least residue of $N/a \pmod{b}$. We will show that if we can write $N = am + bn$ for integers $m, n \geq 0$, then there is such a representation of N with $m = j$: Since $am \equiv N \equiv aj \pmod{b}$, therefore $m \equiv j \pmod{b}$ and so $m \geq j$. We may write $m = j + bi$ for some integer i, and so $N = a(j + bi) + n = aj + b(n + ai) = aj + bk$ for some $k \geq n \geq 0$, as desired.

We deduce that

$$\mathcal{P}(a,b) = \bigcup_{j=0}^{b-1} \{N \in \mathbb{Z} : N \equiv aj \pmod{b} \text{ and } N \geq aj\}.$$

Therefore

$$\mathcal{E}(a,b) = \bigcup_{j=0}^{b-1} \{N \in \mathbb{Z} : N \equiv aj \pmod{b} \text{ and } 0 \leq N < aj\}.$$

The largest element of the jth set in $\mathcal{E}(a,b)$ is $aj - b$ (as this is ≥ 0). Hence the largest element of $\mathcal{E}(a,b)$ is $a(b-1) - b = ab - a - b$. □

Exercise 3.25.1. Let a and b be positive integers with $g = \gcd(a,b)$. Prove that $ab/g - a - b$ is the largest integer, divisible by g, that is not represented as $am + bn$ with $m, n \geq 0$.

Exercise 3.25.2. Let a and b be positive coprime integers. Suppose that $r = ma + nb$ for some integers m, n in the ranges $1 \leq m \leq b - 1$ and $1 \leq n \leq a - 1$.
 (a) Prove that r is the smallest integer in $\mathcal{P}(a,b)$ which is $\equiv r \pmod{ab}$.

123

(b) What is the smallest integer $s \in \mathcal{P}(a,b)$ which is $\equiv -r \pmod{ab}$?
(c) Show that exactly one of r and s is $< ab$, and deduce that if $1 \leq N \leq ab$ where $a \nmid N$ and $b \nmid N$, then exactly one of N and $ab - N$ belongs to $\mathcal{P}(a,b)$.
(d) Show that there are exactly $\frac{1}{2}(a-1)(b-1)$ elements of $\mathcal{E}(a,b)$.

Exercise 3.25.3. (This continues from exercise 1.16.2.) An integer n is *powerful* if p^2 divides n whenever prime p divides n. Prove that we can write any powerful integer n as $n = a^2 b^3$ where a and b are integers.

Exercise 3.25.4. Let a and b be positive coprime integers and select r and s in the ranges $1 \leq r \leq b-1$ and $1 \leq s \leq a-1$ so that $ar \equiv 1 \pmod{b}$ and $bs \equiv 1 \pmod{a}$. Prove that there are $\frac{N}{ab} + 1 - \{\frac{rN}{b}\} - \{\frac{sN}{a}\}$ representations of N as $N = ma + nb$ with integers $m, n \geq 0$.

Exercise 3.25.5. Suppose that $a_1, \ldots, a_k$ are positive integers for which $\gcd(a_1, \ldots, a_k) = 1$. Let
$$\mathcal{P}(a_1, \ldots, a_k) := \{a_1 n_1 + \cdots + a_k n_k : n_1, \ldots, n_k \in \mathbb{Z}, \ n_1, \ldots, n_k \geq 0\}.$$
(a) Prove that there exists an integer N such that every integer $\geq N$ belongs to $\mathcal{P}(a_1, \ldots, a_k)$.
(b) Show that we may take $N = (k-1) \operatorname{lcm}[a_1, \ldots, a_k]$.

It is not easy to prove the analogy to Theorem 3.12 for sets of three or more integers a_j. In fact, determining a formula for the largest integer that does not belong to $\mathcal{P}(a,b,c)$ (for arbitrary a,b,c) is an open problem.

3.26. Egyptian fractions for $3/b$

In section 1.17 of appendix 1E we showed that every fraction a/b with $(a,b) = 1$ can be written as a sum of at most a distinct fractions of the form $1/n$. We proved that this is best possible for $a = 1$ and 2, and we do so now for $a = 3$:

Theorem 3.13. *Suppose that b is a prime. Then $3/b$ can be written as the sum of two distinct Egyptian fractions if and only if $b \equiv -1 \pmod 3$.*

Proof. Suppose that $3/b = 1/m + 1/n$ with $m < n$. Let $g = \gcd(m,n)$ so that $m = gM, n = gN$ with $\gcd(M,N) = 1$, and therefore $3gMN = (M+N)b$. We have $(M+N, M) = (M+N, N) = (M, N) = 1$ and so MN divides the prime b. As $M < N$ this implies that $M = 1$ and $N = b$, and therefore $b = 3g - 1 \equiv -1 \pmod 3$.

On the other hand if $b = 3k - 1$, then $\frac{3}{3k-1} = \frac{1}{k} + \frac{1}{k(3k-1)}$. $\square$

Exercise 3.26.1. Fix integer $a \geq 3$. Suppose that b is a prime. Prove that a/b can be written as the sum of two distinct Egyptian fractions if and only if $b \equiv -1 \pmod a$.

Exercise 3.26.2. Suppose that a and b are coprime positive integers. Prove that we have a solution to $\frac{a}{b} = \frac{1}{m} + \frac{1}{n}$ with $(m,n) = 1$ if and only if $a^2 - 4b$ is the square of an integer.

3.27. The $3x+1$ conjecture

Start with any positive integer n and transform it according to the following map:
$$n \longrightarrow \begin{cases} 3n+1 & \text{if } n \text{ is odd,} \\ n/2 & \text{if } n \text{ is even.} \end{cases}$$

3.27. The $3x+1$ conjecture

Iterate. The $3x+1$ conjecture states that one eventually gets down to 1. For example

$$99 \to 298 \to 149 \to 448 \to 224 \to 112 \to 56 \to 28 \to 14 \to 7 \to 22 \to 11 \to$$
$$\to 34 \to 17 \to 52 \to 26 \to 13 \to 40 \to 20 \to 10 \to 5 \to 16 \to 8 \to 4 \to 2 \to 1,$$

a long and circuitous route, called the *orbit* of 99. (Note that $1 \to 4 \to 2 \to 1$.) This is what makes the $3x+1$ conjecture so difficult; there seems to be no formula that helps understand where the iterates go in the long run. And, in many examples, the elements of the orbit do not get smaller quickly; indeed they often get quite a bit larger than the number that one started with, before decreasing to 1. One can prove that there are orbits in which the numbers get larger than the original integer by an arbitrary factor.

Before proving this we modify the $3x+1$ algorithm for convenience: Given n odd we transform to $3n+1$, but given even n we divide out as many powers of 2 as possible, all in one go. Thus the above orbit now looks like

$$99 \to 298 \to 149 \to 448 = 2^6 \cdot 7 \to 7 \to 22 \to$$
$$11 \to 34 \to 17 \to 52 \to 13 \to 40 \to 5 \to 16 \to 1.$$

Exercise 3.27.1. Let $x_0 = 2^k m - 1$. Suppose that the iterates of the modified $3x+1$ algorithm go $x_0 \to y_1 \to x_1 \to y_2 \to x_2 \to \dots$.
 (a) Prove that $x_j = 3^j 2^{k-j} m - 1$ for $j = 0, 1, \dots, k$.
 (b) Deduce that there exist integers x_0 such that there is an nth iterate x_n for which x_n/x_0 is arbitrarily large.

Exercise 3.27.2. Prove that $2^k \| N$ if and only if $N \equiv 2^k \pmod{2^{k+1}}$.

To analyze how quickly we expect the numbers in the modified $3x+1$ algorithm to decrease, we now determine the expected size of the largest odd part of $3n+1$.

The probability that, for a random odd integer n, we have $3n + 1 \equiv 2^k \pmod{2^{k+1}}$ is $1/2^k$ (since we are guaranteed that $3n+1$ is even) by exercise 3.27.2. Hence the expected size of the largest odd part of $3n+1$ is

$$\sum_{k \geq 1} \frac{1}{2^k} \cdot \frac{3n+1}{2^k} = (3n+1) \sum_{k \geq 1} \frac{1}{4^k} = n + \frac{1}{3}.$$

Of course it is never near to this size (the possibilities are $\frac{3n+1}{2}, \frac{3n+1}{4}, \frac{3n+1}{8}, \dots$) but this is the average of those possibilities.

Similar probabilistic models suggest that if an orbit starts from n, then all numbers in the orbit are $< Cn^2$ for some large constant C. The orbit that begins with $n = 1980976057694848447$ contains a number which is almost $25n^2$.

It is not impossible that there exists some integer $n > 1$, such that after some iterate we return to n, and therefore the $3x+1$ conjecture is false. If such an example exists, then it is known that the cycle has period at least 10 billion!

Further reading on these open problems

[1] Ronald L. Graham, *Paul Erdős and Egyptian fractions*, Bolyai Soc. Math. Stud., **25** (2013), 289–309.

[2] Jeffrey C. Lagarias, *The $3x+1$ problem and its generalizations*, Amer. Math. Monthly **92** (1985), 3–23.

[3] Alfonsin Ramirez, *The Diophantine Frobenius problem*, Oxford Lecture Series in Mathematics and its Applications **30**, Oxford University Press, Oxford, 2005.

Chapter 4

Multiplicative functions

In the previous chapter we discussed $\tau(n)$, which counts the number of divisors of n. We discovered that $\tau(n)$ is a multiplicative function, which allowed us to calculate its value fairly easily. *Multiplicative functions*, so called since

$$f(mn) = f(m)f(n) \text{ for all pairwise coprime, positive integers } m \text{ and } n,$$

play a central role in number theory. (Moreover f is *totally multiplicative*, or *completely multiplicative*, if $f(mn) = f(m)f(n)$ for all integers $m, n \geq 1$.) Thus the divisor function, $\tau(n)$, is multiplicative but not totally multiplicative, since $\tau(p^a) = a+1$, and so $\tau(p^2) = 3$ is not equal to $\tau(p)^2 = 2^2$. Common examples of totally multiplicative functions include $f(n) = 1$, $f(n) = n$, and $f(n) = n^s$ for a fixed complex number s. Also Liouville's function $\lambda(n)$ which equals -1 to the power of the total number of prime factors of n, counting repetitions of the same prime factor. For example $\lambda(2) = \lambda(3) = \lambda(12) = \lambda(32) = -1$ and $\lambda(4) = \lambda(6) = \lambda(10) = \lambda(60) = 1$.

What makes multiplicative functions central to number theory is that one can evaluate a multiplicative function $f(n)$ in terms of the $f(p^e)$ for the prime powers p^e dividing n.

Exercise 4.0.1. Show that if f is multiplicative and $n = \prod_{p \text{ prime}} p^{e_p}$, then

$$f(n) = \prod_{p \text{ prime}} f(p^{e_p}).$$

Deduce that if f is totally multiplicative, then $f(n) = \prod_p f(p)^{e_p}$.

Exercise 4.0.2. Prove that if f is a multiplicative function, then either $f(n) = 0$ for all $n \geq 1$ or $f(1) = 1$.

Exercise 4.0.3. Prove that if f and g are multiplicative functions, then so is h, where $h(n) = f(n)g(n)$ for all $n \geq 1$.

Exercise 4.0.4. Prove that if f is completely multiplicative and $d|n$, then $f(d)$ divides $f(n)$.

127

Exercise 4.0.5. Prove that if f is multiplicative and a and b are any two positive integers, then
$$f((a,b))f([a,b]) = f(a)f(b).$$

In this chapter we will focus on two multiplicative functions of great interest.

4.1. Euler's ϕ-function

There are
$$\phi(n) := \#\{m : 1 \leq m \leq n \text{ and } (m,n) = 1\}$$
elements in any reduced system of residues mod n. Obviously $\phi(1) = 1$.

Lemma 4.1.1. *$\phi(n)$ is a multiplicative function.*

Proof. Suppose that $n = mr$ where $(m,r) = 1$. By the Chinese Remainder Theorem (Theorem 3.6) there is a natural bijection between the integers $a \pmod{n}$ with $(a,n) = 1$ and the pairs of integers $(b \pmod{m}, c \pmod{r})$ with $(b,m) = (c,r) = 1$. Since there are $\phi(m)\phi(r)$ such pairs (b,c) we deduce that $\phi(n) = \phi(m)\phi(r)$. □

Hence to evaluate $\phi(n)$ for all n we simply need to evaluate it on the prime powers, by exercise 4.0.1. This is straightforward because $(m, p^e) = 1$ if and only if $(m, p) = 1$; and $(m, p) = 1$ is not satisfied if and only if p divides m. Therefore
$$\begin{aligned}\phi(p^e) &= \#\{m : 1 \leq m \leq p^e \text{ and } (m,p) = 1\} \\ &= \#\{m : 1 \leq m \leq p^e\} - \#\{m : 1 \leq m \leq p^e \text{ and } p|m\} \\ &= p^e - p^{e-1}\end{aligned}$$
by exercise 1.7.6(c). We deduce the following:

Theorem 4.1. *If $n = \prod_{p \text{ prime}} p^{e_p}$, then*
$$\phi(n) = \prod_{\substack{p \text{ prime} \\ p|n}} (p^{e_p} - p^{e_p - 1}) = \prod_{\substack{p \text{ prime} \\ p|n}} p^{e_p}\left(1 - \frac{1}{p}\right) = n \prod_{\substack{p \text{ prime} \\ p|n}} \left(1 - \frac{1}{p}\right).$$

Example. $\phi(60) = 60 \cdot \left(1 - \frac{1}{2}\right)\left(1 - \frac{1}{3}\right)\left(1 - \frac{1}{5}\right) = 16$, the least positive residues being
$$1, 7, 11, 13, 17, 19, 23, 29, 31, 37, 41, 43, 47, 49, 53, \text{ and } 59.$$

We give an alternative proof of Theorem 4.1, based on the inclusion-exclusion principle, in section 4.5.

Studying the values taken by $\phi(n)$, one makes a surprising observation:

Proposition 4.1.1. *We have $\sum_{d|n} \phi(d) = n$.*

Example. For $n = 30$, we have
$$\begin{aligned}\phi(1) + \phi(2) + \phi(3) + \phi(5) + \phi(6) &+ \phi(10) + \phi(15) + \phi(30) \\ = 1 + 1 + 2 + 4 + 2 + 4 + 8 + 8 &= 30.\end{aligned}$$

Proof. Given any integer m with $1 \leq m \leq n$, let $d = n/(m,n)$, which divides n. Then $(m,n) = n/d$ so one can write $m = an/d$ with $(a,d) = 1$ and $1 \leq a \leq d$. Now, for each divisor d of n the number of integers m for which $(m,n) = n/d$ equals the number of integers a for which $(a,d) = 1$ and $1 \leq a \leq d$, which is $\phi(d)$ by definition. We have therefore shown that

$$n = \#\{m : 1 \leq m \leq n\} = \sum_{d|n} \#\{m : 1 \leq m \leq n \text{ and } (m,n) = n/d\}$$

$$= \sum_{d|n} \#\{m : m = a(n/d), \ 1 \leq a \leq d \text{ and } (a,d) = 1\}$$

$$= \sum_{d|n} \#\{a : 1 \leq a \leq d \text{ and } (a,d) = 1\} = \sum_{d|n} \phi(d),$$

which is the result claimed. □

Exercise 4.1.1. Prove that if $d|n$, then $\phi(d)$ divides $\phi(n)$.

Exercise 4.1.2. Prove that if n is odd and $\phi(n) \equiv 2 \pmod{4}$, then n has exactly one prime factor (perhaps repeated several times).

Exercise 4.1.3. Prove that $\sum_{1 \leq m \leq n, \ (m,n)=1} m = n\phi(n)/2$ and $\prod_{d|n} d = n^{\tau(n)/2}$.

Exercise 4.1.4. (a) Prove that $\phi(n^2) = n\phi(n)$.
 (b) Prove that if $\phi(n)|n-1$, then n is squarefree.
 (c) Find all integers n for which $\phi(n)$ is odd.

Exercise 4.1.5.[†] Suppose that n has exactly k prime factors, each of which is $> k$. Prove that $\phi(n) \geq n/2$.

4.2. Perfect numbers. *"The whole is equal to the sum of its parts."*

The number 6 is a *perfect number* since it is the sum of its *proper* divisors (the *proper* divisors of m are those divisors d of m for which $1 \leq d < m$); that is,

$$6 = 1 + 2 + 3.$$

> Six is a number perfect in itself, and not because God created all things in six days; rather, the converse is true. God created all things in six days because the number is perfect.
> — from *The City of God* by SAINT AUGUSTINE (354–430)

The next perfect number is $28 = 1 + 2 + 4 + 7 + 14$ which is the number of days in a lunar month. However the next, $496 = 1 + 2 + 4 + 8 + 16 + 31 + 62 + 124 + 248$, appears to have little obvious cosmic relevance. Nonetheless, we will be interested in trying to classify all perfect numbers. To create an equation we will add n to both sides to obtain that n is perfect if and only if

$$2n = \sigma(n), \text{ where } \sigma(n) := \sum_{d|n} d.$$

Exercise 4.2.1. Show that $\sigma(n) = \sum_{d|n} n/d$, and so deduce that n is perfect if and only if $\sum_{d|n} \frac{1}{d} = 2$.

Exercise 4.2.2. (a) Prove that each divisor d of ab can be written as ℓm where $\ell | a$ and $m | b$.
(b) Show that if $(a, b) = 1$, then there is a unique such pair ℓ, m for each divisor d.

By this last exercise we see that if $(a, b) = 1$, then

$$\sigma(ab) = \sum_{d | ab} d = \sum_{\ell | a,\ m | b} \ell m = \sum_{\ell | a} \ell \cdot \sum_{m | b} m = \sigma(a)\sigma(b),$$

proving that σ is a multiplicative function. Now

$$\sigma(p^k) = 1 + p + p^2 + \cdots + p^k = \frac{p^{k+1} - 1}{p - 1}$$

by definition, and so

$$\boxed{\text{If } n = \prod_p p^{k_p}, \text{ then } \sigma(n) = \prod_p \frac{p^{k_p+1} - 1}{p - 1}.}$$

For example $\sigma(2^5 \cdot 3^3 \cdot 5^2 \cdot 7) = \frac{2^6-1}{2-1} \cdot \frac{3^4-1}{3-1} \cdot \frac{5^3-1}{5-1} \cdot \frac{7^2-1}{7-1}$.

Euclid observed that the first perfect numbers factor as $6 = 2 \cdot 3$ where $3 = 2^2 - 1$ is prime, and $28 = 2^2 \cdot 7$ where $7 = 2^3 - 1$ is prime, and then that this pattern persists:

Proposition 4.2.1 (Euclid). *If $2^p - 1$ is a prime number, then $2^{p-1}(2^p - 1)$ is a perfect number.*

The cases $p = 2, 3, 5$ correspond to the Mersenne primes $2^2 - 1 = 3$, $2^3 - 1 = 7$, $2^5 - 1 = 31$ and therefore yield the three smallest perfect numbers 6, 28, 496 (and the next smallest examples are given by $p = 7$ and $p = 13$).

Proof. Since σ is multiplicative we have, for $n = 2^{p-1}(2^p - 1)$,

$$\sigma(n) = \sigma(2^{p-1}) \cdot \sigma(2^p - 1) = \frac{2^p - 1}{2 - 1} \cdot (1 + (2^p - 1)) = (2^p - 1) \cdot 2^p = 2n. \quad \square$$

After extensive searching one finds that perfect numbers of the form $2^{p-1}(2^p-1)$ with 2^p-1 prime appear to be the only perfect numbers. Euler succeeded in proving that these are the only *even* perfect numbers, and we believe (but don't know) that there are no *odd* perfect numbers. If there are no odd perfect numbers, as claimed, then we would achieve our goal of classifying all the perfect numbers.

Theorem 4.2 (Euclid). *If n is an even perfect number, then there exists a prime number of the form $2^p - 1$ such that $n = 2^{p-1}(2^p - 1)$.*

In exercise 3.9.1 we showed that if $2^p - 1$ is prime, then p must itself be prime. Now, although $2^2 - 1, 2^3 - 1, 2^5 - 1$, and $2^7 - 1$ are all prime, $2^{11} - 1 = 23 \times 89$ is not, so we do not know for sure whether $2^p - 1$ is prime, even if p is prime. However it is conjectured that there are infinitely many Mersenne primes $M_p = 2^p - 1$,[1] which would imply that there are infinitely many even perfect numbers.

---

[1] It is known that $2^p - 1$ is prime for $p = 2, 3, 5, 7, 13, 17, 19, \ldots, 82589933$, a total of 51 values as of September 2019 (and this last is currently the largest prime explicitly known). There is a long history of the search for Mersenne primes, from the first serious computers to the first great distributed computing project, GIMPS (Great Internet Mersenne Prime Search).

4.2. Perfect numbers. "The whole is equal to the sum of its parts."

Proof. Any even integer can be written as $n = 2^{k-1}m$ where m is odd and $k \geq 2$, so that if n is perfect, then

$$2^k m = 2n = \sigma(n) = \sigma(2^{k-1})\sigma(m) = (2^k - 1)\sigma(m).$$

Now $(2^k - 1, 2^k) = 1$ and so $2^k - 1$ divides m. Writing $m = (2^k - 1)M$ we find that $\sigma(m) = 2^k M = m + M$. That is, $\sigma(m)$, which is the sum of all of the divisors of m, equals the sum of just two of its divisors, namely m and M (and note that these are different integers since $m = (2^k - 1)M \geq (2^2 - 1)M > M$). This implies that m and M are the only divisors of m. The only integers with just two divisors are the primes, so that m is a prime and $M = 1$, and the result follows. □

It is widely believed that the only perfect numbers were those identified by Euclid, that is, that there are no odd perfect numbers. It has been proved that if there is an odd perfect number, then it must be $> 10^{1500}$, and it would have to have more than 100 (not necessarily distinct) prime factors.

Exercise 4.2.3. (a) Prove that if p is odd and k is odd, then $\sigma(p^k)$ is even.
(b)† Deduce that if n is an odd perfect number, then $n = p^\ell m^2$ where p is a prime that does not divide the integer $m \geq 1$ and $p \equiv \ell \equiv 1 \pmod{4}$.

Exercise 4.2.4. Fix integer $m > 1$. Show that there are only finitely many integers n for which $\sigma(n) = m$.

Exercise 4.2.5.† (a) Prove that for all integers $n > 1$ we have the inequalities

$$\prod_{p|n} \frac{p+1}{p} \leq \frac{\sigma(n)}{n} < \prod_{p|n} \frac{p}{p-1}.$$

(b) We have seen that every even perfect number has exactly two distinct prime factors. Prove that every odd perfect number has at least three distinct prime factors.

Additional exercises

Exercise 4.3.1. Suppose that $f(n) = 0$ if n is even, $f(n) = 1$ if $n \equiv 1 \pmod{4}$, and $f(n) = -1$ if $n \equiv -1 \pmod{4}$. Prove that $f(.)$ is a multiplicative function.

Exercise 4.3.2.† Suppose that $r(.)$ is a multiplicative function taking values in $\mathbb{C}$. Let $f(n) = 1$ if $r(n) \neq 0$, and $f(n) = 0$ if $r(n) = 0$. Prove that $f(.)$ is also a multiplicative function.

Exercise 4.3.3.† Suppose that f is a multiplicative function, such that the value of $f(n)$ depends only on the value of $n \pmod{3}$. What are the possibilities for f?

Exercise 4.3.4.‡ Suppose that f is a multiplicative function, such that the value of $f(n)$ depends only on the value of $n \pmod{8}$. What are the possibilities for f?

Exercise 4.3.5. How many of the fractions a/n with $1 \leq a \leq n - 1$ are reduced?

Looking at the values of $\phi(m)$, Carmichael conjectured that for all integers m there exists an integer $n \neq m$ such that $\phi(n) = \phi(m)$.

Exercise 4.3.6.† (a) Find all integers n for which $\phi(2n) = \phi(n)$.
(b) Find all integers n for which $\phi(3n) = \phi(2n)$.
(c) Can you find other classes of m for which Carmichael's conjecture is true?

Carmichael's conjecture is still an open problem but it is known that if it is false, then the smallest counterexample is $> 10^{10^{10}}$.

Exercise 4.3.7.[†] (a) Given a polynomial $f(x) \in \mathbb{Z}[x]$ let $N_f(m)$ denote the number of a (mod m) for which $f(a) \equiv 0$ (mod m). Show that $N_f(m)$ is a multiplicative function.
(b) Be explicit about $N_f(m)$ when $f(x) = x^2 - 1$. (You can use section 3.8.)

Exercise 4.3.8.[‡] Given a polynomial $f(x) \in \mathbb{Z}[x]$ let $R_f(m)$ denote the number of b (mod m) for which there exists a (mod m) with $f(a) \equiv b$ (mod m). Show that $R_f(m)$ is a multiplicative function. Can you be more explicit about $R_f(m)$ for $f(x) = x^2$, the example of exercise 2.5.6?

Exercise 4.3.9. Let $\tau(n)$ denote the number of divisors of n (as in section 3.3), and let $\omega(n)$ and $\Omega(n)$ be the number of prime divisors of n not counting and counting repeated prime factors, respectively. Therefore $\tau(12) = 6, \omega(12) = 2,$ and $\Omega(12) = 3$. Prove that
$$2^{\omega(n)} \leq \tau(n) \leq 2^{\Omega(n)} \text{ for all integers } n \geq 1.$$

Exercise 4.3.10. Let $\sigma_k(n) = \sum_{d|n} d^k$. Prove that $\sigma_k(n)$ is multiplicative.

Exercise 4.3.11. (a) Prove that $\tau(ab) \leq \tau(a)\tau(b)$ for all positive integers a and b, with equality if and only if $(a,b) = 1$,
(b) Prove that $\sigma_k(ab) \leq \sigma_k(a)\sigma_k(b)$ for all positive integers a, b, and k.
(c) Prove that $\sigma_{k+\ell}(n) \leq \sigma_k(n)\sigma_\ell(n)$ for all positive integers k, ℓ, and n.

Exercise 4.3.12. Give closed formulas for (a)[†] $\sum_{m=1}^{n} \gcd(m,n)$ and (b)[‡] $\sum_{m=1}^{n} \text{lcm}(m,n)$ in terms of the prime power factors of n.

Exercise 4.3.13. n is *multiplicatively perfect* if it equals the product of its proper divisors.
(a) Show that n is multiplicatively perfect if and only if $\tau(n) = 4$.
(b) Classify exactly which integers n satisfy this.

The integers m and n are *amicable* if the sum of the proper divisors of m equals n and the sum of the proper divisors of n equals m. For example, 220 and 284 are amicable, as are 1184 and 1210.[2]

Exercise 4.3.14. (a) Show that m and n are amicable if and only if $\sigma(m) = \sigma(n) = m + n$.
(b) Verify Thâbit ibn Qurrah's 9th-century claim that if $p = 3 \times 2^{n-1} - 1$, $q = 3 \times 2^n - 1$, and $r = 9 \times 2^{2n-1} - 1$ are each odd primes, then $2^n pq$ and $2^n r$ are amicable.[3]
(c) Find an example (other than the two given above) using the construction in (b).

An integer n is *abundant* if the sum of its proper divisors is $> n$, for example $n = 12$; and n is *deficient* if the sum of its proper divisors is $< n$, for example $n = 8$. Each positive integer is either deficient, perfect, or abundant, a classification that goes back to antiquity.[4]

Exercise 4.3.15. (a) Prove that every prime number is deficient.
(b) Prove that every multiple of 6 is abundant.
(c) How do these concepts relate to the value of $\sigma(n)/n$?
(d) Prove that every multiple of an abundant number is abundant.
(e)[†] Prove that if n is the product of k distinct primes that are each $> k$, then n is deficient.
(f) Prove that every divisor of a deficient number is deficient.

---

[2]The 14th-century scholar Ibn Khladun claimed: "Experts on talismans assure me that these numbers have a special influence in establishing strong bonds of friendship between individuals ... A bond so close that they cannot be separated. The author of the Ghaia, and other great masters in this art, swear that they have seen this happen again and again."

[3]This was rediscovered by Descartes in the 17th century.

[4]Specifically a book by Nichomachus from A.D. 100. Another interesting reference is the 10th-century German nun Hrotsvitha who depicts the heroine of her play "*Sapientia*" challenging Emperor Hadrian while he is persecuting Christians, to surmise the ages of her children from information about this classification and the number of Olympic games that each has been alive for!

Carl André is a controversial minimalist artist, his most infamous work being his *Equivalent I–VIII* series exhibited at several of the world's leading museums. Each of the eight sculptures involves 120 bricks arranged in a different rectangular formation. In *Equivalent VIII*, at the Tate Modern in London, the bricks are stacked 2 deep, 6 wide, and 10 long. (See http://thesingleroad.blogspot.co.uk/2011/01/test-post.html for a photo of the original eight formations.)

Exercise 4.3.16. (a) How many different 2-deep, 120-brick, rectangular formations are there?
  (b) What if there must be at least three bricks along the width and along the length?

Appendix 4A. More multiplicative functions

4.4. Summing multiplicative functions

We have already seen that the functions 1, n, $\phi(n)$, $\sigma(n)$, and $\tau(n)$ are all multiplicative. In Proposition 4.1.1 we saw the surprising connection that n is the sum of the multiplicative function $\phi(d)$, summed over the divisors d of n. Similarly $\tau(n)$ is the sum of 1, and $\sigma(n)$ is the sum of d, summed over the divisors d of n. This suggests that there might be a general such phenomenon.

Theorem 4.3. *For any given multiplicative function f, the function*

$$F(n) := \sum_{d|n} f(d)$$

is also multiplicative.

Proof. Suppose that $n = ab$ with $(a,b) = 1$. In exercise 4.2.2 we showed that the divisors of n can be written as ℓm where $\ell|a$ and $m|b$. Note that $(\ell, m) = 1$ since (ℓ, m) divides $(a, b) = 1$ and so $f(\ell m) = f(\ell)f(m)$. Therefore

$$F(ab) = \sum_{d|ab} f(d) = \sum_{\substack{\ell|a \\ m|b}} f(\ell m) = \sum_{\ell|a} f(\ell) \sum_{m|b} f(m) = F(a)F(b),$$

as desired. □

It is worth noting that if we write $m = n/d$, then Theorem 4.3 becomes

$$F(n) := \sum_{m|n} f(n/m).$$

Above we have the examples $\{F(n), f(d)\} = \{n, \phi(d)\}$, $\{\tau(n), 1\}$, $\{\sigma(n), d\}$; but what about for other $F(n)$? For $F(n) = 1$ we have $1 = \sum_{d|n} \delta(d)$ where

4.5. Inclusion-exclusion and the Möbius function

$\delta(d) = 1$ if $d = 1$, and $= 0$ otherwise. Finding f when $F(n) = \delta(n)$ looks more complicated. This leads us to two questions: For every multiplicative F, does there exist a multiplicative f for which $F(n) := \sum_{d|n} f(d)$? And, if so, is f unique? To answer these questions we begin by defining another multiplicative function which arises in a rather different context.

Exercise 4.4.1. Prove that $\frac{n}{\phi(n)} = \sum_{\substack{d|n \\ d \text{ squarefree}}} \frac{1}{\phi(d)}$.

4.5. Inclusion-exclusion and the Möbius function

In the proof of Theorem 4.1 we saw that if $n = p^a$ is a prime power, then $\phi(n)$ is the total number of integers up to n, minus the number of those that are divisible by p. This leads to the formula

$$\phi(n) = n - \frac{n}{p} = n\left(1 - \frac{1}{p}\right).$$

Similarly if $n = p^a q^b$, then we wish to count the number of positive integers up to n that are not divisible by either p or q. To do so we take the n integers up to n, subtract the n/p that are divisible by p and the n/q that are divisible by q. This is not quite right as we have twice subtracted the n/pq integers divisible by both p and q, and so we need to add them back in. This leads to the formula

$$\phi(n) = n - \frac{n}{p} - \frac{n}{q} + \frac{n}{pq} = n\left(1 - \frac{1}{p}\right)\left(1 - \frac{1}{q}\right).$$

This argument generalizes to arbitrary n, though we need to keep track of the terms of the form $\pm n/d$. In our examples so far, we see that each such d is a divisor of n, but the term n/d only has a non-zero coefficient if d is squarefree. When d is squarefree the coefficient is given by $(-1)^{\omega(d)}$ where

$$\omega(d) := \sum_{p \text{ prime}, \, p|d} 1$$

is the number of distinct prime factors of d. One therefore deduces that the coefficient of n/d is always given by the *Möbius* function, $\mu(d)$, a multiplicative function defined by

$$\mu(p) = -1, \text{ with } \mu(p^k) = 0 \text{ for all } k \geq 2, \text{ for every prime } p.$$

For example $\mu(1) = 1$, $\mu(2) = \mu(3) = -1$, $\mu(4) = 0$, $\mu(6) = \mu(10) = 1$, and $\mu(1001) = -1$ as $1001 = 7 \times 11 \times 13$.

The argument for general n uses the *inclusion-exclusion principle*, which we formulate here to fit well with the topic of multiplicative functions.

Corollary 4.5.1. *We have*

$$\sum_{d|n} \mu(d) = \begin{cases} 1 & \text{if } n = 1, \\ 0 & \text{otherwise.} \end{cases}$$

Proof. The result for $n = 1$ is trivial. If n is a prime power p^a with $a \geq 2$, then $\sum_{d|p^a} \mu(d) = 1 + (-1) + 0 + \cdots + 0 = 0$ by definition.

The result for general n then follows from Theorem 4.3. □

Exercise 4.5.1. (a) Show that if m is squarefree, then
$$(1+x)^{\omega(m)} = \sum_{d|m} x^{\omega(d)}.$$

(b) Deduce Corollary 4.5.1.

A proof of Theorem 4.1 using the inclusion-exclusion principle. We want a function that counts 1 if $(a,n) = 1$ and 0 otherwise. This counting function can be given by Corollary 4.5.1:
$$\sum_{d|a\ \&\ d|n} \mu(d) = \begin{cases} 1 & \text{if } (a,n) = 1, \\ 0 & \text{otherwise.} \end{cases}$$

Therefore
$$\phi(n) = \sum_{a=1}^{n} \begin{cases} 1 & \text{if } (a,n) = 1, \\ 0 & \text{otherwise.} \end{cases}$$
$$= \sum_{a=1}^{n} \sum_{d|a\ \&\ d|n} \mu(d)$$
$$= \sum_{d|n} \mu(d) \sum_{\substack{1 \leq a \leq n \\ d|a}} 1 = \sum_{d|n} \mu(d) \frac{n}{d} = n \sum_{d|n} \frac{\mu(d)}{d}.$$

The last line comes from first swapping the order of summation and then using exercise 1.7.6(c) as $[n/d] = n/d$ since each d divides n. Exercise 4.5.2 completes the proof. □

Exercise 4.5.2. Prove that for any positive integer n we have
$$\sum_{d|n} \frac{\mu(d)}{d} = \prod_{\substack{p \text{ prime} \\ p|n}} \left(1 - \frac{1}{p}\right).$$

Exercise 4.5.3. Prove that $\mu(n)^2$ is the characteristic function for the squarefree integers, and deduce that $\frac{n}{\phi(n)} = \sum_{d|n} \frac{\mu(d)^2}{\phi(d)}$.

4.6. Convolutions and the Möbius inversion formula

In the proof of Theorem 4.1 in the last section we saw that
$$\phi(n) = \sum_{d|n} \mu(d) \frac{n}{d}.$$

If we let $r = n/d$, then the sum is over all factorizations of n into two positive integers $n = dr$, and so
$$\phi(n) = \sum_{\substack{d,r \geq 1 \\ n=dr}} \mu(d) r.$$

This can be compared to Proposition 4.1.1 which yielded
$$n = \sum_{d|n} \phi(d) = \sum_{\substack{d,r \geq 1 \\ n=dr}} \phi(d) 1(r),$$

4.6. Convolutions and the Möbius inversion formula

where $1(r)$ is the function that is always 1 (which is a multiplicative function). Something similar happens for the sum of any function f defined on the positive integers.

> **Theorem 4.4** (The Möbius inversion formula). *For any two arithmetic functions f and g we have*
> $$g(n) = \sum_{ab=n} f(b) \text{ for all integers } n \geq 1$$
> *if and only if*
> $$f(m) = \sum_{cd=m} \mu(c)g(d) \text{ for all integers } m \geq 1.$$

This can be rewritten as
$$g(n) = \sum_{d|n} f(d) \text{ for all } n \geq 1 \text{ if and only if } f(m) = \sum_{d|m} \mu(m/d)g(d) \text{ for all } m \geq 1.$$

Proof. If $f(m) = \sum_{cd=m} \mu(c)g(d)$ for all integers $m \geq 1$, then
$$\sum_{ab=n} f(b) = \sum_{ab=n} \sum_{cd=b} \mu(c)g(d) = \sum_{acd=n} \mu(c)g(d) = \sum_{d|n} g(d) \cdot \sum_{ac=n/d} \mu(c) = g(n),$$
since this last sum is 0 unless $n/d = 1$, that is, unless $d = n$. Similarly if $g(n) = \sum_{ab=n} f(b)$ for all integers $n \geq 1$, then
$$\sum_{cd=m} \mu(c)g(d) = \sum_{cd=m} \mu(c) \sum_{ab=d} f(b) = \sum_{abc=m} \mu(c)f(b) = \sum_{b|m} f(b) \sum_{ac=\frac{m}{b}} \mu(c) = f(m),$$
as desired. □

In the discussion above we saw several examples of the *convolution* $f * g$ of two multiplicative functions f and g, which we define by
$$(f * g)(n) := \sum_{ab=n} f(a)g(b).$$

Note that $f * g = g * f$. We saw that if $I(n) = n$, then $\phi * 1 = I$ and $\mu * I = \phi$, as well as $1 * \mu = \delta$.

Exercise 4.6.1. Prove that $\delta * f = f$ for all f, $\tau = 1 * 1$, and $\sigma(n) = 1 * I$.

Proposition 4.6.1. *For any two multiplicative functions f and g, the convolution $f * g$ is also multiplicative.*

Exercise 4.6.2. Prove that if $ab = mn$, then there exist integers r, s, t, u with $a = rs$, $b = tu$, $m = rt$, $n = su$ with $(s, t) = 1$.

Proof. Suppose that $(m, n) = 1$. For $h = f * g$, we have
$$h(mn) = \sum_{ab=mn} f(a)g(b).$$

We use exercise 4.6.2 and note that (r,s) and (t,u) both divide $(m,n) = 1$ and so both equal 1. Therefore $f(a) = f(rs) = f(r)f(s)$ and $g(b) = g(tu) = g(t)g(u)$. This implies that

$$h(mn) = \sum_{\substack{rt=m,\\su=n}} f(rs)g(tu) = \sum_{rt=m} f(r)g(t) \sum_{su=n} f(s)g(u) = h(m)h(n). \quad \square$$

In this new language, Theorem 4.3, which states that $1 * f$ is multiplicative whenever f is, is the special case $g = 1$ of Proposition 4.6.1. Corollary 4.5.1 states that $1 * \mu = \delta$. The Möbius inversion formula states that $F = 1 * f$ if and only if $f = \mu * F$. It is also easy to prove the Möbius inversion formula with this notation since if $F = 1 * f$, then $\mu * F = \mu * 1 * f = \delta * f = f$; and if $f = \mu * F$, then $1 * f = 1 * \mu * F = \delta * F = F$.

Exercise 4.6.3. Prove that $(\mu * \sigma)(n) = n$ for all integers $n \geq 1$.

Exercise 4.6.4. (a) Show that $(a * f) + (b * f) = (a + b) * f$.
 (b) Let $f(n) \geq 0$ for all integers $n \geq 1$. Prove that $(1 * f)(n) + (\mu * f)(n) \geq 2f(n)$ for all integers $n \geq 1$.
 (c) Prove that $\sigma(n) + \phi(n) \geq 2n$ for all integers $n \geq 1$.

Exercise 4.6.5. Suppose that $g(n) = \prod_{d|n} f(d)$. Deduce that $f(n) = \prod_{d|n} g(d)^{\mu(n/d)}$.

4.7. The Liouville function

The number of prime factors of a given integer $n = \prod_{i=1}^{k} p_i^{e_i}$ can be interpreted in two different ways:

$$\omega(n) := \sum_{p|n} 1 = \#\{\text{distinct primes that divide } n\} = k$$

and

$$\Omega(n) := \sum_{\substack{p \text{ prime, } k \geq 1 \\ p^k|n}} 1 = \#\{\text{distinct prime powers that divide } n\} = \sum_{i=1}^{k} e_i.$$

In other words, $\Omega(n)$ counts the number of primes when one factors n into primes without using exponents, so $\Omega(12) = 3$ as $12 = 2 \times 2 \times 3$, while $\omega(n)$ counts the number of primes when one factors n into primes using exponents, so $\omega(12) = 2$ as $12 = 2^2 \cdot 3$. For other examples, $\omega(27) = 1$ with $\Omega(27) = 3$, and $\omega(36) = 2$ with $\Omega(36) = 4$, while $\Omega(105) = \omega(105) = 3$.

Another interesting multiplicative function is Liouville's function, defined at the start of this chapter by $\lambda(n) = (-1)^{\Omega(n)}$ so that, for example, $\lambda(12) = (-1)^3 = -1$. We notice that λ is the totally multiplicative function that agrees with μ on the squarefree integers. Liouville's function feels, intuitively, more natural, but Möbius's function fits better with the theory.

Exercise 4.7.1. Prove that $\Omega(n) \geq \omega(n)$ for all integers $n \geq 1$, with equality if and only if n is squarefree.

Exercise 4.7.2. Prove that $\lambda * \mu^2 = \delta$.

Exercise 4.7.3. (a) Prove that
$$\sum_{d|n} \lambda(d) = \begin{cases} 1 & \text{if } n \text{ is a square,} \\ 0 & \text{otherwise.} \end{cases}$$

(b)† By summing the formula in (a) over all positive integers $n \leq N$, deduce that for all integers $N \geq 1$ we have
$$\sum_{n \geq 1} \lambda(n) \left[\frac{N}{n}\right] = [\sqrt{N}].$$

Additional exercises

Exercise 4.8.1. Prove that $\mu(n)\mu(n+1)\mu(n+2)\mu(n+3) = 0$ for all integers $n \geq 1$.

Exercise 4.8.2. Prove that $\phi(n) + \sigma(n) = 2n$ if and only if $n = 1$ or n is a prime.

Exercise 4.8.3. (a) By summing the formula in Corollary 4.5.1 over all positive integers $n \leq N$, deduce that
$$\sum_{n \geq 1} \mu(n) \left[\frac{N}{n}\right] = 1 \quad \text{for all } N \geq 1.$$

(b)† Deduce that
$$\left|\sum_{n \leq N} \frac{\mu(n)}{n}\right| \leq 1 \quad \text{for all } N \geq 1.$$

It is a much deeper problem to prove that $\sum_{n \leq N} \mu(n)/n$ tends to a limit as $N \to \infty$.

Exercise 4.8.4. (a) Prove that if f is an arithmetic function, then
$$\sum_{n \geq 1} f(n) \frac{x^n}{1-x^n} = \sum_{m \geq 1} (1 * f)(m) x^m,$$
without worrying about convergence.

(b) Write out explicitly the example $f = \mu$ as well as some other common multiplicative functions.

Appendix 4B. Dirichlet series and multiplicative functions

4.9. Dirichlet series

In section 4.2 we started looking at finite sums of multiplicative functions. Now we study the (infinite) sum of a given multiplicative function $g(n)$, summing over *all* positive integers n. The Fundamental Theorem of Arithmetic gives each positive integer n a unique factorization $\prod_p p^{n_p}$, and this allows us to factor $g(n)$ as $\prod_p g(p^{n_p})$. Proceeding as in the proof of Theorem 4.3 and putting convergence issues to one side for now, we obtain

$$\sum_{n \geq 1} g(n) = \prod_{p \text{ prime}} (1 + g(p) + g(p^2) + \cdots).$$

The product over the primes is known as the *Euler product*. In all of the examples of multiplicative functions given above, both sides of this equation diverge. However one can take $g(n) = f(n)/n^s$ for well-chosen s, so that convergence is no longer a problem. (Note that if $f(n)$ is multiplicative, then $f(n)/n^s$ is multiplicative for any $s \in \mathbb{C}$ by exercise 4.0.3.) In this case we define the *Dirichlet series*

$$F(s) := \sum_{n \geq 1} \frac{f(n)}{n^s} = \prod_{p \text{ prime}} \left(1 + \frac{f(p)}{p^s} + \frac{f(p^2)}{p^{2s}} + \cdots \right),$$

which is only really well-defined for complex numbers s for which the infinite sum is absolutely convergent.[5] If f is totally multiplicative, then the sum for each p is

---

[5] One can think of a Dirichlet series in two ways: first as a function in its own right, in which case questions about convergence are important (since the function needs to be well-defined), and second as a formal sum which allows us to work directly with the values of $f(n)$ so as to focus on properties of $f(n)$ and we do not consider $F(s)$ as s varies. Such formal manipulations of $F(s)$ connect to a lot of combinatorial ideas just as generating functions allow us to better understand sequences.

a geometric series and so

$$(4.9.1) \qquad F(s) := \sum_{n \geq 1} \frac{f(n)}{n^s} = \prod_{p \text{ prime}} \left(1 - \frac{f(p)}{p^s}\right)^{-1}.$$

If an infinite sum is absolutely convergent, then one can rearrange the terms in the sum with impunity. Now $|n^s| = n^{\text{Re}(s)}$ so if we let $\sigma = \text{Re}(s)$, then

$$\sum_{n \geq 1} \frac{|f(n)|}{|n^s|} = \sum_{n \geq 1} \frac{|f(n)|}{n^\sigma}.$$

For multiplicative functions of interest the $|f(n)|$ are typically bounded by a constant times a power of n, which implies that there exists some real number σ_0 such that $F(s)$ is absolutely convergent precisely when $\text{Re}(s) \geq \sigma_0$ or $> \sigma_0$. (We will work out the details in the special case where each $f(n) = 1$ in section 5.12 of appendix 5B.)

Exercise 4.9.1.[†] (a) Prove that if there exists a constant $c > 0$ and $\tau \in \mathbb{R}$ such that $|f(n)| < cn^\tau$ for all integers $n \geq 1$, then $F(s)$ is absolutely convergent for $s = \sigma + it$ with $\sigma > 1 + \tau$.
(b) Prove that if $F(s)$ is absolutely convergent for $\text{Re}(s) = \sigma$, then there exists a constant $c > 0$ such that $|f(n)| < cn^\sigma$ for all integers $n \geq 1$.

Some interesting examples of Dirichlet series include the *Riemann zeta-function*

$$\zeta(s) := \sum_{n \geq 1} \frac{1}{n^s}, \quad \text{the Dirichlet series with } f(n) = 1 \text{ for each } n \geq 1,$$

and its inverse

$$\zeta(s)^{-1} = \sum_{n \geq 1} \frac{\mu(n)}{n^s}, \quad \text{the Dirichlet series with } f(n) = \mu(n) \text{ for each } n \geq 1.$$

We will prove that this really is the inverse of $\zeta(s)$ in the next section.

4.10. Multiplication of Dirichlet series

Given two multiplicative functions f and g, let $h = f * g$ be the convolution of f and g, as defined in appendix 4A. If we multiply the associated Dirichlet series for f and g, say

$$F(s) = \sum_{a \geq 1} \frac{f(a)}{a^s}, \quad G(s) = \sum_{b \geq 1} \frac{g(b)}{b^s}, \quad \text{and} \quad H(s) = \sum_{n \geq 1} \frac{h(n)}{n^s},$$

then, grouping together terms where $ab = n$, we have

$$F(s)G(s) = \sum_{a,b \geq 1} \frac{f(a)g(b)}{(ab)^s} = \sum_{n \geq 1} \left(\sum_{ab=n} f(a)g(b)\right) \cdot \frac{1}{n^s}$$

$$= \sum_{n \geq 1} (f * g)(n) \cdot \frac{1}{n^s} = \sum_{n \geq 1} \frac{(f * g)(n)}{n^s} = H(s).$$

Define $\mathbf{1}(s) = 1$, the Dirichlet series with nth coefficient $\delta(n)$. Corollary 4.5.1 yields that $\mu * 1 = \delta$ and so

$$\sum_{n \geq 1} \frac{\mu(n)}{n^s} \cdot \zeta(s) = \mathbf{1}$$

and so we justify our claim for the Dirichlet series of $\zeta(s)^{-1}$. On the other hand by considering the coefficient of $1/n^s$ in the identity $\zeta(s) \cdot \zeta(s)^{-1} = \mathbf{1}(s)$, we can deduce Corollary 4.5.1. The Möbius inversion formula can be restated as the more transparent equivalence

$$G(s) = \zeta(s) F(s) \text{ if and only if } F(s) = \zeta(s)^{-1} G(s).$$

Exercise 4.10.1. By studying the Euler products or otherwise, prove the following:
(a) $\sum_{n \geq 1} n/n^s = \zeta(s-1)$;
(b) $\sum_{n \geq 1} \tau(n)/n^s = \zeta(s)^2$;
(c) $\sum_{n \geq 1} \sigma(n)/n^s = \zeta(s)\zeta(s-1)$ and $\sum_{n \geq 1} \sigma_k(n)/n^s = \zeta(s)\zeta(s-k)$ for all integers $k \geq 1$;
(d) $\sum_{n \geq 1} \mu(n)^2/n^s = \zeta(s)/\zeta(2s)$;
(e) $\sum_{n \geq 1} \phi(n)/n^s = \zeta(s-1)/\zeta(s)$;
(f) $\sum_{n \geq 1} \lambda(n)/n^s = \zeta(2s)/\zeta(s)$.

Exercise 4.10.2. (a) Describe the identity in Proposition 4.1.1 in terms of Dirichlet series.
(b) Reprove exercise 4.10.1(d) and (e) by multiplying through by denominators.
(c) Give a formula for the coefficients of $F(s)\zeta(s)$ in terms of the values of the $f(n)$.
(d) Suppose that $f(p)$ is totally multiplicative with $f(2) = 0$, $f(p) = 1$ if $p \equiv 1 \pmod 3$, and $f(p) = -1$ if $p \equiv -1 \pmod 3$. Describe the coefficient of $1/n^s$ in $F(s)\zeta(s)$ in terms of the prime power factors of n.

4.11. Other Dirichlet series of interest

Taking the logarithm of the Euler product for the Riemann zeta-function,

$$\zeta(s) = \prod_{p \text{ prime}} \left(1 - \frac{1}{p^s}\right)^{-1},$$

yields, since $-\log(1-t)$ has Taylor series $t + \frac{t^2}{2} + \frac{t^3}{3} + \cdots$ whenever $|t| < 1$,

$$\log \zeta(s) = -\sum_{p \text{ prime}} \log\left(1 - \frac{1}{p^s}\right) = \sum_{\substack{p \text{ prime} \\ m \geq 1}} \frac{1}{m p^{ms}},$$

taking $t = 1/p^s$ for every prime p. Differentiating we obtain

$$-\frac{\zeta'(s)}{\zeta(s)} = \sum_{\substack{p \text{ prime} \\ m \geq 1}} \frac{\log p}{p^{ms}}.$$

The right-hand side is a Dirichlet series in which the only non-zero Dirichlet coefficients occur for $1/n^s$ where n is a prime power. This provides us with a completely different way to recognize primes from any that we have seen before, and it was the genius of Riemann to see that this observation could be manipulated into a technique for counting primes. For this reason we rewrite this as

$$-\frac{\zeta'(s)}{\zeta(s)} = \sum_{n \geq 1} \frac{\Lambda(n)}{n^s}$$

4.11. Other Dirichlet series of interest

where we define the *von Mangoldt function* as

$$\Lambda(n) = \begin{cases} \log p & \text{if } n = p^m, \text{ a prime power,} \\ 0 & \text{otherwise.} \end{cases}$$

We say that Λ is *supported* on the prime powers, since it is only when n is a prime power that $\Lambda(n)$ can be non-zero.

Writing $\log n$ as a sum of $\log p$ over all prime powers p^k dividing n, we obtain the identity

$$\log n = \sum_{d|n} \Lambda(d).$$

This can be rewritten as $\log = 1 * \Lambda$ and so this identity can also be obtained by taking the coefficient of $1/n^s$ in $-\zeta'(s) = \zeta(s) \cdot (-\zeta'(s)/\zeta(s))$. If instead we write

$$-\frac{\zeta'(s)}{\zeta(s)} = \zeta(s)^{-1} \cdot (-\zeta'(s)),$$

then the coefficient of $1/n^s$ yields the identity

(4.11.1) $$\Lambda(n) = \sum_{d|n} \mu(d) \log(n/d).$$

This identity relates the characteristic function for the prime numbers with the convolution of the multiplicative function $\mu(\cdot)$ and the smooth function $\log(\cdot)$. It can be used to prove the prime number theorem (see [**GS**]), a subject which will be discussed in the next chapter.

These ideas generalize very nicely to *any* multiplicative function f. The pth Euler factor of the Dirichlet series, $F(s)$, for f, is always invertible, since it is a power series in $1/p^s$ with leading term 1:

Exercise 4.11.1.† (a) Suppose that $a(t) = 1 + a_1 t + a_2 t^2 + \cdots$. Prove that there exists an inverse $b(t) = 1 + b_1 t + b_2 t^2 + \cdots$ for which $a(t)b(t) = 1$.
(b) Deduce that $\left(1 + \frac{f(p)}{p^s} + \frac{f(p^2)}{p^{2s}} + \cdots\right)$, the pth Euler factor of $F(s)$, is invertible.

Let $G(s) = \sum_{n\geq 1} g(n)/n^s$ be the product over all primes p of these inverse Euler factors, and so $F(s)G(s) = 1$ which implies that $f * g = \delta$. Now we define the coefficients, $\Lambda_f(n)$, of the logarithmic derivative of F by

$$-\frac{F'(s)}{F(s)} = \sum_{n\geq 1} \frac{\Lambda_f(n)}{n^s}.$$

Working with this series leads to lots of intriguing identities.

Exercise 4.11.2. (a) Prove that $\Lambda_f(n) + \Lambda_g(n) = 0$ for all integers $n \geq 1$.
(b) Use the identity $-F'(s) = F(s) \cdot (-F'(s)/F(s))$ to prove that

$$f(n)\log n = \sum_{ab=n} f(a)\Lambda_f(b) \quad \text{for all integers } n \geq 1.$$

(c) Use the identity $-F'(s)/F(s) = G(s) \cdot (-F'(s))$ to prove that

$$\Lambda_f(n) = \sum_{ab=n} g(a)f(b)\log b \quad \text{for all integers } n \geq 1.$$

(d) Deduce (4.11.1).

Appendix 4C. Irreducible polynomials modulo p

The Fundamental Theorem of Algebra (section 3.22 of appendix 3F) states that every monic polynomial can be factored into monic irreducible polynomials in a unique way, which is completely analogous to how positive integers are factored into primes in the Fundamental Theorem of Arithmetic. In appendix 4B we associated the primes with $\zeta(s)$, the generating function for the integers. In this appendix we associate the irreducible monic polynomials mod p with the generating function for the monic polynomials mod p.

4.12. Irreducible polynomials modulo p

The monic polynomials mod p of degree d take the form

$$x^d + a_{d-1}x^{d-1} + \cdots + a_1 x + a_0 \text{ such that each } a_j \in \mathbb{Z}/p\mathbb{Z}.$$

(The notation $\mathbb{Z}/p\mathbb{Z}$ was introduced in appendix 2A, to stand for the residue classes mod p.) There are p possible values for each a_j, and there are d different a_j to be selected, so there are a total of p^d monic polynomials mod p, of degree d.

Exercise 4.12.1. Suppose that $h(x)$ is a given polynomial mod p of degree d. Prove that there are exactly p^{m-d} monic polynomials in $\mathbb{F}_p[x]$ of degree m that are divisible by $h(x)$, provided $m \geq d$.

We will determine N_d, the number of monic *irreducible* polynomials mod p, of degree d. This is surprisingly straightforward using the Möbius inversion formula:[6] It is most elegant if we define the analogy to the von Mangoldt function by

$$\Lambda(g(x)) = \begin{cases} \deg p(x) & \text{if } g(x) = p(x)^k \text{ with } p(x) \text{ irreducible and integer } k \geq 1, \\ 0 & \text{otherwise.} \end{cases}$$

---

[6]Though there is no known analogous argument known for counting the primes themselves.

4.12. Irreducible polynomials modulo p

We are going to compute the degree of the product of all of the monic polynomials mod p, of degree m, in two different ways: Firstly, since there are p^m such polynomials and since they each have degree m, the total degree is mp^m. On the other hand, the degree of each such polynomial $h(x)$ equals the sum of the degrees of all of its irreducible monic factors (counting each factor the number of times it occurs in the factorization); that is,

$$\deg h(x) = \sum_{\substack{p(x) \text{ monic irreducible} \\ k \geq 1,\ p(x)^k \text{ divides } h(x)}} \deg p(x)$$

$$= \sum_{\substack{g(x) \text{ monic} \\ g(x) \text{ divides } h(x)}} \Lambda(g(x)).$$

Therefore, summing this over all polynomials $h(x)$ of degree m we obtain

$$mp^m = \sum_{h(x) \text{ monic of degree } m} \deg h(x) = \sum_{g(x) \text{ monic}} \Lambda(g(x)) \sum_{\substack{h(x) \text{ monic of degree } m \\ g(x) \text{ divides } h(x)}} 1.$$

If $g(x)$ has degree d, then the last term is 0 unless $d \leq m$, in which case exercise 4.12.1 yields that

$$mp^m = \sum_{d=1}^{m} p^{m-d} \sum_{g(x) \text{ monic of degree } d} \Lambda(g(x)).$$

Subtracting p times this identity for $m-1$ from this identity for m, we obtain

(4.12.1) $$\boxed{\sum_{g(x) \text{ monic of degree } m} \Lambda(g(x)) = p^m.}$$

The terms on the left-hand side are given by $g(x) = p(x)^k$ for each factorization $m = dk$ as $p(x)$ runs over the irreducible polynomials of degree d. Therefore

(4.12.2) $$\sum_{d \mid m} dN_d = p^m,$$

and then, by the Möbius inversion formula (given in section 4.6 of appendix 4A, with $f(d) = dN_d$), we deduce that

(4.12.3) $$\boxed{mN_m = \sum_{ab=m} \mu(a)p^b = \sum_{d \mid m} \mu(d) p^{m/d}.}$$

This is an *exact formula* for the number of irreducible polynomials of degree m. The largest term in the sum on the right side comes from $d = 1$, the term being p^m, so the number of irreducible polynomials of degree m is roughly p^m/m. The second largest term has $d = 2$ (when m is even), so equals $-p^{m/2}$ and otherwise is smaller. Therefore

$$|mN_m - p^m| \leq \sum_{d \mid m,\ d \neq 1} p^{m/d} \leq \sum_{k \leq [m/2]} p^k \leq 2p^{[m/2]}.$$

One can then deduce that $N_m > 0$ for all prime powers p^m; that is, there exists an irreducible polynomial mod p of every degree ≥ 1. This is useful since, as will be explained in section 7.25 of appendix 7E, an irreducible polynomial mod p of degree n can be used to construct a finite field with p^n elements.

An alternative approach uses generating functions. We define

$$F(t) := \sum_{h(x) \text{ monic}} t^{\deg h(x)} = \sum_{m \geq 1} \sum_{h(x) \text{ monic of degree } m} t^m = \sum_{m \geq 1} p^m t^m = \frac{1}{1-pt}.$$

On the other hand

$$t^{\deg h(x)} = \prod_{\substack{p(x) \text{ monic irreducible} \\ k \geq 1, \ p(x)^k \| h(x)}} t^{\deg(p(x)^k)} = \prod_{\substack{p(x) \text{ monic irreducible} \\ k \geq 1, \ p(x)^k \| h(x)}} t^{k \deg(p(x))}$$

and so, summing $t^{\deg h(x)}$ over all monic polynomials $h(x)$, we obtain (using the Fundamental Theorem of Arithmetic for polynomials to ensure unique factorization)

$$F(t) = \prod_{p(x) \text{ monic irreducible}} \left(1 + t^{\deg(p(x))} + t^{2\deg(p(x))} + \cdots \right)$$

$$= \prod_{d \geq 1} \prod_{p(x) \text{ monic irreducible, degree } d} (1-t^d)^{-1} = \prod_{d \geq 1} (1-t^d)^{-N_d}.$$

We have therefore proved the remarkable identity

$$F(t) = \boxed{\frac{1}{1-pt} = \prod_{d \geq 1} (1-t^d)^{-N_d}}$$

which we have obtained much like we obtained Dirichlet series and their Euler products in appendix 4B. Now we take the logarithmic derivative to obtain

$$\frac{-F'(t)}{F(t)} = \frac{p}{1-pt} = \sum_{d \geq 1} \frac{dN_d t^{d-1}}{1-t^d}.$$

Multiplying through by t and expanding we have

$$\sum_{n \geq 1} p^n t^n = \frac{pt}{1-pt} = \sum_{d \geq 1} dN_d \frac{t^d}{1-t^d} = \sum_{d \geq 1} dN_d \sum_{m \geq 1} t^{dm} = \sum_{n \geq 1} \left(\sum_{d | n} dN_d \right) t^n.$$

Comparing the coefficients of t^n on both sides we again obtain the identity (4.12.2), which then leads to our exact formula for the number of irreducible polynomials of degree d.

Exercise 4.12.2. Prove that $N_m \leq p^m/m$.

Appendix 4D. The harmonic sum and the divisor function

The number of divisors of integers, $\tau(n)$, varies considerably with n: The data begins

$$\tau(1) = 1,\ \tau(2) = 2,\ \tau(3) = 2,\ \tau(4) = 3,\ \tau(5) = 2,\ \tau(6) = 4,\ \tau(7) = 2,$$
$$\tau(8) = 4,\ \tau(9) = 3,\ \tau(10) = 4,\ \tau(11) = 2,\ \tau(12) = 6,\ldots,$$

and sampling a bit further along:

$$\tau(101) = 2,\ \tau(102) = 8,\ \tau(103) = 2,\ \tau(104) = 8,\ \tau(105) = 8,$$
$$\tau(106) = 4,\ \tau(107) = 2,\ \tau(108) = 12,\ \tau(109) = 2,\ldots.$$

The values of $\tau(n)$ bounce around as n grows, often a power of 2, without seeming to be close to some smooth function.[7] On the other hand, it is not difficult to show that the "average" of $\tau(n)$ up to x is well-approximated by a smooth, familiar function, namely $\log x$:

4.13. The average number of divisors

The average of $\tau(n)$ up to x is given by, using the definition of $\tau(n)$,

$$\frac{1}{x} \sum_{1 \leq n \leq x} \tau(n) = \frac{1}{x} \sum_{n \leq x} \sum_{\substack{d \leq x \\ d|n}} 1 = \frac{1}{x} \sum_{d \leq x} \sum_{\substack{n \leq x \\ d|n}} 1 = \frac{1}{x} \sum_{d \leq x} \left[\frac{x}{d}\right].$$

The first equality follows as $\tau(n) = \sum_{d \leq x,\ d|n} 1$, the second by swapping the order of summation, and the third by exercise 1.7.6. Now $[x/d]$ is an awkward quantity to work with so we replace it by something easier to work with. From the inequalities

---

[7] Often one wishes to say that a given complicated mathematical function "is well-approximated" by a function that is "smooth" and simple to describe.

$x/d - 1 < [x/d] \leq x/d$ we therefore obtain

$$\frac{1}{x} \sum_{d \leq x} \left(\frac{x}{d} - 1\right) < \frac{1}{x} \sum_{1 \leq n \leq x} \tau(n) \leq \frac{1}{x} \sum_{d \leq x} \frac{x}{d},$$

which simplifies to

(4.13.1) $$\sum_{d \leq x} \frac{1}{d} - 1 < \frac{1}{x} \sum_{1 \leq n \leq x} \tau(n) \leq \sum_{d \leq x} \frac{1}{d}.$$

The upper and lower bounds here differ by just 1; it's amazing that the average of $\tau(n)$ can be nailed down so precisely, given how much $\tau(n)$ varies. Next we need a smooth function that provides a good estimate for $\sum_{d \leq x} 1/d$ as x varies.

4.14. The harmonic sum

We wish to obtain good approximations to the value of the *harmonic sums*

$$L_N := \frac{1}{1} + \frac{1}{2} + \cdots + \frac{1}{N},$$

as N grows. As an example we now describe L_4 in terms of the area of the shaded region in the graph below.

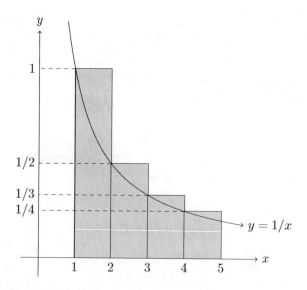

Figure 4.1. The harmonic sum, L_4, presented as an area, bounded below by the curve $y = 1/x$.

The first shaded rectangle, with corners at $(1,0)$, $(2,0)$, $(2,1)$, and $(1,1)$, has area $\frac{1}{1}$; the next with corners at $(2,0)$, $(3,0)$, $(3,\frac{1}{2})$, and $(2,\frac{1}{2})$, has area $\frac{1}{2}$, etc. So the shaded region has area $L_4 = \frac{1}{1} + \frac{1}{2} + \frac{1}{3} + \frac{1}{4}$. The shaded region contains the curve $y = 1/x$ for x in the range $1 \leq x \leq 5$, and so the shaded region has area which is

4.14. The harmonic sum

larger than the area under this curve. Precisely, $L_4 \geq \int_1^5 \frac{dt}{t} = \log 5$. This argument generalizes to obtain that, for any integer $N \geq 1$,

$$L_N = \int_1^{N+1} \frac{dx}{[x]} \geq \int_1^{N+1} \frac{dx}{x} = \log(N+1) > \log N.$$

To obtain an upper bound on L_N we shift our boxes one to the left.

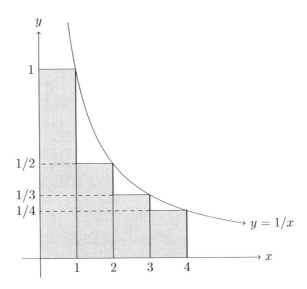

Figure 4.2. The harmonic sum, L_4, presented as an area, bounded above by the curve $y = 1/x$.

Now the area under the curve $y = 1/x$ for x in the range $0 \leq x \leq 4$ contains the shaded region and so provides an upper bound for L_4. However the area under $y = 1/x$ for x in the range $0 \leq x \leq 1$ is ∞, so this does not yield a useful upper bound on L_4. The trick is to treat the first term in the sum, 1, separately. Therefore $L_4 = \frac{1}{1} + \frac{1}{2} + \frac{1}{3} + \frac{1}{4} \leq \frac{1}{1} + \int_1^4 \frac{dt}{t} = 1 + \log 4$, which is not much bigger than $\log 5$. In general, as $1/[x+1] < 1/x$ for all $x \geq 0$, we have

$$L_N = 1 + \int_1^N \frac{dx}{[x+1]} < 1 + \int_1^N \frac{dx}{x} = 1 + \log N.$$

Therefore $\log N$ is a very accurate approximation to L_N (as $0 < L_N - \log N \leq 1$ for all N) but we can and will do better! Inserting these bounds for the harmonic sum into (4.13.1) we obtain

$$\log x - 1 < \frac{1}{x} \sum_{1 \leq n \leq x} \tau(n) \leq \log x + 1,$$

for any number $x \geq 1$. Therefore the mean value of $\tau(n)$, over all $n \leq x$, is $\log x$ with an error of at most 1, a far better estimate than we might have guessed was feasible given how this section started. We will obtain even better bounds in section 4.15 of appendix 4D.

In the next exercise we will prove that $L_N - \log N$ tends to a limit as N goes to infinity; that is,
$$\lim_{N \to \infty} \left(\frac{1}{1} + \frac{1}{2} + \cdots + \frac{1}{N} - \log N \right)$$
exists. This limit is usually denoted by γ and is called the *Euler-Mascheroni constant*. It can be shown that $\gamma = .5772156649\ldots$.[8]

Exercise 4.14.1.[†] Let $x_n = 1/1 + 1/2 + 1/3 + \cdots + 1/n - \log n$ for each integer $n \geq 1$.
(a) By modifying the argument above, show that if $n > m$, then
$$0 \leq x_m - x_n \leq \log(1 + 1/m) - \log(1 + 1/n) < 1/m.$$
Therefore $(x_n)_{n \geq 1}$ is a Cauchy sequence and converges to a limit, as desired.
(b) Deduce that $0 \leq \frac{1}{1} + \frac{1}{2} + \cdots + \frac{1}{N} - \log N - \gamma \leq \frac{1}{N}$.
(c)[‡] Prove that
$$\gamma = 1 - \int_1^\infty \frac{\{t\}}{t^2}\, dt.$$

Inserting this estimate into (4.13.1) does not give a much better approximation for the average of $\tau(n)$; the problem is that there is a difference of 1 between the two sides of (4.13.1), which we will improve in the next section.

In the next exercise we obtain an analogous estimate for $N!$.

Exercise 4.14.2.[†] Notice that $\log x$ is a monotone increasing function for $x \geq 1$. Assume that $N \geq M$.
(a) Justify the lower bound $\log n \geq \int_{n-1}^n \log t\, dt$ and deduce that $N!/M! \geq (N/e)^N/(M/e)^M$.
(b) Prove an analogous upper bound on $\log n$ and deduce that $N!/M! \leq (N/e)^{N+1}/(M/e)^{M+1}$.
(c) Deduce that $1/\sqrt{N} \leq N! \Big/ \left(\frac{N}{e}\right)^N \sqrt{e^2 N} \leq \sqrt{N}$.

4.15. Dirichlet's hyperbola trick

In the argument in section 4.13 of appendix 4D we noted that d is a divisor of exactly $[x/d]$ integers up to x and then approximated this using the inequalities $x/d - 1 < [x/d] \leq x/d$. This difference of 1 between the upper and lower bounds gives rise to the difference 1 between the upper and lower bounds in (4.13.1). The difference 1 is negligible compared to the "main term" x/d when x/d is large but seems more relevant when x/d is small, that is, if d is close to x. For example, if $x/2 < d \leq x$, the actual value of $[x/d]$ is 1, and our bounds differ by that much, a source of much of the error term in our approximation to the average. So can we somehow avoid this source of most of the error in our summation technique?

Dirichlet observed if $d|n$ and d is close to n, then its complementary divisor, n/d, must be small. Inspired by this observation, Dirichlet counted the number of divisors using only the smaller number of each pair. Therefore, writing $n = ab$, we have
$$\tau(n) = \sum_{\substack{a,b \geq 1 \\ ab = n}} 1 = \sum_{\substack{b \geq a \geq 1 \\ ab = n}} 1 + \sum_{\substack{a > b \geq 1 \\ ab = n}} 1,$$

---

[8] One might reasonably ask whether γ can be described other than as this limit? Although there are other descriptions of γ (usually as an infinite integral, as in exercise 4.14.1(c)), none are easy to work with, so we do not even know whether γ is rational, a longstanding open question. However γ is a fundamental constant appearing in all sorts of important settings in number theory, so we would dearly like to understand it better.

4.15. Dirichlet's hyperbola trick

so that

$$\frac{1}{x}\sum_{1\leq n\leq x}\tau(n) = \frac{1}{x}\sum_{\substack{b\geq a\geq 1\\ab\leq x}}1 + \frac{1}{x}\sum_{\substack{a>b\geq 1\\ab\leq x}}1$$

$$= \frac{1}{x}\sum_{1\leq a\leq \sqrt{x}}\sum_{a\leq b\leq x/a}1 + \frac{1}{x}\sum_{1\leq b<\sqrt{x}}\sum_{b<a\leq x/b}1$$

$$= \frac{1}{x}\sum_{1\leq a\leq \sqrt{x}}([x/a]+1-a) + \frac{1}{x}\sum_{1\leq b<\sqrt{x}}([x/b]-b).$$

Now that a (and b) is $\leq \sqrt{x}$ the approximation x/a for $[x/a]$ is far better. Writing $N=[\sqrt{x}]$ and using the upper bound $[x/c]\leq x/c$ for $c=a$ and b, we obtain

$$\frac{1}{x}\sum_{1\leq n\leq x}\tau(n) \leq \frac{1}{x}\sum_{1\leq a\leq N}\left(\frac{x}{a}+1-a\right) + \frac{1}{x}\sum_{1\leq b<N}\left(\frac{x}{b}-b\right)$$

$$\leq 2\sum_{1\leq d\leq N}\frac{1}{d} - \frac{1}{x}\sum_{1\leq d\leq N}(2d-1)$$

$$\leq 2\left(\log N + \gamma + \frac{1}{N}\right) - \frac{N^2}{x},$$

by exercise 4.14.1(b). Using the lower bound $[x/c]>x/c-1$ the analogous argument yields

$$\frac{1}{x}\sum_{1\leq n\leq x}\tau(n) > 2\left(\log N + \gamma - \frac{1}{N}\right) - \frac{(N-2)^2}{x}.$$

Therefore, if x is sufficiently large, then

$$\left|\frac{1}{x}\sum_{1\leq n\leq x}\tau(n) - (\log x + 2\gamma - 1)\right| \leq \frac{5}{\sqrt{x}},$$

a tiny error term. Getting the main term, $\log x + 2\gamma - 1$, was not difficult, but getting the upper bound on the error term is not so easy.[9] How much can the bounds on the error term be improved? Calculations suggest bounds like $< c/x^{2/3}$ or even better. What is the best possible? Can we get a good lower bound on the error term? These are fundamental questions (for this and other problems) that intrigue analytic number theorists.

People like Euler and Gauss were not as interested in the explicit average, as much as determining what the summed function grows like. Thus they would write something like

$$\tau(n) \text{ grows, on average, like } \log n + 2\gamma.$$

---

[9] Although both the upper and lower bounds are obviously at most some multiple of $1/\sqrt{x}$, getting a sharp bound on that multiple is not easy. And who really cares? Typically in these situations we aim to obtain a multiple of $1/\sqrt{x}$ that is not difficult to prove rather than aim for a multiple that takes a lot of work to establish yet knowledge of this improvement adds little to our understanding.

To see that this is true, we observe that exercise 4.14.2 implies, after the last displayed equation,
$$\left|\sum_{1\leq n\leq x} (\tau(n) - \log n - 2\gamma)\right| \leq 5\sqrt{x}.$$
In the final exercise we show how to obtain an even better estimate for $N!$.

Exercise 4.15.1.[‡] Let $M_N := \log 1 + \log 2 + \cdots + \log N$ and $x_N := M_N - (N + \frac{1}{2})(\log N - 1)$.
 (a) Prove that there exists a constant $c_1 > 0$ such that for all integers $n \geq 1$ we have
$$\left|\int_{n-\frac{1}{2}}^{n+\frac{1}{2}} \log t \, dt - \log n\right| \leq \frac{c_1}{n^2}.$$
 (b) Deduce that if $M > N$, then $|x_M - x_N| < c_2/N$ for some constant $c_2 > 0$.
 (c) Prove that $(x_N)_{N\geq 1}$ tends to a limit.
 (d) Deduce that there exists a constant $c_0 > 0$ such that
$$\lim_{N\to\infty} N! \Big/ \left(\frac{N}{e}\right)^N \sqrt{c_0 N} = 1.$$
Stirling showed that $c_0 = 2\pi$ (this amazingly accurate approximation to $N!$ is known as *Stirling's formula*).

Appendix 4E. Cyclotomic polynomials

We give here one final application of the Möbius inversion formula to construct a family of polynomials that are particularly important in number theory, as we will see in later sections.

4.16. Cyclotomic polynomials

Every non-zero complex number can be written in the form $re^{2i\pi\theta}$ where r is a positive real number and $\theta \in \mathbb{R}/\mathbb{Z}$. If this is a root of $x^m - 1$, then $r^m e^{2i\pi m\theta} = 1$. Taking absolute values we see that $r^m = 1$, and therefore $r = 1$. Since $e^{2i\pi\tau} = 1$ if and only if $\tau \in \mathbb{Z}$, we have $e^{2i\pi m\theta} = 1$ if and only if $m\theta \in \mathbb{Z}$, that is,

$$\theta = 0 \text{ or } \tfrac{1}{m} \text{ or } \tfrac{2}{m} \text{ or } \ldots \text{ or } \tfrac{m-1}{m} \text{ (in } \mathbb{R}/\mathbb{Z}\text{)}.$$

The first of these yields the root 1; the second gives the root $\zeta_m = e^{2i\pi/m}$. The others are the powers of ζ_m; that is,

$$1, \zeta_m, \zeta_m^2, \ldots, \zeta_m^{m-1} \text{ are the } m \text{ distinct roots of } x^m - 1.$$

These are the *m*th *roots of unity*.

Exercise 4.16.1. (a) Show that $\zeta_m^i = \zeta_m^j$ if and only if $i \equiv j \pmod{m}$.
(b) Prove that $x^m - 1 = (x-1)(x-\zeta_m)(x-\zeta_m^2)\cdots(x-\zeta_m^{m-1})$.
(c) Deduce that $x^m - y^m = (x-y)(x-\zeta_m y)\cdots(x-\zeta_m^{m-1}y)$.
(d) Deduce that if m is odd, then $x^m + y^m = (x+y)(x+\zeta_m y)\cdots(x+\zeta_m^{m-1}y)$.

If α is an *m*th root of unity, but not an *r*th root of unity for any r, $1 \leq r \leq m-1$, then α is a *primitive m*th root of unity. The *cyclotomic polynomials* are defined as

$$\phi_m(x) := \prod_{\substack{\alpha \text{ a primitive} \\ m\text{th root of unity}}} (x - \alpha).$$

Since every root of $\phi_m(x)$ is a root of $x^m - 1$, we deduce that $\phi_m(x)$ divides $x^m - 1$. In fact $\phi_m(x)$ divides $x^n - 1$ if and only if m divides n: If m divides n, this follows since $\phi_m(x)$ divides $x^m - 1$, which divides $x^n - 1$. In the other direction, if α is an nth root of unity, then $\alpha^n = 1$, and if α is a primitive mth root of unity, then $\alpha^m = 1$. Now select integers u, v such that $um + vn = \gcd(m, n)$, and so $\alpha^{\gcd(m,n)} = (\alpha^m)^u (\alpha^n)^v = 1$. Now $\gcd(m, n) \leq m$ and therefore $\gcd(m, n) = m$ by the minimality of m, and so m divides n.

The polynomials $\phi_m(x)$ all have distinct roots and so $\prod_{m|n} \phi_m(x)$ divides $x^n - 1$. On the other hand the roots of $x^n - 1$ are all nth roots of unity and hence are each a primitive mth root of unity for some m dividing n. Since $x^n - 1$ has no repeated roots (or else the root would also be a root of its derivative, nx^{n-1}), we deduce that
$$x^n - 1 = \prod_{m|n} \phi_m(x).$$

Exercise 4.16.2. (a) Show that $\sum_{m|n} \deg \phi_m = n$ for all integers $n \geq 1$.
(b) Deduce that $\phi_m(x)$ has degree $\phi(m)$ for each $m \geq 1$.

Exercise 4.16.3. (a) Show that if m divides n, then $(\zeta_n^k)^m = 1$ if and only if n/m divides k.
(b) Deduce that if m divides n, then ζ_n^k is a primitive mth root of unity if and only if there exists an integer r, coprime with m, for which $k = (n/m)r$.
(c) Show that the set of primitive mth roots of unity is $\{\zeta_m^j : 1 \leq j \leq m$ and $(j, m) = 1\}$.
(d) Deduce that $\phi_m(x)$ has degree $\phi(m)$.

Exercise 4.16.4. By using the Möbius inversion formula, prove that
$$\phi_m(t) = \prod_{d|m} (t^d - 1)^{\mu(m/d)}.$$

Exercise 4.16.5. (a) Prove that $\prod_{m|n, m>1} \phi_m(1) = n$.
(b)† Deduce that if $m > 1$, then $\log \phi_m(1) = \Lambda(m)$, the von Mangoldt function of appendix 4B.

Chapter 5

The distribution of prime numbers

Once one begins to determine which integers are primes, one quickly finds that there are many. Are there infinitely many? One notices that the primes seem to make up a decreasing proportion of the positive integers. Can we explain this? Can we give a formula for how many primes there are up to a given point? Or at least give a good estimate?

When we write out the primes there seem to be patterns, though the patterns rarely persist for long. Can we find patterns that do persist? Is there a formula that describes all of the primes? Or at least some of them?

Is it possible to recognize prime numbers quickly and easily?

These questions motivate different parts of this chapter and of chapter 10.

5.1. Proofs that there are infinitely many primes

The first known proof appears in Euclid's *Elements*, Book IX, Proposition 20:

Theorem 5.1. *There are infinitely many primes.*

Proof #1.[1] Suppose that there are only finitely many primes, which we will denote by $2 = p_1 < p_2 = 3 < \cdots < p_k$. What are the prime factors of $p_1 p_2 \ldots p_k + 1$? Since this number is > 1 it must have a prime factor by the Fundamental Theorem of Arithmetic, and this must be p_j for some j, $1 \leq j \leq k$, since *all* primes are contained amongst $p_1, p_2, \ldots, p_k$. But then p_j divides both $p_1 p_2 \ldots p_k$ and $p_1 p_2 \ldots p_k + 1$, and hence p_j divides their difference, 1, by exercise 1.1.1(c), which is impossible. □

---

[1] Not until relatively recently has there been mathematical notation to describe a collection of objects, for example, $p_1, p_2, \ldots, p_k$. Neither Euclid nor Fermat had subscripts or "..." or "etc." (Gauss used "&c"). So instead the reader had to infer from the context how many objects the author meant. In Euclid's *Elements*, he writes that he assumes α, β, γ denote all of the prime numbers and then gives, in terms of ideas, the same proof as here. The reader had to understand that in writing "α, β, γ", Euclid meant an arbitrary number of primes, not just three!

There are many variants on Euclid's proof. For example:

Exercise 5.1.1 (Proof #2). Suppose that there are only finitely many primes, the largest of which is $n > 2$. Show that this is impossible by considering the prime factors of $n! - 1$.

Other variants include Furstenberg's curious proof using point-set topology (see appendix 5F). These all boil down to showing that there exists an integer $q > 1$ that is not divisible by any of a given finite set of primes $p_1, \ldots, p_k$. If $m = p_1 p_2 \cdots p_k$, then we wish to show there exists an integer $q > 1$ with $(q, m) = 1$, and there are $\phi(m) - 1$ such integers up to m. There is therefore such an integer by the formula in Theorem 4.1 once $m > 2$.

Exercise 5.1.2. Prove that there are infinitely many composite numbers.

Euclid's proof that there are infinitely many primes is a "proof by contradiction", showing that it is impossible that there are finitely many, and so does not suggest how one might find infinitely many. We can use the following constructive technique to determine infinitely many primes:

Lemma 5.1.1. *Suppose that $a_1 < a_2 < \cdots$ is an infinite sequence of pairwise coprime positive integers, and let p_n be a prime factor of a_n for each $n \geq 2$. Then $p_2, p_3, \ldots$ is an infinite sequence of distinct primes.*

Proof. If $p_m = p_n$ with $1 < m < n$, then p_m divides both a_m and a_n and so divides $(a_m, a_n) = 1$, which is impossible. □

By Lemma 5.1.1 we need only find an infinite sequence of pairwise coprime positive integers to obtain infinitely many primes. This can be achieved by modifying Euclid's construction. We define the sequence

$$a_1 = 2, \ a_2 = 3 \text{ and then } \ a_n = a_1 a_2 \ldots a_{n-1} + 1 \text{ for each } n \geq 2.$$

Now if $m < n$, then $a_n \equiv 1 \pmod{a_m}$ and so $(a_m, a_n) = (a_m, 1) = 1$ by exercise 2.1.5, as desired. Therefore, by Lemma 5.1.1, we can take a prime factor p_n of each a_n with $n > 1$ to obtain an infinite sequence of prime numbers.

Fermat conjectured that the integers $F_n = 2^{2^n} + 1$ are primes for all $n \geq 0$. His claim starts off correct: $3, 5, 17, 257, 65537$ are all prime, but his conjecture is false for $F_5 = 641 \times 6700417$, as Euler famously noted. It is an open question as to whether there are infinitely many primes of the form F_n.[2] Using the identity

(5.1.1) $$F_n = F_1 F_2 \ldots F_{n-1} + 2 \text{ for each } n \geq 1$$

we see that if $m < n$, then $F_n \equiv 2 \pmod{F_m}$ so that $(F_m, F_n) = (F_m, 2) = 1$, the last equality since each F_m is odd. Therefore, by Lemma 5.1.1, we can take a prime factor p_n of each F_n to obtain an infinite sequence of prime numbers.[3]

These proofs that there are infinitely many primes will be generalized using dynamical systems in appendix 5H.

---

[2] The only Fermat numbers known to be primes have $n \leq 4$. We know that the F_n are composite for $5 \leq n \leq 30$ and for many other n besides. It is always a significant moment when a Fermat number is factored for the first time. It could be that all F_n with $n > 4$ are composite or they might all be prime from some sufficiently large n onwards or some might be prime and some composite. Currently, we have no way of knowing which is true.

[3] This proof that there are infinitely many primes first appeared in a letter from Goldbach to Euler in July 1730.

Exercise 5.1.3. Prove (5.1.1).

Exercise 5.1.4. Suppose that $p_1 = 2 < p_2 = 3 < \cdots$ is the sequence of prime numbers. Use the fact that every Fermat number has a distinct prime divisor to prove that $p_n \leq 2^{2^n} + 1$. What can one deduce about the number of primes up to x?

Exercise 5.1.5. (a) Show that if m is not a power of 2, then $2^m + 1$ is composite by showing that $2^a + 1$ divides $2^{ab} + 1$ whenever b is odd.
(b) Deduce that if $2^m + 1$ is prime, then there exists an integer n such that $m = 2^n$; that is, if $2^m + 1$ is prime, then it is a Fermat number $F_n = 2^{2^n} + 1$. (This also follows from exercise 3.9.3(b).)

Another interesting sequence is the *Mersenne numbers*,[4] which take the form $M_n = 2^n - 1$. After exercise 3.9.2 we observed that if M_n is prime, then n is prime and, in our discussion of perfect numbers (section 4.2) we observed that M_2, M_3, M_5, and M_7 are each prime but $M_{11} = 23 \times 89$ is not. The Lucas-Lehmer test provides a relatively quick and elegant way to test whether a given M_p is prime (see Corollary 10.10.1 in appendix 10C).

5.2. Distinguishing primes

We can determine whether a given integer n is prime in practice, by proving that it is not composite: If a given integer n is composite, then we can write it as ab, two integers both > 1. If we suppose that $a \leq b$, then $a^2 \leq ab = n$ and so $a \leq \sqrt{n}$. Hence n must be divisible by some integer a in the range $1 < a \leq \sqrt{n}$. Therefore we can test divide n by every integer a in this range, and we either discover a factor of n or, if not, we know that n must be prime. This process is called *trial division* and is too slow, in practice, except for relatively small integers n. We can slightly improve this algorithm by noting that if p is a prime dividing a, then p divides n, so we only need to test divide by the primes up to $\sqrt{n}$. This is still very slow, in practice.[5] We discuss more practical techniques in chapter 10.

Trial division is a very slow way of recognizing whether an *individual* integer is prime, but it can be organized to be a highly efficient way to determine *all* of the primes up to some given point, as observed by Eratosthenes around 200 B.C.[6]

The *sieve of Eratosthenes* provides an efficient method for finding all of the primes up to x. For example to find all the primes up to 100, we begin by writing down every integer between 2 and 100 and then deleting every composite even number; that is, one deletes (or *sieves out*) every second integer up to x after 2.

---

[4]In 1640, France was home to the great philosophers and mathematicians of the age, such as Descartes, Desargues, Fermat, and Pascal. From 1630 on, Father Marin Mersenne wrote letters to all of these luminaries, posting challenges and persuading them all to think about perfect numbers.

[5]How slow is "slow"? If we could test divide by one prime per second, for a year, with no rest, then we could determine the primality of 17-digit numbers but not 18-digit numbers. If we used the world's fastest computer in 2019, we could test divide 53-, but not 54-, digit numbers. In chapter 10 we will encounter much better methods that can test such a number for primality, in moments.

[6]Eratosthenes lived in Cyrene in ancient Greece, from 276 to 195 B.C. He created the grid system of latitude and longitude to draw an accurate map of the world incorporating parallels and meridians. He was the first to calculate the circumference of the earth, the tilt of the earth's axis, and the distance from the earth to the sun (and so invented the leap day). He even attempted to assign dates to what was then ancient history (like the conquest of Troy) using available evidence.

	2	3	5	7	9
	11	13	15	17	19
	21	23	25	27	29
	31	33	35	37	39
	41	43	45	47	49
	51	53	55	57	59
	61	63	65	67	69
	71	73	75	77	79
	81	83	85	87	89
	91	93	95	97	99

Deleting every even number > 2, between 2 and 100

The first undeleted integer > 2 is 3; one then deletes every composite integer divisible by 3; that is, one sieves out every third integer up to x after 3. The next undeleted integer is 5 and one sieves out every fifth integer after 5, and then every seventh integer after 7.

2	3	5	7			2	3	5	7	
11	13		17	19		11	13		17	19
	23	25		29			23			29
31		35	37			31			37	
41	43		47	49		41	43		47	
	53	55		59			53			59
61		65	67			61			67	
71	73		77	79		71	73			79
	83	85		89			83			89
91		95	97						97	

Then delete remaining integers > 3 and > 5 that are divisible by 5
that are divisible by 3   and > 7 that are divisible by 7.

The sieve of Eratosthenes enables us to find all of the primes up to 100.

What's left are the primes up to 100. To obtain the primes up to any given limit x, one keeps on going like this, finding the next undeleted integer, call it p, which must be prime since it is undeleted, and then deleting every pth integer beyond p and up to x. We stop once $p > \sqrt{x}$ and then the undeleted integers are the primes $\leq x$. There are about $x \log \log x$ steps[7] in this algorithm, so it is a remarkably efficient way to find all the primes up to some given x,[8] but not for finding any particular prime.

Exercise 5.2.1. Use this method to find all of the primes up to 200.

The number of integers left after one removes the multiples of 2 is roughly $\frac{1}{2} \cdot x$, since about half of the integers up to x are divisible by 2. After one then removes

---

[7] How should one think about an expression like $\log \log x$? It goes to ∞ as x does, but it is a *very* slow growing function of x. For example, if $x = 10^{100}$, far more than the current estimate for the number of atoms in the universe, then $\log \log x < 5\frac{1}{2}$. Dan Shanks once wrote that "$\log \log x$ goes to infinity with great dignity."

[8] In practice, this algorithm determines which of the first x integers are prime in no more than $6x$ steps.

5.3. Primes in certain arithmetic progressions

the multiples of 3, one expects that there are about $\frac{2}{3} \cdot \frac{1}{2} \cdot x$ integers left, since about a third of the odd integers up to x are divisible by 3. In general removing the multiples of p removes, we expect, about $1/p$ of the integers in our set and so leaves a proportion $1 - \frac{1}{p}$. Therefore we *expect* that the number of integers left unsieved in the sieve of Eratosthenes, up to x, after sieving by the primes up to y, is about
$$x \prod_{\substack{p \leq y \\ p \text{ prime}}} \left(1 - \frac{1}{p}\right).$$

The product $\prod_{p \leq y}(1 - \frac{1}{p})$ is well-approximated by $e^{-\gamma}/\log y$, where γ is the Euler-Mascheroni constant discussed in section 4.14 of appendix 4D.[9] The logarithm, used here and elsewhere in this book, is the natural logarithm.

When we take $y = \sqrt{x}$, then only 1 and the primes up to x should be left in the sieve of Eratosthenes, and so one might guess from this analysis of sieve methods that the number of primes up to x is approximately

(5.2.1) $$2e^{-\gamma} \frac{x}{\log x}.$$

This guess is not correct; the constant is off,[10] as we will discuss in section 5.4.

5.3. Primes in certain arithmetic progressions

How are the primes split between the arithmetic progressions modulo 3? Or modulo 4? Or modulo any given integer m? Evidently every integer in the arithmetic progression 0 (mod 3) (that is, integers of the form $3k$) is divisible by 3, so the only prime in that arithmetic progression is 3 itself. There are no such divisibility restrictions for the arithmetic progressions 1 (mod 3) and 2 (mod 3) and if we partition the primes up to 100 into these arithmetic progressions, we find:

Primes $\equiv 1 \pmod 3$: $7, 13, 19, 31, 37, 43, 61, 67, 73, 79, 97, \ldots$

Primes $\equiv 2 \pmod 3$: $2, 5, 11, 17, 23, 29, 41, 47, 53, 59, 71, 83, 89, \ldots$

There seem to be a lot of primes in each arithmetic progression, and they seem to be roughly equally split between the two. Let's see what we can prove. First let's deal, in general, with the analogy to the case 0 (mod 3). This includes not only 0 (mod m) but also cases like 2 (mod 4):

Exercise 5.3.1. (a) Prove that any integer $\equiv a \pmod m$ is divisible by (a, m).
  (b) Deduce that if $(a, m) > 1$ and if there is a prime $\equiv a \pmod m$, then that prime is (a, m).
  (c) Give examples of arithmetic progressions which contain exactly one prime and examples which contain none.
  (d) Show that the arithmetic progression 2 (mod 6) contains infinitely many prime powers.

Therefore all but finitely many primes are distributed among the $\phi(m)$ arithmetic progressions $a \pmod m$ with $(a, m) = 1$. How are they distributed? If the $m = 3$ case is anything to go by, it appears that there are infinitely many in each

---

[9]This is a fact that is beyond the scope of this book but will be discussed in [**Graa**]. In fact $e^{-\gamma} = .56145948\ldots$.
  [10]Though not by much. The correct constant is 1 whereas $2e^{-\gamma} = 1.12291896\ldots$.

such arithmetic progression, and maybe even roughly equal numbers of primes in each up to any given point.

We will prove that there are infinitely many primes in each of the two feasible residue classes mod 3 (see Theorems 5.2 and 7.7).

Theorem 5.2. *There are infinitely many primes $\equiv -1$ (mod 3).*

Proof. Suppose that there are only finitely many primes $\equiv -1$ (mod 3), say $p_1, p_2, \ldots, p_k$. The integer $N = 3p_1p_2 \ldots p_k - 1$ must have a prime factor $q \equiv -1$ (mod 3), by exercise 5.3.2. However q divides both N and $N+1$ (since it must be one of the primes p_i), and hence q divides their difference, 1, which is impossible. □

Exercise 5.3.2. Use exercise 3.1.4(a) to show that if $n \equiv -1$ (mod 3), then there exists a prime factor p of n which is $\equiv -1$ (mod 3).

In 1837 Dirichlet showed that whenever $(a, q) = 1$ there are infinitely many primes $\equiv a$ (mod q). (We discuss this deep result in sections 8.17 of appendix 8D and 13.7.) In fact there are roughly equally many primes in each of these arithmetic progressions mod q. For example, half the primes are $\equiv 1$ (mod 3) and half are $\equiv -1$ (mod 3), as our data above suggested. Roughly 1% of the primes are $\equiv 69$ (mod 101) and indeed there are roughly 1% of the primes in each arithmetic progression a mod 101 with $1 \leq a \leq 100$. This is a deep result and will be discussed at length in our book [**Graa**].

Exercise 5.3.3. Prove that there are infinitely many primes $\equiv -1$ (mod 4).

Exercise 5.3.4. Prove that there are infinitely many primes $\equiv 5$ (mod 6).

Exercise 5.3.5.[†] Prove that at least two of the arithmetic progressions mod 8 contain infinitely many primes.

Exercise 5.8.6 generalizes these results considerably, using similar ideas.

5.4. How many primes are there up to x?

When people started to develop large tables of primes, perhaps looking for a pattern, they discovered no patterns but did find that the proportion of integers that are prime is gradually diminishing (which will be proved in section 5.13 of appendix 5B). In 1808 Legendre quantified this, suggesting that there are roughly $\frac{x}{\log x}$ primes up to x.[11] A few years earlier, aged 15 or 16, Gauss had already made a much better guess, based on studying tables of primes:

> In 1792 or 1793 ... I turned my attention to the decreasing frequency of primes ... counting the primes in intervals of length 1000. I soon recognized that behind all of the fluctuations, this frequency is on average inversely proportional to the logarithm ....
> — from a letter to ENCKE by K. F. GAUSS (Christmas Eve, 1849)

---

[11]And even the more precise assertion that there exists a constant B such that $\pi(x)$, the number of primes up to x, is well-approximated by $x/(\log x - B)$ for large enough x. This turns out to be true with $B = 1$, though this was not the value for B suggested by Legendre (who presumably made a guess based on data for small values of x).

5.4. How many primes are there up to x?

His observation may be best phrased as

About 1 in $\log x$ of the integers near x are prime,

which is (subtly) different from Legendre's assertion: Gauss's observation suggests that a good approximation to the number of primes up to x is $\sum_{n=2}^{x} \frac{1}{\log n}$. As $\frac{1}{\log t}$ does not vary much for t between n and $n+1$, Gauss deduced that $\pi(x)$ should be well-approximated by

$$(5.4.1) \qquad \int_2^x \frac{dt}{\log t}.$$

We denote this quantity by $\text{Li}(x)$ and call it *the logarithmic integral*.[12] The logarithm here is again the natural logarithm. Here is a comparison of Gauss's prediction with the actual count of primes up to various values of x:

x	$\pi(x) = \#\{\text{primes} \leq x\}$	Overcount: $\text{Li}(x) - \pi(x)$
10^3	168	10
10^4	1229	17
10^5	9592	38
10^6	78498	130
10^7	664579	339
10^8	5761455	754
10^9	50847534	1701
10^{10}	455052511	3104
10^{11}	4118054813	11588
10^{12}	37607912018	38263
10^{13}	346065536839	108971
10^{14}	3204941750802	314890
10^{15}	29844570422669	1052619
10^{16}	279238341033925	3214632
10^{17}	2623557157654233	7956589
10^{18}	24739954287740860	21949555
10^{19}	234057667276344607	99877775
10^{20}	2220819602560918840	222744644
10^{21}	21127269486018731928	597394254
10^{22}	201467286689315906290	1932355208
10^{23}	1925320391606803968923	7250186216
10^{24}	18435599767349200867866	17146907278
10^{25}	176846309399143769411680	55160980939

Primes up to various x and the overcount in Gauss's prediction.

Gauss's prediction is amazingly accurate. From the data, Gauss's prediction seems to overcount by a small amount, for all $x \geq 8$.[13] To quantify this "small amount", we observe that the last column (representing the overcount) is always about half the width of the central column (representing the number of primes up to x), so these data suggest that the difference is no bigger than a small multiple of $\sqrt{x}$.

---

[12] Some authors begin the integral defining $\text{Li}(x)$ at $x = 0$. This adds complication since the integrand equals ∞ at $x = 1$; nonetheless this can be handled, and the difference between the two definitions is then the constant $1.045163\ldots$, which has little relevance to our discussion.

[13] It is not true that $\text{Li}(x) > \pi(x)$ for all $x > 2$ but the first counterexample is far beyond where we can hope to calculate. Understanding how we know this is well beyond the scope of this book, but see [**Graa**].

This might be optimistic but, at the very least, the ratio of $\pi(x)$, the number of primes up to x, to Gauss's guess, $\text{Li}(x)$, should tend to 1 as $x \to \infty$; that is,

$$\pi(x) \,/\, \text{Li}(x) \to 1 \quad \text{as} \quad x \to \infty.$$

In exercise 5.8.11 we show that $\text{Li}(x) \,/\, \frac{x}{\log x} \to 1$ as $x \to \infty$, and combining these last two limits, we deduce that

$$\pi(x) \,/\, \frac{x}{\log x} \to 1 \quad \text{as} \quad x \to \infty.$$

The notation of limits is cumbersome; it is easier to write

(5.4.2) $$\boxed{\pi(x) \sim \frac{x}{\log x}}$$

as $x \to \infty$, "$\pi(x)$ *is asymptotic to* $x/\log x$".[14] This is different by a multiplicative constant from (5.2.1), our guesstimate based on the sieve of Eratosthenes. Our data makes it seem more likely that the constant 1 given here, rather than the $2e^{-\gamma}$ given in (5.2.1), is the correct constant.

The asymptotic (5.4.2) is called the *prime number theorem*. Its proof came in 1896, more than 100 years after Gauss's guess, involving several remarkable developments. It was a high point of 19th-century mathematics and there is still no straightforward approach. The main reason is that the prime number theorem can be shown to be equivalent to a statement about zeros of the analytic continuation of a function (the Riemann zeta-function which we discuss in appendices 4B, 5B, and 5D), which seems preposterous at first sight. Although proofs can be given that avoid mentioning these zeros, they are still lurking somewhere just beneath the surface.[15] A proof of the prime number theorem is beyond the scope of this book (but see [**Graa**] and [**GS**]).

Exercise 5.4.1.[†] Assume the prime number theorem.
  (a) Show that there are infinitely many primes whose leading digit is a "1". How about leading digit "7"?
  (b) Show that for all $\epsilon > 0$, if x is sufficiently large, then there are primes between x and $x + \epsilon x$.
  (c) Deduce that $\mathbb{R}_{\geq 0}$ is the set of limit points of the set $\{p/q : p, q \text{ primes}\}$.
  (d) Let $a_1, \ldots, a_d$ be any sequence of digits, that is, integers between 0 and 9, with $a_1 \neq 0$. Show that there are infinitely many primes whose first (leading) d digits are $a_1, \ldots, a_d$.

Exercise 5.4.2.[†] Let $p_1 = 2 < p_2 = 3 < \cdots$ be the sequence of primes. Assume the prime number theorem and prove that

$$p_n \sim n \log n \quad \text{as} \quad n \to \infty.$$

Exercise 5.4.3.[†] (a) Show that the sum of primes and prime powers $\leq x$ is $\sim x^2/(2\log x)$.
  (b) Deduce that if the sum equals N, then $x \sim \sqrt{N \log N}$.

---

[14] In general, $A(x) \sim B(x)$, that is, $A(x)$ is asymptotic to $B(x)$, is equivalent to $\lim_{x \to \infty} A(x)/B(x) = 1$. It does *not* mean that "$A(x)$ is approximately equal to $B(x)$", which has no strict mathematical meaning, rather that for any $\epsilon > 0$, no matter how small, one has

$$(1-\epsilon)B(x) < A(x) < (1+\epsilon)B(x)$$

once x is sufficiently large (where how large is "sufficiently large" depends on ϵ). This definition concerns the ratio $A(x)/B(x)$, *not* their difference $A(x) - B(x)$. Therefore $n^2 + 1 \sim n^2$ and $n^2 + n^2/\log n \sim n^2$ are equally true, even though the first is a better approximation to n^2 than the second ([**Sha85**], p. 16),

[15] Including the so-called "elementary proof" of the prime number theorem.

Primes in arithmetic progressions. As we mentioned in section 5.3, Dirichlet showed in 1837 that if $(a,q) = 1$, then there are infinitely many primes $p \equiv a \pmod{q}$. Dirichlet's proof was combined in 1896 with the proof of the prime number theorem to establish that

$$\#\{p \leq x : p \text{ is prime}, p \equiv a \pmod{q}\} \sim \frac{\pi(x)}{\phi(q)} \sim \frac{x}{\phi(q) \log x}.$$

The factor "$1/\phi(q)$" emerges as there are $\phi(q)$ reduced residues a modulo q.

Exercise 5.4.4.‡ Use the prime number theorem in arithmetic progressions to prove that for any integers $a_1, \ldots, a_d, b_0, \ldots, b_d \in \{0, \ldots, 9\}$, with $a_1 \neq 0$ and $b_0 = 1, 3, 7,$ or 9, there are infinitely many primes whose first d digits are $a_1, \ldots, a_d$ and whose last d digits are $b_d, \ldots, b_0$.

5.5. Bounds on the number of primes

The first quantitative lower bound proven on the number of primes is due to Euler in the mid-18th century who showed that

$$\sum_{p \text{ prime}} \frac{1}{p} \text{ diverges,}$$

as we will prove in section 5.12 of appendix 5B. This gives some idea of how numerous the primes are in comparison to other sequences of integers. For example $\sum_{n \geq 1} \frac{1}{n^2}$ converges, so the primes are, in this sense, more numerous than the squares. This implies that there are arbitrarily large values of x for which $\pi(x) > \sqrt{x}$.

Exercise 5.5.1.† Do better than this using Euler's result.
 (a) Prove that $\sum_{n \geq 1} \frac{1}{n(\log n)^2}$ converges.
 (b) Deduce that there are arbitrarily large x for which $\pi(x) > x/(\log x)^2$.

Next we will prove upper and lower bounds for the number of primes up to x, of the form
$$(5.5.1) \qquad c_1 \frac{x}{\log x} \leq \pi(x) \leq c_2 \frac{x}{\log x}$$

for some constants $0 < c_1 < 1 < c_2$, for all sufficiently large x. The prime number theorem is equivalent to being able to take $c_1 = 1 - \epsilon$ and $c_2 = 1 + \epsilon$ for any fixed $\epsilon > 0$ in (5.5.1). Instead we will prove Chebyshev's weaker 1850 result that one can take any $c_1 < \log 2$ and any $c_2 > \log 4$ in (5.5.1).

Theorem 5.3. *For all integers $n \geq 2$ we have*

$$(\log 2) \frac{n}{\log n} - 1 \leq \pi(n) \leq (\log 4) \frac{n}{\log n} + 4 \frac{n}{(\log n)^2}.$$

Exercise 5.5.2. Fix $\epsilon > 0$ arbitrarily small. Deduce Chebyshev's bounds (5.5.1) with $c_1 = \log 2 - \epsilon$ and $c_2 = \log 4 + \epsilon$, for all sufficiently large x, from Theorem 5.3.

Proof. The binomial theorem states that $(x+y)^N = \sum_{j=0}^{N} \binom{N}{j} x^j y^{N-j}$. Taking $x = y = 1$ we get

$$(5.5.2) \qquad \sum_{j=0}^{N} \binom{N}{j} = 2^N.$$

Lemma 5.5.1. *The product of the primes up to N is $\leq 4^{N-1}$ for all $N \geq 1$.*

Proof. Each prime in $[n+1, 2n]$ appears in the numerator of the binomial coefficient $\binom{2n-1}{n}$ but not in the denominator, and so their product divides $\binom{2n-1}{n}$. Now if $N = 2n - 1$ is odd, then $\binom{2n-1}{n-1} = \binom{2n-1}{n}$ so the value appears twice in the sum in (5.5.2). Therefore

$$(5.5.3) \qquad \prod_{\substack{n < p \leq 2n \\ p \text{ prime}}} p \leq \binom{2n-1}{n} < \frac{1}{2} \sum_{j=0}^{2n-1} \binom{2n-1}{j} = 2^{2n-2} = 4^{n-1}.$$

We now prove the claimed result by induction on $N \geq 1$. The result is straightforward for $N = 1, 2$ by calculation. If $N = 2n$ or $2n - 1$, then the product of the primes up to N is at most the product of the primes up to n times the product of the primes in $[n+1, 2n]$. The first product is $\leq 4^{n-1}$ by the induction hypothesis, and the second $< 4^{n-1}$ by the previous paragraph. Combining these two upper bounds gives the upper bound $\leq 4^{2n-2} \leq 4^{N-1}$, as claimed. $\square$

If we take logarithms in (5.5.3), we obtain

$$(5.5.4) \qquad \sum_{\substack{p \text{ prime} \\ n < p \leq 2n}} \log p < (n-1) \log 4.$$

As each term of the left side is $> \log n$ we deduce that

$$(5.5.5) \qquad \pi(2n) - \pi(n) = \#\{p \text{ prime} : n < p \leq 2n\} \leq \frac{n-1}{\log n} \cdot \log 4.$$

We now use this to deduce the upper bound claimed in Theorem 5.3. We verify the bound by calculations for all $N \leq 100$ and then proceed by induction for $N \geq 101$. If $N = 2n$ or $2n - 1$ (so that $n \geq 51$), then by the induction hypothesis and (5.5.5)

$$\pi(N) \leq \pi(2n) = \pi(n) + (\pi(2n) - \pi(n)) \leq (\log 4) \frac{2n-1}{\log n} + 4 \frac{n}{(\log n)^2},$$

and for all $n \geq 51$ this is

$$< (\log 4) \frac{2n-1}{\log 2n} + 4 \frac{2n-1}{(\log 2n)^2} < (\log 4) \frac{N}{\log N} + 4 \frac{N}{(\log N)^2},$$

as a careful calculation reveals. This yields the upper bound claimed in Theorem 5.3.

To obtain the lower bound claimed in Theorem 5.3 we begin by observing that the largest binomial coefficient $\binom{n}{m}$ occurs with $m = [n/2]$. All the other binomial coefficients are smaller, as is $\binom{n}{0} + \binom{n}{n}$, so that

$$2^n = \left(\binom{n}{0} + \binom{n}{n}\right) + \sum_{m=1}^{n-1} \binom{n}{m} \leq n \binom{n}{[n/2]},$$

by (5.5.2). Now, all prime factors of any $\binom{n}{[n/2]}$ are $\leq n$, and in fact if p^{e_p} divides $\binom{n}{[n/2]}$, then $p^{e_p} \leq n$ by Corollary 3.10.1 to Kummer's Theorem. Therefore

$$2^n \leq n \binom{n}{[n/2]} \leq n \prod_{\substack{p \text{ prime} \\ p \leq n}} p^{e_p} \leq n^{\pi(n)+1}.$$

Taking logarithms we deduce the claimed result. □

Exercise 5.5.3. Use exercise 3.10.3 and the last displayed equation to prove that

(5.5.6) $$\operatorname{lcm}[m: m \leq n] \geq \frac{2^n}{n}.$$

5.6. Gaps between primes

Let $p_1 = 2 < p_2 = 3 < \cdots$ be the sequence of primes. We are interested in the possible gaps, $p_{n+1} - p_n$, between primes.

The prime number theorem tells us that there are about $x/\log x$ primes up to x, so that the average gap between primes $\leq x$ is about $\log x$: If $N = \pi(x)$, then p_N is the largest prime $\leq x$, and $p_N \sim x$ by exercise 5.4.1(b). This implies that the average gap between consecutive primes up to x is

$$\frac{1}{N-1} \sum_{n=1}^{N-1} (p_{n+1} - p_n) = \frac{p_N - 2}{N-1} \sim \frac{x}{x/\log x} = \log x,$$

by the prime number theorem.

Are there gaps between consecutive primes that are much smaller than the average? Much larger than the average? What is the largest that gaps between primes can be, and what is the smallest?

Exercise 5.6.1. (a) Prove that there are gaps between primes $\leq x$ that are at least as large as the average gap between primes up to x.
(b) Prove that there are gaps between primes $\leq x$ that are no bigger than the average gap between primes up to x.

Legendre conjectured that there are always primes between consecutive squares, that is, that there are primes in the interval $(n^2, (n+1)^2)$ for every integer n.

Exercise 5.6.2. (a) Show that if every interval $(x, x + 2\sqrt{x})$ contains a prime, then there are always primes between consecutive squares.
(b) Show that if there are always primes between consecutive squares, then every interval $(x, x + 4\sqrt{x} + 3]$ contains a prime.

At present we do not know how to prove every interval $(x, x + C\sqrt{x})$ contains primes, for any given $C > 0$. However it has been proven, by Baker, Harman, and Pintz, that any interval $(x, x + cx^{\frac{1}{2}+\frac{1}{40}}]$ contains a prime for c sufficiently large.

Exercise 5.6.3. Deduce from this that there is a prime between any consecutive, sufficiently large, cubes.

There is a simple way to construct a long interval which contains no primes:

Proposition 5.6.1. *For any integer m the interval $m!+2, m!+3, \ldots, m!+m$ contains no primes. Therefore if p_n is the largest prime $\leq m!+1$, then $p_{n+1}-p_n \geq m$.*

Proof. If $2 \leq j \leq m$, then j is included in the product for $m!$, and so j divides $m!+j$. Therefore $m!+j$ is composite as it is $> j$. Now $p_{n+1} \neq m!+j$ for each such j and so $p_{n+1} \geq m!+m+1 \geq p_n + m$. □

The gaps between primes constructed in this way are not quite as large as the average gaps. However one can extend this idea, creating a long interval of integers which each have a small prime factor, to prove that

$$\limsup_{n \to \infty} \frac{p_{n+1}-p_n}{\log p_n} = \infty.$$

Proving this is again beyond the scope of this book but a proof can be found in [**Graa**].

What about small gaps between primes?

Exercise 5.6.4. Prove that 2 and 3 are the only two primes that differ by 1.

There are plenty of pairs of primes that differ by two, namely 3 and 5, 5 and 7, 11 and 13, 17 and 19, etc., seemingly infinitely many, and this *twin prime conjecture* that there are infinitely many prime twins $p, p+2$ remains an open problem. Until recently, very little was proved about short gaps between primes, but that changed in 2009, when Goldston, Pintz, and Yildirim (see [**1**]) showed that

$$\liminf_{n \to \infty} \frac{p_{n+1}-p_n}{\log p_n} = 0.$$

In 2013, Yitang Zhang, until then a practically unknown mathematician,[16] showed that there are infinitely many pairs of primes that differ by at most a bounded amount. More precisely there exists a constant B such that there are infinitely many pairs of distinct primes that differ by at most B. This was soon improved by Maynard and Tao, though by a different method, so that we now know there are infinitely many pairs of consecutive primes p_n, p_{n+1} such that

$$p_{n+1}-p_n \leq 246.$$

This is not quite the twin prime conjecture, but it is a very exciting development. (See [**2**] for a discussion.)

The proofs of Maynard and of Tao yield a further great result: For any integer $m \geq 3$ there are infinitely many intervals of length 2^{14m} which contain m primes. That is, there are infinitely many m-tuples of consecutive primes $p_n, p_{n+1}, \ldots, p_{n+m-1}$ such that

$$p_{n+m-1}-p_n \leq 2^{14m}.$$

---

[16]See the movie *Counting from Infinity* (Zala Films, 2015) for an account of his fascinating story.

Further reading on hot topics in this section

[1] K. Soundararajan, *Small gaps between prime numbers: The work of Goldston-Pintz-Yıldırım*, Bull. Amer. Math. Soc. (N.S.) 44 (2007), 1–18.

[2] Andrew Granville, *Primes in intervals of bounded length*, Bull. Amer. Math. Soc. (N.S.) 52 (2015), 171–222.

5.7. *Formulas for primes*

Are there polynomials (of degree ≥ 1) that only yield prime values? That is, is $f(n)$ prime for every integer n? The example $6n + 5$ begins by taking the prime values $5, 11, 17, 23, 29$ before getting to $35 = 5 \times 7$. Continuing on, we get more primes $41, 47, 53, 59$ till we hit $65 = 5 \times 13$, another multiple of 5. So every fifth term of the arithmetic progression seems to be divisible by 5, which we verify as $6(5k) + 5 = 5(6k + 1)$. More generally $qn + a$ is a multiple of a whenever n is a multiple of a, since $q(ak) + a = a(qk + 1)$. A famous example of a polynomial that takes lots of prime values is $f(x) = x^2 + x + 41$. Indeed $f(n)$ is prime for $0 \leq n \leq 39$. However $f(40) = 41^2$ and $f(41k) = 41(41k^2 + k + 1)$. Therefore $f(41k)$ is composite for each integer k for which $41k^2 + k + 1 \neq -1, 0$, or 1.

We will develop this argument to work for all polynomials, but we will need the following result, which is a consequence of the Fundamental Theorem of Algebra and is proved in Theorem 3.11 of section 3.22 in appendix 3F.

Lemma 5.7.1. *A non-zero degree d polynomial has no more than d distinct roots in $\mathbb{C}$.*

The main consequence that we need is the following:

Corollary 5.7.1. *Suppose that $f(x) \in \mathbb{Z}[x]$ has degree $d \geq 1$. For any integer $B \geq 1$, there are no more than $(2B + 1)d$ integers n for which $|f(n)| \leq B$.*

Proof. If n is an integer, then so is $f(n)$, and therefore if $|f(n)| \leq B$, then $f(n) = m$ for some integer m with $|m| \leq B$. Therefore n is a root of one of the $2B + 1$ polynomials $f(x) - m$, each of which has no more than d roots by Lemma 5.7.1, and so the result follows. □

Proposition 5.7.1. *If $f(x) \in \mathbb{Z}[x]$ has degree $d \geq 1$, then there are infinitely many integers n for which $|f(n)|$ is composite.*

Proof. By Corollary 5.7.1 there are no more than $3d$ integers n for which $f(n) = -1, 0$, or 1, so there exists an integer a in the range $0 \leq a \leq 3d$ for which $|f(a)| > 1$. Let $m := |f(a)| > 1$. Now $km + a \equiv a \pmod{m}$ and so, by Corollary 2.3.1, we have
$$f(km + a) \equiv f(a) \equiv 0 \pmod{m}.$$
There are at most $3d$ values of k for which $km + a$ is a root of one of $f(x) - m$, $f(x)$, or $f(x) + m$, by Corollary 5.7.1. For any other k we have that $|f(km+a)| \neq 0$ or m, in which case $|f(km + a)|$ is divisible by m and $|f(km + a)| > m$, so that $|f(km + a)|$ is composite. □

Exercise 5.7.1. Show that if $f(x, y) \in \mathbb{Z}[x, y]$ has degree $d \geq 1$, then there are infinitely many pairs of integers m, n for which $|f(m, n)|$ is composite.

Nine of the first ten values of the polynomial $6n+5$ are primes. The polynomial $n^2 + n + 41$, discovered by Euler in 1772, is prime for $n = 0, 1, 2, \ldots, 39$ and the square of a prime for $n = 40$. However, in the proof of Proposition 5.7.1, we saw that $n^2 + n + 41$ is composite whenever n is a positive multiple of 41. See section 12.5 for more on such prime rich polynomials.

We discuss other places to look for primes in section 5.21 of appendix 5G.

It is not difficult to show that if a polynomial f takes on infinitely many prime values, then f must be irreducible. The next result indicates how many prime values f needs to take before we *know* that f is irreducible.

Theorem 5.4. *If $f(x) \in \mathbb{Z}[x]$ has degree $d \geq 1$ and $|f(n)|$ is prime for $\geq 2d+1$ integers n, then $f(x)$ is irreducible.*

Proof. Suppose that f is reducible; that is, $f = gh$ for polynomials $g(x), h(x) \in \mathbb{Z}[x]$. If $|f(n)| = p$, a prime, then $g(n)h(n) = p$ or $-p$. Therefore one of $g(n)$ and $h(n)$ equals p or $-p$, the other 1 or -1. In particular n is a root of $(g(x) - 1)(h(x) - 1)(g(x) + 1)(h(x) + 1)$, a polynomial of degree $2d$. This has no more than $2d$ roots by Lemma 5.7.1, and so $|f(n)|$ can be prime for no more than $2d$ integers n. □

This is often more than we need, as we see in the following beautiful result:

Theorem 5.5. *Write a given prime p in base 10 as $p = a_0 + a_1 10 + \cdots + a_d 10^d$ (with each $a_i \in \{0, 1, 2, \ldots, 9\}$ and $a_d \neq 0$). Then $a_0 + a_1 x + \cdots + a_d x^d$ is an irreducible polynomial.*

Proof. Let $f(x) = a_0 x + \cdots + a_d x^d$ and suppose that $f = gh$. As $g(10)h(10)$ is prime, one of $g(10)$ and $h(10)$ equals 1 or -1. We will suppose that it is g (swapping g and h if necessary). As $g(x) \in \mathbb{Z}[x]$ it can be written in the form $g(x) = c \prod_{j=1}^{D} (x - \alpha_j)$ with $c \in \mathbb{Z}$, and so $\prod_{j=1}^{D} |10 - \alpha_j| \leq |g(10)| = 1$. Therefore there is a root α of $g(x)$ for which $|\alpha - 10| \leq 1$. This implies that $\text{Re}(\alpha) \in [9, 11]$ and so $\text{Re}(1/\alpha) > 0$ and $|\alpha| \geq 9$.

As $f(\alpha) = 0$ we deduce that

$$0 = \text{Re}\left(\frac{f(\alpha)}{\alpha^d}\right) = a_d + a_{d-1}\text{Re}\left(\frac{1}{\alpha}\right) + \sum_{i=2}^{d} a_{d-i}\text{Re}\left(\frac{1}{\alpha^i}\right).$$

As discussed above $\text{Re}(1/\alpha) > 0$ and so $a_{d-1}\text{Re}(1/\alpha) \geq 0$. On the other hand, $\text{Re}(1/\alpha^i)$ might be negative and so $a_{d-i}\text{Re}(1/\alpha^i) \geq -9/|\alpha|^i$. Therefore

$$0 \geq 1 + 0 - 9\sum_{i=2}^{d} \frac{1}{|\alpha|^i},$$

which implies that

$$1 < 9\sum_{i \geq 2} \frac{1}{|\alpha|^i} = \frac{9}{|\alpha|(|\alpha|-1)} \leq \frac{1}{8}$$

as $|\alpha| \geq 9$, which yields a contradiction. □

Exercise 5.7.2. Prove an analogous result for primes written in an arbitrary base $b \geq 3$.

Questions about primes   169

Exercise 5.7.3.† Suppose that $f(x) = a_0x + \cdots + a_dx^d \in \mathbb{Z}[x]$ with each $|a_i| \leq A$ and $a_d \neq 0$. Prove that if $f(n)$ is prime for some integer $n \geq A + 2$, then $f(x)$ is irreducible.

There are many books on the distribution of primes. My favorites for beginners are [**TMF00**] which explains the key ideas behind the prime number theorem and other important results in an accessible way, and [**Rib91**] which is more recreational but full of good stuff. The introductory book [**HW08**] proves quite a few of the easier theorems in the subject.

Additional exercises

Exercise 5.8.1. Let m be the product of the primes ≤ 1000. Prove that if n is an integer between 10^3 and 10^6, then n is prime if and only if $(n, m) = 1$.

Exercise 5.8.2. Show that if $p > 3$ and $q = p + 2$ are twin primes, then $p + q$ is divisible by 12.

Exercise 5.8.3. Show that there are infinitely many integers n for which each of $n, n+1, \ldots, n+1000$ is composite.

Exercise 5.8.4. Fix integer $m > 1$. Show that there are infinitely many integers n for which $\tau(n) = m$.

Exercise 5.8.5.† Fix integer $k > 1$. Prove that there are infinitely many integers n for which $\mu(n) = \mu(n+1) = \cdots = \mu(n+k)$.

Exercise 5.8.6. Let H be a proper subgroup[17] of $(\mathbb{Z}/m\mathbb{Z})^*$.
  (a) Show that if a is coprime to m and q is a given non-zero integer, then there are infinitely many integers $n \equiv a \pmod{m}$ such that $(n, q) = 1$.
  (b) Prove that if n is an integer coprime to m but which is not in a residue class of H, then n has a prime factor which is not in a residue class of H.
  (c) Deduce there are infinitely many primes which do not belong to any residue class of H.

Exercise 5.8.7.† Suppose that for any coprime integers a and q there exists *at least one* prime $\equiv a \pmod{q}$. Deduce that for any coprime integers A and Q, there are *infinitely many* primes $\equiv A \pmod{Q}$.

Exercise 5.8.8. Prove that there are infinitely many primes p for which there exists an integer a such that $a^3 - a + 1 \equiv 0 \pmod{p}$.

Exercise 5.8.9. Prove that for any $f(x) \in \mathbb{Z}[x]$ of degree ≥ 1, there are infinitely many primes p for which there exists an integer a such that p divides $f(a)$.

Exercise 5.8.10. Let $\mathcal{L}(n) = \text{lcm}[1, 2, \ldots, n]$.
  (a) Show that $\mathcal{L}(n)$ divides $\mathcal{L}(n+1)$ for all $n \geq 1$.
  (b) Express $\mathcal{L}(n)$ as a function of the prime powers $\leq n$.
  (c) Prove that for any integer k there exist integers n for which $\mathcal{L}(n) = \mathcal{L}(n+1) = \cdots = \mathcal{L}(n+k)$.
  (d)‡ Prove that if k is sufficiently large, then there is such an integer n which is $< 3^k$.

Exercise 5.8.11.† Prove that
$$\text{Li}(x) \Big/ \frac{x}{\log x} \to 1 \text{ as } x \to \infty.$$

Exercise 5.8.12. Prove that 1 is the best choice for B when approximating $\text{Li}(x)$ by $x/(\log x - B)$.

Exercise 5.8.13.† Using the Maynard-Tao result, prove that there exists a positive integer $k \leq 246$ for which there are infinitely many prime pairs $p, p + k$.

---

[17] H is a *proper* subgroup of G if it is a subgroup of G but not the whole of G

Exercise 5.8.14. Suppose that a and b are integers for which $g(a) = 1$ and $g(b) = -1$, where $g(x) \in \mathbb{Z}[x]$.
  (a) Prove that $b = a - 2$, $a - 1$, $a + 1$, or $a + 2$.
  (b)† Deduce that there are no more than four integer roots of $(g(x) - 1)(g(x) + 1) = 0$.
  (c)† Show that if $g(x)$ has degree 2 and there are four integer roots of $(g(x) - 1)(g(x) + 1) = 0$, then $g(x) = \pm h(x - A)$ where $h(t) = t^2 - 3t + 1$, with roots A, $A + 1$, $A + 2$, and $A + 3$.
  (d)† Modify the proof of Theorem 5.4 to establish that if $f(x) \in \mathbb{Z}[x]$ has degree $d \geq 6$ and $|f(n)|$ is prime for $\geq d + 3$ integers n, then $f(x)$ is irreducible.

Let $f(x) = h(x)h(x - 4)$, which has degree 4. Note that $|f(n)|$ is prime for the eight values $n = 0, 1, \ldots, 7$, and so there is little room in which to improve (d).

One can show that there are reducible polynomials $f(x) \in \mathbb{Z}[x]$ of arbitrarily large degree d for which $|f(n)|$ takes on at least $d+1$ prime values: Let $p_1 < \cdots < p_m$ be distinct primes. Let $g(x) = \prod_{j=1}^{m}(p_j^2 - x^2)$ and $q = g(1)$. By Dirichlet's Theorem (section 5.3) we know that there are infinitely many primes $p_0 \equiv 1 \pmod{q}$.[18] We select one such prime and write $p_0 = 1 + \ell q$ for some positive integer ℓ. Now let $f(x) = x(1 + \ell g(x))$ which has degree $d := 2m + 1$. We have that $|f(\pm 1)| = 1 + \ell q = p_0$ and $|f(\pm p_j)| = p_j$ for $j = 1, \ldots, m$, so there are $\geq 2m + 2 = d + 1$ integers n for which $|f(n)|$ is prime.

In the next exercise, assuming certain conjectures,[19] we construct reducible polynomials $f(x) \in \mathbb{Z}[x]$ of arbitrarily large degree d for which $|f(n)|$ takes on $d + 2$ prime values. This implies that the result in exercise 5.8.14(d) is "best possible".

Exercise 5.8.15.† Assume that there are infinitely many positive integers n for which $n^2 - 3n + 1$ is prime, and denote these integers by $n_1 < n_2 < \cdots$. Let $g_m(x) := (n_1 - x) \cdots (n_m - x)$. If ℓ is a positive integer for which $1 + \ell g_m(0), 1 + \ell g_m(1), 1 + \ell g_m(2), 1 + \ell g_m(3)$ are simultaneously prime, then prove that the polynomial $f(x) := (x^2 - 3x + 1)(1 + \ell g_m(x))$ has degree $d := m + 2$ and that there are exactly $d + 2$ integers n for which $|f(n)|$ is prime.

---

[18] We will prove this later, in Theorem 7.8.
[19] These conjectures follows from the *Polynomial prime values conjecture* stated in the bonus section of this chapter.

Appendix 5A. Bertrand's postulate and beyond

5.9. Bertrand's postulate

In 1845 Bertrand conjectured, on the basis of calculations up to a million:

Theorem 5.6 (Bertrand's postulate). *For every integer $n \geq 1$, there is a prime number between n and $2n$.*

Bertrand's postulate was proved in 1850 by Chebyshev. We will follow the 19-year-old Erdős's proof, or, as N. J. Fine put it (in the voice of Erdős):

> Chebyshev said it, but I'll say it again:
> There's always a prime between n and $2n$.

Exercise 5.9.1. Show that prime p does not divide $\binom{2n}{n}$ when $2n/3 < p \leq n$.

Proof of Bertrand's postulate. Let p^{e_p} be the exact power of prime p dividing $\binom{2n}{n}$. We know that

- $e_p = 1$ if $n < p \leq 2n$ by Kummer's Theorem (Theorem 3.7),
- $e_p = 0$ if $2n/3 < p \leq n$ by exercise 5.9.1,
- $e_p \leq 1$ if $\sqrt{2n} < p \leq 2n$ by Corollary 3.10.1,
- $p^{e_p} \leq 2n$ if $p \leq 2n$ by Corollary 3.10.1.

Combining these gives

$$\frac{2^{2n}}{2n} \leq \binom{2n}{n} = \prod_{p \leq 2n} p^{e_p} \leq \prod_{n < p \leq 2n} p \prod_{p \leq 2n/3} p \prod_{p \leq \sqrt{2n}} 2n$$

$$\leq \left(\prod_{n < p \leq 2n} p \right) \times 4^{2n/3 - 1} \times (2n)^{(\sqrt{2n}+1)/2},$$

using Lemma 5.5.1 to bound $\prod_{p \leq 2n/3} p$ and the bound $\pi(\sqrt{2n}) \leq \frac{1}{2}(\sqrt{2n}+1)$ (as neither 1 nor any even integer > 2 is prime). Taking logarithms we deduce that

$$\sum_{\substack{p \text{ prime} \\ n < p \leq 2n}} \log p > \frac{\log 4}{3} n - \frac{\sqrt{2n}+3}{2} \log(2n).$$

This implies that

(5.9.1) $$\sum_{\substack{p \text{ prime} \\ n < p \leq 2n}} \log p \geq \frac{1}{3} n$$

for all $n \geq 2349$, which implies Bertrand's postulate in this range. (This lower bound should be compared to the upper bound (5.5.4).)

If $1 \leq n \leq 5000$, then the interval $(n, 2n]$ contains at least one of the primes 2, 3, 5, 7, 13, 23, 43, 83, 163, 317, 631, 1259, 2503, and 5003. □

Exercise 5.9.2. Use Bertrand's postulate to prove that there are infinitely many primes with first digit "1".

Exercise 5.9.3. Use Bertrand's postulate to show, by induction, that every integer $n > 6$ can be written as the sum of distinct primes.

Exercise 5.9.4. Goldbach conjectured that every even integer ≥ 6 can be written as the sum of two primes. Deduce Bertrand's postulate from Goldbach's conjecture.

Exercise 5.9.5. Use Bertrand's postulate to prove that $\frac{1}{n+1} + \frac{1}{n+2} + \cdots + \frac{1}{2n}$ is never an integer.

Exercise 5.9.6. Prove that for every $n \geq 1$ one can partition the set of integers $\{1, 2, \ldots, 2n\}$ into pairs $\{a_1, b_1\}, \ldots, \{a_n, b_n\}$ such that each sum $a_j + b_j$ is a prime.

Exercise 5.9.7.[†] (a) Prove that prime p divides $\binom{2n}{n}$ when $n/2 < p \leq 2n/3$.
 (b) Prove that the product of the primes in $(3m, 12m]$ divides $\binom{12m}{6m}\binom{6m}{4m}$.
 (c)[†] Deduce that we can take any constant $c_2 > \frac{2}{9}\log(432)$ in (5.5.1).
   (Note that $\frac{2}{9}\log(432) = 1.3485\ldots < \log 4 = 1.3862\ldots$)
 (d) Now deduce Bertrand's postulate for all sufficiently large x from (5.5.1).

5.10. The theorem of Sylvester and Schur

Bertrand's postulate can be rephrased to state that at least one of the integers $k+1, k+2, \ldots, 2k$ has a prime factor $> k$. This can be generalized as follows:

Theorem 5.7 (Sylvester-Schur Theorem). *For any integers $n \geq k \geq 1$, at least one of the integers $n+1, n+2, \ldots, n+k$ is divisible by a prime $p > k$.*

Proposition 5.10.1. *If, for given integers $n \geq k \geq 1$, we have*

(5.10.1) $$\binom{n+k}{k} > (n+k)^{\pi(k)},$$

then at least one of the integers $n+1, n+2, \ldots, n+k$ is divisible by a prime $p > k$. If (5.10.1) holds for $n_1(k)$, then it holds for all $n \geq n_1(k)$.

5.10. The theorem of Sylvester and Schur

Proof. If the prime factors of $n+1, n+2, \ldots, n+k$ are all $\leq k$, then all of the prime factors p of $\binom{n+k}{k}$ are $\leq k$. If $p^e \| \binom{n+k}{k}$, then $p^e \leq n+k$ by Corollary 3.10.1. Therefore

$$(5.10.2) \qquad \binom{n+k}{k} \leq \prod_{p \leq k}(n+k) = (n+k)^{\pi(k)},$$

contradicting (5.10.1). This proves the first part of the result.

We prove the second part by induction on $n \geq n_1(k)$ using the following result.

Exercise 5.10.1. Prove that $\left(1 + \frac{1}{x+k}\right)^k \leq \left(1 + \frac{k}{x+1}\right)$ for all $x \geq k \geq 1$.

The result holds for $n = n_1(k)$, so now suppose that (5.10.1) holds for some given n. Then

$$\binom{n+1+k}{k} = \left(1 + \frac{k}{n+1}\right)\binom{n+k}{k} > \left(1 + \frac{1}{n+k}\right)^k (n+k)^{\pi(k)} > (n+1+k)^{\pi(k)},$$

by exercise 5.10.1 and the induction hypothesis, and so (5.10.1) holds for $n+1$. The result follows. $\square$

Proof of the Sylvester-Schur Theorem for all $k \leq 1500$. Calculations give some value for $n_1(k)$ in Proposition 5.10.1 for all $k \leq 1500$, and so the Sylvester-Schur Theorem follows for these k and all $n \geq n_1(k)$ by Proposition 5.10.1. Now $n_1(k) = k$ for $202 \leq k \leq 1500$, and $k \leq n_1(k) \leq k+17$ for all $k \leq 201$. We verify the theorem for $k \leq n \leq k+16$ with $k \leq 201$, case by case. $\square$

A just failed proof of the Sylvester-Schur Theorem. Calculations suggest that $\binom{2k}{k} > (2k)^{\pi(k)}$ for all $k \geq 202$. If so, the Sylvester-Schur Theorem follows for all $k \geq 202$ by Proposition 5.10.1. However we just failed to prove this inequality as a consequence of the upper bound in Theorem 5.3. If one combines the upper bound on $\pi(k/4)$ from Theorem 5.3, together with exercise 5.9.7(b), then we can prove that $\binom{2k}{k} > (2k)^{\pi(k)}$ for all sufficiently large k. However "sufficiently large" here is likely to be extremely large. $\square$

Exercise 5.10.2. Prove that if $\pi(k) < \frac{k \log 4}{\log(2k)} - 1$ for all integers $k \geq 1$, then Theorem 5.7 holds for all $n \geq k \geq 1$.

Proof of the Sylvester-Schur Theorem for all $k > 1500$. If (5.10.1) holds, then the result follows from Proposition 5.10.1. Hence we may assume that (5.10.2) holds. Now, $\pi(k) < k/3$ (which can be proved by accounting for divisibility by 2 and 3), and $\frac{n+k-j}{k-j} > \frac{n+k}{k}$ for $j = 0, \ldots, k-1$ so that $\binom{n+k}{k} \geq (\frac{n+k}{k})^k$. Therefore (5.10.2) implies that

$$\left(\frac{n+k}{k}\right)^k \leq \binom{n+k}{k} \leq (n+k)^{\pi(k)} \leq (n+k)^{k/3},$$

which in turn implies that

$$n + k \leq k^{3/2}; \text{ that is, } n \leq k^{3/2} - k.$$

Next we note that if $p > (n+k)^{1/2}$ and $p^e \| \binom{n+k}{k}$ so that $p^e \leq n+k$, then $e = 0$ or 1. Therefore we can refine (5.10.2) to

(5.10.3) $$\binom{n+k}{k} \leq \prod_{p \leq (n+k)^{1/2}} (n+k) \prod_{p \leq k} p = k^{\frac{1}{3}k^{3/4}} 4^{k-1},$$

by (5.5.4), as $\pi((n+k)^{1/2}) \leq \frac{1}{3}(n+k)^{1/2} \leq \frac{1}{3}k^{3/4}$.

Now if $n \geq 3k$, then, by exercise 4.14.2 of appendix 4D,

$$\frac{(4^4/3^3)^k}{ek} \leq \binom{4k}{k} \leq \binom{n+k}{k} \leq k^{\frac{1}{2}k^{3/4}} 4^{k-1}$$

which is false for all $k \geq 1$. Therefore $n + k \leq 4k$, and so if $n + k > \frac{5}{2}k$, then our inequality becomes

$$\frac{(5^5/3^3 2^2)^{k/2}}{ek} \leq \binom{5k/2}{k} \leq \binom{n+k}{k} \leq (4k)^{k^{1/2}} 4^{k-1}.$$

This is false for all $k \geq 780$.

Finally for the range $k \leq n \leq 3k/2$ if prime p is in the range $(n+k)/3 < p \leq k$, then $2p$ is the only multiple of p that appears in $(n+1) \cdots (n+k)$ and so p does not divide $\binom{n+k}{k}$. Therefore

$$\binom{2k}{k} \leq \binom{n+k}{k} \leq \prod_{p \leq (n+k)^{1/2}} (n+k) \prod_{p \leq (n+k)/3} p \leq \prod_{p \leq (n+k)^{1/2}} (3k)^{\pi(2k^{1/2})} \prod_{p \leq 5k/6} p,$$

which implies that

$$\frac{4^k}{ek} \leq (4k)^{k^{1/2}} 4^{5k/6-1}$$

which is false for all $k \geq 1471$. □

Exercise 5.10.3. (a) Use Bertrand's postulate and the Sylvester-Schur Theorem to show that if $1 \leq r < s$, then there is a prime p that divides exactly one of the integers $r+1, \ldots, s$.
(b) Deduce that if $1 \leq r < s$, then $\frac{1}{r+1} + \cdots + \frac{1}{s}$ is never an integer.

Bonus read: A review of prime problems

5.11. Prime problems

In this bonus section we will discuss various natural sequences that are expected to contain infinitely many primes, highlighting recent progress.

> *Mathematicians have tried in vain to discover some order in the sequence of the prime numbers and we have every reason to believe that there are some mysteries that the human mind shall never penetrate.*
> — LEONHARD EULER (1740)

Prime values of polynomials in one variable

In section 5.6 we mentioned the twin prime conjecture, that there are infinitely many pairs of primes that differ by 2. What about other pairs? Obviously there can be no more than one pair of primes that differ by an odd integer k (as one of the two integers must be divisible by 2), but when the difference is an even integer k there is no such obstruction. Calculations then suggest that:

For all even integers $2m > 0$ there are infinitely many pairs of primes that differ by $2m$. That is, there are infinitely many prime pairs $p, p + 2m$.

Here we asked for simultaneous prime values of two *monic* linear polynomials x and $x + 2m$. What if we select polynomials with different leading coefficients, like x and $2x + 1$? Such *prime pairs* come up naturally in Sophie Germain's Theorem 7.11 (of section 7.27 in appendix 7F) and calculations support the guess that there are many (like 3 and 7; 5 and 11; 11 and 23; 23 and 47;...). We therefore conjecture:

There are infinitely many pairs of primes $p, 2p + 1$.

One can generalize this to other pairs of linear polynomials but we might again have the problem that at least one is even, as with $p, 3p + 1$.

Exercise 5.11.1. Give conditions on integers a, b, c, d with $a, c > 0$, assuming that $(a, b) = (c, d) = 1$, which guarantee that there are infinitely many integers n for which $an + b$ and $cn + d$ are different and both positive and odd. We conjecture, under these conditions that:

There are infinitely many pairs of primes $am + b, cm + d$.

For triples of linear forms and even k-tuplets of linear forms, there are more exceptional cases. For example, the three polynomials $n, n + 2, n + 4$ can all simultaneously take odd values but, for each integer n, one of them is divisible by 3. We call 3 a *fixed prime divisor*, which plays the same role as 2 in the example $n, n + k$ with k odd. In general we need that a given set of linear forms $a_1 x + b_1, a_2 x + b_2, \ldots, a_k x + b_k$ with integer coefficients is *admissible*; that is, there is no fixed prime divisor p. Specifically, for each prime p, there exists an integer n_p for which none of the $a_j n_p + b_j$ is divisible by p, which implies that p does not divide $a_j n + b_j$ for $1 \leq j \leq k$ for every integer $n \equiv n_p \pmod{p}$. This leads us to

The prime k-tuplets conjecture. *Let $a_1 x + b_1, \ldots, a_k x + b_k$ be an admissible set of k linear polynomials with integer coefficients, such that each a_j is positive. Then there are infinitely many positive integers m for which*

$$a_1 m + b_1, \ldots, a_k m + b_k \text{ are all prime.}$$

Exercise 5.11.2.† Assuming the prime k-tuplets conjecture deduce that there are infinitely many pairs of *consecutive* primes $p, p + 100$.

Exercise 5.11.3.† Assuming the prime k-tuplets conjecture deduce that there are infinitely many triples of *consecutive* primes in an arithmetic progression.

Exercise 5.11.4.† Assuming the prime k-tuplets conjecture deduce that there are infinitely many quadruples of *consecutive* primes formed of two pairs of prime twins.

Exercise 5.11.5.† Let $a_{n+1} = 2a_n + 1$ for all $n \geq 0$. Fix an arbitrarily large integer N. Use the prime k-tuplets conjecture to show that we can choose a_0 so that $a_0, a_1, \ldots, a_N$ are all primes.

Exercise 5.11.6. Show that the set of linear polynomials $a_1 m + 1, a_2 m + 1, \ldots, a_k m + 1$, with each a_j positive, is admissible.

There is more on prime k-tuplets of linear polynomials in appendix 5E.

What about other polynomials? For example, the polynomial $n^2 + 1$ takes prime values $2, 5, 17, 37, 101, \ldots$ seemingly on forever, so we conjecture that:

There are infinitely many primes of the form $n^2 + 1$.

The polynomial $x^2 + 2x$ cannot be prime for many integer values since it is reducible (recall Theorem 5.4 and exercise 5.8.14(c)). This is a different reason (from the fixed prime factors above) for a polynomial not to take more than finitely many prime values. These are the only reasons known for a polynomial not to take infinitely many prime values and, if neither of them holds, then we believe that the polynomial does take on infinitely many prime values. More precisely:

Polynomial prime values conjecture. *Let $f_1(x), \ldots, f_k(x) \in \mathbb{Z}[x]$, each irreducible, with positive leading coefficients. If $f_1 \cdots f_k$ has no fixed prime divisor, then:*

There are infinitely many integers m for which $f_1(m), \ldots, f_k(m)$ are all prime.

To be precise, if $f_1, \ldots, f_k$ have "no fixed prime divisor" then we mean that for every prime p there exists an integer n_p such that $f_1(n_p) \cdots f_k(n_p)$ is not divisible by p. The polynomial prime values conjecture specialized to linear polynomials is the prime k-tuplets conjecture.[20]

Exercise 5.11.7. Prove that the only prime pair $p, p^2 + 2$ is $3, 11$.

Exercise 5.11.8. (a) Prove that if $f_1 \cdots f_k$ has no fixed prime divisor, then, for each prime p, there are infinitely many integers n such that $f_1(n) \cdots f_k(n)$ is not divisible by p.
(b)[†] Show that if $p > \deg(f_1(x) \cdots f_k(x))$ and p does not divide $f_1(x) \cdots f_k(x)$, then n_p exists.
(c) Prove that if $f_j(x) = x + h_j$ for given integers $h_1, \ldots, h_k$, then n_p exists for a given prime p if and only if $\#\{\text{distinct } h_j \pmod{p}\} < p$.

The only case of the polynomial prime values conjecture that has been proved is when $k = 1$ with $f_1(.)$ is linear. The hypothesis ensures that $f(x) = qx + a$ with $q \geq 1$ and $(a, q) = 1$. This is Dirichlet's Theorem (that there are infinitely many primes $\equiv a \pmod{q}$ whenever $(a, q) = 1$, which we discuss in sections 8.17 of appendix 8D and 13.7).

Distinguishing primes and P_k's from other integers. The Möbius function was introduced in section 4.5, and in Corollary 4.5.1 we saw that the sum

$$\sum_{d|n} \mu(d)$$

is non-zero only if $n = 1$ and so allows us to distinguish the integer 1 from all other positive integers. In section 4.11 of appendix 4B we saw that if the sum

$$\sum_{d|n} \mu(d) \log(n/d)$$

is non-zero, then n has exactly one prime factor and so allows us to distinguish primes and prime powers from all other positive integers. A positive integer is called a "P_k" if it has no more than k distinct prime factors. In the next exercise we will see how an analogous sum allows us to distinguish P_k's.

Exercise 5.11.9.[†] (a)[‡] Let $x_0, \ldots, x_m$ be variables. Prove that if $m > k \geq 0$, then

$$\sum_{S \subset \{1,2,\ldots,m\}} (-1)^{|S|} \left(x_0 + \sum_{j \in S} x_j\right)^k = 0.$$

(b) Deduce that if n has more than k different prime factors, then

$$\sum_{d|n} \mu(d)(\log(n/d))^k = 0.$$

(c)[‡] What value does this take when n has exactly k different prime factors?

Exercise 5.11.10. Show that if each prime factor of n is $> n^{1/3}$, then n is either prime or the product of two primes.

Prime values of polynomials in several variables

One can ask for prime values of polynomials in two or more variables, for example, primes of the form $m^2 + n^2$ or the form $a^2 + b^2 + 1$ or more complicated polynomials of mixed degree like $4a^3 + 27b^2$. What is known?

---

[20]This conjecture was first formulated by Andrzej Schinzel in 1958. He called it "*Hypothesis H*" in that paper, and the name has stuck.

The proof of the prime number theorem can be adapted to many situations, for example to primes of the form $m^2 + n^2$ or the form $2u^2 + 2uv + 3v^2$ or indeed the prime values of any irreducible binary quadratic form (which are discussed in chapters 9 and 12) without a fixed prime divisor. The proof for $m^2 + n^2$ uses the fact that $m^2 + n^2 = (m+in)(m-in)$, the *norm* of $m + in$. One can develop this to prove that any such *norm form* (the appropriate generalization[21] of $m^2 + n^2$ to higher degree) takes on infinitely many prime values as long as it has no fixed prime factor. A norm form is always a degree d polynomial in d variables.

One can then ask for prime values of norm forms in which we fix some of the variables (perhaps to 0). For example, if $m = 1$ in $m^2 + n^2$, we are back to the open question about prime values of $n^2 + 1$. *However* in 2002 Heath-Brown was able to prove that $a^3 + 2b^3$ takes on infinitely prime values and then extended this, with Moroz, to any irreducible cubic form in two variables. In 2018, Maynard proved such a result for a family of norm forms[22] in $3m$ variables of degree $4m$ (or less).

These results on norm forms were all inspired by Friedlander and Iwaniec's 1998 breakthrough in which they took n to be a square in $m^2 + n^2$ (and therefore found prime values of $u^2 + v^4$), following Fouvry and Iwaniec's 1997 paper in which they took n to be prime (and therefore obtained infinitely many prime pairs $p, m^2 + p^2$). This was the first example in which the polynomial in question is *sparse* in that the number of integer values it takes up to x is roughly x^c for some $c < 1$. The current record sparsity is $c = \frac{2}{3}$ from the work of Heath-Brown and Moroz. In 2017, Heath-Brown and Xiannan Li went beyond the Fouvry-Iwaniec and Friedlander-Iwaniec results by showing that there are infinitely many prime pairs $p, m^2 + p^4$.

In every case we expect that the proportion of values of the polynomial up to x which are prime is about $c/\log x$, where c is a constant which depends on how often each prime divides values of the polynomial.

Back in 1974, Iwaniec had shown how versatile sieve methods could be by showing that any quadratic polynomial in two variables (which is irreducible and has no fixed prime divisor) takes on infinitely many prime values, for example, $m^2 + n^2 + 1$. We will see this result put to good use in appendix 12G when tiling a circle with smaller circles.

What about the prime values of more than one polynomial in several variables? We can generalize our conjectures as follows:

Multivariable polynomial prime values conjecture. *Let $f_1(x_1, \ldots, x_n), \ldots, f_k(x_1, \ldots, x_n) \in \mathbb{Z}[x_1, \ldots, x_n]$, each of which is irreducible. Suppose that there are infinitely many n-tuplets of integers $m_1, \ldots, m_n$ for which each $f_j(m_1, \ldots, m_n)$ is positive. If $f_1 \cdots f_k$ has no fixed prime divisor, then there are*

Infinitely many n-tuplets of integers $m_1, \ldots, m_n$ for which
$f_1(m_1, \ldots, m_n), \ldots, f_k(m_1, \ldots, m_n)$ are all prime.

In 1939, van der Corput showed that there are infinitely many three-term arithmetic progressions of primes, which can be written as

$$a, a+d, a+2d,$$

---

[21]More precisely the norm of $\sum_i x_i \omega_i$ where the ω_i are a basis for the ring of integers of some number field of degree d and the x_i are the variables.

[22]The norm of $\sum_{i=1}^{3m} x_i \omega^i$ where the field, of degree $4m$, is generated by ω over $\mathbb{Q}$.

three degree-one polynomials in two variables. For a long time, methods seemed inadequate to extend this to length four arithmetic progressions, but this was resolved in 2008 by Green and Tao, who proved that for any fixed integer $k \geq 3$ there are infinitely many prime k-tuplets of the form

$$a, a+d, a+2d, \ldots, a+(k-1)d.$$

The methods used were quite new to the search for prime numbers and this has led to widespread interest. In 2012, along with Ziegler, they were able to prove a very general result for linear polynomials, which is as good as one can hope for, given that there has been no progress directly on the prime k-tuplets conjecture:

Until we prove the twin prime conjecture we will be unable to prove the multivariable polynomial prime values conjecture, in full generality, even for linear polynomials, since two of the polynomials might differ by two, for example if $x+3y$ and $x+3y+2$ are in our set. More generally, without progress on the prime k-tuplets conjecture, we must avoid any linear relation between two of our polynomials.

Theorem 5.8 (The Green-Tao-Ziegler Theorem). *Suppose that $f_1(\mathbf{x}), \ldots, f_k(\mathbf{x})$ are linear polynomials which satisfy the hypothesis of the multivariable polynomial prime values conjecture. Moreover assume that if $1 \leq i < j \leq k$, there do not exist integers a, b, c, not all zero, for which $af_i + bf_j = c$. Then there are infinitely many $\mathbf{m} \in \mathbb{Z}^n$ for which $f_1(\mathbf{m}), \ldots, f_k(\mathbf{m})$ are all prime.*

We will discuss applications of the Green-Tao-Ziegler Theorem in appendix 5E.

It is not difficult to show that there are infinitely many primes of the form $b^2 - 4ac$, the discriminant of an arbitrary quadratic polynomial. However we do not know how to prove that there are infinitely many primes of the form $4a^3 + 27b^2$, the discriminant of the cubic polynomial $x^3 + ax + b$. Proving this would have a significant impact on our understanding of various questions about degree 3 Diophantine equations.

Exercise 5.11.11. Let $g(x) = 1 + \prod_{j=1}^{k}(x-j)$. Prove that there exist integers a and b such that the reducible polynomial $f(x) = (ax+b)g(x)$ is prime when $x = n$ for $1 \leq n \leq k$. Compare this to the result in exercise 5.8.14(c) (with $d = k+1$).

Goldbach's conjecture and variants

Goldbach's 1742 conjecture is the statement that every even integer ≥ 4 can be written as the sum of two primes. It is still an open question though it has now been verified for all even numbers $\leq 4 \times 10^{18}$.

Great problems motivate mathematicians to think of new techniques, which can have great influence on the subject, even if they fail to resolve the original question. For example, although there have been few plausible ideas for proving Goldbach's conjecture, it has motivated some of the development of sieve theory, and there are some beautiful results on modifications of the original problem. The most famous are:

In 1975 Montgomery and Vaughan showed that if there are any exceptions to Goldbach's conjecture (that is, even integers n that are not the sum of two primes), then there are very few of them.

In 1973 Jingrun Chen showed that every sufficiently large even integer is the sum of a prime and an integer that is the product of at most two primes. Here "sufficiently large" means enormous.

In 1934 I. M. Vinogradov proved that every sufficiently large odd integer is the sum of three primes. The "sufficiently large" has recently been removed: Harald Helfgott, with computational assistance from David Platt, proved that every odd integer > 1 is the sum of at most three primes.

Exercise 5.11.12. Show that the Goldbach conjecture is equivalent to the statement that every integer > 1 is the sum of at most three primes.[23]

Other questions

Before this chapter we asked if there are infinitely many primes of the form $2^p - 1$ (Mersenne primes) or of the form $2^{2^n} + 1$ (Fermat primes). We can ask other questions in this vein, for example prime values of second-order linear recurrences which start $0, 1$ (like the Fibonacci numbers) or their companion sequences (see exercise 3.9.3) or prime values of high-order linear recurrence sequences.

Mersenne primes written in binary look like $111\ldots 111$, and so are palindromic. Some people have been interested in primes of the form $\frac{1}{9}(10^n - 1)$ which equal $111\ldots 111$ in base 10 and so are palindromic. We are unable to prove there are infinitely many Mersenne primes, so how about the easier question, are there infinitely many palindromic primes when written in binary or in decimal or indeed in any other base? Also open.

We saw earlier that it is not difficult to show that there are infinitely many primes with the first few digits given. But how about missing digits? Can one find infinitely many primes which have no 7 in their decimal expansion or no 9 or no consecutive digits 123? These questions are all answered in a remarkable recent paper of Maynard [4].

Let M be a given n-by-n matrix. The (i, j)th entry of $M, M^2, \ldots$ can all be described by an nth-order linear recurrence sequence. To see this think of the powers of $\begin{pmatrix} 2 & 0 \\ 0 & 1 \end{pmatrix}$. We have already asked whether the trace can take infinitely many prime values. A recent question of interest is to take two (or more) such matrices M and N say, and then look at the entries of all "words" created by M and N, for example $M^a N^b M^c \cdots N^z$, and ask whether the entries are infinitely often prime (see section 9.15 of appendix 9D and appendix 12G for a beautiful example).

Guides to conjectures and the Green-Tao Theorem

[1] David Conlon, Jacob Fox, and Yufei Zhao, *The Green-Tao theorem: An exposition*, EMS Surv. Math. Sci. **1** (2014), 249–282.

---

[23]This was in fact the form in which Goldbach made his conjecture. Goldbach was a friend of Euler, arguably the greatest mathematician of the 18th century, and would often send Euler mathematical questions. In one letter Goldbach asked whether every integer > 1 is the sum of at most three primes, and Euler observed that this is equivalent to showing that every even number ≥ 4 is the sum of two primes. Why then does Goldbach get credit for this conjecture that he did not make? Perhaps because "*Euler is rich, and Goldbach is poor.*"

[2] G. H. Hardy and J. E. Littlewood, *Some problems of 'Partitio Numerorum'; III: On the expression of a number as a sum of primes*, Acta Math. **44** (1923), 1–70.

[3] Bryna Kra, *The Green-Tao theorem on arithmetic progressions in the primes: An ergodic point of view*, Bull. Amer. Math. Soc. **43** (2006), 3–23.

[4] James Maynard, *Small gaps between primes*, Annals Math. **181** (2015), 383–413.

[5] A. Schinzel and W. Sierpiński, *Sur certaines hypothèses concernant les nombres premiers*, Acta Arith. **4** (1958), 185–208; erratum **5** (1958), 259.

Appendix 5B. An important proof of infinitely many primes

5.12. Euler's proof of the infinitude of primes

In the 17th century Euler gave a different proof that there are infinitely many primes, one which would prove highly influential in what was to come later. Suppose again that the list of primes is $p_1 < p_2 < \cdots < p_k$. Euler observed that the fundamental theorem of arithmetic implies that the sets $\{n \geq 1 : n \text{ is a positive integer}\}$ and $\{p_1^{a_1} p_2^{a_2} \ldots p_k^{a_k} : a_1, a_2, \ldots, a_k \geq 0\}$ contain the same numbers. Therefore a sum involving the elements of the first set must equal the analogous sum involving the elements of the second set. In particular,

$$\sum_{\substack{n \geq 1 \\ n \text{ a positive integer}}} \frac{1}{n^s} = \sum_{a_1, a_2, \ldots, a_k \geq 0} \frac{1}{(p_1^{a_1} p_2^{a_2} \ldots p_k^{a_k})^s}$$

$$= \left(\sum_{a_1 \geq 0} \frac{1}{(p_1^{a_1})^s} \right) \left(\sum_{a_2 \geq 0} \frac{1}{(p_2^{a_2})^s} \right) \cdots \left(\sum_{a_k \geq 0} \frac{1}{(p_k^{a_k})^s} \right)$$

$$= \prod_{j=1}^{k} \left(1 - \frac{1}{p_j^s} \right)^{-1}.$$

The last equality holds because each sum in the second-to-last line is over a geometric progression. Euler proved that if we take $s = 1$, then the left-hand side becomes $\sum_{n \geq 1} \frac{1}{n}$ which diverges, as we saw in section 4.14 of appendix 4D, and the right-hand side equals a finite product of rational numbers, $\prod_{i=1}^{k} \frac{p_i}{p_i - 1}$, which is itself a rational (and so finite) number. This contradiction implies that there cannot be finitely many primes.

5.12. Euler's proof of the infinitude of primes

What is wonderful about Euler's formula is that something like it holds without assumption about the number of primes. So

$$\sum_{\substack{n \geq 1 \\ n \text{ a positive integer}}} \frac{1}{n^s} = \prod_{p \text{ prime}} \left(1 - \frac{1}{p^s}\right)^{-1} \tag{5.12.1}$$

holds whenever both sides are absolutely convergent, which we will now show occurs when $\text{Re}(s) > 1$. (This is the most special case of the general Dirichlet series studied in section 4.9 of appendix 4B, though we will now be more careful about these convergence issues.) For the left-hand side, if $s = \sigma + it$ with $\sigma > 1$, then the sum of the absolute values of the terms converge as

$$\sum_{n \geq 1} \left|\frac{1}{n^s}\right| = \sum_{n \geq 1} \frac{1}{n^\sigma} \leq 1 + \int_1^\infty \frac{dt}{t^\sigma} = 1 + \left[\frac{t^{1-\sigma}}{1-\sigma}\right]_1^\infty = 1 + \frac{1}{\sigma - 1} = \frac{\sigma}{\sigma - 1}.$$

Here we have used that $n^{-\sigma} < \int_{n-1}^n t^{-\sigma} dt$ since $t^{-\sigma}$ is a decreasing function in t (much as in the discussion in section 4.14 of appendix 4D of the more difficult case when $\sigma = 1$).

A product $\prod_{j \geq 1}(1 - a_j)$ with each $|a_j| < 1$ converges absolutely if $\sum_{j \geq 1} |a_j|$ converges. As each $|p^{-s}| = |p^{-\sigma}| < p^{-1} < 1$, we deduce that the Euler product in (5.12.1) converges absolutely as

$$\sum_{p \text{ prime}} \left|\frac{1}{p^s}\right| = \sum_{p \text{ prime}} \frac{1}{p^\sigma} \leq \sum_{n \geq 1} \frac{1}{n^\sigma} \leq \frac{\sigma}{\sigma - 1},$$

by the last displayed equation.

We have just seen that (5.12.1) makes sense when s is to the right of the horizontal line in the complex plane going through the point 1. Like Euler, we want to be able to interpret what happens to (5.12.1) when $s = 1$. To not fall afoul of convergence issues we need to take the limit of both sides as $s \to 1^+$, since (5.12.1) holds for real values of $s > 1$. Now $1/n^\sigma > \int_n^{n+1} dt/t^\sigma$ for each n, as $1/t^\sigma$ is a decreasing function and the integral is over an interval of length 1, and so

$$\sum_{n \geq 1} \frac{1}{n^\sigma} \geq \int_1^\infty \frac{dt}{t^\sigma} = \frac{1}{\sigma - 1}.$$

Then (5.12.1) implies that

$$\prod_{p \text{ prime}} \left(1 - \frac{1}{p^\sigma}\right) \leq \sigma - 1,$$

so that

$$\prod_{p \text{ prime}} \left(1 - \frac{1}{p}\right) = \lim_{\sigma \to 1^+} \prod_{p \text{ prime}} \left(1 - \frac{1}{p^\sigma}\right) \leq \lim_{\sigma \to 1^+} (\sigma - 1) = 0. \tag{5.12.2}$$

Upon taking logarithms this implies Euler's famous result that

$$\sum_{p \text{ prime}} \frac{1}{p} \text{ diverges.} \tag{5.12.3}$$

An explicit estimate for the sum in (5.12.3). It is useful to estimate the sum of $1/p$ over the primes $p \leq x$. One can do so using the (difficult to prove) prime number theorem, but one can also obtain the following good estimate from more elementary methods (see chapter 4 of [**Graa**]): There exists a constant c such that

$$(5.12.4) \qquad \lim_{x \to \infty} \left(\sum_{\substack{p \text{ prime} \\ p \leq x}} \frac{1}{p} - \log \log x \right) = c.$$

The difference between $\log \prod_{p \leq x}(1 - \frac{1}{p})^{-1}$ and the first term in its approximation, $\sum_{p \leq x} \frac{1}{p}$, is a sum of positive terms, which are $\leq \sum_p \frac{1}{p(p-1)} \leq \sum_{n \geq 2} \frac{1}{n(n-1)} = 1$. The difference therefore converges to a limit as $x \to \infty$. Exponentiating we deduce that there exists a constant γ such that

$$\prod_{p \leq x} \left(1 - \frac{1}{p}\right)^{-1} \sim \frac{e^{-\gamma}}{\log x},$$

an explicit improvement on (5.12.2). (This estimate appeared in section 5.2.)

Another use of (5.12.4) is to deduce that the number of steps in the sieve of Eratosthenes (as in section 5.2) is

$$\frac{x}{2} + \frac{x}{3} + \cdots + \frac{x}{p_k} = x \sum_{p \leq \sqrt{x}} \frac{1}{p} < x(\log \log x + C)$$

for any constant $C > c - \log 2$, once x is sufficiently large, where p_k is the largest prime $\leq \sqrt{x}$.

Another derivation of (5.12.1). One begins with $\sum_{n \geq 1} \frac{1}{n^s}$, the sum of $1/n^s$ over all integers $n \geq 1$. Now suppose that we wish to remove the even integers from this sum. Their contribution to this sum is

$$\sum_{\substack{n \geq 1 \\ n \text{ even}}} \frac{1}{n^s} = \sum_{m \geq 1} \frac{1}{(2m)^s} = \frac{1}{2^s} \sum_{m \geq 1} \frac{1}{m^s}$$

writing even n as $2m$, and hence

$$\sum_{\substack{n \geq 1 \\ (n,2)=1}} \frac{1}{n^s} = \sum_{n \geq 1} \frac{1}{n^s} - \sum_{\substack{n \geq 1 \\ n \text{ even}}} \frac{1}{n^s} = \left(1 - \frac{1}{2^s}\right) \sum_{n \geq 1} \frac{1}{n^s}.$$

If we wish to remove the multiples of 3, we can proceed similarly, to obtain

$$\sum_{\substack{n \geq 1 \\ (n,2\cdot 3)=1}} \frac{1}{n^s} = \left(1 - \frac{1}{2^s}\right)\left(1 - \frac{1}{3^s}\right) \sum_{n \geq 1} \frac{1}{n^s},$$

and for arbitrary y, letting $m = \prod_{p \leq y} p$,

$$\sum_{\substack{n \geq 1 \\ (n,m)=1}} \frac{1}{n^s} = \prod_{p \leq y}\left(1 - \frac{1}{p^s}\right) \cdot \sum_{n \geq 1} \frac{1}{n^s}.$$

5.13. The sieve of Eratosthenes and estimates for the primes up to x

As $y \to \infty$, the left side becomes the sum over all integers $n \geq 1$ which do not have any prime factors: The only such integer is $n = 1$ so the left-hand side becomes $1/1^s = 1$. Hence

$$\prod_{p \text{ prime}} \left(1 - \frac{1}{p^s}\right) \cdot \sum_{n \geq 1} \frac{1}{n^s} = 1,$$

an alternative formulation of (5.12.1). The advantage of this proof is that we see what happens when we "sieve" by various primes, that is, when we remove the integers from our set that are divisible by given primes.

Exercise 5.12.1. Show that if $\text{Re}(s) > 1$, then

$$\left(1 - \frac{1}{2^{s-1}}\right) \sum_{n \geq 1} \frac{1}{n^s} = \sum_{\substack{n \geq 1 \\ n \text{ odd}}} \frac{1}{n^s} - \sum_{\substack{n \geq 1 \\ n \text{ even}}} \frac{1}{n^s}.$$

Reference on Euler's many contributions

[1] Raymond Ayoub, *Euler and the zeta function*, Amer. Math. Monthly **81** (1974), 1067–1086.

5.13. The sieve of Eratosthenes and estimates for the primes up to x

Fix $\epsilon > 0$ to be an arbitrarily small positive constant. By (5.12.2) we know that there exists y such that

$$\prod_{p \leq y} \left(1 - \frac{1}{p}\right) < \frac{\epsilon}{3}.$$

Let m be the product of the primes $\leq y$, and select x to be sufficiently large, where $x > 3y/\epsilon$ is large enough. Any prime $> y$ must be coprime to m, the product of the primes $\leq y$, and so

$$\pi(x) - \pi(y) = \#\{\text{primes } p \in (y, x]\} \leq \sum_{\substack{n \leq x \\ (n,m)=1}} 1.$$

Obviously $\pi(y) \leq y$. We let $k = [x/m]$, so that $km \leq x < 2km$, and therefore, writing $n = jm + i$,

$$\sum_{\substack{n \leq x \\ (n,m)=1}} 1 \leq \sum_{\substack{n \leq 2km \\ (n,m)=1}} 1 = \sum_{j=0}^{2k-1} \sum_{\substack{1 \leq i \leq m \\ (jm+i,m)=1}} 1 = \sum_{j=0}^{2k-1} \sum_{\substack{1 \leq i \leq m \\ (i,m)=1}} 1 = 2k\phi(m).$$

We deduce that

$$\pi(x) \leq y + 2k\phi(m) < \frac{\epsilon}{3}x + 2km \prod_{p \leq y}\left(1 - \frac{1}{p}\right) < \frac{\epsilon}{3}x + 2x \cdot \frac{\epsilon}{3} = \epsilon x.$$

Since this holds for any $\epsilon > 0$ we deduce that

$$(5.13.1) \qquad \lim_{x \to \infty} \frac{1}{x} \#\{p \leq x : \ p \text{ prime}\} \to 0;$$

that is, a vanishing proportion of the integers are prime.

5.14. Riemann's plan for Gauss's prediction, I

In 1859, Riemann wrote a nine-page memoir that was to shape the future of number theory. This was his only paper in number theory, yet its ideas guided the study of the distribution of prime numbers from then on. Riemann proposed a plan to prove Gauss's guesstimate for the number of primes up to x, discussed in section 5.4. This involved moving the question from number theory to analysis via the theory of analytic continuation.

We define the Riemann zeta-function $\zeta(s)$ by

$$\zeta(s) := \sum_{n \geq 1} \frac{1}{n^s},$$

an infinite sum whose value is well-defined when the sum is absolutely convergent, that is, for $\text{Re}(s) > 1$ as discussed in section 5.12. Our starting point is Euler's formula (5.12.1), which connects the Riemann zeta-function to the set of primes:

$$\zeta(s) = \prod_{p \text{ prime}} \left(1 - \frac{1}{p^s}\right)^{-1}.$$

Taking logarithms of both sides, we get

$$\log \zeta(s) = \sum_{\substack{p \text{ prime} \\ m \geq 1}} \frac{1}{m(p^m)^s}.$$

The coefficient of $1/n^s$ is 0, unless n is a power of a prime p, say p^k, in which case the coefficient is $1/k$. In other words, if we let $a_n = 1/k$ if $n = p^k$ is a prime power, and 0 otherwise, be the characteristic function for the prime powers, then

$$\log \zeta(s) = \sum_{n \geq 1} \frac{a_n}{n^s}.$$

What we need now is a way to extract $\sum_{n \leq x} a_n$, the sum of the coefficients up to x of such a Dirichlet series, and this, Riemann realized, was provided by an idea of Perron, as long as one could analytically continue the Dirichlet series well to the left of the line $\text{Re}(s) = 1$. If one can do this, and this is a big "if", then one gets a formula for the number of prime powers up to x, in terms of the zeros of the analytic continuation of $\log \zeta(s)$, something we will discuss more in section 5.16 of appendix 5D.

Appendix 5C. What should be true about primes?

Gauss suggested (see section 5.4) that the density of primes *near* x should be about $1/\log x$, so that in an interval of length y around x, we expect that

(5.15.1) $\qquad \#\{p \text{ prime} : x < p \leq x+y\} \text{ is about } \dfrac{y}{\log x}.$

This makes no sense if y, say, is $\frac{1}{2}\log x$ as you can't have half a prime. Does it make sense for larger y? And is there a way to interpret Gauss's suggestion to better understand $\pi(x+y) - \pi(x)$ when $y = \frac{1}{2}\log x$?

5.15. The Gauss-Cramér model for the primes

The great probabilist Cramér interpreted Gauss's suggestion as the following model for the sequence of primes: Let $X_3, X_4, \ldots$ be an infinite sequence of independent random variables such that

$$\text{Prob}(X_n = 1) = \frac{1}{\log n} \quad \text{and} \quad \text{Prob}(X_n = 0) = 1 - \frac{1}{\log n}.$$

This defines a probability space on all infinite sequences of 0's and 1's. The indicator function for the odd primes is such a sequence:

$$\begin{array}{lcccccccccccccc}
 & 1 & 0 & 1 & 0 & 1 & 0 & 0 & 0 & 1 & 0 & 1 & 0 & 0 & 0 & 1 & 0\ldots \\
\text{corresponding to} & 3 & & 5 & & 7 & & & & 11 & & 13 & & & & 17 &
\end{array}$$

Cramér proposed that one should think of this as a "typical" sequence of 0's and 1's, under this probability measure, so that anything that can be said with probability 1 for the space of such sequences should be true for the sequence of primes. First notice that the expectation for the sum up to N, which we would guess would be a good approximation for the number of primes up to N, is

$$\mathbb{E}\left(\sum_{n=3}^{N} X_n\right) = \sum_{n=3}^{N} \mathbb{E}(X_n) = \sum_{n=3}^{N} \frac{1}{\log n}$$

which equals Li(x) within an error of 1, so that works! But it is more important to ask whether the sum is usually near to its expected value? We will prove that, for any fixed $\epsilon > 0$, then with probability $\to 1$ as $N \to \infty$, that

(5.15.2) $$\left|\sum_{n=3}^{N} X_n - \int_{3}^{N} \frac{dt}{\log t}\right| \leq N^{\frac{1}{2}+\epsilon}.$$

This fits very well with what we observed about the data for the sequence of primes. Moreover in the next appendix we will see that this being true about the primes is equivalent to the famous "Riemann Hypothesis". To prove (5.15.2) one calculates the *variance*, a key quantity in probability theory:

$$\mathbb{E}\left(\sum_{n=3}^{N} X_n - \sum_{n=3}^{N} \frac{1}{\log n}\right)^2 = \mathbb{E}\left(\sum_{n=3}^{N} X_n\right)^2 - \left(\sum_{n=3}^{N} \frac{1}{\log n}\right)^2.$$

The first term here equals

$$\sum_{n=3}^{N} \mathbb{E}(X_n^2) + \sum_{m=3}^{N} \mathbb{E}(X_m) \sum_{\substack{n=3 \\ m \neq n}}^{N} \mathbb{E}(X_n) = \sum_{n=3}^{N} \frac{1}{\log n} + \sum_{3 \leq m \neq n \leq N} \frac{1}{\log m} \cdot \frac{1}{\log n},$$

and therefore

$$\mathbb{E}\left(\sum_{n=3}^{N} X_n - \sum_{n=3}^{N} \frac{1}{\log n}\right)^2 = \sum_{n=3}^{N} \left(\frac{1}{\log n} - \frac{1}{(\log n)^2}\right),$$

which can be shown to be $< 2N/\log N$. Now, for any random variable Y, we have

$$\mathbb{E}(Y^2) = \int_{-\infty}^{\infty} t^2 \cdot d\mathbb{P}(Y > t) \geq T^2 \cdot \mathbb{P}(|Y| > T).$$

(Here, $\mathbb{P}$("event") means the probability that the "event" takes place.) Therefore, taking $Y = \sum_{n=3}^{N} X_n - \text{Li}(N)$ we deduce from the above that (5.15.2) fails with probability $< N^{-2\epsilon}$ once N is sufficiently large.

By modifying this argument one can prove that, for short intervals, for any fixed $\epsilon > 0$ we have, with probability $\to 1$ as $x \to \infty$,

$$\left|\sum_{n=x}^{x+y} X_n - \int_{x}^{x+y} \frac{dt}{\log t}\right| \leq y^{\frac{1}{2}+\epsilon},$$

and so (5.15.1) should be about right provided $y \geq (\log x)^{2+\epsilon}$, a rather short interval. However we are very far from proving anything like this for the primes themselves. A more careful analysis of this kind even suggests that

$$\pi(x+y) - \pi(x) \sim \frac{y}{\log x}$$

for 100% of the intervals $(x, x+y]$ with $x \leq X$ where $y = y(x)$ is a function of x for which $y/\log x \to \infty$ as $x \to \infty$. The "100%" here does not mean the same thing as "all". This is 100% in the sense that this estimate is true for a certain proportion of intervals up to X, and this proportion tends to 100% as $X \to \infty$.

Short intervals

For intervals of length $\frac{1}{2} \log x$, our calculation should be interpreted as meaning that at least half of such intervals contain no primes, a reasonable proportion of such intervals contain one or two primes, and very few contain a large number of primes, all so the average is $\frac{1}{2}$. This situation has been considered by probability theorists and gives rise to a *Poisson distribution*. It suggests that for any fixed real number $t > 0$ and any fixed integer $k \geq 0$ with probability that tends to 1 as $X \to \infty$,

$$\mathbb{P}\{(x, x + t \log x] \text{ where } x \leq X, \text{ contains exactly } k \text{ primes}\} \to \frac{e^{-t} t^k}{k!}.$$

This probability is maximized at the integers k closest to t, as one might guess.

One can also turn to the Gauss-Cramér model to guess how big y needs to be, in terms of x, to *guarantee* a prime in $(x, x + y]$ for every x. That is tantamount to asking for the largest gaps between consecutive primes. Now, if x and y are integers, then

$$\mathbb{P}(X_n = 0 \text{ for all } n \in (x, x+y]) = \prod_{n=x+1}^{x+y} \mathbb{P}(X_n = 0) = \prod_{n=x+1}^{x+y} \left(1 - \frac{1}{\log n}\right).$$

If y is not very large, then $\log n$ is very close to $\log x$, and so we can approximate this product by

$$\left(1 - \frac{1}{\log x}\right)^y \text{ which is roughly } e^{-y/\log x}.$$

Therefore if $X < x \leq 2X$ and y is "small", then the probability that $X_n = 1$ for some $n \in (x, x+y]$ is roughly $1 - e^{-y/\log X}$. In this context we want this to be very close to 1 (which implies that y is substantially larger than $\log X$). There are X intervals of the form $(x, x+y]$ with x an integer for which $X < x \leq 2X$, and so the probability that all of them contain at least one prime is therefore roughly

$$\left(1 - e^{-y/\log X}\right)^X \text{ which is roughly } e^{-X e^{-y/\log X}}.$$

This quantity is very close to 1 if y is a little bit bigger than $(\log X)^2$, and it is very close to 0 if y is a little bit smaller than $(\log X)^2$. Cramér developed this into a formal proof that with probability 1 as $X \to \infty$, if $n_1 < n_2 < n_3 < \cdots$ are the indices n for which $X_n = 1$, then

$$\max_{k:\ n_k \leq X} (n_{k+1} - n_k) \sim (\log X)^2.$$

Therefore Cramér conjectured that the largest gap between consecutive primes $p_n < p_{n+1}$ should be about $\log^2 p_n$. In other words, if $p_1 = 2 < p_2 = 3 < \cdots$ is the sequence of prime numbers, then

$$\max_{p_n \leq x} (p_{n+1} - p_n) \sim (\log x)^2.$$

Here are the record-breaking gaps, in terms of the ratio of these two quantities:

p_n	$p_{n+1} - p_n$	$(p_{n+1} - p_n)/\log^2 p_n$
113	14	.6264
1327	34	.6576
31397	72	.6715
370261	112	.6812
2010733	148	.7026
20831323	210	.7395
25056082087	456	.7953
2614941710599	652	.7975
19581334192423	766	.8178
218209405436543	906	.8311
1693182318746371	1132	.9206

(Known) record-breaking gaps between primes

We see from the data that the constant is slowly creeping upwards but will it ever reach 1? And will it go beyond? We don't know.

Twin primes

The Gauss-Cramér model suggests that the number of prime pairs $p, p+k$ with $p \leq N$ is about

$$\mathbb{E}\left(\sum_{n=3}^{N} X_n X_{n+k}\right) = \sum_{n=3}^{N} \mathbb{E}(X_n)\,\mathbb{E}(X_{n+k}) = \sum_{n=3}^{N} \frac{1}{(\log n)\log(n+k)}$$

which is $\sim \frac{N}{(\log N)^2}$. One gets the same number whether $k=1$ (when there is just one such prime pair) or $k=2$ (when we expect infinitely many), so the model gives an incorrect prediction, because it does not take account of divisibility by 2. We can modify the Gauss-Cramér model to take account of divisibility by "small" primes. When asking questions about primes around x we will take $X_n = 0$ if n has a prime factor $\leq z$ for an appropriate "small" number z. This new model will not yield the same prediction for the number of primes up to x, as above, unless, when n has no prime factor $\leq z$, we now have

$$\text{Prob}(X_n = 1) = \frac{\kappa}{\log n} \quad \text{and} \quad \text{Prob}(X_n = 0) = 1 - \frac{\kappa}{\log n},$$

where $\kappa = \kappa(z) = \prod_{p \leq z} \frac{p}{p-1}$. With this modified model we avoid pairs $X_n = X_{n+k} = 1$ with k odd and small, and we are even able to predict that the number of prime pairs $p, p+k \leq N$ with k even should be about

$$\mathbb{E}\left(\sum_{n=3}^{N} X_n X_{n+k}\right) = \sum_{\substack{n=3 \\ (n(n+k),m)=1}}^{N} \frac{\kappa}{\log n} \cdot \frac{\kappa}{\log(n+k)} \sim C_k \frac{N}{(\log N)^2}$$

where $m = \prod_{p \leq z} p$ and $C_k = 2 \prod_{p \text{ an odd prime}} \left(1 - \frac{1}{(p-1)^2}\right) \cdot \prod_{p | k,\ p \text{ odd}} \frac{p-1}{p-2}$.

We can also use this model to predict the number of pairs of primes p, q for which $p + q = N$ in Goldbach's conjecture, which should be about

$$\mathbb{E}\left(\sum_{n=3}^{N-3} X_n X_{N-n}\right) = \sum_{\substack{n=3 \\ (n(N-n),m)=1}}^{N-3} \frac{\kappa}{\log n} \cdot \frac{\kappa}{\log(N-n)} \sim C_N \frac{N}{(\log N)^2}.$$

Computations suggest that these predictions are about right.

Appendix 5D. Working with Riemann's zeta-function

In section 5.4 the data led us to believe that $\pi(x)$, the number of primes up to x, is best approximated by the complicated integral $\text{Li}(x) = \int_2^x \frac{dt}{\log t}$. This is not an easy function to work with (see exercises 5.8.11 and 5.18.2). If, instead, we count the primes with the weight $\log p$ attached to prime p, then the expected value of

$$\sum_{p \leq x} \log p \quad \text{is} \quad \sum_{n=2}^{x} \log n \cdot \frac{1}{\log n} = x - 1,$$

a far easier function to manipulate. If one can estimate the weighted sum well, then one can deduce good estimates for $\pi(x)$ by a mathematical technique called *partial summation* (see the book [**Graa**]).

5.16. Riemann's plan for Gauss's prediction

Riemann came up with the most surprising plan to try to verify Gauss's prediction, by using complex analysis, that is, the theory of calculus in the complex plane.[24] Riemann begins with an extraordinary identity (due to Perron): For any "reasonably behaved" sequence of complex numbers $\{a_n\}_{n \geq 1}$ we have

$$\sum_{n \leq x} a_n = \frac{1}{2i\pi} \int_{2-i\infty}^{2+i\infty} \sum_{n \geq 1} \frac{a_n}{n^s} \frac{x^s}{s} ds.$$

Here the integral is along the straight line going from $2 - i\infty$ to $2 + i\infty$. The key point to appreciate is that in order to evaluate the finite sum on the left, we use the infinite integral on the right, which involves a Dirichlet series which is an infinite sum, an identity that seems only to (vastly) complicate matters. *However* the integral on the right can be approached with the tools of calculus and sometimes

---

[24]This is not the place to discuss this intriguing theory in detail but we will use from it, and elucidate, the ideas that we need to describe Riemann's plan.

5.16. Riemann's plan for Gauss's prediction

this yields a rich dividend. If we wish to count the primes with the log weight suggested above, then the integrand on the right-hand side of our identity involves the Dirichlet series
$$\sum_{p \text{ prime}} \frac{\log p}{p^s},$$
where the only non-zero coefficients of $1/n^s$ occur when n is a prime. This is more or less $-\frac{\zeta'(s)}{\zeta(s)}$ which was discussed in section 4.11 of appendix 4B, the only difference being that this also has non-zero coefficients when n is a prime power p^m with $m \geq 2$ (extra terms which will have little impact on our discussion.) So $-\frac{\zeta'(s)}{\zeta(s)}$ is fit for Riemann's purpose.[25]

The sum in the Dirichlet series for $\zeta(s) = \sum_{n \geq 1} 1/n^s$ is absolutely convergent only when $\text{Re}(s) > 1$. The theory of analytic continuation allows one to extend the function to the whole complex plane so that $(s-1)\zeta(s)$ is differentiable, arbitrarily often, everywhere in $\mathbb{C}$.[26] It is too complicated to go into all of the details here,[27] but it leads to Riemann's amazing exact formula

$$(5.16.1) \qquad \sum_{\substack{p \text{ prime} \\ m \geq 1 \\ p^m \leq x}} \log p = x - \sum_{\rho: \zeta(\rho)=0} \frac{x^\rho}{\rho} - \frac{\zeta'(0)}{\zeta(0)},$$

relating the number of prime powers up to x to the zeros, ρ, of the analytic continuation of $\zeta(s)$. This is a surprising thing to do. We take a nice elementary question, the count of the primes up to x, and relate it to some sum over the (infinite set of) zeros of an analytic continuation. What is great is that the estimate x, expected for the left-hand side, pops out of the calculation. But in order to show that this is really a good estimate for our sum over primes, we will need to establish that the sum over the zeros ρ on the right-hand side of (5.16.1) is small compared to x.

How big is each $\frac{x^\rho}{\rho}$ term? The key point is that $|x^\rho| = x^{\text{Re}(\rho)}$ so the smaller the real parts of the zeros ρ, the smaller the x^ρ/ρ terms in comparison to x, especially as x grows. Indeed, by establishing that each $\text{Re}(\rho) < 1$, Hadamard and de la Vallée Poussin both proved the prime number theorem in 1896. Riemann himself calculated a few zeros of $\zeta(s)$ and all the real parts seemed to be $1/2$, which led him to his famous conjecture:[28]

Conjecture 5.1 (The Riemann Hypothesis).
If $\zeta(\rho) = 0$ with $0 \leq \text{Re}(\rho) \leq 1$, then $\text{Re}(\rho) = \frac{1}{2}$.

---

[25]Riemann himself used $\log \zeta(s)$, which also only has non-zero coefficients when n is a prime power. Today people prefer to use $-\zeta'(s)/\zeta(s)$ since it makes the calculus significantly easier.

[26]Euler was able to determine the values of $\zeta(0), \zeta(-1), \zeta(2), \ldots$; these values have no obvious connection to the series that defines $\zeta(s)$ when $\text{Re}(s) > 1$. For example, if $s = -1$, one has the series, as $1/n^{-1} = n$ for each integer $n \geq 1$,
$$1 + 2 + 3 + \cdots, \quad \text{which is evidently not the same as} \quad \zeta(-1) = -\frac{1}{12}.$$
In fact $\zeta(1-n) = -B_n/n$ where B_n is the nth Bernoulli number, for each integer $n \geq 2$.

[27]Though see [**Dav80**] for a wonderful introduction, expanded on in my book [**Graa**].

[28]Riemann wrote: "It is very probable that all roots are [of the form $\frac{1}{2} + it$ with t] real. Certainly one would wish for a stricter proof ... I have meanwhile temporarily put aside the search for this after some fleeting futile attempts."

If the Riemann Hypothesis is true, then each $|x^\rho| = x^{1/2}$, from which one can deduce that if $x \geq 3$, then

$$\left|\sum_{p \leq x} \log p - x\right| \leq x^{1/2}(\log x)^2.$$

This in turn implies that

$$|\pi(x) - \text{Li}(x)| \leq x^{1/2} \log x.$$

The implications go two ways. If these bounds are true, then the Riemann Hypothesis is true!

We still do not know whether the Riemann Hypothesis is true, but the available evidence points towards it. For example we know that the 10^{13} zeros nearest to, but not on, the x-axis all lie on the line $\text{Re}(s) = \frac{1}{2}$. The Riemann Hypothesis is a famously difficult question and appears on every list of famous problems in mathematics.[29] It is said that the person who resolves the Riemann Hypothesis will become immortal, literally.[30]

Exercise 5.16.1. Prove that $\zeta(s) = 0$ has no zeros ρ with $\text{Re}(\rho) > 1$.

The key to all of this is Riemann's extraordinary identity (5.16.1). Where does it come from? How should we think about it? We will address these questions in the next section.

5.17. Understanding the zeros

Fourier analysis is built on the idea that most functions on $[0, 1)$ can be given as an (infinite) sum of sines and cosines. The first interesting example is given by

$$x - \frac{1}{2} = -2 \sum_{n=1}^{\infty} \frac{\sin(2\pi n x)}{2\pi n},$$

which can be shown to hold whenever $0 < x < 1$. This representation of $x - \frac{1}{2}$ by an infinite sum is not something one can calculate term by term in practice, and so one hopes to approximate $x - \frac{1}{2}$ by the sum of the first few terms of the series. The first term $-\frac{1}{\pi}\sin(2\pi x)$ does not provide a good approximation but even the sum of the first two, $-\frac{1}{\pi}\sin(2\pi x) - \frac{1}{2\pi}\sin(4\pi x)$, gives the general shape of $x - \frac{1}{2}$. By the time we add the sine terms for n from 1 to 100, we get a very good approximation.

---

[29]Sometimes with big financial rewards for resolving it, though there are easier ways to make a million dollars!

[30]Those who have made the biggest leaps in our understanding, Hadamard and de la Vallée Poussin, lived to 97 and 95, respectively. Most important contributors have lived to a ripe old age.

5.17. Understanding the zeros

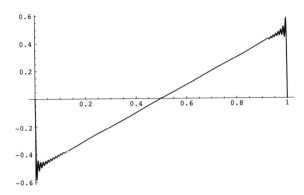

In our formula for $x - \frac{1}{2}$, the numbers $2\pi n$ (inside the "$\sin(2\pi nx)$") are the *frequencies* of the various component waves (controlling how fast they go through their cycles), while the coefficients $-2/2\pi n$ are their *amplitudes* (controlling how far up and down they go).

Riemann's formula (5.16.1) is a bit too technical to fully appreciate. However it can be simply, albeit rather surprisingly, rephrased colloquially in the following terms:

The primes can be counted as a sum of waves.

Moreover it will be significantly easier to understand, not to say more elegant,[31] if we assume that the Riemann Hypothesis is true. With this assumption we can write the difference between $\pi(x)$ and $\text{Li}(x)$ as a (suitably weighted) sum of sine waves:

$$(5.17.1) \quad \frac{\int_2^x \frac{dt}{\log t} - \#\{\text{primes} \leq x\}}{\sqrt{x}/\log x} \approx 1 + 2 \sum_{\substack{\text{All real numbers } \gamma > 0 \\ \text{such that } \frac{1}{2} + i\gamma \\ \text{is a zero of } \zeta(s)}} \frac{\sin(\gamma \log x)}{\gamma}.$$

The numerator of the left-hand side of this formula is the error term when comparing the Gauss prediction $\text{Li}(x)$ with the actual count $\pi(x)$ for the number of primes up to x. We saw earlier that the overcounts seemed to be roughly the size of the square root of x, so the denominator $\sqrt{x}/\log x$ appears to be an appropriate thing to divide through by. The right side of the formula bears much in common with our formula for $x - \frac{1}{2}$. It is a sum of sine functions, with the numbers γ employed in two different ways in place of $2\pi n$: Each γ is used inside the sine (as the "frequency"), and the reciprocal of each γ forms the coefficient of the sine (as the "amplitude"). We even get the same factor of 2 in each formula. However, the numbers γ here are much more subtle than the straightforward numbers $2\pi n$ in the corresponding formula for $x - \frac{1}{2}$.

---

[31] The aesthete believes the Riemann Hypothesis because its consequences, like this formula, are too beautiful and well-fitting not to be true!

We can obtain approximations for the number of primes up to x by evaluating the right-hand side of this formula by summing over the first ten or hundred zeros $\frac{1}{2}+i\gamma$ of $\zeta(s)=0$ (or more accurately, using the formula in (5.16.1)). Here we can order the zeros by the size of the γ. The first few zeros give

$$\gamma_1 = 14.135\ldots;\quad \gamma_2 = 21.022\ldots;\quad \gamma_3 = 25.011\ldots;\quad \gamma_4 = 30.425\ldots.$$

(These do not seem to be rationals nor numbers that can easily be identified in some other context.) For example we approximate the left-hand side of (5.16.1) for x up to 20 by using the first ten zeros, then the first 100, and finally the first 1000 zeros.

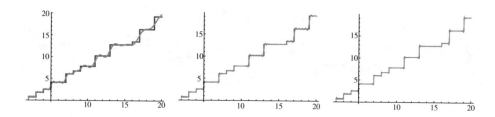

Figure 5.1. Using Riemann's zeros to approximate the count of primes

The step function in these graphs is the actual function on the left-hand side of (5.16.1), the more complicated function, constructed of waves, is the sum of the first few terms of the right-hand side. By the time we use a thousand zeros the function and its approximation are indistinguishable to the naked eye.

Riemann's paper and the subsequent observations gave birth to the subject known today as *analytic number theory*. There are many books on the subject, including [**Graa**], in which we will develop these and other ideas so as to be better able to count primes.

An elementary proof

Riemann's formula (5.16.1) shows that estimates for the number of primes up to x and understanding the location of the zeros of $\zeta(s)$ are more or less tautologous questions. This led to a widespread belief that there could not be an "elementary proof" of the prime number theorem, one which avoids the zeros of $\zeta(s)$. It therefore came as a great shock when, in 1949, Selberg and Erdős gave elementary (but complicated) proofs. At the heart is Selberg's elementary estimate for the number of (suitably weighted) integers up to x that are the product of at most two primes:

$$\left| \sum_{\substack{p \leq x \\ p\text{ prime}}} (\log p)^2 + \sum_{\substack{pq \leq x \\ p,q\text{ primes}}} (\log p)(\log q) - 2x\log x \right| \leq Cx,$$

for some constant $C > 0$.

5.18. Reformulations of the Riemann Hypothesis

Each of the following estimates are equivalent to the Riemann Hypothesis. Fix $\epsilon > 0$. If N is sufficiently large, then

- $|\log(\text{lcm}[1, 2, \ldots, N]) - N| \leq N^{1/2+\epsilon}$,
- $\left|\sum_{n \leq N} \mu(n)\right| < N^{1/2+\epsilon}$,
- $|\#\{n \leq N : \Omega(n) \text{ is even}\} - \#\{n \leq N : \Omega(n) \text{ is odd}\}| < N^{1/2+\epsilon}$,

where $\Omega(n)$ denotes the number of prime factors of n, counting multiplicities. The connections can be seen through the following exercise:

Exercise 5.18.1. (a) Prove that $\log(\text{lcm}[1, 2, \ldots, N]) = \sum_{p^m \leq N} \log p$.
(b)† Use (4.11.1) to show that $\sum_{p^m \leq N} \log p = \sum_{ab \leq N} \mu(b) \log a$.
(c) Express $\mu(n)$ in terms of $\Omega(n)$ and $\omega(n)$.

A key difficulty in counting primes lies in their definition, given in terms of what they are *not* (they *are not* the product of two integers > 1) rather than in terms of what they are. The advantage of the formulation in exercise 5.18.1(b) in terms of sums of the Möbius function is that this is a multiplicative function, built up constructively and so lends itself more naturally to elementary manipulations. This leads to an alternative approach to analytic number theory without deep complex analysis (see [**GS**]).

Exercise 5.18.2. For any integer $m \geq 1$:
(a) Prove that there exists a constant c_m such that if $x \geq 2$, then
$$\int_2^x \frac{dt}{(\log t)^m} = \frac{x}{(\log x)^m} - c_m + m \int_2^x \frac{dt}{(\log t)^{m+1}}.$$
(b) Prove that there exists a constant C_m such that if $x \geq 2$, then
$$\text{Li}(x) = \sum_{k=0}^{m-1} \frac{k!x}{(\log x)^{k+1}} - C_m + m! \int_2^x \frac{dt}{(\log t)^{m+1}}.$$
(c)† Prove that there exists a constant κ_m such that if $x \geq 3$, then
$$0 \leq \int_2^x \frac{dt}{(\log t)^{m+1}} \leq \frac{\kappa_m x}{(\log x)^{m+1}}.$$

Exercise 5.18.3. (a) Prove that $\overline{n^\rho} = n^{\overline{\rho}}$ for any integer n and $\rho \in \mathbb{C}$.
(b) Explain why if $\zeta(\rho) = 0$, then $\zeta(\overline{\rho}) = 0$.
(c) Show that if $\rho = \frac{1}{2} + i\gamma$, then
$$\frac{x^\rho}{\rho} + \frac{x^{\overline{\rho}}}{\overline{\rho}} = x^{1/2} \cdot \frac{\cos(\gamma \log x) + 2\gamma \sin(\gamma \log x)}{\frac{1}{4} + \gamma^2}.$$
(d) Show that if γ is large, then the expression in (c) is roughly $x^{1/2} \cdot \frac{2\sin(\gamma \log x)}{\gamma}$.
This exercise explains how (5.16.1) yields the approximation (5.17.1).

Primes and complex analysis

[1] Brian J. Conrey, *The Riemann hypothesis*, Notices Amer. Math. Soc. **50** (2003), 341–353.

Appendix 5E. Prime patterns: Consequences of the Green-Tao Theorem

In 2008 Green and Tao proved that there are infinitely many k-term arithmetic progressions of primes, that is, non-zero integers a, d such that

$$a, a+d, \ldots, a+(k-1)d$$

are all prime. The smallest arithmetic progression of ten primes is given by

$$199, 409, 619, 829, 1039, 1249, 1459, 1669, 1879, 2089,$$

which we can write as $199 + 210n$, $0 \leq n \leq 9$.

Length k	Arithmetic Progression ($0 \leq n \leq k-1$)	Last Term
3	$3 + 2n$	7
4	$5 + 6n$	23
5	$5 + 6n$	29
6	$7 + 30n$	157
7	$7 + 150n$	907
8	$199 + 210n$	1669
9	$199 + 210n$	1879
10	$199 + 210n$	2089
11	$110437 + 13860n$	249037
12	$110437 + 13860n$	262897
13	$4943 + 60060n$	725663
14	$31385539 + 420420n$	36850999
15	$115453391 + 4144140n$	173471351
16	$53297929 + 9699690n$	198793279
17	$3430751869 + 87297210n$	4827507229
18	$4808316343 + 717777060n$	17010526363
19	$8297644387 + 4180566390n$	83547839407
20	$214861583621 + 18846497670n$	572945039351
21	$5749146449311 + 26004868890n$	6269243827111

The k-term arithmetic progression of primes with smallest last term.

Despite there being arbitrarily long arithmetic progression of primes, it is not easy to find long ones. The longest explicitly known is the first 26 terms of

$$3486107472997423 + 371891575525470n.$$

Green and Tao proved that the smallest k-term arithmetic progressions of primes are all

$$< 2^{2^{2^{2^{2^{2^{2^{2^{100k}}}}}}}},$$

although we might guess that $\leq k! + 1$ is true, for each $k \geq 3$.

5.19. Generalized arithmetic progressions of primes

There are squares filled with primes, which are in arithmetic progression, when one looks along any row or any column like

5	17	29
47	59	71
89	101	113

29	41	53
59	71	83
89	101	113

503	1721	2939	4157
863	2081	3299	4517
1223	2441	3659	4877
1583	2801	4019	5237

Are there such squares of arbitrary size? The entries of such an N-by-N square can be parametrized as

$$a + mb + nc \quad \text{with } 0 \leq m, n \leq N - 1.$$

This is a 2-dimensional generalization of an arithmetic progression. How about three dimensions?

47	383	719
179	431	683
311	479	647

149	401	653
173	347	521
197	293	389

251	419	587
167	263	359
83	107	131

The three layers of a 3-by-3-by-3 Balog cube of primes.

The arithmetic progressions in this cube run in three directions, along each row and along each column and up through the layers. For example the top left entries of each layer, 47, 149, 251, are in arithmetic progression, as are the primes in the center of each layer, 431, 347, 263.

What about cubes of primes of higher dimension? The $(n_1, n_2, \ldots, n_d)$ entry of an N-by-N-by-$\cdots$-by-N cube, in d-dimensions, where each n_j lies in the range $0 \leq n_j \leq N - 1$ is given by

$$a + n_1 b_1 + n_2 b_2 + \cdots + n_d b_d,$$

for some integers $a, b_1, \ldots, b_d$. If we let each $b_i = N^{i-1} q$ for some integer q, then the entries are

$$a + nq \quad \text{where } 0 \leq n \leq k - 1$$

writing $n = n_1 + n_2 N + n_3 N^2 + \cdots + n_d N^{d-1}$ in base N.

The Green-Tao Theorem states there exist k-term arithmetic progressions of primes, $a + nq$, $0 \leq n \leq k - 1$, and so this gives rise to an N-by-$\cdots$-by-N Balog cube of primes by the above construction.

Consecutive prime values of a polynomial

Arithmetic progressions $a + nd$, $n = 1, 2, \ldots$, can be viewed as the values of a degree-one polynomial. Hence the Green-Tao Theorem can be rephrased as stating that for any k there are infinitely many different degree-one polynomials such that their first k values are prime.

How about degree-two polynomials? A famous example is the infamous quadratic polynomial $n^2 + n + 41$, which is prime for $n = 0, 1, \ldots, 39$ (which we will study in detail in section 12.5); there is no other monic quadratic polynomial that has such a long run of primes compared to the size of its coefficients. Other examples of quadratic polynomials known to have a long run of initial prime values are $36n^2 - 810n + 2753$ and $36n^2 - 2358n + 36809$, both of which give distinct primes for $n = 0, 1, \ldots, 44$. Are there quadratic polynomials whose first k values are prime, for any given integer $k \geq 1$? And if so, are there degree d polynomials whose first k values are prime? We resolve these questions by using the Green-Tao Theorem:

We begin with a k^d-term arithmetic progression of primes,

$$a + jb \text{ for every integer } j \text{ in the range } 0 \leq j \leq k^d - 1.$$

Then $a + bn^d$ is prime for every integer n in the range $0 \leq n \leq k - 1$. This restated gives:

The first k values of the polynomial $bx^d + a$ are all prime.

This technique does not yield monic polynomials, since we cannot control the value of b. To achieve this we need the extra power of the Green-Tao-Ziegler Theorem (Theorem 5.8). Consider the set of k linear forms

$$b + a + 1, \ b + 2a + 2^2, \ b + 3a + 3^2, \ \ldots, \ b + ka + k^d.$$

Exercise 5.19.1. Show that if $i \neq j$ are integers and a and b are variables, then there do not exist integers u, v, w, not all zero, for which $u(b + ia + i^2) + v(b + ja + j^2) = w$.

We can show there is no fixed prime divisor: For each prime p, select $a \equiv a_p \equiv -1 \pmod{p}$ so that $b + ja + j^2 \equiv b + j^2 - j \pmod{p}$. Now

$$S_p := \{j - j^2 \pmod{p} : j \in \mathbb{Z}\} = \{j - j^2 \pmod{p} : j = 0, \ldots, p - 1\}$$

by Corollary 2.3.1. Since $j - j^2 \equiv 0 \pmod{p}$ for $j \equiv 0$ and $1 \pmod{p}$, the set S_p contains $\leq p - 1$ distinct elements, and let $b \equiv b_p \pmod{p}$ where b_p is a residue that is not in S_p. Therefore $b + an + n^2 \not\equiv 0 \pmod{p}$ for all integers n, $1 \leq n \leq k$.

We can therefore deduce from the Green-Tao-Ziegler Theorem that there exist infinitely many pairs of integers a and b such that:

The first k values of the polynomial $x^2 + ax + b$ are all prime.

Exercise 5.19.2. Prove that there exist infinitely many pairs of integers a and b such that the first k values of the polynomial $x^d + ax + b$ are all prime.

Magic squares of primes

We discussed constructing magic squares in section 1.13 of appendix 1C. Here are two small examples, whose entries are primes.

17	89	71
113	59	5
47	29	101

41	89	83
113	71	29
59	53	101

Examples of 3-by-3 magic squares of primes.

Do you recognize the primes involved? Do you notice any similarities with the examples of 3-by-3 squares of primes above?

Exercise 5.19.3. Prove that every 3-by-3 square of integers in arithmetic progressions along each row and column can be rearranged to form a 3-by-3 magic square and vice versa.

37	83	97	41
53	61	71	73
89	67	59	43
79	47	31	101

41	71	103	61
97	79	47	53
37	67	83	89
101	59	43	73

Examples of 4-by-4 magic squares of primes.

It has long been known that there are n-by-n normal magic squares for any $n \geq 3$. We will take one, whose entries are $m_{i,j}$, $1 \leq i, j \leq n$, the distinct integers between 1 and n^2. The square with (i,j)th entry $a + m_{i,j}b$ is also an n-by-n magic square (exercise 1.13.1). The Green-Tao Theorem implies that there are infinitely many pairs of integers a, b for which all of the integers $a + \ell b$, $1 \leq \ell \leq n^2$, are prime; in particular, each $a + m_{i,j}b$ is prime. This then yields infinitely many n-by-n magic squares of primes for all integers $n \geq 3$.

Primes as averages

Balog showed that there exist arbitrarily large sets A of distinct primes such that for any $a, b \in A$ the average $\frac{a+b}{2}$ is also prime (and all of these averages are distinct):

Exercise 5.19.4. Prove that the averages of any two distinct elements of the set $2, 2^2, 2^3, \ldots, 2^m$ are distinct.

The Green-Tao Theorem implies that there are infinitely many pairs of integers a, b for which all of the integers $a + nb$, $1 \leq n \leq k = 2^m$, are prime, and so let

$$A = \{a + 2b, a + 4b, a + 8b, \ldots, a + 2^k b\}.$$

Exercise 5.19.5. Prove that the averages of any two distinct elements of A are distinct and prime.

Exercise 5.19.6.‡ Prove that there exist arbitrarily large sets A of primes such that the average of any subset of A yields a distinct prime (e.g $\{7, 19, 67\}$, $\{5, 17, 89, 1277\}$ and $\{209173, 322573, 536773, 1217893, 2484733\}$).

Appendix 5F. A panoply of prime proofs

The main idea in Euclid's proof is, given a list of primes $p_1, \ldots, p_k$, to find an integer $q > 1$ that is coprime to $p_1 \cdots p_k$ and so its prime factors are not in that list. Euclid took $q = p_1 \cdots p_k + 1$ but he might have taken q to be $p_1 \cdots p_k - 1$, or $mp_1 \cdots p_k + 1$ for any integer $m \neq 0$.

We could also split our list of primes into any two subsets, $\mathcal{M} \cup \mathcal{N}$, and let $q = m + n$ where $m = \prod_{p \in \mathcal{M}} p$ and $n = \prod_{\ell \in \mathcal{N}} \ell$.

Exercise 5.20.1. Show that $(q, mn) = 1$ and deduce that q has a prime factor not on our list.

One could take $q = |m - n|$, as long as this is not 1.[32] One can have more than two summands: If $N = p_1 \cdots p_k$, let $q = \sum_{i=1}^{k} N/p_i$. Now p_j divides N/p_i whenever $i \neq j$, so that $(q, p_j) = (N/p_j, p_j) = 1$.

In 2017 Meštrović gave a nice new proof: Suppose that q is the product of the finite set of odd primes, so that $q \geq 5$. Now $q - 2$ is odd but cannot be divisible by an odd prime (or else it would divide the difference, 2, between q and $q - 2$), and therefore $q - 2 = 1$; that is, $q = 3$, a contradiction.

Furstenberg's (point-set) topological proof

One of the most elegant ways to present Euclid's idea is in Furstenberg's extraordinary proof using basic notions of point-set topology:

Define a topology on the set of integers $\mathbb{Z}$ in which a set S is open if it is empty or if for every $a \in S$ there is an arithmetic progression

$$\mathbb{Z}(a, m) := \{a + nm : n \in \mathbb{Z}\},$$

---

[32]There are a couple of examples known where $m - n = 1$. Most famously, at least for baseball afficionados, Babe Ruth's record of $714 = 2 \times 3 \times 7 \times 17$ home runs was overtaken by Hank Aaron's $715 = 5 \times 11 \times 13$.

with $m \neq 0$, which is a subset of S. Obviously each $\mathbb{Z}(a, m)$ is open, and it is also closed since
$$\mathbb{Z}(a, m) = \mathbb{Z} \setminus \bigcup_{b:\ 0 \leq b \leq m-1,\ b \neq a} \mathbb{Z}(b, m).$$
If there are only finitely many primes p, then $A = \bigcup_p \mathbb{Z}(0, p)$ is also closed, and so $\mathbb{Z} \setminus A = \{-1, 1\}$ is open, but this is false since $\{-1, 1\}$ is finite and so cannot contain any arithmetic progression $\mathbb{Z}(1, m)$, as this would contain infinitely many integers. This contradiction implies that there are infinitely many primes. □

Some love the surprising sparse elegance of this proof. However, others dislike the way it obscures what is really going on.

An analytic proof

We count the positive integers up to x whose prime factors come only from a given set of primes $\mathcal{P} = \{p_1 < p_2 < \cdots < p_k\}$. These integers all take the form

(5.20.1) $\qquad p_1^{e_1} p_2^{e_2} \cdots p_k^{e_k}$ for some integers e_j, each ≥ 0.

We are going to count the integers up to $x = 2^m - 1$, for an arbitrary integer $m \geq 1$: For each j, the prime $p_j \geq 2$, and every other $p_i^{e_i} \geq 1$, so that
$$2^{e_j} \leq p_j^{e_j} \leq p_1^{e_1} p_2^{e_2} \cdots p_k^{e_k} \leq 2^m - 1.$$
This implies that e_j is at most $m - 1$, and so there are at most m possibilities for the integer e_j, the integers from 0 through to $m - 1$. Therefore the number of integers of the form (5.20.1), up to $2^m - 1$, is
$$\leq \prod_{j=1}^k \#\{\text{integers } e_j : 0 \leq e_j \leq m - 1\} = m^k.$$
Now if $\mathcal{P}$ (which contains $2, 3, 5, 7, 11$ so that $k \geq 5$) is the set of all primes, then every positive integer is of the form (5.20.1) and therefore the last equation implies that $2^m - 1 \leq m^k$ for all integers m. We select $m = 2^k$, so that $2^k \leq k^2$, which is false since $k \geq 5$. We deduce that there cannot be finitely many primes. □

This proof highlights counting arguments, the basis of analytic number theory.

A proof by irrationality

Euler exhibited the inspiring identity
$$\frac{\pi}{4} = 1 - \frac{1}{3} + \frac{1}{5} - \frac{1}{7} + \frac{1}{9} - \cdots.$$
Let $\delta(n) = 1$ or -1 as $n \equiv 1$ or $-1 \pmod 4$, and let $\delta(n) = 0$ if $n \equiv 0 \pmod 2$. This is a multiplicative function and so, by (4.9.1), we have
$$\frac{\pi}{4} = \sum_{n \geq 1} \frac{\delta(n)}{n} = \prod_{\substack{p \text{ prime} \\ p \equiv 1 \pmod 4}} \left(1 - \frac{1}{p}\right)^{-1} \cdot \prod_{\substack{p \text{ prime} \\ p \equiv 3 \pmod 4}} \left(1 + \frac{1}{p}\right)^{-1}.$$
It is well known that π (and so $\pi/4$) is irrational, but under the assumption that there are only finitely many primes, the right-hand side is a finite product of rational numbers so is itself rational, a contradiction.

Appendix 5G. Searching for primes and prime formulas

5.21. Searching for prime formulas

Proposition 5.7.1 proves that there is no (non-constant) polynomial that takes only prime values, and exercise 5.7.1 says the same thing for polynomials in more than one variable. But perhaps there is a more exotic formula than mere polynomials, which yields only primes? Earlier we discussed the Fermat numbers, $2^{2^n}+1$, which Fermat had mistakenly believed to all be prime, but maybe there is some other formula? One intriguing possibility stems from the fact that

$$2^2-1,\ 2^{2^2-1}-1,\ 2^{2^{2^2-1}-1}-1,\ \text{and}\ 2^{2^{2^{2^2-1}-1}-1}-1$$

are all prime. Could every term in this sequence be prime? No one knows and the next example is so large that no one will be able to determine whether or not it is prime in the foreseeable future. (Draw lessons on the power of computation from this example!)

With a little imagination it is not so difficult to develop formulas that easily yield all of the primes. For example if $p_1 = 2 < p_2 = 3 < \cdots$ is the sequence of primes, then define

$$\alpha = \sum_{m\geq 1} \frac{p_m}{10^{m^2}} = .200300005000000700000000011\ldots.$$

One can read off the primes from the decimal expansion of α, the mth prime coming from the few digits to the left of the m^2th digit, or, more formally,

$$p_m = [10^{m^2}\alpha] - 10^{2m-1}[10^{(m-1)^2}\alpha].$$

Is α truly interesting? If one could easily describe α, other than by the definition that we gave, then it might provide an easy way to determine the primes. But with its artificial definition it does not seem like it can be used in any practical way. There are other such constructions.

In a rather different vein, Matijasevič, while working on Hilbert's tenth problem, discovered that there exist polynomials f in many variables, such that the set of positive values taken by f when each variable is set to be a non-negative integer is precisely the set of primes.[33] One can find many different polynomials for the primes; we will give one with 26 variables of degree 21. (One can cut the degree to as low as 5 at the cost of having an enormous number of variables. No one knows the minimum possible degree nor the minimum possible number of variables): Our polynomial is $k + 2$ times

$$1 - (n + l + v - y)^2 - (2n + p + q + z - e)^2 - (wz + h + j - q)^2$$
$$- ((gk + 2g + k + 1)(h + j) + h - z)^2 - (z + pl(a - p) + t(2ap - p^2 - 1) - pm)^2$$
$$- (p + l(a - n - 1) + b(2an + 2a - n^2 - 2n - 2) - m)^2 - ((a^2 - 1)l^2 + 1 - m^2)^2$$
$$- (q + y(a - p - 1) + s(2ap + 2a - p^2 - 2p - 2) - x)^2 - ((a^2 - 1)y^2 + 1 - x^2)^2$$
$$- (16(k + 1)^3(k + 2)(n + 1)^2 + 1 - f^2)^2 - (e^3(e + 2)(a + 1)^2 + 1 - o^2)^2$$
$$- (16r^2 y^4 (a^2 - 1) + 1 - u^2)^2 - (ai + k + 1 - l - i)^2$$
$$- (((a + u^2(u^2 - a))^2 - 1)(n + 4dy)^2 + 1 - (x + cu)^2)^2.$$

Stare at this for a while and try to figure out how it works: The key is to determine when the displayed polynomial takes positive values. Note that it is equal to 1 minus a sum of squares so, if the polynomial is positive, with $k + 2 > 0$, then the second factor must equal 1 and therefore each of the squares must equal 0, so that

$$n + l + v - y = 2n + p + q + z - e = wz + h + j - q = \cdots = 0.$$

Understanding much beyond this seems difficult, and it seems that the only way to appreciate this polynomial is to understand its derivation; see [1]. In the current state of knowledge it seems that this absolutely extraordinary and beautiful polynomial is entirely useless in helping us better understand the distribution of primes!

5.22. Conway's prime producing machine

Begin with the integer 2 and multiply it by the first fraction in the list

$$\frac{17}{91}, \frac{78}{85}, \frac{19}{51}, \frac{23}{38}, \frac{29}{33}, \frac{77}{29}, \frac{95}{23}, \frac{77}{19}, \frac{1}{17}, \frac{11}{13}, \frac{13}{11}, \frac{15}{14}, \frac{15}{2}, \frac{55}{1}$$

for which the product is an integer. So we have $2 \times \frac{15}{2}$, giving 15. Repeat the process with this product, 15, and continue over and over again. One obtains the integers

$$2, 15, 825, 725, 1925, 2275, 425, 390, 330, 290, 770, \ldots.$$

Other than the first 2, the powers of 2 on the list are 2^2, 2^3, 2^5, 2^7, 2^{11}, 2^{13}, 2^{17} .... In other words, if 2^k appears, then k is prime, and one can even show that every prime power of 2 appears. This is an extraordinary way to find the primes. It is a challenge to determine why this works. (Hint: Study how the exponent of each prime dividing our integer varies as it is multiplied with each fraction in the list.)

---

[33] One can also construct such polynomials so as to yield the set of Fibonacci numbers or the set of Fermat primes or the set of Mersenne primes or the set of even perfect numbers (see section 4.2) and indeed any *Diophantine* set (see [1]).

5.23. Ulam's spiral

Ulam's idea was to write the integers in a spiral starting from 1, marking the prime numbers and looking for patterns. Here is the spiral up to 151:

```
                        151  150  149  148  147  146  145
         111  110  109  108  107  106  105  104  103  102  101  144
         112   73   72   71   70   69   68   67   66   65  100  143
         113   74   43   42   41   40   39   38   37   64   99  142
         114   75   44   21   20   19   18   17   36   63   98  141
         115   76   45   22    7    6    5   16   35   62   97  140
         116   77   46   23    8    1    4   15   34   61   96  139
         117   78   47   24    9    2    3   14   33   60   95  138
         118   79   48   25   10   11   12   13   32   59   94  137
         119   80   49   26   27   28   29   30   31   58   93  136
         120   81   50   51   52   53   54   55   56   57   92  135
         121   82   83   84   85   86   87   88   89   90   91  134
         122  123  124  125  126  127  128  129  130  131  132  133
```

Figure 5.2. Ulam's spiral up to 151 with the primes encircled.

Ulam spirals capturing many more integers can be found online.

Ulam observed that there are diagonals with lots of primes, for example the rising diagonal

$$79, 47, 23, 7, 19, 39, 67, 103, 147.$$

Starting from the 7, the numbers on the descending diagonal from 7 to the left are the values of the polynomial $4n^2 + 4n - 1$ with $n = 1, 2, 3, \ldots$, while the numbers on the rising diagonal from 7 to the right are the values of the polynomial $4n^2 + 3$ with $n = 1, 2, 3, \ldots$.

Likewise the descending diagonal from 11 contains primes $11, 29, 89, 131$; this diagonal contains the values of the polynomial $4n^2 + 6n + 1$ with $n = 1, 2, 3, \ldots$.

In an extended diagram one observes many prime-rich lines, for example a line which corresponds to the values of the polynomial $4x^2 + 170x + 1847 = X^2 + X + 41$ where $X = 2x + 42$, as well as a line which corresponds to the values of $4x^2 + 4x + 59$.

To better understand this diagram, let $U(0,0) = 1$, and let $U(x,y)$ be the integer that is x to the right and y up from 1; for example, $U(3,2) = 36$.

Exercise 5.23.1. Prove that we have

$$U(x,y) = \begin{cases} 4x^2 - x + 1 + y & \text{if } -x \leq y \leq x \text{ with } x \geq 0, \\ 4y^2 + y + 1 - x & \text{if } -y \leq x \leq y \text{ with } y \geq 0, \\ 4x^2 - 3x + 1 - y & \text{if } -|x| \leq y \leq |x| \text{ with } x \leq 0, \\ 4y^2 + 3y + 1 + x & \text{if } -|y| < x \leq |y| \text{ with } y \leq 0. \end{cases}$$

A line that hits infinitely many points with integer coordinates takes the form $ay = bx + c$ with a, b, c integers with $(a, b) = 1$. By Theorem 3.5, the solutions all take the form $x = r + na$, $y = s + nb$, as n varies. For a line in the right quadrant, $-a \leq b \leq a$, so we are looking for prime values of the polynomial

$$f(n) = 4x^2 - x + 1 + y = An^2 + Bn + C \text{ with } n \geq 0,$$

where $A = (2a)^2$, $B = 8ar - a + b$, $C = 4r^2 - r + 1 + s$, which has discriminant $(b-a)^2 - 16a(a+c)$. Therefore we are looking for a quadratic polynomial with many prime values.

The examples we found earlier were on the diagonal so that $a = 1, b = \pm 1$. As $a = 1$ we can take $r = 0$ and $s = c$ so that $A = 4, B = b - 1, C = c + 1$. For the polynomial f to take odd values we need C odd, so c even, and therefore the polynomial is

$$\text{either} \quad 4n^2 + C \quad \text{or} \quad 4n^2 - 2n + C.$$

We will discuss "prime-rich" quadratic polynomials in chapter 12.

Exercise 5.23.2. Let three consecutive values of a quadratic polynomial f be $f(n-1) = u, f(n) = v, f(n+1) = w$. Prove that f has discriminant $\left(\frac{u-4v+w}{2}\right)^2 - uw$.

5.24. Mills's formula

Although Legendre's conjecture that there is always a prime between consecutive squares is unresolved, we do know that there is a prime between every pair of consecutive cubes $> N_0$ (exercise 5.6.3). In 1947, Mills deduced from this that there exists a constant τ (now called *Mills's constant*), such that

$$\lfloor \tau^{3^n} \rfloor \text{ is a prime } q_n \text{ for every integer } n \geq 0.$$

To prove this, let $q_0 > N_0$ be prime and then select q_{n+1} to be any prime for which

$$q_n^3 < q_{n+1} < (q_n + 1)^3 \text{ for all } n \geq 0.$$

If $\ell_n = q_n^{1/3^n}$ and $u_n = (q_n + 1)^{1/3^n}$, then, taking 3^{n+1}th roots in the line above,

$$\ell_n < \ell_{n+1} < \cdots < u_{n+1} \leq u_n.$$

The ℓ_n are therefore an increasing bounded sequence so must tend to a limit; call it τ. Then $\ell_n < \tau \leq u_n$ for all n and so $q_n < \tau^{3^n} \leq q_n + 1$. We cannot have equality in the upper bound or else τ^{3^n} is an integer, and so τ^{3^m} is an integer for all $m \geq n$, implying that $q_m = \tau^{3^m} - 1$ for all $m \geq n$, and in particular

$$q_{n+1} = \tau^{3^{n+1}} - 1 = q_n(q_n^2 + 3q_n + 3)$$

so is not prime. Therefore $q_n < \tau^{3^n} < q_n + 1$ for all n and so $q_n = \lfloor \tau^{3^n} \rfloor$, as claimed.

Further reading on primes in surprising places

[1] James P. Jones, Daihachiro Sato, Hideo Wada, and Douglas Wiens, *Diophantine representation of the set of prime numbers*, Amer. Math. Monthly **83** (1976), 449–464.

[2] M. L. Stein, S. M Ulam, and M. B. Wells, *Mathematical notes: A visual display of some properties of the distribution of primes*, Amer. Math. Monthly **71** (1964), 516–520.

Appendix 5H. Dynamical systems and infinitely many primes

We will show that various different polynomial dynamical systems each give rise to a different proof that there are infinitely many primes.

5.25. A simpler formulation

The sequences $(a_n)_{n\geq 0}$ and $(F_n)_{n\geq 0}$, used to prove that there are infinitely many primes, can be determined by multiplying all of the terms of the sequence so far together and adding a constant. We can rewrite this as follows:

$$a_{n+1} = a_0 a_1 \cdots a_{n-1} \cdot a_n + 1 = (a_n - 1)a_n + 1 = f(a_n),$$

where $f(x) = x^2 - x + 1$. Similarly the F_n-values can be determined by

$$F_{n+1} = (2^{2^n} + 1)(2^{2^n} - 1) + 2 = F_n(F_n - 2) + 1 = f(F_n),$$

where $f(x) = x^2 - 2x + 2$. So they are both examples of sequences $(x_n)_{n\geq 0}$ for which

$$x_{n+1} = f(x_n)$$

for some polynomial $f(x) \in \mathbb{Z}[x]$. The terms of the sequence are all given by the recursive formula, so that

$$x_n = \underbrace{f(f(\ldots f(x_0)))}_{n \text{ times}} = f^n(x_0),$$

where the notation f^n denotes the polynomial obtained by composing f with itself n times (which is definitely *not* the nth power of f). Any such sequence $(x_n)_{n\geq 0}$ is called the *orbit* of x_0 under the map f, since the sequence is completely determined once one knows x_0 and f. We sometimes write the orbit as $x_0 \to x_1 \to x_2 \to \cdots$.

5.26. Different starting points

The key to the proof that the a_n's are pairwise coprime is that $a_n \equiv 1 \pmod{a_m}$ whenever $n > m \geq 0$, and the key to the proof that the F_n's are pairwise coprime is that $F_n \equiv 2 \pmod{F_m}$ whenever $n > m \geq 0$. These congruences are not difficult to deduce using Corollary 2.3.1. For $f(x) = x^2 - x + 1$, the orbit of 0 under the map f is $0 \to 1 \to 1 \to \cdots$. Therefore if $n > m \geq 0$, then we have

$$a_n = f^{n-m}(a_m) \equiv f^{n-m}(0) = 1 \pmod{a_m},$$

and so $(a_n, a_m) = (1, a_m) = 1$. Similarly if $f(x) = x^2 - 2x + 2$, then the orbit of 0 under the map f is $0 \to 2 \to 2 \to \cdots$ and so $F_n = f^{n-m}(F_m) \equiv f^{n-m}(0) = 2 \pmod{F_m}$, which implies that if $n > m$, then $(F_n, F_m) = (2, F_m) = 1$ as each F_m is odd.

This reformulation of two of the best-known proofs of the infinitude of primes hints at the possibility of a more general approach.

Exercise 5.25.1. Show that if $f_m(a) = a$, then $f_{m+n}(a) = f_n(a)$ for all $n \geq 0$.

5.26. Different starting points

We will use Lemma 5.1.1 to prove that there are infinitely many primes. This requires finding an infinite sequence of integers $a_1 < a_2 < \cdots$ that are pairwise coprime.

The orbit of 2 under the map $x \to x^2 - x + 1$ is an infinite sequence of pairwise coprime integers, $2 \to 3 \to 7 \to 43 \to 1807 \to \cdots$. What about other orbits? The orbit $4 \to 13 \to 157 \to 24493 \to \cdots$ is also an infinite sequence of pairwise coprime integers, as is the orbit $5 \to 21 \to 421 \to \cdots$. The same proof as before yields that no two integers in a given orbit have a common factor.

The orbit of 3 under the map $x \to x^2 - 2x + 2$ yielded the Fermat numbers $3 \to 5 \to 17 \to 257 \to \cdots$, but starting at 4 we get $4 \to 10 \to 82 \to 6562 \to \cdots$. These are obviously not pairwise prime as every number in the orbit is even, but if we divide through by 2, then we get

$$2 \to 5 \to 41 \to 3281 \to \cdots$$

which are pairwise coprime. To prove this, note that $x_0 = 4$ and $x_{n+1} = x_n^2 - 2x_n + 2$ for all $n \geq 0$, and so the above proof yields that if $m < n$, then $(x_m, x_n) = (x_m, 2) = 2$. Therefore, taking $a_n = x_n/2$ for every n we deduce that $(a_m, a_n) = (x_m/2, x_n/2) = (x_m/2, 1) = 1$. This same idea works for every orbit under this map: We get an infinite sequence of pairwise coprime integers by dividing through by 1 or 2, depending on whether x_0 is odd or even.

Exercise 5.26.1. Perform a similar analysis of the map $x \to x^2 - 2$ beginning by studying the orbit of 0. (The orbit of 4 under this map is shown, in the Lucas-Lehmer test (Corollary 10.10.1), to provide an efficient way to test whether a given Mersenne number is prime.)

However, things can be more complicated: Consider the orbit of $x_0 = 3$ under the map $x \to x^2 - 6x - 1$. We have

$$3 \to -10 \to 159 \to 24326 \to 591608319 \to \cdots.$$

Here x_n is divisible by 3 if n is even, and it is divisible by 2 if n is odd. If we let $a_n = x_n/3$ when n is even, and $a_n = x_n/2$ when n is odd, then one can show the

terms of the resulting sequence,
$$1 \to -5 \to 53 \to 12163 \to 197202773 \to \cdots,$$
are indeed pairwise coprime.

Another surprising example is given by the orbit of 6 under the map $x \to 7 + x^5(x-1)(x-7)$. Reducing the elements of the orbit mod 7 we find that
$$6 \to 5 \to 4 \to 3 \to 2 \to 1 \to 0 \to 0 \to \cdots \pmod 7,$$
as $x^5(x-1)(x-7) \equiv x^6(x-1) \pmod 7$, which is $\equiv x-1 \pmod 7$ if $x \not\equiv 0 \pmod 7$ by Fermat's little theorem. So x_n is divisible by 7 for every $n \geq 6$, but for no smaller n, and to obtain the pairwise coprime a_n we let $a_n = x_n$ for $n \leq 5$ and $a_n = x_n/7$ once $n \geq 6$.

It starts to look as though it might become complicated to formulate how to define a sequence $(a_n)_{n\geq 0}$ of pairwise coprime integers in general; certainly a case-by-case description is unappealing. However, there is a simpler way to obtain the a_n: In these last two examples we have $a_n = x_n/(x_n, 6)$ and then $a_n = x_n/(x_n, 7)$, respectively, for all $n \geq 0$, a description that will generalize well.

5.27. Dynamical systems and the infinitude of primes

One models evolution by determining the future development of the object of study from its current state (more sophisticated models incorporate the possibility of random mutations). This gives rise to *dynamical systems*, a rich and bountiful area of study. One simple model is that the state of the object at time n is denoted by x_n, and given an initial state x_0, one can find subsequent states via a map $x_n \to f(x_n) = x_{n+1}$ for some given function $f(.)$. Orbits of linear polynomials $f(.)$ are easy to understand,[34] but quadratic polynomials can give rise to evolution that is very far from what one might naively guess (the reader might look into the extraordinary Mandelbrot set).

This is the set-up above, where $f(x)$ is a polynomial with integer coefficients and x_0 an integer. It will be useful to use dynamical systems terminology.

If $f^n(\alpha) = \alpha$ for some integer $n \geq 1$, then (the orbit of) α is *periodic*, and the smallest such n is the *exact period* length for α. The orbit begins with the *cycle*
$$\alpha, f(\alpha), \ldots, f^{n-1}(\alpha)$$
of distinct values and repeats itself, so that $f^n(\alpha) = \alpha, f^{n+1}(\alpha) = f(\alpha), \ldots$, and, in general $f^{n+k}(\alpha) = f^k(\alpha)$ for all $k \geq 0$.

The number α is *preperiodic* if $f^m(\alpha)$ is periodic for some $m \geq 0$ and is *strictly preperiodic* if α is preperiodic but not itself periodic. In all of our examples so far, 0 has been strictly preperiodic. In fact, if any two elements of the orbit of α are equal, say $f^{m+n}(\alpha) = f^m(\alpha)$, then $f^{k+n}(\alpha) = f^{k-m}(f^{m+n}(\alpha)) = f^{k-m}(f^m(\alpha)) = f^k(\alpha)$ for all $k \geq m$, so that α is preperiodic.

Finally, α has a *wandering orbit* if it is not preperiodic, that is, if its orbit never repeats itself so that the $\{f^m(\alpha)\}_{m\geq 0}$ are all distinct. Therefore, we wish to start only with integers x_0 that have wandering orbits.

---

[34] One can verify: If $f(t) = at + b$, then $x_n = x_0 + nb$ if $a = 1$, and $x_n = a^n x_0 + \frac{a^n - 1}{a-1} b$ if $a \neq 1$.

Our general result for constructing infinitely many primes from orbits of a polynomial map is

Theorem 5.9. *Suppose that $f(x) \in \mathbb{Z}[x]$ and that 0 is a strictly preperiodic point of the map $x \to f(x)$. Let $\ell(f) = \mathrm{lcm}[f(0), f^2(0)]$. For any integer x_0 that has a wandering orbit, $(x_n)_{n \geq 0}$, let*
$$a_n = \frac{x_n}{\gcd(x_n, \ell(f))} \quad \text{for all} \quad n \geq 0.$$
Then the $(a_n)_{n \geq 0}$ are an infinite sequence of pairwise coprime integers.

One can prove that $|a_n| > 1$ for all $n \geq 3$ (see [**1**]), and so all such a_n have a private prime factor.[35] The example $f(x) = 3 - x(x-3)^2$ with $0 \to 3 \to 3$ has a wandering orbit $2 \to 1 \to -1 \to \cdots$, so that if $x_0 = 2$, then $x_1 = 1$ and $x_2 = -1$. Therefore we cannot in general improve the lower bound, $n \geq 3$.

We have $\ell(f) = 1, 2, 6$, and 7, respectively, for the four polynomials $f(x)$ of section 5.26 of appendix 5H.

In exercise 5.28.2 we will determine which polynomials f satisfy the hypothesis of Theorem 5.9, that is, those f for which 0 is a strictly preperiodic point.

Proof. Suppose that $k = n - m > 0$. Then, by Corollary 2.3.1,
$$x_n = f^k(x_m) \equiv f^k(0) \pmod{x_m},$$
and so $\gcd(x_m, x_n)$ divides $\gcd(x_m, f^k(0))$, which divides $f^k(0)$. But this divides
$$L(f) := \mathrm{lcm}[f^k(0) : k \geq 1],$$
which is the lcm of a finite number of nonzero integers, as 0 is preperiodic. Therefore (x_m, x_n) divides $L(f)$, and so (x_m, x_n) divides both $(x_m, L(f))$ and $(x_n, L(f))$. This implies that $A_m := x_m/(x_m, L(f))$ divides $x_m/(x_m, x_n)$ and A_n divides $x_n/(x_m, x_n)$. But $x_m/(x_m, x_n)$ and $x_n/(x_m, x_n)$ are pairwise coprime which therefore implies that $(A_m, A_n) = 1$.

We will show in exercise 5.28.4(b) that $L(f) = \ell(f)$; that is, the lcm of all the elements of the orbit of 0 is the same as the lcm of the first two terms. This then implies that $a_n = A_n$ for all n. □

5.28. Polynomial maps for which 0 is strictly preperiodic

We have already seen the examples $x^2 - x + 1$ for which $0 \to 1 \to 1$ and $x^2 - 6x - 1$ for which $0 \to -1 \to 6 \to -1$, where 0 is preperiodic with period length one and two, respectively. In fact exact periods cannot be any larger:

Lemma 5.28.1. *Let $f(x) \in \mathbb{Z}[x]$. If the orbit of a_0 is periodic, then its exact period length is either one or two.*

Proof. Let N be the exact period length so that $a_N = a_0$, and then $a_{N+k} = f^k(a_N) = f^k(a_0) = a_k$ for all $k \geq 1$. Now assume that $N > 1$ so that $a_1 \neq a_0$. Corollary 2.3.1 implies that $a_{n+1} - a_n$ divides $f(a_{n+1}) - f(a_n) = a_{n+2} - a_{n+1}$ for

---
[35]p is a *private prime factor* of a_n if p divides a_n but no other a_m.

all $n \geq 0$. Therefore,

$a_1 - a_0$ divides $a_2 - a_1$, which divides $a_3 - a_2, \ldots$, which divides $a_N - a_{N-1} = a_0 - a_{N-1}$; and this divides $a_1 - a_N = a_1 - a_0$,

the nonzero number we started with. Therefore each $a_{j+1} - a_j = \pm(a_1 - a_0)$ by exercise 1.1.1(b). This implies that there must be some $j \geq 1$ for which $a_{j+1} - a_j = -(a_j - a_{j-1})$, or else each $a_{j+1} - a_j = a_1 - a_0$ and so

$$0 = a_N - a_0 = \sum_{j=0}^{N-1}(a_{j+1} - a_j) = \sum_{j=0}^{N-1}(a_1 - a_0) = N(a_1 - a_0) \neq 0,$$

a contradiction. Therefore $a_{j+1} = a_{j-1}$, and we deduce that

$$a_2 = a_{N+2} = f^{N+1-j}(a_{j+1}) = f^{N+1-j}(a_{j-1}) = a_N = a_0,$$

as desired. □

One can show that there are just four possibilities for the period length and the preperiod length of the orbit of 0 (yielding the polynomials which satisfy the hypothesis of Theorem 5.9). We now give examples of each type. For some non-zero integer a, the orbit of 0 is one of the following:

$\boxed{0 \to a \to a \to \cdots}$ for $x \to x^2 - ax + a$ (e.g., $(a_n)_{n\geq 0}$ and $(F_n)_{n\geq 0}$); or

$\boxed{0 \to -a \to a \to a \to \cdots}$ for $a = 1$ or 2, with $x \to 2x^2 - 1$ or $x^2 - 2$, respectively; or

$\boxed{0 \to -1 \to a \to -1 \to \cdots}$ with $f(x) = x^2 - ax - 1$; or

$\boxed{0 \to 1 \to 2 \to -1 \to 2 \to -1 \to \cdots}$ with $f(x) = 1 + x + x^2 - x^3$; and

all other possible orbits are obtained from these, by using the observation that if $0 \to a \to b \to \cdots$ is an orbit for $x \to f(x)$, then $0 \to -a \to -b \to \cdots$ is an orbit for $x \to g(x)$ where $g(x) = -f(-x)$.

Exercise 5.28.1. Suppose that $f(x)$ has an orbit $x_0 \to x_1 \to \cdots$. Let $g(x) = f(x+a) - a$. Prove that $g(x)$ has an orbit $x_0 - a \to x_1 - a \to x_2 - a \to \cdots$.

Exercise 5.28.2. Find every $f(x) \in \mathbb{Z}[x]$ with each of the four orbits above. (As an example, $f_0(x) = a$ gives $0 \to a \to a$, so $f(x)$ is another with this orbit if and only if 0 and a are roots of $f(x) - f_0(x)$; that is, $f(x) - f_0(x) = x(x-a)g(x)$ for some $g(x) \in \mathbb{Z}[x]$.)

Exercise 5.28.3.‡ Prove that the four orbits above are the only possible ones.

Exercise 5.28.4. Let $f(x) \in \mathbb{Z}[x]$ and deduce from our classification of possible orbits:
(a) 0 is strictly preperiodic if and only if $f^2(0) = f^4(0) \neq 0$;
(b) $L(f) = \text{lcm}[f(0), f^2(0)] =: \ell(f)$ (as claimed in the proof of Theorem 5.9);
(c) x_0 has a wandering orbit if and only if $x_2 \neq x_4$.

Exercise 5.28.5. Suppose that $u_0 \in \mathbb{C}$ has period p under the map $x \to f(x)$ where $f(x) \in \mathbb{Z}[x]$, so that u is a root of the polynomial $f^p(x) - x$. Prove that if f is monic, then $\frac{u_j - u_i}{u_1 - u_0}$ is a unit for all $0 \leq i < j \leq p-1$.

We have been iterating the map $n \to f(n)$ where $f(t) \in \mathbb{Z}[t]$. If one allows $f(t) \in \mathbb{Q}[t]$, then it is an open question as to the possible period lengths in the integers. Even the simplest case, $f(x) = x^2 + c$, with $c \in \mathbb{Q}$, is not only open but leads to the magnificent world of dynamical systems (see [2]). We would like to know what primes divide the numerators when iterating, starting from a given integer.

References for this chapter

[1] Andrew Granville, *Using dynamical systems to construct infinitely many primes*, Amer. Math. Monthly **125** (2018), 483–496 (preprint).

[2] J. H. Silverman, *The arithmetic of dynamical systems*, Graduate Texts in Mathematics **241**, Springer, New York, 2007.

Chapter 6

Diophantine problems

Diophantine equations are polynomial equations in which we study the integer or rational solutions. They are named after Diophantus (who lived in Alexandria in the third century A.D.) who wrote up his understanding of such equations in his thirteen volume *Arithmetica* (though only six part-volumes survive today). This work was largely forgotten until interest was revived by Bachet's 1621 translation of *Arithmetica* into Latin.[1]

6.1. The Pythagorean equation

Right-angled triangles with sides $3, 4, 5$ and $5, 12, 13$, etc, were known to the ancient Babylonians. We wish to determine all right-angled triangles with integer sides, which amounts to finding all solutions in positive integers x, y, z to the Pythagorean equation

$$x^2 + y^2 = z^2.$$

Note that $z > x, y > 0$ as x, y, and z are all positive. We can reduce the problem, without loss of generality, so as to work with some convenient assumptions:

- That x, y, and z are pairwise coprime, by dividing through by their gcd, as in exercise 1.7.8.

- That x is even and y is odd, and therefore that z is odd: First note that x and y cannot both be even, since x, y, and z are pairwise coprime; nor both odd, by exercise 2.5.6(b). Hence one of x and y is even, the other odd, and we interchange them, if necessary, to ensure that x is even and y is odd.

Under these assumptions we reorganize the equation and factor to get

$$(z-y)(z+y) = z^2 - y^2 = x^2.$$

---

[1] Translations of various ancient Greek texts into Latin helped inspire the Renaissance.

We now prove that $(z-y, z+y) = 2$: We observe that $(z-y, z+y)$ divides $(z+y) - (z-y) = 2y$ and $(z+y) + (z-y) = 2z$, and that $(2y, 2z) = 2(y, z) = 2$. Therefore $(z-y, z+y)$ divides 2, and so equals 2 as $z-y$ and $z+y$ are both even.

Therefore, since $(z-y)(z+y) = x^2$ and $(z-y, z+y) = 2$, there exist integers r, s such that

$$z - y = 2s^2 \quad \text{and} \quad z + y = 2r^2; \quad \text{or} \quad z - y = -2s^2 \quad \text{and} \quad z + y = -2r^2,$$

by exercise 3.3.7(c). The second case is impossible since r^2, y, and z are all positive. From the first case we deduce that

$$x = 2rs, \quad y = r^2 - s^2, \quad \text{and} \quad z = r^2 + s^2.$$

To ensure that x, y, and z are pairwise coprime we need $(r, s) = 1$ and $r + s$ odd. If we now multiply back in any common factors, we get the general solution

(6.1.1) $\boxed{x = 2grs, \quad y = g(r^2 - s^2), \quad \text{and} \quad z = g(r^2 + s^2).}$

If we want an actual triangle, then the side lengths should all be positive so we may assume that $g > 0$ and $r > s > 0$, as well as $(r, s) = 1$ and r and s having different parities.[2] The reader should verify that the integers x, y, and z given by this parametrization always satisfy the Pythagorean equation.

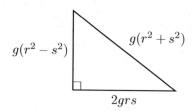

Figure 6.1. Parameterization of all integer-sided right-angled triangles.

One can also give a nice geometric proof of the parametrization in (6.1.1). We start with a reformulation of the question.

Exercise 6.1.1. Prove that the integer solutions to $x^2 + y^2 = z^2$ with $z > 0$ and $(x, y, z) = 1$ are in 1-to-1 correspondence with the rational solutions u, v to $u^2 + v^2 = 1$.

Where else does a line going through $(1, 0)$ intersect the circle $x^2 + y^2 = 1$? Unless the line is vertical it will hit the unit circle in exactly one other point, which we will denote by (u, v). Note that $u < 1$. If the line has slope t, then $t = v/(u-1)$ is rational if u and v are.

---

[2] That is, one is even, the other is odd.

6.1. The Pythagorean equation

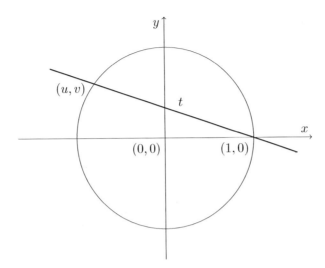

Figure 6.2. A line through $(1,0)$ on the circle $x^2 + y^2 = 1$.

In the other direction, the line through $(1,0)$ of slope t is $y = t(x-1)$ which intersects $x^2 + y^2 = 1$ where $1 - x^2 = y^2 = t^2(x-1)^2$, so that either $x = 1$ and $y = 0$, or we have $1 + x = t^2(1-x)$, which yields the point (u, v) with

$$u = \frac{t^2 - 1}{t^2 + 1} \quad \text{and} \quad v = \frac{-2t}{t^2 + 1}.$$

These are both rational if t is. We have therefore proved that $u, v \in \mathbb{Q}$ if and only if $t \in \mathbb{Q}$. In other words the line of slope t through $(1,0)$ hits the unit circle again at another rational point if and only if t is rational, and then we can classify those points in terms of t. Therefore, writing $t = -r/s$ where $(r, s) = 1$, we have

$$u = \frac{r^2 - s^2}{r^2 + s^2} \quad \text{and} \quad v = \frac{2rs}{r^2 + s^2},$$

the same parametrization to the Pythagorean equation as in (6.1.1) when we clear out denominators.

Exercise 6.1.2.† Find a formula for all the rational points on the curve $x^2 - y^2 = 3$.

Exercise 6.1.3. We call $\{a, b, c\}$ a *primitive Pythagorean triple* if a, b, and c are pairwise coprime integers for which $a^2 + b^2 = c^2$.
  (a) Prove that, in a primitive Pythagorean triple, the difference in length between the hypotenuse and each of the other sides is either a square or twice a square.
  (b) Can one find primitive Pythagorean triples in which the hypotenuse is three units longer than one of the other sides? Either give an example or prove that it is impossible.
  (c)† One can find primitive Pythagorean triples in which the hypotenuse is one unit longer than one of the other sides, e.g., $\{3, 4, 5\}$, $\{5, 12, 13\}$, $\{7, 24, 25\}$, $\{9, 40, 41\}$, $\{11, 60, 61\}$. Parametrize all such solutions.
  (d)† One can find primitive Pythagorean triples in which the hypotenuse is two units longer than one of the other sides, e.g., $\{3, 4, 5\}$, $\{8, 15, 17\}$, $\{12, 35, 37\}$, $\{16, 63, 65\}$, $\{20, 99, 101\}$. Parametrize all such solutions.

Exercise 6.1.4. (a) Prove that the side lengths of a primitive Pythagorean triple are $\not\equiv 2$ (mod 4).
  (b) Given integer $n > 1$ with $n \not\equiv 2$ (mod 4), explicitly give a primitive Pythagorean triple which has n as a side length.

Exercise 6.1.5.[†] Prove that there are infinitely many triples of coprime squares in arithmetic progressions.

Around 1637, Pierre de Fermat was studying the proof of (6.1.1) in his copy of Bachet's translation of Diophantus's *Arithmetica*. In the margin he wrote:

> *I have discovered a truly marvellous proof that it is impossible to separate a cube into two cubes, or a fourth power into two fourth powers, or in general, any power higher than the second into two like powers. This margin is too narrow to contain it.*
>
> —Pierre de Fermat (1637), in his copy of *Arithmetica*

In other words, Fermat claimed that for every integer $n \geq 3$ there do not exist positive integers x, y, z for which

$$x^n + y^n = z^n.$$

This is known as "Fermat's Last Theorem". Fermat did not subsequently mention this problem or his truly marvellous proof elsewhere, and the proof has not, to date, been rediscovered, despite many efforts.[3] Fermat did show that there are no solutions when $n = 4$ and we will present his proof in section 6.4, as well as some consequences for more general exponents n in Fermat's Last Theorem.

6.2. No solutions to a Diophantine equation through descent

Some Diophantine equations can be shown to have no solutions by starting with a purported smallest solution and finding an even smaller one, thereby establishing a contradiction. Such a *proof by descent* can be achieved in various different ways.

No solutions through prime divisibility

For some equations one can perform descent by considering the divisibility of the variables by various primes. We now give such a proof that $\sqrt{2}$ is irrational.

Proof of Proposition 3.4.1 by 2-divisibility. [$\sqrt{2}$ *is irrational.*] Let us recall that if $\sqrt{2}$ is rational, then we can write it as a/b so that $a^2 = 2b^2$. Let us suppose that (b, a) gives the smallest solution to $y^2 = 2x^2$ in positive integers. Now 2 divides $2b^2 = a^2$ so that $2|a$. Writing $a = 2A$, thus $b^2 = 2A^2$, and so $2|b$. Writing $b = 2B$ we obtain a solution $A^2 = 2B^2$ where A and B are half the size of a and b, contradicting the assumption that (b, a) is minimal. □

Exercise 6.2.1. Show that there are no non-zero integer solutions to $x^3 + 3y^3 + 9z^3 = 0$.

---

[3] Fermat wrote several important thoughts about number theory on his personal copy of *Arithmetica*, without proof. When he died his son, Samuel, made these available by republishing *Arithmetica* with his father's annotations. This is the *last* of those claims to have been fully understood.

No solutions through geometric descent

Proof of Proposition 3.4.1 by geometric descent. Again assume that $\sqrt{2} = a/b$ with a and b positive integers, where a is minimal. Hence $a^2 = 2b^2$ which gives rise to the smallest isosceles, right-angled triangle, OPQ with integer side lengths $\overline{OP} = \overline{OQ} = b$, $\overline{PQ} = a$ and angles $P\hat{O}Q = 90°$, $P\hat{Q}O = Q\hat{P}O = 45°$. Now mark a point R which is b units along PQ from Q and then drop a perpendicular to meet OP at the point S so that SR is perpendicular to PQ. Then $R\hat{P}S = Q\hat{P}O = 45°$, and so $R\hat{S}P = 180° - 90° - 45° = 45°$ by considering the angles in the triangle RSP. Therefore RSP is a smaller isosceles, right-angled triangle than OPQ. Moreover we have side lengths $\overline{RS} = \overline{PR} = a - b$. To establish our contradiction we need to show that the hypoteneuse, PS, also has integer length.

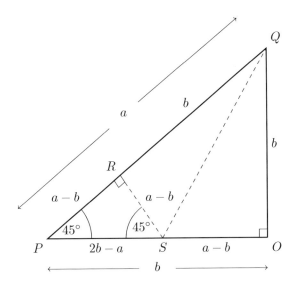

Figure 6.3. No solutions through geometric descent.

The two triangles, OQS and RQS, are congruent, since they both contain a right-angle opposite SQ and adjacent to a side of length b (OQ and RQ, respectively). Therefore $\overline{OS} = \overline{SR} = a - b$ and so $\overline{PS} = \overline{OP} - \overline{OS} = b - (a - b) = 2b - a$. Hence RSP is a smaller isosceles, right-angled triangle than OPQ with integer side lengths, contradicting the assumed minimality of OPQ. □

One can write this proof more algebraically: As $a^2 = 2b^2$, so $a > b > a/2$. Now

$$(2b - a)^2 = a^2 - 4ab + 2b^2 + 2b^2 = a^2 - 4ab + 2b^2 + a^2 = 2(a - b)^2.$$

However $0 < 2b - a < a$, contradicting the minimality of a.

Proof of Proposition 3.4.2 by an analogous descent. [*If d is an integer for which $\sqrt{d}$ is rational, then $\sqrt{d}$ is an integer.*] If $\sqrt{d}$ is rational, then we can write it as a/b so that $a^2 = db^2$. Let us suppose that (b, a) gives the smallest solution to $y^2 = dx^2$ in positive integers. Let r be the smallest integer $\geq db/a$, so that $\frac{db}{a} + 1 > r \geq \frac{db}{a}$, and therefore $a > ra - db \geq 0$. Then

$$(ra - db)^2 = da^2 - 2rdab + d^2b^2 + (r^2 - d)a^2$$
$$= da^2 - 2rdab + d^2b^2 + (r^2 - d)db^2 = d(rb - a)^2.$$

However $0 \leq ra - db < a$, contradicting the minimality of a, unless $ra - db = 0$. In this case $r^2 = d \cdot db^2/a^2 = d$. $\square$

6.3. Fermat's "infinite descent"

Fermat proved that there are no right-angled triangles with all integer sides whose area is a square (see exercise 6.3.1 below). In so doing he developed the important technique of "infinite descent", which we now exhibit in two related questions. (The reader can read the proof of only one of the two following similar theorems. They both lead to the same Corollary 6.4.1.)

Theorem 6.1. *There are no solutions in non-zero integers x, y, z to*

$$x^4 + y^4 = z^2.$$

Proof. Assume that there is a solution and let x, y, z be the solution in positive integers with z minimal. We may assume that $\gcd(x, y) = 1$ or else we can divide the equation through by the fourth power of $\gcd(x, y)$ to obtain a smaller solution. Here we have

$$(x^2)^2 + (y^2)^2 = z^2 \quad \text{with} \quad \gcd(x^2, y^2) = 1,$$

and so, by (6.1.1), there exist integers r, s with $(r, s) = 1$ and $r + s$ odd such that

$$x^2 = 2rs, \quad y^2 = r^2 - s^2, \quad \text{and} \quad z = r^2 + s^2$$

(swapping the roles of x and y if necessary to ensure that x is even). Now r and s have the same sign since $rs = x^2/2$, so we may assume they are both > 0 (multiplying each by -1 if necessary). Now $s^2 + y^2 = r^2$ with y odd and $(r, s) = 1$ and so, by (6.1.1), there exist integers a, b with $(a, b) = 1$ and $a + b$ odd such that

$$s = 2ab, \quad y = a^2 - b^2, \quad \text{and} \quad r = a^2 + b^2,$$

and so

$$x^2 = 2rs = 4ab(a^2 + b^2).$$

Now a and b have the same sign since $ab = s/2 > 0$, and therefore we may assume they are both > 0 (multiplying each by -1 if necessary).

Now a, b, and $a^2 + b^2$ are pairwise coprime positive integers whose product is a square so they must each be squares by exercise 3.3.7(b). Write $a = u^2$, $b = v^2$, and $a^2 + b^2 = w^2$ for some positive integers u, v, w. Therefore

$$u^4 + v^4 = a^2 + b^2 = w^2$$

yields another solution to the original equation. We wish to compare this to the

solution (x, y, z) we started with. We find that
$$w \leq w^2 = a^2 + b^2 = r < r^2 + s^2 = z,$$
contradicting the minimality of z. □

Theorem 6.2. *There are no solutions in positive integers x, y, z to*
$$x^4 - y^4 = z^2.$$

Proof. If there is a solution, take the one with x minimal. We may assume $(x, y) = 1$ or else we divide through by the fourth power of the common factor.

We begin by noting that
$$(y^2)^2 + z^2 = (x^2)^2 \text{ with } \gcd(x^2, y^2) = 1.$$
If z is even, then, by (6.1.1), there exist integers X, Y with $(X, Y) = 1$, of opposite parity, for which
$$x^2 = X^2 + Y^2 \text{ and } y^2 = X^2 - Y^2, \text{ so that } X^4 - Y^4 = (xy)^2.$$
Now $X^2 < x^2$, contradicting the minimality of x.

Therefore z is odd. By (6.1.1) there exist integers r, s with $(r, s) = 1$, of opposite parity, for which
$$x^2 = r^2 + s^2 \text{ and } y^2 = 2rs.$$
Now r and s have the same sign since $rs = y^2/2 > 0$, and therefore we may assume they are both > 0 (multiplying each by -1 if necessary). From the equation $2rs = y^2$ we deduce that $r = 2R^2, s = Z^2$ for some integers R, Z (swapping the roles of r and s, if necessary). From (6.1.1) applied to the equation $r^2 + s^2 = x^2$, there exist integers u, v with $(u, v) = 1$, of opposite parity, for which $r = 2uv$ and $s = u^2 - v^2$. Now $uv = r/2 = R^2$, so we may assume they are both positive (multiplying each by -1 if necessary), and so $u = m^2, v = n^2$ for some integers m, n. Therefore
$$m^4 - n^4 = u^2 - v^2 = s = Z^2.$$
Now $m^2 < (mn)^2 = uv = r/2 < x/2$, contradicting the minimality of x. □

Exercise 6.3.1 (Fermat, 1659).
(a)† Prove that there is no right-angled, integer-sided, triangle whose area is a square.
(b) Deduce that there is no right-angled, rational-sided, triangle whose area is 1.
(c) Deduce that there are no integer solutions to $x^4 + 4y^4 = z^2$.

In appendix 6B we will see an alternative proof of these results using classical Greek geometry.

6.4. Fermat's Last Theorem

Fermat's Last Theorem is the assertion that for every integer $n \geq 3$ there do not exist positive integers x, y, z for which
$$x^n + y^n = z^n.$$

Corollary 6.4.1 (Fermat). *There are no solutions in non-zero integers x, y, z to*
$$x^4 + y^4 = z^4.$$

Exercise 6.4.1. Prove this using Theorem 6.1 or Theorem 6.2.

We deduce that Fermat's Last Theorem holds for all exponents $n \geq 3$ if it holds for all odd prime exponents:

Proposition 6.4.1. *If Fermat's Last Theorem is false, then there exists an odd prime p and pairwise coprime non-zero integers x, y, z such that*
$$x^p + y^p + z^p = 0.$$

Proof. Suppose that $x^n + y^n = z^n$ with $x, y, z > 0$ and $n \geq 3$. If two of x, y, and z have a common factor, then it must divide the third and so we can divide out the common factor. Hence we may assume that x, y, z are pairwise coprime positive integers. Now any integer $n \geq 3$ has a factor m which is either $= 4$ or is an odd prime (see exercise 3.1.3(b)). Hence, if $n = dm$, then $(x^d)^m + (y^d)^m = (z^d)^m$, so we get a solution to Fermat's Last Theorem with exponent m. We can rule out $m = 4$ by Corollary 6.4.1. Therefore $m = p$ is an odd prime and we have the desired solution $(x^d)^p + (y^d)^p + (-z^d)^p = 0$. □

A brief history of equation solving

There have been many attempts to prove Fermat's Last Theorem, inspiring the development of much great mathematics, for example, ideal theory (see appendices 3D and 12B). We will discuss one beautiful advance due to Sophie Germain from the beginning of the 19th century (see section 7.27 of appendix 7F).

In 1994 Andrew Wiles proved Fermat's Last Theorem, developing ideas of Frey, Ribet, and Serre involving modular forms, a subject far removed from the original question. The proof is extraordinarily deep, involving some of the most profound themes in arithmetic geometry.[4] If the whole proof were written in the leisurely style of, say, this book, it would probably take a couple of thousand pages. This could not be the proof that Fermat believed that he had—could Fermat have been correct? Could there be a short, elementary, marvelous proof still waiting to be found? Or will Fermat's claim always remain a mystery?

To some extent one can measure the difficulty of solving Diophantine equations (especially rational solutions to equations with two variables) by their degree.[5] The first three chapters of this book focus on linear (degree-one) equations, culminating in section 3.6. Much of the rest of this book provides tools for studying degree-two (quadratic) equations; see chapters 8 and 9, sections 11.2 and 11.3, and chapter 12. Degree-three (cubic) equations give rise to elliptic curves; many of the key questions about elliptic curves lay shrouded in mystery and so they are intensively researched in number theory today (see chapter 17). In 1983 Gerd Faltings showed that higher-degree Diophantine equations only have finitely many rational solutions (though not how to find those solutions).

For higher-degree equations perhaps the most interesting cases are Diophantine equations with varying degree, like the Fermat equation. Another famous example is *Catalan's conjecture*: The positive integer powers are

$$1, 4, 8, 9, 16, 25, 27, 32, 36, 49, 64, \ldots$$

---

[4]See our sequel [**Grab**] for some discussion of the ideas involved in the proof.

[5]A better but more sophisticated invariant is the *genus*, which requires quite a bit of algebraic geometry to define and is beyond the scope of this book.

A brief history of equation solving

which seem to get wider spread out as they get larger. Only two of the numbers in our list, 8 and 9, differ by 1, and Catalan conjectured that this is the only example of powers differing by 1. That is, the only integer solution to

$$x^p - y^q = 1 \text{ with } x, y \neq 0 \text{ and } p, q \geq 2,$$

is $3^2 - 2^3 = 1$. This was shown to be true by Preda Mihăilescu in 2002.

Combining these two famous equations leads to the *Fermat-Catalan equation*

$$x^p + y^q = z^r \text{ where } (x, y, z) = 1 \text{ and } \frac{1}{p} + \frac{1}{q} + \frac{1}{r} < 1.$$

We insist that $(x, y, z) = 1$ because one can find "trivial" solutions like $2^k + 2^k = 2^{k+1}$ in many cases (see exercise 6.5.8 for more examples). Obviously there are solutions when one of p, q, r is 1, so we insist they are all ≥ 2. One can find solutions when two of the exponents equal 2, and so the peculiar looking condition $\frac{1}{p} + \frac{1}{q} + \frac{1}{r} < 1$ turns out to be the correct one. We do know of ten solutions:

$$1 + 2^3 = 3^2, \quad 2^5 + 7^2 = 3^4, \quad 7^3 + 13^2 = 2^9, \quad 2^7 + 17^3 = 71^2, \quad 3^5 + 11^4 = 122^2,$$

$$17^7 + 76271^3 = 21063928^2, \quad 1414^3 + 2213459^2 = 65^7, \quad 9262^3 + 15312283^2 = 113^7,$$

$$43^8 + 96222^3 = 30042907^2, \quad 33^8 + 1549034^2 = 15613^3.$$

It is conjectured that there are only finitely many solutions x^p, y^q, z^r to the Fermat-Catalan equation; perhaps these ten are all the solutions. All of our ten solutions have an exponent equal to 2. So one might further conjecture that there are no solutions to the Fermat-Catalan equation with p, q, r all > 2. These are open questions and mathematicians are making headway. Henri Darmon and I proved in 1995 that there are only finitely many solutions for each *fixed* triple p, q, r. Today we know that for various infinite families exponent triples p, q, r, the Fermat-Catalan equation has no solutions: For example when $p = q$ and $\frac{1}{p} + \frac{1}{q} + \frac{1}{r} < 1$ there are no solutions if r is divisible by 2 or by 3 or by p, or if p is even and r is divisible by 5, etc. (see [**1**] for the state of the art).

Now that Fermat's Last Theorem has been proved, what can take its place as the "holy grail" of Diophantine equations? The *abc*-conjecture is clearly an important problem that would have profound effects on equations and even in other areas of number theory. In appendix 6A we will discuss its analogy for polynomials and then discuss the *abc*-conjecture itself and its influence on other equations, in section 11.5.

References for this chapter

[1] Michael Bennett, Imin Chen, Sander Dahmen, Soroosh Yazdani, *Generalized Fermat equations: A miscellany*, Int. J. Number Theory **11** (2015), 1–28.

[2] John J. Watkins, chapter 5 of *Number theory, a historical approach*, Princeton University Press, 2014.

Additional exercises

Exercise 6.5.1. Find all rational-sided right-angled triangles in which the area equals the perimeter. Prove that $5, 12, 13$ and $6, 8, 10$ are the only such integer-sided triangles.

Exercise 6.5.2.[†] Let n be an integer > 2 that is $\not\equiv 2 \pmod{4}$. Prove that there are $2^{\omega(n)-1}$ distinct primitive Pythagorean triangles in which n is the length of a side which is not the hypotenuse, where $\omega(n)$ counts the number of distinct prime factors of n.

Exercise 6.5.3.[†] Find a 1-to-1 correspondence between pairs of integers $b, c > 0$ for which $x^2 - bx - c$ and $x^2 - bx + c$ are both factorable over $\mathbb{Z}$, and right-angled triangles in which all three sides are integers.

Exercise 6.5.4. Prove that if $f(x) \in \mathbb{Z}[x]$ is a quadratic polynomial for which $f(x)$ and $f(x) + 1$ both have integer roots, then $f(x) + 1$ is the square of a linear polynomial. (Try substituting the roots of $f(x)$ into $f(x) + 1$ and studying divisibilities of the differences of the roots.)

Exercise 6.5.5.[†] We wish to show that $\alpha = \frac{\sqrt{5}+1}{2}$ is irrational. Suppose it is rational, so that $\alpha = p/q$ with $(p, q) = 1$. Now α satisfies the equation $x^2 = x + 1$, so dividing through by x we have $x = (1 + x)/x$, and so $\alpha = (p + q)/p$. Prove that p/q cannot equal $(p + q)/p$ and therefore establish a contradiction.

Exercise 6.5.6.[†] Generalize the proof in the last exercise, to prove that if α is a rational root of $x^2 - ax - b \in \mathbb{Z}[x]$, then α is an integer which divides b.

Exercise 6.5.7.[‡] Prove that $2n$ is the length of the perimeter of a right-angled integer-sided triangle if and only if there exist divisors d_1, d_2 of n for which $d_1 < d_2 < 2d_1$.

Exercise 6.5.8. Suppose that integers p, q, r are given. For any integers a and b let $c = a^p + b^q$. If we multiply this through by c^n, where n is divisible by p and q, then $(ac^{n/p})^p + (bc^{n/q})^q = c^{n+1}$. Determine conditions on p, q, and r under which we find an integer n such that c^{n+1} is an rth power (and therefore find an integer solution to $x^p + y^q = z^r$, albeit with $(x, y, z) > 1$).

Exercise 6.5.9. Calculations show that every integer in $[129, 300]$ is the sum of distinct squares. Deduce that every integer > 128 is the sum of distinct squares. (In exercise 2.5.6(f) we showed that there are infinitely many integers that cannot be written as the sum of three squares. In appendix 12E we will show that every integer is the sum of four squares.)

Exercise 6.5.10. Prove that there are infinitely many integers that cannot be written as the sum of three cubes.

Exercise 6.5.11.[‡] Calculations show that every integer in $[12759, 30000]$ is the sum of distinct cubes of positive integers. Deduce that every integer > 12758 is the sum of distinct cubes of positive integers. (In 2015 Siksek showed that every integer > 454 is the sum of at most seven positive cubes. It is believed, but not proven, that every sufficiently large integer is the sum of at most four positive cubes.)

Exercise 6.5.12. Verify the identity $6x = (x+1)^3 + (x-1)^3 - 2x^3$. Deduce that every prime is the sum of no more than five cubes of integers (which can be positive or negative).

Exercise 6.5.13. (a) Prove that $n^4 \equiv 0$ or $1 \pmod{16}$ for all integers n.
  Let N be divisible by 16.
 (b) Show that if N is the sum of 15 fourth powers, then each of those fourth powers is even.
 (c) Deduce that N is the sum of 15 fourth powers if and only if $N/16$ is the sum of 15 fourth powers.
 (d) Prove that 31 is not the sum of 15 fourth powers but is the sum of 16 fourth powers.
 (e) Deduce that there are infinitely positive integers N that are not the sum of 15 fourth powers.

  (In 2005, Deshouillers, Kawada, and Wooley showed that every integer > 13792 can be written as the sum of 16 fourth powers.)

In 1770 Waring asked whether for all integers k there exists an integer $g(k)$ such that every positive integer is the sum of at most $g(k)$ kth powers of positive integers. This was proved by Hilbert in 1909 but it is still a challenge to evaluate the smallest possible $g(k)$ for each k. We discuss this further in appendix 17D.

Appendix 6A. Polynomial solutions of Diophantine equations

6.6. Fermat's Last Theorem in $\mathbb{C}[t]$

The notation $\mathbb{C}[t]$ denotes polynomials whose coefficients are complex numbers. In section 6.1 we saw that all integer solutions to $x^2+y^2=z^2$ are derived from letting t be a rational number in the polynomial solution

$$(t^2-1)^2 + (2t)^2 = (t^2+1)^2.$$

We now prove that there are no "genuine" polynomial solutions to Fermat's equation

(6.6.1) $$x^p + y^p = z^p$$

with exponent p larger than 2 (where by *genuine* we mean that $(x(t), y(t), z(t))$ is not a polynomial multiple of a solution of (6.6.1) in complex numbers).

Proposition 6.6.1. *There are no genuine polynomial solutions $x(t), y(t), z(t) \in \mathbb{C}[t]$ to $x(t)^p + y(t)^p = z(t)^p$ with $p \geq 3$.*

Proof. Assume that there is a solution with x, y, and z all non-zero to (6.6.1) where $p \geq 3$. We may assume that x, y, and z have no common (polynomial) factor or else we can divide out by that factor (and that they are pairwise coprime by the same argument as in section 6.1). Our first step will be to differentiate (6.6.1) to get

$$px^{p-1}x' + py^{p-1}y' = pz^{p-1}z'$$

and after dividing out the common factor p, this leaves us with

(6.6.2) $$x^{p-1}x' + y^{p-1}y' = z^{p-1}z'.$$

We now have two linear equations (6.6.1) and (6.6.2) (thinking of x^{p-1}, y^{p-1}, and z^{p-1} as our variables), which suggests we use linear algebra to eliminate a variable: Multiply (6.6.1) by y' and (6.6.2) by y, and subtract, to get

$$x^{p-1}(xy' - yx') = x^{p-1}(xy' - yx') + y^{p-1}(yy' - yy') = z^{p-1}(zy' - yz').$$

Therefore x^{p-1} divides $z^{p-1}(zy' - yz')$, but since x and z have no common factors, this implies that

(6.6.3) $\qquad\qquad x^{p-1}$ divides $zy' - yz'$.

This is a little surprising, for if $zy' - yz'$ is non-zero, then a high power of x divides $zy' - yz'$, something that does not seem consistent with (6.6.1).

Now, if $zy' - yz' = 0$, then $(y/z)' = 0$ and so y is a constant multiple of z, contradicting our statement that y and z have no common factor. Therefore (6.6.3) implies, taking degrees of both sides, that

$$(p-1)\ \text{degree}(x) \leq \text{degree}(zy' - yz') \leq \text{degree}(y) + \text{degree}(z) - 1,$$

since $\text{degree}(y') = \text{degree}(y) - 1$ and $\text{degree}(z') = \text{degree}(z) - 1$. Adding $\text{degree}(x)$ to both sides gives

(6.6.4) $\qquad p\ \text{degree}(x) < \text{degree}(x) + \text{degree}(y) + \text{degree}(z).$

The right side of (6.6.4) is symmetric in x, y, and z. The left side is a function of x simply because of the order in which we chose to do things above. We could just as easily have derived the same statement with y or z in place of x on the left side of (6.6.4), so that

$$p\ \text{degree}(y) < \text{degree}(x) + \text{degree}(y) + \text{degree}(z)$$
$$\text{and}\quad p\ \text{degree}(z) < \text{degree}(x) + \text{degree}(y) + \text{degree}(z).$$

Adding these last three equations together and then dividing out by $\text{degree}(x) + \text{degree}(y) + \text{degree}(z)$ implies

$$p < 3,$$

and so Fermat's Last Theorem is proved, at least for polynomials. □

That Fermat's Last Theorem is not difficult to prove for polynomials is an old result, going back certainly as far as Liouville in 1851.

Exercise 6.6.1. Prove that all solutions to $x(t)^2 + y(t)^2 = z(t)^2$ in polynomials are a scalar multiple of some solution of the form $(r(t)^2 - s(t)^2)^2 + (2r(t)s(t))^2 = (r(t)^2 + s(t)^2)^2$.

6.7. $a + b = c$ in $\mathbb{C}[t]$

We now intend to extend the idea in our proof of Fermat's Last Theorem for polynomials to as wide a range of questions as possible. It takes a certain genius to generalize to something far simpler than the original. But what could possibly be more simply stated, yet more general, than Fermat's Last Theorem? It was Richard C. Mason (1983) who gave us that insight: *Look for solutions to*

$$a + b = c.$$

We will just follow through the above proof of Fermat's Last Theorem for polynomials (Proposition 6.6.1) and see where it leads: Start by assuming, with no loss

6.7. $a+b=c$ in $\mathbb{C}[t]$

of generality, that a, b, and c are all non-zero polynomials without common factors (or else all three share the common factor and we can divide it out). Then we differentiate to get

$$a' + b' = c'.$$

Next we need to do linear algebra. It is not quite so obvious how to proceed analogously, but what we do learn in a linear algebra course is to put our coefficients in a matrix and solutions follow if the determinant is non-zero. This suggests defining

$$\Delta(t) := \begin{vmatrix} a(t) & b(t) \\ a'(t) & b'(t) \end{vmatrix}.$$

Then if we add the first column to the second, we get

$$\Delta(t) = \begin{vmatrix} a(t) & c(t) \\ a'(t) & c'(t) \end{vmatrix},$$

and similarly

$$\Delta(t) = \begin{vmatrix} c(t) & b(t) \\ c'(t) & b'(t) \end{vmatrix}$$

by adding the second column to the first, a beautiful symmetry.

We note that $\Delta(t) \neq 0$, or else $ab' - a'b = 0$ so b is a scalar multiple of a (with the same argument as above), contradicting our hypothesis.

To find the appropriate analogy to (6.6.3), we consider the power to which the factors of a (as well as b and c) divide our determinant: Let α be a root of $a(t)$, and suppose that $(t-\alpha)^e$ is the highest power of $(t-\alpha)$ which divides $a(t)$ (we write $(t-\alpha)^e \| a(t)$). Now we can write $a(t) = U(t)(t-\alpha)^e$ where $U(t)$ is a polynomial that is not divisible by $(t-\alpha)$, so that $a'(t) = (t-\alpha)^{e-1} V(t)$ where $V(t) := U'(t)(t-\alpha) + eU(t)$. Now $(t-\alpha, V(t)) = (t-\alpha, eU(t)) = 1$, and so $(t-\alpha)^{e-1} \| a'(t)$. Therefore

$$\Delta(t) = a(t)b'(t) - a'(t)b(t) = (t-\alpha)^{e-1} W(t)$$

where $W(t) := U(t)(t-\alpha)b'(t) - V(t)b(t)$ and $(t-\alpha, W(t)) = (t-\alpha, V(t)b(t)) = 1$ as $t-\alpha$ does not divide $b(t)$ or $V(t)$. Therefore we have proved that

$$(t-\alpha)^{e-1} \| \Delta(t).$$

This implies that $(t-\alpha)^e$ divides $\Delta(t)(t-\alpha)$. Multiplying all such $(t-\alpha)^e$ together we obtain (since they are pairwise coprime) that

$$a(t) \text{ divides } \Delta(t) \prod_{a(\alpha)=0} (t-\alpha).$$

In fact $a(t)$ only appears on the left side of this equation because we studied the linear factors of a; analogous statements for $b(t)$ and $c(t)$ are also true, and since $a(t), b(t), c(t)$ have no common roots, we can combine those statements to read

(6.7.1) $$a(t)b(t)c(t) \text{ divides } \Delta(t) \prod_{(abc)(\alpha)=0} (t-\alpha).$$

The next step is to take the degrees of both sides of (6.7.1). The degree of $\prod_{(abc)(\alpha)=0}(t-\alpha)$ is precisely the total number of distinct roots of $a(t)b(t)c(t)$.

Therefore
$$\deg(a) + \deg(b) + \deg(c) \leq \deg(\Delta) + \#\{\alpha \in \mathbb{C} : (abc)(\alpha) = 0\}.$$
Now, using the three different representations of Δ above, we have
$$\deg(\Delta) \leq \begin{cases} \deg(a) + \deg(b) - 1, \\ \deg(a) + \deg(c) - 1, \\ \deg(c) + \deg(b) - 1. \end{cases}$$
Inserting all this into the previous inequality we get
$$\deg(a), \deg(b), \deg(c) < \#\{\alpha \in \mathbb{C} : (abc)(\alpha) = 0\}.$$
Put another way, this result can be read as:

Theorem 6.3 (The *abc* Theorem for Polynomials). *If $a(t), b(t), c(t) \in \mathbb{C}[t]$ do not have any common roots and provide a genuine polynomial solution to $a(t) + b(t) = c(t)$, then the maximum of the degrees of $a(t), b(t), c(t)$ is less than the number of distinct roots of $a(t)b(t)c(t) = 0$.*

This is a "best possible" result in that we can find infinitely many examples where there is exactly one more zero of $a(t)b(t)c(t) = 0$ than the largest of the degrees, for example the familiar identity
$$(2t)^2 + (t^2 - 1)^2 = (t^2 + 1)^2;$$
or the rather less interesting
$$t^n + 1 = (t^n + 1).$$

Exercise 6.7.1. Let a, b, and c be given non-zero integers, and suppose $n, p, q, r > 1$.
 (a) Prove that there are no genuine polynomial solutions $x(t), y(t), z(t)$ to $ax^n + by^n = cz^n$ with $n \geq 3$.
 (b) Prove that if there is a genuine polynomial solution $x(t), y(t), z(t)$ to $ax^p + by^q = cz^r$ in which $x, y,$ and z have no common root, then $\frac{1}{p} + \frac{1}{q} + \frac{1}{r} > 1$.
 (c) Deduce in (b) that this implies that at least one of p, q, and r must equal 2.
 (d) One can find solutions in (b) if one allows common factors, for example $x^3 + y^3 = z^4$ where $x = t(t^3 + 1)$ and $y = z = t^3 + 1$. Generalize this construction to as many other sets of exponents p, q, r as you can. (Try to go beyond the construction in exercise 6.5.8.)

Exercise 6.7.2. Let a and b be given non-zero integers, $p, q > 1$, and $x(t), y(t) \in \mathbb{C}[t]$. Let D be the maximum of the degrees of x^p and y^q, and assume that $ax^p + by^q \neq 0$.
 (a) Prove that the degree of $ax^p + by^q$ is $> D(1 - \frac{1}{p} - \frac{1}{q})$.
 (b)† Prove that if $g = (p, q) > 1$, then the degree of $ax^p + by^q$ is $\geq D/g$.
 (c) Deduce that the degree of $ax^p + by^q$ is always $> D/6$.
  (This is "best possible" in the case $(t^2 + 2)^3 - (t^3 + 3t)^2 = 3t^2 + 8$.)

Appendix 6B. No Pythagorean triangle of square area via Euclidean geometry

In this appendix we use Euclidean geometry to show that there is no integer-sided right-angled triangle whose area is a square,[6] rather than the algebraic methods of section 6.3. In this proof one sees algebra and geometry working together, foreshadowing a theme one frequently encounters as one studies advanced mathematics.

An algebraic proof, by descent

We will suppose that there are integer-sided right-angled triangles with square area and establish a contradiction. We take the integer-sided right-angled triangle ABC whose area is a square with smallest hypotenuse. By (6.1.1) its sides have lengths

$$AB = 2MN, \quad AC = N^2 - M^2, \quad \text{and} \quad BC = M^2 + N^2,$$

where M and N are coprime positive integers of different parities. The area of this triangle, $MN(N-M)(N+M)$, is a square by hypothesis. We now prove that the factors M, N, $N-M$, and $N+M$ are pairwise coprime:

We have $(M, N \pm M) = (N \pm M, M) = (N, M) = 1$. Finally let $g = (N-M, N+M)$. Then g is odd since $N-M$ is odd. Moreover g divides both $(N+M) - (N-M) = 2M$ and $(N+M) + (N-M) = 2N$, so that g divides $(2M, 2N) = 2(M, N) = 2$. The only odd positive integer g dividing 2 is $g = 1$.

We have proved that the product $MN(N-M)(N+M)$ is a square and that the factors M, N, $N-M$, and $N+M$ are pairwise coprime. Since they are all

---

[6]This proof is due to a student, Stephanie Chan, working with me in London in 2017.

positive integers, this implies that each of M, N, $N - M$, and $N + M$ has to be a square. We write

$$M = m^2, \quad N = n^2, \quad N - M = p^2, \quad \text{and} \quad N + M = q^2$$

for integers m, n, p, q, where $p \equiv q \pmod{2}$.

Now we take the triangle UVW with integer side lengths

$$UV = \frac{q-p}{2}, \quad VW = \frac{q+p}{2}, \quad \text{and} \quad UW = n.$$

This is a right-angled triangle since $(\frac{q-p}{2})^2 + (\frac{q+p}{2})^2 = \frac{q^2+p^2}{2} = N = n^2$, and it has area $\frac{1}{2} \cdot \frac{q-p}{2} \cdot \frac{q+p}{2} = \frac{q^2-p^2}{8} = \frac{M}{4} = (\frac{m}{2})^2$, which is a square. However its hypotenuse, n, is $< N < M^2 + N^2$, which contradicts the minimality of ABC.

Remarkably, Stephanie Chan showed that a scalar multiple of the triangle UVW could be constructed using a ruler and compass:

6.8. A geometric viewpoint

We set $T = M/N$, so that T, $1 - T = (N - M)/N$, and $1 + T = (N + M)/N$ are all squares. We now construct an integer-sided right-angled triangle $A'B'C'$ from ABC whose area is a square, such that when we divide the sides by a common factor, we will obtain UVW.

Let 2β be the angle BCA, so that $\tan 2\beta = \frac{AB}{AC} = \frac{2T}{1-T^2}$ and one can deduce $\tan \beta = T$ from the usual double-angle formulas. Draw a circle centred at C through B, and then extend the line AC to a point C' where it meets the circle. The geometry implies that the angle $BC'A$ is β.

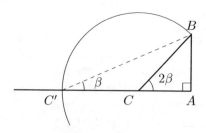

Figure 6.4. Halving the angle 2β of the initial triangle ABC.

Now rotate the line AC' about A by drawing a circle centered at A through C'. Draw a line parallel to AC passing through B, which meets the circle in two points; let B' denote the intersection of this line and circle which is furthest from C'.

6.8. A geometric viewpoint

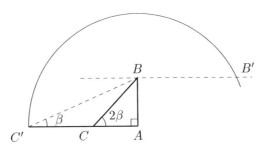

Figure 6.5. Obtaining a new point B' at the intersection of the line and the circle.

Extending the line AC' we can drop a perpendicular from B' to this line and denote the intersection by A'. Our new triangle is $A'B'C'$. Let α denote the angle $B'C'A'$.

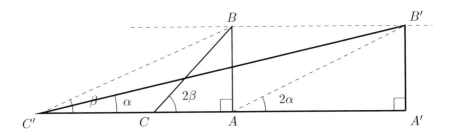

Figure 6.6. Forming the new triangle ABC.

Now $AB' = AC'$ as they are both arcs of the same circle, and $AB = A'B'$ by construction. Therefore

$$\sin 2\alpha = \frac{A'B'}{AB'} = \frac{AB}{AC'} = \tan\beta = T,$$

and so $\cos 2\alpha = \sqrt{1-T^2}$. Since $1-T$ and $1-T$ are both squares, therefore $\sqrt{1+T} > \sqrt{1-T} > 0$ are both rational numbers. Moreover

$$(\sqrt{1+T} + \sqrt{1-T})^2 = 2 + 2\sqrt{1-T^2} = 2 + 2\cos 2\alpha = (2\cos\alpha)^2.$$

Taking square roots, we deduce that $\cos\alpha = \frac{1}{2}(\sqrt{1+T} + \sqrt{1-T})$, as they are both positive, and so $\cos\alpha$ is a rational number. Similarly $\sin\alpha = \frac{1}{2}(\sqrt{1+T} - \sqrt{1-T})$ is also a rational number.

Now $A'B' = AB$ is an integer by the hypothesis. Therefore

$$B'C' = \frac{A'B'}{\sin\alpha} \quad \text{and then} \quad A'C' = B'C' \cdot \cos\alpha$$

are both rational; that is, $A'B'C'$ is rational-sided. The area of this new triangle is

$$\frac{1}{2} \cdot A'B' \cdot A'C' = \frac{1}{2}(B'C'\sin\alpha)(B'C'\cos\alpha) = \frac{1}{4}(B'C')^2 \sin 2\alpha = \frac{T}{4}(B'C')^2,$$

which is a square, as T is a square. The side lengths of our new triangle are

(6.8.1) $\qquad A'B' = 2MN, \quad A'C' = 2N(N + \sqrt{N^2 - M^2}),$ and

(6.8.2) $\qquad\qquad B'C' = 2N(\sqrt{N(N+M)} + \sqrt{N(N-M)}).$

At the beginning of this appendix we had $M = m^2$, $N = n^2$, $N - M = p^2$, and $N + M = q^2$ for integers m, n, p, q, and so

$$A'B' = n^2(q^2 - p^2), \quad A'C' = n^2(p+q)^2, \quad \text{and} \quad B'C' = 2n^3(p+q).$$

If we divide each side length through by $2n^2(q+p)$, then we obtain the triangle UVW from the previous section and establish the required contradiction. $\square$

Exercise 6.8.1. Suppose that ABC is an integer-sided right-angled triangle. Insert a circle inside the triangle which is tangent to each of the sides of the triangle. Prove that the circle's radius is an integer.

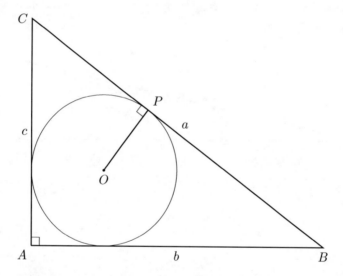

Figure 6.7. A circle inscribed inside a right-angled triangle

Appendix 6C. Can a binomial coefficient be a square?

Can $\binom{n+k}{k}$ with $n \geq k$ be a square? We will show that if so, then $k = 0, 1, 2,$ or 3.

6.9. Small k

Evidently $\binom{n}{0} = 1$ is always a square. Moreover $\binom{n+1}{1}$ is a square if and only if $n = m^2 - 1$ for some integer $m \geq 1$.

Lemma 6.9.1. *The positive integers n for which $\binom{n+2}{2}$ is a square are given exactly by the integer solutions to*
$$x^2 - 2y^2 = \pm 1.$$
When the sign is -1 let $n = x^2 - 1$, and when the sign is $+1$ let $n = x^2 - 2$ so that $\binom{n+2}{2} = (xy)^2$ in either case.

The integer solutions to this equation, the Pell equation, will be discussed at length in section 11.2.

Proof. If $\binom{n+2}{2} = w^2$, then $(n+2)(n+1) = 2w^2$. Now $(n+2, n+1) = 1$ so either $n + 2 = x^2$ and $n + 1 = 2y^2$, or $n + 2 = 2y^2$ and $n + 1 = x^2$, yielding our result. □

One finds an infinite set of solutions by induction: We begin with $\binom{9}{2} = 6^2$, which corresponds to $3^2 - 2 \cdot 2^2 = 1$ above. Now given any solution to $\binom{N}{2} = m^2$ we obtain a larger solution
$$\binom{(2N-1)^2}{2} = (4N-2)^2 \cdot \binom{N}{2} = ((4N-2)m)^2.$$
We perform this operation repeatedly to find an infinite set of solutions.

For $k = 3$ we write $t = n + 2 \in \mathbb{Z}$ so that if $\binom{n+3}{3} = m^2$, then
$$t^3 - t = 6m^2,$$

giving the right-angled triangle with side lengths $(2t/m, (t^2-1)/m, (t^2+1)/m)$ of area 6. We will determine the infinitely many rational-sided right-angled triangles of area 6 in appendix 17B, obtaining the integral solution $\binom{50}{3} = 140^2$ (as well as $\binom{3}{3}$ and $\binom{4}{3}$) to our problem, but no others.

6.10. Larger k

For larger k we can prove there are no solutions using our understanding of prime divisors of binomial coefficients:

Theorem 6.4. *There are no integers $n \geq k \geq 4$ for which $\binom{n+k}{k}$ is the square of an integer.*

Proof. We write each $n+j = a_j m_j^2$ with each a_j squarefree. If $\binom{n+k}{k} = m^2$, then

(6.10.1) $$a_1 \cdots a_k (m_1 \cdots m_k)^2 = k! m^2.$$

We first show that if p is a prime dividing a_i, then $p \leq k$: Otherwise $p > k$, and so the power of p dividing $k! m^2$ is the same as the power of p dividing m^2, which is even. Therefore the power of p dividing $a_1 \cdots a_k$ is even by (6.10.1), and so p divides an even number of the a_j, since each a_j is squarefree. Since p divides a_i, this implies that p divides a_j for some $j \neq i$. Therefore p divides $(n+j) - (n+i) = j-i$ and so $p \leq |j-i| < k$, a contradiction.

By the theorem of Sylvester and Schur (Theorem 5.7) we know that some $n+j$ has a prime factor $p > k$ and so p divides m_j. Therefore $n+k \geq n+j = a_j m_j^2 \geq m_j^2 \geq p^2 \geq (k+1)^2$; that is, $n \geq k^2 + k + 1$.

Let $g = (a_1 \cdots a_k, k!)$. Dividing g out from both sides of (6.10.1), we deduce that $a_1 \cdots a_k / g$ is a square. Now the power of prime p dividing $a_1 \cdots a_k$ is $\leq \#\{j : p|n+j\}$ as each a_j is squarefree, and this equals $[\frac{n+k}{p}] - [\frac{n}{p}]$. The power of p dividing $k!$ is $\geq [\frac{k}{p}]$. Therefore the power of p dividing $a_1 \cdots a_k / g$ is

$$\leq \left[\frac{n+k}{p}\right] - \left[\frac{n}{p}\right] - \left[\frac{k}{p}\right] \leq 1.$$

But this must be even as $a_1 \cdots a_k / g$ is a square, and so equals 0. Therefore $a_1 \cdots a_k / g = 1$; that is, $g = a_1 \cdots a_k$ divides $k!$.

If the a_j are distinct positive integers, then the smallest must be ≥ 1, the next smallest ≥ 2, etc. Therefore $k! \leq a_1 \cdots a_k \leq k!$, and so the numbers $a_1, \ldots, a_k$ must be the numbers $1, 2, \ldots, k$ arranged in some order. Therefore if $k \geq 4$, then some a_j equals 4, contradicting that the a_j are squarefree.

Therefore two a_j are the same, say $a_i = a_j$ where $i < j$. Then

$$k > j - i = (n+j) - (n+i) = a_i(m_j^2 - m_i^2) \geq a_i((m_i+1)^2 - m_i^2)$$
$$> 2a_i m_i \geq 2\sqrt{n+i} > 2\sqrt{k^2+k}$$

which is impossible. □

Exercise 6.10.1.[‡] Show that $\binom{n+k}{k}$ cannot be an ℓth power for any $\ell \geq 3$.

Chapter 7

Power residues

We begin by calculating the least residues of the small powers of each given residue mod m, to look for interesting patterns:

a^0	a	a^2
1	0	0
1	1	1

Least power residues (mod 2).

a^0	a	a^2	a^3	a^4	a^5
1	0	0	0	0	0
1	1	1	1	1	1
1	2	1	2	1	2

Least power residues (mod 3).[1]

In these small examples, the columns soon settle into repeating patterns as we go from left to right: For example, in the mod 3 case, the columns alternate between $0, 1, 1$ and $0, 1, 2$. How about for slightly larger moduli?

a^0	a	a^2	a^3	a^4	a^5
1	0	0	0	0	0
1	1	1	1	1	1
1	2	0	0	0	0
1	3	1	3	1	3

Least power residues (mod 4).

a^0	a	a^2	a^3	a^4	a^5
1	0	0	0	0	0
1	1	1	1	1	1
1	2	4	3	1	2
1	3	4	2	1	3
1	4	1	4	1	4

Least power residues (mod 5).

---

[1] Why did we take 0^0 to be 1 (mod m) for $m = 2, 3, 4$, and 5? In mathematics we create symbols and protocols (like taking powers) to represent numbers and actions on those numbers, and then we need to be able to interpret all combinations of those symbols and protocols. Occasionally some of those combinations do not have an immediate interpretation, for example 0^0. So how do we deal with this? Usually mathematicians develop a convenient interpretation that allows that not-well-defined use of a protocol to nonetheless be consistent with the many appropriate uses of the protocol. Therefore, for example, we let 0^0 be 1, because it is true that $a^0 = 1$ for *every* non-zero number a, so it makes sense (and is often convenient) to define this to also be so for $a = 0$.

Perhaps the best known dilemma of this sort comes in asking whether ∞ is a number. The correct answer is "No, it is a symbol" (representing an upper bound on the set of real numbers) but it is certainly convenient to treat it as a number in many situations.

Again the patterns repeat, every second power mod 4, and every fourth power mod 5. Our goal in this chapter is to understand the power residues, and in particular when we get these repeated patterns.

7.1. Generating the multiplicative group of residues

We begin by verifying that for each coprime pair of integers a and m, the power residues do repeat periodically:

Lemma 7.1.1. *For any integer a, with $(a, m) = 1$, there exists an integer k, $1 \leq k \leq \phi(m)$, for which $a^k \equiv 1 \pmod{m}$.*

Proof. Each term of the sequence $1, a, a^2, a^3, \ldots$ is coprime with m by exercise 3.3.5. But then each is congruent to some element from any given reduced set of residues mod m (which has size $\phi(m)$). Therefore, by the pigeonhole principle, there exist i and j with $0 \leq i < j \leq \phi(m)$ for which $a^i \equiv a^j \pmod{m}$.

Next we divide both sides of this equation by a^i. To justify doing this we observe that $(a^i, m) = 1$ (as $(a, m) = 1$) and so we can use Corollary 3.5.1 to obtain our result with $k = j - i$, so that $1 \leq k \leq \phi(m)$. □

Exercise 7.1.1. (a) Show that for *any* integers a and $m \geq 2$, there exist integers i and k, with $0 \leq i \leq m - 1$ and $1 \leq k \leq m - i$ such that $a^{n+k} \equiv a^n \pmod{m}$ for every $n \geq i$.
(b) For each integer $m \geq 2$ determine an integer a such that $a \not\equiv 1 \pmod{m}$ but $a^2 \equiv a \pmod{m}$. (This explains why we need the hypothesis that $(a, m) = 1$ in Lemma 7.1.1.)

Another proof of Corollary 3.5.2. [*If $(a, m) = 1$, then a has an inverse* mod m.] Let $r = a^{k-1}$ so that $ar = a^k \equiv 1 \pmod{m}$. □

Examples. In the geometric progression $2, 4, 8, \ldots$, the first term $\equiv 1 \pmod{13}$ is $2^{12} = 4096$. The first term $\equiv 1 \pmod{23}$ is $2^{11} = 2048$. Similarly $5^6 = 15625 \equiv 1 \pmod{7}$ but $5^5 \equiv 1 \pmod{11}$. We see that in some cases the power needed is as big as $\phi(p) = p - 1$, the bound given by Lemma 7.1.1, but not always.

If $a^k \equiv 1 \pmod{m}$, then $a^{k+j} \equiv a^j \pmod{m}$ for all $j \geq 0$, and so the geometric progression $a^0, a^1, a^2, \ldots$ modulo m has period k. Thus if $u \equiv v \pmod{k}$, then $a^u \equiv a^v \pmod{m}$. Therefore one can easily determine the residues of powers $\pmod{m}$. For example, to compute $3^{1000} \pmod{13}$, first note that $3^3 \equiv 1 \pmod{13}$. Now $1000 \equiv 1 \pmod{3}$, and so $3^{1000} \equiv 3^1 = 3 \pmod{13}$.

If $(a, m) = 1$, then let $\text{ord}_m(a)$, the *order of a* $\pmod{m}$, denote the smallest positive integer k for which $a^k \equiv 1 \pmod{m}$. We know that there must be such an integer, by Lemma 7.1.1. We have $\text{ord}_3(2) = \text{ord}_4(3) = 2$, $\text{ord}_5(2) = \text{ord}_5(3) = 4$ (from the tables above), and $\text{ord}_{13}(2) = 12$, $\text{ord}_{23}(2) = 11$, $\text{ord}_7(5) = 6$, and $\text{ord}_{11}(5) = 5$ from the examples above. The powers of 3 $\pmod{16}$ are $1, 3, 9, 3^3 \equiv 11, 3^4 \equiv 1, 3, 9, 11, 1, 3, 9, 11, 1, \ldots$ so that the residues are periodic with period $\text{ord}_{16}(3) = 4$.

Lemma 7.1.2. *Suppose that a and m are coprime integers with $m \geq 1$. Then n is an integer for which $a^n \equiv 1 \pmod{m}$ if and only if $\text{ord}_m(a)$ divides n.*

7.2. Fermat's Little Theorem

Proof. Let $k := \text{ord}_m(a)$ so that $a^k \equiv 1 \pmod{m}$. Suppose that n is an integer for which $a^n \equiv 1 \pmod{m}$. There exist integers q and r such that $n = qk+r$ where $0 \leq r \leq k-1$. Hence $a^r = a^n/(a^k)^q \equiv 1/1^q \equiv 1 \pmod{m}$. Therefore $r=0$ by the minimality of k (from the definition of order), and so k divides n as claimed.

In the other direction, if k divides n, then $a^n = (a^k)^{n/k} \equiv 1 \pmod{m}$. □

Exercise 7.1.2. Let $k := \text{ord}_m(a)$ where $(a,m) = 1$.
(a) Show that $1, a, a^2, \ldots, a^{k-1}$ are distinct $\pmod{m}$.
(b) Deduce that $a^j \equiv a^i \pmod{m}$ if and only if $j \equiv i \pmod{k}$.

We see that $\text{ord}_m(a)$ is the smallest period of the sequence $1, a, a^2, \ldots \pmod{m}$.

We wish to understand the possible values of $\text{ord}_m(a)$, especially for fixed m, as a varies over integers coprime to m. We begin by taking $m = p$ prime. The theory for composite m can be deduced from an understanding of the prime power modulus case, using the Chinese Remainder Theorem as determined in detail in section 7.18 of appendix 7B.

Theorem 7.1. *If p is a prime and p does not divide a, then $\text{ord}_p(a)$ divides $p-1$.*

Proof. Let $k := \text{ord}_p(a)$ and $A = \{1, a, a^2, \ldots, a^{k-1} \pmod{p}\}$. For any non-zero b $\pmod{p}$ define the set $bA = \{b\alpha \pmod{p} : \alpha \in A\}$.

Let b and b' be any two reduced residues mod p. We now show that either bA and $b'A$ are disjoint or they are equal: If they have an element, c, in common, then there exists $0 \leq i, j \leq k-1$ such that $ba^i \equiv c \equiv b'a^j \pmod{p}$. Therefore $b' \equiv ba^h$ $\pmod{p}$ where h is the least non-negative residue of $i-j \pmod{k}$. Hence

$$b'a^\ell \equiv \begin{cases} ba^{h+\ell} \pmod{p} & \text{if } 0 \leq \ell \leq k-1-h, \\ ba^{h+\ell-k} \pmod{p} & \text{if } k-h \leq \ell \leq k-1, \end{cases}$$

which implies that $b'A \subset bA$. Since the two sets are finite and of the same size they must be identical.

Since any two sets of the form bA are either identical or disjoint, we deduce that they partition the non-zero elements mod p. That is, the reduced residues $1, \ldots, p-1 \pmod{p}$ may be partitioned into disjoint *cosets* bA, of A, each of which has size $|A|$; and therefore $|A| = k$ divides $p-1$. □

To highlight this proof let $a = 5$ and $p = 13$ so that $A = \{1, 5, 5^2 \equiv 12, 5^3 \equiv 8 \pmod{13}\}$. Then the *cosets* A, $2A \equiv \{2, 10, 11, 3 \pmod{13}\}$, and $4A \equiv \{4, 7, 9, 6 \pmod{13}\}$ partition the reduced residues mod 13, and therefore $3|A| = 12$. Also note that $7A \equiv \{7, 9, 6, 4 \pmod{13}\} = 4A$, as claimed, the same residues but in a rotated order.

7.2. Fermat's Little Theorem

Theorem 7.1 limits the possible values of $\text{ord}_p(a)$. The beauty of the proof of Theorem 7.1, which is taken from Gauss's *Disquisitiones Arithmeticae*, is that it works in any finite group, as we will see in Proposition 7.22.1 of appendix 7D.[2] This

---
[2] What is especially remarkable is that Gauss produced this surprising proof before anyone had thought up the abstract notion of a group!

result leads us directly to one of the great results of elementary number theory, first observed by Fermat in a letter to Frénicle on October 18, 1640:

Theorem 7.2 (Fermat's "Little" Theorem). *If p is a prime and a is an integer that is not divisible by p, then*

$$p \text{ divides } a^{p-1} - 1.$$

Proof. We know that $\mathrm{ord}_p(a)$ divides $p-1$ by Theorem 7.1, and therefore $a^{p-1} \equiv 1 \pmod{p}$ by Lemma 7.1.2. □

Here is a useful reformulation of Fermat's "Little" Theorem:

Fermat's Little Theorem, v2. *If p is a prime and a is a positive integer, then*

$$p \text{ divides } a^p - a.$$

Exercise 7.2.1. Prove that our two versions of Fermat's Little Theorem are *equivalent* to each other (that is, easily imply one another).

We now present several different proofs of Fermat's "Little" Theorem and then a surprising proof in appendix 7A.

"Sets of reduced residues" proof. In exercise 3.5.2 we proved that $\{a \cdot 1, a \cdot 2, \ldots, a \cdot (p-1)\}$ form a reduced set of residues mod p. The residues of these integers mod p are therefore the same as the residues of $\{1, 2, \ldots, p-1\}$ although in a different order. Since the two sets are the same mod p, the products of the elements of each set are equal mod p, and so

$$(a \cdot 1)(a \cdot 2) \cdots (a \cdot (p-1)) \equiv 1 \cdot 2 \cdots (p-1) \pmod{p};$$

that is,

$$a^{p-1} \cdot (p-1)! \equiv (p-1)! \pmod{p}.$$

As $(p, (p-1)!) = 1$, we can divide the $(p-1)!$ out from both sides to obtain the desired

$$a^{p-1} \equiv 1 \pmod{p}.$$
□

Euler's 1741 proof. We shall show that $a^p - a$ is divisible by p for every integer $a \geq 1$. We proceed by induction on a: For $a = 1$ we have $1^{p-1} - 1 = 0$, and so the result is trivial. Otherwise, by the binomial theorem,

$$(a+1)^p - a^p - 1 = \sum_{i=1}^{p-1} \binom{p}{i} a^i \equiv 0 \pmod{p},$$

as p divides the numerator but not the denominator of $\binom{p}{i}$ for each $i, 1 \leq i \leq p-1$ (as in exercise 2.5.8). Reorganizing we obtain

$$(a+1)^p - (a+1) \equiv (a^p + 1) - (a+1) \equiv a^p - a \equiv 0 \pmod{p},$$

the last congruence following from the induction hypothesis. □

7.2. Fermat's Little Theorem

Combinatorial proof. The numerator, but not the denominator, of the multinomial coefficient $\binom{p}{i,j,k,\ldots}$ is divisible by p unless one of $i, j, k, \ldots$ equals p and the others equal 0. In this case the multinomial coefficient equals 1. Therefore, by the multinomial theorem,[3]

$$(a+b+c+\cdots)^p \equiv a^p + b^p + c^p + \cdots \pmod{p}.$$

Taking $a = b = c = \cdots = 1$ gives $\ell^p \equiv \ell \pmod{p}$ for all integers $\ell \geq 1$. □

Another proof of Theorem 7.1. Theorem 7.1 follows from Fermat's Little Theorem and Lemma 7.1.2 with $m = p$ and $n = p - 1$. (This is not a circular argument as our last three proofs of Fermat's Little Theorem do not use Theorem 7.1.) □

We can use Fermat's Little Theorem to help quickly determine large powers in modular arithmetic. For example for $2^{1000001} \pmod{31}$, we have $2^{30} \equiv 1 \pmod{31}$ by Fermat's Little Theorem, and so, as $1000001 \equiv 11 \pmod{30}$, we obtain $2^{1000001} \equiv 2^{11} \pmod{31}$ and it remains to do the final calculation. However, using the order makes this calculation significantly easier: Since $\text{ord}_{31}(2) = 5$ we have $2^5 \equiv 1 \pmod{31}$ and therefore, as $1000001 \equiv 1 \pmod 5$, we obtain $2^{1000001} \equiv 2^1 \equiv 2 \pmod{31}$.

It is worth stating the converse to Fermat's Little Theorem:

Corollary 7.2.1. *If $(a, n) = 1$ and $a^{n-1} \not\equiv 1 \pmod n$, then n is composite.*

For example $(2, 15) = 1$ and $2^4 = 16 \equiv 1 \pmod{15}$ so that $2^{14} \equiv 2^2 \equiv 4 \pmod{15}$. Hence 15 is a composite number. The surprise here is that we have proved that 15 is composite without having to factor 15. Indeed whenever Corollary 7.2.1 is applicable we will not have to factor n to show that it is composite. This is important because we do not know a fast way to factor an arbitrarily large integer n, but one can compute rapidly with Corollary 7.2.1 (as discussed in section 7.13 of appendix 7A). We will discuss such compositeness tests in section 7.6.

Exercise 7.2.2. Prove that for any $m > 1$ if $(a, m) = 1$, then $\text{ord}_m(a)$ divides $\phi(m)$ (by an analogous proof to that of Theorem 7.1).

Theorem 7.3 (Euler's Theorem). *For any $m > 1$ if $(a, m) = 1$, then $a^{\phi(m)} \equiv 1 \pmod m$.*

Proof. By definition $a^{\text{ord}_m(a)} \equiv 1 \pmod m$. By exercise 7.2.2 there exists an integer k for which $\phi(m) = k \, \text{ord}_m(a)$ and so $a^{\phi(m)} = (a^{\text{ord}_m(a)})^k \equiv 1 \pmod m$. □

This result and proof generalizes even further, to any finite group, as we will see in Corollary 7.23.1 of appendix 7D.

Exercise 7.2.3. Prove Euler's Theorem using the idea in the "sets of reduced residues" proof of Fermat's Little Theorem, given above.

Exercise 7.2.4. Determine the last decimal digit of 3^{8643}.

---

[3] For the reader who has seen it before.

7.3. Special primes and orders

We now look at prime divisors of the Mersenne and Fermat numbers using our results on orders.

Exercise 7.3.1. Show that if p is prime and q is a prime dividing $2^p - 1$, then $\text{ord}_q(2) = p$.

Hence, by exercise 7.3.1, if q divides $2^p - 1$, then $p = \text{ord}_2(q)$ divides $q - 1$ by Theorem 7.1; that is, $q \equiv 1 \pmod{p}$.

Another proof that there are infinitely many primes. If p is the largest prime, let q be a prime factor of $2^p - 1$. We have just seen that p divides $q - 1$, so that $p \leq q - 1 < q$. This contradicts the assumption that p is the largest prime. □

Exercise 7.3.2.† Show that if prime p divides $F_n = 2^{2^n} + 1$, then $\text{ord}_p(2) = 2^{n+1}$. Deduce that $p \equiv 1 \pmod{2^{n+1}}$.

Theorem 7.4. *Fix $k \geq 2$. There are infinitely many primes $\equiv 1 \pmod{2^k}$.*

Proof. If p_n is a prime factor of $F_n = 2^{2^n} + 1$, then $p_n \equiv 1 \pmod{2^k}$ for all $n \geq k - 1$, by exercise 7.3.2. We saw that the p_n are all distinct in section 5.1. □

7.4. Further observations

Lemma 5.7.1, a weak form of the Fundamental Theorem of Algebra (Theorem 3.11), states that any polynomial in $\mathbb{C}[x]$ of degree d has at most d roots. An analogous result can be proved for polynomials mod p.

Proposition 7.4.1 (Lagrange). *Suppose that p is a prime and that $f(x)$ is a non-zero polynomial with coefficients in $\mathbb{Z}/p\mathbb{Z}$ of degree d. Then $f(x)$ has no more than d roots mod p (counted with multiplicity).*

Proof. By induction on $d \geq 0$. This is trivial for $d = 0$. For $d \geq 1$ we will suppose that $f(a) \equiv 0 \mod p$. Then write $f(x) = \sum_{i=0}^{d} f_i x^i$ and define

$$g(x) = \frac{f(x) - f(a)}{x - a} = \sum_{i=0}^{d} f_i \frac{x^i - a^i}{x - a} = \sum_{i=0}^{d} f_i (x^{i-1} + ax^{i-2} + \cdots + a^{i-1}),$$

a polynomial of degree $d - 1$ with leading coefficient f_d (so is non-zero). Therefore $g(x)$ has no more than $d - 1$ roots mod p, by the induction hypothesis. Now

$$f(x) = f(x) - f(a) = (x - a)g(x)$$

and so if $f(b) \equiv 0 \pmod{p}$, then $(b - a)g(b) \equiv 0 \pmod{p}$. Either $b \equiv a \pmod{p}$ or $g(b) \equiv 0 \pmod{p}$, and so f has no more than $1 + (d - 1) = d$ roots mod p. □

Fermat's Little Theorem implies that $1, 2, 3, \ldots, p-1$ are $p-1$ distinct roots of $x^{p-1} - 1 \pmod{p}$, and are therefore all the roots, by Proposition 7.4.1. Therefore the polynomials $x^{p-1} - 1$ and $(x - 1)(x - 2) \cdots (x - (p - 1))$ mod p are the same up to a multiplicative constant. Since they are both monic they must be identical; that is,

(7.4.1) $\qquad x^{p-1} - 1 \equiv (x - 1)(x - 2) \cdots (x - (p - 1)) \pmod{p}$,

which implies that
$$x^p - x \equiv x(x-1)(x-2)\cdots(x-(p-1)) \pmod{p}.$$

Theorem 7.5 (Wilson's Theorem). *For any prime p we have*
$$\boxed{(p-1)! \equiv -1 \pmod{p}.}$$

Proof. Take $x = 0$ in (7.4.1), and note that $(-1)^{p-1} \equiv 1 \pmod{p}$. □

Gauss's proof of Wilson's Theorem. Let S be the set of pairs (a,b) for which $1 \leq a < b < p$ and $ab \equiv 1 \pmod{p}$; that is, every residue is paired up with its inverse unless it equals its inverse. Now if $a \equiv a^{-1} \pmod{p}$, then $a^2 \equiv 1 \pmod{p}$, in which case $a \equiv 1$ or $p-1 \pmod{p}$ by Lemma 3.8.1. Therefore
$$1 \cdot 2 \cdots (p-1) = 1 \cdot (p-1) \cdot \prod_{(a,b) \in S} ab \equiv 1 \cdot (-1) \cdot \prod_{(a,b) \in S} 1 \equiv -1 \pmod{p}. \quad \square$$

Example. For $p = 13$ we have
$$12! = 12(2 \times 7)(3 \times 9)(4 \times 10)(5 \times 8)(6 \times 11) \equiv -1 \cdot 1 \cdot 1 \cdot 1 \cdot 1 \cdot 1 \equiv -1 \pmod{13}.$$

Exercise 7.4.1. (a) Show that if $n > 4$ is composite, then n divides $(n-1)!$.
(b) Show that $n \geq 2$ is prime if and only if n divides $(n-1)! + 1$.

Combining Wilson's Theorem with the last exercise we have an indirect primality test for integers $n > 2$: Compute $(n-1)! \pmod{n}$. If it is $\equiv -1 \pmod{n}$, then n is prime; if it is $\equiv 0 \pmod{n}$, then n is composite. Note however that in determining $(n-1)!$ we need to do $n-2$ multiplications, so that this primality test takes far more steps than trial division (see section 5.2)!

Exercise 7.4.2. (a) Use the idea in Gauss's proof of Wilson's Theorem to show that
$$\prod_{\substack{1 \leq a \leq n \\ (a,n)=1}} a \equiv \prod_{\substack{1 \leq b \leq n \\ b^2 \equiv 1 \pmod{n}}} b \pmod{n}.$$
(b) Evaluate this product using exercise 3.8.3 or by pairing b with $n - b$.

Exercise 7.4.3. (a) Show that $\binom{p-1}{(p-1)/2} \equiv (-1)^{(p-1)/2} \pmod{p}$.
(b) Deduce that if $p \equiv 3 \pmod{4}$, then $\left(\frac{p-1}{2}\right)! \equiv 1$ or $-1 \pmod{p}$.
(c) Deduce that if $p \equiv 1 \pmod{4}$, then $\left(\frac{p-1}{2}\right)!$ is a root of $x^2 \equiv -1 \pmod{p}$.[4]

7.5. The number of elements of a given order, and primitive roots

In Theorem 7.1 we saw that the order modulo p of any integer a which is coprime to p must be an integer which divides $p - 1$. In this section we show that for each divisor m of $p - 1$, there are residue classes mod p of order m.

---

[4]This explicitly provides a square root of $-1 \pmod{p}$ which is interesting, as there is no easy way in general to determine square roots mod p. However we do not know how to rapidly calculate the least residue of $\left(\frac{p-1}{2}\right)! \pmod{p}$.

Example. For the primes $p = 13$ and $p = 19$ we have

Order	$a \pmod{13}$
1	1
2	12
3	3, 9
4	5, 8
6	4, 10
12	2, 6, 7, 11

Order	$a \pmod{19}$
1	1
2	18
3	7, 11
6	8, 12
9	4, 5, 6, 9, 16, 17
18	2, 3, 10, 13, 14, 15

How many residues are there of each order? From these examples we might guess the following result.

Theorem 7.6. *If m divides $p-1$, then there are exactly $\phi(m)$ elements $a \pmod{p}$ of order m.*

A *primitive root* a mod p is a reduced residue mod p of order $p-1$. The least residues of the powers

$$1, a, a^2, a^3, \ldots, a^{p-2} \pmod{p}$$

are distinct reduced residues by exercise 7.1.2 and so must equal

$$1, 2, \ldots, p-1$$

in some order. Therefore every reduced residue is congruent to some power a^j (mod p) of a, and the power j can be reduced mod $p-1$. For example, 2, 3, 10, 13, 14, and 15 are the primitive roots mod 19. We can verify that the powers of 3 mod 19 yield a reduced set of residues:

$$1, 3, 3^2, 3^3, 3^4, 3^5, 3^6, 3^7, 3^8, 3^9, 3^{10}, 3^{11}, 3^{12}, 3^{13}, 3^{14}, 3^{15}, 3^{16}, 3^{17}, 3^{18}, \ldots$$
$$\equiv 1, 3, 9, 8, 5, -4, 7, 2, 6, -1, -3, -9, -8, -5, 4, -7, -2, -6, 1, \ldots \pmod{19},$$

respectively, so 3 is a primitive root mod p. Taking $m = p-1$ in Theorem 7.6 we obtain the following:

Corollary 7.5.1. *For every prime p there exists a primitive root* mod p. *In fact there are $\phi(p-1)$ distinct primitive roots* mod p.

To prove Theorem 7.6 it helps to first establish the following lemma:

Lemma 7.5.1. *If m divides $p-1$, then there are exactly m elements $a \pmod{p}$ for which $a^m \equiv 1 \pmod{p}$.*

Proof. We saw in (7.4.1) that

$$x^{p-1} - 1 = (x^m - 1)(x^{p-1-m} + x^{p-1-2m} + \cdots + x^m + 1)$$

factors into distinct linear factors mod p, and therefore $x^m - 1$ does so also. □

The residue $a \pmod{p}$ is counted in Lemma 7.5.1 if and only if the order of a divides m. Now we prove Theorem 7.6 which counts the number of residue classes $a \pmod{p}$ whose order is exactly m.

7.5. The number of elements of a given order, and primitive roots

Proof of Theorem 7.6. Let $\psi(d)$ denote the number of elements $a \pmod p$ of order d. The set of roots of $x^m - 1 \pmod p$ is precisely the union of the sets of residue classes mod p of order d, over each d dividing m, so that

$$\sum_{d \mid m} \psi(d) = m \tag{7.5.1}$$

for all positive integers m dividing $p - 1$, by Lemma 7.5.1. We now prove that $\psi(m) = \phi(m)$ for all m dividing $p - 1$, by induction on m. The only element of order 1 is 1 $\pmod p$, so that $\psi(1) = 1 = \phi(1)$. For $m > 1$ we have $\psi(d) = \phi(d)$ for all $d < m$ that divide m, by the induction hypothesis. Therefore

$$\psi(m) = m - \sum_{\substack{d \mid m \\ d < m}} \psi(d) = m - \sum_{\substack{d \mid m \\ d < m}} \phi(d) = \phi(m),$$

the last equality following from Proposition 4.1.1. The result follows. □

Although there are many primitive roots mod p ($\phi(p-1)$ of them by Theorem 7.6), it is not obvious how to always find one rapidly. In section 7.15 of appendix 7B we will present Gauss's practical algorithm for finding primitive roots (as well as special cases in exercises 8.9.20, 8.9.21, and 8.9.22).

It is believed that 2 is a primitive root mod p for infinitely many primes p though this remains an open question. Artin's *primitive root conjecture* states that every prime q is a primitive root mod p for infinitely many primes p. This is known to be true for all, but at most two, primes.[5] Gauss himself conjectured that 10 is a primitive root mod p for infinitely many primes p and this is also an open question. Any integer m, which is neither a perfect square nor -1, is conjectured to be a primitive root mod p for infinitely many primes p.

Corollary 7.5.2. *For every prime p and every integer k, we have*

$$1^k + 2^k + \cdots + (p-1)^k \equiv \begin{cases} 0 & \text{if } p-1 \nmid k \\ -1 & \text{if } p-1 \mid k \end{cases} \pmod p.$$

Proof. Let $S_k := 1^k + 2^k + \cdots + (p-1)^k$. If $p-1$ divides k, then each $j^k \equiv 1 \pmod p$ by Fermat's Little Theorem and so $S_k \equiv 1 + \cdots + 1 = p - 1 \pmod p$, as claimed. So, henceforth assume that $p - 1$ does not divide k.

Let a be a primitive root mod p, so that $a^k \not\equiv 1 \pmod p$ since $p - 1$ does not divide k. The integers $\{a \cdot 1, a \cdot 2, \ldots, a \cdot (p-1)\}$ form a reduced set of residues mod p and so are a rearrangement of the residues of $\{1, 2, \ldots, p-1\}$ mod p. Therefore any symmetric function of these two sets of integers residues are congruent mod p (as we saw in the "Sets of reduced residues" proof of Fermat's Little Theorem); in particular,

$$S_k \equiv \sum_{j=1}^{p-1} (aj)^k = a^k S_k \pmod p.$$

Therefore $(a^k - 1)S_k \equiv 0 \pmod p$ but $a^k \not\equiv 1 \pmod p$ and so $S_k \equiv 0 \pmod p$. □

---

[5]This result is strangely formulated because of the nature of what was proved (by Heath-Brown [2], improving a result of Gupta and Murty, see [3])—that in any set of three distinct primes q_1, q_2, q_3, at least one is a primitive root mod p for infinitely many primes p. Therefore there cannot be three exceptions to the conjecture, and we believe that there are none.

Near the beginning of this section we noted that if a is a primitive root (mod p), then every reduced residue is congruent to some power a^j (mod p). This property is extremely useful for it allows us to treat multiplication as addition of exponents in the same way that the introduction of logarithms simplifies usual multiplication. We will discuss this further in section 7.16 of appendix 7B.

Exercise 7.5.1. Write each reduced residue mod p as a power of the primitive root a, and use this to evaluate $1^k + 2^k + \cdots + (p-1)^k$ (mod p) as a function of a and k. Use this to give another proof of Corollary 7.5.2.

Exercise 7.5.2. Let g be a primitive root modulo odd prime p.
(a) Prove that $g^a \equiv 1$ (mod p) if and only if $p-1$ divides a.
(b) Show that $g^{(p-1)/2} \equiv -1$ (mod p).

In order to determine the order of an element mod n, one can use the following result:

Proposition 7.5.1. *Suppose that a and n are coprime integers. Then d is the order of a (mod n) if and only if $a^d \equiv 1$ (mod n) and $a^{d/q} \not\equiv 1$ (mod n) for every prime q dividing d.*

Proof. If d is the order of a (mod n), then $a^d \equiv 1$ (mod n) and $a^{d/q} \not\equiv 1$ (mod n) by the definition of order, since $d/q < d$.

On the other hand let $m := \mathrm{ord}_n(a)$. By Lemma 7.1.2 we know that m divides d but does not divide d/q for any prime q dividing d. Therefore q does not divide d/m for any prime q dividing d, so there cannot be any primes q that divide d/m. This implies that $d/m = 1$ and so $\mathrm{ord}_n(a) = m = d$. □

We deduce an important practical way to recognize primitive roots mod p:

Corollary 7.5.3. *Suppose that p is a prime that does not divide integer a. Then a is a primitive root (mod p) if and only if*

$$a^{(p-1)/q} \not\equiv 1 \pmod{p}$$

for all primes q dividing $p - 1$.

Proof. By definition a is a primitive root (mod p) if and only if $m := \mathrm{ord}_p(a) = p - 1$. The result follows from Proposition 7.5.1. □

Exercise 7.5.3. Find all residues of order 5 mod 31, given that $2^5 \equiv 1$ (mod 31).

Exercise 7.5.4. (a) Prove that 2 is a primitive root (mod 13).
(b) *Use this* to determine all of the other primitive roots (mod 13).

Exercise 7.5.5. Let g be a primitive root modulo odd prime p.
(a) Prove that if m divides $p-1$, then g^m has order $\frac{p-1}{m}$.
(b)† Prove that g^k (mod p) is a primitive root mod p if and only if $(k, p-1) = 1$.
(c) Deduce that there are $\phi(p-1)$ primitive roots mod p.

7.6. Testing for composites, pseudoprimes, and Carmichael numbers

In the converse to Fermat's Little Theorem, Corollary 7.2.1, we saw that if an integer n *does not divide* $a^{n-1} - 1$ for some integer a coprime to n, then n is composite. For example, taking $a = 2$ we calculate that

$$2^{1000} \equiv 562 \pmod{1001},$$

so we know that 1001 is composite. We might ask whether this always works. In other words:

Is it true that *if n is composite, then n does not divide $2^n - 2$?*

For, if so, we have a very nice way to distinguish primes from composites. Unfortunately the answer is "no" since, for example,

$$2^{340} \equiv 1 \pmod{341},$$

but $341 = 11 \times 31$. We call 341 a *base-2 pseudoprime*. Note though that

$$3^{340} \equiv 56 \pmod{341},$$

and so the converse to Fermat's Little Theorem, with $a = 3$, implies that 341 is composite.

Are there composites n for which $2^{n-1} \equiv 3^{n-1} \equiv 1 \pmod{n}$? Or $2^{n-1} \equiv 3^{n-1} \equiv 5^{n-1} \equiv 1 \pmod{n}$? Or, even *Carmichael numbers*, composite numbers that "masquerade" as primes in that $a^{n-1} \equiv 1 \pmod{n}$ for all integers a coprime to n? A quick computer search finds the smallest example: $561 = 3 \cdot 11 \cdot 17$. The next few Carmichael numbers are $1105 = 5 \cdot 13 \cdot 17$, then $1729 = 7 \cdot 13 \cdot 19$, etc.

Exercise 7.6.1. Show that squarefree n is a Carmichael number if and only if n is composite and divides $a^n - a$ for all integers a.

Carmichael numbers are a nuisance, masquerading as primes like this (and so preventing a quick and easy, surefire primality test). Calculations reveal that Carmichael numbers are rare, but in 1994 Alford, Pomerance, and I [1] proved that there are infinitely many of them. Here is a more elegant way to recognize Carmichael numbers:

Lemma 7.6.1. *A positive integer n is a Carmichael number if and only if n is squarefree and composite and $p - 1$ divides $n - 1$ for every prime p dividing n.*

Proof. Suppose that n is squarefree and composite and $p - 1$ divides $n - 1$ for every prime p dividing n. If $(a, n) = 1$ and prime p divides n, then $\text{ord}_p(a)$ divides $p - 1$ by Theorem 7.1, which divides $n - 1$, and so $a^{n-1} \equiv 1 \pmod{p}$ by Lemma 7.1.2. Therefore $a^{n-1} \equiv 1 \pmod{n}$ by the Chinese Remainder Theorem as n is squarefree, and so it is a Carmichael number.

Now suppose that n is a Carmichael number. If prime p divides n, then $a^{n-1} \equiv 1 \pmod{p}$ for all integers a coprime to n. In particular, if a is a primitive root mod p, then $p - 1 = \text{ord}_p(a)$ divides $n - 1$ by Lemma 7.1.2.

Now assume that $p^e \| n$ with $e \geq 2$. We note that $(1 + p)^k \equiv 1 + kp \pmod{p^2}$ for all integers $k \geq 1$, by the binomial theorem, so that $\text{ord}_{p^2}(1 + p) = p$. Select $a \equiv 1 + p \pmod{p^e}$ with $a \equiv 1 \pmod{n/p^e}$ so that $(a, n) = 1$. As $p | n$ we have

$1 \equiv (1+p)^n \equiv a^n \equiv a \equiv 1+p \pmod{p^2}$, a contradiction. Therefore n must be squarefree. □

Lemma 7.6.1 imples that $561 = 3 \cdot 11 \cdot 17$ is a Carmichael number as $2, 10$, and 16 divide 560.

Exercise 7.6.2. Show that if n is a Carmichael number, then it is odd.

Exercise 7.6.3.[†] Show that if n is a Carmichael number, then it has at least three prime factors.

Exercise 7.6.4. Prove that if $6m+1$, $12m+1$, and $18m+1$ are all primes, then their product is a Carmichael number. (It is an open problem whether there exist infinitely many such prime triples, though it is not difficult to find examples, like $7 \times 13 \times 19$ and $37 \times 73 \times 109$.)

7.7. Divisibility tests, again

In section 2.4 we found simple tests for the divisibility of integers by 7, 9, 11, and 13, promising to return to this theme later. The key to these earlier tests was that $10 \equiv 1 \pmod{9}$ and $10^3 \equiv -1 \pmod{7 \cdot 11 \cdot 13}$; that is, $\text{ord}_9(10) = 1$ and $\text{ord}_7(10)$, $\text{ord}_{11}(10)$, and $\text{ord}_{13}(10)$ divide 6. For all primes $p \neq 2$ or 5 we know that $k := \text{ord}_p(10)$ is an integer dividing $p-1$. Hence if $n = \sum_{j=0}^{d} n_j 10^j$, then

$$n = \sum_{m \geq 0} \left(\sum_{i=0}^{k-1} n_{km+i} 10^i \right) (10^k)^m \equiv \sum_{m \geq 0} \left(\sum_{i=0}^{k-1} n_{km+i} 10^i \right) \pmod{p},$$

since if $j = km + i$, then $10^j \equiv 10^i \pmod{p}$. In the displayed equation we have cut up the integer n, written in decimal, into blocks of digits of length k and added these blocks together, which is clearly an efficient way to test for divisibility. The length of these blocks, k, is always $\leq p-1$ no matter what the size of n. Therefore we can, in practice, quickly test whether n is divisible by p, once we know the p-divisibility of every integer $< 10^k$ ($\leq 10^{p-1}$).

If $k = 2\ell$ is even, we can do a little better (as we did with $p = 7$, 11, and 13) since $10^\ell \equiv -1 \pmod{p}$, namely that

$$n = \sum_{j=0}^{d} n_j 10^j \equiv \sum_{m \geq 0} \left(\sum_{i=0}^{\ell-1} n_{km+i} 10^i - \sum_{i=0}^{\ell-1} n_{km+\ell+i} 10^i \right) \pmod{p},$$

thus breaking n up into blocks of length $\ell = k/2$.

7.8. The decimal expansion of fractions

The fraction $\frac{1}{3} = .3333\ldots$ is given by a recurring digit 3, so we write it as $.\overline{3}$. More interesting to us are the set of fractions

$$\frac{1}{7} = .\overline{142857}, \quad \frac{2}{7} = .\overline{285714}, \quad \frac{3}{7} = .\overline{428571},$$

$$\frac{4}{7} = .\overline{571428}, \quad \frac{5}{7} = .\overline{714285}, \quad \frac{6}{7} = .\overline{857142}.$$

These decimal expansions of the six fractions $\frac{a}{7}$, $1 \leq a \leq 6$, are each periodic of period length 6, and each contains the same six digits in the same order but starting at a different place. Starting with the period for $1/7$ we find that we go through the fractions $a/7$ with $a = 1, 3, 2, 6, 4, 5$ when we rotate the period one step at a time.

7.8. The decimal expansion of fractions

Do you recognize this sequence of numbers? These are the least positive residues of $10^0, 10^1, 10^2, 10^3, 10^4, 10^5 \pmod{7}$. To prove this, we begin by noting that since $10^6 \equiv 1 \pmod 7$, we have that

$$\frac{10^6 - 1}{7} = 142857 \text{ is an integer,}$$

which is ≤ 6 digits long. Putting the $1/7$ on the other side and dividing through by 10^6, we obtain

$$\frac{1}{7} = \frac{142857}{10^6} + \frac{10^{-6}}{7} = .142857 + \frac{1}{10^6} \cdot \frac{1}{7}.$$

Substituting this expression in for the last term, divided by 10^6, we obtain

$$\frac{1}{7} = .142857 + \frac{.142857}{10^6} + \frac{1}{10^{12}} \cdot \frac{1}{7} = \cdots = .\overline{142857},$$

the final equality by repeating this process infinitely often. Now if we multiply this through by 10, we obtain

$$\frac{10}{7} = 1.\overline{428571}, \text{ so that } \frac{3}{7} = \frac{10}{7} - 1 = .\overline{428571},$$

and similarly, as $10^2 \equiv 2 \pmod 7$,

$$\frac{2}{7} = \frac{10^2}{7} - \left[\frac{10^2}{7}\right] = .\overline{285714}.$$

We obtain all the other decimal expansions analogously.

What happens when we multiply $1/7$ through by 10^k? For example, if $k = 4$, then

$$\frac{10^4}{7} = 1428.\overline{571428} = 1428 + \frac{4}{7}.$$

The part after the decimal point is always $\{\frac{10^k}{7}\}$ which equals $\frac{\ell}{7}$ where ℓ is the least positive residue of $10^k \pmod 7$ (as in exercise 1.7.4(b)). We can now give two results.

Proposition 7.8.1. *Suppose that m is an integer that is coprime to 10. If $1 \leq a < m$, then the decimal expansion of the period for a/m is periodic with period of length $\text{ord}_m(10)$. This is the minimal period length if $(a,m) = 1$.*

Proof sketch. We proceed analogously to the above. Let $n = \text{ord}_m(10)$ and $r = (10^n - 1)a/m$, so that r is a positive integer $< 10^n$. Let $\mathbf{r}$ be the sequence of digits that give the integer r. The same argument as above gives that

$$\frac{a}{m} = \frac{r}{10^n} + \frac{1}{10^n} \cdot \frac{a}{m} = \frac{r}{10^n} + \frac{r}{10^{2n}} + \frac{1}{10^{2n}} \cdot \frac{a}{m} = \cdots = \overline{\mathbf{r}}.$$

On the other hand, if this equation holds and the decimal expansion has period n, then $(10^n - 1)a/m = (10^n - 1).\overline{\mathbf{r}} = \mathbf{r}.\overline{\mathbf{r}} - .\overline{\mathbf{r}} = r$. In other words, $(10^n - 1)a/m$ is the integer r, so that $10^n \equiv 1 \pmod m$ if $(a,m) = 1$. □

Exercise 7.8.1.[†] Suppose that p is an odd prime for which 10 is a primitive root. Let a_k be the least residue of $10^k \pmod p$, and suppose that $a_k/p = .\overline{r_k}$ where $1 \leq r_k < 10^{p-1}$. Prove that r_k is obtained from r_1, by removing the leading k digits and concatenating them on to the end.

Exercise 7.8.2. Prove that the decimal expansion of every rational number is eventually periodic. (One can see why we need "eventually" with the example $\frac{1}{30} = .03333\ldots$.)

7.9. Primes in arithmetic progressions, revisited

We can use the ideas in this chapter to prove that there are infinitely many primes in certain arithmetic progressions $1 \pmod{m}$.

Theorem 7.7. *There are infinitely many primes $\equiv 1 \pmod{3}$.*

Proof. Suppose there are only finitely many primes $\equiv 1 \pmod 3$, say $p_1, p_2, \ldots, p_k$. Let $a = 3p_1p_2 \cdots p_k$, and let q be a prime dividing $a^2 + a + 1$. Now $q \neq 3$ as $a^2 + a + 1 \equiv 1 \pmod 3$. Moreover q divides $a^3 - 1 = (a-1)(a^2 + a + 1)$, but not $a - 1$ (or else $0 \equiv a^2 + a + 1 \equiv 1 + 1 + 1 \equiv 3 \pmod q$ but $q \neq 3$). Therefore $\text{ord}_q(a) = 3$ and so $q \equiv 1 \pmod 3$ by Theorem 7.1. Hence $q = p_j$ for some j, so that q divides a as well as $a^2 + a + 1$, and thus q divides $(a^2 + a + 1) - a(a+1) = 1$, which is impossible. □

This, together with Theorem 5.2, proves that there are infinitely many primes in both of the residue classes $1 \pmod 3$ and $2 \pmod 3$, as predicted from the data at the start of section 5.3.

Exercise 7.9.1. Generalize this argument to primes that are 1 (mod 4), to primes that are 1 (mod 5), and to primes that are 1 (mod 6).

In order to generalize this argument to proving the existence of primes $\equiv 1 \pmod m$ for every integer $m \geq 3$, including composite m, we need to replace the polynomial $a^2 + a + 1$ by one that recognizes when a has order m. Evidently this must be a divisor of the polynomial $a^m - 1$; indeed $a^m - 1$ divided through by all of the factors corresponding to orders which are proper divisors of m. This discussion leads us to define the *cyclotomic polynomials* $\phi_n(t) \in \mathbb{Z}[t]$, inductively, by the requirement

$$(7.9.1) \qquad t^m - 1 = \prod_{d \mid m} \phi_d(t) \qquad \text{for all } m \geq 1,$$

with each $\phi_d(t)$ monic (see also appendix 4E). Therefore $\phi_1(t) = t - 1$, $\phi_2(t) = t + 1$, $\phi_3(t) = t^2 + t + 1$, $\phi_4(t) = t^2 + 1$, $\phi_5(t) = t^4 + t^3 + t^2 + t + 1, \ldots$.

Theorem 7.8. *For any integer $m \geq 2$, there are infinitely many primes $\equiv 1 \pmod m$.*

Proof. Suppose that $p_1, \ldots, p_k$ are all the primes that are $\equiv 1 \pmod m$ and let $a = mp_1 \cdots p_k$. Let q be a prime divisor of $\phi_m(a)$, which divides $a^m - 1$, so that $a^m \equiv 1 \pmod q$. This implies that (q, a) divides $(a^m - 1, a) = 1$ and so $(q, a) = 1$. In particular q is not a p_j and does not divide m.

Let $d = \text{ord}_q(a)$ so that $q \equiv 1 \pmod d$ by Theorem 7.1. Moreover d divides m as $a^m \equiv 1 \pmod q$. But q is not a p_j and so $q \not\equiv 1 \pmod m$, which implies that $d \neq m$, and therefore $d < m$.

Now $\phi_m(x)$ divides $\frac{x^m - 1}{x^d - 1}$ by definition. Substituting in $x = a$ we deduce that q divides both $\frac{a^m - 1}{a^d - 1}$ and $a^d - 1$, so that

$$0 \equiv \frac{a^m - 1}{a^d - 1} = \sum_{j=0}^{m/d-1} (a^d)^j \equiv \sum_{j=0}^{m/d-1} 1 = m/d \pmod q.$$

This implies that q divides m/d, and therefore divides m, which contradicts what we proved above. □

References for this chapter

[1] W. Red Alford, Andrew Granville, and Carl Pomerance, *There are infinitely many Carmichael numbers*, Ann. of Math. **139** (1994), 703–722.

[2] D. R. Heath-Brown, *Artin's conjecture for primitive roots*, Quart. J. Math. Oxford Ser. **37** (1986), 27–38.

[3] M. Ram Murty, *Artin's conjecture for primitive roots*, Math. Intelligencer **10** (1988), 59–67.

Additional exercises

Exercise 7.10.1. Prove that we can write any polynomial $f(x) \bmod p$ of degree $\leq p-1$ as

$$f(x) \equiv \sum_{a=0}^{p-1} f(a)(1-(x-a)^{p-1}) \pmod{p}.$$

Exercise 7.10.2.[†] Prove that if $f(x) \in \mathbb{Z}[x]$ is monic and has degree d and if prime p divides $f(0), f(1), \ldots, f(d)$, then $p \leq d$ and p divides $f(n)$ for all integers n.

Exercise 7.10.3. We will find all powers of 2 and 3 that differ by 1, a special case of Catalan's conjecture mentioned in section 6.4.
  (a) What are the powers of 3 (mod 8)? What are the powers of 2 (mod 8)?
  (b) Show that if $2^n - 3^m \equiv 1 \pmod{8}$ for some positive integers m, n, then $n = 1$ or 2.
  (c) Deduce that the only solutions to $2^n - 3^m = 1$ are $4 - 3 = 2 - 1 = 1$.
  (d) Prove that if $3^m - 2^n = 1$ with m odd, then $m = n = 1$.
  (e) Prove that if $3^{2k} - 2^n = 1$, then both $3^k - 1$ and $3^k + 1$ are powers of 2, and that this is only possible if $k = 1$. We deduce that the only solutions to $3^m - 2^n = 1$ are $3 - 2 = 9 - 8 = 1$.

(This is the proof of Levi ben Gershon from around 1320.)

Exercise 7.10.4.[†] Show that if $\binom{n}{3}$ with $n > 3$ has no more than one prime factor which is > 3, then $n = 3, 4, 5, 6, 8, 9, 10$, or 18. (Use exercise 7.10.3.)

Exercise 7.10.5. (a) Prove that if $a > 1$, then the order of $a \bmod N := a^q - 1$ is exactly q.
  Now let q be a prime.
  (b) Deduce that if prime p divides $a^q - 1$ but not $a - 1$, then p is a prime $\equiv 1 \pmod{q}$.
  (c) Prove that $(\frac{a^q-1}{a-1}, a-1) = (q, a-1)$.
  (d)[†] Prove that there are infinitely many primes $\equiv 1 \pmod{q}$.

Exercise 7.10.6. Let p be an odd prime, and let x, y, and z be pairwise coprime, positive integers.
  (a)[†] Prove that if p divides $z - y$, then $\frac{z^p - y^p}{z - y} \equiv p \pmod{p^2}$.
  (b) Show that if $x^p + y^p = z^p$, then there exists an integer r for which $z - y = r^p$ or $z - y = p^{p-1}r^p$.

(This problem continues on from exercise 3.9.7.)

Exercise 7.10.7. Deduce Theorem 7.6 from (7.5.1) using the Möbius inversion formula (Theorem 4.4).

Exercise 7.10.8. Let p be a prime. Prove that every quadratic non-residue (mod p) is a primitive root if and only if p is a Fermat prime.

Exercise 7.10.9. Suppose that g is a primitive root modulo odd prime p. Prove that $-g$ is also a primitive root mod p if and only if $p \equiv 1 \pmod{4}$.

Exercise 7.10.10. (a) Show that the number of primes up to N equals, exactly,
$$\sum_{2 \leq n \leq N} \frac{n}{n-1} \cdot \left\{ \frac{(n-1)!}{n} \right\} - \frac{2}{3}.$$
(Here $\{t\}$ is the fractional part of t, defined as in exercise 1.7.4(b).)
(b) Suppose that $n > 1$. Show that n and $n+2$ are both odd primes if and only if $n(n+2)$ divides $4((n-1)!+1)+n$.

Exercise 7.10.11. Prove that if $f(x) \in \mathbb{Z}[x]$ has degree $\leq p-2$, then $\sum_{a=0}^{p-1} f(a) \equiv 0 \pmod{p}$.

Exercise 7.10.12.† Let p be an odd prime and k be an odd integer which is $\not\equiv 1 \pmod{p-1}$. Prove that $1^k + 2^k + \cdots + (p-1)^k \equiv 0 \pmod{p^2}$.

Exercise 7.10.13.† Let $a_{n+1} = 2a_n + 1$ for all $n \geq 0$. Can we choose a_0 so that this sequence consists entirely of primes?

We define n to be a *base-b pseudoprime* if n is composite and $b^{n-1} \equiv 1 \pmod{n}$.

Exercise 7.10.14. Show that if n is not prime, then it a base-b pseudoprime if and only if $\text{ord}_{p^k}(b)$ divides $n-1$ for every prime power p^k dividing n.

Exercise 7.10.15. Suppose that n is a squarefree, composite integer.
(a) Show that $\#\{a \pmod{p} : a^{n-1} \equiv 1 \pmod{p}\} = (p-1, n-1)$.
(b) Show that there are $\prod_{p|n}(p-1, n-1)$ reduced residue classes $b \pmod{n}$ for which n is a base-b pseudoprime.

Exercise 7.10.16. (a) Prove that if n is composite, then $\{b \pmod{n} : n$ is a base-b pseudoprime$\}$ is a subgroup of the reduced residues mod n.
(b)† Prove that if n is not a Carmichael number, then it is not a base-b pseudoprime for at least half of the reduced residues $b \pmod{n}$.
(c)† Suppose that p and $2p-1$ are both prime and let $n = p(2p-1)$. Prove that
$$\#\{b \pmod{n} : n \text{ is a base-}b \text{ pseudoprime}\} = \frac{1}{2}\phi(n).$$

Exercise 7.10.17. (a) Show that if p is prime, then the Mersenne number $2^p - 1$ is either a prime or a base-2 pseudoprime.
(b) Show that every Fermat number $2^{2^n} + 1$ is either a prime or a base-2 pseudoprime.
(c) Show that p^2 divides $2^{p-1} - 1$ if and only if p^2 is a base-2 pseudoprime.

None of these criteria guarantee that there are infinitely many base-2 pseudoprimes. However this is provable:

Exercise 7.10.18.† Prove that there are infinitely many base-2 pseudoprimes by proving and developing one of the following two observations:
- Start with 341, and show that if n is a base-2 pseudoprime, then so is $N := 2^n - 1$.
- Prove that if $p > 3$ is prime, then $(4^p - 1)/3$ is a base-2 pseudoprime.

Can you generalize either of these proofs to other bases?

Exercise 7.10.19. Let a, b, c be pairwise coprime positive integers. Prove that there exists a (unique) residue class $m_0 \pmod{abc}$ such that if $m \equiv m_0 \pmod{abc}$ and if $am+1$, $bm+1$, and $cm+1$ are all primes, then their product is a Carmichael number (for example, $a=1, b=2, c=3$ in exercise 7.6.4 with $m_0 = 0$).

Exercise 7.10.20. Let D be a finite set of at least two distinct positive integers, the elements of which sum to n. Suppose that d divides n for every $d \in D$. Prove that if there exists an integer m for which $p_d := dnm + 1$ is prime for every $d \in D$, then $\prod_{d \in D} p_d$ is a Carmichael number. (In particular note the case in which n is perfect and D is the set of proper divisors of n. The perfect number 6, for example, gives rise to the triple $6m+1, 12m+1, 18m+1$, which we explored in exercise 7.6.4.)

Exercise 7.10.21. (a) Prove that .010010000100... is irrational. (Here we put a "1" two digits after the decimal point, then 3 digits later, then 5 digits later, etc., with all the other digits being 0, the spacings between the "1"'s being $p-1$ for each consecutive prime p.)

(b)† Develop this idea to find a large class of irrationals.

Appendix 7A. Card shuffling and Fermat's Little Theorem

In this appendix we will define order in terms of card shuffling, give a combinatorial proof of Fermat's Little Theorem, and discuss quick calculations of powers mod n.

7.11. Card shuffling and orders modulo n

The cards in a 52-card deck can be arranged in $52! \approx 8 \times 10^{67}$ different orders. Between card games we shuffle the cards to make the order of the cards unpredictable. But what if someone can shuffle "perfectly"? How unpredictable will the order of the cards then be? Let's analyze this by carefully figuring out what happens in a "perfect shuffle". There are several ways of shuffling cards, the most common being the *riffle shuffle*. In a riffle shuffle one splits the deck in two, places the two halves in either hand, and then drops the cards, using one's thumbs, in order to more or less interlace the cards from the two decks.

One begins with a deck of 52 cards and, to facilitate our discussion, we will call the top card, card 1, the next card down, card 2, etc. If one performs a *perfect riffle shuffle*, one cuts the cards into two 26 card halves, one half with the cards 1 through 26, the other half with the cards 27 through 52. An "out-shuffle" then interlaces the two halves so that the new order of the cards becomes (from the top) cards

$$1, 27, 2, 28, 3, 29, 4, 30, \ldots.$$

That is, cards $1, 2, 3, \ldots, 26$ go to positions $1, 3, 5, \ldots, 51$, and cards $27, 28, \ldots, 52$ go to positions $2, 4, \ldots, 52$, respectively. We can give formulas for each half:

$$k \to \begin{cases} 2k - 1 & \text{for } 1 \leq k \leq 26, \\ 2k - 52 & \text{for } 27 \leq k \leq 52. \end{cases}$$

7.11. Card shuffling and orders modulo n

These coalesce into one formula $k \to 2k - 1 \pmod{51}$ for all $k, 1 \le k \le 52$. The top and bottom cards do not move, that is, $1 \to 1$ and $52 \to 52$, so we focus on understanding the permutation of the other fifty cards:

Any shuffle induces a *permutation* σ on $\{1, \ldots, 52\}$.[6] For the out-shuffle, $\sigma(1) = 1, \sigma(52) = 52$, and

$\sigma(1 + m)$ is the least positive residue of $1 + 2m \pmod{51}$ for $1 \le m \le 50$.

To determine what happens after two or more out-shuffles, we simply compute the function $\sigma^k(.) \; (= \underbrace{\sigma(\sigma(\ldots \sigma(.))))}_{k \text{ times}}$. Evidently $\sigma^k(1) = 1, \sigma^k(52) = 52$, and then

$\sigma^k(1 + m)$ is the least positive residue of $1 + 2^k m \pmod{51}$ for $1 \le m \le 50$.

Now $2^8 \equiv 1 \pmod{51}$, and so $\sigma^8(1 + m) \equiv 1 + m \pmod{51}$ for all m. Therefore eight perfect out-shuffles return the deck to its original state—so much for the 52! possible orderings!

Eight more perfect out-shuffles will also return the deck to its original state, a total of 16 perfect out-shuffles, and also 24 or 32 or 40, etc. Indeed any multiple of 8. So we see that the order of 2 (mod 51) is 8 and that $2^r \equiv 1 \pmod{51}$ if and only if r is divisible by 8. This shows, we hope, why the notion of order is interesting and exhibits one of the key results (Lemma 7.1.2) about orders.

Exercise 7.11.1.[†] An "in-shuffle" is the riffle shuffle that interlaces the cards the other way; that is, after one shuffle, the order becomes cards $27, 1, 28, 2, 29, \ldots, 52, 26$. Analyze this in an analogous way to the above, and determine how many "in-shuffles" it takes to get the cards back into their original order.

Exercise 7.11.2.[†] What happens when one performs riffle shuffles on n-card decks, with n even?

Exercise 7.11.3.[‡] Suppose that the dealer alternates between in-shuffles and out-shuffles. How many such pairs of shuffles does it take to get the deck of cards back into their original order?

Persi Diaconis is a Stanford mathematics professor who left home at the age of fourteen to learn from sleight-of-hand legend Dai Vernon.[7] It is said that Diaconis can shuffle to obtain any permutation of a deck of playing cards. We are interested in the highest possible order of a shuffle. To analyze this question, remember that a shuffle can be reinterpreted as a permutation σ on $\{1, \ldots, n\}$ (where $n = 52$ for a usual deck). One way to explicitly write down a permutation is to track the orbit of each number. For example, for the permutation σ on 5 elements given by

$$\sigma(1) = 4, \; \sigma(2) = 5, \; \sigma(3) = 1, \; \sigma(4) = 3, \; \sigma(5) = 2,$$

1 gets mapped to 4, which gets mapped to 3, and 3 gets mapped back to 1, whereas 2 gets mapped to 5 and 5 gets mapped back to 2, so we can write

$$\sigma = (1, 4, 3)(2, 5).$$

Each of $(1, 4, 3)$ and $(2, 5)$ is a *cycle*, and cycles cannot be decomposed any further. Any permutation can be decomposed into cycles in a unique way, the analogy of the Fundamental Theorem of Arithmetic, for permutations. What is the order of σ? Now $\sigma^n = (1, 4, 3)^n (2, 5)^n$, so that $\sigma^n(1) = 1$, $\sigma^n(4) = 4$, and $\sigma^n(3) = 3$ if

---

[6]That is, $\sigma : \{1, \ldots, 52\} \to \{1, \ldots, 52\}$ such that the $\sigma(i)$ are all distinct (and so σ has an inverse).
[7]Check out this story, and these larger-than-life characters, on Wikipedia.

and only if 3 divides n, while $\sigma^n(2) = 2$ and $\sigma^n(5) = 5$ if and only if 2 divides n. Therefore σ^n is the identity if and only if 6 divides n, and so σ has order 6.

Exercise 7.11.4. Suppose that σ is a permutation on $\{1, \ldots, n\}$ and that $\sigma = C_1 \cdots C_k$ where $C_1, \ldots, C_k$ are disjoint cycles.
 (a) Show that the order of σ equals the least common multiple of the lengths of the cycles C_j, $1 \leq j \leq k$.
 (b) Use this to find the order of the permutation corresponding to an out-shuffle.
 (c) Prove that if $n_1, \ldots, n_k$ are any set of positive integers for which $n_1 + \cdots + n_k = n$, then there exists a permutation $\sigma = C_1 \cdots C_k$ on $\{1, \ldots, n\}$, where each C_j has length n_j.
 (d) Deduce that the maximum order, $m(n)$, of a permutation σ on $\{1, \ldots, n\}$ is given by
$$\max \operatorname{lcm}[n_1, \ldots, n_k] \text{ over all integers } n_1, \ldots, n_k \geq 1 \text{ for which } n_1 + \cdots + n_k = n.$$

Our goal is to determine $m(52)$, the highest order of any shuffle that Diaconis can perform on a regular deck of 52 playing cards. However it is unclear how to determine $m(n)$ systematically. Working through the possibilities for small n, using exercise 7.11.4, we find that

$$\begin{aligned}
m(5) &= 6 & \text{obtained from} & & 6 &= 3 \cdot 2 & \text{and} & & 5 &= 3 + 2, \\
m(6) &= 6 & \text{obtained from} & & 6 &= 3 \cdot 2 \cdot 1 & \text{and} & & 6 &= 3 + 2 + 1, \\
m(7) &= 12 & \text{obtained from} & & 12 &= 4 \cdot 3 & \text{and} & & 7 &= 4 + 3, \\
m(8) &= 12 & \text{obtained from} & & 12 &= 4 \cdot 3 \cdot 1 & \text{and} & & 8 &= 4 + 3 + 1, \\
m(9) &= 20 & \text{obtained from} & & 20 &= 5 \cdot 4 & \text{and} & & 9 &= 5 + 4, \\
m(10) &= 30 & \text{obtained from} & & 30 &= 5 \cdot 3 \cdot 2 & \text{and} & & 10 &= 5 + 3 + 2, \\
m(11) &= 30 & \text{obtained from} & & 30 &= 6 \cdot 5 & \text{and} & & 11 &= 6 + 5, \\
m(12) &= 60 & \text{obtained from} & & 60 &= 5 \cdot 4 \cdot 3 & \text{and} & & 12 &= 5 + 4 + 3.
\end{aligned}$$

No obvious pattern jumps out (at least to the author) from this data, though one observes one technical issue:

Exercise 7.11.5.[†] Show that there is a permutation $\sigma = C_1 \cdots C_k$ on $\{1, \ldots, n\}$ of order $m(n)$ in which the length of each cycle is either 1 or a power of a distinct prime.

Exercise 7.11.6.[†] Use the previous exercise to determine $m(52)$.

Exercise 7.11.7.[‡] Use exercise 5.4.3 to prove that $\log m(n) \sim \sqrt{n \log n}$.

7.12. The "necklace proof" of Fermat's Little Theorem

Little Sophie has a necklace-making kit, which comes with wires that each accommodate p beads, and unlimited supplies of beads of a different colors. How many genuinely *different* necklaces can be Sophie make? Two necklaces are *equivalent* if they can be obtained from each other by a rotation; otherwise they are different; and so Sophie is asking for the number of equivalence classes of sequences of length p where each entry is selected from a possible colors.

Suppose we have a necklace with the jth bead having color $c(j)$ for $1 \leq j \leq p$. One can rotate the necklace in p different ways: If we rotate the necklace k places for some k in the range $0 \leq k \leq p-1$, then the jth bead will have color $c(j+k)$ for $1 \leq j \leq p$, where $c(.)$ is taken to be a function of period p. If two of these equivalent necklaces are identical, then $c(j+k) = c(j+\ell)$ for all j, for some $0 \leq k < \ell \leq p-1$. Then $c(n+d) = c(n)$ for all n, where $d = \ell - k \in [1, p-1]$, and so $c(md) = c(0)$ for all m; that is, all of the beads in the necklace have the same color.

Therefore we have proved that, other than the a necklaces made of beads of the same color which each belong to an equivalence class of size 1, all other necklaces belong to equivalence classes of size p. Since there are a^p possible sequences of length p with a possible colors for each entry, and a sequences that all have the same color, the total number of equivalence classes (different necklaces) is

$$a + \frac{a^p - a}{p}.$$

In particular, we have established that p divides $a^p - a$ for all a, as desired.[8]

Exercise 7.12.1. Let p be prime. Let X denote a finite set and $f : X \to X$ where $f^p = i$, the identity map. (Here f^p means composing f with itself p times.) Let $X_{\text{fixed}} := \{x \in X : f(x) = x\}$.
  (a) Prove that $|X| \equiv |X_{\text{fixed}}| \pmod{p}$.
      Let G be a finite multiplicative group and $X = \{(x_1, \ldots, x_p) \in G^p : x_1 \cdots x_p = 1\}$.
  (b)[†] Deduce that $\#\{g \in G : g \text{ has order } p\} \equiv |G|^{p-1} - 1 \pmod{p}$.
  (c) Deduce that if p divides the order of finite group G, then G contains an element of order p. Combined with Lagrange's Theorem, Corollary 7.23.1 of appendix 7D, this is an "if and only if" criterion.

Exercise 7.12.2. Let p be a given prime.
  (a) Use (4.12.3) of appendix 4C to determine the number of irreducible polynomials mod p of prime degree q.
  (b) Deduce that $q^p \equiv q \pmod{p}$ for every prime q.
  (c) Deduce Fermat's Little Theorem.

More combinatorics and number theory

[1] Melvin Hausner, *Applications of a simple counting technique*, Amer. Math. Monthly **90** (1983), 127–129.

7.13. Taking powers efficiently

How can one raise a residue class mod m to the nth power "quickly", when n is very large? In 1785 Legendre computed high powers mod p by *fast exponentiation*: To determine $5^{65} \pmod{161}$, we write 65 in base 2, that is, $65 = 2^6 + 2^1$, so that $5^{65} = 5^{2^6} \cdot 5^{2^1}$. Let $f_0 = 5$ and $f_1 \equiv f_0^2 \equiv 5^2 \equiv 25 \pmod{161}$. Next let $f_2 \equiv f_0^4 \equiv f_1^2 \equiv 25^2 \equiv 142 \pmod{161}$, and then $f_3 \equiv f_0^8 \equiv f_2^2 \equiv 142^2 \equiv 39 \pmod{161}$. We continue computing $f_k \equiv f_0^{2^k} \equiv f_{k-1}^2 \pmod{161}$ by successive squaring: $f_4 \equiv 72$, $f_5 \equiv 32$, $f_6 \equiv 58 \pmod{161}$ and so $5^{65} = 5^{64+1} \equiv f_6 \cdot f_0 \equiv 58 \cdot 5 \equiv 129 \pmod{161}$. We have determined the value of $5^{65} \pmod{161}$ in seven multiplications mod 161, as opposed to 64 multiplications by the more obvious algorithm.

In general to compute $a^n \pmod{m}$ quickly: Define

$$f_0 = a \text{ and then } f_j \equiv f_{j-1}^2 \pmod{m} \text{ for } j = 1, 2, \ldots, j_1,$$

where j_1 is the largest integer for which $2^{j_1} \leq n$. Writing n in binary, say as $n = 2^{j_1} + 2^{j_2} + \cdots + 2^{j_\ell}$ with $j_1 > j_2 > \cdots > j_\ell \geq 0$, let $g_1 = f_{j_1}$ and then

---

[8] We've seen that Fermat's Little Theorem arises in many different contexts. Even its earliest discoverers got there for different reasons: Fermat, Euler, and Lagrange were led to Fermat's Little Theorem by the search for perfect numbers, whereas Gauss was led to it by studying the periods in the decimal expansion of fractions (as in section 7.8). It seems to be a universal truth, rather than simply an ad hoc discovery.

$g_i \equiv g_{i-1} f_{j_i} \mod m$ for $i = 2, 3, \ldots, \ell$. Therefore

$$g_\ell \equiv f_{j_1} \cdot f_{j_2} \cdots f_{j_\ell} \equiv a^{2^{j_1} + 2^{j_2} + \cdots + 2^{j_\ell}} = a^n \pmod{m}.$$

This involves $j_\ell + \ell - 1 \leq 2 j_\ell \leq \frac{2 \log n}{\log 2}$ multiplications mod m as opposed to n multiplications mod m by the more obvious algorithm.

One can often use fewer multiplications. For example, for $31 = 1 + 2 + 4 + 8 + 16$ the above uses 8 multiplications, but we can use just 7 multiplications if, instead, we determine $a^{31} \pmod{m}$ by computing $a^2 \equiv a \cdot a$; $a^3 \equiv a^2 \cdot a$; $a^6 \equiv a^3 \cdot a^3$; $a^{12} \equiv a^6 \cdot a^6$; $a^{24} \equiv a^{12} \cdot a^{12}$; $a^{30} \equiv a^{24} \cdot a^6 \pmod{m}$; and finally $a^{31} \equiv a^{30} \cdot a \pmod{m}$.

These exponents form an *addition chain*, a sequence of integers $e_1 = 1 < e_2 < \cdots < e_k$ where, for all $k > 1$, we have $e_k = e_i + e_j$ for some $i, j \in \{1, \ldots, k-1\}$. In the example above, the binary digits of 31 led to the addition chain $1, 2, 3, 4, 7, 8, 15, 16, 31$, but the addition chain $1, 2, 3, 6, 12, 24, 30, 31$ is shorter.

For most exponents n, there is an addition chain which is substantially shorter than $j_\ell + \ell - 1$, though never less than half that size. There are many open questions about addition chains. The best known is Scholz's conjecture that the shortest addition chain for $2^n - 1$ has length $\leq n - 1$ plus the length of the shortest addition chain for n. For much more on addition chains, see Knuth's classic book [**Knu98**].

7.14. Running time: The desirability of polynomial time algorithms

In this section we discuss how to measure how fast an algorithm is. The inputs into the algorithm in the previous section for calculating $a^n \pmod{m}$ are the integers a and m, with $1 \leq a \leq m$, and the exponent n. We will suppose that m has d digits (so that d is proportional to $\log m$). The usual algorithms for adding and subtracting integers with d digits take about $2d$ steps, whereas the usual algorithm for multiplication takes about d^2 steps.[9]

Exercise 7.14.1. Justify that multiplying two residues mod m together and reducing mod m takes no more than $2d^2$ steps.

The algorithm described in the previous section involves about $c \log n$ multiplications of two residues mod m, for some constant $c > 0$, and so the total number of steps is proportional to

$$(\log m)^2 \log n.$$

Is this good? Given any mathematical problem, the cost (measured by the number of steps) of an algorithm to resolve the question must include the time taken to read the input data, which can be measured by the number of digits, D, in the input. In this case the input is the numbers a, m, and n, so that D is proportional to $\log m + \log n$. Now if a and m are fixed and we allow n to grow, then the algorithm takes CD steps for some constant $C > 0$, which is C times as long as it takes to read the input. You cannot hope to do much better than that. On the other hand, if m and n are roughly the same size, then the algorithm takes time proportional

---

[9]Since we have to multiply each pair of digits together, one from each of the given numbers.

to D^3. We still regard this as fast—any algorithm whose speed is bounded by a polynomial in D is a *polynomial time algorithm* and is considered to be pretty fast.

It is important to distinguish between a mathematical problem and an algorithm used for resolving it. There can be many choices of algorithm and one wants a fast one. However, we might only know a slowish algorithm which, even though it may seem clever, does not necessarily mean that there is no fast algorithm.

Let P be the class of problems that can be resolved by an algorithm that runs in polynomial time. Few mathematical problems belong to P and the key question is whether we can identify which problems. We'll discuss P in section 10.4.

Exercise 7.14.2. Prove that the Euclidean algorithm works in polynomial time.

Appendix 7B. Orders and primitive roots

> The problem of finding primitive roots is one of the deepest mysteries of numbers.
>
> — from *Opuscula Analytica* 1, 152 (1783) by L. EULER

It is easy to determine whether a given integer g is a primitive root mod p by using Corollary 7.5.3. Moreover, given one primitive root one can find them all since the set of all $\phi(p-1)$ primitive roots is given by

$$\{g^j \pmod{p} : 1 \leq j \leq p-1 \text{ and } (j, p-1) = 1\}.$$

The proportion of reduced residues that are primitive roots, $\frac{\phi(p-1)}{p-1}$, is rarely small. Therefore to find a primitive root we can select and test random elements mod p, and we should quickly be lucky. However Gauss described a search method that is more efficient than this, stemming from a different description of the primitive roots.

7.15. Constructing primitive roots modulo p

Gauss's efficient algorithm for constructing primitive roots mod p stems from a different description of primitive roots.

Proposition 7.15.1. *Suppose that $p - 1 = \prod_q q^b$. The set of primitive roots mod p is precisely the set*

$$\left\{ \prod_{q|p-1} A_q : A_q \text{ has order } q^b \pmod{p} \right\}.$$

To prove this we need the following:

Lemma 7.15.1 (Legendre). *If $\operatorname{ord}_m(a) = k$ and $\operatorname{ord}_m(b) = \ell$ where $(k, \ell) = 1$, then $\operatorname{ord}_m(ab) = k\ell$.*

7.15. Constructing primitive roots modulo p

Proof. Since $(ab)^{k\ell} = (a^k)^\ell (b^\ell)^k \equiv 1^\ell 1^k \equiv 1 \pmod{m}$, we see that $\text{ord}_m(ab) | k\ell$, so we may write $\text{ord}_m(ab) = k_1 \ell_1$ where $k_1 | k$ and $\ell_1 | \ell$ (by exercise 4.2.2). Now

$$a^{k_1 \ell} \equiv a^{k_1 \ell} (b^\ell)^{k_1} = ((ab)^{k_1 \ell_1})^{\ell/\ell_1} \equiv 1 \pmod{m},$$

so that $k | k_1 \ell$ by Lemma 7.1.2. As $(k, \ell) = 1$ we deduce that $k | k_1$ and so $k_1 = k$. Analogously we have $\ell_1 = \ell$ and the result follows. □

Proof of Proposition 7.15.1. Lemma 7.15.1 implies that each $\prod_{q|p-1} A_q$ is a primitive root mod p.

These products are all distinct for if $\prod_{q|p-1} A_q \equiv \prod_{q|p-1} B_q \pmod{p}$, then, raising this to the power ℓ where $\ell \equiv 0 \pmod{(p-1)/q^b}$ and $\equiv 1 \pmod{q^b}$, we see that each $A_q \equiv (\prod_{q|p-1} A_q)^\ell \equiv (\prod_{q|p-1} B_q)^\ell \equiv B_q \pmod{p}$ for each prime $q|p-1$.

Finally, by Theorem 7.6 we know that there are $\phi(q^b)$ such A_q, and therefore a total of $\prod_{q^b \| p-1} \phi(q^b) = \phi(p-1)$ different such products. That is, they give all of the $\phi(p-1)$ primitive roots. □

Proposition 7.15.1 provides a satisfactory way to construct primitive roots provided we can find the A_q of order q^b:

Lemma 7.15.2. *Suppose that $a^{(p-1)/q} \not\equiv 1 \pmod{p}$. If q^b divides $p-1$, then $A_q := a^{(p-1)/q^b} \pmod{p}$ has order q^b mod p.*

Proof. Now $A_q^{q^b} \equiv a^{p-1} \equiv 1 \pmod{p}$ and $A_q^{q^{b-1}} \equiv a^{(p-1)/q} \not\equiv 1 \pmod{p}$. Therefore $\text{ord}_p(A)$ divides q^b and not q^{b-1}, and so $\text{ord}_p(A) = q^b$. □

GAUSS'S ALGORITHM to find primitive roots: For each prime power $q^b \| p - 1$:

- Find an integer a_q for which $a_q^{(p-1)/q} \not\equiv 1 \pmod{p}$.
- Let $A_q \equiv a_q^{(p-1)/q^b} \pmod{p}$.

Then $\prod_{q|p-1} A_q$ is a primitive root $\pmod{p}$.

How do we find appropriate a_q? We know that there are exactly $\frac{p-1}{q}$ roots mod p of $x^{(p-1)/q} \equiv 1 \pmod{p}$, that is, a proportion $\frac{1}{q}$ of the reduced residues mod p. For the remaining $a \pmod{p}$ we have $a^{(p-1)/q} \not\equiv 1 \pmod{p}$ as desired, that is, for a proportion $1 - \frac{1}{q}$ of the reduced residues mod p.

One can try to find a_q by trying $2, 3, 5, 7, \ldots$ until one finds an appropriate number, but there are no guarantees that this will succeed in a reasonable time period. However if we select values of a at random, then the probability that we fail to find an appropriate a_q after k tries is $1/q^k$, which is negligible for $k > 20$.

Finding elements of order n. For any integer n dividing $p-1$ the residue $h \equiv g^{(p-1)/n} \pmod{p}$ provides a solution to $x^n \equiv 1 \pmod{p}$, where g is a primitive root, perhaps found by the method described above. Moreover $\text{ord}_p(h) = n$. The set of solutions

$$\{u \pmod{p} : u^n \equiv 1 \pmod{p}\} = \{h^j \pmod{p} : 1 \leq j \leq n\}.$$

An alternative way to find a residue h mod p of order n is to modify Gauss's algorithm: One only needs to determine the a_q for the primes q dividing n and then one can take
$$h \equiv \prod_{\substack{q^e \| n,\ e \geq 1 \\ q^b \| p-1}} A_q^{q^{b-e}} \pmod{p}.$$

7.16. Indices / Discrete Logarithms

Suppose that g is a given primitive root (mod p). If $b \equiv g^e \pmod{p}$, then e is the *index*, or *discrete logarithm*, of b in base g, denoted $\text{ind}_p(b)$. This value is only determined mod $p - 1$. It is a challenging open problem to determine the discrete logarithm of a given residue in a given base in a short amount of time.

Exercise 7.16.1. (a) Show that $\text{ind}_p(ab) \equiv \text{ind}_p(a) + \text{ind}_p(b) \pmod{p-1}$.
(b) Show that $\text{ind}_p(1) = 0$ and $\text{ind}_p(-1) = (p-1)/2$, irrespective of the base used.
(c) Show that $\text{ind}_p(a^n) \equiv n\ \text{ind}_p(a) \pmod{p-1}$.

There are several parameters that go into the definition of index. The one that appears to be of some concern at first sight is the choice of primitive root to use as a base. The next result shows that there is little difference between the choice of one base and another.

Exercise 7.16.2. Suppose that g and h are two primitive roots mod p, where $h \equiv g^\ell \pmod{p}$.
(1) Show that $(\ell, p-1) = 1$.
(2) Show that the index with respect to g is ℓ times the index with respect to h, mod p.
(3) Prove that there exists an integer m for which $g \equiv h^m \pmod{p}$.

We have described residues in terms of index and in terms of order. What is the link between the two?

Proposition 7.16.1. *For any reduced residue a mod p we have*
$$\text{ord}_p(a) \cdot (p-1, \text{ind}_p(a)) = p - 1.$$

Proof. Let g be a primitive root with $k = \text{ind}_p(a)$ and let $m = \text{ord}_p(a)$. This means that m is the smallest integer for which $g^{km} \equiv a^m \equiv 1 \pmod{p}$, that is, the smallest integer for which $p - 1$ divides km, by Lemma 7.1.2. Therefore $(p-1)/(p-1,k)$ divides m by Corollary 3.2.1, and the smallest such integer m is $(p-1)/(p-1,k)$. □

Exercise 7.16.3. (a) Suppose that k divides $p-1$. Show that a is a kth power mod p if and only if k divides $\text{ind}_p(a)$.
(b) Show that if a has order m mod p, then $\{a^k \pmod{p} : 1 \leq k \leq m,\ (k,m) = 1\}$ is the set of residues mod p of order m.

7.17. Primitive roots modulo prime powers

A *primitive root* mod m (whether m is prime or not) is a residue $g \pmod{m}$ whose powers, $1, g, g^2, \ldots, g^{\phi(m)-1} \pmod{m}$, give all of the $\phi(m)$ reduced residues mod m. We will show that there are primitive roots mod m if and only if m is a prime power other than 2^k with $k \geq 3$.

7.17. Primitive roots modulo prime powers

Proposition 7.17.1. *Given integer a coprime to odd prime p, let $\ell = \text{ord}_p(a)$. Therefore we can write $a^\ell = 1 + p^j m$ for some integer $j \geq 1$ where $p \nmid m$. Then, for all $k \geq j$ we have $\text{ord}_{p^k}(a) = p^{k-j}\ell$.*
(If $p = 2$, the same result holds for $k \geq 2$ if we now define $\ell = \text{ord}_4(a) = 1$ or 2.)

The proof of Proposition 7.17.1 depends on the following "lifting lemma".

Lemma 7.17.1. *Let p^k be a prime power and $a = b + p^k m$ where $p \nmid bm$ and $k \geq 1$.*
(a) *If $a^r \equiv b^r \pmod{p^{k+1}}$, then p divides r.*
(b) *If $p^k > 2$, then $a^p = b^p + p^{k+1}M$ for some integer M that is not divisible by p.*

Proof. Using the binomial theorem we have

$$a^r = (b + p^k m)^r = \sum_{j=0}^{r} \binom{r}{j} b^{r-j} (p^k m)^j$$

$$\equiv b^r + r b^{r-1} m p^k + \binom{r}{2} b^{r-2} m^2 p^{2k} \pmod{p^{k+2}},$$

since subsequent terms are divisible by p^{3k} which is divisible by p^{k+2}. We deduce that $a^r \equiv b^r \pmod{p^{k+1}}$ if and only if p divides $r b^{r-1} m$, and so p divides r, as $p \nmid bm$. This gives (a).

For (b) we let $r = p$. If $k \geq 2$ or $p \neq 2$, then $\binom{p}{2} p^{2k} \equiv 0 \pmod{p^{k+2}}$, so that $a^p \equiv b^p + b^{p-1} m p^{k+1} \pmod{p^{k+2}}$, and the result follows with the integer $M \equiv b^{p-1} m \equiv m \pmod{p}$. □

Proof of Proposition 7.17.1. We begin by proving that $a^{p^{k-j}\ell} = 1 + p^k m_k$ for some integer m_k that is not divisible by p, for all $k \geq j$ by induction. It is true by hypothesis for $k = j$. Then we use Lemma 7.17.1(b), with a replaced by $a^{p^{k-j}\ell}$ and $b = 1$, to deduce the claim for $k + 1$.

Now we prove that $\text{ord}_{p^k}(a) = p^{k-j}\ell$ for all $k \geq j$, by induction.

Let $r = \text{ord}_{p^j}(a)$. Then r divides ℓ, and $a^r \equiv 1 \pmod{p^j}$ implies $a^r \equiv 1 \pmod{p}$ and so ℓ divides r. Therefore $r = \text{ord}_{p^j}(a) = \ell$.

Now suppose that the result is true mod p^k and let $r = \text{ord}_{p^{k+1}}(a)$. Now $a^r \equiv 1 \pmod{p^{k+1}}$ implies that $a^r \equiv 1 \pmod{p^k}$ and so $p^{k-j}\ell$ divides r, say $r = R p^{k-j}\ell$. Now $a^{p^{k-j}\ell} = 1 + p^k m_k \not\equiv 1 \pmod{p^{k+1}}$ and so $R \neq 1$. Then Lemma 7.17.1(a), with a replaced by $a^{p^{k-j}\ell}$ and $b = 1$, shows that R must be divisible by p, and therefore $R = p$ as $a^{p^{k+1-j}\ell} = 1 + p^{k+1} m_{k+1}$. □

We define *Carmichael's λ-function* $\lambda(m)$ to be the maximal order of a reduced residue mod m, so that $\lambda(p) = p - 1$ for all primes p, because there are primitive roots for all primes p.

Corollary 7.17.1. *If p^k is an odd prime power, then there is a primitive root g (mod p^k), so that $\lambda(p^k) = \phi(p^k)$ and each $\text{ord}_{p^k}(a)$ divides $\lambda(p^k)$. Moreover if g is* a primitive root mod p^2, then it is a primitive root mod p^k for all $k \geq 1$. We also have $\lambda(2) = 1$, $\lambda(4) = \lambda(8) = 2$, and $\lambda(2^k) = \phi(2^{k-1}) = \text{ord}_{2^k}(3)$ for all $k \geq 3$.

Proof. Let a be as in the hypothesis of Proposition 7.17.1. If p is an odd prime, then ℓ divides $p - 1$ and so $\text{ord}_{p^k}(a) = p^{k-j}\ell \leq p^{k-1}(p - 1) = \phi(p^k)$. To obtain equality we need $\ell = p - 1$, that is, a is a primitive root mod p, and $j = 1$, that is, p^2 does not divide $a^{p-1} - 1$.

There exists a primitive root g mod p by Corollary 7.5.1. If $g^{p-1} \not\equiv 1 \pmod{p^2}$, then let $a = g$. Otherwise let $a = g + p$, so that $a^{p-1} \not\equiv g^{p-1} \equiv 1 \pmod{p^2}$ by Lemma 7.17.1(a). Either way, $\ell = p - 1$ and $a^{p-1} \not\equiv 1 \pmod{p^2}$, so $\text{ord}_{p^k}(a) = \phi(p^k)$, the maximum possible. We deduce that $\lambda(p^k) = \phi(p^k)$ and then for every reduced residue class b, we know that $\text{ord}_{p^k}(b)$ divides $\phi(p^k) = \lambda(p^k)$.

For $p = 2$, we have $a^2 \equiv 1 \pmod 8$ for all odd a, and $3^2 = 1 + 8$. As in the above argument, we deduce that $\lambda(2^k) = \text{ord}_{2^k}(3) = \phi(2^{k-1})$ for all $k \geq 3$. □

Corollary 7.17.2. *If prime power $p^k = 2, 4$ or is odd, then*

(7.17.1) $$(\mathbb{Z}/p^k\mathbb{Z})^* \cong \mathbb{Z}/\phi(p^k)\mathbb{Z}.$$

Otherwise $p = 2$ and $k \geq 3$ in which case

(7.17.2) $$(\mathbb{Z}/2^k\mathbb{Z})^* \cong \mathbb{Z}/2^{k-2}\mathbb{Z} \oplus \mathbb{Z}/2\mathbb{Z}.$$

Proof. If $p^k = 2, 4$ or is odd, then Corollary 7.17.1 states that $\lambda(p^k) = \phi(p^k)$, so all of the reduced residues must be a power of the primitive root. If $p = 2$ and $k \geq 3$, then we saw that 3 has order 2^{k-2} in the proof of Corollary 7.17.1. Reducing mod 8, we see that -1 cannot be a power of 3 mod 2^k, so the residues

$$\{3^a(-1)^b : 0 \leq a \leq 2^{k-2} - 1 \text{ and } 0 \leq b \leq 1\}$$

are distinct and therefore give all the reduced residues mod 2^k. □

Exercise 7.17.1. Use Euler's Theorem and Lemma 7.1.2 to prove that $\lambda(m)$ divides $\phi(m)$. Prove that there is a primitive root mod m if and only if $\lambda(m) = \phi(m)$.

Exercise 7.17.2.[†] Suppose that p^k is a prime power dividing n and let $m = \text{ord}_p(2)$. Prove that $\text{ord}_{p^k}(2)$ divides $n - 1$ if and only if m divides $(p - 1, n - 1)$ and p^k divides $2^m - 1$.

Exercise 7.17.3. Let $x_n = a^n - b^n$ and suppose that $m = m_p$ is the smallest positive integer for which prime p divides x_m. In exercise 1.7.24 we proved that $p|x_n$ if and and only if $m|n$. Now suppose that $p^k \| x_m$, and $m|n$ so that $x_m | x_n$. Prove that the power of p that divides x_n/x_m equals the power of p that divides n/m. (This also follows from exercise 7.33.2(c).)

Exercise 7.17.4.[†] Suppose that $q^n - p^m = 1$ where p is prime, with $m \geq 1$ and $n \geq 2$.
 (a) Prove that if $m = 1$, then $q = 2$, n is prime, and p is a Mersenne prime.
 (b) Prove that $q - 1 = p^k$ for some integer $k \geq 0$.
  For now assume that $k \geq 1$ (and use Lemma 7.17.1 throughout).
 (c) Prove that n is a power of p.
 (d) Prove that if $p^k > 2$, then $q^p - 1 = p^{k+1}$, which is impossible.
 (e) Deduce that $p = 2$ and $q = 3$, so that $9 - 8 = 1$ by exercise 7.10.3.
  Now we may assume that $m \geq 2$ and $k = 0$ so that $q = 2$.

(f) Suppose that r divides m with m/r odd. Prove that $p^r + 1 = 2^j$ for some integer $j \geq 1$.
(g) Deduce that $m = r$ and therefore that m is a power of 2.
(h) Deduce that $p^m + 1 \equiv 2 \pmod{8}$ so that $n = 1$, which is impossible.

Therefore the only solution to $q^n - p^m = 1$ with p prime and $m, n \geq 2$ is $3^2 - 2^3 = 1$.

Exercise 7.17.5.† Prove that $(1+x)^{p^r} \equiv (1+x^p)^{p^{r-1}} \pmod{p^r}$ for all prime powers p^r.

7.18. Orders modulo composites

We now show that if m has two or more distinct prime factors, then there is no primitive root mod m. Moreover we determine the structure of the multiplicative group of reduced residues mod m. We can understand orders of reduced residues modulo a composite number m by breaking m up into its prime power factors:

Proposition 7.18.1. *If $(a, m) = 1$, then $\mathrm{ord}_m(a) = \mathrm{lcm}[\mathrm{ord}_{p^e}(a) : p^e \| m]$ divides $\lambda(m) = \mathrm{lcm}[\lambda(p^e) : p^e \| m]$.*

Proof. By induction on the number of distinct prime factors of m. It holds when m is a prime power p^e, by Corollary 7.17.1. We now assume that $m = rs$ where r and s are coprime integers for which the result is proved.

Let $k = \mathrm{ord}_m(a)$ so that $a^k \equiv 1 \pmod{m}$. Therefore $a^k \equiv 1 \pmod{r}$ and $a^k \equiv 1 \pmod{s}$, so that $[\mathrm{ord}_r(a), \mathrm{ord}_s(a)]$ divides $k = \mathrm{ord}_m(a)$ by Lemma 7.1.2.

On the other hand let $h = [\mathrm{ord}_r(a), \mathrm{ord}_s(a)]$ so that $a^h \equiv 1 \pmod{r}$ and $a^h \equiv 1 \pmod{s}$. Therefore $a^h \equiv 1 \pmod{m}$ by the Chinese Remainder Theorem, and therefore $\mathrm{ord}_m(a)$ divides $h = [\mathrm{ord}_r(a), \mathrm{ord}_s(a)]$ by Lemma 7.1.2.

Combining the last two paragraphs yields that $\mathrm{ord}_m(a) = [\mathrm{ord}_r(a), \mathrm{ord}_s(a)]$.

By the induction hypothesis, $\mathrm{ord}_r(a)$ divides $\lambda(r)$ and $\mathrm{ord}_s(a)$ divides $\lambda(s)$, so that $\mathrm{ord}_m(a) = [\mathrm{ord}_r(a), \mathrm{ord}_s(a)]$ divides $[\lambda(r), \lambda(s)]$. Therefore $\lambda(m)$ divides $[\lambda(r), \lambda(s)]$.

On the other hand select $b \pmod{r}$ so that $\mathrm{ord}_r(b) = \lambda(r)$ and select $c \pmod{s}$ so that $\mathrm{ord}_s(c) = \lambda(s)$. Now we select $a \pmod{m}$ so that $a \equiv b \pmod{r}$ and $a \equiv c \pmod{s}$, using the Chinese Remainder Theorem. Therefore we have $\mathrm{ord}_m(a) = [\mathrm{ord}_r(b), \mathrm{ord}_s(c)] = [\lambda(r), \lambda(s)]$. Combining this with the last paragraph we see that this must be the maximum possible order mod m; that is, $\lambda(m) = [\lambda(r), \lambda(s)]$. □

Exercise 7.18.1.† Prove that $\lambda(m) < \phi(m)$ if m is divisible by $4p$ or pq for odd primes $p < q$.

Corollary 7.18.1. *There is a primitive root mod m if and only if $m = 2$ or 4 or p^k or $2p^k$ where p is an odd prime.*

Proof. If $\lambda(m) = \phi(m)$, then m is of the form 2^k or p^k or $2p^k$ for some odd prime p and integer $k \geq 1$ by exercise 7.18.1. In Corollary 7.17.1 we saw that $\lambda(m) < \phi(m)$ if $m = 2^k$ with $k \geq 3$, and we showed that $\lambda(m) = \phi(m)$ if $m = 2$ or 4 or p^k where p is an odd prime. Finally $\lambda(2p^k) = \mathrm{lcm}[\lambda(p^k), \lambda(2)] = \lambda(p^k) = \phi(2p^k)$. □

Exercise 7.18.2. Determine $\lambda(65520)$.

Exercise 7.18.3. Prove that $a^{\lambda(m)} \equiv 1 \pmod{m}$ for all integers a coprime to m.

Exercise 7.18.4. Show that composite n is a Carmichael number if and only if $\lambda(n)$ divides $n - 1$.

Exercise 7.18.5. Let $N_m(n) = \#\{x \pmod{n} : x^m \equiv 1 \pmod{n}\}$ for some given integer $m \geq 2$.
 (a) Prove that $N_m(n)$ is a multiplicative function of n.
 (b)† Prove that $x^m \equiv 1 \pmod{n}$ if and only if $x^g \equiv 1 \pmod{n}$ where $g = (m, \lambda(n))$.
 (c) Deduce that $N_m(n) = N_g(n)$ where $g = (m, \lambda(n))$.
 (d) Use Theorem 7.6 to determine $N_m(p)$ for every prime p.

Exercise 7.18.6. Prove that 2 is a primitive root mod 3^m for all $m \geq 1$, and mod 5^n for all $n \geq 1$, and mod 11^r for all $r \geq 1$.

Appendix 7C. Finding nth roots modulo prime powers

7.19. nth roots modulo p

Given n, a, and p with $(a,p) = 1$ we are interested in finding all of the solutions x (mod p) to

(7.19.1) $$x^n \equiv a \pmod{p}.$$

The question is equivalent to one in which the exponent is a divisor of $p-1$.

Theorem 7.9. *Suppose that $(a,p) = 1$ and let $g = (n, p-1)$. The solutions x (mod p) to $x^n \equiv a$ (mod p) are in 1-to-1 correspondence with the solutions y (mod p) to $y^g \equiv a$ (mod p).*

Proof. Given x let $y \equiv x^{n/g}$ (mod p) so that $y^g \equiv x^n \equiv a$ (mod p).

On the other hand, suppose that $g = un + v(p-1)$ for some integers u and v. Given y let $x \equiv y^u$ (mod p), and so $x^n \equiv (y^u)^n (y^{p-1})^v = y^g \equiv a$ (mod p). □

We can therefore restrict our attention to the case that n divides $p-1$. We can provide easily verified conditions for there to be a solution to (7.19.1), and given one solution we can find them all quickly.

Proposition 7.19.1. *Suppose that $(a,p) = 1$ and n divides $p-1$.*
  (a) *There are solutions x (mod p) of (7.19.1) if and only if $a^{(p-1)/n} \equiv 1$ (mod p).*
  (b) *Given one solution x_0, the set of all solutions mod p to (7.19.1) is given by $x_0 u$ (mod p) as u runs through the n roots of $u^n \equiv 1$ (mod p).*

Proof. (a) If there is a solution to (7.19.1), then $a^{(p-1)/n} \equiv (x^n)^{(p-1)/n} = x^{p-1} \equiv 1$ (mod p). On the other hand if $a^{(p-1)/n} \equiv 1$ (mod p), then $\text{ord}_p(a)$ divides $(p-1)/n$ and so n divides $\text{ind}_p(a)$ by Proposition 7.16.1. Writing $\text{ind}_p(a) = nk$, we let $x \equiv g^k$ so that $x^n \equiv (g^k)^n = g^{kn} \equiv a$ (mod p).

(b) Given a solution x (mod p) to (7.19.1), let $u \equiv x/x_0$ so that $u^n \equiv x^n/x_0^n \equiv a/a \equiv 1$ (mod p). On the other hand if $x \equiv x_0 u$ (mod p), then $x^n \equiv x_0^n u^n \equiv a \cdot 1 \equiv a$ (mod p). □

In section 7.15 of appendix 7B we modified Gauss's efficient algorithm to construct all of the nth roots of 1 mod p. Therefore to quickly determine all of the solutions to (7.19.1), we are left with the task of finding an initial solution x_0.

Exercise 7.19.1. Show that the solutions x (mod m) to $x^n \equiv a$ (mod m) are in 1-to-1 correspondence with the solutions y (mod m) to $y^g \equiv a$ (mod m) where $g = (n, \lambda(m))$.

Exercise 7.19.2. Prove that if odd prime p does not divide a, then

$$\#\{x \pmod{p} : x^4 \equiv a \pmod{p}\} = \begin{cases} 0 \text{ or } 4 & \text{if } p \equiv 1 \pmod 4, \\ 0 \text{ or } 2 & \text{if } p \equiv 3 \pmod 4. \end{cases}$$

7.20. Lifting solutions

Gauss discovered that if an equation has solutions mod p, then one can often use those solutions to determine solutions to the same equation mod p^k.

Proposition 7.20.1. *Suppose that p does not divide a and that $u^n \equiv a$ (mod p). If p does not divide n, then, for each integer $k \geq 2$, there exists a unique congruence class b (mod p^k) such that $b^n \equiv a$ (mod p^k) and $b \equiv u$ (mod p).*

Proof. We prove this by induction on $k \geq 2$. We may assume that there exists a unique congruence class b (mod p^{k-1}) such that $b^n \equiv a$ (mod p^{k-1}) and $b \equiv u$ (mod p). Therefore if $B^n \equiv a$ (mod p^k) and $B \equiv u$ (mod p), then $B^n \equiv a$ (mod p^{k-1}) and so $B \equiv b$ (mod p^{k-1}). Writing $B = b + mp^{k-1}$ we have

$$B^n = (b + mp^{k-1})^n \equiv b^n + nmp^{k-1}b^{n-1} \pmod{p^k}$$

which is $\equiv a$ (mod p^k) if and only if

$$m \equiv \frac{a - b^n}{np^{k-1}b^{n-1}} \equiv \frac{u}{an} \cdot \frac{a - b^n}{p^{k-1}} \pmod{p},$$

as $ub^{n-1} \equiv u^n \equiv a$ (mod p). Since p^{k-1} divides $a - b^n$, and $(an, p) = 1$, the quantity on the right-hand side is a congruence class mod p and therefore gives m (mod p) uniquely. □

Exercise 7.20.1. Show that if prime $p \nmid an$, then the number of solutions x (mod p^k) to $x^n \equiv a$ (mod p^k) does not depend on k.

We will focus on the case $n = 2$ in Proposition 8.8.1. Proposition 7.20.1 may be generalized to more-or-less arbitrary polynomial solutions, as we will see in Proposition 16.3.1.

We will also need the analogous result when p divides n, or at least when $n = p$. The difficulty lies in the fact that every residue class a (mod p) has a pth root mod p, namely $a^p \equiv a$ (mod p) by Fermat's Little Theorem, but only p residue classes a (mod p^2) have a pth root. To see this, note that if $a \equiv b$ (mod p), then $a^p \equiv b^p$ (mod p^2) by Lemma 7.17.1.

Proposition 7.20.2. *Let p be a prime, with $\kappa = 2$ if p is odd, and $\kappa = 3$ if $p = 2$.*

Suppose that p does not divide a and that $u^p \equiv a \pmod{p^\kappa}$. For each integer $k \geq \kappa$, there exists a unique congruence class $b \pmod{p^{k-1}}$ such that $b^p \equiv a \pmod{p^k}$ and $b \equiv u \pmod{p^{\kappa-1}}$.

Proof. We prove this by induction on $k \geq \kappa$. This is immediate for $k = \kappa$. For $k \geq \kappa + 1$ we may assume that there exists a unique congruence class $b \pmod{p^{k-2}}$ such that $b^p \equiv a \pmod{p^{k-1}}$ and $b \equiv u \pmod{p^{\kappa-1}}$ by the induction hypothesis. Therefore if $B^p \equiv a \pmod{p^k}$ and $B \equiv u \pmod{p^{\kappa-1}}$, then $B^p \equiv a \pmod{p^{k-1}}$ and so $B \equiv b \pmod{p^{k-2}}$. Writing $B = b + mp^{k-2}$ we have

$$B^p = (b + mp^{k-2})^p \equiv b^p + mp^{k-1}b^{p-1} \equiv b^p + mp^{k-1} \pmod{p^k}$$

as $b^{p-1} \equiv 1 \pmod{p}$ since $(b, p) = 1$. This is $\equiv a \pmod{p^k}$ if and only if

$$m \equiv \frac{a - b^p}{p^{k-1}} \pmod{p},$$

which yields a unique residue class as p^{k-1} divides $a - b^p$. This implies that $B = b + mp^{k-2}$ occupies a unique residue class mod p^{k-1}. □

7.21. Finding nth roots quickly

Suppose that there is a solution to $x^2 \equiv a \pmod{p}$. This implies that $a^{(p-1)/2} \equiv 1 \pmod{p}$ by Proposition 7.19.1. We will try to find a solution to $x^2 \equiv a \pmod{p}$ by taking $x \equiv a^k \pmod{p}$ for some integer k. This works if and only if $a^{2k} \equiv a \pmod{p}$. This holds for any such a provided $\frac{p-1}{2}$ divides $2k - 1$, which can only hold if $\frac{p-1}{2}$ is odd, that is, $p \equiv 3 \pmod{4}$, in which case we can take $k = \frac{p+1}{4}$ (see exercise 8.2.4). We now rework this in a more general setting.

Proposition 7.21.1. *Suppose that n divides $p-1$ and that a is an nth power mod p. Assume that $(n, \frac{p-1}{n}) = 1$ so that there exists an integer k for which $nk \equiv 1 \pmod{\frac{p-1}{n}}$. If $x \equiv a^k \pmod{p}$, then $x^n \equiv a \pmod{p}$.*

Proof. Since a is an nth power mod p we know that $a^{\frac{p-1}{n}} \equiv 1 \pmod{p}$, and therefore $x^n \equiv a^{nk} \equiv a \pmod{p}$. □

When $(n, \frac{p-1}{n}) > 1$ we are unable to find a solution of $x^n \equiv a \pmod{p}$ but we can show that the problem is equivalent to one where a comes from a restricted set. For example, if $n = 2$ and $p \equiv 5 \pmod 8$, then $(2, \frac{p-1}{4}) = 1$ so there exist integers k and odd m for which $2k + m\frac{p-1}{4} = 1$. Therefore $(a^k)^2 \equiv ab^m \pmod{p}$ where $b \equiv a^{-\frac{p-1}{4}} \pmod{p}$. Now $b^2 \equiv a^{\frac{p-1}{2}} \equiv 1 \pmod{p}$ and so $b \equiv \pm 1 \pmod{p}$. If $a^{\frac{p-1}{4}} \equiv b \equiv 1 \pmod{p}$, then $(a^k)^2 \equiv a \pmod{p}$. If $a^{\frac{p-1}{4}} \equiv b \equiv -1 \pmod{p}$ and $y^2 \equiv -1 \pmod{p}$, then $(ya^k)^2 \equiv (-1)a(-1)^m \equiv a \pmod{p}$. So we have reduced the difficulty to finding the square root of $-1 \pmod{p}$, a seemingly easier question than the original. Gauss generalized this argument as follows:

Proposition 7.21.2. *Suppose that n divides $p-1$ and that a is an nth power mod p. Let N be the smallest positive integer such that $(n, \frac{p-1}{N}) = 1$ (so that n divides N, and N has the same prime factors as n). There exists a solution $b \pmod{p}$*

to $b^{N/n} \equiv 1 \pmod{p}$ *such that the solutions* $x \pmod{p}$ *to* $x^n \equiv a \pmod{p}$ *are in 1-to-1 correspondence with the solutions* $y \pmod{p}$ *to* $y^n \equiv b \pmod{p}$.

Proof. Let $b \equiv a^{\frac{p-1}{N}} \pmod{p}$ so that $b^{N/n} \equiv a^{\frac{p-1}{n}} \equiv 1 \pmod{p}$. If $x^n \equiv a \pmod{p}$, let $y \equiv x^{\frac{p-1}{N}} \pmod{p}$ so that $y^n \equiv (x^n)^{\frac{p-1}{N}} \equiv a^{\frac{p-1}{N}} \equiv b \pmod{p}$.

Now suppose that $y^n \equiv b \pmod{p}$. Select integers m and k for which $kn + m\frac{p-1}{N} = 1$ and let $x \equiv y^m a^k \pmod{p}$. Therefore $a^{kn} = a^{1-m\frac{p-1}{N}} = a(a^{\frac{p-1}{N}})^{-m} \equiv ab^{-m} \pmod{p}$ and so

$$x^n \equiv (y^m a^k)^n \equiv (y^n)^m \cdot a^{kn} \equiv b^m \cdot ab^{-m} \equiv a \pmod{p}. \qquad \square$$

Exercise 7.21.1. Show that N is the largest divisor of $p-1$ with exactly the same prime factors as n.

Example. To solve $x^2 \equiv a \pmod{29}$ where $a^{14} \equiv 1 \pmod{29}$, we have $n = 2$ and $N = 4$, and we take $b \equiv a^7 \pmod{29}$, so that $b^2 \equiv 1 \pmod{29}$ and therefore $b \equiv 1$ or $-1 \pmod{29}$. By Proposition 7.21.2 we need to solve $y^2 \equiv b \pmod{29}$. If $b \equiv 1 \pmod{29}$, then $y \equiv 1$ or $-1 \pmod{29}$. If $b \equiv -1 \pmod{29}$, then $y \equiv -12$ or $12 \pmod{29}$ (which requires some calculation).

Now $2k + 7m = 1$ has the solution $m = 1, k = -3$, so that $x \equiv ya^{-3} \pmod{29}$. So if $a^7 \equiv 1 \pmod{29}$, then $x \equiv \pm a^4 \pmod{29}$; and if $a^7 \equiv -1 \pmod{29}$, then $x \equiv \pm 12a^4 \pmod{29}$

Example. Solve $x^3 \equiv 31 \pmod{37}$. Here $p - 1 = 36$, $n = 3$, $N = 9$, and $3k + 4m = 1$ so we can take $m = 1, k = -1$. Now $b \equiv 31^4 \equiv 1 \pmod{37}$, and so we need to solve $y^3 \equiv 1 \pmod{37}$. Now 37 divides 111 which divides 999 so that $10^3 \equiv 1 \pmod{37}$ and therefore $y \equiv 1, 10,$ or $10^2 \equiv -11 \pmod{37}$. Therefore $x \equiv y \cdot 31^{-1} \equiv -31y \pmod{37} \equiv 6, 23,$ or $8 \pmod{37}$.

Exercise 7.21.2. Determine the square roots of 3 (mod 37) using the technique above.

Appendix 7D. Orders for finite groups

We will see in this appendix that the key results of this chapter, like Fermat's Little Theorem and Wilson's Theorem, can be formulated and proved for general finite groups.

7.22. Cosets of general groups

Suppose that G is a given finite group, not necessarily commutative (see section 0.11 of appendix 0D for definitions). We explored the cosets of additive groups in appendix 2A, and now we look at the cosets in an arbitrary group. If H is a subgroup of G, then we define a *left coset* to be the set $a * H = \{a * h : h \in H\}$ for any $a \in G$. Right cosets are of the form $H * a = \{h * a : h \in H\}$, and these are indistinguishable if G is a commutative group.

Proposition 7.22.1. *Let H be a subgroup of G. The left cosets of H in G are disjoint. Moreover if G is finite, then they partition G, so that $|H|$ divides $|G|$.*

This generalizes Theorem 7.1 and Proposition 2.7.1 and their proofs:

Proof. We begin by showing that $a*H$ and $b*H$ are either disjoint or identical. For if they are not disjoint, then there exist elements $h_1, h_2 \in H$ such that $a*h_1 = b*h_2$. Therefore $b = b * h_2 * (h_2)^{-1} = a * h_1 * (h_2)^{-1}$ so that $b = a * h_0 \in a * H$ where $h_0 := h_1 * (h_2)^{-1} \in H$ since H is closed under the operation "$*$". If $g \in b * H$, then $g = b * h$ for some $h \in H$, and so $g = (a * h_0) * h = a * (h_0 * h) \in a * H$ by associativity and the closure of H. Hence $b * H \subset a * H$. By an analogous proof we have $a * H \subset b * H$, and hence $a * H = b * H$.

If G is finite, let $a_1 * H, \ldots, a_k * H$ be a maximal set of disjoint cosets of H inside G. Their union must equal G or else there exists $a \in G$ which is in none of these cosets. But then the coset $a * H$ is disjoint from these cosets (by the previous paragraph), which contradicts maximality (since $a*H, a_1*H, \ldots, a_k*H$ would be a larger set of disjoint cosets of H). □

Exercise 7.22.1. Show that every element of G belongs to a unique coset of H.

One can prove the analogous result about right cosets, with the analogous proof. It is tempting to guess that $a*H$ is equal to $H*a$ but that is not true in general; see section 7.24 of this appendix for more on that.

7.23. Lagrange and Wilson

Given the group operation $*$, we denote $g*g$ by g^2, and $g*g*g$ by g^3, etc. An element a has *order* m in G if m is the least positive integer for which $a^m = 1$ (if such an integer exists).

Exercise 7.23.1. Prove that if a has order m in G, then $H := \{1, a, a^2, \ldots, a^{m-1}\}$ is a subgroup of G.

Theorem 7.1 was used to prove Fermat's Little Theorem. Exercise 7.23.1 implies the following generalization of Fermat's Little Theorem:

Corollary 7.23.1 (Lagrange's Theorem). *For any element a of any finite group G we have $a^{|G|} = 1$.*

Proof. Let m be the order of a in G, that is, the least positive integer for which $a^m = 1$. Then $H := \{1, a, a^2, \ldots, a^{m-1}\}$ is a subgroup of G by exercise 7.23.1. We deduce that $m = |H|$ divides $|G|$ by Proposition 7.22.1, and so

$$a^{|G|} = (a^m)^{|G|/m} = 1^{|G|/m} = 1.$$ □

To deduce Euler's Theorem let $G = (\mathbb{Z}/m\mathbb{Z})^*$ so that $|G| = \phi(m)$.

Exercise 7.23.2. Let p be a prime which does not divide the order of the finite group G.
 (a) Prove that G contains no elements of order p.
 (b) Let $X = \{(x_1, \ldots, x_p) \in G^p : x_1 \cdots x_p = 1\}$, and then use exercise 7.12.1(a) to prove that $|G|^{p-1} \equiv 1 \pmod{p}$.
 (c) Deduce Fermat's Little Theorem by applying (b) to the cyclic groups of order a for $1 \leq a \leq p-1$.

Theorem 7.10 (Wilson's Theorem for finite abelian groups). *The product of the elements of any given finite abelian group G equals 1 unless the group contains exactly one element, ℓ, of order two, in which case the product equals ℓ.*

Proof. We partition the elements of G of order > 2 into subsets of size two, each element with its inverse. The elements in each such subset are distinct for if $y = x$ and $xy = 1$, then $x^2 = 1$; that is, x has order 1 or 2. Now the two elements of each of these subsets each multiply together to give 1, and so the product of all of the elements of order > 2 multiply together to give 1. Therefore we have proved that

$$\prod_{g \in G} g = \prod_{h \in H} h \text{ where } H := \{g \in G : g^2 = 1\}.$$

We can see that H is a subgroup of G, for if $a, b \in H$, then $(ab)^2 = a(ba)b = a(ab)b = a^2b^2 = 1 \cdot 1 = 1$. If H only has one element, 1, or two elements, 1 and ℓ

(the element of order two), then the result follows immediately. If H has at least two elements of order two, call them ℓ and m. Then $L := \{1, \ell, m, \ell m\}$ is a subgroup of H, and the product of the elements of each coset xL of H is

$$x \cdot x\ell \cdot xm \cdot x\ell m = x^4 \ell^2 m^2 = 1.$$

Therefore the product over all of the elements of H, which is the union of the elements of all of the cosets of L, is 1. $\square$

7.24. Normal subgroups

A group G is *cyclic* if there exists an element $g \in G$ such that $G = \{g^m : m = 0, 1, \ldots\}$, and we say that G is *generated* by g. The additive group $\mathbb{Z}/m\mathbb{Z}$ is cyclic; the elements of the group are precisely the multiples of the *generator* 1. The primitive roots of appendix 7B are precisely the generators of the multiplicative group $(\mathbb{Z}/m\mathbb{Z})^*$, when it is cyclic.

Exercise 7.24.1. Prove that every finite cyclic group is isomorphic to some $\mathbb{Z}/n\mathbb{Z}$.

Exercise 7.24.2. Prove that if $|G|$ is a prime, then G is cyclic.

Exercise 7.24.3. Show that the product of the elements in a finite cyclic group G is 1 if $|G|$ is odd, and equals the (unique) element of G of order two if m is even.

If H is a subgroup of G for which $a * H = H * a$ for all $a \in G$, then we call H a *normal subgroup* of G, denoted $H \triangleleft G$.[10] Normal subgroups are useful not only because the left and right cosets are the same, but also because one can then make sense of the quotient group G/H since then

$$(a * H) * (b * H) = a * (H * b) * H = a * (b * H) * H = (a * b) * H.$$

There are two "trivial" normal subgroups $\{1\}$ and G of any group G (which reminds one of the "trivial" factors of any integer n). If there are no other normal subgroups of G, other than $\{1\}$ and the group itself, then we call G a *simple group* (which is analogous to the definition of a prime number).

For any non-simple finite group G there is a maximal proper normal subgroup H of G, and one can prove that the quotient group G/H is simple. But then either H is simple or it has a maximal proper normal subgroup L, say, such that H/L is simple. We can iterate this construction, and in the case of a finite group we have

$$H_k = \{1\} \triangleleft H_{k-1} \triangleleft H_{k-2} \triangleleft \cdots \triangleleft H_1 \triangleleft G = H_0.$$

This process is finite since the $|H_j|$ are strictly decreasing. The quotient groups

$$H_0/H_1, H_1/H_2, \ldots, H_{k-1}/H_k$$

are all simple groups. There may be more than one maximal proper normal subgroup, and so we might find ourselves with different possible sequences of quotient groups. However the *Jordan-Hölder Theorem* states that the eventual list of finite simple groups is unique, other than the order in which they arise. This is analogous to what happens when we factor large integers, and so the Jordan-Hölder Theorem is a wonderful and surprising generalization of the Fundamental Theorem

---

[10] The name "normal" is somewhat misleading since most subgroups of most groups are not normal!

of Arithmetic. To properly understand this would take us too far afield of number theory.

Exercise 7.24.4. Prove that if G is a finite, simple, abelian group, then G is isomorphic to the additive, cyclic group $\mathbb{Z}/p\mathbb{Z}$, where p is a prime.

Exercise 7.24.5.[†] By taking $G = \mathbb{Z}/n\mathbb{Z}$ deduce the Fundamental Theorem of Arithmetic from the Jordan-Hölder Theorem.

Feit and Thompson showed that, otherwise, every non-cyclic finite simple group has even order.

Appendix 7E. Constructing finite fields

7.25. Classification of finite fields

We have already explored the finite fields $\mathbb{F}_p = \{0, 1, \ldots, p-1\}$, though in the guise of the additive and multiplicative groups of residues mod p. In this section we determine all the other possible finite fields. To do so we need two field properties:

A field $\mathbb{F}$ has no zero divisors; that is, if $ab = 0$ with $a, b \in \mathbb{F}$, then either $a = 0$ or $b = 0$. This holds since the non-zero elements of the field form a multiplicative group and so are closed under multiplication. Moreover if $\mathbb{F}$ is a finite field, then $|\mathbb{F}| \cdot 1 = 0$ by applying Lagrange's Theorem (Corollary 7.23.1) to the additive group of $\mathbb{F}$. These properties makes it surprisingly easy to determine all finite fields.

Exercise 7.25.1. Let $\mathbb{F}$ be a finite field.
  (a) Show that if prime q divides $|\mathbb{F}|$, then either $q \cdot 1 = 0$ or $|\mathbb{F}|/q \cdot 1 = 0$. Use an induction hypothesis to deduce that there exists a prime p such that $p \cdot 1 = 0$ in $\mathbb{F}$.
  (b) Show that this prime p is unique.
  (c) Begin with a non-zero element $a_1 \in \mathbb{F}$. If $a_2 \notin \{n_1 a_1 : n_1 \in \mathbb{F}_p\}$, then show that $\{n_1 a_1 + n_2 a_2 : n_1, n_2 \in \mathbb{F}_p\}$ has p^2 distinct elements.
  (d) Deduce by induction that there exist $a_1, \ldots, a_r \in \mathbb{F}$ for some integer $r \geq 1$ such that
  $$\mathbb{F} = \{n_1 a_1 + n_2 a_2 + \cdots + n_r a_r : n_1, \ldots, n_r \in \mathbb{F}_p\},$$
  the elements $n_1 a_1 + n_2 a_2 + \cdots + n_r a_r$ being all distinct, and so $\mathbb{F}$ has p^r elements.

We have shown that all finite fields must have p^r elements for some prime power p^r. We know that there are fields of size p, the rings $\mathbb{Z}/p\mathbb{Z}$ often denoted $\mathbb{F}_p$, and we ask whether there are fields of size p^r for every $r \geq 2$?

The easiest way to construct a finite field of p^r elements is to use a root α of a polynomial $f(x)$ of degree r which is irreducible in $\mathbb{F}_p[x]$. (We showed that such polynomials exist in section 4.12 of appendix 4C; it is much more challenging to actually determine an example of such a polynomial.) Then we can represent the elements of the finite field as

$$a_0 + a_1\alpha + \cdots + a_{r-1}\alpha^{r-1} \text{ where we take the } a_i \in \mathbb{F}_p.$$

Exercise 7.25.2. Verify that this indeed gives a field with p^r distinct elements.

We will write $q = p^r$ for convenience. The given proof of Proposition 7.4.1, suitably modified, works in any finite field, so we may state the following:

Proposition 7.25.1 (Lagrange). *Let q be a prime power and $\mathbb{F}$ be a field with q elements. Any non-zero polynomial $f(x) \in \mathbb{F}[x]$ of degree d has no more than d roots in $\mathbb{F}$.*

The multiplicative group of $\mathbb{F}$ has $q-1$ elements so that $a^{q-1} = 1$ for all $a \in \mathbb{F}^*$ by Lagrange's Theorem, Corollary 7.23.1. Lagrange's Proposition, Proposition 7.25.1, therefore implies that these are all of the roots of the polynomial $x^{q-1} - 1$ and so

(7.25.1) $$\boxed{x^{q-1} - 1 = \prod_{\substack{a \in \mathbb{F} \\ a \neq 0}} (x - a).}$$

Therefore the finite field of q elements is unique (up to isomorphism)! We denote it by $\mathbb{F}_q$.

Exercise 7.25.3. (a) Prove that if $d | p^r - 1$ for some integer $r \geq 1$, then there are precisely $\phi(d)$ elements in $\mathbb{F}_{p^r}$ of order d.
(b) Deduce that $\mathbb{F}_q^*$ is a cyclic group (and therefore has a generator/primitive root) for any prime power q.

Exercise 7.25.4. Show that the finite field of p^2 elements is not isomorphic to the integers mod p^2 (that is, $\mathbb{F}_{p^2} \not\cong \mathbb{Z}/p^2\mathbb{Z}$).

Let $\mathcal{N}_m$ be the set of irreducible, monic polynomials in $\mathbb{F}_p[x]$ of degree m, not including the polynomial x, and let $N_m = |\mathcal{N}_m|$. Above we showed that if $f(x) \in \mathcal{N}_r$ and $f(\alpha) = 0$, then we can construct the unique field with $q = p^r$ elements, $\mathbb{F}_q$, by taking all polynomials in α with coefficients in $\mathbb{F}_p$; in other words $\mathbb{F}_q \cong \mathbb{F}_p[\alpha]$. This implies that $\alpha \in \mathbb{F}_q$, so that $\alpha^{q-1} - 1 = 0$, and therefore α is a root of $x^{q-1} - 1$. We deduce that $f(x)$ is a factor of $x^{q-1} - 1$, and as the polynomials of $\mathcal{N}_m$ cannot share any roots (as they are irreducible), we deduce that their product

$$\Phi_r(x) := \prod_{f(x) \in \mathcal{N}_r} f(x) \text{ divides } x^{p^r - 1} - 1.$$

Now if d divides r, then $p^d - 1$ divides $p^r - 1$, and so $x^{p^d - 1} - 1$ divides $x^{p^r - 1} - 1$. We deduce that $\Phi_d(x)$ divides $x^{p^r - 1} - 1$. Since the elements of the different $\mathcal{N}_d$ must be different, we deduce that

$$\prod_{d | r} \Phi_d(x) \text{ divides } x^{p^r - 1} - 1.$$

The degree of $\Phi_m(x)$ is mN_m by definition, and so

the degree of $\prod_{d|r} \Phi_d(x)$ equals $\sum_{d|r} dN_d = p^r - 1$

7.25. Classification of finite fields

by (4.12.2).[11] This is the same as the degree of $x^{p^r-1}-1$, and so the two polynomials are equal, as one divides the other, they both have the same degree, and they are both monic. Moreover if we multiply through by x, then we obtain that

(7.25.2) $$\boxed{x^{p^r} - x = \prod_{\substack{f(x) \in \mathbb{F}_p[x] \\ f(x) \text{ monic, irreducible,} \\ \text{of degree dividing } r}} f(x).}$$

We complete this survey of finite fields by showing that if $f(x) \in \mathbb{F}_p[x]$ is an irreducible polynomial of degree r, with root $\alpha \in \mathbb{F}_q$ (where $q = p^r$), then[12]

$$f(x) = (x-\alpha)(x-\alpha^p)(x-\alpha^{p^2})\cdots(x-\alpha^{p^{r-1}}).$$

Proof. Write $f(x) = b_0 + b_1 x + \cdots + b_r x^r \in \mathbb{F}_p[x]$ so that
$$b_0 + b_1 \alpha + \cdots + b_r \alpha^r = f(\alpha) = 0.$$
Taking the pth power of both sides and using the Child's binomial theorem, we obtain
$$b_0^p + b_1^p \alpha^p + \cdots + b_r^p \alpha^{pr} = (b_0 + b_1\alpha + \cdots + b_r \alpha^r)^p = 0^p = 0 \text{ in } \mathbb{F}_p[\alpha].$$
But each $b_j^p = b_j$ by Fermat's Little Theorem, and so
$$f(\alpha^p) = b_0 + b_1 \alpha^p + \cdots + b_r(\alpha^p)^r = 0.$$
One can repeat this to show that α^{p^2} is also a root of $f(x)$, and then $\alpha^{p^3},\ldots,\alpha^{p^{r-1}}$. Since $\alpha \in \mathbb{F}_q$ we know that $\alpha^{p^r} = \alpha^q = \alpha$. We claim that
$$\alpha,\ \alpha^p,\ \alpha^{p^2},\ \alpha^{p^3},\ldots,\alpha^{p^{r-1}}$$
are the distinct roots of f, for if $\alpha^{p^i} = \alpha^{p^j}$ with $0 \le i < j \le r-1$, then, by taking the p^{r-i}th power of both sides, we obtain $\alpha^{p^d} = \alpha$ where $d = j-i$ so that $1 \le d \le r-1$. This implies that $f(x)$ divides $x^{p^d} - x$ which contradicts the formula in (7.25.2) with r replaced by d, as $f(x)$ is irreducible of degree r. The result then follows from Proposition 7.25.1. □

Exercise 7.25.5. Fix prime p. Suppose that the sequence $(u_n)_{n\ge 1}$ of integers satisfies the linear recurrence $u_{n+d} \equiv a_{d-1} u_{n+d-1} + \cdots + a_0 u_n \pmod{p}$, where d is minimal.
(a) Suppose that $u_{n+D} \equiv b_{D-1} u_{n+D-1} + \cdots + b_0 u_n \pmod{p}$. Prove that either the $u_n \equiv 0 \pmod{p}$ for all n or $f(x) := x^d - a_{d-1}x^{d-1} - \cdots - a_1 x - a_0$ divides $x^D - b_{D-1} x^{D-1} - \cdots - b_0 \bmod p$.
(b) Prove that if $f(x) \pmod{p}$ is irreducible and $u_j \not\equiv 0 \pmod{p}$ for some j, $0 \le j \le d-1$, then $(u_n)_{n\ge 1}$ is periodic mod p with period $p^d - 1$.

This implies that the upper bound in exercise 2.5.21 is best possible.

Further reading on arithmetic associated with finite fields

[1] Jean-Pierre Serre, *On a theorem of Jordan*, Bull. Amer. Math. Soc. **40** (2003), 429–440.
[2] Neal Koblitz, *Why study equations over finite fields?*, Math. Mag. **55** (1982), 144–149.

---

[11] The value of N_1 here is one less than in section 4.12 of appendix 4C, since here we exclude the polynomial x from our count. Otherwise the N_d are all the same.

[12] The map $\alpha \to \alpha^p$ of roots is usually credited to Frobenius from the end of the 19th century. However this was discussed by Gauss, almost a hundred years earlier, in one of the chapters of his *Disquisitiones* that was discarded by the printer and not published until after Gauss's death.

7.26. The product of linear forms in $\mathbb{F}_q$

Any line in the (x,y)-plane in $\mathbb{F}_q$ that goes through the origin takes the form $ax + by = 0$ for some $a, b \in \mathbb{F}_q$, not both 0. These lines are not all distinct since, for example, $2x + 6y = 0$ gives the same line as $x + 3y = 0$, so we need to take account of scalars, by noting that distinct lines have different $a : b$ ratios. If $a \neq 0$, then any such ratio is proportional to a ratio of the form $1 : r$ for some $r \in \mathbb{F}_q$ (and these are all different); if $a = 0$, then any such ratio is proportional to $0 : 1$. Therefore there are $q + 1$ distinct such lines. The product of these, up to a scalar, is therefore

$$\prod_{a \in \mathbb{F}_q} (x - ay) \cdot y \equiv x(x^{q-1} - y^{q-1})y = x^q y - y^q x,$$

which we prove by replacing x with x/y in (7.25.1) and multiplying through by y^{q-1}.

The set of ratios $a : b$ is really a 1-dimensional set, which we call *projective space*, denoted $\mathbb{P}^1(\mathbb{F}_q)$. Therefore we can rewrite the above as

(7.26.1) $\qquad \prod_{(a,b) \in \mathbb{P}^1(\mathbb{F}_q)} (ax + by)$ is proportional to $xy^q - yx^q$.

The right-hand side is a polynomial which equals the determinant of the matrix

$$\begin{pmatrix} x & y \\ x^q & y^q \end{pmatrix}.$$

We can determine the factorization of this determinant directly because for any $a, b \in \mathbb{F}_q$, we have

$$\begin{pmatrix} x & y \\ x^q & y^q \end{pmatrix} \begin{pmatrix} a & 0 \\ b & 1 \end{pmatrix} = \begin{pmatrix} ax + by & y \\ ax^q + by^q & y^q \end{pmatrix} = \begin{pmatrix} ax + by & y \\ (ax + by)^q & y^q \end{pmatrix}$$

since $(ax + by)^q = (ax)^q + (by)^q = ax^q + by^q$ by the Child's binomial theorem (as q is a power of p) and then Fermat's Little Theorem. Therefore $ax + by$ divides each term in the first column on this matrix, and so $ax + by$ divides the determinant for each such pair a, b. Now each of the factors in the product of the left-hand side of (7.26.1) divides the right-hand side and so their product does as they are coprime in $\mathbb{Z}[x, y]$.[13] By counting degrees we see that the two sides must therefore be equal, up to a scalar multiple.

This latter argument generalizes surprisingly easily. The matrix

$$\begin{pmatrix} x & y & z \\ x^q & y^q & z^q \\ x^{q^2} & y^{q^2} & z^{q^2} \end{pmatrix}$$

looks a bit like the Vandermonde matrix though now the changing powers are in the second exponent, not the first. (We could write the top row as x^{q^0}, y^{q^0}, z^{q^0}.) As above we can alter the first column by multiplying through by an arbitrary column

---

[13]It is more conventional to consider the polynomials with $a \neq 0$ to belong to $\mathbb{F}_p(y)[x]$ and to establish their coprimality using field properties.

7.26. The product of linear forms in $\mathbb{F}_q$

matrix (while leaving the other two columns alone)

$$\begin{pmatrix} x & y & z \\ x^q & y^q & z^q \\ x^{q^2} & y^{q^2} & z^{q^2} \end{pmatrix} \begin{pmatrix} a \\ b \\ c \end{pmatrix} = \begin{pmatrix} ax + by + cz \\ ax^q + by^q + cz^q \\ ax^{q^2} + by^{q^2} + cz^{q^2} \end{pmatrix} \equiv \begin{pmatrix} ax + by + cz \\ (ax + by + cz)^q \\ (ax + by + cz)^{q^2} \end{pmatrix},$$

and so $ax + by + cz$ is a factor of the determinant. Therefore the product of all such (distinct) factors divides the determinant. In this case we need 2-dimensional projective space, $\mathbb{P}^2(\mathbb{F}_q)$, the set of (a, b, c) that are distinct up to a scalar multiple. By a similar analysis one can show that this set contains $p^2 + p + 1$ elements, which is the degree of the determinant of our matrix (this can be seen by considering the contribution to the degree from each row, given the evident fact that there is no cancellation when we expand the usual 3! terms of the determinant). Therefore the determinant of our matrix is equal in $\mathbb{F}_q$ to a scalar multiple times

$$\prod_{(a,b,c) \in \mathbb{P}^2(\mathbb{F}_q)} (ax + by + cz).$$

Exercise 7.26.1.[†] Generalize this to the appropriate n-by-n determinant in $\mathbb{F}_q$, with proof.

Appendix 7F. Sophie Germain and Fermat's Last Theorem

Sophie Germain, born in 1776 to a wealthy Parisian family, studied and researched mathematics under the pseudonym Antoine LeBlanc as women were not accepted in universities at that time.[14] She wrote to and worked with Lagrange and Legendre on the leading questions of the day, though she was cautious about revealing her true identity. Reading *Disquisitiones* in 1804, she wrote to Gauss, sharing her ideas and became a regular correspondent.

In 1807 Napoleon's troops occupied much of Prussia where Gauss lived. Germain was concerned that Gauss might suffer the fate of Archimedes.[15] Being wealthy and well-connected, she asked General Pernety, a family friend, to ensure Gauss's safety. Gauss appreciated but did not understand such concern for his well-being from a woman he did not know! Three months later, Germain disclosed her true identity to Gauss. He replied:

> But how to describe to you my admiration and astonishment at seeing my esteemed correspondent, Monsieur LeBlanc, metamorphose himself into this illustrious personage who gives such a brilliant example of what I would find it difficult to believe. A taste for the abstract sciences in general and above all the mysteries of numbers is excessively rare: one is not astonished at it: the enchanting charms of this sublime science reveal only to those who have the courage to go deeply into it. But when a person of the sex which, according to our customs and prejudices, must encounter infinitely more difficulties than men to familiarize herself with these thorny researches, succeeds nevertheless in surmounting these obstacles and penetrating the most obscure parts of them, then without doubt she must have the noblest courage, quite extraordinary talents and superior genius. Indeed nothing could prove to me in so flattering and less equivocal manner that the attractions of this science, which has enriched my life with so many joys, are not chimerical, than the predilection with which you have honored it.
>
> — excerpt from a letter from C. F. GAUSS to SOPHIE GERMAIN (1807)

---

[14] The École Polytechnique, which opened when Germain was 18, did not accept women, but she was able to study by correspondence.

[15] Archimedes died in 212 B.C. when Romans captured his home city of Syracuse. Archimedes had just drawn a mathematical diagram in the sand when a Roman soldier asked him to get up to be arrested. Archimedes ignored him, wishing to finish what he was doing. The Roman soldier lost patience and ran Archimedes through with his sword.

7.27. Fermat's Last Theorem and Sophie Germain

The first strong result for general exponents in Fermat's Last Theorem was:

Theorem 7.11 (Sophie Germain's Theorem). *Suppose that p is an odd prime and $q = 2p + 1$ is also prime. If $x, y,$ and z are integers for which $x^p + y^p + z^p = 0$, then p divides $x, y,$ or z.*

We will need the following simple lemma.

Lemma 7.27.1. *Suppose that p is an odd prime for which $q = 2p+1$ is also prime. If a, b, c are coprime integers with $a^p + b^p + c^p \equiv 0 \pmod{q}$, then q divides abc.*

Proof. If n is not divisible by q, then $n^p = n^{\frac{q-1}{2}} \equiv -1$ or $1 \pmod{q}$. Therefore if q does not divide abc, then $a^p, b^p, c^p \equiv -1$ or $1 \pmod{q}$, and so $a^p + b^p + c^p \equiv -3, -1, 1,$ or $3 \pmod{q}$. This is impossible as $q = 2p+1 > 3$. □

Proof of Sophie Germain's Theorem. Assume that there is a solution in which p does not divide x, y, z. In the proof of Proposition 6.4.1 we saw that we may assume x, y, z are pairwise coprime so, by Lemma 7.27.1, exactly one of x, y, z is divisible by q: Let us suppose that q divides x, since we may rearrange x, y, z.

By exercise 7.10.6(b) there exist integers a, b, c, d such that

$$z + y = a^p, \; z + x = b^p, \; x + y = c^p, \text{ and } \frac{y^p + z^p}{y + z} = d^p \text{ where } x = -ad,$$

as $p \nmid xyz$. Now $a^p = z + y \equiv (z+x) + (x+y) \equiv b^p + c^p \pmod{q}$ as $q|x$, and so we see that q divides at least one of a, b, c by the lemma. However q does not divide b since $(q,b)|(x, z+x) = (x,z) = 1$ as $q|x$ and $b|z+x$; and similarly q does not divide c. Hence q divides a; that is, $-z \equiv y \pmod{q}$. But then

$$d^p = \frac{y^p + z^p}{y + z} = \sum_{j=0}^{p-1}(-z)^{p-1-j}y^j \equiv \sum_{j=0}^{p-1} y^{p-1-j}y^j = py^{p-1} \pmod{q}.$$

Therefore, as $y \equiv x + y = c^p \pmod{q}$ and $q - 1 = 2p$, we deduce that

$$4 \equiv 4d^{2p} = (2d^p)^2 \equiv (2py^{p-1})^2 = (-1)^2(c^{2p})^{p-1} \equiv 1 \pmod{q},$$

which is impossible as $q > 3$. Therefore p divides $x, y,$ or z. □

The *first case of Fermat's Last Theorem* is the claim that there do not exist integers x, y, z not divisible by p for which $x^p + y^p + z^p = 0$. Sophie Germain's Theorem implies that if there are infinitely many pairs of primes of the form $p, q = 2p + 1$, then there are infinitely many primes p for which the first case of Fermat's Last Theorem is true. However it is still an open question as to whether there are infinitely many *Sophie Germain prime pairs*, that is, those of the form $p, q = 2p+1$.

Subsequently Germain's idea was developed to show that if $m \equiv 2$ or $4 \pmod 6$, then there exists a constant $N_m \neq 0$ such that if p and $q = mp + 1$ are primes for which $q \nmid N_m$, then the first case of Fermat's Last Theorem is true for exponent p. Adleman, Fouvry, and Heath-Brown used this to show that the first case of Fermat's Last Theorem is true for infinitely many prime exponents (long before Wiles fully proved Fermat's Last Theorem).

Appendix 7G. Primes of the form $2^n + k$

7.28. Covering sets of congruences

Are there many primes of the form $k \cdot 2^n \pm 1$ or $k \pm 2^n$ or $2^n \pm k$ for given integer k? In this generality it seems like a more difficult question than asking about primes of the form $2^n \pm 1$, but Erdős showed, ingeniously, how these questions can be resolved for certain integers k:

Let $F_n = 2^{2^n} + 1$ be the Fermat numbers; remember that F_0, F_1, F_2, F_3, F_4 are prime and $F_5 = 641 \times 6700417$. Let $k_0 \pmod{F_0 F_1 F_2 F_3 F_4 F_5}$ be defined by $k_0 \equiv 1 \pmod{641 F_0 F_1 F_2 F_3 F_4}$ and $k_0 \equiv -1 \pmod{6700417}$. For any positive integer k such that $k \equiv k_0 \pmod{F_0 F_1 F_2 F_3 F_4 F_5}$ we have:

- if $n \equiv 1 \pmod{2}$, then $k \cdot 2^n + 1 \equiv 1 \cdot 2^1 + 1 = F_0 \equiv 0 \pmod{F_0}$;
- if $n \equiv 2 \pmod{4}$, then $k \cdot 2^n + 1 \equiv 1 \cdot 2^2 + 1 = F_1 \equiv 0 \pmod{F_1}$;
- if $n \equiv 4 \pmod{8}$, then $k \cdot 2^n + 1 \equiv 1 \cdot 2^{2^2} + 1 = F_2 \equiv 0 \pmod{F_2}$;
- if $n \equiv 8 \pmod{16}$, then $k \cdot 2^n + 1 \equiv 1 \cdot 2^{2^3} + 1 = F_3 \equiv 0 \pmod{F_3}$;
- if $n \equiv 16 \pmod{32}$, then $k \cdot 2^n + 1 \equiv 1 \cdot 2^{2^4} + 1 = F_4 \equiv 0 \pmod{F_4}$;
- if $n \equiv 32 \pmod{64}$, then $k \cdot 2^n + 1 \equiv 1 \cdot 2^{2^5} + 1 = F_5 \equiv 0 \pmod{641}$; and
- if $n \equiv 0 \pmod{64}$, then $k \cdot 2^n + 1 \equiv -1 \cdot 2^0 + 1 = 0 \pmod{6700417}$.

Every integer n belongs to one of these arithmetic progressions (these are called a *covering system* of congruences), and so we have exhibited a prime factor of $k \cdot 2^n + 1$ for every integer n. Therefore $k \cdot 2^n + 1$ is composite unless it equals that prime factor, which is impossible as each $k \cdot 2^n + 1$ is too large. We deduce that for a positive proportion of integers ℓ (that is, the positive integers $\ell \equiv k_0 \pmod{F_6 - 2}$), there is no prime p for which $(p-1)/\ell$ is a power of 2.

Exercise 7.28.1. Deduce that $k \cdot 2^n + 1$ is composite for every integer $n \geq 0$ (with k as defined above).

Exercise 7.28.2. Prove that there exist infinitely many integers k for which $2^n + k$ is composite for every integer $n \geq 0$. (That is, there is no prime p equal to k plus a power of 2.)

Exercise 7.28.3. Let k be as above. Let x_n be a second-order linear recurrence sequence for which $x_n = 3x_{n-1} - 2x_{n-2}$ for all $n \geq 2$. Show that x_n is composite for all $n \geq 0$ if (a) $x_0 = k+1$ and $x_1 = k+2$ or (b) $x_0 = k+1$ and $x_1 = 2k+1$.

Exercise 7.28.4. Let ℓ be any positive integer for which $\ell \equiv -k \pmod{F_6 - 2}$. Prove that $\ell \cdot 2^n - 1$ and $|2^n - \ell|$ are composite for every integer $n \geq 0$. Deduce that a positive proportion of odd integers m cannot be written in the form $p + 2^n$ with p prime.

Exercise 7.28.5. Prove that $13 \cdot 20^k + 1$ is not prime for any $k \geq 1$.

John Selfridge showed that at least one of the primes $3, 5, 7, 13, 19, 37,$ and 73 divides $78557 \cdot 2^n + 1$ for every integer $n \geq 0$. This is the smallest k known for which $k \cdot 2^n + 1$ is always composite. It is an open problem as to whether this is the smallest such k.

Exercise 7.28.6.[‡] Prove that if $n \geq 3$, then $F_n - 2 = F_0 F_1 \ldots F_{n-1}$ cannot be written in the form $p + 2^k + 2^\ell$ where p is prime and $k > \ell \geq 0$.

For $n = 2$ there are the two solutions $F_2 - 2 = 17 - 2 = 5 + 2^3 + 2 = 3 + 2^3 + 2^2$, as well as $F_1 - 2 = 3 = 2 + 1$.

We suspect that there are infinitely many primes in the sequence $2^n - 1$ with $n \geq 1$, the Mersenne primes. And we now have many examples of integers k for which every $2^n + k$ is composite, those integers k for which we can create a covering system. So what about other integers k? For example, is $2^n - 3$ infinitely often prime? Calculations suggest there might be infinitely many primes of the form $2^n - 3$, though we cannot compute very far since the numbers grow so rapidly. The most optimistic conjecture would be: *Fix integer k.*

- *Either there is a finite set of primes P such that for every positive integer n, the number $2^n + k$ is divisible by some element of P (and so is not prime).*
- *Or there are infinitely many positive integers n for which $2^n + k$ is prime.*

7.29. Covering systems for the Fibonacci numbers

We will show that 727413 cannot be written as a prime plus or minus a Fibonacci number, by using covering systems to show the following.

Theorem 7.12. *If N is an integer $\equiv 93687$ or $103377 \pmod{M}$, where $M = 312018 = 2 \cdot 3 \cdot 7 \cdot 17 \cdot 19 \cdot 23$, then $(N + F_k, M) > 1$ for all integers $k \geq 0$.*

Here we let F_k be the kth Fibonacci number. In the covering systems of section 7.28, the key idea is that if 2 has order $m \pmod p$, then the sequence $1, 2, 2^2, 2^3, \ldots \pmod{p}$ has period m. Therefore the zeros in the sequence $2^n - 1 \pmod{p}$ occur every mth value of n, where m is the period of this sequence. However for the Fibonacci numbers the frequency of the $0 \pmod{p}$ is not necessarily the same as the period of the Fibonacci numbers mod p. For example, the period of the

Fibonacci numbers mod 3 is

$$0, 1, 1, 2, 0, 2, 2, 1,$$

which has length 8 whereas 3 divides F_n whenever 4 divides n. Notice that 1 and 2 both appear three times in the period, which is more often than 0 appears; we will exploit this, and similar repetitions, in our construction:

A Fibonacci covering system

If $N \equiv 93687 \pmod{M}$, then the divisibilities lead to a covering system but with several congruence classes to each modulus:

$2 | N + F_k$ if and only if $k \equiv 1$ or $2 \pmod{3}$,

$3 | N + F_k$ if and only if $k \equiv 0 \pmod{4}$,

$7 | N + F_k$ if and only if $k \equiv 1, 2, 6$, or $15 \pmod{16}$,

$17 | N + F_k$ if and only if $k \equiv 0 \pmod{9}$,

$19 | N + F_k$ if and only if $k \equiv 3, 8$, or $15 \pmod{18}$,

$23 | N + F_k$ if and only if $k \equiv 30$ or $42 \pmod{48}$.

This proves our theorem for $N \equiv 93687 \pmod{M}$, and a similar analysis reveals the proof for $N \equiv 103377 \pmod{M}$.

From our theorem we deduce that 727413 cannot be written as $p - F_k$ with p prime, for if so, then $(p, M) = (727413 + F_k, M) > 1$, as $727413 \equiv 103377 \pmod{M}$, and so p divides M. This is impossible or else $23 \geq p = 727413 + F_k > 23$, a contradiction. We also note that 727413 cannot be written as $p + F_k$ with p prime, simply by showing, by computations, that $727413 - F_k$ is composite, whenever $F_k < 727413$.

Calculations suggest that 6851 and 7293 might not have a representation as a prime plus or minus a Fibonacci number, but there seems little chance of finding a covering system for either. The smallest such integer known is 135225 thanks to Ismailescu and Shim.

7.30. The theory of covering systems

A set of congruence classes

$$a_1 \pmod{m_1}, \ldots, a_k \pmod{m_k}$$

is called a *covering system* if every integer belongs to one of the congruence classes. For example, one might take all the congruence classes mod m, and we have seen several examples above. For example $2^{j-1} \pmod{2^j}$ for $1 \leq j \leq J$ and $0 \pmod{2^J}$. With Selfridge's construction one can use

1 mod 2, 1 mod 3, 2 mod 4, 0 mod 9, 8 mod 12, 6 mod 18, and 12 mod 36.

We have seen in this appendix some beautiful applications of covering systems to showing that there are no primes in certain sequences. One can also study covering systems for their own sake. The most precise is an *exact covering system* in which every integer belongs to exactly one of the congruence classes. If we take

7.30. The theory of covering systems

each a_i to be the least residue mod m_i, then our covering system is exact if and only if

$$\sum_{i=1}^{k} \frac{x^{a_i}}{1-x^{m_i}} = \frac{1}{1-x}.$$

Exercise 7.30.1. (a) Prove this.
  (b)[†] Deduce that if $m_1 < m_2 < \cdots < m_k$ in an exact covering system, then either $m_k = 1$ or $m_k = m_{k-1}$.

A *distinct covering system* is one in which the m_i are distinct, so we may write $m_1 < m_2 < \cdots < m_k$. In 2015, Bob Hough proved that $m_1 \leq 10^{16}$, a famous old question of Erdős (whether or not the minimum modulus could be unbounded).[16] This leaves the Erdős-Selfridge question as to whether there is a distinct covering system in which the moduli are all odd. Schinzel showed that, if so, then for any $f(x) \in \mathbb{Z}[x]$ with $f(0) \neq 0$ and $f(1) \neq -1$, the polynomials $x^n + f(x)$ are irreducible for all n in some arithmetic progression, an intriguing connection.

Interesting articles on covering systems

[1] Bob Hough, *Solution of the minimum modulus problem for covering systems*, Ann. of Math. (2) 181 (2015), 361–382.

[2] Dan Ismailescu and Peter C. Shim, *On numbers that cannot be expressed as a plus-minus weighted sum of a Fibonacci number and a prime*, Integers 14 (2014), Paper No. A65, 12 pp.

[3] A. Schinzel, *Reducibility of polynomials and covering systems of congruences*, Acta Arith. 13 (1967/1968), 91–101.

---

[16] This upper bound has recently been improved to $m_1 < 616000$ by Balister, Bollobas, Morris, Sahasrabudhe, and Tiba.

Appendix 7H. Further congruences

In this chapter we saw some extraordinary congruences mod p, like Fermat's Little Theorem and Wilson's Theorem. In this appendix we seek to develop some of these elegant congruences to higher powers of p.

7.31. Fermat quotients

Fermat's Little Theorem tells us that p always divides $a^p - a$, but what about divisibility by p^2? Experiments indicate that this seems to happen rarely, but it does happen; for example

$$p^2 \text{ divides } 2^p - 2 \text{ for } p = 1093 \text{ and for } p = 3511.$$

These are the only two examples known, despite extensive searching (up to 7.2×10^{16}), and we have no idea whether there are any more examples. We define the *Fermat quotient*

$$q_p(a) := \frac{a^{p-1} - 1}{p}$$

and note that $q_p(2) \equiv 0 \pmod{p}$ if and only if p^2 divides $2^p - 2$. It turns out that we can find some interesting congruences for the Fermat quotients mod p.

By the binomial theorem and exercise 2.5.9(b) we have

$$\mathcal{L}_p(1-x) := \frac{(1-x)^p - 1 + x^p}{p} = \frac{1}{p} \sum_{m=1}^{p-1} \binom{p}{m}(-x)^m \equiv -\sum_{m=1}^{p-1} \frac{x^m}{m} \pmod{p}.$$

For example, for $x = -1$ and then for $x = 2$, we obtain

$$\frac{2^p - 2}{p} \equiv -\sum_{m=1}^{p-1} \frac{(-1)^m}{m} \equiv -\sum_{m=1}^{p-1} \frac{2^m}{m} \pmod{p}.$$

The first congruence implies that

$$2q_p(2) \equiv \sum_{\substack{1 \leq m \leq p-1 \\ m \text{ odd}}} \frac{1}{m} - \sum_{\substack{1 \leq m \leq p-1 \\ m \text{ even}}} \frac{1}{m}$$

(7.31.1)
$$\equiv \sum_{1 \leq m \leq p-1} \frac{1}{m} - 2 \sum_{\substack{1 \leq m \leq p-1 \\ m \text{ even}}} \frac{1}{m} \equiv -\sum_{n=1}^{\frac{p-1}{2}} \frac{1}{n} \pmod{p},$$

by adding $\sum_{1 \leq m \leq p-1, \, m \text{ even}} 1/m$ to both sums, writing $m = 2n$ in the second sum, and then by Corollary 7.5.2 with $k = -1$.

By their definitions, we have that

$$\mathcal{L}_p(1-x) = x\, q_p(x) + (1-x)\, q_p(1-x).$$

The Taylor expansion of $\mathcal{L}_p(1-x)$ is a truncation of the Taylor expansion for the logarithm function,

$$\log(1-x) = -\sum_{m \geq 1} \frac{x^m}{m}.$$

We discuss an appropriate way to extend $\mathcal{L}_p(1-x)$ in this direction in section 16.6. There is another connection since

(7.31.2) $\qquad q_p(a) + q_p(b) \equiv q_p(ab) \pmod{p}.$

To see this, note that

$$1 + p q_p(ab) = (ab)^{p-1} = a^{p-1} b^{p-1} = (1 + p q_p(a))(1 + p q_p(b))$$
$$\equiv 1 + p(q_p(a) + q_p(b)) \pmod{p^2},$$

subtract 1, and divide by p.

Binomial coefficients

We have seen that $(-1)^j \binom{p-1}{j} \equiv 1 \pmod{p}$. What about mod p^2? In that case

$$(-1)^j \binom{p-1}{j} = \prod_{i=1}^{j} \frac{i-p}{i} = \prod_{i=1}^{j} \left(1 - \frac{p}{i}\right) \equiv 1 - p \sum_{i=1}^{j} \frac{1}{i} \pmod{p^2}.$$

If $j = \frac{p-1}{2}$, we can use (7.31.1) and (7.31.2) to deduce that

$$(-1)^{\frac{p-1}{2}} \binom{p-1}{\frac{p-1}{2}} \equiv 1 + 2p q_p(2) \equiv 1 + p q_p(4) \equiv 4^{p-1} \pmod{p^2}.$$

In 1894 Morley took this one step further by proving that

$$(-1)^{\frac{p-1}{2}} \binom{p-1}{(p-1)/2} \equiv 4^{p-1} \pmod{p^3}$$

and that this holds mod p^4 if and only if p divides the Bernoulli number B_{p-3}.

Exercise 7.31.1.   (a) Show that $\binom{2p}{p} = 2\prod_{n=1}^{p-1}\left(1 + \frac{p}{n}\right)$.

(b) By expanding the product, deduce that for odd primes p, we have

$$\frac{1}{2}\binom{2p}{p} \equiv 1 + p\sum_{n=1}^{p-1}\frac{1}{n} + \frac{p^2}{2}\left(\left(\sum_{n=1}^{p-1}\frac{1}{n}\right)^2 - \sum_{n=1}^{p-1}\frac{1}{n^2}\right) \pmod{p^3}.$$

(c) Use Corollary 7.5.2 and exercise 7.10.12 to deduce that $\binom{2p}{p} \equiv 2 \pmod{p^3}$ for all primes $p > 3$.

(d)‡ Prove *Wolstenholme's Theorem* that, for any prime $p > 3$ and any integers $n \geq m \geq 0$,

$$\binom{np}{mp} \equiv \binom{n}{m} \pmod{p^3}.$$

(The difference is divisible by p^4 if and only if p divides the Bernoulli number B_{p-3}.)

Bernoulli numbers modulo p

In section 0.6 of appendix 0B we introduced the Bernoulli numbers and showed that they are rational numbers; moreover $B_1 = -\frac{1}{2}$ and $B_n = 0$ if n is odd and > 1. It is important to know the prime factors of the denominators of the B_{2n}:

Theorem 7.13 (The Von Staudt-Clausen Theorem). *Prime p divides the denominator of B_{2n} if and only if $p-1$ divides $2n$. The denominator of B_{2n} is always squarefree; in fact*

$$(7.31.3) \qquad B_{2n} + \sum_{\substack{p \text{ prime} \\ p-1 \mid 2n}} \frac{1}{p} \quad \text{is an integer for all } n \geq 1.$$

Proof. For each prime p, we will verify that $pB_{2n} \equiv 0 \pmod{p}$ if $p-1 \nmid 2n$, and $pB_{2n} \equiv p-1 \pmod{p}$ if $p-1 \mid 2n$, which implies the theorem. We will prove this by induction on $n \geq 1$. It is evidently true for $n = 1$ as $B_1 = -\frac{1}{2}$, so now assume $n \geq 2$, and the result holds for all $m \leq n-1$. By Theorem 0.1 and (0.6.1) with $k = 2n + 1$ we have

$$(7.31.4) \quad \sum_{j=0}^{p-1} j^{2n} = \frac{1}{2n+1}(B_{2n+1}(p) - B_{2n+1}) = \frac{1}{2n+1}\sum_{r=0}^{2n}\binom{2n+1}{r} B_r p^{2n+1-r}.$$

Our goal is to evaluate both sides of this equation mod p. We begin with the right-hand side. The $r = 1$ term gives $-p^{2n}/2$ which is $\equiv 0 \pmod{p}$. We are left with the terms with $r = 2m$, $0 \leq m \leq n$ (as $B_r = 0$ when r is odd and > 1). For each such term, $\frac{1}{2n+1}\binom{2n+1}{2m} = \frac{1}{2m}\binom{2n}{2m-1}$; as each binomial coefficient is an integer, this is a rational number whose denominator divides both $2n+1$ and $2m$ and therefore divides $(2n+1, 2m)$. Let $p^e \| (2n+1, 2m)$ so that the denominator of $\frac{p^e}{2n+1}\binom{2n+1}{2m}$ is not divisible by p. Now p^e divides $2n+1-2m$ and so if $e \geq 2$, then $2n - 1 - 2m \geq p^e - 2 \geq 2^e - 2 \geq e$; if $e = 0$ or 1, then $2n - 1 - 2m \geq 1 \geq e$, if $m \leq n - 1$. By the induction hypothesis $p^2 B_{2m} \equiv 0 \pmod{p}$ for all $m \leq n - 1$. Collecting all this information together gives that if $0 \leq m \leq n - 1$, then

$$\frac{1}{2n+1}\binom{2n+1}{2m} B_{2m} p^{2n+1-2m} = \frac{p^e}{2n+1}\binom{2n+1}{2m} \cdot p^2 B_{2m} \cdot \frac{p^{2n-1-2m}}{p^e} \equiv 0 \pmod{p}.$$

Substituting these congruences into (7.31.4), we obtain

$$pB_{2n} \equiv \sum_{j=0}^{p-1} j^{2n} \equiv \begin{cases} p-1 & \text{if } p-1 \text{ divides } 2n \\ 0 & \text{otherwise} \end{cases} \pmod{p}$$

by Corollary 7.5.2, and the result follows. □

Sums of powers of integers modulo p^2

Assume, for convenience, that $p > 3$. By Theorem 0.1 and (0.6.1) we have

(7.31.5) $$\sum_{n=0}^{p-1} n^{k-1} = \frac{1}{k} \sum_{r=0}^{k-1} \binom{k}{r} B_r p^{k-r}.$$

If p does not divide (r,k), then it does not divide the denominator of $\frac{1}{k}\binom{k}{r}$, and so the rth term is then $\equiv 0 \pmod{p^3}$ whenever $k - r \geq 4$, by the von Staudt-Clausen Theorem. Otherwise p^ℓ divides $\frac{1}{k}\binom{k}{r} B_r p^{k-r}$ where $\ell = k - r - 1 - e$ and $p^e \| (k,r)$ with $e \geq 1$. Now p^e divides $k - r > 0$ and so $\ell \geq p^e - e - 1 \geq 5^e - e - 1 \geq 3$. Therefore if $m = k - 1 \geq 2$, then

$$\sum_{n=0}^{p-1} n^m \equiv B_m p + \frac{m}{2} B_{m-1} p^2 + \frac{m(m-1)}{6} B_{m-2} p^3 \pmod{p^3}.$$

A similar analysis would allow us to extend this to modulo any power of p. We deduce from this that if $p > 3$ and $m \geq 2$, then

(7.31.6) $$\sum_{n=0}^{p-1} n^m \equiv \begin{cases} pB_m \pmod{p^2} & \text{if } m \not\equiv 1 \pmod{p-1}, \\ -mp/2 \pmod{p^2} & \text{if } m \equiv 1 \pmod{p-1}, \end{cases}$$

since $\frac{m}{2} B_{m-1} p^2 \equiv 0 \pmod{p^2}$ unless $p-1$ divides $m-1$, by the von Staudt-Clausen Theorem, in which case $B_m = 0$ and $pB_{m-1} \equiv -1 \pmod{p}$.

This improves on Corollary 7.5.2. Note that (7.31.6) is $\equiv 0 \pmod{p^2}$ if m is odd and $m \not\equiv 1 \pmod{p-1}$, though this is more easily proved in exercise 7.10.12. It is also $\equiv 0 \pmod{p^2}$ when the exponent $m = p$.

Exercise 7.31.2. Prove that

$$\sum_{n=1}^{p-1} n^k q_p(n) \equiv \begin{cases} 1/2 \pmod{p} & \text{if } k = 1, \\ B_{p-1+k} - B_k \pmod{p} & \text{if } 2 \leq k \leq p-1. \end{cases}$$

The Wilson quotient

Wilson's Theorem states that p divides $(p-1)! + 1$. We define the *Wilson quotient*,

$$w_p = \frac{(p-1)! + 1}{p}.$$

By the additivity property, (7.31.2), of Fermat quotients we obtain

$$\sum_{n=1}^{p-1} q_p(n) = q_p((p-1)!) = \frac{(-1+pw_p)^{p-1} - 1}{p} \equiv -(p-1)w_p \equiv w_p \pmod{p}.$$

Therefore
$$p - 1 + pw_p \equiv \sum_{n=1}^{p-1}(1 + pq_p(n)) = \sum_{n=1}^{p-1} n^{p-1} \equiv p B_{p-1} \pmod{p^2},$$

by (7.31.6) (and remember that the Von Staudt-Clausen Theorem implies that $pB_{p-1} \equiv -1 \pmod{p}$). We obtain the surprising connection:

(7.31.7) $$w_p \equiv \frac{pB_{p-1} - (p-1)}{p} \pmod{p}.$$

Exercise 7.31.3. Show that if $p > 3$, then $\prod_{j=1}^{p-1}(ap+j) \equiv (p-1)! \pmod{p^3}$ for every integer a.

Beyond Fermat's Little Theorem

Let p be an odd prime. We have seen that $\prod_{n=0}^{p-1}(x-n) \equiv x^p - x \pmod{p}$, but what about mod p^2? We define

$$\mathcal{I}_p := -\log\left(\prod_{n=1}^{p-1}(1-nx)\right) = \sum_{k \geq 1}\left(\sum_{n=1}^{p-1} n^k\right)\frac{x^k}{k}.$$

We will restrict our attention to $k \leq p-1$ (as the original polynomial has degree $p-1$), and so (7.31.6) implies that

$$\mathcal{I}_p \equiv \sum_{k=1}^{p-1} pB_k \frac{x^k}{k} \pmod{(p^2, x^p)}.$$

Therefore $\mathcal{I}_p \equiv x^{p-1} \pmod{(p, x^p)}$ and so $\mathcal{I}_p^2 \equiv 0 \pmod{(p^2, x^p)}$. Therefore

$$\prod_{n=1}^{p-1}(1-nx) \equiv e^{-\mathcal{I}_p} \equiv 1 - \mathcal{I}_p \equiv 1 - \sum_{k=1}^{p-1} pB_k \frac{x^k}{k} \pmod{(p^2, x^p)}.$$

Since the left-hand side has degree $p-1$, the two sides are congruent mod p^2. Replacing x by $1/x$, multiplying through by x^p, and using (7.31.7), we deduce that

(7.31.8) $$\prod_{n=0}^{p-1}(x-n) \equiv x^p - x + p\left(w_p x + \sum_{m=2}^{p-1} B_{p-m}\frac{x^m}{m}\right) \pmod{p^2}.$$

Exercise 7.31.4. Prove that
$$\prod_{n=0}^{p-1}(x-n) - \prod_{n=0}^{p-1}(x+n) \equiv px^{p-1} \pmod{p^2}$$
in two ways. First deduce it from (7.31.8) and then prove it by substituting in $x = 0, 1, \ldots, p-1$.

Reference for this section

[1] Emma Lehmer *On Congruences Involving Bernoulli Numbers and the Quotients of Fermat and Wilson*, Ann. of Math. **39** (1938), 350–360.

7.32. Frequency of p-divisibility

Fermat quotients

We do not know much about the value of $q_p(2)$ (mod p). Our best guess is that if we do not understand a sequence of mathematical objects and do not see any algebraic structure to prejudice it one way or another, then it must be "randomly distributed", at least with the correct interpretation of what "random" means. In this case, we guess that $q_p(2)$ is as likely to be in any given residue class mod p as any other, in particular 0 (mod p), and therefore the "probability" that $q_p(2) \equiv 0$ (mod p) is roughly $1/p$. This is a meaningless statement but can be reinterpreted much like the Gauss-Cramér model for primes discussed in section 5.15 of appendix 5C. So let $X_2, X_3, X_5, X_7, \ldots$ be a sequence of independent random variables indexed by the primes, with $\text{Prob}(X_p = 1) = \frac{1}{p}$ and $\text{Prob}(X_p = 0) = 1 - \frac{1}{p}$. We think of the sequence $x_2, x_3, \ldots$ with $x_p = 1$ whenever p divides $q_p(2)$, and $x_p = 0$ otherwise, as being typical of the sequences in this probability space. Therefore we expect that

$$\#\{p \leq x : p \text{ divides } q_p(2)\} \text{ is roughly } \mathbb{E}\left(\sum_{p \leq x} X_p\right) = \sum_{p \leq x} \mathbb{E}(X_p) = \sum_{p \leq x} \frac{1}{p}$$

which $\to \infty$ by (5.12.3), growing like $\log \log x$ by (5.12.4). This suggests there are infinitely many primes p for which p divides $q_p(2)$ but they should be extremely rare. There are just two examples known with $p \leq 7.2 \times 10^{16}$, so when should we expect to find the next example? Our best guess is that it should be $\leq x$ where $\log \log x - \log \log(7.2 \times 10^{16})$ is roughly 1; that is, $x \approx 10^{43}$. This is so large that we may never be able to compute another example, even if examples exist as frequently as expected!

Bernoulli numbers

How frequently are the Bernoulli numbers B_{2n} divisible by a given prime p, with $2 \leq 2n < p - 1$?[17] Again we have little idea but reams of data suggest that the B_{2n} (mod p) are distributed much like random numbers mod p. If this is so, then we can guess that for a random large prime p and a random n, the probability that p divides (the numerator of) B_{2n} is $1/p$. If these probabilities are "independent", then the "probability" that p divides none of these numerators is

$$\left(1 - \frac{1}{p}\right)^{\frac{p-3}{2}} \longrightarrow_{p \to \infty} e^{-1/2} = 0.6065306597\ldots.$$

Calculations with Bernoulli numbers suggest that this is about right, though we have absolutely no idea how to prove this. This analysis also suggests that the number of B_{2n} divisible by p should behave like a Poisson variable with parameter $\frac{1}{2}$. In other words we can guess that for each $m \geq 0$, the proportion of primes p for which $p|B_{2n}$ for exactly m values of $2n$ with $2 \leq 2n \leq p - 3$ is $\frac{e^{-1/2}}{2^m \cdot m!}$.

---

[17] In a famous early paper in algebraic number theory Kummer showed that a class number related to cyclotomic fields is divisible by p if and only if one of these Bernoulli numbers is divisible by p.

Appendix 7I. Primitive prime factors of recurrence sequences

7.33. Primitive prime factors

Prime p is a *primitive prime factor* of $a^m - 1$ if p divides $a^m - 1$ but does not divide $a^r - 1$ for any $1 \leq r \leq m - 1$. In other words a has order m mod p and so $p \equiv 1 \pmod{m}$, by Theorem 7.1. Moreover p divides $\phi_m(a)$ but not $\phi_r(a)$ for any $1 \leq r \leq m - 1$. (Here the $\phi_n(x)$ are the cyclotomic polynomials defined in section 4.16 of appendix 4E and used in section 7.9.)

Proposition 7.33.1. *If a has order m mod p, then $p | \phi_n(a)$ if and only if n/m is a power of p. Moreover if $n \neq m$, then $p^2 \nmid \phi_n(a)$, except when $a \equiv 3 \pmod 4$ with $m = 1$ and $n = p = 2$.*

Proof. Now $(m, p) = 1$ as m divides $p - 1$. If $p | \phi_n(a)$, which divides $a^n - 1$, then $m | n$ by Lemma 7.1.2. We now prove that n/m is a power of p. If not, then there exists a prime $q \neq p$ which divides n/m. Then m divides n/q so p divides $a^{n/q} - 1$, which divides $a^n - 1$. By Lemma 7.17.1(a) we deduce that $\frac{a^n-1}{a^{n/q}-1}$ is not divisible by p. But $\phi_n(a)$ is a factor of $\frac{a^n-1}{a^{n/q}-1}$, and so it cannot be divisible by p, a contradiction.

Now let $n = mp^\ell$. By Proposition 7.17.1 we have that p divides $\frac{a^n-1}{a^{n/p}-1}$ but not p^2 (except perhaps if $p = 2$, and 4 divides $a^{n/2} + 1$). Moreover

$$\frac{a^n - 1}{a^{n/p} - 1} = \frac{a^{mp^\ell} - 1}{a^{mp^{\ell-1}} - 1} = \prod_{d|m} \phi_{dp^\ell}(a)$$

and $m \nmid dp^\ell$ for $d < m$, so p divides $\phi_n(a)$ but not p^2 (except perhaps if $p = 2$, $n/2$ is odd, and $a \equiv 3 \pmod 4$). In this last case, a must have order 1 (mod 2), and so $m = 1, n = 2$. □

Corollary 7.33.1. *Let n be an integer > 2. If prime p divides $\phi_n(a)/(n, \phi_n(a))$, then a has order n mod p. In particular if $|\phi_n(a)| > n$, then there exists a prime $p \equiv 1 \pmod n$ that is a primitive prime factor of $a^n - 1$.*

Proof. If a has order m mod p where $m < n$, and p divides $\phi_n(a)$, then $n = mp^k$ for some $k \geq 1$, by Proposition 7.33.1. Moreover, p^2 does not divide $\phi_n(a)$ and so p divides $(n, \phi_n(a))$, so that p does not divide $\phi_n(a)/(n, \phi_n(a))$. Therefore a has order n (mod p).

Now $(n, \phi_n(a)) \leq n$ and so if $|\phi_n(a)| > n$, then $|\phi_n(a)|/(n, \phi_n(a)) > 1$. Therefore such a prime p exists and must be $\equiv 1 \pmod n$. □

Exercise 7.33.1. (a) Use exercise 4.16.4 to prove that if $|a| \geq 2$, then $|\phi_n(a)| \geq \alpha \, |a|^{\phi(n)}$ where $\alpha = \alpha(a) := \prod_{k \geq 1}(1 - 1/|a|^k)$. Note that $\alpha(a) \geq \alpha(2) = .288788\ldots$.
(b)† Deduce that $a^n - 1$ has a primitive prime factor for every integer $a \neq -1, 0, 1$ and $n \geq 1$, except for the special cases $n = 1$, $a = 2$; or $n = 2$, $a = -1 \pm 2^k$ for some integer k; or $n = 3$, $a = -2$; or $n = 6$, $a = 2$.

We have therefore proved the following:

Corollary 7.33.2. *If a is an integer > 1, then $a^n - 1$ has a primitive prime factor for all $n \geq 1$, except for $2^1 - 1$, $(2^k - 1)^2 - 1$, and $2^6 - 1$. Moreover $a^n + 1$ has a primitive prime factor for all $n \geq 1$ except $2^3 + 1$.*

Proof. The first part follows from exercise 7.33.1(b). Now $a^n + 1$ has the same primitive prime factors as $a^{2n} - 1$, so the only possible exceptions correspond to the even exponent cases in the first part, leading to just the one example. □

One can also deduce Theorem 7.8.

Prime power divisibility of second-order linear recurrence sequences

The next exercise gives a generalization of Proposition 7.17.1 to second-order linear recurrence sequences: Let $\{x_n : n \geq 0\}$ be the second-order linear recurrence sequence given by

$$x_n = ax_{n-1} + bx_{n-2} \text{ for all } n \geq 2 \text{ with } x_0 = 0 \text{ and } x_1 = 1.$$

Exercise 2.5.19 implies that, for any given $m \geq 2$, x_m divides $x_{2m}, x_{3m}, \ldots$. Let $X_k := x_{km}/x_m$ be the sequence of quotients. Applying exercise 0.7.3(c) (with m replaced by $(k-2)m$ and n by m), X_k starts $X_0 = 0$, $X_1 = 1$ and then

$$X_k = y_m X_{k-1} + z_m X_{k-2} \quad \text{for all} \quad k \geq 2,$$

where

$$x^2 - y_m x - z_m = (x - \alpha^m)(x - \beta^m).$$

Exercise 7.33.2.‡ Suppose that $(a, b) = 1$. Let Δ_m denote the discriminant of the minimum polynomial for $(X_k)_{k \geq 0}$.
(a) Prove that y_m, z_m, and Δ_m are pairwise coprime.
(b) Show that $\Delta_m = \Delta^2 x_m^2$.

(c) Let m_d be the smallest positive integer for which $d|x_m$. Now let $m = m_p$ for some prime p, and suppose that $p^k \| x_m$, so that $m_p = m_{p^2} = \cdots = m_{p^k}$. Prove that $m_{p^{k+1}} = pm_p$ and, in general, $m_{p^{k+j}} = p^j m_p$ for all $j \geq 0$.

Exercise 7.33.3.[†] Suppose that $(a, b) = 1$ and that odd prime p divides Δ.
(a) Prove that $x_n \equiv n(a/2)^{n-1} \pmod{\Delta}$, and deduce that $m_p = p$.
(b) Show that if $p > 3$, then $x_n \equiv n(a/2)^{n-1} + \frac{n(n-1)(n-2)}{24} n^2 \Delta (a/2)^{n-3} \pmod{\Delta^2}$ for all $n \geq 3$.
(c) Deduce that p divides x_p but p^2 does not.

7.34. Closed form identities and sums of powers

In section 0.2 we began our study of closed form expressions for sums of kth powers and observed the remarkable identity

$$s_3(N) = s_1(N)^2$$

for all $N \geq 1$ where $s_k(N) = 1^k + 2^k + \cdots + N^k$. Moreover we asked whether there are any other such identities. We now prove that there are not.

Corollary 7.34.1. *Suppose there is an identity* $r \prod_{i=1}^m s_{j_i}(N) = R \prod_{I=1}^M s_{J_I}(N)$ *for some integers* $r, R, j_1, \ldots, j_m, J_1, \ldots, J_M \geq 1$, *where all the* j_i *are different from the* J_I, *and* $(r, R) = 1$. *Then the identity must be of the form* $s_3(N)^\ell = s_1(N)^{2\ell}$ *for some integer* $\ell \geq 1$.

Proof. By changing the order if necessary we may assume that $J_M > J_I, j_i$ for all $I < M$ and $i \leq m$. By Theorem 0.1 and (0.6.1), the main term of $s_k(N)$ is $N^{k+1}/(k+1)$. Therefore, comparing the coefficient of the leading term on both sides of our identity and multiplying through by denominators, we deduce that

$$r \prod_{I=1}^M (J_I + 1) = R \prod_{i=1}^m (j_i + 1),$$

which implies that r divides $\prod_{i=1}^m (j_i + 1)$.

Let $N = 2$ in our identity and suppose that p is a primitive prime factor of $s_{J_M}(2) = 2^{J_M} + 1$. Then p divides $r \prod_{i=1}^m s_{j_i}(2)$, but p cannot divide any $s_{j_i}(2)$, and so p divides r, which divides $\prod_{i=1}^m (j_i + 1)$. But $p \equiv 1 \pmod{2J_M}$ so that $p \geq 2J_M + 1$, but this yields a contradiction since it implies that p cannot divide any $j_i + 1$. Hence $2^{J_M} + 1$ does not have a primitive prime factor which implies that $J_M = 3$ by Corollary 7.33.2.

Therefore any identity involves only $s_1(N)$, $s_2(N)$, and $s_3(N)$. Now $s_2(N)$ has a factor $2N - 1$ which is coprime with $N(N - 1)$ and so $s_2(N)$ cannot be involved in any identity. Comparing powers of $N(N - 1)/2$ we deduce that only possible identities take the form $s_3(N)^\ell = s_1(N)^{2\ell}$ as claimed. □

7.35. Primitive prime factors and second-order linear recurrence sequences

The Bang-Zsigmondy Theorem (1892) states that if a and b are coprime positive integers, then $a^n - b^n$ has a primitive prime divisor for all $n > 1$ except for the special cases $n = 6$, $a = 2$, $b = 1$, and $n = 2$, $a + b = 2^k$. It can be proved

7.35. Primitive prime factors and second-order linear recurrence sequences

by elaborating on the ideas in section 7.33. One can also prove such results for second-order linear recurrence sequences that begin $0, 1$:

Let $x_0 = 0$ and $x_1 = 1$, and let $x_n = ax_{n-1} + bx_{n-2}$ for all $n \geq 2$. Let $\Delta := a^2 + 4b$.

When $\Delta > 0$, then it is easy to show that the x_n grow exponentially fast (which is the key to the proof in exercise 7.33.1), since $x_n = (\alpha^n - \beta^n)/(\alpha - \beta)$ with $|\alpha| > 1, |\beta|$. Using this, Carmichael showed in 1913 that if $\Delta > 0$, then x_n has a primitive prime factor for each $n \neq 1, 2,$ or 6 except for $F_{12} = 144$ where F_n is the Fibonacci sequence, and for $F_n'' = (-1)^{n-1} F_n$.

The case with $\Delta < 0$ is much more difficult. Nonetheless, in 1974 Schinzel [3] succeeded in showing that x_n has a primitive prime factor once $n > n_0$, for some sufficiently large n_0, other than in the periodic case $a = \pm 1$, $b = -1$. Determining the smallest possible value of n_0 has required great efforts culminating in the beautiful work of Bilu, Hanrot, and Voutier [1] who proved that $n_0 = 30$ is best possible. Indeed, they show that such examples occur only for $n = 5, 7, 8, 10, 12, 13, 18, 30$: If $a = 1$, $b = -2$, then $x_5, x_8, x_{12}, x_{13}, x_{18}, x_{30}$ have no primitive prime factors; if $a = 1$, $b = -5$, then $x_7 = 1$; if $a = 2$, $b = -3$, then x_{10} has no primitive prime factors; there are a handful of other examples besides, all with $n \leq 12$.

It is conjectured that each x_n with n sufficiently large (say $n \geq n_0$) has a primitive prime factor which divides x_n to the power 1. This would have several interesting applications, for example, to showing that no Fibonacci number after 12^2 is a square and that, for any integer d, the equation $F_n = dx^2$ has at most one solution with $n \geq n_0$.

Exercise 7.35.1. Assume that if $n \geq n_0$, then x_n has a primitive prime factor which divides x_n to the power 1 (that is, p divides x_n but not p^2).
(a) Show that the equation $x_n = y^2$ has no solutions with $n \geq n_0$.
(b) Show that, for any integer $d \geq 1$, the equation $x_n = dy^2$ has at most one solution with $n \geq n_0$.
(c) Show that there are no finite sequences $n_0 \leq n_1 < n_2 < \cdots < n_r$ such that $x_{n_1} x_{n_2} \ldots x_{n_r}$ is a square.
(d) Make the same deductions assuming that if $n > n_0$, then x_n has a primitive prime factor which divides x_n to exactly an odd power.

Recently it was shown in [2] that the assumption in (d) is valid for $x_n = 2^n - 1$ with $n_0 = 6$, and even for any second-order linear recurrence sequence with $(a, b) = 1$, $b \equiv 2 \pmod 4$, $\Delta > 0$, and $n_0 = 6$. It is an open question whether this result can be extended to other recurrence sequences, the most interesting being the Fibonacci numbers.

It seems laborious that we have to make the assumption that x_n has a primitive prime factor which divides it to exactly an odd power, rather than to the power 1, but it is feasible (though unlikely) that this latter assumption is untrue:

Lemma 7.35.1. *Assume that p^2 divides $2^{p-1} - 1$ for all primes $p > n_0$. If q is a primitive prime factor of $x_n = 2^n - 1$ with $n \geq n_0$, then q^2 divides x_n. (The result in [2] then guarantees that some such q divides x_n to an odd power > 1.)*

In section 7.32 of appendix 7H, we discussed that we believe that it is extremely rare that p^2 divides $2^{p-1} - 1$; however we cannot prove that it does not happen for all sufficiently large p, as in the hypothesis of the lemma.

Proof. If q is a primitive prime factor of $x_n = 2^n - 1$, then $\text{ord}_q(2) = n$ which divides $q - 1$, and so $q \geq n + 1 > n_0$. By the hypothesis q^2 divides $2^{q-1} - 1$. Therefore n divides $q - 1$, and so $2^n - 1$ is divisible by q to the same power as $2^{q-1} - 1$, by Lemma 7.17.1(a). Therefore q^2 divides x_n. □

References for this chapter

[1] Y. Bilu, G. Hanrot, and P. Voutier, *Existence of primitive divisors of Lucas and Lehmer numbers*, J. Reine Angew. Math. **539** (2001), 75–122.

[2] Andrew Granville, *Primitive prime factors in second-order linear recurrence sequences*, Acta Arithmetica **155** (2012), 431–452.

[3] A. Schinzel, *Primitive divisors of the expression $A^n - B^n$ in algebraic number fields*, J. Reine Angew. Math. **268/269** (1974), 27–33.

Chapter 8

Quadratic residues

In this chapter we will develop an understanding of the squares mod n, in particular how many there are and how to quickly identify whether a given residue is a square mod n. We mostly discuss the squares modulo primes and from there understand the squares mod prime powers via "lifting", and modulo composites through the Chinese Remainder Theorem.

8.1. Squares modulo prime p

There are two types of squares mod p. We always have $0^2 \equiv 0 \pmod{p}$. Then there are the "*quadratic residues* (mod p)", which are the non-zero residues a (mod p) which are congruent to a square modulo p. All other residue classes are "*quadratic non-residues*". If there is no ambiguity, we simply say "residues" and "non-residues". In the next table we list the quadratic residues modulo each of the primes between 5 and 17.

Modulus	Quadratic residues
5	1, 4
7	1, 2, 4
11	1, 3, 4, 5, 9
13	1, 3, 4, 9, 10, 12
17	1, 2, 4, 8, 9, 13, 15, 16

Exercise 8.1.1. (a) Prove that 337 is not a square (that is, the square of an integer) by reducing it mod 5.
  (b) Prove that 391 is not a square by reducing it mod 7.
  (c) Prove that there do not exist integers x and y for which $x^2 - 3y^2 = -1$, by reducing any solution mod 3.

In each row of our table there seem to be $\frac{p-1}{2}$ quadratic residues mod p:

Lemma 8.1.1. *The distinct quadratic residues* mod p *are given by* $1^2, 2^2, \ldots, \left(\frac{p-1}{2}\right)^2$ (mod p).

Proof. If $r^2 \equiv s^2 \pmod{p}$ with $1 \leq s < r \leq p-1$, then $p \mid r^2 - s^2 = (r-s)(r+s)$ and so p divides either $r - s$ or $r + s$. Now $0 < r - s < p$ and so p does not divide

295

$r - s$. Therefore p divides $r + s$, and $0 < r + s < 2p$, so we must have $r + s = p$. Hence the residues of $1^2, 2^2, \ldots, \left(\frac{p-1}{2}\right)^2$ (mod p) are distinct, and if $s = p - r$, then $s^2 \equiv (-r)^2 \equiv r^2$ (mod p). This implies our result. □

Define the *Legendre symbol* as follows: For each odd prime p let

$$\left(\frac{a}{p}\right) = \begin{cases} 0 & \text{if } a \equiv 0 \pmod{p}, \\ 1 & \text{if } a \text{ is a quadratic residue mod } p, \\ -1 & \text{if } a \text{ is a quadratic non-residue mod } p. \end{cases}$$

Exercise 8.1.2. (a) Prove that if $a \equiv b \pmod{p}$, then $\left(\frac{a}{p}\right) = \left(\frac{b}{p}\right)$.
(b) Prove that $\sum_{a=0}^{p-1} \left(\frac{a}{p}\right) = 0$.

Corollary 8.1.1. *There are exactly $1 + \left(\frac{a}{p}\right)$ residues classes b (mod p) for which $b^2 \equiv a \pmod{p}$.*

Proof. If a is a quadratic non-residue, there are no solutions. For $a = 0$ if $b^2 \equiv 0 \pmod{p}$, then $b \equiv 0 \pmod{p}$ so there is just one solution. If a is a quadratic residue, then, by definition, there exists b such that $b^2 \equiv a \pmod{p}$, and then there are the two solutions $(p - b)^2 \equiv b^2 \equiv a \pmod{p}$ and no others, by the proof in Lemma 8.1.1 (or by Proposition 7.4.1). We have therefore proved

$$\#\{b \pmod{p} : b^2 \equiv a \pmod{p}\} = \begin{cases} 1 & \text{if } a \equiv 0 \pmod{p}, \\ 2 & \text{if } a \text{ is a quadratic residue mod } p, \\ 0 & \text{if } a \text{ is a quadratic non-residue mod } p. \end{cases}$$

This equals $1 + \left(\frac{a}{p}\right)$, looking above at the definition of the Legendre symbol. □

Theorem 8.1. *We have $\left(\frac{ab}{p}\right) = \left(\frac{a}{p}\right)\left(\frac{b}{p}\right)$ for any integers a, b. That is:*

(i) *The product of two quadratic residues* (mod p) *is a quadratic residue.*

(ii) *The product of a quadratic residue and a non-residue is itself a non-residue.*

(iii) *The product of two quadratic non-residues* (mod p) *is a quadratic residue.*

Proof (Gauss). (i) If $a \equiv A^2$ and $b \equiv B^2$, then $ab \equiv (AB)^2 \pmod{p}$.

Let $R := \{r \pmod{p} : (r/p) = 1\}$ be the set of quadratic residues mod p. We saw that if $(a/p) = 1$, then $(ar/p) = 1$ for all $r \in R$. In other words, $ar \in R$; that is, $aR \subset R$. The elements of aR are distinct, so that $|aR| = |R|$, and therefore $aR = R$.

(ii) Let $N = \{n \pmod{p} : (n/p) = -1\}$ be the set of quadratic non-residues mod p, so that $N \cup R$ partitions the reduced residues mod p. By exercise 3.5.2, we deduce that $aR \cup aN$ also partitions the reduced residues mod p, and therefore $aN = N$ since $aR = R$. That is, the elements of the set $\{an : (n/p) = -1\}$ are all quadratic non-residues mod p.

By Lemma 8.1.1, we know that $|R| = \frac{p-1}{2}$, and hence $|N| = \frac{p-1}{2}$ since $N \cup R$ partitions the $p - 1$ reduced residues mod p.

(iii) In (ii) we saw that if $(n/p) = -1$ and $(a/p) = 1$, then $(na/p) = -1$. Hence $nR \subset N$ and, as $|nR| = |R| = \frac{p-1}{2} = |N|$, we deduce that $nR = N$. But $nR \cup nN$ partitions the reduced residues mod p, and so $nN = R$. That is, the elements of the set $\{nb : (b/p) = -1\}$ are all quadratic residues mod p. □

Exercise 8.1.3. Suppose that prime p does not divide ab.
(a) Prove that $\left(\frac{a/b}{p}\right) = \left(\frac{ab}{p}\right)$.
(b) Prove that there are non-zero residues x and y (mod p) for which $ax^2 + by^2 \equiv 0$ (mod p) if and only if $\left(\frac{-ab}{p}\right) = 1$.

Exercise 8.1.4. Prove that if odd prime p divides $b^2 - 4ac$ but neither a nor c, then $\left(\frac{a}{p}\right) = \left(\frac{c}{p}\right)$.

Exercise 8.1.5. Let p be a prime > 3. Prove that if there is no residue x (mod p) for which $x^2 \equiv 2$ (mod p), and no residue y (mod p) for which $y^2 \equiv 3$ (mod p), then *there is* a residue z (mod p) for which $z^2 \equiv 6$ (mod p).

We deduce from Theorem 8.1 that $\left(\frac{\cdot}{p}\right)$ is a multiplicative function. Therefore if we have a factorization of a into prime factors as $a = \pm q_1^{e_1} q_2^{e_2} \ldots q_k^{e_k}$, and $(a,p) = 1$, then[1]

$$\left(\frac{a}{p}\right) = \left(\frac{\pm 1}{p}\right) \prod_{i=1}^{k} \left(\frac{q_i}{p}\right)^{e_i} = \left(\frac{\pm 1}{p}\right) \prod_{\substack{i=1 \\ e_i \text{ odd}}}^{k} \left(\frac{q_i}{p}\right),$$

since $(q/p)^2 = 1$ whenever $p \nmid q$ as this implies that $\left(\frac{q_i}{p}\right)^{e_i} = 1$ if e_i is even, and $\left(\frac{q_i}{p}\right)^{e_i} = \left(\frac{q_i}{p}\right)$ if e_i is odd. Therefore, in order to determine $\left(\frac{a}{p}\right)$ for all integers a, it is only necessary to know the values of $\left(\frac{-1}{p}\right)$, and of $\left(\frac{q}{p}\right)$ for all primes q.

Exercise 8.1.6. One can write each non-zero residue mod p as a power of a primitive root g.
(a) Prove that the quadratic residues are precisely those residues that are an even power of g, and the quadratic non-residues are those that are an odd power.
(b) Deduce that $\left(\frac{g}{p}\right) = -1$.

Exercise 8.1.7. (a) Show that if n is odd and p divides $a^n - 1$, then $\left(\frac{a}{p}\right) = 1$.
(b) Show that if n is prime and p divides $a^n - 1$, but $a \not\equiv 1$ (mod p), then $p \equiv 1$ (mod n).
(c) Give an example to show that (b) can be false if we only assume that n is odd.

Exercise 8.1.8. (a) Prove that, for every prime $p \neq 2, 5$, at least one of 2, 5, and 10 is a quadratic residue mod p.
(b)† Prove that, for every prime $p > 5$, there are two consecutive positive integers that are both quadratic residues mod p and are both ≤ 10.

8.2. The quadratic character of a residue

Fermat's Little Theorem (Theorem 7.2) states that the $(p-1)$st power of any reduced residue mod p is congruent to 1 (mod p). Are there other patterns to be found among the lower powers?

---
[1] Each of "$\pm$" and "± 1" is to be read as "either '+' or '−'". We deal with these two cases together since the proofs are entirely analogous, taking care throughout to be consistent with the choice of sign.

a	a^2	a^3	a^4
1	1	1	1
2	-1	-2	1
-2	-1	2	1
-1	1	-1	1

The powers of a mod 5

a	a^2	a^3	a^4	a^5	a^6
1	1	1	1	1	1
2	-3	1	2	-3	1
3	2	-1	-3	-2	1
-3	2	1	-3	2	1
-2	-3	-1	2	3	1
-1	1	-1	1	-1	1

The powers of a mod 7

As expected the $(p-1)$st column is all 1's, but there is another pattern that emerges: The entries in the "middle" column, that is, the a^2 column mod 5 and the a^3 column mod 7, are all -1's and 1's. This column represents the least residues of numbers of the form $a^{\frac{p-1}{2}}$ (mod p), and it appears that these are all -1's and 1's. Can we decide which are $+1$ and which are -1? For $p=5$ we see that $1^2 \equiv 4^2 \equiv 1$ (mod 5) and $2^2 \equiv 3^2 \equiv -1$ (mod 5); recall that 1 and 4 are the quadratic residues mod 5. For $p=7$ we see that $1^3 \equiv 2^3 \equiv 4^3 \equiv 1$ (mod 7) and $3^3 \equiv 5^3 \equiv 6^3 \equiv -1$ (mod 7); recall that 1, 2, and 4 are the quadratic residues mod 7. So we have observed a pattern: The ath entry in the middle column is $+1$ if a is a quadratic residues mod p, and it is -1 if a is a quadratic residues mod p; in either case it equals the value of the Legendre symbol, $\left(\frac{a}{p}\right)$. This observation was proved by Euler in 1732.

Theorem 8.2 (Euler's criterion). *We have $a^{\frac{p-1}{2}} \equiv \left(\frac{a}{p}\right)$ (mod p) for all primes p and integers a.*

Proof #1. If $\left(\frac{a}{p}\right) = 1$, then there exists b such that $b^2 \equiv a$ (mod p) so that $a^{\frac{p-1}{2}} \equiv b^{p-1} \equiv 1$ (mod p), by Fermat's Little Theorem.

If $\left(\frac{a}{p}\right) = -1$, then we proceed as in Gauss's proof of Wilson's Theorem though pairing up the residues slightly differently. Let

$$S = \{(r,s): 1 \leq r < s \leq p-1, \ rs \equiv a \pmod{p}\}.$$

Note that if $rs \equiv a$ (mod p), then $r \not\equiv s$ (mod p), or else $a \equiv r^2$ (mod p), contradicting that $\left(\frac{a}{p}\right) = -1$. Therefore each integer m, $1 \leq m \leq p-1$, appears exactly once, in exactly one pair in S. We deduce that

$$(p-1)! = \prod_{(r,s) \in S} rs \equiv a^{|S|} = a^{\frac{p-1}{2}} \pmod{p},$$

and the result follows from Wilson's Theorem. □

For example, for $p = 13, a = 2$ we have

$$-1 \equiv 12! = (1 \cdot 2)(3 \cdot 5)(4 \cdot 7)(6 \cdot 9)(8 \cdot 10)(11 \cdot 12) \equiv 2^6 \pmod{13}.$$

Exercise 8.2.1.[†] Prove Euler's criterion for $(a/p) = 1$, by evaluating $(p-1)!$ (mod p) as in the second part of proof #1, but now taking account of the solutions r (mod p) to $r^2 \equiv a$ (mod p).

8.2. The quadratic character of a residue

Proof #2 of Euler's criterion. We began Proof #1 by showing that if $\left(\frac{a}{p}\right) = 1$, then $a^{\frac{p-1}{2}} \equiv 1 \equiv \left(\frac{a}{p}\right) \pmod{p}$. This implies that a is a root of $x^{\frac{p-1}{2}} - 1 \pmod{p}$. By Lemma 8.1.1 there are exactly $\frac{p-1}{2}$ quadratic residues mod p, and we now know that these are all roots of $x^{\frac{p-1}{2}} - 1 \pmod{p}$ and are therefore *all* of the roots of $x^{\frac{p-1}{2}} - 1 \pmod{p}$. That is,

$$(8.2.1) \qquad x^{\frac{p-1}{2}} - 1 \equiv \prod_{\substack{1 \leq a \leq p \\ (a/p)=1}} (x - a) \pmod{p}.$$

In (7.4.1) we noted that

$$x^{p-1} - 1 \equiv (x-1)(x-2)\cdots(x-(p-1)) \pmod{p};$$

that is, the $p-1$ roots of $x^{p-1} - 1 = (x^{\frac{p-1}{2}} - 1)(x^{\frac{p-1}{2}} + 1) \pmod{p}$ are precisely the reduced residues mod p, each occurring exactly once. Since the set of reduced residues mod p is the union of the set of quadratic residues and the set of quadratic non-residues, we can divide this last equation through by (8.2.1), to obtain

$$(8.2.2) \qquad x^{\frac{p-1}{2}} + 1 \equiv \prod_{\substack{1 \leq b \leq p \\ (b/p)=-1}} (x - b) \pmod{p}.$$

This implies that if b is a quadratic non-residue mod p, then $b^{\frac{p-1}{2}} + 1 \equiv 0 \pmod{p}$; that is, $b^{\frac{p-1}{2}} \equiv -1 = \left(\frac{b}{p}\right) \pmod{p}$. □

We can use Euler's criterion to determine the value of Legendre symbols as follows: $\left(\frac{3}{13}\right) = 1$ since $3^6 = 27^2 \equiv 1^2 \equiv 1 \pmod{13}$, and $\left(\frac{2}{13}\right) = -1$ since $2^6 = 64 \equiv -1 \pmod{13}$.

Exercise 8.2.2. Let p be an odd prime. Explain how one can determine the integer $\left(\frac{a}{p}\right)$ by knowing $a^{\frac{p-1}{2}} \pmod{p}$. (Euler's criterion gives a congruence, but here we are asking for the value of the integer $\left(\frac{a}{p}\right)$.)

Exercise 8.2.3. Use Euler's criterion to reprove Theorem 8.1.

Proof #3 of Euler's criterion. Let g be a primitive root mod p. We have $g^{\frac{p-1}{2}} \equiv -1 \pmod{p}$ by exercise 7.5.2. Suppose that $a \equiv g^r \pmod{p}$ for some integer r, so that $a^{\frac{p-1}{2}} \equiv (g^r)^{\frac{p-1}{2}} = (g^{\frac{p-1}{2}})^r \equiv (-1)^r \pmod{p}$. If a is a quadratic residue mod p, then r is even by exercise 8.1.6, and so $a^{\frac{p-1}{2}} \equiv (-1)^r \equiv 1 \pmod{p}$. If a is a quadratic non-residue mod p, then r is odd, and so $a^{\frac{p-1}{2}} \equiv (-1)^r \equiv -1 \pmod{p}$. □

Square roots and non-squares modulo p**.** How can we tell whether a reduced residue $a \pmod{p}$ is a square mod p? One idea is to try to find the square root, but it is not clear how to go about this efficiently (for example, try to find the square root of 77 (mod 101)). One consequence of Euler's criterion is that one does not have to try to find the square root to determine whether a given residue class is a square mod p. Indeed one can determine whether a is a square mod p by calculating

$a^{\frac{p-1}{2}}$ (mod p). This might look like it will be equally difficult, but we have shown in section 7.13 of appendix 7A that one can calculate a high power of a mod p quite efficiently.

There are some special cases in which *one can determine* a square root of a (mod p) quite easily. For example, when $p \equiv 3$ (mod 4):

Exercise 8.2.4. Let p be a prime $\equiv 3$ (mod 4). Show that if $\left(\frac{a}{p}\right) = 1$ and $b \equiv a^{\frac{p+1}{4}}$ (mod p), then $b^2 \equiv a$ (mod p). (This idea is explored further in section 7.21 of appendix 7C.)

However if $p \equiv 1$ (mod 4), then it is not so easy to determine a square root. For example, -1 is a square mod p (as we will prove in the next section) but we do not know a simple practical way to quickly determine a square root of -1 (mod p).

How can one quickly find a quadratic non-residue mod p? One would think it would be easy, as half of the residues mod p are quadratic non-residues, but there is no simple way to guarantee finding one quickly. In practice it is most efficient to select numbers in $[1, p-1]$ at random, independently. The probability that any given selection is a quadratic residue is $\frac{1}{2}$; so the probability that every one of the first k choices is a quadratic residue is $1/2^k$. Therefore, the probability that none of the first 20 selections is a quadratic non-residue mod p is less than one in a million. Moreover it is easy to verify whether each selection is a quadratic residue mod p, using Euler's criterion. This algorithm will almost always rapidly determine a quadratic non-residue mod p, but one might just be terribly unlucky and the algorithm might fail.

It is useful to determine for which primes p a given small integer a is a quadratic residue (mod p). We study this for $a = -1, 2$, and -2 in the next few sections.

8.3. The residue -1

Theorem 8.3. *If p is an odd prime, then -1 is a quadratic residue* (mod p) *if and only if $p \equiv 1$* (mod 4).

We will give five proofs of this result (even though we don't need more than one!) to highlight how the various ideas in the book dovetail in this key result. It is worth recalling that in exercise 7.4.3(c) we showed that if $p \equiv 1$ (mod 4), then $\left(\frac{p-1}{2}\right)!$ is a square root of -1 (mod p). We developed more efficient ways of finding a square root of -1 (mod p) in section 7.21 of appendix 7C.

Proof #1. Euler's criterion implies that $\left(\frac{-1}{p}\right) \equiv (-1)^{\frac{p-1}{2}}$ (mod p). Since each side of the congruence is -1 or 1, and p, which is > 2, divides their difference, they must be equal and so $\left(\frac{-1}{p}\right) = (-1)^{\frac{p-1}{2}}$, and the result follows. □

Proof #2. In exercise 7.5.2 we saw that $-1 \equiv g^{(p-1)/2}$ (mod p) for any primitive root g modulo p. Now if $-1 \equiv (g^k)^2$ (mod p) for some integer k, then $\frac{p-1}{2} \equiv 2k$ (mod $p-1$), and there exists such an integer k if and only if $\frac{p-1}{2}$ is even. □

Proof #3. The number of quadratic non-residues (mod p) is $\frac{p-1}{2}$, and so, by Wilson's Theorem, we have

$$\left(\frac{-1}{p}\right) = \left(\frac{(p-1)!}{p}\right) = \prod_{a \,(\text{mod } p)} \left(\frac{a}{p}\right) = (-1)^{\frac{p-1}{2}}. \qquad \square$$

Proof #4. If a is a quadratic residue, then so is $1/a$ (mod p). Therefore we may "pair up" the quadratic residues (mod p), except those for which $a \equiv 1/a$ (mod p). The only solutions to $a \equiv 1/a$ (mod p) (that is, $a^2 \equiv 1$ (mod p)) are $a \equiv 1$ and -1 (mod p). Therefore the product of the quadratic residues mod p is congruent to $-(-1/p)$. On the other hand the roots of $x^{\frac{p-1}{2}} - 1$ (mod p) are precisely the quadratic residues mod p, and so, taking $x = 0$ in (8.2.1), the product of the quadratic residues mod p is congruent to $(-1)(-1)^{\frac{p-1}{2}}$ (mod p). Comparing these yields that $(-1/p) \equiv (-1)^{\frac{p-1}{2}}$ (mod p), and the result follows. $\square$

Proof #5. (Euler) The first part of Proof #4 implies that

$$\frac{p-1}{2} = \#\{a \,(\text{mod } p) : a \text{ is a quadratic residue (mod } p)\}$$

has the same parity as

$$\#\{a \in \{1, -1\} : a \text{ is a quadratic residue (mod } p)\} = \frac{1}{2}\left(3 + \left(\frac{-1}{p}\right)\right).$$

Multiplying through by 2 yields $p \equiv \left(\frac{-1}{p}\right)$ (mod 4), and the result follows. $\square$

Theorem 8.3 implies that if $p \equiv 1$ (mod 4), then $\left(\frac{-r}{p}\right) = \left(\frac{r}{p}\right)$; and if $p \equiv -1$ (mod 4), then $\left(\frac{-r}{p}\right) = -\left(\frac{r}{p}\right)$.

Exercise 8.3.1. Let p be a prime $\equiv 3$ (mod 4), which does not divide integer a. Prove that either there exists x (mod p) for which $x^2 \equiv a$ (mod p) or there exists y (mod p) for which $y^2 \equiv -a$ (mod p), but not both.

Exercise 8.3.2. (a) Prove that every prime factor p of $4n^2 + 1$ satisfies $p \equiv 1$ (mod 4).
(b) Deduce that there are infinitely many primes $\equiv 1$ (mod 4).

8.4. The residue 2

Calculations reveal that the odd primes $p < 100$ for which $\left(\frac{2}{p}\right) = 1$ are

$$p = 7, 17, 23, 31, 41, 47, 71, 73, 79, 89, \text{ and } 97.$$

These are exactly the primes < 100 that are $\equiv \pm 1$ (mod 8). This observation is established as fact as follows:

Theorem 8.4. *If p is an odd prime, then*

$$\left(\frac{2}{p}\right) = \begin{cases} 1 & \text{if } p \equiv 1 \text{ or } -1 \pmod{8}, \\ -1 & \text{if } p \equiv 3 \text{ or } -3 \pmod{8}. \end{cases}$$

Proof. We will evaluate the product
$$S := \prod_{\substack{1 \leq m \leq p-1 \\ m \text{ even}}} m \pmod{p}$$
in two different ways. First note that each m in the product can be written as $2k$ with $1 \leq k \leq \frac{p-1}{2}$, and so
$$S = \prod_{k=1}^{\frac{p-1}{2}} (2k) = 2^{\frac{p-1}{2}} \left(\frac{p-1}{2}\right)!.$$
One can also rewrite each m in the product as $p - n$ where n is odd; and if m is in the range $\frac{p+1}{2} \leq m \leq p - 1$, then $1 \leq n \leq \frac{p-1}{2}$. Therefore
$$S = \prod_{\substack{1 \leq m \leq \frac{p-1}{2} \\ m \text{ even}}} m \cdot \prod_{\substack{1 \leq n \leq \frac{p-1}{2} \\ n \text{ odd}}} (p - n).$$
Let's suppose there are r such values of n, and note that each $p - n \equiv -n \pmod{p}$. Therefore
$$S \equiv \prod_{\substack{1 \leq m \leq \frac{p-1}{2} \\ m \text{ even}}} m \cdot \prod_{\substack{1 \leq n \leq \frac{p-1}{2} \\ n \text{ odd}}} (-n) = (-1)^r \left(\frac{p-1}{2}\right)! \pmod{p}.$$
Comparing the two ways that we have evaluated S, and dividing through by $\left(\frac{p-1}{2}\right)!$, we find that
$$2^{\frac{p-1}{2}} \equiv (-1)^r \pmod{p}.$$
The result follows from Euler's criterion and verifying that r is even if $p \equiv \pm 1 \pmod{8}$, while r is odd if $p \equiv \pm 3 \pmod{8}$ (see exercise 8.4.1). □

Exercise 8.4.1. For any odd integer q, let r denote the number of positive odd integers $\leq \frac{q-1}{2}$. Prove that r is even if $q \equiv \pm 1 \pmod{8}$, while r is odd if $q \equiv \pm 3 \pmod{8}$.

Gauss's Lemma (Theorem 8.6 in appendix 8A) cleverly generalizes this proof of Theorem 8.4 to classify the values of $\left(\frac{a}{p}\right)$ for any fixed integer a.

Calculations reveal that the odd primes $p < 100$ for which $\left(\frac{-2}{p}\right) = 1$ are
$$p = 3, 11, 17, 19, 41, 43, 59, 67, 73, 83, 89, \text{ and } 97.$$
These are exactly the primes < 100 that are $\equiv 1$ or $3 \pmod{8}$. This observation is established as fact by combining Theorems 8.3 and 8.4, which allow us to evaluate $\left(\frac{-2}{p}\right)$ by taking $\left(\frac{-2}{p}\right) = \left(\frac{-1}{p}\right)\left(\frac{2}{p}\right)$ for every odd prime p.

Exercise 8.4.2. Prove that if p is an odd prime, then
$$\left(\frac{-2}{p}\right) = \begin{cases} 1 & \text{if } p \equiv 1 \text{ or } 3 \pmod{8}, \\ -1 & \text{if } p \equiv 5 \text{ or } 7 \pmod{8}. \end{cases}$$

Exercise 8.4.3. Prove that if 2 is a primitive root mod p, then $p \equiv 3$ or $5 \pmod{8}$.

Exercise 8.4.4.† (a) Prove that if prime $p | M_n := 2^n - 1$ where $n > 2$ is prime, then $p \equiv 1 \pmod{n}$ and $p \equiv \pm 1 \pmod{8}$.
(b) Prove that if $p = 2n + 1$ is prime, then $p | 2^n - 1$ if and only if $p \equiv \pm 1 \pmod{8}$.
(c) Prove that if $p = 2n + 1$ is prime, then $p | 2^n + 1$ if and only if $p \equiv \pm 3 \pmod{8}$.

8.5. The law of quadratic reciprocity

(d) Prove that if q and $p = 2q + 1$ are both prime, then p divides $2^q - 1$ if and only if $q \equiv 3 \pmod 4$.

(e) Factor $2^{11} - 1 = 2047$.

Exercise 8.4.5.[†] In exercise 7.3.2 we proved that if prime p divides $2^{2^k} + 1$, then $p \equiv 1 \pmod{2^{k+1}}$. Now show that $p \equiv 1 \pmod{2^{k+2}}$ if $k \geq 2$.[2]

8.5. The law of quadratic reciprocity

We have already seen that if p is an odd prime, then

$$\left(\frac{-1}{p}\right) = \begin{cases} 1 & \text{if } p \equiv 1 \pmod 4, \\ -1 & \text{if } p \equiv -1 \pmod 4 \end{cases}$$

and

$$\left(\frac{2}{p}\right) = \begin{cases} 1 & \text{if } p \equiv 1 \text{ or } -1 \pmod 8, \\ -1 & \text{if } p \equiv 3 \text{ or } -3 \pmod 8. \end{cases}$$

To be able to evaluate arbitrary Legendre symbols we will also need the *law of quadratic reciprocity*.

Theorem 8.5 (The law of quadratic reciprocity). *If p and q are given distinct odd primes, then*

$$\left(\frac{p}{q}\right)\left(\frac{q}{p}\right) = \begin{cases} 1 & \text{if } p \equiv 1 \pmod 4 \text{ or } q \equiv 1 \pmod 4, \\ -1 & \text{if } p \equiv q \equiv -1 \pmod 4. \end{cases}$$

These rules, taken together, allow us to rapidly evaluate any Legendre symbol. For example, to evaluate (m/p), we first reduce m mod p, so that $(m/p) = (n/p)$ where $n \equiv m \pmod p$ and $|n| < p$. Next we factor n and, by the multiplicativity of the Legendre symbol, we can evaluate (n/p) in terms of $(-1/p), (2/p)$ and the (q/p) for those primes q dividing n. We can easily determine the values of $(-1/p)$ and $(2/p)$ from determining $p \pmod 8$, and then we need to evaluate each (q/p) where $q \leq |n| < p$. We do this by the law of quadratic reciprocity since $(q/p) = \pm(p/q)$ depending only on the values of p and q mod 4.[3] We repeat the procedure on each (p/q). Clearly this process will quickly finish as the numbers involved are always getting smaller. Let us work through some examples.

$$\left(\frac{111}{71}\right) = \left(\frac{40}{71}\right) = \left(\frac{2}{71}\right)^3 \left(\frac{5}{71}\right) \quad \text{as } 111 \equiv 40 \pmod{71} \text{ and } 40 = 2^3 \cdot 5,$$

$$= 1^3 \cdot 1 \cdot \left(\frac{71}{5}\right) \quad \text{as } 71 \equiv -1 \pmod 8 \text{ and } 5 \equiv 1 \pmod 4,$$

$$= \left(\frac{1}{5}\right) = 1 \quad \text{as } 71 \equiv 1 \pmod 5.$$

---

[2]We can use this to "demystify" Euler's factorization of F_5: Exercise 8.4.5 implies that any prime factor p of F_5 must be of the form $128m + 1$. This is divisible by 3, 5, and 3 for $m = 1, 3,$ and 4, respectively, so is not prime. If $m = 2$, then $p = F_4$ which we proved is coprime with F_5 in section 5.1. Finally, if $m = 5$, then $p = 541$ is a prime factor of F_5.

[3]Note that if $\left(\frac{p}{q}\right)\left(\frac{q}{p}\right) = \eta \, (= \pm 1)$ by the law of quadratic reciprocity, then $\left(\frac{q}{p}\right) = \eta \left(\frac{p}{q}\right)$.

There is more than one way to proceed with these rules:

$$\left(\frac{111}{71}\right) = \left(\frac{-1}{71}\right)\left(\frac{31}{71}\right) \qquad \text{as } 111 \equiv -31 \pmod{71},$$

$$= (-1) \cdot (-1) \cdot \left(\frac{71}{31}\right) \qquad \text{as } 71 \equiv 31 \equiv -1 \pmod{4},$$

$$= \left(\frac{9}{31}\right) = \left(\frac{3}{31}\right)^2 = 1 \qquad \text{as } 71 \equiv 9 \equiv 3^2 \pmod{31}.$$

A slightly larger example is

$$\left(\frac{869}{311}\right) = \left(\frac{247}{311}\right) = \left(\frac{13}{311}\right)\left(\frac{19}{311}\right) = 1 \cdot \left(\frac{311}{13}\right) \cdot (-1) \cdot \left(\frac{311}{19}\right)$$

$$= -\left(\frac{-1}{13}\right)\left(\frac{7}{19}\right) = -1 \cdot 1 \cdot (-1)\left(\frac{19}{7}\right) = \left(\frac{-2}{7}\right) = -1.$$

Although longer, each step is straightforward except when we factored $247 = 13 \times 19$ (a factorization which is not obvious for most of us, and imagine how difficult factoring might be when we are dealing with much larger numbers). Indeed, this is an efficient procedure provided that one is capable of factoring the numbers n that arise. Although this may be the case for small examples, it is not practical for large examples. We can bypass this potential difficulty by using the Jacobi symbol, a generalization of the Legendre symbol, which we will discuss in section 8.7.

In the next subsection we will prove the law of quadratic reciprocity, justifying the algorithm used above to determine the value of any given Legendre symbol.

The law of quadratic reciprocity is easily used to determine various other rules. For example, when is 3 a square mod p? This is the same as asking when $(3/p) = 1$. Now by quadratic reciprocity we have two cases:

- If $p \equiv 1 \pmod{4}$, then $(3/p) = (p/3)$, and $(p/3) = 1$ when $p \equiv 1 \pmod{3}$, so we have $(3/p) = 1$ when $p \equiv 1 \pmod{12}$ (using the Chinese Remainder Theorem).
- If $p \equiv -1 \pmod{4}$, then $(3/p) = -(p/3)$, and $(p/3) = -1$ when $p \equiv -1 \pmod{3}$, so we have $(3/p) = 1$ when $p \equiv -1 \pmod{12}$ (using the Chinese Remainder Theorem).

We have therefore proved that $(3/p) = 1$ if and only if $p \equiv 1$ or $-1 \pmod{12}$.

Exercise 8.5.1. Determine (a) $\left(\frac{13}{31}\right)$; (b) $\left(\frac{323}{31}\right)$; (c) $\left(\frac{377}{233}\right)$; (d) $\left(\frac{13}{71}\right)$; (e) $\left(\frac{-104}{131}\right)$.

Exercise 8.5.2. (a) Show that if prime $p \equiv 1 \pmod{5}$, then 5 is a quadratic residue mod p.
(b) Show that if prime $p \equiv 3 \pmod{5}$, then 5 is a quadratic non-residue mod p.
(c) Determine all odd primes p for which $(5/p) = -1$.

Exercise 8.5.3. Prove that if $p := 2^n - 1$ is prime with $n > 2$, then $(3/p) = -1$.

Exercise 8.5.4.[†] Suppose that $F_m = 2^{2^m} + 1$ with $m \geq 2$ is prime. Prove that $3^{\frac{F_m-1}{2}} \equiv 5^{\frac{F_m-1}{2}} \equiv -1 \pmod{F_m}$.

Exercise 8.5.5.[†] (a) Determine all odd primes p for which $(7/p) = 1$.
(b) Find all primes p such that there exists $x \pmod{p}$ for which $2x^2 - 2x - 3 \equiv 0 \pmod{p}$.

8.6. Proof of the law of quadratic reciprocity

Exercise 8.5.6. Show that if p and $q = p + 2$ are "twin primes", then p is a quadratic residue mod q if and only if q is a quadratic residue mod p.

Exercise 8.5.7. Prove that $(-3/p) = (p/3)$ for all primes p.

8.6. Proof of the law of quadratic reciprocity

Suppose that $p < q$ are odd primes, and let $n = pq$. Given residue classes a (mod p) and b (mod q) there exists a unique residue class r (mod n) for which $r \equiv a$ (mod p) and $\equiv b$ (mod q), by the Chinese Remainder Theorem. Let $r(a, b)$ be the least residue of r mod n in absolute value and let $m(a, b) = |r(a, b)|$, so that $1 \leq m(a, b) \leq n/2$, and $m(a, b) = r(a, b)$ or $-r(a, b)$. We claim that

$$\left\{ m(a,b) : 1 \leq a \leq p-1 \text{ and } 1 \leq b \leq \frac{q-1}{2} \right\} = \left\{ m : 1 \leq m \leq \frac{n}{2} \text{ with } (m,n)=1 \right\},$$

since the two sets both have $\phi(n)/2$ elements, each such $m(a,b) \in [1, \frac{n}{2}]$ with $(m, n) = 1$, and the $m(a, b)$ are distinct. This last assertion holds or else if $m(a, b) = m(a', b')$, then $r(a, b) \equiv \pm r(a', b')$ (mod n), so that $b \equiv \pm b'$ (mod q). As $1 \leq b, b' \leq \frac{q-1}{2}$ this implies that $b = b'$ so that the sign is "+", and therefore $a \equiv a'$ (mod p) implying that $a = a'$.

Since each $m(a, b) = \pm r(a, b)$, we deduce that there exists $\sigma = -1$ or 1 such that

$$(8.6.1) \qquad \sigma \prod_{\substack{1 \leq a < p-1 \\ 1 \leq b \leq \frac{q-1}{2}}} r(a,b) = \prod_{\substack{1 \leq a < p-1 \\ 1 \leq b \leq \frac{q-1}{2}}} m(a,b) = \prod_{\substack{1 \leq m < n/2 \\ (m,n)=1}} m.$$

We will calculate the two sides in this identity, mod p and mod q, and compare.

As $r(a, b) \equiv a$ (mod p) the product on the left-hand side of (8.6.1) is

$$\prod_{\substack{1 \leq a < p-1 \\ 1 \leq b \leq \frac{q-1}{2}}} r(a,b) \equiv \prod_{1 \leq b \leq \frac{q-1}{2}} \prod_{1 \leq a < p-1} a = (p-1)!^{\frac{q-1}{2}} \equiv (-1)^{\frac{q-1}{2}} \pmod{p},$$

using Wilson's Theorem. We rewrite the right-hand side of (8.6.1), multiplying top and bottom by the integers $m \in [1, \frac{n}{2})$ that are divisible by q, to obtain

$$\prod_{\substack{1 \leq m < n/2 \\ (m,p)=1}} m \bigg/ \prod_{\substack{1 \leq m < n/2 \\ q | m}} m.$$

We partition the m's in the numerator into intervals of length p, because

$$\prod_{\substack{ip \leq m < (i+1)p \\ (m,p)=1}} m = \prod_{j=1}^{p-1} (ip+j) \equiv \prod_{j=1}^{p-1} j \equiv (p-1)! \equiv -1 \pmod{p},$$

by Wilson's Theorem. Applying this for $0 \leq i \leq \frac{q-3}{2}$ we get a contribution of $(-1)^{\frac{q-1}{2}}$ to the numerator. The remaining integers in the numerator contribute

$$\prod_{\substack{\frac{q-1}{2} \cdot p \leq m < \frac{n}{2} \\ (m,p)=1}} m = \prod_{j=1}^{(p-1)/2} \left(\frac{q-1}{2} p + j \right) \equiv \prod_{j=1}^{(p-1)/2} j \equiv \left(\frac{p-1}{2} \right)! \pmod{p}.$$

On the other hand the m's in the denominator can be written as qk with $1 \leq k \leq \frac{p-1}{2}$, and so

$$\prod_{\substack{1 \leq m < n/2 \\ q \mid m}} m = \prod_{1 \leq k \leq \frac{p-1}{2}} qk = q^{(p-1)/2} \left(\frac{p-1}{2}\right)! \equiv \left(\frac{q}{p}\right)\left(\frac{p-1}{2}\right)! \pmod{p},$$

by Euler's criterion. Cancelling the $\left(\frac{p-1}{2}\right)!$ from the numerator and denominator, we deduce that the right-hand side of (8.6.1) is $\equiv (-1)^{\frac{q-1}{2}} \left(\frac{q}{p}\right) \pmod{p}$. Comparing our calculation of the left- and right-hand sides of (8.6.1) mod p, we obtain

(8.6.2) $$\sigma(-1)^{\frac{q-1}{2}} \equiv (-1)^{\frac{q-1}{2}} \left(\frac{q}{p}\right) \pmod{p}.$$

Since both sides are 1 or -1 and are congruent mod p, they must be equal and so we deduce that

$$\sigma = \left(\frac{q}{p}\right).$$

Next we reduce (8.6.1) mod q. For the right-hand side we proceed entirely analogously to how we did mod p, with the roles of p and q reversed and so obtain $(-1)^{\frac{p-1}{2}} \left(\frac{p}{q}\right) \pmod{q}$.

For the left-hand side of (8.6.1) mod q, we note that each $r(a,b) \equiv b \pmod{q}$, so that

$$\prod_{\substack{1 \leq a < p-1 \\ 1 \leq b \leq \frac{q-1}{2}}} r(a,b) \equiv \prod_{1 \leq a < p-1} \prod_{1 \leq b \leq \frac{q-1}{2}} b = \left(\left(\frac{q-1}{2}\right)!\right)^{p-1} \pmod{q}.$$

In exercise 7.4.3 we saw $\binom{q-1}{(q-1)/2} \equiv (-1)^{\frac{q-1}{2}} \pmod{q}$,[4] and therefore

$$\left(\left(\frac{q-1}{2}\right)!\right)^2 \equiv (-1)^{\frac{q-1}{2}}(q-1)! \equiv -(-1)^{\frac{q-1}{2}} \pmod{q},$$

by Wilson's Theorem. Therefore

$$\prod_{\substack{1 \leq a < p-1 \\ 1 \leq b \leq \frac{q-1}{2}}} r(a,b) \equiv \left(\left(\left(\frac{q-1}{2}\right)!\right)^2\right)^{\frac{p-1}{2}} \equiv (-1)^{\frac{p-1}{2}} \cdot (-1)^{\frac{p-1}{2} \cdot \frac{q-1}{2}} \pmod{q}.$$

Substituting this and the above into (8.6.1) we obtain

(8.6.3) $$\left(\frac{q}{p}\right)(-1)^{\frac{p-1}{2}} \cdot (-1)^{\frac{p-1}{2} \cdot \frac{q-1}{2}} \equiv (-1)^{\frac{p-1}{2}} \left(\frac{p}{q}\right) \pmod{q}.$$

Again both sides are 1 or -1 and are congruent mod q, so must be equal. Multiplying both sides through by $(-1)^{\frac{p-1}{2}} \left(\frac{q}{p}\right)$ implies that

$$\left(\frac{p}{q}\right)\left(\frac{q}{p}\right) = (-1)^{\frac{p-1}{2} \cdot \frac{q-1}{2}}.$$

---

[4] See the solution to exercise 7.4.3 at the end of the book for a proof.

From here we work through the four cases for p and q mod 4 and deduce the law of quadratic reciprocity (Theorem 8.5). □

There are many proofs of the law of quadratic reciprocity, 246 at the last count (see the list at http://www.rzuser.uni-heidelberg.de/~hb3/fchrono.html). In this chapter's appendices we present two of the best: the original proof due to Gauss and an elegant proof due to Eisenstein. We also discuss two other proofs in the exercises and then two sophisticated but shorter proofs in chapter 14.

8.7. The Jacobi symbol

The *Jacobi symbol* is defined as follows: If m is a positive odd integer, we write $m = \prod_p p^{e_p}$, where the p are distinct odd primes, and then

$$\left(\frac{a}{m}\right) = \prod_p \left(\frac{a}{p}\right)^{e_p}.$$

This is defined only for odd m, not for even m.

If a is a square modulo m, then, by the Chinese Remainder Theorem, a is a square modulo every prime p dividing m; that is, $(a/p) = 0$ or 1 for all $p|m$ and so $(a/m) = 0$ or 1. However the converse is not always true; for example, 2 is not a square mod 15 as

$$\left(\frac{2}{3}\right) = \left(\frac{2}{5}\right) = -1, \text{ even though this implies that } \left(\frac{2}{15}\right) = \left(\frac{2}{3}\right)\left(\frac{2}{5}\right) = 1.$$

Exercise 8.7.1. Suppose that m is an odd positive integer.
(a) Prove that $\left(\frac{a}{m}\right) = \left(\frac{b}{m}\right)$ whenever $a \equiv b \pmod{m}$.
(b) Prove that $\left(\frac{ab}{m}\right) = \left(\frac{a}{m}\right)\left(\frac{b}{m}\right)$.
(c) Prove that if $\left(\frac{a}{m}\right) = -1$, then a is not a square mod m.
(d) Prove that $\left(\frac{a}{m}\right) = 0$ if and only if $(a, m) > 1$.

Exercise 8.7.2. (a) Prove that $\sum_{a=0}^{m-1} \left(\frac{a}{m}\right) = 0$ for every non-square odd integer $m \geq 2$.
(b) For how many residues a mod m do we have $(a/m) = 1$?
(c) For how many residues a mod m do we have $(a/m) = -1$?

Exercise 8.7.3. Show that if $n \geq 1$, then $\left(\frac{n}{4n-1}\right) = 1$.

Theorems 8.3, 8.4, and 8.5 can all be extended to the Jacobi symbol (as we will prove at the end of this section): If m and n are odd, coprime integers > 1, then

(8.7.1) $$\left(\frac{-1}{n}\right) = \begin{cases} 1 & \text{if } n \equiv 1 \pmod{4}, \\ -1 & \text{if } n \equiv -1 \pmod{4}, \end{cases}$$

(8.7.2) $$\left(\frac{2}{n}\right) = \begin{cases} 1 & \text{if } n \equiv 1 \text{ or } -1 \pmod{8}, \\ -1 & \text{if } n \equiv 3 \text{ or } -3 \pmod{8}, \end{cases}$$

and the law of quadratic reciprocity

(8.7.3) $$\left(\frac{m}{n}\right)\left(\frac{n}{m}\right) = (-1)^{\frac{m-1}{2} \cdot \frac{n-1}{2}}.$$

We can use these three rules to easily evaluate (m/n) for any odd coprime integers m and n. One begins by selecting $M \equiv m \pmod{n}$ as conveniently as possible, usually with $|M| < n$. Then we factor $M = \pm 2^k \ell$ where ℓ is an odd positive integer $< n$, so that $\left(\frac{m}{n}\right) = \left(\frac{M}{n}\right) = \left(\frac{\pm 1}{n}\right)\left(\frac{2}{n}\right)^k \left(\frac{\ell}{n}\right)$. We can evaluate the first two Jacobi symbols using the first two rules above (which depend only on the value of $n \pmod 8$), and then we know that $\left(\frac{\ell}{n}\right) = \pm \left(\frac{n}{\ell}\right)$ by the third rule. To evaluate $\left(\frac{n}{\ell}\right)$ we repeat this process, but now with a smaller pair of numbers, so that the algorithm will terminate after finitely many steps.

This algorithm only involves dividing out powers of 2 and a possible minus sign, so it goes fast and avoids serious factoring; in fact it is guaranteed to go at least as fast as the Euclidean algorithm since it involves very similar steps.[5] Here is a first straightforward example using the Jacobi symbol, instead of the Legendre symbol:

$$\left(\frac{106}{71}\right) = \left(\frac{35}{71}\right) = -\left(\frac{71}{35}\right) = -\left(\frac{1}{35}\right) = -1.$$

(Note that (71/35) is not the Legendre symbol as 35 is not prime, but it is a Jacobi symbol.) Now let's revisit the example $\left(\frac{869}{311}\right)$ from section 8.5 and avoid factoring 247:

$$\left(\frac{869}{311}\right) = \left(\frac{247}{311}\right) = (-1)\left(\frac{311}{247}\right) = -\left(\frac{64}{247}\right) = -1.$$

We did not need to factor 247, and each step of the algorithm was straightforward.

Exercise 8.7.4. Determine (a) $\left(\frac{13}{27}\right)$; (b) $\left(\frac{323}{225}\right)$; (c) $\left(\frac{233}{377}\right)$; (d) $\left(\frac{-104}{135}\right)$.

Proof of (8.7.1), (8.7.2), **and** (8.7.3). We proceed by induction on the number of prime factors of m and n. The results follows when m and n have one prime factor by Theorems 8.3, 8.4, and 8.5, respectively. Otherwise we write $n = ap$ for some prime p dividing n (swapping the roles of m and n if necessary).

Exercise 8.7.5. Prove that $\frac{a-1}{2} + \frac{b-1}{2} \equiv \frac{ab-1}{2} \pmod 2$ for any odd integers a, b.

Equation (8.7.1) can be rephrased as $\left(\frac{-1}{n}\right) = (-1)^{\frac{n-1}{2}}$. By induction, using the multiplicativity of the denominator of the Jacobi symbol,

$$\left(\frac{-1}{n}\right) = \left(\frac{-1}{ap}\right) = \left(\frac{-1}{a}\right)\left(\frac{-1}{p}\right) = (-1)^{\frac{a-1}{2}+\frac{p-1}{2}} = (-1)^{\frac{ap-1}{2}} = (-1)^{\frac{n-1}{2}}$$

by exercise 8.7.5.

Similarly by induction and multiplicativity of the numerator and denominator,

$$\left(\frac{m}{n}\right)\left(\frac{n}{m}\right) = \left(\frac{m}{ap}\right)\left(\frac{ap}{m}\right) = \left(\frac{m}{a}\right)\left(\frac{m}{p}\right) \cdot \left(\frac{a}{m}\right)\left(\frac{p}{m}\right) = \left(\frac{m}{a}\right)\left(\frac{a}{m}\right) \cdot \left(\frac{m}{p}\right)\left(\frac{p}{m}\right)$$

$$= (-1)^{\frac{m-1}{2} \cdot \frac{a-1}{2} + \frac{m-1}{2} \cdot \frac{p-1}{2}} = (-1)^{\frac{m-1}{2} \cdot \frac{ap-1}{2}} = (-1)^{\frac{m-1}{2} \cdot \frac{n-1}{2}}$$

by exercise 8.7.5.

---

[5] As in the "speeded up" version of the Euclidean algorithm, given in section 1.11 of appendix 1B.

If $\left(\frac{2}{a}\right) = \left(\frac{2}{p}\right)$, then $a \equiv \pm p \pmod 8$, so that $n = ap \equiv \pm 1 \pmod 8$, and therefore $\left(\frac{2}{n}\right) = \left(\frac{2}{a}\right)\left(\frac{2}{p}\right) = (\pm 1)^2 = 1$. If $\left(\frac{2}{a}\right) = -\left(\frac{2}{p}\right)$, then $a \equiv \pm 3p \pmod 8$, so that $n = ap \equiv \pm 3 \pmod 8$, and therefore $\left(\frac{2}{n}\right) = \left(\frac{2}{a}\right)\left(\frac{2}{p}\right) = (1)(-1) = -1$. □

Gauss gave a different proof of (8.7.2), tying the question directly into finding solutions to quadratic equations. This foreshadows Gauss's proof of the full law of quadratic reciprocity, which we will give in appendix 8C.

Gauss's induction step for integers $n \equiv \pm 3 \pmod 8$. We suppose that (8.7.2) is true for all odd integers $m < n$ and that $n \equiv \pm 3 \pmod 8$. If $n = ab$ is composite with $1 < a, b < n$, then $\left(\frac{2}{n}\right) = \left(\frac{2}{a}\right)\left(\frac{2}{b}\right)$ and the result for n follows by applying the induction hypothesis with $m = a$ and with $m = b$.

Therefore we may suppose that $n = p$ is prime and assume that $\left(\frac{2}{p}\right) = 1$. Let a be the smallest odd positive integer for which $a^2 \equiv 2 \pmod p$ so that $1 \leq a \leq p-1$ (for if b is the smallest positive integer for which $b^2 \equiv 2 \pmod p$, then let $a = b$ if b is odd, and $a = p - b$ if b is even), and write $a^2 - 2 = pr$. Evidently $pr \equiv a^2 - 2 \equiv -1 \pmod 8$ and so $r \equiv p^2 r \equiv p(pr) \equiv -p \equiv \pm 3 \pmod 8$. Now $a^2 \equiv 2 \pmod r$ and so $\left(\frac{2}{r}\right) = 1$ with $r = \frac{a^2-2}{p} < p$ and $r \equiv \pm 3 \pmod 8$. This contradicts the induction hypothesis, and so our assumption is wrong. Therefore $\left(\frac{2}{p}\right) = -1$. □

Exercise 8.7.6. Prove an analogous induction step for integers $n \equiv 5$ or $7 \pmod 8$ when establishing the value of $\left(\frac{-2}{n}\right)$.

Exercise 8.7.7 (A useful reformulation of the law of quadratic reciprocity). For a given odd, squarefree integer $n > 1$ let $n^* = \left(\frac{-1}{n}\right) n$. Prove that $n^* \equiv 1 \pmod 4$ and that we have $\left(\frac{m}{n}\right) = \left(\frac{n^*}{m}\right)$ for all odd integers $m > 1$.

8.8. The squares modulo m

To determine the squares mod m, that is, the residues $a \pmod m$ for which there exists $b \pmod m$ with $b^2 \equiv a \pmod m$, we may use the Chinese Remainder Theorem: We know that a is a square mod m if and only if a is a square modulo every prime power factor of m. So it is sufficient to understand the squares modulo every prime power.

Above we have understood the squares modulo every prime p. We now "lift" these squares to determine the squares modulo every prime power, p^k. Let's begin by studying the squares mod p^2:

The squares mod 9 are 0, 1, 4, and 7 mod 9 (these are the least residues of $0^2, 1^2, \ldots, 8^2 \pmod 9$, excluding repetitions). The non-zero residues, 1, 4, and 7 are all $\equiv 1 \pmod 3$; in fact they are all of the residue classes $a \pmod 9$ for which $a \equiv 1 \pmod 3$. We have seen that $1 \pmod 3$ is the only quadratic residue mod 3.

Similarly mod 25 we have the squares

$$0, 1, 4, 9, 16, 11, 24, 14, 6, 21, \text{ and } 19 \pmod{25}.$$

The non-zero squares here are 1, 6, 11, 16, and 21 (mod 25), the residue classes a (mod 25) for which $a \equiv 1$ (mod 5), and 4, 9, 14, 19, and 24 (mod 25), the residue classes a (mod 25) for which $a \equiv 4$ (mod 5). Moreover 1 and 4 (mod 5) are the quadratic residues mod 5.

A pattern begins to emerge. Define a to be a *quadratic residue* (mod m) if $(a, m) = 1$ and there exists b (mod m) for which $b^2 \equiv a$ (mod m).

Proposition 8.8.1. *Let p be a prime. If r is a quadratic residue mod p^k, then r is a quadratic residue mod p^{k+1} whenever $k \geq 1$, except perhaps when $p^k = 2$ or 4.*

Proof. There exists an integer x for which $x^2 \equiv r$ (mod p^k), and $(x, p) = 1$ as $(r, p) = 1$. We let n be that integer for which $x^2 = r + np^k$.

Now if p is odd, then, for any integer j, we have
$$(x - jp^k)^2 = x^2 - 2jxp^k + j^2 p^{2k} \equiv r + (n - 2jx)p^k \pmod{p^{k+1}}.$$
This is $\equiv r \pmod{p^{k+1}}$ if and only if $2jx \equiv n \pmod{p}$, which holds if and only if $j \equiv n/2x \pmod{p}$ (as $(2x, p) = 1$). Therefore r is a square mod p^{k+1}, and our proof yields that there is a unique X (mod p^{k+1}) for which $X \equiv x$ (mod p^k) and $X^2 \equiv r$ (mod p^{k+1}), namely $X \equiv x - jp^k$ (mod p^{k+1}) where $j \equiv n/2x$ (mod p).

If $p = 2$, then $x^2 = r + n \cdot 2^k$ and x is odd so that $x^2 - nx2^k \equiv r$ (mod 2^{k+1}). Therefore
$$(x - n2^{k-1})^2 = x^2 - nx2^k + n^2 2^{2k-2} \equiv r \pmod{2^{k+1}},$$
provided the exponent $2k - 2 \geq k + 1$; that is, $k \geq 3$. □

Exercise 8.8.1. Deduce that an integer r is a quadratic residue mod p^k if and only if r is a quadratic residue mod p, when p is odd, and if and only if $r \equiv 1$ (mod $\gcd(2^k, 8)$) when $p = 2$.

This implies that exactly half of the reduced residue classes mod p^k are quadratic residues, when p is odd, and exactly one quarter when $p = 2$ and $k \geq 3$.

Using the Chinese Remainder Theorem we therefore deduce from exercise 8.8.1 the following:

Corollary 8.8.1. *Suppose that $(a, m) = 1$. Then a is a square mod m if and only if $\left(\frac{a}{p}\right) = 1$ for every odd prime p dividing m, and $a \equiv 1$ (mod $\gcd(m, 8)$).*

Exercise 8.8.2. Suppose that $(a, n) = 1$ and that $b^2 \equiv a$ (mod n). Prove that the set of solutions x (mod n) to $x^2 \equiv a$ (mod n) is given by the values br (mod n) as r runs through the solutions to $r^2 \equiv 1$ (mod n). (Determining the square roots of 1 (mod n) is discussed in section 3.8.)

Additional exercises

Exercise 8.9.1. Let p be an odd prime where $p \nmid a$. Show that the congruence $ax^2 + bx + c \equiv 0$ (mod p) has a solution x (mod p) if and only if $b^2 - 4ac$ is a square mod p.

Exercise 8.9.2.[†] Prove that m^2 and $m^2 + 1$ are both squares mod p, for m equal to at least one of a, $a + 1$, or $a^2 + a + 1$, for any integer a. (This generalizes exercise 8.1.8(a).)

Exercise 8.9.3. The polynomial $x^4 - 4x^2 + 1$ is irreducible over $\mathbb{Q}[x]$ by Theorem 3.4.
(a) Prove that $x^4 - 4x^2 + 1$ can be factored mod p as $(x^2 - \alpha)(x^2 - \beta)$ or $(x^2 - ax + 1)(x^2 + ax + 1)$ or $(x^2 - ax + 1)(x^2 + ax + 1)$ if 3 or 6 or 2 is a square mod p, respectively.

(b) Deduce that $x^4 - 4x^2 + 1 \pmod{p}$ is reducible for every prime p.
(c)† Prove that every quadratic polynomial of the form $x^4 + ax^2 + b^2$ factors into two quadratics mod p, for every prime p.

Exercise 8.9.4. Prove that if $p \equiv 1 \pmod 4$, then $x^4 + 4$ factors into four linear factors mod p.

Exercise 8.9.5. Let $f(.)$ be the totally multiplicative function for which $f(3) = 1$ and $f(p) = \left(\frac{p}{3}\right)$ if $p \ne 3$.
(a) Give a formula for $f(n)$ for an arbitrary integer n.
(b)† For any given large constant B, suppose that p is a prime for which $(q/p) = f(q)$ for every prime $q \le B$. Show that there are no three consecutive squares mod p that are all $\le B$.

This shows that the result in exercise 8.1.8(b) cannot be extended to three consecutive integers provided the hypothesis in (b) holds. This hypothesis will be justified in exercise 8.17.2 of appendix 8D.

Exercise 8.9.6. Show that if $\left(\frac{n}{p}\right) = -1$, then $\sum_{d|n} \left(\frac{d}{p}\right) = 0$.

Exercise 8.9.7. Suppose that a and b are integers and $\{x_n : n \ge 0\}$ is the second-order linear recurrence sequence given by (0.1.2) with $x_0 = 0$ and $x_1 = 1$. Using exercise 0.4.10(b) prove that if odd prime p divides some x_n with n odd, then $(-b/p) = 1$. Deduce that if $(-b/p) = -1$ and p divides x_n, then n is even.

Exercise 8.9.8. (a) Suppose that p^k is an odd prime power. Prove that there are $1 + \left(\frac{a}{p}\right)$ residue classes $b \pmod{p^k}$ for which $b^2 \equiv a \pmod{p^k}$.
(b) Suppose that n is an odd positive integer. Prove that there are $\prod_{p \text{ prime: } p|n} \left(1 + \left(\frac{a}{p}\right)\right)$ residue classes $b \pmod n$ for which $b^2 \equiv a \pmod n$.
(c) Show that this equals $\sum_{d|n} \left(\frac{a}{d}\right)$ where the sum is restricted to squarefree integers d.

Exercise 8.9.9.† Let p be a given odd prime.
(a) Prove that for every $m \pmod p$ there exist a and b mod p such that $a^2 + b^2 \equiv m \pmod p$.
(b) Deduce that there are three squares, not all divisible by p, whose sum is divisible by p.
(c) Generalize this argument to show that if a, b, and c are not divisible by p, then there are at least p solutions $x, y, z \pmod p$ to $ax^2 + by^2 + cz^2 \equiv 0 \pmod p$.

Exercise 8.9.10.† Let m be a squarefree integer $\ne 1$, and let a be an odd positive integer.
(a) Prove that the Jacobi symbol $\left(\frac{4m}{a}\right)$ is a periodic function of a of period dividing $4m$.
(b) Show that the Jacobi symbol $\left(\frac{12}{a}\right)$ has minimal period 12.
(c) Prove that if m is odd and $(a, 2m) = 1$, then $\left(\frac{4m}{a+2m}\right) = \left(\frac{-1}{m}\right)\left(\frac{4m}{a}\right)$.
Now suppose that $m \equiv 3 \pmod 4$.
(d) Prove that there exists an integer r for which $\left(\frac{4m}{r}\right) = -1$.
(e) Prove that $\sum_{a=1}^{4m} \left(\frac{4m}{a}\right) = 0$.

Exercise 8.9.11. (This extends exercise 8.2.4.)
(a) Let $n = pq$ where p and q are distinct primes $\equiv 3 \pmod 4$, and $m = \frac{1}{2}(\frac{p-1}{2} \cdot \frac{q-1}{2} + 1)$. Show that if $\left(\frac{a}{p}\right) = \left(\frac{a}{q}\right) = 1$ and $b \equiv a^m \pmod n$, then $b^2 \equiv a \pmod n$.
(b) Any odd prime p can be written uniquely in the form $p = 1 + 2^k m$ where m is odd and $k \ge 1$. Prove that if a is a 2^kth power mod p and $b \equiv a^{\frac{m+1}{2}} \pmod p$, then $b^2 \equiv a \pmod p$.

If prime $p \equiv 1 \pmod 4$ and $(a/p) = 1$ but a is not a fourth power mod p, then we do not know how to use this idea to find a square root of $a \pmod p$. Known methods in this case are considerably more complicated (see, e.g., [**CP05**]).

Exercise 8.9.12. Suppose that p is a prime $\equiv 3 \pmod 4$ and $\left(\frac{b}{p}\right) = 1$. Prove that there are exactly two solutions $x \pmod p$ to $x^4 \equiv b \pmod p$.

Exercise 8.9.13.[†] Show that if p is a prime which divides $m^2 - 15$ for some integer m, then either $p = 2, 3,$ or 5, or $p \equiv \pm 1, \pm 7, \pm 11,$ or $\pm 17 \pmod{60}$.

Exercise 8.9.14.[†] Show that if p is a prime $\equiv 1 \pmod 4$, then -1 is a fourth power $\pmod p$ if and only if 2 is a square mod p.

Exercise 8.9.15.[†] If $(a, n) = 1$, then multiplication by $a \pmod n$ generates a permutation of the reduced residues mod n. For example for $3 \pmod 7$ we get the permutation $\sigma_{3,7} := (1, 3, 2, -1, -3, -2)$, whereas for $2 \pmod 7$ we get the permutation $\sigma_{2,7} := (1, 2, 4)(3, 6, 5)$. Prove that if p is prime and $(a, p) = 1$, then the signature[6] of the permutation

$$\epsilon(\sigma_{a,p}) = \left(\frac{a}{p}\right).$$

Exercise 8.9.16. (a) Prove that $\left(\frac{2^n-1}{2^m-1}\right) = 0$ if $(m, n) > 1$.

(b) Suppose that $n = mq + r$ where $n \geq m \geq r \geq 2$. Prove that $\left(\frac{2^n-1}{2^m-1}\right) = -\left(\frac{2^m-1}{2^r-1}\right)$.

(c)[‡] Prove that if $n/m = [a_0, a_1, \ldots, a_k]$ with $(n, m) = 1$ and $a_k \geq 2$, then $\left(\frac{2^n-1}{2^m-1}\right) = (-1)^{k+1}$.

Infinitely many primes.

Exercise 8.9.17.[†] Fix odd, squarefree integer $n > 1$. Prove that there are infinitely many primes p for which $(p/n) = -1$.

Exercise 8.9.18.[†] Let n be a squarefree integer.
(a) By considering the prime divisors of $m^2 - n$, for well-chosen values of m, prove that there are infinitely many primes p for which $(n/p) = 1$.
(b) Deduce that there are infinitely many primes $\equiv 1 \pmod 3$.
(c) Refine this to deduce that there are infinitely many primes $\equiv 7 \pmod{12}$.
(d) Prove that there are infinitely many primes $\equiv 11 \pmod{12}$.
(e) Prove that there are infinitely many primes $\equiv 5 \pmod 8$.
(f) Prove that there are infinitely many primes $\equiv 7 \pmod 8$.
(g) Prove that there are infinitely many primes $\equiv 3 \pmod 8$.
(h) Prove that there are infinitely many primes $\equiv 5 \pmod{12}$.

Exercise 8.9.19.[†] Fix odd, squarefree integer $n > 1$. Using exercises 8.9.18(a) and 8.7.7 prove that there are infinitely many primes p for which $(p/n) = 1$.

In Ram Murty's undergraduate thesis (1976, Carleton University, Ottawa) he defined a *Euclidean proof* that there are infinitely many primes $\equiv a \pmod q$ to be one in which we use a polynomial all of whose prime divisors either divide q or are $\equiv 1$ or $a \pmod q$. Several of the proofs for the different arithmetic progressions in the last three questions can be formulated in this way. We gave such a proof for $a = 1$ in Theorem 7.8. Murty went on to show that there is a Euclidean proof that there are infinitely many primes $\equiv a \pmod q$ if and only if $a^2 \equiv 1 \pmod q$ (as in all our examples here). To prove that there are infinitely many primes $\equiv 2$ or $\equiv 3 \pmod 5$, or $5 \pmod 7$, etc., we will have to develop other techniques.

---

[6] Any permutation can be described by a sequence of transpositions (swaps) of pairs of elements. Although the sequence, and even the number of swaps in such a sequence is not unique, the parity of the number of swaps is. This is called the *signature* of the permutation and is given by -1 or 1 (for an odd or even number of transpositions, respectively).

Further reading on Euclidean proofs

[1] M. Ram Murty and N. Thain, *Primes in certain arithmetic progressions*, Funct. Approx. Comment. Math. **35** (2006), 249–259.

Primitive roots for specially chosen primes.

Exercise 8.9.20.† Suppose that q and $p = 2q + 1$ are odd (Sophie Germain twin) primes.
(a) Show that if $p \equiv 3 \pmod 8$, then 2 is a primitive root mod p (e.g., 11, 59, 83, 107, ...).
(b) Show that if $p \equiv 7 \pmod 8$, then -2 is a primitive root mod p.
(c) Prove that -3 is a primitive root mod p, but 3 is not.

Exercise 8.9.21.† Suppose that q and $p = 4q + 1$ are odd primes. Prove that $2, -2, 3$, and -3 are all primitive roots mod p.

Exercise 8.9.22.† Suppose that the Fermat number $F_m = 2^{2^m} + 1$ is prime with $m \geq 1$. Prove that if $(q/F_m) = -1$, then q is a primitive root mod F_m. (We deduce that 3 and 5 (for $m > 1$) are primitive roots mod F_m by exercise 8.5.4.)

Alternate proofs of the value of $(2/n)$.

Exercise 8.9.23. Let p be a prime $\equiv 1 \pmod 4$ so that there exists a reduced residue $r \pmod p$ such that $r^2 \equiv -1 \pmod p$.
(a) By expanding $(r+1)^2 \pmod p$ prove that 2 is a square mod p if and only if r is a square mod p.
(b) Prove that r is a square mod p if and only if there is an element of order 8 mod p.
(c) Use Theorem 7.6 to deduce that 2 is a square mod p if and only if $p \equiv 1 \pmod 8$.

Exercise 8.9.24 (Proof of (8.7.2)). By induction on odd $n \geq 1$. By the law of quadratic reciprocity, as stated in (8.7.3), we have

$$\left(\frac{2}{n}\right) = \left(\frac{-1}{n}\right)\left(\frac{n-2}{n}\right) = \left(\frac{-1}{n}\right)\left(\frac{n}{n-2}\right) = \left(\frac{-1}{n}\right)\left(\frac{2}{n-2}\right),$$

as one of n and $n - 2$ is $\equiv 1 \pmod 4$. Complete the proof.

Exercise 8.9.25. Every odd prime p may be written in the form $p = 4k + \sigma$ with $\sigma = \left(\frac{-1}{p}\right)$. We will show that $\left(\frac{2}{p}\right) = (-1)^k$ which implies Theorem 8.4. Let $m = 2k + \sigma$ so that $2m = p + \sigma$. Verify that

$$\left(\frac{2\sigma}{p}\right) = \left(\frac{2p+2\sigma}{p}\right) = \left(\frac{4m}{p}\right) = \left(\frac{m}{p}\right) = \left(\frac{\sigma p}{m}\right) = \left(\frac{2\sigma m - 1}{m}\right) = \left(\frac{-1}{m}\right)$$

and deduce the result from here.

Further proofs of the law of quadratic reciprocity.

Exercise 8.9.26.† (a) In the mid-18th century, Euler conjectured that if $m > n$ are coprime, odd, positive integers, then $\left(\frac{a}{m}\right) = \left(\frac{a}{n}\right)$ where $m - n = 4a$ if $m \equiv n \pmod 4$, and $m + n = 4a$ otherwise. Use the law of quadratic reciprocity to prove Euler's conjecture.
(b) Use Euler's conjecture to prove (8.7.3), the law of quadratic reciprocity.

Scholze (1938) proved Euler's conjecture using Gauss's Lemma (Theorem 8.6) and so gave a different proof of the law of quadratic reciprocity.

Exercise 8.9.27.‡ Finally we present my own variation of Rousseau's proof of quadratic reciprocity, as a series of (challenging) exercises. Let $p < q$ be odd primes, and let $n = pq$. Let $A = \prod_{1 \leq m < n/2,\ (m,n)=1} m$. In the proof given of Theorem 8.5 in section 8.6, we showed that $A \equiv \left(\frac{-1}{q}\right)\left(\frac{q}{p}\right) \pmod p$ and, analogously, $A \equiv \left(\frac{-1}{p}\right)\left(\frac{p}{q}\right) \pmod q$. We now evaluate $A \pmod n$ much as in Gauss's proof of Wilson's Theorem, where we paired up each residue with its inverse: Let S be the set of (unordered) pairs $\{a, b\} \in [1, \frac{n}{2})$ for which $ab \equiv 1$ or $-1 \pmod n$.

(a) Prove that the residues a and b are distinct unless $a^2 \equiv 1$ or -1 (mod n).
(b) Prove that if $a^2 \equiv 1$ (mod n), then $a \equiv 1, -1, r,$ and $-r$ (mod n) for some $r \not\equiv \pm 1$ (mod n).
(c) Prove that the product of the integers $a \in [1, \frac{n}{2})$ with $a^2 \equiv 1$ (mod n) is $\equiv \pm r$ (mod n).
(d) Prove that if $b^2 \equiv -1$ (mod n), then $p \equiv q \equiv 1$ (mod 4). In this case:
  - Deduce that the product of the integers $b \in [1, \frac{n}{2})$ for which $b^2 \equiv -1$ (mod n) is $\equiv \pm r$ (mod n).
  - Deduce that $A \equiv \pm 1$ (mod n).
  - Combine the above to show that $\left(\frac{-1}{q}\right)\left(\frac{q}{p}\right) = \left(\frac{-1}{p}\right)\left(\frac{p}{q}\right)$.
(e) If at least one of p and q is $\equiv 3$ (mod 4):
  - Deduce that $A \equiv \pm r$ (mod n).
  - Combine the above to show that $\left(\frac{-1}{q}\right)\left(\frac{q}{p}\right) = -\left(\frac{-1}{p}\right)\left(\frac{p}{q}\right)$.
(f) Deduce Theorem 8.5.

Appendix 8A. Eisenstein's proof of quadratic reciprocity

8.10. Eisenstein's elegant proof, 1844

A lemma of Gauss gives a complicated but useful formula to determine (a/p):

Theorem 8.6 (Gauss's Lemma). *Given an integer a which is not divisible by odd prime p, define r_n to be the absolutely least residue of an (mod p), and then define the set $\mathcal{N} := \{1 \leq n \leq \frac{p-1}{2} : r_n < 0\}$. Then $\left(\frac{a}{p}\right) = (-1)^{|\mathcal{N}|}$.*

For example, if $a = 3$ and $p = 7$, then $r_1 = 3, r_2 = -1, r_3 = 2$ so that $\mathcal{N} = \{2\}$ and therefore $\left(\frac{3}{7}\right) = (-1)^1 = -1$.

Proof. For each $m, 1 \leq m \leq \frac{p-1}{2}$, there is exactly one integer n, $1 \leq n \leq \frac{p-1}{2}$, such that $r_n = m$ or $-m$ (mod p) (for if $an \equiv \pm an'$ (mod p), then $p | a(n \mp n')$, and so $p | n \mp n'$, which is possible in this range only if $n = n'$). Therefore

$$\left(\frac{p-1}{2}\right)! = \prod_{1 \leq m \leq \frac{p-1}{2}} m = \prod_{\substack{1 \leq n \leq \frac{p-1}{2} \\ n \notin \mathcal{N}}} r_n \cdot \prod_{\substack{1 \leq n \leq \frac{p-1}{2} \\ n \in \mathcal{N}}} (-r_n)$$

$$\equiv \prod_{\substack{1 \leq n \leq \frac{p-1}{2} \\ n \notin \mathcal{N}}} (an) \cdot \prod_{\substack{1 \leq n \leq \frac{p-1}{2} \\ n \in \mathcal{N}}} (-an) = a^{\frac{p-1}{2}} (-1)^{|\mathcal{N}|} \cdot \left(\frac{p-1}{2}\right)! \pmod{p}.$$

Cancelling out the $\left(\frac{p-1}{2}\right)!$ from both sides, the result follows from Euler's criterion. □

This proof is a clever generalization of the proof of Theorem 8.4.

Exercise 8.10.1.[†] Use Gauss's Lemma to determine the values of (a) $(-1/p)$ and of (b) $(3/p)$, for all primes $p > 3$.

Exercise 8.10.2.[†] Let r be the absolutely least residue of $N \pmod{p}$. Prove that the least non-negative residue of $N \pmod{p}$ is given by

$$N - p\left[\frac{N}{p}\right] = \begin{cases} r & \text{if } r \geq 0, \\ p + r & \text{if } r < 0. \end{cases}$$

Corollary 8.10.1. *If p is a prime > 2 and a is an odd integer not divisible by p, then*

(8.10.1) $$\left(\frac{a}{p}\right) = (-1)^{\sum_{n=1}^{\frac{p-1}{2}} \left[\frac{an}{p}\right]}.$$

Proof. (Gauss) By exercise 8.10.2 we have

(8.10.2) $$\sum_{n=1}^{\frac{p-1}{2}} \left(an - p\left[\frac{an}{p}\right]\right) = \sum_{\substack{n=1 \\ n \notin \mathcal{N}}}^{\frac{p-1}{2}} r_n + \sum_{\substack{n=1 \\ n \in \mathcal{N}}}^{\frac{p-1}{2}} (p + r_n) = \sum_{n=1}^{\frac{p-1}{2}} r_n + p|\mathcal{N}|.$$

In the proof of Gauss's Lemma we saw that for each $m, 1 \leq m \leq \frac{p-1}{2}$, there is exactly one integer n, $1 \leq n \leq \frac{p-1}{2}$, such that $r_n = m$ or $-m$, and so $r_n \equiv m \pmod{2}$. Therefore, as a and p are odd, (8.10.2) implies that

$$|\mathcal{N}| \equiv \sum_{n=1}^{\frac{p-1}{2}} \left[\frac{an}{p}\right] \pmod{2} \quad \text{as} \quad \sum_{n=1}^{\frac{p-1}{2}} r_n \equiv \sum_{m=1}^{\frac{p-1}{2}} m \equiv a \sum_{n=1}^{\frac{p-1}{2}} n \pmod{2}.$$

We now deduce (8.10.1) from Gauss's Lemma. □

The exponent $\sum_{n=1}^{\frac{p-1}{2}} \left[\frac{an}{p}\right]$ on the right-hand side of (8.10.1) looks excessively complicated. However it arises in a different context that is easier to work with:

Lemma 8.10.1. *Suppose that a and b are odd, coprime positive integers. There are*

$$\sum_{n=1}^{\frac{b-1}{2}} \left[\frac{an}{b}\right]$$

lattice points $(n, m) \in \mathbb{Z}^2$ for which $bm < an$ with $0 < n < b/2$.

Proof. We seek the number of lattice points (n, m) inside the triangle bounded by the lines $y = 0$, $x = \frac{b}{2}$, and $by = ax$. For such a lattice point, n can be any

8.10. Eisenstein's elegant proof, 1844

integer in the range $1 \leq n \leq \frac{b-1}{2}$. For a given value of n, the triangle contains the lattice points (n, m) where m is any integer in the range $0 < m < \frac{an}{b}$. These are the lattice points in the shaded rectangle in Figure 8.1.

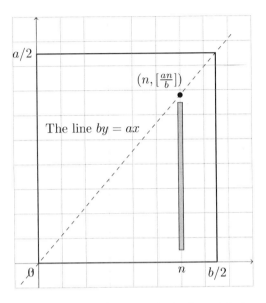

Figure 8.1. The shaded rectangle covers the lattice points (n, m) with $1 \leq m \leq [\frac{an}{b}]$.

Evidently m ranges from 1 to $[\frac{an}{b}]$, and so there are $[\frac{an}{b}]$ such lattice points. Summing this up over the possible values of n gives the lemma. □

Corollary 8.10.2. *If a and b are odd coprime positive integers, then*

$$\sum_{n=1}^{\frac{b-1}{2}} \left[\frac{an}{b}\right] + \sum_{m=1}^{\frac{a-1}{2}} \left[\frac{bm}{a}\right] = \frac{(a-1)(b-1)}{2}.$$

Proof. The idea is to split the triangle

$$R := \left\{(x,y):\ 0 < x < \frac{b}{2}\ \text{and}\ 0 < y < \frac{a}{2}\right\}$$

into two parts: the points in R on or below the line $by = ax$, that is, in the region

$$A := \{(x,y):\ 0 < x < b/2\ \text{and}\ 0 < y \leq ax/b\};$$

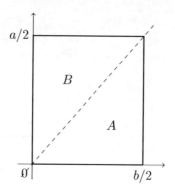

Figure 8.2. Splitting the rectangle R into two parts.

and the points in R above the line $by = ax$, that is, in the region
$$B := \{(x,y) : 0 < x < by/a \text{ and } 0 < y < a/2\}.$$

We count the lattice points (that is, the points with integer coordinates) in R and then in A and B together. To begin with
$$R \cap \mathbb{Z}^2 = \left\{(n,m) \in \mathbb{Z}^2 : 1 \leq n \leq \frac{b-1}{2} \text{ and } 1 \leq m \leq \frac{a-1}{2}\right\},$$
so that $|R \cap \mathbb{Z}^2| = \frac{a-1}{2} \cdot \frac{b-1}{2}$.

Since there are no lattice points in R on the line $by = ax$, as $(a,b) = 1$, therefore
$$A \cap \mathbb{Z}^2 = \{(n,m) \in \mathbb{Z}^2 : 0 < n < b/2 \text{ and } bm < an\},$$
and so $|A \cap \mathbb{Z}^2| = \sum_{n=1}^{\frac{b-1}{2}} \left[\frac{an}{b}\right]$ by Lemma 8.10.1. Similarly
$$B \cap \mathbb{Z}^2 = \{(n,m) \in \mathbb{Z}^2 : 0 < m < a/2 \text{ and } an < bm\},$$
and so $|B \cap \mathbb{Z}^2| = \sum_{m=1}^{\frac{a-1}{2}} \left[\frac{bm}{a}\right]$ by Lemma 8.10.1 (with the roles of a and b interchanged). The result then follows from the observation that $A \cap \mathbb{Z}^2$ and $B \cap \mathbb{Z}^2$ partition $R \cap \mathbb{Z}^2$. □

Eisenstein's proof of the law of quadratic reciprocity. By Corollary 8.10.1 with $a = q$, and then with the roles of p and q reversed, and then by Corollary 8.10.2, we deduce the desired law of quadratic reciprocity:
$$\left(\frac{q}{p}\right)\left(\frac{p}{q}\right) = (-1)^{\sum_{n=1}^{\frac{p-1}{2}}\left[\frac{qn}{p}\right]} \cdot (-1)^{\sum_{m=1}^{\frac{q-1}{2}}\left[\frac{pm}{q}\right]} = (-1)^{\frac{p-1}{2} \cdot \frac{q-1}{2}}.$$
□

Appendix 8B. Small quadratic non-residues

Given a prime p we wish to find as small an integer q as possible which is a quadratic non-residue mod p. This has become a central issue in several important questions in number theory. We discuss it at some length in this appendix and present some sophisticated techniques for finding small q.

The number 1 is always a quadratic residue mod p, as are $4, 9, 16, \ldots$. If 2 and 3 are quadratic non-residues, then $2 \cdot 3 = 6$ is a quadratic residue, by Theorem 8.1(iii). Hence at least one of 2, 3, and 6 is a quadratic residue, and this kind of reasoning implies that one is always guaranteed lots of small quadratic residues. How about small quadratic non-residues mod p? Since half the residues are quadratic non-residues one might expect to find lots of them, but a priori one is only guaranteed to find one that is $\leq \frac{p+1}{2}$. Can one do better? This is an important question in number theory and one where the best results known are surprisingly weak.

Exercise 8.11.1. Prove that the smallest quadratic non-residue mod p must be a prime.

We can therefore restrict our attention to determining the smallest prime q for which $(q/p) = -1$. The analogous question for quadratic residues is of interest (remember that finding the smallest integer that is a quadratic residue is of limited interest since we always have $1, 4, 9, \ldots$, and at least one of 2, 3, and 6). However this turns out to be more difficult than one might guess as it ties in with deep algebraic issues (which are discussed in section 12.5). To give one surprising example, one finds that every one of the dozen primes $q \leq 37$ is a quadratic non-residue mod 163. Therefore 41 is the smallest prime which is a quadratic residue mod 163.

8.11. The least quadratic non-residue modulo p

Theorem 8.7. *For every odd prime p there exists a prime $q < \sqrt{p} + 1$ for which $\left(\frac{q}{p}\right) = -1$.*

Proof. Let q be the least quadratic non-residue (mod p). By exercise 8.11.1 we know that q is prime. Let m be the least integer $> p/q$ so that $p/q < m < p/q + 1$ and therefore $0 < a := mq - p < q$. Therefore a is a quadratic residue (mod p) and so $\left(\frac{m}{p}\right) = \left(\frac{a}{p}\right)\left(\frac{q}{p}\right) = 1 \cdot (-1) = -1$. This implies that m is a quadratic non-residue and so $q \leq m < p/q + 1$ which implies that $q^2 < p + q$, and therefore $(q-1)^2 < q^2 - q < p$, and the result follows by taking square roots. □

This is better than the similar result proved in exercise 9.7.3(b), though by very different methods.

8.12. The smallest prime q for which p is a quadratic non-residue modulo q

In Gauss's original proof of quadratic reciprocity he needed to show that each prime $p \equiv 1 \pmod{4}$ is a quadratic non-residue modulo some prime $q < p$ (as part of an induction argument). This result follows from Theorem 8.7, by quadratic reciprocity, but we need to avoid using quadratic reciprocity in a proof of quadratic reciprocity! We now present Gauss's most ingenious argument.

Theorem 8.8. *For every prime $p \equiv 1 \pmod{8}$ there exists a prime $q < 2\sqrt{p} + 1$ such that $\left(\frac{p}{q}\right) = -1$.*

Proof. (Gauss) Let $m = [\sqrt{p}]$ and consider the product $(p-1^2)(p-2^2)\cdots(p-m^2)$, under the assumption that $\left(\frac{p}{q}\right) = 1$ for all odd primes $q \leq 2m+1$. Now since $p \equiv 1 \pmod{8}$ and $\left(\frac{p}{q}\right) = 1$ for all odd primes $q \leq 2m+1$, we deduce that for any given prime power q^n with $q \leq 2m+1$, there exists a residue a_q (mod q^n) such that $p \equiv a_q^2 \pmod{q^n}$, by exercise 8.8.1. Since this is true for each $q \leq 2m+1$ and since $(2m+1)!$ is divisible only by powers of primes $q \leq 2m+1$, we use the Chinese Remainder Theorem to construct an integer A for which $p \equiv A^2 \pmod{(2m+1)!}$. Now $(p, (2m+1)!) = 1$ and so $(A, (2m+1)!) = 1$, which implies that A is invertible mod $(2m+1)!$. Therefore

$$(p-1^2)(p-2^2)\cdots(p-m^2) \equiv (A^2-1^2)(A^2-2^2)\cdots(A^2-m^2)$$
$$\equiv \frac{(A+m)!}{(A-m-1)!} \cdot \frac{1}{A} \pmod{(2m+1)!}.$$

Now $\binom{A+m}{2m+1}$ is an integer, and so

$$\frac{(A+m)!}{(A-m-1)!} \cdot \frac{1}{A} = \frac{1}{A} \cdot (2m+1)! \binom{A+m}{2m+1} \equiv 0 \pmod{(2m+1)!}.$$

Therefore $(2m+1)!$ divides $(p-1^2)(p-2^2)\cdots(p-m^2)$. However $p < (m+1)^2$ and so

$$(2m+1)! \leq (p-1^2)(p-2^2)\cdots(p-m^2)$$
$$< ((m+1)^2-1^2)((m+1)^2-2^2)\cdots((m+1)^2-m^2) = \frac{(2m+1)!}{m+1},$$

giving a contradiction. □

The following proof is also due to Gauss.

Theorem 8.9. *For every prime $p \equiv 5 \pmod{8}$ there exists a prime $q < \sqrt{8p}$ such that $\left(\frac{p}{q}\right) = -1$.*

Proof. (Gauss) Suppose that $p \equiv 5 \pmod 8$. Let a be the largest integer $< \sqrt{p/2}$, so that $a > (p/2)^{1/2} - 1$. Now $p - 2a^2 \equiv 3$ or $5 \pmod 8$ and so has a prime divisor $q \equiv 3$ or $5 \pmod 8$ (by exercise 3.1.4(b)). But then, by Theorem 8.4, we have $\left(\frac{2}{q}\right) = -1$ and so $\left(\frac{p}{q}\right) = \left(\frac{2a^2}{q}\right) = -1$. Finally
$$q \leq p - 2a^2 < p - 2(\sqrt{p/2} - 1)^2 = 2(\sqrt{2p} - 1) < \sqrt{8p}. \qquad \square$$

It is hard to resist giving another proof of this type.

Theorem 8.10. *If $p \equiv 1 \pmod 4$, $p > 17$, there exists a prime $q < 4\sqrt{p} + 4$ with $q \equiv 3 \pmod 4$ and $\left(\frac{-p}{q}\right) = -1$.*

Proof. Let $2a$ be that even integer immediately greater than $\sqrt{p}$, so that $4a^2 - p \equiv 3 \pmod 4$. Let q be a prime divisor of $4a^2 - p$ which is $\equiv 3 \pmod 4$ so that $p \equiv 4a^2 \pmod q$ and hence $\left(\frac{p}{q}\right) = 1$. But then $\left(\frac{-p}{q}\right) = -1$ as $q \equiv 3 \pmod 4$. Also $2a < \sqrt{p} + 2$ and so $q \leq 4a^2 - p < (\sqrt{p} + 2)^2 - p = 4(\sqrt{p} + 1)$. $\qquad \square$

8.13. Character sums and the least quadratic non-residue

If there are N quadratic residues and R quadratic non-residues mod p up to x, where $1 \leq x < p$, then $N + R = x$ and $N - R$ is the value of the *character sum*
$$S(x) := \sum_{n \leq x} \left(\frac{n}{p}\right).$$
As each $|\left(\frac{n}{p}\right)| \leq 1$, we trivially have $|S(x)| \leq x$ and would like to obtain improvements on this upper bound when possible. For if $S(x) = 0$, then $N = R = \frac{1}{2}x$; and if $S(x)$ is small compared to x, then roughly half the residues are quadratic residues, and roughly half are quadratic non-residues; more precisely
$$N = \tfrac{1}{2}(x + S(x)) \quad \text{and} \quad R = \tfrac{1}{2}(x - S(x)).$$

Now $S(p) = S(p-1) = 0$ as there are $\frac{p-1}{2}$ quadratic residues mod p, and the same number of non-residues. The sum, $S(p)$, sums the value of $\left(\frac{n}{p}\right)$ as n runs through a complete set of residues mod p and so is a *complete character sum*. In contrast, any sum $S(x)$ with $x < p$ is known as an *incomplete character sum*.

If $(-1/p) = 1$, then $(n/p) = ((p-n)/p)$ and so $\sum_{n < p/2}(n/p) = \sum_{p/2 < n < p}(n/p)$. This implies $S(\frac{p-1}{2}) = \frac{1}{2}S(p-1) = 0$. On the other hand, if $(-1/p) = -1$, then $S(\frac{p-1}{2}) \equiv \sum_{n \leq \frac{p-1}{2}} 1 = \frac{p-1}{2} \equiv 1 \pmod 2$ and so is non-zero. Its value relates to questions about quadratic forms we will discuss in section 12.15 of appendix 12D.

Exercise 8.13.1. (a) Prove that $S(x)$ is periodic with period p.
(b) Prove that $|S(x)| \leq \frac{p-1}{2}$ for all integers x.

The non-zero values of $\left(\frac{n}{p}\right)$ are 1 and -1. If we imagine that the sums of $\left(\frac{n}{p}\right)$ are distributed much like sums of independent random variables taking values 1 and -1 with equal probability, then we might expect that the maximum value of $|S(x)|$ is about $\sqrt{x}\log p$; and thus $|S(x)|$ should be no more than $\sqrt{p}\log p$ as we vary over x. This is confirmed by the Pólya-Vinogradov Theorem (1919) which states for any integers M and $N \geq 1$ one has

$$\left|\sum_{n=M+1}^{M+N}\left(\frac{n}{p}\right)\right| \leq \sqrt{p}\log p.$$

This improves the trivial upper bound $\leq N$, if N is large enough. However if $N \leq \sqrt{p}$, then the Pólya-Vinogradov bound is worse than the trivial bound, and it would be useful to obtain an improvement. At the very least we might hope that for any given $\epsilon > 0$, we have

$$\left|\sum_{n=1}^{x}\left(\frac{n}{p}\right)\right| \leq \epsilon x,$$

once x is sufficiently large (as a function of p and ϵ). We believe that this is true if, for instance, $x = p^\delta$ for any fixed $\delta > 0$. However the best result known is due to Burgess (1962), who showed that this holds when $x = p^{1/4+\delta}$. One can deduce from this that the least quadratic non-residue mod prime p is $< p^{1/4}$, and by a more involved argument is $< p^{1/4\sqrt{e}}$ once p is sufficiently large. Burgess's result has not been significantly improved in a long time and falls far short of Vinogradov's conjecture that the least quadratic non-residue is $< p^\epsilon$ for all sufficiently large p. These questions will be discussed in detail in [**GS**].

What is the true size of the least quadratic non-residue mod p? It is believed that there is a quadratic non-residue mod p that is $\leq 2(\log p)^2$, but we cannot prove this (though we can also deduce this from the Generalized Riemann Hypothesis). Brave souls have even argued that the least quadratic non-residue might always be $< c\log p\log\log p$ for a suitable constant $c > 0$. One can show that there are primes p for which the least quadratic non-residue is at least this large.

To find a quadratic non-residue we might try out $2, 3, 4, \ldots$ until we find one; we can test whether each given integer a is a quadratic non-residue by calculating the Legendre symbol (a/p) (using the algorithm described in section 8.5). We very much expect to find one with $a \leq 2(\log p)^2$, which should be a quick calculation. If we fail, then there is an extraordinary bonus: It would imply that one of the most famous conjectures in mathematics, the Generalized Riemann Hypothesis, is false!

Since half the reduced residues are quadratic residues and half are quadratic non-residues mod p, if we pick a reduced residue mod p at random, then the probability that it is a quadratic non-residue is $\frac{1}{2}$. If it was a quadratic residue, then we can try again, and if we get another quadratic residue, then try again, etc. In other words, pick integers $a_1, a_2, \ldots, a_k, \ldots$ from $\{1, 2, 3, \ldots, p-1\}$ at random until we find a quadratic non-residue mod p. The probability that none of $a_1, a_2, \ldots, a_k$ are quadratic non-residues mod p is $1/2^k$. With $k = 100$ it is inconceivable that we could fail! This test is expected (but not certain) to run in polynomial time (see section 7.14 of appendix 7A) and so is said to be a *random polynomial time* algorithm.

Appendix 8C. The first proof of quadratic reciprocity

8.14. Gauss's original proof of the law of quadratic reciprocity

Gauss gave four proofs of the law of quadratic reciprocity, and there are now literally hundreds of proofs. None of the proofs are easy. For an elementary textbook like this, one wishes to avoid any deeper ideas, which considerably cuts down the number of choices. The one that has been long preferred stems from an idea of Eisenstein and is discussed in section 8.10. It ends up with an elegant lattice point counting argument though the intermediate steps are difficult to follow and motivate. Gauss's very first proof was long and complicated yet elementary and the motivation is quite clear. Subsequent authors[7] have shortened Gauss's proof and we present a version of that proof here.

We define $\left(\frac{m}{-n}\right) = \left(\frac{m}{n}\right)$ and prove (8.7.3) extended to all odd integers, negative as well as positive. If (8.7.3) holds for (m, n) with given values of $|m|$ and $|n|$, then (8.7.3) holds for $(\pm m, \pm n)$ since

$$\left(\frac{-m}{n}\right)\left(\frac{n}{-m}\right) = \left(\frac{-1}{n}\right) \cdot \left(\frac{m}{n}\right)\left(\frac{n}{m}\right) = (-1)^{\frac{n-1}{2}} \cdot (-1)^{\frac{m-1}{2} \cdot \frac{n-1}{2}} = (-1)^{\frac{-m-1}{2} \cdot \frac{n-1}{2}}.$$

We assume that $|m| > |n| > 1$ and prove the result by induction on $|m|$, followed by induction on $|n|$. If $m = ab$ is composite with $1 < a, b < m$, then

$$\left(\frac{m}{n}\right)\left(\frac{n}{m}\right) = \left(\frac{a}{n}\right)\left(\frac{n}{a}\right) \cdot \left(\frac{b}{n}\right)\left(\frac{n}{b}\right) = (-1)^{\frac{a-1}{2} \cdot \frac{n-1}{2}} \cdot (-1)^{\frac{b-1}{2} \cdot \frac{n-1}{2}} = (-1)^{\frac{m-1}{2} \cdot \frac{n-1}{2}},$$

by exercise 8.7.5. The analogous proof works if n is composite. We are therefore left with the case that $|n| = p < |m| = q$ are odd primes. The proof is modeled on

---

[7] A preprint may be found at http://www.math.cornell.edu/~web401/steve.gauss17gon.pdf, entitled *The mathematics of Gauss* by David Savitt.

Gauss's induction step for evaluating $(\frac{2}{n})$ for integers $n \equiv \pm 3 \pmod 8$ given at the end of section 8.7. There are two cases here:

- When $(\frac{p}{q}) = 1$ or $(\frac{-p}{q}) = 1$, let $\ell = p$ or $-p$, respectively, so that $(\frac{\ell}{q}) = 1$, and we will prove the result with p replaced by ℓ: There exists an even integer e, $1 \le e \le q-1$, such that $e^2 \equiv \ell \pmod q$, and therefore there exists an integer s for which
$$e^2 = \ell + qs.$$

  - If p does not divide s, then $(\frac{\ell}{s}) = 1$ as $e^2 \equiv \ell \pmod s$, and $(\frac{qs}{\ell}) = 1$ as $e^2 \equiv qs \pmod \ell$. Moreover $|s| = \left|\frac{e^2-\ell}{q}\right| < \frac{(q-1)^2+q}{q} < q$ as $p < q$, so the reciprocity law works for the pair ℓ, s by the induction hypothesis. As $(\frac{\ell}{q}) = 1$, we therefore deduce
$$\left(\frac{\ell}{q}\right)\left(\frac{q}{\ell}\right) = 1 \cdot \left(\frac{q}{\ell}\right) = \left(\frac{q}{\ell}\right) \cdot \left(\frac{qs}{\ell}\right) \cdot \left(\frac{\ell}{s}\right) = \left(\frac{s}{\ell}\right) \cdot \left(\frac{\ell}{s}\right) = (-1)^{\frac{\ell-1}{2} \cdot \frac{s-1}{2}}$$
  by the induction hypothesis. This yields the result if $\ell \equiv 1 \pmod 4$. On the other hand if $\ell \equiv -1 \pmod 4$, then $qs = e^2 - \ell \equiv 1 \pmod 4$, that is, $s \equiv q \pmod 4$, and the result follows.

  - If p divides s, we write $s = \ell S$, $e = \ell E$ to obtain $\ell E^2 = 1 + qS$, and so $(\frac{\ell}{S}) = (\frac{-qS}{\ell}) = 1$. Therefore
$$\left(\frac{\ell}{q}\right)\left(\frac{q}{\ell}\right) = 1 \cdot \left(\frac{q}{\ell}\right) \cdot \left(\frac{-qS}{\ell}\right) \cdot \left(\frac{\ell}{S}\right) = \left(\frac{-S}{\ell}\right) \cdot \left(\frac{\ell}{-S}\right) = (-1)^{\frac{\ell-1}{2} \cdot \frac{S+1}{2}}$$
  and the result follows since $S \equiv -q \pmod 4$.

- When $(\frac{p}{q}) = (\frac{-p}{q}) = -1$, we have $(\frac{-1}{q}) = 1$ so that $q \equiv 1 \pmod 4$. Therefore there exists a prime $r < q$ such that $(\frac{q}{r}) = -1$ by Theorems 8.8 and 8.9. If $r = p$, then the result follows, so now assume that $r \neq p$. Moreover we must have $(\frac{r}{q}) = -1$ or else, since we have already proved the reciprocity law when $(\frac{r}{q}) = 1$, this would imply that $(\frac{q}{r}) = 1$ as $q \equiv 1 \pmod 4$, a contradiction.

  Therefore $(\frac{pr}{q}) = 1$ and so there exists an even integer e, $1 \le e \le q-1$, such that $e^2 \equiv pr \pmod q$, which implies that there exists an integer s with
$$e^2 = pr + qs.$$
Now $|s| = \left|\frac{e^2-pr}{q}\right| \le \left|\frac{\max\{(q-1)^2, pr\}}{q}\right| < q$, so the reciprocity law works for any two of r, p, and s, by the induction hypothesis.

  We proceed much as above but now there are four possibilities for $d = (pr, qs) = (pr, s)$, which we handle all at once: Since d is squarefree and $d | pr + qs = e^2$, hence $d | e$. We write $e = dE$, $pr = dL$, and $s = dS$ so that $dE^2 = L + qS$, and dE^2, L, qS are pairwise coprime. But then
$$\left(\frac{-LqS}{d}\right) = \left(\frac{dqS}{L}\right) = \left(\frac{dL}{q}\right) = \left(\frac{dL}{S}\right) = 1,$$
from the equations above. Multiplying these all together and reorganizing, and then using that $pr = dL$, we obtain
$$\left(\frac{-L}{d}\right)\left(\frac{d}{-L}\right) = \left(\frac{qS}{dL}\right)\left(\frac{dL}{qS}\right) = \left(\frac{qS}{pr}\right)\left(\frac{pr}{qS}\right).$$

8.14. Gauss's original proof of the law of quadratic reciprocity

After some rearrangement this becomes

$$\left(\frac{p}{q}\right)\left(\frac{q}{p}\right) = \left(\frac{-L}{d}\right)\left(\frac{d}{-L}\right) \cdot \left(\frac{S}{p}\right)\left(\frac{p}{S}\right) \cdot \left(\frac{S}{r}\right)\left(\frac{r}{S}\right)$$

since $\left(\frac{q}{r}\right)\left(\frac{r}{q}\right) = 1$ by the choice of r. We use the induction hypothesis for the pairs $(r, S), (p, S)$, and $(-L, d)$ (even when one of L and d is ± 1) to obtain

$$\left(\frac{p}{q}\right)\left(\frac{q}{p}\right) = 1 \cdot (-1)^{\frac{L+1}{2} \cdot \frac{d-1}{2} + \frac{p-1}{2} \cdot \frac{S-1}{2} + \frac{r-1}{2} \cdot \frac{S-1}{2}}.$$

Now $S = -qL \equiv -L \pmod 4$ as 2 divides E and $q \equiv 1 \pmod 4$, and so the exponent here (the power of (-1)) is

$$\equiv \frac{L+1}{2}\left(\frac{d-1}{2} + \frac{p-1}{2} + \frac{r-1}{2}\right) \equiv \frac{L+1}{2} \cdot \frac{dpr-1}{2} \pmod 2,$$

by exercise 8.7.5. Now $dpr = d^2 L \equiv L \pmod 4$, so the above exponent is $\equiv \frac{L+1}{2} \cdot \frac{L-1}{2} \equiv 0 \pmod 2$, and the result follows as $q \equiv 1 \pmod 4$. □

Appendix 8D. Dirichlet characters and primes in arithmetic progressions

In this appendix we introduce Dirichlet characters, an important and natural generalization of the Legendre and Jacobi symbols. Dirichlet characters are used in many parts of number theory, and we shall highlight this by sketching Dirichlet's proof that there are infinitely many primes in any arithmetic progression $a \pmod{q}$ with $(a, q) = 1$. First we begin by reviewing some of what we already know about Legendre and Jacobi symbols, from a slightly different perspective.

8.15. The Legendre symbol and a certain quotient group

Let p be an odd prime and $G_p := (\mathbb{Z}/p\mathbb{Z})^*$ be the multiplicative group of reduced residues mod p. The subgroup of quadratic residues,

$$H_p := G_p^2 = \{g^2 : g \in G_p\} = \left\{ a \pmod{p} : \left(\frac{a}{p}\right) = 1 \right\},$$

is of size $(p-1)/2$. This subgroup partitions G_p into two cosets, H_p and bH_p, where $(b/p) = -1$. If $a \in H_p$, then $(a/p) = 1$, and if $a \in bH_p$, then $(a/p) = -1$. Hence the Legendre symbol distinguishes between the two equivalence classes in G_p/H_p. Now G_p/H_p is isomorphic to $\mathbb{Z}/2\mathbb{Z}$, which can be written in the multiplicative form with representatives -1 and 1, that is, the values given by the Legendre symbol. So one can view the Legendre symbol as the quotient map $G_p \to G_p/H_p$ given by $a \to \left(\frac{a}{p}\right)$. Such quotients are called *characters*.

If $n = p_1 p_2 \cdots p_k$ where the p_i are distinct odd primes, then, by the Chinese Remainder Theorem, the multiplicative group of reduced residues mod n is given by

$$G_n := (\mathbb{Z}/n\mathbb{Z})^* \cong (\mathbb{Z}/p_1\mathbb{Z})^* \oplus (\mathbb{Z}/p_2\mathbb{Z})^* \oplus \cdots \oplus (\mathbb{Z}/p_k\mathbb{Z})^*.$$

This has subgroup $H_n := H_{p_1} \oplus H_{p_2} \oplus \cdots \oplus H_{p_k}$, with quotient group
$$G_n/H_n \cong G_{p_1}/H_{p_1} \oplus G_{p_2}/H_{p_2} \oplus \cdots \oplus G_{p_k}/H_{p_k},$$
so that $G_n/H_n \cong (\mathbb{Z}/2\mathbb{Z})^k$. The quotient map $G_n \to G_n/H_n$ can therefore be presented as
$$a \pmod{n} \longrightarrow \left(\left(\frac{a}{p_1}\right), \left(\frac{a}{p_2}\right), \ldots, \left(\frac{a}{p_k}\right) \right).$$
Now G_n/H_n has the subgroup
$$J_n = \{(s_1, \ldots, s_n) \in G_n/H_n : s_1 \cdots s_n = 1\}$$
which partitions G_n/H_n into two cosets, J_n and bJ_n, where b is a reduced residue for which $(b/n) = -1$ (one can find such a b, for instance where $(b/p_1) = -1$ and $(b/p_j) = 1$ for all $j \geq 2$, using the Chinese Remainder Theorem). If $a \in J_n$, then the Jacobi symbol $(a/n) = 1$, and if $a \in bJ_n$, then $(a/n) = -1$.

The Jacobi symbol is defined to be a homomorphism from $\mathbb{Z}/n\mathbb{Z} \to \mathbb{C}$, taking values inside the multiplicative closed set $\{-1, 0, 1\}$. However, we used it as if it is a map $\mathbb{Z} \to \mathbb{C}$; this can be justified for if $a \in \mathbb{Z}$, let $\bar{a}$ be the residue class of a (mod n) and then define $(a/n) = (\bar{a}/n)$. We can view this map as the composition of the maps $\mathbb{Z} \to \mathbb{Z}/n\mathbb{Z} \to \mathbb{C}$, the first being reduction mod n, the second the Jacobi symbol. In the next section we see how Dirichlet generalized these ideas.

Exercise 8.15.1.[†] Suppose that A is a subset of G_p of size $\frac{p-1}{2}$ that is closed under multiplication. Prove that $A = H_p$.

8.16. Dirichlet characters

The Jacobi symbol restricted to the reduced residues mod n is an example of a *group homomorphism* $\chi : (\mathbb{Z}/n\mathbb{Z})^* \to \mathbb{C}^*$. This means that
$$\chi(ab) = \chi(a)\chi(b) \text{ for all } a, b \in (\mathbb{Z}/n\mathbb{Z})^*.$$

Exercise 8.16.1. Deduce that either $\chi(1) = 1$ or $\chi(a) = 0$ for all $a \in (\mathbb{Z}/n\mathbb{Z})^*$.

Let m be the smallest integer for which $a^m \equiv 1 \pmod{n}$ for all reduced residues $a \pmod{n}$. Then $\chi(a)^m = \chi(a^m) = \chi(1) = 1$, and so the image of χ is a subset of $\{z \in \mathbb{C} : z^m = 1\}$, the set of mth roots of unity. One can define χ on all of $\mathbb{Z}/n\mathbb{Z}$ simply by taking $\chi(a) = 0$ if $(a, n) > 1$, and then extend it to $\mathbb{Z}$, by defining $\chi(a) = \chi(\bar{a})$ where $\bar{a}$ is the residue class of a (mod n). Such χ are called *Dirichlet characters*, or simply *characters*. Dirichlet characters play a central role in advanced number theory. We say that χ is a Dirichlet character of modulus n if $\chi(a) = \chi(a+n)$ for all integers a; $\chi(a) = 0$ if $(a, n) > 1$; and $\chi : (\mathbb{Z}/n\mathbb{Z})^* \to \mathbb{C}^*$ is a group homomorphism. Define $X(n)$ to be the set of Dirichlet characters mod n.

Exercise 8.16.2.[†] Given $\chi, \psi \in X(n)$, define $\chi\psi$ by $(\chi\psi)(a) = \chi(a)\psi(a)$ for all integers a, and then prove that $\chi\psi \in X(n)$. Prove that $\overline{\chi}$ defined by $\overline{\chi}(a) = \overline{\chi(a)}$ is a character. Prove that $X(n)$ forms a group under multiplication.

If $n = p$ is a prime, let g be a primitive root mod p. For any reduced residue $a \pmod{p}$ there exists an integer k for which $a \equiv g^k \pmod{p}$, and so if $\chi \in X(p)$, then $\chi(a) = \chi(g^k) = \chi(g)^k$; that is, the values of $\chi(a)$ are all determined from the value of $\chi(g)$. The only restriction on the value of $\chi(g)$ is that $\chi(g)^{p-1} = \chi(g^{p-1}) = \chi(1) = 1$, and therefore $\chi(g)$ can be any $(p-1)$st root of unity. There are therefore $p-1$ distinct characters mod p, one for each $(p-1)$st root of unity.

For example, if $p = 2$, we have just one character, and so this must be the character χ for which $\chi(a) = 1$ whenever a is odd. In fact, for every modulus n one has the *principal character* (mod n), denoted χ_0, for which $\chi_0(a) = 1$ whenever $(a, n) = 1$. If $p = 3$, then there is one other character, and so it must be the Legendre symbol $\left(\frac{a}{3}\right)$. For $p = 5$ there are four characters, the principal character, the Legendre symbol $\left(\frac{a}{5}\right)$, and two others. Now 2 is a primitive root mod 5, and so the characters are determined by the value of $\chi(2)$, which must be a fourth root of unity. For the two characters we already know about, $\chi(2) = 1$ and -1, respectively, so the other two must have $\chi(2) = i$ and $-i$, respectively. We see that these two characters take on complex values. In general χ takes only real values if and only if $\chi(g) \in \mathbb{R}$ (in which case we call χ a *real character*). But $\chi(g)$ is a $(p-1)$st root of unity and so either $\chi(g) = 1$ in which case $\chi = \chi_0$, or $\chi(g) = -1$ in which case $\chi(a) = \left(\frac{a}{p}\right)$, the Legendre symbol mod p.

The same argument determines $X(n)$ whenever $(\mathbb{Z}/n\mathbb{Z})^*$ has a primitive root, that is, when $n = 1, 2, 4, p^e$, or $2p^e$ for some odd prime p and $e \geq 1$, as we saw in Corollary 7.18.1. In this case $\chi(g)$ is a $\phi(n)$th root of unity, and its value determines χ. For example, if $n = 4$, there are two characters, one of which is the principal character, which is the same as the principal character mod 2. Since -1 is a primitive root mod 4, the other has $\chi(-1) = -1$ and $\chi(1) = 1$, and therefore $\chi(a) = \left(\frac{-4}{a}\right)$. By definition, characters mod p^e are also characters mod p^f for all $f \geq e$. Since there can be just two real characters mod p^e, they must therefore be χ_0 and $\left(\frac{\cdot}{p}\right)$.

In general if $(\mathbb{Z}/n\mathbb{Z})^*$ is generated by elements $g_1, \ldots, g_r$ of orders $m_1, \ldots, m_r$, then

(8.16.1) $$(\mathbb{Z}/n\mathbb{Z})^* = \{g_1^{a_1} \cdots g_r^{a_r},\ 0 \leq a_j \leq m_j - 1 \text{ for each } j\};$$

and so $\chi(a) = \chi(g_1)^{a_1} \cdots \chi(g_r)^{a_r}$. That is, χ is completely determined by the values of $\chi(g_1), \ldots, \chi(g_r)$, where $\chi(g_j)$ is an arbitrary m_jth root of unity.

Exercise 8.16.3. (a) † Use the Chinese Remainder Theorem to show that if $n = \prod_p p^{e_p}$, then
$$X(n) = \left\{\prod_p \chi_p :\ \chi_p \in X(p^{e_p})\right\}.$$
(b) Deduce that $X(n)$ has $\phi(n)$ elements.
(c) Show that if $\chi \in X(n)$, then $\chi^{\lambda(n)} = \chi_0$ where $\lambda(n)$ is Carmichael's function (as defined in section 7.17 of appendix 7B).
(d) Show that if $\chi \in X(n)$ is non-principal, then there exists a (mod n) such that $\chi(a) \neq 0$ or 1.
(e) Prove that if p is an odd prime and $m \not\equiv 0$ or 1 (mod p), then there exists $\chi \in X(p)$ such that $\chi(m) \neq 1$.
(f) Prove that if $(m, n) = 1$ and $m \not\equiv 1$ (mod n), then there exists $\chi \in X(n)$ such that $\chi(m) \neq 1$.

Exercise 8.16.3(a) shows that we can determine the characters mod n by determining the characters modulo the prime power divisors of n. After our discussion above, it remains to understand $X(2^e)$ for $e \geq 3$. In the proof of Corollary 7.17.2 we saw that $(\mathbb{Z}/2^e\mathbb{Z})^*$ is generated by -1 and 3, which have orders 2 and 2^{e-2}, respectively. Therefore there are four real characters mod 2^e with $e \geq 3$, namely χ_0, $\left(\frac{-4}{\cdot}\right)$, and we have determined the other two, $\left(\frac{2}{\cdot}\right)$ in (8.7.2), and $\left(\frac{-2}{\cdot}\right)$ when multiplying this with (8.7.1).

We saw that $\left(\frac{\cdot}{p}\right)$ is a character mod p^2 as well as mod p. Given a character it makes sense to determine the smallest modulus for the character, in this case p, which we call the *conductor* of $\left(\frac{\cdot}{p}\right)$. There is another way to trivially construct a character of larger modulus from a given character: Given a character $\chi \pmod{m}$, we can multiply it by the principal characters $\chi_0 \pmod{n}$ to obtain a new character $\chi\chi_0 \pmod{[m,n]}$. We call this new character an *induced* character. Any character that cannot be induced from another is called *primitive*. If χ is induced from a primitive character of modulus n, then χ has *conductor* n.

Since $\left(\frac{3}{5}\right) = -\left(\frac{3}{11}\right)$, we see that $\left(\frac{3}{\cdot}\right)$ cannot have period 3 or even 6. In fact it has period 12, since $\left(\frac{3}{\cdot}\right) = \left(\frac{-1}{\cdot}\right)\left(\frac{\cdot}{3}\right)$, the product of characters of moduli 4 and 3, respectively. More generally if q is an odd, squarefree integer, then $\left(\frac{q}{\cdot}\right)$ has modulus q if $q \equiv 1 \pmod 4$. However, if $q \equiv 3 \pmod 4$, then $\left(\frac{q}{\cdot}\right) = \left(\frac{-1}{\cdot}\right)\left(\frac{\cdot}{q}\right)$. The first is a character of period 4, and so $\left(\frac{q}{\cdot}\right)$ has period $4q$. The character mod $4q$ is then given by $\left(\frac{-4}{\cdot}\right)\left(\frac{\cdot}{q}\right)$, using exercise 8.16.3.

Exercise 8.16.4.[‡] (a) Prove that if n is a squarefree odd integer, then the Jacobi symbol $\left(\frac{\cdot}{n}\right)$ is primitive.
(b) Prove that the real, primitive characters are given by 1, and for each squarefree, odd n:
- The Jacobi symbol $\left(\frac{\cdot}{n}\right)$ of modulus n, if $n > 1$.
- The symbol $\left(\frac{2n}{\cdot}\right)$ of modulus $8|n|$.
- The symbol $\left(\frac{4n}{\cdot}\right)$ of modulus $4|n|$, if $n \equiv 3 \pmod 4$.

We will now show that if χ is a non-principal character mod n, then the sum

(8.16.2) $$\boxed{\sum_{m=1}^{n} \chi(m) = 0.}$$

First note that as $\chi(m) = 0$ whenever $(m,n) > 1$, we get the same value when we restrict the sum to reduced residues $m \pmod n$, a sum which we denote by S. If we multiply each term in S through by $\chi(g)$ where $(g,n) = 1$, then we get the sum of $\chi(g)\chi(m) = \chi(gm)$ over the reduced residues $m \pmod n$, which is just a rearrangement of S (as $\{gm \pmod n : (m,n) = 1\}$ is a rearrangement of the reduced residues mod n). Therefore $\chi(g)S = S$. However, as χ is a non-principal character, we can select g so that $\chi(g) \neq 1$ by exercise 8.16.3(d), and therefore $S = 0$.

Exercise 8.16.5. Reprove (8.16.2) using the representation of the reduced residues given in (8.16.1).

Finally we prove the "dual" estimate to (8.16.2):[8] For every reduced residue m (mod n) we have

(8.16.3) $$\boxed{\frac{1}{\phi(n)} \sum_{\chi \in X(n)} \chi(m) = \begin{cases} 1 & \text{if } m \equiv 1 \ (\text{mod } n), \\ 0 & \text{otherwise.} \end{cases}}$$

This is immediate when $m = 1$. For all other reduced residues m (mod n) there exists a character χ_1 (mod n) for which $\chi_1(m) \neq 1$ (exercise 8.16.3). If we multiply our sum through by $\chi_1(m)$, then the χ-term becomes $\chi_1(m)\chi(m) = (\chi_1\chi)(m)$ and $\{\chi_1\chi : \chi \in X(n)\} = X(n)$ as $X(n)$ forms a group under multiplication. Hence if the sum is S, then $\chi_1(m)S = S$, which implies that $S = 0$ as $\chi_1(m) \neq 1$, as claimed.

8.17. Dirichlet series and primes in arithmetic progressions

In 1837 Dirichlet proved that for any given integer $q \geq 2$ and any integer a with $(a, q) = 1$, there are infinitely many primes $\equiv a$ (mod q). We have seen how to prove this for specific a and q (or unions of a (mod q)) throughout this book (see, e.g., exercise 8.9.18), but these elementary methods seem incapable of reproving Dirichlet's result. Dirichlet built an extraordinary mechanism to prove his result, which became the basis for much of modern number theory.[9]

In section 4.9 of appendix 4B we introduced the Riemann zeta-function and discussed the convergence of more general series. Dirichlet introduced the *Dirichlet L-functions*, in which the coefficient of $1/n^s$ is now $\chi(n)$, a Dirichlet character:

$$L(s, \chi) := \sum_{n \geq 1} \frac{\chi(n)}{n^s}.$$

Since each $|\chi(n)| \leq 1$, this is absolutely convergent whenever $\text{Re}(s) > 1$ (which can be proved by modifying the proof we gave for $\zeta(s)$ is section 5.12 of appendix 5B), and so the infinite sum is well-defined in this domain. It will require something more to define it for values of s on the line $\text{Re}(s) = 1$, or further to the left in the complex plane, and that something is provided by (8.16.2). The idea is that if χ has modulus q, then we can sum q terms at a time, which will allow us to make sense of the definition of $L(s, \chi)$ in a wider region. We will develop this in section 13.7 of appendix 13B. We will therefore be able to show that, for each non-principal character χ,

$$\boxed{\sum_{n \leq x} \frac{\chi(n)}{n} \text{ converges to } L(1, \chi) \text{ as } x \to \infty.}$$

---

[8] A careful study of the proofs in this section reveals that we use little more than the fact that $(\mathbb{Z}/n\mathbb{Z})^*$ is a multiplicative group to construct the set of characters $X(n)$. This theory, suitably modified, can be elegantly extended to all groups, even non-abelian groups. These ideas lead far beyond the scope of this book.

[9] The creativity needed to resolve a major question in number theory often yields tools that are so potent that they apply to a vast number of situations of which the original author-creator had no inkling.

8.17. Dirichlet series and primes in arithmetic progressions

One can generalize (5.12.1) to obtain

$$(8.17.1) \qquad \sum_{\substack{n \geq 1 \\ n \text{ a positive integer}}} \frac{\chi(n)}{n^s} = \prod_{p \text{ prime}} \left(1 - \frac{\chi(p)}{p^s}\right)^{-1}$$

when $\operatorname{Re}(s) > 1$. We have just seen that the left-hand side converges at $s = 1$ provided the terms are taken in order. This does not automatically tell us anything about the convergence of the infinite product on the right-hand side, at $s = 1$, but one can show that if $L(1, \chi) \neq 0$ (more on that in section 12.15 of appendix 12D when χ is a real non-principal character), then one can take logarithms in (8.17.1) to deduce that

$$(8.17.2) \qquad \boxed{\sum_{\substack{p \text{ prime} \\ p \leq x}} \frac{\chi(p)}{p} \text{ converges as } x \to \infty.}$$

Dirichlet attacked primes in arithmetic progressions by using (8.16.3) to identify primes p for which $p \equiv a \pmod{q}$. The trick is to rewrite this condition as $pa^{-1} \equiv 1 \pmod{q}$, so that $\chi(pa^{-1}) = \overline{\chi}(a)\chi(p)$, and therefore

$$\frac{1}{\phi(q)} \sum_{\chi \in X(q)} \overline{\chi}(a)\chi(p) = \begin{cases} 1 & \text{if } p \equiv a \pmod{q}, \\ 0 & \text{otherwise}. \end{cases}$$

This allowed Dirichlet to note that

$$\sum_{\substack{p \text{ prime}, p \leq x \\ p \equiv a \pmod{q}}} \frac{1}{p} = \sum_{\substack{p \text{ prime} \\ p \leq x}} \frac{1}{p} \cdot \frac{1}{\phi(q)} \sum_{\chi \in X(q)} \overline{\chi}(a)\chi(p) = \frac{1}{\phi(q)} \sum_{\chi \in X(q)} \overline{\chi}(a) \sum_{\substack{p \text{ prime} \\ p \leq x}} \frac{\chi(p)}{p}.$$

Now (8.17.2) implies that all of the terms on the right-hand side with χ non-principal are bounded. The sum in the χ_0-term is the sum of $1/p$ over all primes $p \leq x$ less a finite number of terms (the prime divisors p of q). Therefore

$$\sum_{\substack{p \text{ prime}, p \leq x \\ p \equiv a \pmod{q}}} \frac{1}{p} \text{ equals } \frac{1}{\phi(q)} \sum_{\substack{p \text{ prime}, \\ p \leq x}} \frac{1}{p} + \text{ a bounded quantity}.$$

Now, in Euler's proof of the infinitude of primes, (5.12.3), we saw that the sum $\sum_{p \leq x} 1/p$ diverges to ∞. Therefore

$$\boxed{\sum_{\substack{p \text{ prime}, p \leq x \\ p \equiv a \pmod{q}}} \frac{1}{p} \text{ diverges to } \infty \text{ as } x \to \infty;}$$

that is, there are infinitely many primes $p \equiv a \pmod{q}$. Moreover we can even deduce that

$$\sum_{\substack{p \text{ prime}, p \leq x \\ p \equiv a \pmod{q}}} \frac{1}{p} \bigg/ \sum_{\substack{p \text{ prime}, p \leq x \\ p \equiv b \pmod{q}}} \frac{1}{p} \to 1 \text{ as } x \to \infty$$

for any integers a, b which are coprime to q.

In these early results we "weight" each prime with $1/p$, which seems unnatural. This can be removed. Indeed, Dirichlet's proof can be combined with the ideas in the proof of the prime number theorem to show that for each $(a, q) = 1$

$$\frac{\#\{p \text{ prime}, \ p \leq x, \text{ and } p \equiv a \pmod{q}\}}{\#\{p \text{ prime}, \ p \leq x\}} \to \frac{1}{\phi(q)} \text{ as } x \to \infty.$$

In other words, if x is sufficiently large, there are roughly the same number of primes in each arithmetic progression $a \pmod{q}$ with $(a, q) = 1$. The proof of this will be a central theme in [**Graa**].

In the next two exercises one applies Dirichlet's Theorem on primes in arithmetic progressions.

Exercise 8.17.1. (a) Prove that there are infinitely many integers n for which $\mu(n) + \mu(n+1) = -1$.
(b) Prove that there are infinitely many integers n for which $\mu(n) + \mu(n+1) = 1$.

Exercise 8.17.2.[‡] (a) Given $\sigma \in \{-1, 1\}$ and prime q, show that there are infinitely many primes p for which $\left(\frac{q}{p}\right) = \sigma$.
(b) Given $\sigma_1, \ldots, \sigma_k \in \{-1, 1\}$ and primes $q_1, \ldots, q_k$, show that there are infinitely many primes p for which $\left(\frac{q_j}{p}\right) = \sigma_j$ for $j = 1, 2, \ldots, k$.
(c) Prove that for any given integer B there are infinitely many primes p such that there are no integers $n, 1 \leq n \leq B$, for which $\left(\frac{n}{p}\right) = \left(\frac{n+1}{p}\right) = \left(\frac{n+2}{p}\right)$.

We will return to the question of finding strings of consecutive quadratic residues with more sophisticated tools in section 14.6.

Uniformity questions

Above we noted that, following the work of Dirichlet and the proof of the prime number theorem, one knows that the primes are more or less equally distributed among the $\phi(q)$ reduced residue classes mod q; that is, if $(a, q) = 1$, then

$$(8.17.3) \qquad \#\{p \text{ prime}, \ p \leq x, \ p \equiv a \pmod{q}\} \sim \frac{\pi(x)}{\phi(q)}$$

as $x \to \infty$. The important issue in many applications is to estimate how large x needs to be for (8.17.3) to hold. To be precise how large does x need to be for the ratio of the two sides to be guaranteed to be between $1 - \eta$ and $1 + \eta$ for some very small $\eta > 0$? Calculations suggest that (8.17.3) holds once $x \geq q^{1+\epsilon}$, but there are no ideas around as to how to prove this. The best that (suitable) generalizations of the Riemann Hypothesis can offer is a range like $x \geq q^{2+\epsilon}$. The results that have been proved unconditionally are a lot weaker: that (8.17.3) holds in the range $x > \exp(q^\epsilon)$.

Appendix 8E. Quadratic reciprocity and recurrence sequences

8.18. The Fibonacci numbers modulo p

Using the binomial theorem, for each $n \geq 0$ we have

$$\frac{(x+y)^n - (x-y)^n}{2y} = \frac{1}{2y}\sum_{j=0}^{n}\binom{n}{j}x^{n-j}(y^j - (-y)^j) = \sum_{\substack{1\leq j\leq n \\ j \text{ odd}}}\binom{n}{j}x^{n-j}y^{j-1}.$$

Writing $j = 1 + 2k$, with $x = 1/2$ and $y = \sqrt{5}/2$ we deduce that

$$F_n = \frac{1}{2^{n-1}}\sum_{0\leq k\leq \frac{n-1}{2}}\binom{n}{2k+1}5^k.$$

Now if $n = p$, where p is prime, then p divides each $\binom{p}{2k+1}$ except when $k = \frac{p-1}{2}$. Therefore

$$F_p \equiv 5^{\frac{p-1}{2}} \equiv \left(\frac{5}{p}\right) \pmod{p},$$

using Euler's criterion. If $n = p+1$, then p divides each $\binom{p+1}{2k+1}$ except when $k = 0$ and $k = \frac{p-1}{2}$, so that

$$F_{p+1} \equiv \frac{1}{2^p}(p+1)(1 + 5^{\frac{p-1}{2}}) \equiv \frac{1}{2}\left(1 + \left(\frac{5}{p}\right)\right) \pmod{p}.$$

Therefore if $\left(\frac{5}{p}\right) = 1$, then $F_p \equiv F_{p+1} \equiv 1 \pmod{p}$, and $F_{p-1} = F_{p+1} - F_p \equiv 0 \pmod{p}$. Therefore $F_{p-1} \equiv F_0 \pmod{p}$, $F_p \equiv F_1 \pmod{p}$, and $F_{p+1} \equiv F_2 \pmod{p}$.

Exercise 8.18.1. Deduce that if $(5/p) = 1$, then $F_{n+p-1} \equiv F_n \pmod{p}$ for all $n \geq 0$.

On the other hand if $(5/p) = -1$, then $F_p \equiv -1 \pmod{p}$ and $F_{p+1} \equiv 0 \pmod{p}$, so that $F_{p+2} \equiv F_{p+3} \equiv -1 \pmod{p}$.

Exercise 8.18.2. Deduce that if $(5/p) = -1$, then $F_{n+2p+2} \equiv F_n \pmod{p}$ for all $n \geq 0$.

Exercise 8.18.3.† Let $x_0 = 0$, $x_1 = 1$, and $x_{n+2} = ax_{n+1} + bx_n$ for all $n \geq 0$, and let $\Delta := a^2 + 4b$. Suppose that prime p does not divide $a\Delta$.
   (a) Show that if $(\Delta/p) = 1$, then $x_{n+p-1} \equiv x_n \pmod{p}$ for all $n \geq 0$.
   (b) Show that if $(\Delta/p) = -1$, then $x_{n+p+1} \equiv -bx_n \pmod{p}$ for all $n \geq 0$.
   (c) Deduce that there exists a positive integer $d \leq p^2 - 1$ such that $x_{n+d} \equiv x_n \pmod{p}$ for all $n \geq 0$.

Exercise 8.18.4.† Let $x_0 = 0$, $x_1 = 1$, and $x_{n+2} = ax_{n+1} + bx_n$ for all $n \geq 0$, and let $\Delta := a^2 + 4b$.
   (a) Show that $x_n \equiv n(a/2)^{n-1} \pmod{\Delta}$ for all $n \geq 1$.
   (b) Deduce that if p divides Δ, then $x_{n+p} \equiv \frac{a}{2}x_n \pmod{p}$ for all $n \geq 0$.
   (c) Prove that if $p|a$, then $x_{n+p-1} \equiv \left(\frac{b}{p}\right)x_n \pmod{p}$ for all $n \geq 0$.

We can combine the previous two exercises to get the following useful general result (checking the $p = 2$ case carefully):

Corollary 8.18.1. *Let $x_0 = 0$, $x_1 = 1$, and $x_{n+2} = ax_{n+1} + bx_n$ for all $n \geq 0$. Let $\Delta := a^2 + 4b$. For every prime p we have*

$$x_{p - \left(\frac{\Delta}{p}\right)} \equiv 0 \pmod{p},$$

except if $p = 2$, in the cases for which a is odd and b is even.

Exercise 8.18.5.† Let $(x_n)_{n \geq 0}$ be as in exercise 8.18.3 with $(a, b) = 1$, and let p be a prime.
   (a) Prove that $(x_n, b) = 1$ for all $n \geq 0$.
   (b) Prove that $(x_n, x_{n+1}) = 1$ for all $n \geq 0$.
   (c) By considering the possible pairs $\{x_n, x_{n+1}\} \pmod{p}$, prove that there exists a positive integer $d \leq p^2 - 1$ such that $x_{n+d} \equiv x_n \pmod{p}$ for all $n \geq 0$.

Exercise 8.18.6.† Let r be the smallest integer ≥ 1 for which given prime p divides F_r.
   (a) Using exercises 0.4.10 and 2.5.19(c) to show that $F_{kr} \equiv rF_k F_{k+1}^{r-1} \pmod{F_k^2}$ and $F_{kr+1} \equiv F_{k+1}^r \pmod{F_k^2}$.
   (b) Suppose that $F_n \pmod{p}$ has period k. Deduce that $F_n \pmod{p^2}$ has period k or kp.
   (c) Deduce that the period of $F_n \pmod{p^2}$ divides $p(p-1)$ if $(5/p) = 1$, and it divides $2p(p+1)$ if $(5/p) = -1$.

8.19. General second-order linear recurrence sequences modulo p

Let $x_0 = 0$, $x_1 = 1$, and $x_n = ax_{n-1} + bx_{n-2}$ for all $n \geq 2$. In this section we think of $x_n \in \mathbb{Z}[a, b]$, a polynomial in a and b, and let $\Delta = a^2 + 4b$. We shall study the values of $x_n \pmod{p}$ in $\mathbb{Z}[a, b]$.

Theorem 8.11. *If p is an odd prime, then $x_p \equiv \Delta^{(p-1)/2} \pmod{p}$,*

$$x_{p-1} \equiv -2a \prod_{\substack{r \pmod{p} \\ (r/p)=1 \\ r \not\equiv 1 \pmod{p}}} (a^2 - r\Delta) \pmod{p},$$

and

$$x_{p+1} \equiv \frac{a}{2} \prod_{\substack{r \pmod{p} \\ (r/p)=-1}} (a^2 - r\Delta) \pmod{p}.$$

We deduce that
$$\Delta x_{p-1}x_{p+1} \equiv a^{2p} - a^2 b^{p-1} \equiv \prod_{k \pmod p} (a^2 - kb) \pmod p.$$

Proof. As above we expand
$$2^n x_n = \frac{1}{\sqrt{\Delta}}\left((a+\sqrt{\Delta})^n - (a-\sqrt{\Delta})^n\right)$$
using the binomial theorem, so that
$$2^p x_p = \sum_{k=1}^{p} \binom{p}{k} a^{p-k} \Delta^{(k-1)/2}(1-(-1)^k) \equiv 2\Delta^{(p-1)/2} \pmod p$$
as p divides the binomial coefficients $\binom{p}{k}$ for $1 \leq k \leq p-1$, and the first result follows by dividing through by 2 and using Fermat's Little Theorem. We also use the binomial expansion of the pth power to study x_{p-1} and x_{p+1} (mod p):

$$2^{p+1}x_{p+1} = \frac{1}{\sqrt{\Delta}}\left((a+\sqrt{\Delta})(a+\sqrt{\Delta})^p - (a-\sqrt{\Delta})(a-\sqrt{\Delta})^p\right)$$
$$= \sum_{k=1}^{p}\binom{p}{k}a^{p+1-k}\Delta^{(k-1)/2}(1-(-1)^k) + \sum_{k=0}^{p}\binom{p}{k}a^{p-k}\Delta^{k/2}(1+(-1)^k)$$
$$\equiv 2a\Delta^{(p-1)/2} + 2a^p \equiv 2a((a^2)^{(p-1)/2} + \Delta^{(p-1)/2}) \pmod p;$$

and multiplying through by $-4b = (a+\sqrt{\Delta})(a-\sqrt{\Delta})$ we also have
$$-2^{p+1}bx_{p-1} = \frac{1}{\sqrt{\Delta}}\left((a-\sqrt{\Delta})(a+\sqrt{\Delta})^p - (a+\sqrt{\Delta})(a-\sqrt{\Delta})^p\right)$$
$$= \sum_{k=1}^{p}\binom{p}{k}a^{p+1-k}\Delta^{(k-1)/2}(1-(-1)^k) - \sum_{k=0}^{p}\binom{p}{k}a^{p-k}\Delta^{k/2}(1+(-1)^k)$$
$$\equiv 2a\Delta^{(p-1)/2} - 2a^p \equiv -2a((a^2)^{(p-1)/2} - \Delta^{(p-1)/2}) \pmod p.$$

The next two parts of the result now follow from (8.2.1) and (8.2.2). The final part comes from multiplying together these last two displayed equations to obtain
$$4^p b \Delta x_{p-1} x_{p+1} \equiv (a^2)^p \Delta - a^2 \Delta^p$$
$$\equiv a^{2p}(a^2 + 4b) - a^2(a^{2p} + 4b^p) = 4(a^{2p}b - a^2 b^p) \pmod p,$$
and the result follows by dividing through by $4b$ (if p does not divide b).

In the case that p divides b we have $x_n \equiv a^{n-1} \pmod p$ for all $n \geq 1$, and so $\Delta x_{p-1}x_{p+1} \equiv a^2 \cdot a^{p-2} \cdot a^p \equiv a^{2p} \pmod p$. □

8.20. Prime values in recurrence sequences

It is conjectured that there are infinitely many Mersenne primes $2^p - 1$ as well as Fibonacci primes F_p. There are 33 known Fibonacci primes. The first few are $F_3 = 2$, $F_4 = 3$, $F_5 = 5$, $F_7 = 13$, $F_{11} = 89$, $F_{13} = 233$, $F_{17} = 1597$, $F_{23} = 28657, \ldots$. Notice that $F_{19} = 4181 = 37 \times 113$ is composite.

Exercise 8.20.1. Assume that p and $q = 2p + 1$ are both prime. Deduce that q divides $2^p - 1$ whenever $p \equiv 3 \pmod 4$, and so $2^p - 1$ is not a Mersenne prime.

In chapter 3 we applied exercise 3.9.2 to the Mersenne numbers $M_n = 2^n - 1$, with $m = 1$, so that if M_n is prime, then n is prime; and to Fibonacci numbers with $m = 2$ so that if F_n is prime, then n is prime or $n = 4$. We can streamline the search for other prime factors of Mersenne numbers by using exercise 8.4.4:

"The hundred-year-old prime": Euler proved that the Mersenne number $M_{31} = 2147483647$ is prime, as follows: If M_{31} is composite and q is the smallest prime dividing $2^{31} - 1$, then $31 | q - 1$ and $q \equiv \pm 1 \pmod 8$, by exercise 8.4.4(a). Hence $q \equiv 1$ or $63 \pmod{248}$ with $q < \sqrt{M_{31}} < 2^{31/2} < 46341$. There are 84 such primes q (as Euler discovered by looking at tables of primes) and then he verified that none of them divide M_{31}. Hence M_{31} is prime. This remained the largest prime known for over a hundred years. As of January 1, 2019, nine of the ten largest known prime numbers are Mersenne primes (the largest being $M_{82589933}$).

Should we believe that there are infinitely many Mersenne primes? A standard heuristic goes as follows: From Gauss's remarks (see sections 5.4 and 5.15 of appendix 5C) we believe that a randomly chosen integer around x is prime with probability $1/\log x$. Obviously the M_p are actual given numbers and are not randomly chosen, but if we suppose that their primality is about as likely as that of a random number, then we would guess that the number of primes M_p with $p \leq x$ is roughly

$$\sum_{p \leq x} \frac{1}{\log(2^p - 1)} \approx \frac{1}{\log 2} \sum_{p \leq x} \frac{1}{p}.$$

Now this $\to \infty$ by (5.12.3) and even grows like $\log \log x$, as in (5.12.4). However this heuristic is not supported by the data: There are 51 known Mersenne primes and M_p is known to be composite for around $2\frac{1}{2}$ million primes p. The heuristic would predict no more than three Mersenne primes up to this point!

We can modify the heuristic to take account of the fact that the prime factors of $2^p - 1$ are all $\equiv 1 \pmod p$ and in particular are all $> p$. The probability that an integer around (sufficiently large) x, that is not divisible by any prime $\leq p$, is prime is around $\frac{c \log p}{\log x}$ for some constant $c > 0$. This alters the sum in our heuristic to

$$c \sum_{p \leq x} \frac{\log p}{p} \approx c \log x.$$

(This last approximation can be deduced from (5.12.4).) This seems much more compatible with the known data; indeed one might even guess that c is around 2.55.

Chapter 9

Quadratic equations

Can we tell whether a given large integer is the sum of two squares of integers (other than by summing every possible pair of smaller squares)? How about the values of other quadratics? We will show, in this chapter, how we can understand a lot about solutions to quadratic equations in integers, by understanding the solutions to those quadratic equations modulo p, for every prime p. We begin by studying the values taken by $x^2 + y^2$ when we substitute integers in for x and y, then $ax^2 + by^2$ for arbitrary integer coefficients a, b, and then finally the general *binary quadratic form*, $ax^2 + bxy + cy^2$.

9.1. Sums of two squares

The list of integers that are the sum of two squares of integers begins:

$$0, 1, 2, 4, 5, 8, 9, 10, 13, 16, 17, 18, 20, 25, 26, 29, 32, 34, 36, 37, 40, 41, 45, 49, 50, \ldots.$$

Is there a pattern? Can we easily determine whether a given integer is the sum of two squares by any means other than trying to find two squares that sum to it? No pattern emerges easily from the list above so we begin focusing on the primes that appear in this list, namely

$$2 = 1^2 + 1^2,\ 5 = 1^2 + 2^2,\ 13 = 2^2 + 3^2,\ 17 = 1^2 + 4^2,\ 29 = 5^2 + 2^2,\ 37 = 1^2 + 6^2, \ldots.$$

What do the odd primes in the list, $5, 13, 17, 29, 37, 41, 53, 61, 73, 89, 97, \ldots$ have in common? The only easy-to-spot pattern is that the differences between consecutive odd primes in our list, $13-5, 17-13, 29-17, \ldots$ are all multiples of 4, which implies that they are *all* $\equiv 1 \pmod 4$.

Proposition 9.1.1. *If p is an odd prime that is the sum of two squares, then $p \equiv 1$* (mod 4).

Proof. If $p = a^2 + b^2$, then $p \nmid a$, or else $p | p - a^2 = b^2$ so that $p | b$ and $p^2 | a^2 + b^2 = p$, which is impossible. Similarly $p \nmid b$. Now $a^2 \equiv -b^2 \pmod{p}$ so that

$$1 = \left(\frac{a}{p}\right)^2 = \left(\frac{-1}{p}\right)\left(\frac{b}{p}\right)^2 = \left(\frac{-1}{p}\right),$$

and therefore $p \equiv 1 \pmod 4$ by Theorem 8.3. □

Exercise 9.1.1. Prove that any odd integer n that can be written as the sum of two squares must be $\equiv 1 \pmod 4$. Deduce Proposition 9.1.1.

Exercise 9.1.2. Prove that if prime p divides $a^2 + b^2$, then either $p = 2$ or p divides (a, b) or $p \equiv 1 \pmod 4$.

Remarkably this is an "if and only if" condition:

Theorem 9.1. *Every prime $p \equiv 1 \pmod 4$ can be written as the sum of two squares (of integers).*

Proof. Since $p \equiv 1 \pmod 4$ we know that there exists an integer b such that $b^2 \equiv -1 \pmod p$. Consider now the set of integers

$$\{j + kb : \ 0 \leq j, k \leq [\sqrt{p}]\}.$$

The number of pairs of integers j, k used in the construction of this set is $([\sqrt{p}] + 1)^2 > p$, and so by the pigeonhole principle, two of the numbers in the set must be congruent mod p; say that

$$j + kb \equiv J + Kb \pmod p$$

where $0 \leq j, k, J, K \leq [\sqrt{p}]$ and $\{j, k\} \neq \{J, K\}$. Let $r = j - J$ and $s = K - k$ so that

$$r \equiv bs \pmod p$$

where $|r|, |s| \leq [\sqrt{p}] < \sqrt{p}$ and r and s are not both 0. Now

$$r^2 + s^2 \equiv (bs)^2 + s^2 = s^2(b^2 + 1) \equiv 0 \pmod p,$$

and $0 < r^2 + s^2 < \sqrt{p}^2 + \sqrt{p}^2 = 2p$. The only multiple of p between 0 and $2p$ is p, and therefore $r^2 + s^2 = p$. □

We will use the identity

(9.1.1) $$(a^2 + b^2)(c^2 + d^2) = (ac - bd)^2 + (ad + bc)^2$$

to determine which composite integers can be written as the sum of two squares. Theorem 9.1 tells us that any prime $p \equiv 1 \pmod 4$ can be written as the sum of two squares; for example $5 = 1^2 + 2^2$ and $13 = 2^2 + 3^2$. Then (9.1.1) yields that $65 = 4^2 + 7^2$; if we write instead $13 = 3^2 + 2^2$, then we obtain $65 = 1^2 + 8^2$. Indeed any integer that is the product of two distinct primes $\equiv 1 \pmod 4$ can be written as the sum of two squares like this, and even in two different ways. We will discuss the number of representations further in appendix 9C.

Exercise 9.1.3. Find four distinct representations of $1105 = 5 \times 13 \times 17$ as a sum of two squares.

Exercise 9.1.4. Prove that if $n = n_1 \cdots n_k$ where $n_1, \ldots, n_k$ are each the sum of two squares, then n is the sum of two squares.

9.1. Sums of two squares

Theorem 9.2. *Positive integer n can be written as the sum of two squares of integers if and only if for every prime $p \equiv 3 \pmod 4$ which divides n, the exact power of p dividing n is even.*

Proof. Suppose that $n = a^2 + b^2$ where $g = (a,b)$, so we can write $a = gA$, $b = gB$, and $n = g^2 N$ for some coprime integers A and B, with $N = A^2 + B^2$. Therefore if p is a prime $\equiv 3 \pmod 4$, then p cannot divide N, by exercise 9.1.2; and so if $p|n$, then $p|g$. Moreover if $p^k \| g$, then $p^{2k} \| n$, as claimed.

On the other hand, if $n = g^2 m$ where m is squarefree, then m has no prime factors $\equiv 3 \pmod 4$ by the hypothesis. Therefore all the prime factors of m can be written as the sum of two squares by Theorem 9.1, and so their product, m, is the sum of two squares by exercise 9.1.4, say $m = u^2 + v^2$. Then $n = (gu)^2 + (gv)^2$. □

Exercise 9.1.5. Prove that if n is squarefree and is the sum of two squares, then every positive divisor of n is also the sum of two squares.

We saw that (9.1.1) is a useful identity. To find such an identity let i be a complex number for which $i^2 = -1$. Then $x^2 + y^2 = (x+iy)(x-iy)$, a factorization into numbers of the form $a + bi$ where a and b are integers. Therefore

$$\begin{aligned}(a^2+b^2) \cdot (c^2+d^2) &= (a+bi)(a-bi) \cdot (c+di)(c-di) \\ &= (a+bi)(c+di) \cdot (a-bi)(c-di) \\ &= ((ac-bd) + (ad+bc)i) \cdot ((ac-bd) - (ad+bc)i) \\ &= (ac-bd)^2 + (ad+bc)^2,\end{aligned}$$

and so we get (9.1.1). A different rearrangement leads to a different identity:

(9.1.2) $(a^2+b^2)(c^2+d^2) = (a+bi)(c-di) \cdot (a-bi)(c+di) = (ac+bd)^2 + (ad-bc)^2$.

Theorem 9.2 has the following surprising corollary:

Exercise 9.1.6. Deduce that positive integer n can be written as the sum of two squares of *rationals* if and only if n can be written as the sum of two squares of integers.

This suggests that we can focus, in this question, on rational solutions. In section 6.1 we saw how to find all solutions to $x^2 + y^2 = 1$ in rationals x, y. How about all rational solutions to $x^2 + y^2 = n$?

Proposition 9.1.2. *Suppose that $n = a^2 + b^2$. Then all solutions in rationals x, y to $x^2 + y^2 = n$ are given by the parametrization*

(9.1.3) $$x = \frac{2brs + a(r^2-s^2)}{r^2+s^2}, \quad y = \frac{2ars + b(s^2-r^2)}{r^2+s^2},$$

where r and s are coprime integers.

Proof. Let x, y be any rationals for which $x^2 + y^2 = n$. Just as in our geometric proof of (6.1.1) we will parametrize these rational points (x,y) by noting that if t is the slope of the line between (a,b) and (x,y), then t is rational, and vice versa. In particular we let $u = x - a$ and $t = (y-b)/u$ when $u \neq 0$, which must both be rational numbers. Then

$$0 = n - n = (a+u)^2 + (b+tu)^2 - (a^2+b^2) = 2u(a+bt) + u^2(1+t^2),$$

so that, as $u \neq 0$, we have
$$u = \frac{-2(a+bt)}{1+t^2} = \frac{2brs - 2as^2}{r^2 + s^2}$$
writing the rational number t as $t = -r/s$ where r and s are coprime integers. Substituting this value of u into $x = a + u$ and $y = b + ut$ gives the claimed parametrization.

If $u = 0$, then $x = a$ so that either $y = b$ or $y = -b$. The line between (a, b) and $(a, -b)$ is the vertical line $x = a$ (corresponding to $r = 1, s = 0$ so that $t = \infty$).

Finally we obtain the initial point (a, b) in this parametrization by taking $r = a, s = b$. This is obtained by taking the slope to be $t = -a/b$, the slope of the tangent line to the curve $x^2 + y^2 = n$ at the point (a, b). $\square$

In Theorem 9.1 we saw that every prime $p \equiv 1 \pmod{4}$ can be written as the sum of two squares. Examples suggest that there is a unique such representation, up to signs and changing the order of the squares, as the reader will now prove:

Exercise 9.1.7.† Suppose that prime $p = a^2 + b^2$.
  (a) Prove that $|a|, |b| < \sqrt{p}$.
  (b) Prove that if $r^2 \equiv -1 \pmod{p}$, then either $r \equiv a/b \pmod{p}$ or $r \equiv b/a \pmod{p}$.
  (c) If prime p divides $c^2 + d^2$ but $p \nmid cd$, show that p divides either $ac - bd$ or $ad - bc$, and deduce that p divides both terms on the right-hand side of either (9.1.1) or (9.1.2), respectively.
  (d) Suppose that $p = a^2 + b^2 = c^2 + d^2$ where $a, b, c, d > 0$. Show that $\{a, b\} = \{c, d\}$.

In other words, we have proved that each prime $\equiv 1 \pmod{4}$ has a unique representation as the sum of two squares, unique up to changing the order of the squares, or their signs.

Exercise 9.1.8.† Prove, using the method of Theorem 9.1, that a squarefree integer n can be written as the sum of two squares if and only if -1 is a square mod n.

9.2. The values of $x^2 + dy^2$

What values does $x^2 + 2y^2$ take? Let's start again with the prime values:
$$2, 3, 11, 17, 19, 41, 43, 59, 67, 73, 83, 89, 97, \ldots.$$
There is no obvious pattern; *but* this list contains exactly the same odd primes that we found in section 8.4 when exploring when $\left(\frac{-2}{p}\right) = 1$. This link is no coincldence for if we suppose that odd prime $p = x^2 + 2y^2$, then p does not divide x or y and so
$$1 = \left(\frac{x}{p}\right)^2 = \left(\frac{x^2}{p}\right) = \left(\frac{-2y^2}{p}\right) = \left(\frac{-2}{p}\right)\left(\frac{y}{p}\right)^2 = \left(\frac{-2}{p}\right).$$
From (8.7.1) and (8.7.2), we know that $\left(\frac{-2}{p}\right) = 1$ if and only if $p \equiv 1$ or $3 \pmod{8}$.

On the other hand if $(-2/p) = 1$, then select $b \pmod{p}$ such that $b^2 \equiv -2 \pmod{p}$. We take $R = 2^{1/4}\sqrt{p}, S = 2^{-1/4}\sqrt{p}$ in exercise 9.7.3, so that there exist integers r and s, not both 0, with $|r| \leq R$ and $|s| \leq S$, for which p divides $r^2 + 2s^2$. Therefore $0 < r^2 + 2s^2 \leq 2^{3/2}p < 3p$, and so $r^2 + 2s^2 = p$ or $2p$. In the latter case, 2 divides $2p - 2s^2 = r^2$ so that $2|r$. Writing $r = 2R$ we have $s^2 + 2R^2 = p$. Hence, either way, p can be written in the form $m^2 + 2n^2$. Therefore we have proved:

Theorem 9.3. *Odd prime p can be written in the form $m^2 + 2n^2$ if and only if $p \equiv 1$ or $3 \pmod{8}$.*

The identity
$$(a^2 + 2b^2)(c^2 + 2d^2) = (ac + 2bd)^2 + 2(ad - bc)^2$$
is analogous to (9.1.1). Using this, one can prove, analogous to the proof for $u^2 + v^2$ in the first half of section 9.1, that positive integer n can be written as $r^2 + 2s^2$ if and only if for every prime $p \equiv 5$ or $7 \pmod 8$ which divides n, the exact power of p dividing n is even.

Can we also modify this proof for values of $x^2 + 3y^2$? Or $x^2 + 5y^2$? We explore this in the following exercises.

Exercise 9.2.1. Fix integer $d \geq 1$. Give an identity showing that the product of two integers of the form $a^2 + db^2$ is also of this form.

Exercise 9.2.2. Which primes are of the form $a^2 + 3b^2$? Which integers?

Exercise 9.2.3. Which primes are of the form $a^2 + 5b^2$? Try listing what primes are represented and compare the list with the set of primes p for which $(-5/p) = 1$.

9.3. Is there a solution to a given quadratic equation?

It is easy to see that there do not exist non-zero integers a, b, c such that $a^2 + 5b^2 = 3c^2$, for, if we take the smallest non-zero solution, then we have
$$a^2 \equiv 3c^2 \pmod 5$$
which implies that $a \equiv c \equiv 0 \pmod 5$ since $(3/5) = -1$, and so $b \equiv 0 \pmod 5$. Therefore $a/5, b/5, c/5$ gives a smaller solution to $x^2 + 5y^2 = 3z^2$, contradicting minimality.

Another proof stems from looking at the equation mod 4 since then $a^2 + b^2 + c^2 \equiv 0 \pmod 4$, and 0 and 1 are the only squares mod 4. Therefore if three squares sum to an integer that is 0 (mod 4), then they must all be even. But then $a/2, b/2, c/2$ gives a smaller solution, contradicting minimality.

So we have now presented two different proofs that there are no non-zero solutions in integers to $a^2 + 5b^2 = 3c^2$, by working with two different moduli.

For all quadratic equations in three or more variables with real solutions, there is never just one prime or prime power modulo which there are no solutions to the given equation—when there is one, there is always a second. And indeed when there is a third proof, then there is always a fourth. A remarkable consequence of the theory (see appendix 9B) is that if a given quadratic equation in three or more variables has non-zero real solutions but no non-zero integer solutions, then there are always exactly *an even number of different primes* p such that the given equation has no non-trivial solutions mod p^k for some $k \geq 1$. Moreover the odd primes involved must divide the coefficients of the equation. On the other hand, if there are no such "mod p^k obstructions", then there must be at least one non-zero integer solution (implying that there must be a real solution!).

In exercise 3.6.4 we proved that there are integer solutions (m, n) to $am + bn = c$ if and only if there are solutions $u, v \pmod b$ to $au + bv \equiv c \pmod b$. Similarly we will show that if $a, b,$ and c are pairwise coprime, positive integers, then there are rational solutions (x, y) to $ax^2 + by^2 = c$ if and only if there are coprime solutions $u, v \pmod{4abc}$ to $au^2 + bv^2 \equiv c \pmod{4abc}$. This is an amazing theorem since to determine whether a quadratic equation has solutions in rationals we need only verify whether it has solutions modulo a finite modulus.

To work on rational solutions (x, y) to $ax^2 + by^2 = c$ it is convenient to develop this into a question about integer solutions and to manipulate the equation to a more convenient form:

(i) We may assume that each of a, b, c is a squarefree integer or else, if, say, $a = p^2 A$, the rational solutions to $ax^2 + by^2 = c$ are in 1-to-1 correspondence with those of $AX^2 + by^2 = c$, taking $X = px$. If b is divisible by a square, we proceed analogously. If $c = q^2 C$, then the rational solutions to $ax^2 + by^2 = c$ are in 1-to-1 correspondence with those of $aX^2 + bY^2 = C$, taking $X = x/q$ and $Y = y/q$.

(ii) We may assume that a, b, c are pairwise coprime or else if, say, $a = pA$ and $b = pB$, then $AX^2 + BY^2 = C$ with $X = px$, $Y = py$, and $C = pc$; and if $a = qA$ and $c = qC$, then $Ax^2 + BY^2 = C$ with $B = bq$ and $Y = y/q$.

(iii) Letting n be the lowest common denominator of the rationals x and y, we write $x = \ell/n$ with $y = m/n$ so that ℓ, m, n are integers with $(\ell, m, n) = 1$ and $a\ell^2 + bm^2 = cn^2$.

(iv) We may assume that $a\ell^2, bm^2, cn^2$ are pairwise coprime. If not, suppose that prime p divides $a\ell^2$ and bm^2, so that p divides $a\ell^2 + bm^2 = cn^2$. Now p can only divide one of a, b, c (since they are pairwise coprime), say, c, and so must divide ℓ^2 and m^2. But then p divides ℓ and m, and so p^2 divides $a\ell^2 + bm^2 = cn^2$. Hence p divides n, as $p^2 \nmid c$, contradicting that $(\ell, m, n) = 1$.

Therefore the correct formulation of our result is as follows:

Theorem 9.4 (The local-global principle for quadratic equations). *Let a, b, and c be given pairwise coprime, squarefree integers. There are solutions in*

Non-zero integers ℓ, m, n to $a\ell^2 + bm^2 + cn^2 = 0$ with $(a\ell^2, bm^2) = 1$

if and only if there are solutions in

Non-zero real numbers λ, μ, ν to $a\lambda^2 + b\mu^2 + c\nu^2 = 0$,

and, for all positive integers r, there exist

Residue classes $u, v, w \pmod{r}$ for which $au^2 + bv^2 + cw^2 \equiv 0 \pmod{r}$,

with $(au^2, bv^2, cw^2, r) = 1$.

Proof $\Longrightarrow$: We may take $\lambda = u = \ell$, $\mu = v = m$, $\nu = w = n$ throughout. □

The proof in the other direction is the difficult part; it follows along the lines of the proof of Theorem 9.1 but is more complicated. In appendix 9a we rephrase that proof in the language of lattices, before completing the proof of the local-global principle.

We can reduce the set of moduli to be considered using the following lemma.

Lemma 9.3.1. *Let a, b, c be given pairwise coprime, squarefree integers. There are residue classes $u, v, w \pmod{r}$ with $(au^2, bv^2, cw^2, r) = 1$ for which*

$$au^2 + bv^2 + cw^2 \equiv 0 \pmod{r}$$

for every positive integer r, if and only if there are such solutions for $r = 8$, and for $r = p$ for every odd prime p dividing abc.

This result implies that, as in exercise 3.6.4, we can restrict our attention in Theorem 9.4 to just one modulus, namely $r = 8|abc|$.

Proof. We can restrict our attention to prime power moduli p^k by the Chinese Reminder Theorem. We will prove that there are such appropriate solutions mod p^k by induction on k: for $k \geq 1$ when p is odd and for $k \geq 3$ when $p = 2$. There are appropriate solutions modulo every odd prime p and modulo 2^3, by the hypothesis for primes p dividing $2abc$, and by exercise 8.9.9 for all odd primes p that do not divide abc.

So now assume we have an appropriate solution mod p^k, so that p does not divide at least one of au^2, bv^2, cw^2, say, au^2 (and an analogous argument works if p does not divide one of the others). Let $R = -a^{-1}(bv^2 + cw^2)$, so that $u^2 \equiv R \pmod{p^k}$ by the induction hypothesis. By Proposition 8.8.1 there exists $U \pmod{p^{k+1}}$ for which $U^2 \equiv R \pmod{p^{k+1}}$ so that $aU^2 + bv^2 + cw^2 \equiv 0 \pmod{p^{k+1}}$ and $(U, p) = 1$. □

Now if $au^2 + bv^2 + cw^2 \equiv 0 \pmod{a}$ with $(a, bv^2, cw^2) = 1$, then $-bc \equiv (cw/v)^2 \pmod{a}$; that is, $-bc$ is a square $\pmod{p}$ for every prime dividing a. Making similar remarks modulo b and c, we find Legendre's formulation of the local-global principle.[1]

Theorem 9.5 (Legendre's local-global principle, 1785). *Let a, b, c be given pairwise coprime, squarefree integers which do not all have the same sign. There are solutions in non-zero integers ℓ, m, n to $a\ell^2 + bm^2 + cn^2 = 0$ if and only if $-ab$ is a square mod $|c|$, $-ac$ is a square mod $|b|$, and $-bc$ is a square mod $|a|$.*

Note that $a\ell^2 + bm^2 + cn^2 = 0$ has solutions in non-zero reals if and only if a, b, c do not all have the same sign.

This principle may be extended to the rational solutions of more or less any quadratic equation: Any quadratic polynomial in n variables can be *diagonalized*; that is, a linear change of variables can change the polynomial into a diagonal quadratic polynomial. We know that in the example $g = ax^2 + bxy + cy^2$ we can let $X = x + by/2a$ and then $g = aX^2 + Dy^2$ where $D = -(b^2 - 4ac)/4a$. In a three-variable example we take the polynomial

$$f = x^2 + 2xy + 3xz + 4y^2 + 5yz + 6z^2 + 7x + 8y + 9z + 10;$$

we let $X = x + y + \frac{3}{2}z + \frac{7}{2}$ replace x to obtain $f = X^2 + 3y^2 + 2yz + \frac{15}{4}z^2 + y - \frac{3}{2}z - \frac{9}{4}$. Then letting $Y = y + \frac{z}{3} + \frac{1}{6}$ we obtain $f = X^2 + 3Y^2 + \frac{41}{12}z^2 - \frac{11}{6}z - \frac{7}{3}$, and if $z = 6Z + \frac{11}{41}$, this becomes

$$F = X^2 + 3Y^2 + 123Z^2 - \frac{423}{164},$$

---

[1] The careful reader will note that we do not seem to have made adequate remarks about the solution modulo powers of 2. However, we noted earlier in this section that if there are solutions in the reals and modulo all but one prime, then there is a solution modulo all powers of this last prime. For more details see appendix 9B.

a diagonal quadratic with no "cross terms" (like XY). Notice that the rational solutions to $F(X,Y,Z) = 0$ are in 1-to-1 correspondence with the rational solutions to $f(x,y,z) = 0$.

Whether or not a given diagonal quadratic with three or more terms has rational solutions can then be resolved by the local-global principle.[2]

Exercise 9.3.1. Given one integer solution to $ax_0^2 + by_0^2 + cz_0^2 = 0$, show that all other integer solutions to $ax^2 + by^2 + cz^2 = 0$ are given by the parametrization
$$x : y : z \,=\, (ar^2 - bs^2)x_0 + 2brsy_0 \,:\, 2arsx_0 - (ar^2 - bs^2)y_0 \,:\, (ar^2 + bs^2)z_0 \,.$$

9.4. Representation of integers by $ax^2 + by^2$ with x, y rational, and beyond

Coprime integer solutions to $au^2 + bv^2 = cw^2$ with $w > 0$ are in 1-to-1 correspondence with the rational solutions to $ax^2 + by^2 = c$, by taking $x = u/w$ and $y = v/w$. Therefore the local-global principle can be restated to give an "if and only if" criterion to determine whether c can be written as $ax^2 + by^2$ with x and y rational. This is most usefully modified as follows:

Corollary 9.4.1. *Suppose that a, b, c are given integers with $(a,b,c) = 1$, and suppose $d = b^2 - 4ac$ is not divisible by the square of any odd prime. For any given squarefree integer N with $(N, d) = 1$, there exist rationals u and v for which $N = au^2 + buv + cv^2$ if and only if the following criteria hold:*

- *N has the same sign as a or c, or $d > 0$;*
- *d is a square mod N;*
- *$\left(\frac{N}{p}\right) = \left(\frac{a}{p}\right)$ for all odd primes p dividing d that do not divide a;*
- *$\left(\frac{N}{p}\right) = \left(\frac{c}{p}\right)$ for all odd primes p dividing both d and a.*

Proof. If $N = au^2 + buv + cv^2$, then we multiply through by $4a$ to obtain $4aN = (2au + bv)^2 - dv^2$; in other words, $aN = U^2 - dV^2$ for some rationals U, V. We may reverse this argument, and so there exist rationals u and v for which $N = au^2 + buv + cv^2$ if and only if there exist rationals U, V for which $aN = U^2 - dV^2$. We now apply Legendre's version of the local-global principle to rational solutions to the equation $aN = u^2 - dv^2$.

We have real solutions if and only if $aN > 0$ or $d > 0$.

Now $U^2 \equiv dV^2 \pmod{aN}$ and so d must be a square mod aN. But $d = b^2 - 4ac \equiv b^2 \pmod{a}$, so we need only verify that d is a square mod N.

If odd prime p divides d, then $aN \equiv u^2 \pmod{p}$, and so $\left(\frac{N}{p}\right) = \left(\frac{a}{p}\right)$ if p does not divide a.

If odd prime p divides both d and a, then it divides b, as it divides $b^2 = d + 4ac$. Therefore p does not divide c as $(a,b,c) = 1$. We then run through the analogous argument with a replaced by c. (For the primes p dividing d, but not $4ac$, our results that $\left(\frac{N}{p}\right) = \left(\frac{a}{p}\right)$ and $\left(\frac{N}{p}\right) = \left(\frac{c}{p}\right)$ are consistent; see exercise 8.1.4.) □

---

[2] Which we have only proved in three variables but is true in three or more variables.

9.5. The failure of the local-global principle for quadratic equations in integers

We have seen how the local-global principle allows us to determine whether there are *rational solutions* x, y to a given equation of the form $ax^2 + by^2 = c$. However we will now show that it does not help when we ask for *integer solutions*. The example
$$x^2 + 23y^2 = 52$$
has rational solutions, like $(\frac{1}{2}, \frac{3}{2}), (\frac{25}{4}, \frac{3}{4}), (\frac{29}{12}, \frac{17}{12}), \ldots$. There are obviously no integer solutions or else $23y^2 \leq x^2 + 23y^2 = 52$ and so $y^2 = 0$ or 1, but then $x^2 = 52 - 23y^2 = 52$ or 29, which are not squares. Since there are rational solutions we know that there are non-trivial solutions to $a^2 + 23b^2 \equiv 52c^2 \pmod{p^k}$ for all prime powers p^k by the local-global principle, but not necessarily to $a^2 + 23b^2 \equiv 52 \pmod{p^k}$. To prove that there are such solutions, we show that solutions exist modulo 8 and all odd prime moduli p, and then we lift these solutions to all prime power moduli p^k, using Proposition 8.8.1.

We have the solutions $2^2 + 23 \cdot 4^2 = 372 \equiv 52 \pmod{8}$, $4^2 + 23 \cdot 1^2 = 39 \equiv 52 \pmod{13}$, and $11^2 + 23 \cdot 0^2 = 121 \equiv 52 \pmod{23}$. For any odd prime p other than 13 or 23, there are $\frac{p+1}{2}$ residues mod p of the form $23y^2$, and $\frac{p+1}{2}$ residues mod p of the form $52 - x^2$, so two of these residues must be equal. Therefore there is a solution to $x^2 + 23y^2 \equiv 52 \pmod{p}$, and evidently one of x and y must be non-zero $\pmod{p}$ (or else p would divide 52).

Therefore we have shown that the local-global principle holds for integer and rational solutions of linear equations, and for rational but not integer solutions of quadratic equations. However it does not even hold for rational solutions of cubic equations: In 1957, Selmer showed that $3x^3 + 4y^3 = 5$ has solutions in the reals, and mod r for all $r \geq 1$, yet has no rational solutions. Further discussion of the failure of the local-global principle for cubic equations can be found in [**Grab**], with a motivating discussion in chapter 7.

9.6. Primes represented by $x^2 + 5y^2$

Calculations reveal that the primes > 5 that are represented by $x^2 + 5y^2$ are
$$29, 41, 61, 89, 101, 109, 149, 181, \ldots.$$
From our explorations of the binary quadratic forms $x^2 + y^2$, $x^2 + 2y^2$, and $x^2 + 3y^2$ we might guess that this should be the set of primes for which $(-5/p) = 1$. However the list of primes for which $(-5/p) = 1$ also includes the primes
$$3, 7, 23, 43, 47, 67, 83, 103, 107, 127, 163, 167, \ldots.$$
What is going on? We quickly see that the primes in the first list end in a 1 or a 9, whereas the primes in the second list end in a 3 or a 7, so there seems to be a further congruence condition that partitions the list. Further examination of the equation $p = x^2 + 5y^2$ makes this evident: Besides $(-5/p) = 1$, we can also deduce that $p \equiv x^2 \pmod{5}$ so that $(p/5) = 1$. Combined with $(-5/p) = 1$, this also yields that $p \equiv 1 \pmod{4}$. These two conditions together give that $p \equiv 1$ or $9 \pmod{20}$,

the primes that we see in the first list, and if $(p/5) = -1$, then we obtain $p \equiv 3$ or $7 \pmod{20}$, the primes that we see in the second list.

Where do the primes in the second list come from? It turns out there is a second, fundamentally different binary quadratic form, $2x^2 + 2xy + 3y^2$, which has the same discriminant -20 as $x^2 + 5y^2$. We first observe that these quadratic forms definitely do not represent the same integers because $2x^2 + 2xy + 3y^2$ represents 3, whereas $x^2 + 5y^2$ evidently does not. A quick calculation reveals that the second list is precisely the set of odd primes represented by $2x^2 + 2xy + 3y^2$. This dichotomy will be explored further in chapter 12, though we observe here that if prime $p = 2x^2 + 2xy + 3y^2$, then $2p = 4x^2 + 4xy + 6y^2 = (2x+y)^2 + 5y^2$; that is, $2p$ can be represented by $a^2 + 5b^2$.

In general if we wish to represent the odd prime p by $x^2 + dy^2$, then $-d$ must be a square mod p. On the other hand, suppose that $-d$ is a square mod p, say $u^2 \equiv -d \pmod{p}$ with $|u| < p/2$.

If $p < 2\sqrt{d}$, then we can write $u^2 + d = ap$, so the binary quadratic form, $pm^2 + 2umn + an^2$, has discriminant $-4d$, the same as $x^2 + dy^2$, and takes the value p when $m = 1, n = 0$.

Now assume that $p > 2\sqrt{d}$. By exercise 9.7.3(a) with $R = d^{1/4}\sqrt{p}, S = d^{-1/4}\sqrt{p}$, there exist integers r and s, not both 0, for which $r \equiv us \pmod{p}$ and so, squaring, $r^2 \equiv -ds^2 \pmod{p}$; that is, $r^2 + ds^2$ is a multiple of p. Moreover we have $0 < r^2 + ds^2 \le R^2 + dS^2 = 2\sqrt{d}p$. Therefore there exists an integer a in the range $1 \le a \le 2\sqrt{d}$ for which

$$r^2 + ds^2 = ap.$$

We may assume that $(r, s) = 1$ for if $g = (r, s)$, then we claim that g^2 divides a, so we can divide r and s through by g. To justify our claim, note that g^2 divides $r^2 + ds^2 = ap$ so if g^2 does not divide a, then p divides g. But then $p^2 \le g^2 \le r^2 + ds^2 = ap$ and so $p \le a \le 2\sqrt{d}$, a contradiction.

Now $(s, a) = 1$ or else if prime q divides a and s, then it divides $ap = -ds^2 = r^2$, and so it divides r, contradicting that $(r, s) = 1$. Let b be an integer for which $b \equiv r/s \pmod{a}$ so that $b^2 \equiv -d \pmod{a}$. We define integers $n = s, m = (r - bs)/a$, and $c = (b^2 + d)/a$. This implies that $am + bn = r$ and so

$$am^2 + 2bmn + cn^2 = \frac{(am+bn)^2 - (b^2-ac)n^2}{a} = \frac{r^2 + ds^2}{a} = p.$$

Therefore, whenever $-d$ is a square mod p, there is a quadratic equation in two variables, with positive leading coefficient $\le 2\sqrt{d}$, and of discriminant $-4d$, which takes the value p. This is the first hint of a general theory: We will study the solutions to quadratic equations in two variables, like this, in detail, in chapter 12.

Additional exercises

Exercise 9.7.1. Let $f(n)$ be the arithmetic function for which $f(n) = 1$ if n can be written as the sum of two squares, and $f(n) = 0$ otherwise. Prove that $f(n)$ is a multiplicative function.

Exercise 9.7.2. Let p be a prime $\equiv 1 \pmod 4$. This exercise yields another proof that p is the sum of two squares.
  (a) Use Theorem 8.3 to prove that there exist integers a and b such that $a^2 + b^2$ is a positive multiple of p.
  (b) Let rp be the smallest such multiple of p. Prove that $r \leq p/2$.
  (c)† Prove that if $r > 1$, then there exists a positive integer $s \leq r/2$ such that $rs = c^2 + d^2$ for some integers c and d, selected so that $ad - bc$ is divisible by r.
  (d) Use (9.1.2) to deduce that if $r > 1$, then sp is a sum of two squares.

This contradicts the minimality of r unless $r = 1$; that is, p is the sum of two squares.

Exercise 9.7.3. Let p be an odd prime.
  (a)† Suppose that $b \pmod p$ is given and that $R, S \geq 1$ such that $RS = p$. Prove that there exist integers r, s with $|r| \leq R, 0 < s \leq S$ such that $b \equiv r/s \pmod p$.
  (b) Prove that there exists an integer m with $|m| < \sqrt{p}$ for which $\left(\frac{m}{p}\right) = -1$.
  (c) Deduce that if $p \equiv 1 \pmod 4$, then there exists an integer n in the range $1 < n < \sqrt{p}$ for which $\left(\frac{n}{p}\right) = -1$.

Exercise 9.7.4. Show that x and y are integers in (9.1.3) if and only if $r^2 + s^2$ divides $2(ar + bs)$, and show that this can only happen if $r^2 + s^2$ divides $2n$.

Exercise 9.7.5. What values of r and s yield the point $(-a, -b)$ in Proposition 9.1.2?

Exercise 9.7.6. Reprove exercise 9.1.8 using Theorem 9.1 and (9.1.1).

Exercise 9.7.7.† $33^2 + 56^2 = 65^2$ and $16^2 + 63^2 = 65^2$ are examples of the side lengths of different primitive Pythagorean triangles with the same hypotenuse. Classify those integers that appear as the hypotenuse of at least two different primitive Pythagorean triangles.

Exercise 9.7.8. Prove that for every integer m there exists an integer n which is the length of the hypoteneuse of at least m different primitive Pythagorean triples. (You may use Theorem 7.4 which implies that there are infinitely many primes $\equiv 1 \pmod 4$.)

Exercise 9.7.9.† Prove that an integer of the form $a^2 + 4b^2$ with $(a, 2b) = 1$ cannot be divisible by any integer of the form $m^2 - 2$ with $m > 1$, or $m^2 + 2$. Conversely prove that an integer of the form $m^2 - 2n^2$ or $m^2 + 2n^2$ with $(m, 2n) = 1$ cannot be divisible by any integer of the form $a^2 + 4$.

Exercise 9.7.10.‡ (Zagier's proof that every prime $\equiv 1 \pmod 4$ is the sum of two squares) Let
$$S := \{(x, y, z) \in \mathbb{N}^3 : p = x^2 + 4yz\}.$$
Define the map $\phi : S \to S$ by
$$\phi : (x, y, z) \to \begin{cases} (x + 2z, z, y - x - z) & \text{if } x < y - z, \\ (2y - x, y, x - y + z) & \text{if } y - z < x < 2y, \\ (x - 2y, x - y + z, y) & \text{if } x > 2y. \end{cases}$$
  (a) Show that ϕ is an *involution*, that is, $\phi^2 = 1$, and verify that each $\phi(S)$ belongs to S.
  (b) Prove that if $\phi(v) = v$, then $v = (1, 1, \frac{p-1}{4})$.
  (c) Deduce that there are an odd number of elements of S (in particular, S is non-empty). Let $\psi : S \to S$ be the involution $\psi(x, y, z) = (x, z, y)$.
  (d) Prove that ψ has a fixed point (x, y, y) so that $z = y$.
  (e) Deduce that $p = x^2 + (2y)^2$ for some integers x, y.

Appendix 9A. Proof of the local-global principle for quadratic equations

In this appendix we will give the difficult part of the proof of the local-global principle for quadratic equations, Theorem 9.4, as discussed at length in section 9.3.

The local-global principle for quadratic equations. *Let a, b, c be given pairwise coprime, squarefree integers. There are solutions in*

non-zero integers ℓ, m, n to $a\ell^2 + bm^2 + cn^2 = 0$ with $(a\ell^2, bm^2) = 1$

if and only if there are solutions in

non-zero real numbers λ, μ, ν to $a\lambda^2 + b\mu^2 + c\nu^2 = 0$,

and, *for all positive integers r, there exist*

residue classes $u, v, w \pmod{r}$ for which $au^2 + bv^2 + cw^2 \equiv 0 \pmod{r}$,

with $(au^2, bv^2, cw^2, r) = 1$.

Our proof depends on an understanding of lattices.

9.8. Lattices and quotients

A *lattice* Λ in $\mathbb{R}^n$ is the set of points obtained by integer linear combinations of n given linearly independent vectors. If the *basis* is $x_1, x_2, \ldots, x_n \in \mathbb{R}^n$, then

$$\Lambda := \{m_1 x_1 + m_2 x_2 + \cdots + m_n x_n : m_1, m_2, \ldots, m_n \in \mathbb{Z}\}.$$

One can see that Λ is an additive group, but it also has some geometry connected to it. The *fundamental domain* of Λ with respect to $x_1, x_2, \ldots, x_n$ is the set

$$P = P(\Lambda) := \{a_1 x_1 + a_2 x_2 + \cdots + a_n x_n : 0 \leq a_i < 1\},$$

9.8. Lattices and quotients

the interior (and part of the boundary) of one of the diamond-shaped cells in Figure 9.1. If $\lambda \in \Lambda$, then $\lambda + P$ gives us another of the diamond shapes, shifted from the original by λ. Therefore the sets $\lambda + P$, $\lambda \in \Lambda$ are disjoint and their union is $\mathbb{R}^n$. Therefore $P(\Lambda)$ is a set of representatives of

$$\mathbb{R}^n/\Lambda,$$

which is often called "$\mathbb{R}^n$ mod Λ".

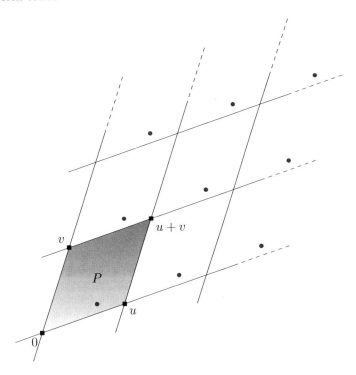

Figure 9.1. Constructing a lattice in $\mathbb{R}^2$, generated by vectors u and v. The shaded grey parallelogram is the fundamental domain $P(\Lambda)$. The dots represent the same point in $\mathbb{R}^2/\Lambda$ repeated in each copy of $P(\Lambda)$; that is, they are the points $P + \lambda$ for each vector $\lambda \in \Lambda$.

In the non-trivial example with $n = 1$, for which $\Lambda = \mathbb{Z}$, we can write every real number z as $m + a$ where $m \in \mathbb{Z}$ and $a \in [0, 1)$, letting $m = [z]$ and $a = \{z\}$. We prefer to think of this as $z = a$ in the ring $\mathbb{R}/\mathbb{Z}$ since their difference, m, is an integer. This generalizes to n dimensions, in which case we can identify $\mathbb{R}^n/\Lambda$ with $(\mathbb{R}/\mathbb{Z})^n$.

The *determinant* $\det(\Lambda)$ of Λ is the volume of P; in fact $\det(\Lambda) = |\det(A)|$, where A is the matrix with column vectors $x_1, x_2, \ldots, x_n$ (written as vectors in $\mathbb{R}^n$). A *convex body* K is a bounded convex open subset[3] of $\mathbb{R}^n$.

---

[3]These are all common terms in geometry. A set $S \subset \mathbb{R}^n$ is *bounded* if it can be contained inside a ball of some finite radius. The set S is *convex* if all the points on the straight line between any two points of S also belong to S. The set S is open if there is a ball around any given point of S, perhaps of very small radius, that also is contained within S.

If $\Lambda \subset \mathbb{Z}^n$, then there are $\det(\Lambda)$ cosets of Λ in $\mathbb{Z}^n$; that is,
$$|\mathbb{Z}^n/\Lambda| = \det(\Lambda).$$
In the proof of Theorem 9.1 we work with the lattice
$$\Lambda := \{(r,s) \in \mathbb{Z}^2 : \ r - ks \equiv 0 \pmod{p}\}$$
(where $k^2 \equiv -1 \pmod{p}$). This lattice is presented there somewhat differently from the definition here, but it can easily be seen that Λ is generated by $(k,1)$ and $(p,0)$, and that $(0,p) = p(k,1) - k(p,0)$. Hence $\det(\Lambda) = p$; in particular we deduce that there are p distinct cosets of Λ within $\mathbb{Z}^2$.

Let S be the set constructed in the proof of Theorem 9.1: S is a convex set of $> p$ elements of $\mathbb{Z}^2$ so that the difference, d, of two of them lies on the lattice Λ. The set S was constructed so that the difference, d, must lie close to the origin. Moreover Λ was constructed so that if $(r,s) \in \Lambda$, then $r^2 + s^2 \equiv 0 \pmod{p}$ (since if $r \equiv ks \pmod{p}$, then $r^2 + s^2 \equiv (ks)^2 + s^2 \equiv (k^2+1)s^2 \equiv 0 \pmod{p}$.)

We will now develop these ideas to give a proof of the local-global principle. In the next section we will modify the last step to make it more elegant.

Proof of the local-global principle. Assume that a, b, and c are squarefree, pairwise coprime integers, with $a,b > 0 > c$ (so that there are non-zero real solutions to $ax^2 + by^2 + cz^2 = 0$), and that there exists a solution to
$$au^2 + bv^2 + cw^2 \equiv 0 \pmod{|abc|},$$
with $(au^2, bv^2, cw^2, abc) = 1$.[4] We may assume that at least two of $a, b, |c|$ are > 1, for the case $a = b = 1$ can be proved directly from Theorem 9.2, while the case $a = 1, c = -1$ is easy as we always have the solution $x = b-1, y = 2, z = b+1$.

Define the lattice
$$\Lambda := \{(x,y,z) \subset \mathbb{Z}^3 : \ aux + bvy + cwz \equiv 0 \pmod{|abc|}\}.$$
We claim that if $(x,y,z) \in \Lambda$, then
$$ax^2 + by^2 + cz^2 \equiv 0 \pmod{|abc|}.$$
We now prove that this holds mod a (and the cases mod b and mod $|c|$ proceed analogously, so that the claim follows using the Chinese Remainder Theorem). Now if $(x,y,z) \in \Lambda$, then $bvy \equiv -cwz \pmod{a}$, and so
$$bv^2 \cdot by^2 = (bvy)^2 \equiv (-cwz)^2 = cw^2 \cdot cz^2 \pmod{a}.$$
Dividing through by $bv^2 \equiv -cw^2 \pmod{a}$, we deduce that $by^2 \equiv -cz^2 \pmod{a}$. Therefore $ax^2 + by^2 + cz^2 \equiv 0 \pmod{a}$, as desired.

In the next exercise we will show that $|\det(\Lambda)| = |abc|$. Let
$$S := \{(i,j,k) : \ 0 \leq i \leq [\sqrt{|bc|}], \ 0 \leq j \leq [\sqrt{|ac|}], \ 0 \leq k \leq [\sqrt{|ab|}]\}.$$
The number of integer points in S is $> \sqrt{|bc|} \cdot \sqrt{|ac|} \cdot \sqrt{|ab|} = |abc| = |\mathbb{Z}^3/\Lambda|$, and so, by the pigeonhole principle, there must be two lattice points in S that differ by

---

[4] Lemma 9.3.1 implies that we should work modulo $8|abc|$ in proving the local-global principle. However, in this first version of our proof, we prefer to not worry about the equation modulo powers of 2. We will revisit this issue in the next section.

non-zero element $(x, y, z) \in \Lambda$. If the two lattice points are (i, j, k) and (I, J, K), then
$$|x| = |i - I| \leq [\sqrt{|bc|}], \quad |y| = |j - J| \leq [\sqrt{|ac|}], \quad |z| = |k - K| \leq [\sqrt{|ab|}].$$
These are all "<" as none of $|bc|, |ac|, |ab|$ are squares, since at least two of $a, b, |c|$ are > 1 and they are pairwise coprime. Therefore $ax^2 + by^2 < 2|abc|$ and $|cz^2| < |abc|$, so that
$$-|abc| < ax^2 + by^2 + cz^2 < 2|abc|.$$
This implies that either $ax^2 + by^2 + cz^2 = 0$ or $ax^2 + by^2 + cz^2 = |abc| = -abc$. We need to eliminate the second case. I know of two ways to do this. The first is inelegant and comes from simply noting that if $ax^2 + by^2 + cz^2 + abc = 0$, then
$$a(xz - by)^2 + b(ax + yz)^2 + c(ab + z^2)^2 = (ab + z^2)(ax^2 + by^2 + cz^2 + abc) = 0.$$
The second involves slightly modifying the definition of Λ, by taking the prime 2 into account more carefully, which we discuss in the next section. □

Exercise 9.8.1. (a) Show that there exist integers U, V, W, coprime with abc, for which $U \equiv u \pmod{bc}$, $V \equiv v \pmod{ac}$, $W \equiv w \pmod{ab}$, so that $aU^2 + bV^2 + cW^2 \equiv 0 \pmod{|abc|}$.
(b) Let U^{-1} be an integer $\equiv 1/U \pmod{abc}$ and W^{-1} be an integer $\equiv 1/W \pmod{abc}$. Show that Λ is generated by the vectors $(1, VU^{-1}, WU^{-1})$, $(0, c, -bVW^{-1})$, and $(0, 0, ab)$.)
(c) Deduce that $\det(\Lambda) = |abc|$.

9.9. A better proof of the local-global principle

The idea is to construct a lattice, based on that in the previous section, but now of determinant $4|abc|$. We begin by defining
$$\Lambda_0 := \{(x, y, z) \subset \mathbb{Z}^3 : aux + bvy + cwz \equiv 0 \pmod{|abc|}\}.$$
If c is even, then let
$$\Lambda := \{(x, y, z) \in \Lambda_0 : y \equiv x \pmod 4 \text{ and } z \equiv wx \pmod 2\}$$
based on the given solution (u, v, w). We construct Λ analogously if a or b is even.

If abc is odd, then one of u, v, w must be even (as $au^2 + bv^2 + cw^2 = 0$), say w. If so, then let
$$\Lambda := \{(x, y, z) \in \Lambda_0 : y \equiv x \pmod 2 \text{ and } z \equiv 0 \pmod 2\},$$
using the given solution (u, v, w). We construct Λ analogously if u or v is even.

Exercise 9.9.1. (a) Prove that if $(x, y, z) \in \Lambda$, then $ax^2 + by^2 + cz^2 \equiv 0 \pmod{4|abc|}$.
(b) Prove that $\det(\Lambda) = 4|abc|$.

Consider the set of integer points
$$S := \{(i, j, k) : 0 \leq i \leq [\sqrt{2|bc|}], \ 0 \leq j \leq [\sqrt{2|ac|}], \ 0 \leq k \leq [2\sqrt{|ab|}]\}.$$
The number of lattice points in S is $> \sqrt{2|bc|} \cdot \sqrt{2|ac|} \cdot 2\sqrt{|ab|} = 4|abc| = |\mathbb{Z}^3/\Lambda|$ by exercise 9.9.1(b), and so, by the pigeonhole principle, there must be two lattice points in S that differ by a non-zero element $(x, y, z) \in \Lambda$. If the two lattice points are (i, j, k) and (I, J, K), then
$$|x| = |i - I| \leq [\sqrt{2|bc|}], \quad |y| = |j - J| \leq [\sqrt{2|ac|}], \quad |z| = |k - K| \leq [2\sqrt{|ab|}].$$

Therefore $ax^2 + by^2 < 4|abc|$ and $|cz^2| < 4|abc|$ (as equality would only be possible if $a = b = 1$), and so
$$|ax^2 + by^2 + cz^2| < 4|abc|.$$
Now, since $(x, y, z) \in \Lambda$, we know that
$$ax^2 + by^2 + cz^2 \equiv 0 \pmod{4|abc|},$$
by exercise 9.9.1(a), and so we must have $ax^2 + by^2 + cz^2 = 0$ as desired. □

A by-product of this proof is that the smallest non-trivial solution satisfies
$$|a\ell^2|,\ |bm^2|,\ |cn^2| \leq 4|abc|.$$
In 1950, Holzer showed that one may replace $4|abc|$ by $|abc|$.

Exercise 9.9.2. Give infinitely many examples in which $\max\{|a\ell^2|, |bm^2|, |cn^2|\} = |abc|$ in the smallest non-trivial solution of $a\ell^2 + bm^2 + cn^2 = 0$.

Appendix 9B. Reformulation of the local-global principle

9.10. The Hilbert symbol

Given pairwise coprime integers a, b, c and prime p we define the *Hilbert symbol*

$$\left(\frac{a,b,c}{p}\right), \text{ which equals } 1, \text{ if}$$

for every $k \geq 1$, there is a solution $u, v, w \pmod{p^k}$ with $(au^2, bv^2, cw^2, p) = 1$ to $au^2 + bv^2 + cw^2 \equiv 0 \pmod{p^k}$, and which equals -1 otherwise. If p is odd, then we need only consider solutions mod p, and if $p = 2$, then we need only consider solutions mod 8, as explained in the proof of Lemma 9.3.1.

We also define

$$\left(\frac{a,b,c}{\infty}\right) := \begin{cases} 1 & \text{if there are real solutions } u,v,w \neq 0 \text{ to } au^2 + bv^2 + cw^2 = 0, \\ -1 & \text{otherwise.} \end{cases}$$

Theorem 9.6. *For any given pairwise coprime, squarefree integers a, b, c we have*

$$\left(\frac{a,b,c}{\infty}\right) \prod_{p \text{ prime}} \left(\frac{a,b,c}{p}\right) = 1.$$

This *Hilbert product theorem* has several remarkable consequences, alluded to in section 9.3. The local-global principle implies that if there are no integer solutions to $au^2 + bv^2 + cw^2 = 0$, then there must be a value of ℓ, either ∞ or a prime p, for which $\left(\frac{a,b,c}{\ell}\right) = -1$. However, Theorem 9.6 implies that the number of such ℓ is always even. This explains the remarks in the second paragraph of section 9.3: If there are real solutions but not integer solutions, then there are an even number of primes p for which $\left(\frac{a,b,c}{p}\right) = -1$, that is, an even number of primes p for which there are no solutions mod p, or mod 8 if $p = 2$. This implies that one can neglect to mention one local criterion in formulating the local-global principle (for if there are no solutions, there must still remain one such local criterion for which there are no solutions). This explains why in Legendre's formulation of the local global principle, one is able to avoid being careful about the role of the prime 2 (which, as we will see in proving Theorem 9.6, is often more complicated than the others).

Proof. The values of the Hilbert symbols remain the same if we rearrange the order of a, b, and c, or if we multiply each of a, b, c through by -1. This means that we may assume that if any of a, b, c are even, then it is a, and that at most one of a, b, c is negative. Therefore $\left(\frac{a,b,c}{\infty}\right) = -(-1)^\eta$, where η is the number of a, b, c that are < 0.

Suppose that p is an odd prime. If $p \nmid abc$, then $\left(\frac{a,b,c}{p}\right) = 1$ by exercise 8.9.9. If odd prime p divides abc, say p divides a, then $\left(\frac{a,b,c}{p}\right) = \left(\frac{-bc}{p}\right)$, and similarly if p divides b or c.

The role of the prime $p = 2$ is complicated. Let $A = a/(2,a)$ so that A, b, c are odd. For convenience we define $\xi = 1$ if $A \equiv b \equiv c \pmod{4}$, and $\xi = -1$ otherwise. A case-by-case analysis then yields that if a is odd, then $\left(\frac{a,b,c}{2}\right) = -\xi$; and if a is even, then $\left(\frac{a,b,c}{2}\right) = 1$ if $b+c \equiv 0$ or $-a \pmod{8}$, and -1 otherwise. One can then verify in every case that

$$\left(\frac{a,b,c}{2}\right) = -\left(\frac{(2,a)}{bc}\right) \cdot \xi.$$

Next, noting that $\left(\frac{a,b,c}{p}\right) = 1$ if p does not divide $2abc$, we obtain

$$\prod_{p \text{ prime}} \left(\frac{a,b,c}{p}\right) = -\left(\frac{(2,a)}{bc}\right) \cdot \xi \cdot \prod_{\substack{p \text{ odd}\\p|a}}\left(\frac{-bc}{p}\right) \cdot \prod_{\substack{p \text{ odd}\\p|b}}\left(\frac{-ac}{p}\right) \cdot \prod_{\substack{p \text{ odd}\\p|c}}\left(\frac{-ab}{p}\right)$$

$$= -\left(\frac{(2,a)}{bc}\right) \cdot \xi \cdot \left(\frac{-bc}{|A|}\right)\left(\frac{-ac}{|b|}\right)\left(\frac{-ab}{|c|}\right)$$

$$= -(-1)^\eta \xi \cdot \left(\frac{-1}{Abc}\right) \cdot \left(\frac{b}{A}\right)\left(\frac{A}{b}\right) \cdot \left(\frac{b}{c}\right)\left(\frac{c}{b}\right) \cdot \left(\frac{c}{A}\right)\left(\frac{A}{c}\right)$$

$$= -(-1)^\eta = \left(\frac{a,b,c}{\infty}\right),$$

and the result follows. To obtain the second-to-last equality we use that $\left(\frac{-1}{Abc}\right) = \left(\frac{-1}{A}\right)\left(\frac{-1}{b}\right)\left(\frac{-1}{c}\right)$, the value of $(-1/\cdot)$, and quadratic reciprocity to obtain that

$$\left(\frac{-1}{A}\right)\left(\frac{-1}{b}\right)\left(\frac{-1}{c}\right) \cdot \left(\frac{b}{A}\right)\left(\frac{A}{b}\right) \cdot \left(\frac{b}{c}\right)\left(\frac{c}{b}\right) \cdot \left(\frac{c}{A}\right)\left(\frac{A}{c}\right)$$

equals (-1) to the power

$$\frac{A-1}{2} + \frac{b-1}{2} + \frac{c-1}{2} + \frac{A-1}{2} \cdot \frac{b-1}{2} + \frac{b-1}{2} \cdot \frac{c-1}{2} + \frac{A-1}{2} \cdot \frac{c-1}{2} \pmod{2}$$

$$= \frac{A+1}{2} \cdot \frac{b+1}{2} \cdot \frac{c+1}{2} - \frac{A-1}{2} \cdot \frac{b-1}{2} \cdot \frac{c-1}{2} - 1 \equiv \begin{cases} 0 & \text{if } A \equiv b \equiv c \pmod{4}, \\ 1 & \text{otherwise,} \end{cases}$$

which equals ξ. $\square$

9.11. The Hasse-Minkowski principle

We have seen that any quadratic polynomial with rational coefficients in no matter how many variables may be diagonalized and then scaled up to become a quadratic form with integer coefficients; call it $f = a_1 x_1^2 + \cdots + a_n x_n^2$. The Hasse-Minkowski principle is a generalization of the local-global principle and states that any diagonal quadratic form with integer coefficients in $n \geq 3$ variables has a non-trivial solution in integers if and only if it does in the reals and modulo every prime power. Moreover the analogy to Theorem 9.6 holds for forms in arbitrarily many variables. This gives rise to the following remarkable result:

Theorem 9.7. *A diagonal quadratic form in five or more variables has a non-trivial solution over the integers if and only if its coefficients do not all have the same sign.*

The proof for this is given in the following exercise.

Exercise 9.11.1. Suppose we are given a quadratic form $a_1 x_1^2 + \cdots + a_n x_n^2$ with each $a_i \in \mathbb{Z}$.
  (a) By changing variables and multiplying through by a suitable constant, show that we may assume each a_i is squarefree.
  (b) Prove that if the a_i do not all have the same sign, then there is a non-trivial real solution to $a_1 x_1^2 + \cdots + a_n x_n^2 = 0$.
  Let p be a given prime.
  (c) By possibly multiplying through by p and changing variables, show that we may assume that for every prime p, no more than $n/2$ of the a_i are divisible by p.
  (d) Deduce that if $n \geq 5$, then there exist integers $m_1, \ldots, m_n$ for which $a_1 m_1^2 + \cdots + a_n m_n^2 \equiv 0 \pmod{p}$, such that there exists some j for which $a_j m_j \not\equiv 0 \pmod{p}$.
  (e) Prove Theorem 9.7 using the Hasse-Minkowski principle.

This is all discussed in detail in Part I of the wonderful book

[1] J.-P. Serre, *A course in arithmetic*, Graduate Texts in Mathematics **7**, Springer-Verlag, 1973.

Appendix 9C. The number of representations

9.12. Distinct representations as sums of two squares

We modify the proof of Theorem 9.1 to obtain the following:

Proposition 9.12.1. *If n is a squarefree positive integer and $m^2 \equiv -1 \pmod{n}$, then there exist coprime integers r, s such that $n = r^2 + s^2$ and $r \equiv ms \pmod{n}$.*

Proof. There are $> n$ integers constructed in the set

$$\{j + km : 0 \leq j, k \leq [\sqrt{n}]\}$$

and so two must be congruent mod n, by the pigeonhole principle. If $j + km \equiv J + Km \pmod{n}$, then let $r = j - J$ and $s = K - k$ so that $r \equiv ms \pmod{n}$ where $|r|, |s| < \sqrt{n}$, and r and s are not both 0. Now

$$r^2 + s^2 \equiv (ms)^2 + s^2 = s^2(m^2 + 1) \equiv 0 \pmod{n}$$

and $0 < r^2 + s^2 < 2n$. The only integer between 0 and $2n$ that is divisible by n is n itself, and so we deduce that $r^2 + s^2 = n$. Moreover $(r, s) = 1$ or else n would be divisible by the square of $(r, s)^2$, contrary to the hypothesis. □

Lemma 9.12.1. *If $n = a^2 + b^2 = c^2 + d^2$ with $(a, b) = (c, d) = 1$ and $ad \equiv bc \pmod{n}$, then either $c = \pm a$ and $d = \pm b$, or $c = \mp b$ and $d = \pm a$.*

Proof. By (9.1.2) we have $(ac + bd)^2 + (ad - bc)^2 = n^2$. Now n^2 divides $(ad - bc)^2$ by the hypothesis, and so divides $(ac + bd)^2$. Therefore if $u = (ac + bd)/n$ and $v = (ad - bc)/n$, then u and v are integers for which $u^2 + v^2 = 1$. This implies that either $u = 0$ or $v = 0$. Now suppose $u = 0$, so that $ad = bc$. Then a divides $ad = bc$ and so a divides c, as $(a, b) = 1$. An analogous argument gives that c divides a, and so $c = \pm a$ and therefore $d = \pm b$. Otherwise $v = 0$ and $ac = -bd$. The analogous argument yields that $d = \pm a$ and $c = \mp b$. □

9.12. Distinct representations as sums of two squares

Let $R(n)$ denote the number of proper representations of n as the sum of two squares.

Theorem 9.8. *If n is squarefree, then*
$$R(n) = 4 \prod_{p \text{ prime: } p|n} \left(1 + \left(\frac{-4}{p}\right)\right) = 4 \sum_{d|n} \left(\frac{-4}{d}\right).$$

Proof. If $n = a^2 + b^2$, then $(a, b) = 1$ as n is squarefree. Therefore $(b, n) = 1$ or else (b, n) divides $n - b^2 = a^2$ which would imply that $(a, b) > 1$. Then $m \equiv a/b \pmod{n}$ is well-defined and satisfies $m^2 \equiv -1 \pmod{n}$. Now for each such m there exists a solution to $n = r^2 + s^2$ with $(r, s) = 1$ and $r \equiv ms \pmod{n}$ by Proposition 9.12.1; and there are in fact exactly four such solutions by Lemma 9.12.1. Hence $R(n)$ equals four times the number of square roots of $-1 \pmod{n}$. The result then follows by exercise 8.9.8(b) and (c). □

A similar discussion extends the result to all n, to obtain that
$$(9.12.1) \qquad R(n) = 4 \sum_{d|n} \left(\frac{-4}{d}\right).$$

We see that the right-hand side equals $4(1 * \left(\frac{-4}{\cdot}\right))(n)$, and so
$$\sum_{n \geq 1} \frac{R(n)}{n^s} = 4\zeta(s) L\left(s, \left(\frac{-4}{\cdot}\right)\right),$$
where $L\left(s, \left(\frac{-4}{\cdot}\right)\right)$ is the Dirichlet L-function that we encountered in section 8.17 of appendix 8D.

Exercise 9.12.1. Give another proof of exercise 9.7.1 using exercise 4.3.2.

Exercise 9.12.2. Prove that $R(n)/4$ equals the number of divisors of n that are $\equiv 1 \pmod{4}$ minus the number of divisors of n that are $\equiv 3 \pmod{4}$.

A different perspective. In $\mathbb{Z}[i]$ there are four units $1, -1, i, -i$.

If p is a prime $\equiv 1 \pmod{4}$, then we can write $p = a^2 + b^2$, which factors in $\mathbb{Z}[i]$ as $(a+bi)(a-bi)$. We can replace $a + bi$ by $x + yi = u(a+bi)$ for any unit u. Note that if $u = i$, then $x = -b, y = a$ and so $x/y \equiv -b/a \equiv a/b \pmod{p}$; in other words $x/y \pmod{p}$ does not vary as we range over the units u (which is essentially another proof of Lemma 9.12.1 for $n = p$). So if we write $P = a + bi$, then we can take p as the norm of $x + yi$ for $x + yi = uP$ or $u\overline{P}$ where u is a unit. Therefore there are $8 = R(p)$ possibilities for x and y.

If $p = 2$, then $1 - i = -i(1+i)$ so we have $R(2) = 4$. Now if $n = p_1 \cdots p_k$ where each $p_j \equiv 1 \pmod{4}$ is the norm of P_j, then the possible representations of n as a sum of two squares, $x^2 + y^2$, correspond to all possible factorizations of $x + iy$ as
$$x + iy = uA_1 A_2 \cdots A_k$$
where u is any unit and each $A_j = P_j$ or $\overline{P_j}$, a total of $4 \cdot 2^k$ possibilities. Moreover if $n = m^2 2^{e_0} p_1^{e_1} \cdots p_k^{e_k}$ where each $p_j \equiv 1 \pmod{4}$, and every prime factor of m is $\equiv 3 \pmod{4}$, then the possible representations of n as a sum of two squares, $x^2 + y^2$, correspond to all
$$x + iy = um(1+i)^{e_0} \cdot P_1^{r_1} \overline{P_1}^{e_1 - r_1} \cdot P_2^{r_2} \overline{P_2}^{e_2 - r_2} \cdots P_k^{r_k} \overline{P_k}^{e_k - r_k}$$

where u is any unit and each r_j lies in range $0 \leq r_j \leq e_j$, a total of $4 \cdot \prod_{j=1}^{k}(e_j+1)$ possibilities. This yields our formula for $R(n)$ above.

The average number of representations. If we sum up $R(n)$ over all $n \leq N$, we are asking for the number of pairs of integers a, b for which $a^2 + b^2 = n \leq N$, in other words the number of integer points $(a,b) \in \mathbb{Z}^2$ inside the circle $x^2 + y^2 \leq N$. We can approach this question by drawing a square of area 1 around each such point (whose corners lie at $(a \pm \frac{1}{2}, b \pm \frac{1}{2})$). These squares are disjoint other than their boundaries, so the area of their union equals the number of pairs a, b. The area contained inside the union of the squares is a good approximation to the area inside the circle of radius $\sqrt{N}$, which has area πN. Therefore the average number of representations of integers up to N as the sum of two squares is

$$\lim_{N \to \infty} \frac{1}{N} \sum_{n \leq N} R(n) = \lim_{N \to \infty} \frac{1}{N} \#\{(a,b) \in \mathbb{Z}^2 : a^2 + b^2 \leq N\},$$

which should be about π.

We can be more precise in estimating $\#\{(a,b) \in \mathbb{Z}^2 : a^2 + b^2 \leq N\}$. We saw above that this equals the area of the union of the unit boxes whose centers lie in the circle $x^2 + y^2 \leq N$. This is close to the area πN of this circle, but there is an error in this estimate, which comes from the area of the unit boxes that straddle the boundary (that is, that are partly inside and partly outside the circle). Since each box has area one, the error is therefore bounded by the number of such boxes.

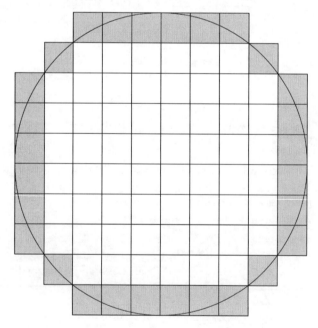

Figure 9.2. A circle approximated by boxes of area one, shading those that straddle the boundary.

9.12. Distinct representations as sums of two squares

To bound the number of such boxes that straddle the boundary, we draw the boxes on the boundary and count the boxes by the number of times neighboring boxes share a vertical edge, plus the number of times they share a horizontal edge. For example to go counterclockwise around the first quadrant of the circle from $(\sqrt{N}, 0)$ to $(0, \sqrt{N})$, we begin with a boundary box with center $(m, 0)$ where m is the nearest integer to $\sqrt{N}$. As we go counterclockwise around the boundary the next box is $(m, 1)$, $(m-1, 0)$, or $(m-1, 1)$. If we are at the box with center (i, j) with $i > 0$ and $j \geq 0$, then the next box is one of the three $(i, j+1)$, $(i-1, j)$, or $(i-1, j+1)$. In particular, the value of $j - i$ transforms monotonically from $-m$ to m, increasing by either 1 or 2 as we move on to the next box. Therefore the number of boundary straddling boxes in the first quadrant is at most $2m + 1 \leq 2\sqrt{N} + 2$. Adding together similar contributions from all four quadrants we have

$$|\#\{(a,b) \in \mathbb{Z}^2 : a^2 + b^2 \leq N\} - \pi N| \leq 8\sqrt{N} + 8.$$

In *Gauss's circle problem* one wishes to improve this error term as much as possible. It is conjectured that for any fixed $\epsilon > 0$, one has the bound $\leq N^{1/4 + \epsilon}$ if N is sufficiently large, and it is known that there are arbitrarily large N for which the error term is considerably larger than $N^{1/4}$ (so one cannot take $\epsilon = 0$).

Appendix 9D. Descent and the quadratics

9.13. Further solutions through linear algebra

We wish to determine all pairs of positive integers $a \geq b$ for which $ab + 1$ divides $a^2 + b^2$. If we have a solution with $a^2 + b^2 = k(ab + 1)$ for some positive integer k, then a is a root of the quadratic polynomial

$$x^2 - kbx + (b^2 - k).$$

If c is the other root, then $a + c = kb$, so $c = kb - a$ is another integer for which $b^2 + c^2 = k(bc + 1)$. Now $c \geq 0$ or else $bc + 1 \leq 0$, in which case $b^2 + c^2 \leq 0$, which is impossible. If $c = 0$, then, looking back at our equations, we see that $k = b^2$ and $a = b^3$. Otherwise $c > 0$ so that $b^2 - k = ac > 0$ and therefore $c = (b^2 - k)/a < b^2/b = b$.

We have proved that $0 \leq c < b$ which means that (b, c) is a smaller solution than the original solution (a, b), with the same quotient. We can iterate this map, $(a, b) \to (b, kb - a)$, to eventually descend, after a finite number of steps, to a basic solution. We only stop descending when $c = 0$, which means that k must be a square, a fact that is far from obvious in the formulation of the problem.

To obtain all solutions, we simply invert our map: Writing $k = m^2$ for some integer $m \geq 1$ we begin with the solution $(m, 0)$ to $m^2 + 0^2 = k(m \cdot 0 + 1)$ and obtain all others by iterating the map

$$(b, c) \to (kb - c, b).$$

This map is better understood through matrices: We have

$$\begin{pmatrix} a \\ b \end{pmatrix} = \begin{pmatrix} kb - c \\ b \end{pmatrix} = \begin{pmatrix} k & -1 \\ 1 & 0 \end{pmatrix} \begin{pmatrix} b \\ c \end{pmatrix},$$

9.15. Apollonian circle packing

the matrix representing a transformation of determinant 1. Therefore, if $k = m^2$, then all solutions to $a^2 + b^2 = k(ab+1)$ in non-negative integers a, b are given by

$$\begin{pmatrix} a \\ b \end{pmatrix} = \begin{pmatrix} m^2 & -1 \\ 1 & 0 \end{pmatrix}^n \begin{pmatrix} m \\ 0 \end{pmatrix},$$

for some integers $m \geq 1$ and $n \geq 0$.

Exercise 9.13.1. Fix integer $k \geq 1$. Let $x_0 = 0$, $x_1 = 1$, and $x_n = kx_{n-1} - x_{n-2}$ for all $n \geq 2$. Prove that all solutions to $a^2 + b^2 = k(ab+1)$ in non-negative integers a, b with $k = m^2$ are given by $a = mx_n$, $b = mx_{n-1}$ for some integer $n \geq 1$.

Exercise 9.13.2. Prove that if A is any 2-by-2 matrix and the vector $u_n = A^n u_0$ for some given matrix u_0, then u_n satisfies a second-order linear recurrence.

Other quadratic equations can also be understood by recursions. Perhaps the most famous is the Markov equation.

9.14. The Markov equation

Here we seek all triples of positive integers x, y, z for which

$$x^2 + y^2 + z^2 = 3xyz.$$

One can find many solutions: $(1,1,1), (1,1,2), (1,2,5), (1,5,13), (2,5,29), \ldots$. If (a, b, c) is a solution, then a is a root of the quadratic, $x^2 - 3bcx + (b^2 + c^2)$, the other root being $3bc - a$, and so we obtain a new solution $(3bc - a, b, c)$. One can perform this same procedure singling out b or c instead of a, and get different solutions. For example, starting from $(1, 2, 5)$ one obtains the solutions $(29, 2, 5)$, $(1, 13, 5)$, and $(1, 2, 1)$, respectively.

If we fix one coordinate, say $z = c$, then we can get a new solution from an old one via the map

$$\begin{pmatrix} a \\ b \end{pmatrix} = \begin{pmatrix} 3c & -1 \\ 1 & 0 \end{pmatrix} \begin{pmatrix} a_0 \\ b_0 \end{pmatrix},$$

and then this map can be repeated arbitrarily often, as in the previous section, to obtain infinitely many solutions.

Despite knowing there are infinitely many, the solutions to the Markov equation remain mysterious. For example, one open question is to determine all of the integers that appear in a Markov triple. The first few are

$$1, 2, 5, 13, 29, 34, 89, 169, 194, 233, 433, 610, 985, 1325, \ldots.$$

It is believed that they are quite sparse.

Exercise 9.14.1.[†] Determine what solutions are obtained from (1,1,1) by using the maps $(x, y) \to (3y - x, y)$ and $(x, y) \to (x, 3x - y)$.

9.15. Apollonian circle packing

To my taste, the most beautiful such problem is the Apollonian circle packing problem.[5] Take three circles that touch each other (for example, take three coins

---
[5] Apollonius lived in Perga, 262–190 B.C.

and push them together):

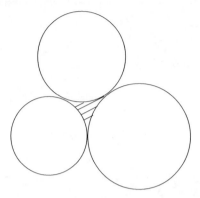

Figure 9.3. Three mutually tangent circles with a shaded crescent shape in between.

In between the circles one has a crescent-type shape (a *hyperbolic triangle*).

There are two circles that are tangent to each of these three circles: Inside that crescent shape one can inscribe a (unique) circle that touches all three of the original circles. There is also a unique circle that contains all of the original cycles and touches each of them.

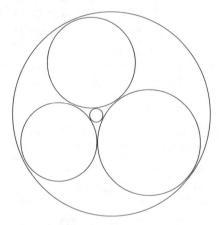

Figure 9.4. Two new circles, each tangent to all three of the old circles.

What is the relationship between the radii of the new circles and the radii of the original circles? Define the *curvature* of a circle to be C/r where r is the radius, for some appropriately selected constant $C > 0$. In 1643 Descartes, in a letter to Princess Elisabeth of Bohemia, noted that if b, c, and d are the curvatures of the original three circles, then the curvatures of the two new circles both satisfy the quadratic equation

$$2(a^2 + b^2 + c^2 + d^2) = (a + b + c + d)^2,$$

9.15. Apollonian circle packing

that is,
$$a^2 + b^2 + c^2 + d^2 - 2(ab + bc + cd + da + ac + bd) = 0.$$

Therefore given b, c, and d, there are two possibilities for a, the roots of the quadratic equation,
$$x^2 - 2(b + c + d)x + (b^2 + c^2 + d^2 - 2bc - 2cd - 2bd) = 0.$$

We select C so that the first three curvatures, b, c, and d are integers with $\gcd(b, c, d) = 1$. We will focus on the case that a is also an integer; for example if we start with $b = c = 2$ and $d = 3$, we have $a^2 - 14a - 15 = 0$ so that $a = -1$ or $a = 15$. Evidently $a = -1$ corresponds to the outer circle,[6] and $a = 15$ the inner one.

If we have one solution (a, b, c, d), then we also have another solution (A, b, c, d) for which
$$A = 2(b + c + d) - a.$$

Moreover if a, b, c, and d are integers, then so is A. We can iterate this (using any of the variables b, c, or d, in place of a) to obtain infinitely many Apollonian circles. These eventually *tile* the whole of the original circle, as each new circle fills in part of the crescent in between three existing circles.

In this example, we take the three most common American coins, a quarter, a nickel, and a dime, which have radii 24, 21, and 18 mms, respectively, to the nearest millimeter. In this case we define the curvature of a circle of radius r mm to be $504/r$, yielding curvatures of 21, 24, and 28, respectively. We proceed by filling in each successive crescent shape with a mutually tangent circle. What emerges is a tiling of the whole outer circle (which has curvature -11) by circles with larger and larger positive integer curvatures.[7]

There are many questions that can be asked: What integers appear as curvatures in a given packing? There are some integers that cannot appear because of congruence restrictions. For example if a, b, c, d are all odd, then all integers that arise as curvatures in this packing will be odd. The conjecture is that all sufficiently large integers that satisfy these congruence constraints, which can all be described mod 24, will appear as curvatures in the given packing. Although this is an open question, we do know that a positive proportion of the integers appear in any such packing, and that there are a

---

[6]The negative sign is intriguing. The mathematics gives a negative integer which surely makes no sense; but can the mathematics lie? A more in-depth analysis indicates that the negative sign should be interpreted as meaning that whereas the *interior* of a circle usually means all points inside the circle, when we have negative curvature the interior is to be interpreted as all points *outside the circle*, going off to ∞. It is best to think of the circles as being drawn on a sphere because, for a circle on a sphere, the circle partitions the sphere into two parts, and there are two choices as to what is the interior and what is the exterior.

[7]Tiled circle defined by U.S. coins, reproduced here with the kind permission of Alex Kontorovich.

surprisingly large number of circles in any packing with curvature $\leq T$, far more than T (so that many circles in the packing have the same curvature, and thus the same radius).[8] Since so many different integers appear as curvatures in any given packing, Peter Sarnak asked (and resolved) whether there are infinitely many pairs of mutually tangent circles, whose curvatures are both prime numbers, the *Apollonian twin prime conjecture*.

This last question is accessible because we see that any given solution $v = (a, b, c, d)$ is mapped to another solution by any permutation of the four elements, as well as the matrix $\begin{pmatrix} -1 & 2 & 2 & 2 \\ 0 & 1 & 0 & 0 \\ 0 & 0 & 1 & 0 \\ 0 & 0 & 0 & 1 \end{pmatrix}$. These (linear) transformations generate a subgroup, G, of $\mathrm{SL}(4, \mathbb{Z})$,[9] and one can proceed by considering *orbits* (that is, the set $\{Av : A \in G\}$ for some starting vector v) under the actions of G.

Sarnak's approach to studying the curvatures brings us back to quadratic equations: In the American coins example, we begin with the circle of curvature 28 that is tangent to the circles of curvature 21 and 24. The circle inside the crescent that is tangent to these three circles has curvature 157. Next we determine the circle that is tangent to those of curvature 21, 24, and 157, and then the next one, always using the two circles of curvature 21 and 24. So if x_n is the curvature of the nth circle in this procedure, then $x_0 = 28$, $x_1 = 157$, and
$$x_{n+1} = 2(21 + 24 + x_n) - x_{n-1} \text{ for all } n \geq 1,$$
by Descartes's equation. We can prove by induction that
$$x_n = 45n^2 + 84n + 28 \text{ for all } n \geq 0;$$
so the circles in our circle problem, tangent to the original circles of curvatures 21 and 24, have curvature x_n for each $n \geq 0$.

Exercise 9.15.1. Suppose that you are given three mutually tangent circles A, B, and C_0 of curvatures a, b, and x_0 in an Apollonian circle packing. For each $n \geq 0$ let C_{n+1} be the circle tangent to the circles A, B, and C_n that lies in the crescent between these three circles, and let x_{n+1} be its curvature. Prove that
$$x_n = (a+b)(n^2 - n) + (x_1 - x_0)n + x_0 \text{ for all } n \geq 0.$$

Peter Sarnak developed this idea further, which we return to in appendix 12G.

Further reading on Apollonian packings

[1] Dana Mackenzie, *A tisket, a tasket, an Apollonian gasket*, American Scientist **98** (Jan–Feb 2010), 10–14.

[2] Peter Sarnak, *Integral Apollonian packings*, Amer. Math. Monthly **118** (2011), 291–306.

---

[8]The total number of circles in any given packing with curvature $\leq T$ is about cT^α where $\alpha = 1.30568\ldots$, and c is a positive constant that depends on the packing.

[9]$\mathrm{SL}(4, \mathbb{Z})$ is the set of 4-by-4 matrices of determinant 1 with integer entries.

Chapter 10

Square roots and factoring

In this chapter we will study the computational side of number theory, which plays an important role in several uses of computers in today's society, particularly when it comes to keeping secrets. We will investigate how to *rapidly* determine whether a given large integer is prime and, if not, how to factor it. The issue of factoring an integer n is closely related to determining square roots mod n:

10.1. Square roots modulo n

How difficult is it to find square roots mod n? The first question to ask is how many square roots does a square have mod n?

Lemma 10.1.1. *If n is an odd integer with k prime factors and A is a square* mod n *with $(A, n) = 1$, then there are exactly 2^k residues* mod n *whose square is $\equiv A$* (mod n).

In particular, all squares mod m, that are coprime to m, have the same number of square roots mod m. We resolved how many square roots 1 (mod n) has in Lemma 3.8.1, and here we modify that proof to better suit the discussion in this chapter. We could have immediately deduced Lemma 10.1.1 for if A is a square mod n, then there exists b (mod n) such that $b^2 \equiv A$ (mod n), and then the solutions to $x^2 \equiv A$ (mod n) are in 1-to-1 correspondence with the solutions to $y^2 \equiv 1$ (mod n) through the invertible transformation $x \equiv by$ (mod n).

Proof. Suppose that $b^2 \equiv A$ (mod n) where $n = p_1^{e_1} p_2^{e_2} \ldots p_k^{e_k}$, and each p_i is odd and distinct. If $x^2 \equiv A$ (mod n), then $n | (x^2 - b^2) = (x-b)(x+b)$ so that p divides $x - b$ or $x + b$ for each prime p dividing n. Now p cannot divide both or else p divides $(x+b) - (x-b) = 2b$ and so $4A \equiv (2b)^2 \equiv 0$ (mod p), which contradicts the fact that $(p, 2A) | (n, 2A) = 1$. So let

$$d = (n, x - b), \text{ and therefore } n/d = (n, x + b),$$

which must be coprime. Then $x \equiv b_d \pmod{n}$ where b_d is that unique residue class mod n for which

(10.1.1) $$b_d \equiv \begin{cases} b \pmod{d}, \\ -b \pmod{n/d}. \end{cases}$$

Note that the b_d are well-defined by the Chinese Remainder Theorem, are distinct, and that $x^2 \equiv b_d^2 \equiv b^2 \equiv A \pmod{n}$ for each d.

The possible values of d are $\prod_{i \in I} p_i^{e_i}$ for each subset I of $\{1, \ldots, k\}$, and therefore there are 2^k possibilities. □

To see how the proof works let's obtain the four square roots of 4 (mod 15) from knowing one square root, 2, and the factorization of 15. These four square roots are given by four pairs of congruences which we solve using the Chinese Remainder Theorem:

2 (mod 1)	and	−2 (mod 15)	which yield	13 (mod 15);
2 (mod 3)	and	−2 (mod 5)	which yield	8 (mod 15);
2 (mod 5)	and	−2 (mod 3)	which yield	7 (mod 15); and
2 (mod 15)	and	−2 (mod 1)	which yield	2 (mod 15).

Consequence. *Let n be an odd integer with at least two different prime factors, and suppose that $b^2 \equiv A \pmod{n}$ with $(A, n) = 1$. Finding square roots of A mod n, other than b and $-b$, is "as difficult as" factoring n into two parts both > 1.*

Sketch of "proof". If we have a factorization $n = d \cdot n/d$, then we select b_d as in (10.1.1) so that $b_d^2 \equiv A \pmod{n}$ but $b_d \not\equiv \pm b \pmod{n}$, as $d, n/d > 1$.

In the other direction, suppose that one has a fast algorithm for rapidly finding arbitrary square roots mod n for odd integers n. In particular given $A \pmod{n}$, the algorithm randomly determines some $x \pmod{n}$ for which $x^2 \equiv A \pmod{n}$; by "random" we mean that each time the "square root finding" algorithm is run it is equally likely to produce any one of the 2^k solutions (as in Lemma 10.1.1). Now define $d = (n, x - b)$ (as in the proof of Lemma 10.1.1) and so we factor n as $d \cdot n/d$. This works provided $d \neq 1$ or n, that is, provided that $x \not\equiv b$ or $-b \pmod{n}$.

Now, the probability that $x \equiv b$ or $-b \pmod{n}$ is $2/2^k$ which is $\leq \frac{1}{2}$ as $k \geq 2$. Therefore the probability of finding a non-trivial factor of n each time the "square root finding" algorithm is run is $\geq \frac{1}{2}$. This does not seem persuasive, but if we run the "square root finding" algorithm 20 times, then the probability that the algorithm gives 1 or n on every run is $\leq \left(\frac{1}{2}\right)^{20}$, which is less than one in a million. So, in practice, we will quickly find a non-trivial factor of n. □

We have shown that finding square roots mod n and factoring n are more or less equally difficult problems.

Exercise 10.1.1. Find all of the square roots of 49 mod $3^2 \cdot 5 \cdot 11$.

10.2. Cryptosystems

Cryptography has been around for as long as the need to communicate secrets at a distance. Julius Caesar, on campaign, communicated military messages by creating *ciphertext* from *plaintext* (the unencrypted message), replacing each letter of the plaintext with that letter which is three letters further on in the alphabet. Thus A becomes D, B becomes E, etc. For example,

```
t h i s i s v e r y i n t e r e s t i n g
```
becomes
```
w k l v l v y h u b l q w h u h v w l q j
```

(Y became B, since we wrap around to the beginning of the alphabet. It is essentially the map $x \to x + 3 \pmod{26}$.) At first sight an enemy might regard $WKLV \ldots WLQJ$ as gibberish even if the message was intercepted. It is easy enough to decrypt the ciphertext, simply by going back three places in the alphabet for each letter, to reconstruct the original message. The enemy could easily do this if (s)he guessed that the *key* is to rotate the letters by three places in the alphabet, or even if they only guessed that one rotates by a fixed number of letters, as there would only be 25 possibilities to try. So in classical cryptography it is essential to keep the key secret, as well as the technique by which the key was created.[1]

One can generalize to arbitrary *substitution ciphers* where one replaces the alphabet by some permutation of the alphabet. There are 26! permutations of our alphabet, which is around 4×10^{26} possibilities, enough one might think to be safe. And it would be if the enemy went through each possibility, one at a time. However the clever *cryptographer* will look for patterns in the ciphertext. In the above short ciphertext we see that L appears four times among the 21 letters, and H, V, W three times each, so it is likely that these letters each represent one of A, E, I, S, T. By looking for multiword combinations (like the ciphertext for THE) one can quickly break any ciphertext of around one hundred letters.

To combat this, armies in the First World War used longer cryptographic keys, rather than of length 1. That is, they would take a word like $ABILITY$ and since A is letter 1 in the alphabet, B is letter 2, and $ILITY$ are letters 9,12,9,20,25, respectively, they would rotate on through the alphabet by $1, 2, 9, 12, 9, -6, -1$ letters to *encrypt* the first seven letters, and then repeat this process on the next seven. For example, we begin with the message, adding the word "ability" as often as is needed:

```
w e n e e d t o m a k e a n e x a m p l e
```
plus
```
a b i l i t y a b i l i t y a b i l i t y
```
becomes
```
x g w q n x s p o j w n u m f z j y y f d
```

This can again be "broken" by statistical analysis, though the longer the key length, the harder it is to do. Of course using a long key on a battlefield would be difficult, so one needed to compromise between security and practicality. A *one-time pad*,

---

[1]*Steganography*, hiding secrets in plain view, is another method for communicating secrets at a distance. In 499 B.C., Histiaeus shaved the head of his most trusted slave, tattooed a message on his bald head, and then sent the slave to Aristagoras, once the slave's hair had grown back. Aristagoras then shaved the slave's head again to recover the secret message telling him to revolt against the Persians. In more recent times, cold war spies reportedly used "microdots" to transmit information, and Al-Qaeda supposedly notified its terrorist cells via messages hidden in images on certain webpages.

where one uses such a long key that one never repeats a pattern, is unbreakable by statistical analysis. This might have been used by spies during the cold war and was perhaps based on the letters in an easily obtained book, so that the spy would not have to possess any obviously incriminating evidence.

During the Second World War the Germans came up with an extraordinary substitution cypher that involved changing several settings on a specially built typewriter (an *Enigma machine*). The number of possibilities was so large that the Germans remained confident that it could not be broken, and they even changed the settings every day so as to ensure that it would be extremely difficult. The Poles managed to obtain an early Enigma machine and their mathematicians determined how it worked. They shared their findings with the Allies so that after a great amount of effort the Allies were able to break German codes quickly enough to be useful, even vital, to their planning and strategy.[2] Early successes led to the Germans becoming more cautious, and thence to horrific decisions having to be made by the Allied leaders to safeguard this most precious secret.[3]

The Allied cryptographers would cut down the number of possibilities (for the settings on the Enigma machine) to a few million, and then their challenge became to build a machine to try out many possibilities very rapidly. Up until then one would have to change, by hand, external settings on the machine to try each possibility; it became a goal to create a machine in which one could change what it was doing, *internally*, by what became known as a *program*, and this stimulated, in part, the creation of the first modern computers.

Exercise 10.2.1. One can also create a cryptosystem using binary addition. For example, our key could be the 20-letter word $k = 10111011101111011001$. Then we could encrypt by using bit-by-bit addition; that is, $0 \oplus 0 = 1 \oplus 1 = 0$ and $0 \oplus 1 = 1 \oplus 0 = 1$. Therefore if the plaintext is $p = 11100010101101000011$, then $c = p \oplus k$, namely

$$\begin{array}{r} 10111\ 01110\ 11110\ 11001 \\ \oplus\ 11100\ 01010\ 11010\ 00011 \\ \hline =\ 01011\ 00100\ 00100\ 11010. \end{array}$$

It is easy to recover the plaintext since $p = c \oplus k$. Prove that one can recover the key if one knows the ciphertext and the plaintext.

10.3. RSA

In the theory of cryptography we always have two (imaginary) people, Alice and Bob, attempting to share a secret over an open communication channel, and the evil Oscar listening in, attempting to figure out what the message says. We will begin by describing a *private key* scheme for exchanging secrets based on the ideas in our number theory course:

Suppose that prime p is given and integers d and e such that $de \equiv 1 \pmod{p-1}$. Alice knows p and e but not d, whereas Bob knows p and d but not e. The numbers

---

[2] As portrayed, rather inaccurately, in the film *The Imitation Game*.
[3] The ability to crack the Enigma code allowed the Allied leaders to save lives. However if they used it so often that every possible life was saved, the Germans would have realized that the Allies had broken the code, and then the Germans were liable to have moved on to a different cryptographic method, which perhaps the Allied codebreakers might have been unable to decipher. Hence the Allied leadership was forced to use its knowledge sparingly so that it would be available in the militarily most advantageous situations. As a consequence, they knowingly sent many sailors to their doom, knowing where the U-boats were waiting in ambush, but being forced not to disclose that information.

10.3. RSA

d and e are kept secret by whoever knows them. Thus if Alice's secret message is M,[4] she *encrypts* M by computing $x \equiv M^e \pmod{p}$. She sends the *ciphertext* x over the open channel. Then Bob *decrypts* by raising x to the dth power mod p, since
$$x^d \equiv (M^e)^d \equiv M^{de} \equiv M \pmod{p}$$
as $de \equiv 1 \pmod{p-1}$. As far as we know, Oscar will discover little by intercepting the encrypted messages x, even if he intercepts many different x, and even if he can occasionally make an astute guess at M. However, if Oscar is able to steal the values of p and e from Alice, he will be able to determine d, since d is the inverse of $e \bmod p - 1$, and this can be determined by the Euclidean algorithm, as discussed in exercise 3.5.5 (see the second proof of Corollary 3.5.2). He is then able to decipher Alice's future secret messages, in the same way as Bob does.

This is the problem with most classical cryptosystems; once one knows the encryption method it is not difficult to determine the decoding method. In 1975 Diffie and Hellman proposed a sensational idea: Can one find a cryptographic scheme in which the encryption method gives no help in determining a decryption method? If one could, one would then have a *public key* cryptographic scheme, which is exactly what is needed in our age of electronic information, in particular allowing people to use passwords in public places (for instance when using an ATM) without fear any lurking Oscar will be able to figure out how to impersonate them.[5]

In 1977 Rivest, Shamir, and Adleman (RSA) realized this ambition, via a minor variation of the above private key cryptosystem:[6] Now let $p \neq q$ be two large primes[7] and $n = pq$. Select integers d and e such that $de \equiv 1 \pmod{\phi(pq)}$. Alice knows pq and e but not d, while Bob knows pq and d. Thus if Alice's secret message is M, the ciphertext is $x \equiv M^e \pmod{pq}$, and Bob decrypts this by taking $x^d \equiv (M^e)^d \equiv M^{de} \equiv M \pmod{pq}$ as $de \equiv 1 \pmod{\phi(pq)}$ using Euler's Theorem.

Now, if Oscar steals the values of pq and e from Alice, will he be able to determine d, the inverse of $e \bmod \phi(pq) = (p-1)(q-1)$? When the modulus was the prime p, Oscar had no difficulty in determining $\phi(p) = p - 1$. Now that the modulus is pq, can Oscar easily determine $(p-1)(q-1)$? If so, then, since he already knows pq, he would be able to determine $pq + 1 - (p-1)(q-1) = p + q$ and hence p and q, since they are the roots of $x^2 - (p+q)x + pq = 0$. In practice, Oscar needs to only know d to factor n (see exercise 5.27 in [**CP05**])[8]. In other words, if Oscar can "break" the RSA algorithm, then he can factor $n = pq$, and vice versa.

We have just shown that breaking RSA is more or less as difficult as factoring. Therefore RSA is a secure cryptographic protocol (when correctly implemented) if and only if n is a difficult integer to factor. But nobody truly knows whether

---

[4] Of course a message is usually in words, but one converts the letters to numbers using some simple substitutions, like "01" for "A", "02" for "B", ... , "26" for "Z", etc., and concatenates these numbers. Thus "cabbie" becomes "030102020905". It is this number that is our message that we denote by M.

[5] When Alice uses a password, a cryptographic protocol might append a *timestamp* to ensure that the encrypted password (plus timestamp) is different with each use, and so Bob will get suspicious if the same timestamp is used again later.

[6] It is now known that (Sir) Clifford Cocks, working for the British secret cryptography agency, GCHQ, had discovered this *RSA algorithm* in 1974, and it had been classified "Top Secret". See https://www.wired.com/1999/04/crypto/ for the story.

[7] We will develop fast methods to find large primes in appendix 10C.

[8] This uses Pollard's $p - 1$ method, which will not be discussed in this book, and is an algorithm that runs in *probabilistic polynomial time*.

factoring is a difficult problem, nor how to select integers that are provably hard to factor. In our current state of knowledge, we do not know any very efficient ways to factor arbitrary large numbers, but that does not necessarily mean that there is no quick way to do so.[9] So why do we put our faith (and secrets and fortunes) in the difficulty of factoring? The security of a cryptographic protocol must evidently be based on the difficulty of resolving *some* mathematical problem,[10] but we do not know how to *prove* that any particular mathematical problem is necessarily difficult to solve.[11] However the problem of factoring efficiently has been studied by many of the greatest minds in history, from Gauss onwards, who have looked for an efficient factoring algorithm and failed. Is this a good basis to have faith in RSA? Probably not, but we have no better. (More on this at the end of section 10.15 of appendix 10F.)

Exercise 10.3.1. Let $n = 11 \times 53$ be an RSA modulus with encryption exponent $e = 7$. Determine d, the decryption exponent, by hand, using the Euclidean algorithm and the Chinese Remainder Theorem.

Exercise 10.3.2. Let $n = 5891$ be an RSA modulus with encryption exponent $e = 29$ and decryption exponent $d = 197$. Use this information to factor n.

10.4. Certificates and the complexity classes P and NP

Algorithms are typically designed to work on any of an arbitrarily large class of examples, and one wishes them to work as fast as possible. If the example is input in ℓ characters, and the function calculated is genuinely a function of all the characters of the input, then one cannot hope to compute the answer any quicker than the length, ℓ, of the input. A *polynomial time algorithm* is one in which the answer is computed in no more than $c\ell^A$ steps, for some constants $c, A > 0$, no matter what the input. These are considered to be quick algorithms. There are many simple problems that can be answered in polynomial time (the set of such problems is denoted by P and was already discussed in section 7.14 of appendix 7A); see section 10.15 of appendix 10F for more details. In modern number theory, because of the intrinsic interest as well as because of the applications to cryptography, we are particularly interested in the running times of factoring and primality testing algorithms.

At the 1903 meeting of the American Mathematical Society, F. N. Cole came to the blackboard and, without saying a word, wrote down

$$2^{67} - 1 = 147573952589676412927 = 193707721 \times 761838257287,$$

long-multiplying the numbers out on the right side of the equation to prove that he was indeed correct. Afterwards he said that figuring this out had taken him "three years of Sundays". The moral of this tale is that although it took Cole a great deal

---

[9]There are some families of numbers that we know are easy to factor (for example, see exercise 10.7.2 for a fast factoring method if p and q are close together) so we need to avoid those when selecting a modulus for RSA.

[10]Here we are talking about cryptographic protocols on computers as we know them today. There is a highly active quest to create *quantum computers*, on which cryptographic protocols are based on a very different set of ideas.

[11]We can *prove* that almost all mathematical problems are "difficult to solve" (see section 10.16 of appendix 10F), but we do not know how to identify *one specific problem* that is provably difficult to solve. This is a notoriously challenging and important open problem.

10.4. Certificates and the complexity classes P and NP

of work and perseverance to find these factors, it did not take him long to justify his result to a room full of mathematicians (and, indeed, to give a proof that he was correct). Thus we see that one can provide a short proof, even if finding that proof takes a long time.

In general one can exhibit factors of a given integer n to give a short proof that n is composite. Such proofs, which can be checked in polynomial time, are called *certificates*. (The set of problems for which the answer can be checked in polynomial time is denoted by NP.) Note that it is not necessary to exhibit factors to give a short proof that a number is composite. Indeed, we already saw in the converse to Fermat's Little Theorem, Corollary 7.2.1, that one can exhibit an integer a coprime to n for which n does not divide $a^{n-1} - 1$ to provide a certificate that n is composite.

What about primality testing? If someone gives you an integer and asserts that it is prime, can you quickly check that this is so? Can they give you better evidence than their say-so that it is a prime number? Can they provide some sort of certificate that gives you all the information you need to quickly verify that the number is indeed a prime? We had hoped (see section 7.6) that we could use the converse of Fermat's Little Theorem to establish a quick primality test, but we saw that Carmichael numbers seem to stop that idea from reaching fruition. Here we are asking for less, for a short certificate for a proof of primality. It is not obvious how to construct such a certificate, certainly not so obvious as with the factoring problem. It turns out that some old remarks of Lucas from the 1870s can be modified for this purpose. We begin with a sure-fire primality test, obtained as a consequence of Proposition 7.5.1.

Corollary 10.4.1. *Suppose that $n > 1$ is a positive integer for which there exists an integer g with $(g, n) = 1$ such that $g^{n-1} \equiv 1 \pmod{n}$ and $g^{(n-1)/q} \not\equiv 1 \pmod{n}$ for every prime q dividing $n - 1$. Then n is a prime.*

Proof. Proposition 7.5.1 implies that g has order $n - 1 \pmod{n}$, so that the $n - 1$ reduced residues $1, g, \ldots, g^{n-1}$ are all distinct mod n. Therefore every integer a in the range $1 \leq a \leq n - 1$ is coprime to n, implying that n is prime. $\square$

We are not suggesting that Corollary 10.4.1 provides a fast primality test. One can probably find g rapidly, if it exists, using Gauss's algorithm which is discussed in section 7.15 of appendix 7B. However the algorithm requires one to completely factor $n - 1$, and we have no particularly fast factoring algorithms. On the other hand, if $n - 1$ has already been factored, then one can proceed rapidly. Indeed we can provide a "certificate" to allow a checker to quickly verify that n is prime, which would consist of

$$g \text{ and } \{q \text{ prime} : q \text{ divides } n - 1\}.$$

The checker would need to verify that $g^{n-1} \equiv 1 \pmod{n}$ whereas $g^{(n-1)/q} \not\equiv 1 \pmod{n}$ for all primes q dividing $n - 1$, something that can be quickly accomplished using fast exponentiation (as explained in section 7.13 of appendix 7A).

There is a problem though: One needs (the additional) certification that each such q is prime. The solution is to iterate the above algorithm; and one can show that no more than $\log n$ odd primes need to be certified prime in the process of proving that n is prime. Thus we have a "short" certificate that n is prime.

At first one might hope that this also provides a quick way to test whether a given integer n is prime. However there are several obstacles. The most important is that we need to factor $n-1$ in creating the certificate. When one is handed the certificate, $n-1$ is already factored, so that is not an obstacle to the use of the certificate; however it is a fundamental impediment to the rapid creation of the certificate (and therefore to using this as a primality test).

Exercise 10.4.1. Assuming only that 2 is prime, provide a certificate that proves that 107 is prime.

Exercise 10.4.2. Let $F_m = 2^{2^m} + 1$ with $m \geq 2$ be a Fermat number.
(a) Prove that if there exists an integer q for which $q^{\frac{F_m-1}{2}} \equiv -1 \pmod{F_m}$, then F_m is prime.
(b) Deduce an "if and only if" condition for the primality of F_m using exercise 8.5.4.

10.5. Polynomial time primality testing

Although the converse to Fermat's Little Theorem does not provide a polynomial time primality test, one can further develop this idea. For example, we know that $a^{\frac{p-1}{2}} \equiv -1$ or $1 \pmod{p}$ by Euler's criterion, and hence if $a^{\frac{n-1}{2}} \not\equiv \pm 1 \pmod{n}$, then n is composite. This identifies even more composite n than Corollary 7.2.1 alone, but not necessarily all n. We develop this idea further in section 10.8 of appendix 10A to find a criterion of this type that is satisfied by all primes but not by any composites. However we are unable to prove that this is indeed a polynomial time primality test without making certain assumptions that are, as yet, unproved.

There have indeed been many ideas for establishing a primality test which is provably polynomial time, but this was not achieved until 2002. This was of particular interest since the proof was given by a professor, Manindra Agrawal, and two undergraduate students, Kayal and Saxena, working together with Agrawal on a summer research project. Their algorithm is based on the following elegant characterization of prime numbers.

Theorem 10.1 (Agrawal, Kayal, and Saxena (AKS)). *For given integer $n \geq 2$, let r be a positive integer $< n$, for which n has order $> 9(\log n)^2$ modulo r. Then n is prime if and only if*

- *n is not a perfect power,*
- *n does not have any prime factor $\leq r$,*
- *$(x+a)^n \equiv x^n + a \mod (n, x^r - 1)$ for each integer a, $1 \leq a \leq 3\sqrt{r} \, \log n$.*

The last equation uses "modular arithmetic" in a way that is new to us, but analogous to what we have seen: $(x+a)^n \equiv x^n + a \mod (n, x^r - 1)$ means that there exist $f(x), g(x) \in \mathbb{Z}[x]$ such that $(x+a)^n - (x^n + a) = nf(x) + (x^r - 1)g(x)$.

At first sight this might seem to be a rather complicated characterization of the prime numbers. However this fits naturally into the historical progression of ideas in this subject (indeed, see appendix 10G for a discussion and a proof), is not so complicated (compared to some other ideas in use), and has the great advantage that it is straightforward to develop into a fast algorithm for proving the primality of large primes. *However*, although the AKS algorithm satisfies the desire to have a rigorously proved polynomial time primality testing algorithm, it is not in practice the fastest algorithm for establishing primality of the largest integers currently being considered.[12]

Exercise 10.5.1. Let p^k be the highest power of prime p that divides n, with $k \geq 1$.
(a) Prove that p^k does not divide $\binom{n}{p}$.
(b) Deduce that n does not divide $\binom{n}{p}$.
(c) Show that if n is composite, then n does not divide all the coefficients of the polynomial $(1+x)^n - x^n - 1$.

Exercise 10.5.2. Use the previous exercise to show:
(a) n is prime if and only if $(x+1)^n \equiv x^n + 1 \pmod{n}$.
(b) If $(n,a) = 1$, then n is prime if and only if $(x+a)^n \equiv x^n + a \pmod{n}$.
(c) Prove that if n is prime, then $(x+a)^n \equiv x^n + a \pmod{n, x^r - 1}$ for any integer a with $(a,n) = 1$ and any $r > 1$.

10.6. Factoring methods

The problem of distinguishing prime numbers from composite numbers and of resolving the latter into their prime factors is known to be one of the most important and useful in arithmetic. It has engaged the industry and wisdom of ancient and modern geometers to such an extent that it would be superfluous to discuss the problem at length. Nevertheless we must confess that all methods that have been proposed thus far are either restricted to very special cases or are so laborious and difficult that even for numbers that do not exceed the limits of tables constructed by estimable workers, they try the patience of even the practiced calculator. And these methods do not apply at all to larger numbers .... It frequently happens that the trained calculator will be sufficiently rewarded by reducing large numbers to their factors so that it will compensate for the time spent. Further, the dignity of the science itself seems to require that every possible means be explored for the solution of a problem so elegant and so celebrated .... It is in the nature of the problem that any method will become more complicated as the numbers get larger. Nevertheless, in the following methods the difficulties increase rather slowly .... The techniques that were previously known would require intolerable labor even for the most indefatigable calculator.
— from article 329 of *Disquisitiones Arithmeticae* (1801) by C. F. GAUSS

The first factoring method, other than trial division, was given by Fermat: His goal was to write a given odd integer n as $x^2 - y^2$, so that $n = (x-y)(x+y)$. He started with m, the smallest integer $\geq \sqrt{n}$, and then looked to see if $m^2 - n$ is a square. If so, say $m^2 - n = r^2$, then $n = (m-r)(m+r)$.

It is not easy to determine (at least by hand) whether a large integer is a square, though most are not. Fermat simplified his algorithm by quickly eliminating non-squares, by testing whether $m^2 - n$ is a square modulo various small primes. If $m^2 - n$ is not a square, then he tested whether $(m+1)^2 - n$ is a square; if that failed, whether $(m+2)^2 - n$ is a square, or $(m+3)^2 - n$, ..., etc. Since Fermat computed by hand he also noted the trick that

$$(m+1)^2 - n = \quad m^2 - n + (2m+1),$$
$$(m+2)^2 - n = (m+1)^2 - n + (2m+3), \text{ etc.,}$$

---
[12] Because other algorithms that we believe, but cannot prove, are polynomial time, run faster.

so that, at each step he only needed to add a relatively small number to the integer he had just tested, and the next add-on is just two larger than the previous one.

For example, Fermat factored $n = 2027651281$ so that $m = 45030$. Then

$45030^2 - n \qquad = \qquad 49619$ which is not a square mod 100;

$45031^2 - n = 49619 + 90061 = 139680$ which is divisible by 2^5, not 2^6;

$45032^2 - n = 139680 + 90063 = 229743$ which is divisible by 3^3, not 3^4;

$45033^2 - n = 229743 + 90065 = 319808$ which is not a square mod 3; etc.

$$\vdots$$

up until $45041^2 - n = 1020^2$, so that

$n = 2027651281 = 45041^2 - 1020^2 = (45041 - 1020) \times (45041 + 1020) = 44021 \times 46061$.

Exercise 10.6.1. Factor 1649 using Fermat's method.

Gauss and other authors further developed Fermat's ideas, most importantly realizing that if $x^2 \equiv y^2 \pmod{n}$ with $x \not\equiv \pm y \pmod{n}$ and $(x, n) = 1$, then

$$\gcd(n, x - y) \cdot \gcd(n, x + y)$$

gives a non-trivial factorization of n.

The issue now becomes to rapidly determine two residues x and $y \pmod{n}$ with $x \not\equiv y$ or $-y \pmod{n}$, such that $x^2 \equiv y^2 \pmod{n}$. Several factoring algorithms work by generating a sequence of integers $a_1, a_2, \ldots$, with each

$$a_i \equiv b_i^2 \pmod{n} \text{ but } a_i \neq b_i^2$$

for some known integer b_i, until some subsequence of the a_i's has product equal to a square, say

$$y^2 = a_{i_1} \cdots a_{i_r}.$$

Then one sets $x^2 = (b_{i_1} \cdots b_{i_r})^2$ to obtain $x^2 \equiv y^2 \pmod{n}$, and there is a good chance that $\gcd(n, x - y)$ is a non-trivial factor of n.

We want to generate the a_i's so that it is not so difficult to find a subsequence whose product is a square; to do so, we need to be able to factor the a_i. This is most easily done by only keeping those a_i that have all of their prime factors $\leq B$, for some appropriately chosen bound B. Suppose that the primes up to B are $p_1, p_2, \ldots, p_k$. If $a_i = p_1^{a_{i,1}} p_2^{a_{i,2}} \cdots p_k^{a_{i,k}}$, then let $v_i = (a_{i,1}, a_{i,2}, \ldots, a_{i,k})$, which is a vector with entries in $\mathbb{Z}$.

Exercise 10.6.2. Show that $\prod_{i \in I} a_i$ is a square if and only if $\sum_{i \in I} v_i \equiv (0, 0, \ldots, 0) \pmod{2}$.

Hence to find a non-trivial subset of the a_i whose product is a square, we simply need to find a non-trivial linear dependency mod 2 amongst the vectors v_i. This is easily achieved through the methods of linear algebra and guaranteed to exist once we have generated more than k such integers a_i.

The quadratic sieve factoring algorithm selects the b_i so that it is easy to find the small prime factors of the a_i, using Corollary 2.3.1. There are other algorithms that attempt to select the b_i so that the a_i are small and therefore more likely to have small prime factors. We discuss some of these in appendix 10B. The best algorithm, the number field sieve, is an analogy to the quadratic sieve algorithm over number fields.

There are many other cryptographic protocols based on ideas from number theory. Some of these will be discussed in the appendices to this chapter.

References: See [CP05] and [Knu98], as well as:

[1] Carl Pomerance, *A tale of two sieves*, Notices Amer. Math. Soc. **43** (1996), 1473–1485.
[2] John D. Dixon *Factorization and primality tests*, Amer. Math. Monthly **91** (1984), 333–352.

Additional exercises

Exercise 10.7.1. Suppose that n is an odd composite integer. Prove that for at least half the pairs x,y with $0 \leq x,y < n$ and $x^2 \equiv y^2 \pmod{n}$, we have $1 < \gcd(x-y, n) < n$.

Exercise 10.7.2. Factor $n = 62749$. Let $m = [\sqrt{n}] + 1 = 251$. Compute $(m+i)^2 \pmod{n}$ for $i = 0, 1, 2, \ldots$ and retain those residues whose prime factors are all ≤ 11. Therefore we have $251^2 \equiv 2^2 \cdot 3^2 \cdot 7$; $253^2 \equiv 2^2 \cdot 3^2 \cdot 5 \cdot 7$; $257^2 \equiv 2^2 \cdot 3 \cdot 5^2 \cdot 11$; $260^2 \equiv 3^2 7^2 \cdot 11$; $268^2 \equiv 3 \cdot 5^2 \cdot 11^2$; $271^2 \equiv 2^2 \cdot 3^5 \cdot 11 \pmod{n}$. Use this information to factor n.

Exercise 10.7.3. Alice is sending Bob messages using RSA with public key modulus $n = 2027651281$ and encryption exponent $e = 66308903$. Oscar recalls that n is the number Fermat factored in section 10.6. Find the decryption exponent for Oscar.

We wish to determine how many different odd primes are involved in the Lucas certificate of section 10.4.

Exercise 10.7.4. Let n be prime and suppose $q_1, \ldots, q_k$ are the odd prime factors of $n-1$.
(a) Prove that the product of these primes, $N_1 := q_1 \cdots q_k$, is $\leq n/2$.
(b)† To certify that $q_1, \ldots, q_k$ are prime we need the set of odd prime factors of $q_1 - 1, \ldots, q_k - 1$. Let's call those primes $p_1, \ldots, p_\ell$. Prove that the product of these primes, $N_2 := p_1 \cdots p_\ell$, is $\leq N_1/2^k$.
(c) Generalize this argument to show that if there are r primes to be certified at the jth stage, then $N_{j+1} \leq N_j/2^r$.
(d)† Prove that if there are m primes that were certified to be prime during all the steps of this argument, then $2^m \leq n$. Explain why this implies that primality testing is in NP.

Exercise 10.7.5.† Suppose n is an odd composite, and $a^{(n-1)/2} \equiv 1$ or $-1 \pmod{n}$ for every a with $(a,n) = 1$. Deduce that $a^{(n-1)/2} \equiv 1 \pmod{n}$ for every a with $(a,n) = 1$ and that n is a Carmichael number.

Appendix 10A. Pseudoprime tests using square roots of 1

In section 7.6 we noted that the converse to Fermat's Little Theorem may be used to give a quick proof that a given integer n is composite: One simply finds an integer a, not divisible by n, for which $a^{n-1} \not\equiv 1 \pmod{n}$ (if this fails, that is, if $a^{n-1} \equiv 1 \pmod{n}$ and n is composite, then n is called a *base-a pseudoprime*). Such a search often works quickly, especially for randomly chosen values of n, but can fail if the tested n have some special structure. For example, it always fails for Carmichael numbers, which have the property that n is a base-a pseudoprime for every a with $(a, n) = 1$. What can we do in these cases? Can we construct a test, based on similar ideas, that is guaranteed to recognize even these composite numbers?

10.8. The difficulty of finding all square roots of 1

Lemma 10.1.1 implies that there are *at least* four distinct square roots of $1 \pmod{n}$, for any odd n which is divisible by at least two distinct primes. This suggests that we might try to prove that a given base-a pseudoprime n is composite by finding a square root of $1 \pmod{n}$ which is neither 1 nor -1. (If we can find such a square root of $1 \pmod{n}$, then we can partially factor n, as discussed in section 10.1.) The issue then becomes: How do we efficiently search for a square root of 1?

This is not difficult: Since n is a base-a pseudoprime, we have

$$\left(a^{\frac{n-1}{2}}\right)^2 = a^{n-1} \equiv 1 \pmod{n},$$

and so $a^{\frac{n-1}{2}} \pmod{n}$ is a square root of $1 \pmod{n}$. By Euler's criterion we know that if p is prime, then $a^{\frac{p-1}{2}} \equiv (a/p) \pmod{p}$, so that $a^{\frac{p-1}{2}} \equiv 1$ or $-1 \pmod{p}$. If n is a base-a pseudoprime (and therefore composite), it is feasible that $a^{\frac{n-1}{2}} \not\equiv (a/n) \pmod{n}$, which would imply that n is composite. If $a^{\frac{n-1}{2}} \pmod{n}$ is neither 1 nor

10.8. The difficulty of finding all square roots of 1

-1, this allows us to factor n into two parts, since

$$n = \gcd(a^{\frac{n-1}{2}} - 1, n) \cdot \gcd(a^{\frac{n-1}{2}} + 1, n).$$

If n is composite and $a^{\frac{n-1}{2}} \equiv (a/n) \pmod{n}$, then we call n a base-a *Euler pseudoprime*.

For example, 1105 is a Carmichael number, and so $2^{1104} \equiv 1 \pmod{1105}$. We take the square root, and determine that $2^{552} \equiv 1 \pmod{1105}$. So this method fails to prove that 1105 is composite, since 1105 is a base-2 Euler pseudoprime. But, wait a minute, 552 is even, so we can take the square root again, and a calculation reveals that $2^{226} \equiv 781 \pmod{1105}$. That is, 781 is a square root of 1 mod 1105, which proves that 1105 is composite. Moreover, since $\gcd(781 - 1, 1105) = 65$ and $\gcd(781 + 1, 1105) = 17$, we can even factor 1105 as 65×17.[13]

This property is even more striking mod 1729. In this case $1728 = 2^6 \cdot 27$ so we can take square roots many times. Indeed, taking successive square roots of 2^{1728} we determine that

$$1 \equiv 2^{1728} \equiv 2^{864} \equiv 2^{432} \equiv 2^{216} \pmod{1729}, \text{ but then } 2^{108} \equiv 1065 \pmod{1729}.$$

This proves that 1729 is composite, and even that

$$1729 = \gcd(1064, 1729) \times \gcd(1066, 1729) = 133 \times 13.$$

This protocol of taking successive square roots can fail to identify that our given pseudoprime is indeed composite; for example, we cannot use 103 to prove that either 561 or 1729 is composite, since

$$103^{35} \equiv 1 \pmod{561}, \text{ and so } 103^{70} \equiv \cdots \equiv 103^{560} \equiv 1 \pmod{561},$$

$$103^{27} \equiv -1 \pmod{1729}, \text{ and so } 103^{54} \equiv \cdots \equiv 103^{1728} \equiv 1 \pmod{1729},$$

but such failures are rare (see exercise 10.8.7).

Suppose that n is a composite integer with $n - 1 = 2^k m$ for some integer $k \geq 1$ with m odd. We call n a base-a *strong pseudoprime* if the sequence of residues

(10.8.1) $\qquad a^{n-1} \pmod{n}, \ a^{(n-1)/2} \pmod{n}, \ldots, \ a^{(n-1)/2^k} \pmod{n}$

is equal to either

$$1, 1, \ldots, 1 \quad \text{or} \quad 1, 1, \ldots, 1, -1, *, \ldots, *$$

where the $*$'s stand for any residue mod n. These are the only two possibilities if n is prime, and so if the sequence of residues in (10.8.1) looks like one of these two possibilities, then this information does not allow us to deduce that n is composite.

On the other hand, if n is a not a base-a strong pseudoprime, then we say that a is a *witness* (to n being composite). To be more precise:

Definition. *Suppose that n is a composite odd integer and $n - 1 = 2^k m$ for some integer $k \geq 1$ with m odd. Assume that n is a base-a pseudoprime; that is, $a^{n-1} \equiv 1 \pmod{n}$. If $a^m \equiv 1 \pmod{n}$ or $a^{m \cdot 2^j} \equiv -1 \pmod{n}$ for some integer $j \geq 0$, then n is a base-a strong pseudoprime. Otherwise a is a witness (to the compositeness of n) and if ℓ is the largest integer for which $a^{m \cdot 2^\ell} \not\equiv -1$ or $1 \pmod{n}$, then $\gcd(a^{m \cdot 2^\ell} - 1, n)$ is a non-trivial factor of n.*

---

[13]We have not factored 1105 into prime factors (since 65 factors further as $65 = 5 \times 13$), but rather into two non-trivial factors.

One can compute high powers modulo n very rapidly using "fast exponentiation" (a technique we discussed in section 7.13 of appendix 7A), so this strong pseudoprime test can be done quickly and easily.

In exercise 10.8.7 we will show that at least three-quarters of the integers a, $1 \leq a \leq n$, with $(a,n) = 1$ are witnesses for n, for each odd composite $n > 9$. So can we find a witness quickly if n is composite?

- The most obvious idea is to try $a = 2, 3, 4, \ldots$ consecutively until we find a witness. It is believed that there is a witness $\leq 2(\log n)^2$, but we cannot prove this (though we can deduce this from a famous conjecture, the Generalized Riemann Hypothesis[14]).

- Pick integers $a_1, a_2, \ldots, a_\ell, \ldots$ from $\{1, 2, 3, \ldots, n-1\}$ at random until we find a witness. By what we wrote above, if n is composite, then the probability that none of $a_1, a_2, \ldots, a_\ell$ are witnesses for n is $\leq 1/4^\ell$. Thus with a hundred or so such tests we get a probability that is so small that it is inconceivable that it could occur in practice; so we believe that any integer n for which none of a hundred randomly chosen a's is a witness is prime. We call such n *"industrial strength primes"* since they have not been proven to be prime, but there is an enormous weight of evidence that they are not composite.

This test is a *random polynomial time* test for compositeness (like our test for finding a quadratic non-residue given at the end of appendix 8B). If n is composite, then the randomized witness test is almost certain to provide a short proof of n's compositeness in 100 runs of the test. On the other hand, if 100 runs of the test do not produce a witness, then we can be almost certain that n is prime, but we cannot be *absolutely* certain since no proof is provided, and therefore we have an industrial strength prime.

In practice the witness test accomplishes Gauss's dream of quickly distinguishing between primes and composites, for either we will quickly get a witness to n being composite or, if not, we can be almost certain that our industrial strength prime is indeed prime. Although this solves the problem in practice, we cannot be absolutely certain that we have distinguished correctly when we claim that n is prime since we have no proof, and mathematicians like proof. Indeed if you claim that industrial strength primes are prime, without proof, then a cynic might not believe that your randomly chosen a are so random or that you are unlucky or . . . . No, what we need is a proof that a number is prime when we think that it is.

Exercise 10.8.1. Find all bases b for which 15 is a base-b Euler pseudoprime.

Exercise 10.8.2.[†] We wish to show that every odd composite n is not a base-b Euler pseudoprime for some integer b, coprime to n. Suppose not, i.e., that n is a base-b Euler pseudoprime for every integer b with $(b,n) = 1$.
  (a) Show that n is a Carmichael number.
  (b) Show that if prime p divides n, then $p-1$ cannot divide $\frac{n-1}{2}$.
  (c) Deduce that $(b/n) \equiv (b/p) \pmod{p}$ for each prime p dividing n.
  (d) Explain why (c) cannot hold for every integer b coprime to n.

---

[14] We discussed the Riemann Hypothesis, and its generalizations, in sections 5.16 and 5.17 of appendix 5D. Suffice to say that this is one of the most famous and difficult open problems of mathematics, so much so that the Clay Mathematics Institute has now offered one million dollars for its resolution (see http://www.claymath.org/millennium-problems/).

10.8. The difficulty of finding all square roots of 1

Exercise 10.8.3. Prove that $F_n = 2^{2^n} + 1$ is either a prime or a base-2 strong pseudoprime.

Exercise 10.8.4. Prove that if n is a base-2 pseudoprime, then $2^n - 1$ is a base-2 strong pseudoprime and a base-2 Euler pseudoprime. Deduce that there are infinitely many base-2 strong pseudoprimes.

Exercise 10.8.5. Pépin showed that one can test Fermat numbers F_m for primality by using just one strong pseudoprime test; i.e., F_m is prime if and only if $3^{(F_m-1)/2} \equiv -1 \pmod{F_m}$.
  (a) Use exercise 8.5.4 to show if F_m is prime, then $3^{(F_m-1)/2} \equiv -1 \pmod{F_m}$.
  (b) In the other direction show that if $3^{(F_m-1)/2} \equiv -1 \pmod{F_m}$, then $\mathrm{ord}_p(3) = 2^{2^m}$ whenever prime $p|F_m$.
  (c) Deduce that $F_m - 1 \leq p - 1$ in (b) and so F_m is prime.

Exercise 10.8.6.† (a) Prove that $A := (4^p + 1)/5$ is composite for all primes $p > 3$.
  (b) Deduce that A is a base-2 strong pseudoprime.

Exercise 10.8.7.‡ How many witnesses are there mod n? Suppose that $n - 1 = 2^k m$ with m odd and $k \geq 1$, and that n has ω distinct prime factors. Let g_p be the largest odd integer dividing $(p - 1, n - 1)$, and let 2^{R+1} be the largest power of 2 dividing $\gcd(p - 1 : p|n)$.
  (a) Prove that $R \leq k - 1$.
  (b) Show that (10.8.1) is $1, 1, \ldots, 1$ if and only if $a^{g_p} \equiv 1 \pmod{p^e}$ for every prime power $p^e \| n$.
  (c) Show that there are $\prod_{p|n} g_p$ such integers $a \pmod n$.
  (d) Show that if (10.8.1) is $1, 1, \ldots, 1, -1, *, \ldots, *$, with r *'s at the end, then $0 \leq r \leq R$, and that this holds if and only if $a^{2^r g_p} \equiv -1 \pmod{p^e}$ for every prime power $p^e \| n$.
  (e) Show that there are $\leq \prod_{p|n} 2^r g_p$ such integers $a \pmod n$.
  (f) Show the number of strong pseudoprimes mod n is
  $$\prod_{p|n}(2^R g_p) \cdot \left(1 + \frac{1}{2^\omega} + \frac{1}{2^{2\omega}} + \cdots + \frac{1}{2^{(R-1)\omega}} + \frac{2}{2^{R\omega}}\right).$$
  (g) Prove that $2^R g_p \leq \frac{p-1}{2}$ and so deduce that the quantity in (f) is $\leq \frac{\phi(n)}{2^{\omega-1}}$, and so is $< \frac{1}{4}\phi(n)$ if $\omega \geq 3$.
  (h) Show that there are $\leq \frac{1}{4}\phi(n)$ reduced residues mod n which are not witnesses, whenever $n \geq 10$ with equality holding if and only if either
  - $n = pq$ where $p = 2m + 1, q = 4m + 1$ are primes with m odd, or
  - $n = pqr$ is a Carmichael number with p, q, r primes each $\equiv 3 \pmod 4$ (e.g., $7 \cdot 19 \cdot 67$).

Appendix 10B. Factoring with squares

An integer n is called *y-smooth* if all of its prime factors are $\leq y$.

In section 10.6 we outlined the main ideas in quadratic sieve-type algorithms. The key question that remains is *how*, explicitly, to select $b_1, b_2, \ldots$ so that if a_i is the least positive residue of $b_i^2 \pmod{n}$, then there is a good chance that a_i is y-smooth, for an appropriately chosen bound y. The idea is to then find a subset, call it I, of the i for which the product $\prod_{i \in I} a_i$ is a square, call it A^2. Then, for $B = \prod_{i \in I} b_i$, we have

$$B^2 = \prod_{i \in I} b_i^2 \equiv \prod_{i \in I} a_i = A^2 \pmod{n},$$

and we hope that either $(B - A, n)$ or $(B + A, n)$ is a proper divisor of n

Here are a few methods to determine such a_i so that they each have a reasonable chance of being y-smooth:

Random squares

Pick the b_i at random in $[1, n]$. We guess that the probability that an a_i is y-smooth is roughly the same as the probability that a random integer $\leq n$ is y-smooth (this is unproven).

Euler's sum of squares method

If we can write $n = u^2 + dv^2 = r^2 + ds^2$ where $d, r, s, u, v > 0$ with $s \neq v$ and $(d, n) = 1$, then we have

$$d(su)^2 = u^2 \cdot ds^2 \equiv (-dv^2)(-r^2) = d(rv)^2 \pmod{n},$$

so that n divides $\gcd(n, su-rv) \cdot \gcd(n, su+rv)$. This gives a factorization of n for if not, then n must divide either $su-rv$ or $su+rv$. Now

$$n^2 = (u^2+dv^2)(r^2+ds^2) = (ur+dsv)^2 + d(us-rv)^2 = (ur-dsv)^2 + d(us+rv)^2.$$

If, say, n divides $su-rv$, then n^2 divides $n^2 - d(us-rv)^2 = (ur+dsv)^2$ and so n divides $ur+dsv$. Dividing through by n we get a solution to $1 = a^2 + db^2$ and so $b=0$ as $d>0$; therefore $us-rv = nb = 0$. Now $(u,v) = (r,s) = 1$ and so $s=v$ which contradicts our assumption. A similar proof holds if n divides $su+rv$.

The Continued fractions method

In section 11.11 of appendix 11B we will see that if p/q is a convergent to $\sqrt{n}$, then $|p^2 - nq^2| < 2\sqrt{n}+1$. Hence above we can take $b_i = p_i$ so that $|a_i| < 2\sqrt{n}+1$. We discussed earlier that for most n the continued fraction for $\sqrt{n}$ has period length about $\sqrt{n}$, so this algorithm gives us many values of a_i, in fact far more than we will typically need. The sizes of p_i and q_i grow exponentially with i which is not good for computations, but since we only need $p_i \bmod n$ we can work mod n when computing the p_i; that is, we simply compute $p_{i+1} \equiv r_{i+1}p_i + p_{i-1} \pmod{n}$ and $q_{i+1} \equiv r_{i+1}q_i + q_{i-1} \pmod{n}$ for each $i \geq 1$, where $\sqrt{d} = [r_0, r_1, \ldots]$. We can determine the r_i as in section 11.9 of appendix 11B (where the r_i here are written as a_i there), so that the numbers involved in the calculation are all $\leq n$.

10.9. Factoring with polynomial values

Let $m = [\sqrt{n}]$, and then let $b_i = m+i$ so that $a_i = (m+i)^2 - n$ if $i \leq \frac{2}{5}m$. Now

$$a_i = i^2 + 2im + (m^2-n) \leq 2im + i^2 \leq 3i\sqrt{n} \quad \text{provided } i \leq \frac{2}{5}m,$$

so that the a_i are not much bigger than $\sqrt{n}$. The probability that a random integer up to $n^{1/2+\epsilon}$ is y-smooth is significantly higher than for a random number up to n.

An important computational issue (that we have not mentioned before) is to determine which of the a_i are y-smooth. In the random squares method one has little option but to test divide to see whether each a_i is y-smooth.[15] In this "factoring with polynomial values" method we can use the formula $a_i = f(i)$, where $f(t) = t^2 + 2mt + (m^2 - n)$ to determine which of the a_i are divisible by q, where q is the power of a prime $\leq y$:

If i is the smallest integer for which q divides a_i, then $0 \leq i \leq q-1$ and q divides a_j whenever $j \equiv i \pmod{q}$, by Corollary 2.3.1. Moreover q divides a_{-2m-i} (since the sum of the two roots of $f(t)$ is $-2m$), and therefore q divides a_j whenever $j \equiv -2m-i \pmod{q}$. To find the smallest $i_p \geq 0$ for which prime p divides a_{i_p} we test divide $a_0, a_1, \ldots$ until we find i_p (if it exists), which must be $< p$. Then to determine the smallest i for which a_i divisible by p^2 or p^3, etc., we use the algorithm suggested by Propositions 7.20.1 and 7.20.2 of section 7.20 in appendix 7C. Therefore we can easily and quickly find all the prime power divisors of the a_i that are powers of primes $p \leq y$. If their product equals a_i, then a_i is y-smooth.

---

[15]Or come up with some other method, but one always has the disadvantage that one has no prior knowledge of the prime factors of the a_i.

This is called the *quadratic sieve* because it reminds one of the sieve of Eratosthenes. There we found primes p by eliminating ("sieving out") every pth value of the polynomial t, starting from $2p$. Here we find y-smooth integers, by sieving through every pth value of the polynomial $f(t)$, starting from i_p, and then from the least residue of $-2m - i_p \pmod{p}$, to determine those a_i divisible by p.

The large prime variation

By the end of the quadratic sieve process we can write each $a_i = r_i s_i$ where s_i is the y-smooth part of a_i and has been completely factored, while all of the prime factors of r_i are $> y$. If $r_i = 1$, then a_i is y-smooth, as desired. It has proved to be useful to also retain r_i if it is itself a prime that is not too much larger than y:

Exercise 10.9.1. Show that if $r_i = r_j$, then $a_i a_j$ is a square times a y-smooth integer.

In practice people also use the *double large prime variation* in which one also keeps r_i if it has no more than two prime factors.

Exercise 10.9.2. Show that if ℓ, p, and q are primes $> y$ with $r_i = \ell p$, $r_j = pq$, and $r_k = \ell q$, then $a_i a_j a_k$ is a square times a y-smooth integer.

The key issue is to determine how fast each of these algorithms (and their variants) factor a typical integer n. It is not always so easy to determine the running time precisely, because we may not know how to analyze how long a particular step in an algorithm will take (for example, in "polynomial values" we are unable to prove how often such numbers are y-smooth, so we make the assumption that they behave much like random numbers of the same size, so as to be able to do the analysis). Also, the best algorithms in practice tend to use some random choice somewhere (like "random squares" above) and so, if our luck is out, then the algorithm could last far longer than expected—nonetheless it suffices to determine an "expected running time". All of these variations of the quadratic sieve algorithm give roughly the same expected running time: If n has d digits (that is, d is the largest integer with $10^d \leq n$), then, with probability close to 1, the quadratic sieve will factor n in around $C^{\sqrt{d \log d}}$ steps; the different variations give rise to different values of C but all with $C > 1$.

None of these is the fastest factoring algorithm known. The fastest known is the *number field sieve*, which is a version of the quadratic sieve algorithm that works in number fields, but exploits the structure of number fields. The details are beyond this book. The number field sieve will factor a typical integer n with d digits in around $C^{d^{1/3}(\log d)^{2/3}}$ steps, with probability close to 1.

Appendix 10C. Identifying primes of a given size

There are many situations in which one requires a prime of a certain size. For example the Goldbach conjecture (that every even integer > 2 is the sum of two primes) is an open question but has been numerically verified for all $n \leq 4 \times 10^{18}$ by Oliveira e Silva in 2013. In Helfgott's proof of the ternary Goldbach conjecture, he began by using deep ideas to show that every odd integer $n > n_0$ is the sum of three primes, where n_0 is a little smaller than 10^{31}. Then Platt and Helfgott showed by calculations that the conjecture holds for all odd n in the range $7 \leq n \leq n_0$. At first sight this might appear to "just be a calculation", but we have little hope of being able to do anything like 10^{31} steps in an algorithm in practice, and so we need a clever idea. The idea is to find a sequence of primes $p_1 < p_2 < \cdots < p_k < p_{k+1} = n_0$ with each $p_{i+1} - p_i < 4 \times 10^{18} - 4$. Given any integer n, $100 < n \leq n_0$, let j be the largest integer with $p_j \leq n - 4$. Therefore $4 \leq n - p_j < 4 + p_{j+1} - p_j < 4 \times 10^{18}$, and so $n - p_j$ is the sum of two primes by Oliveira e Silva's calculations. Therefore n is the sum of three primes.

To make this work we only need to determine a suitable sequence of primes p_i. The difference between consecutive p_j should be around 4×10^{18}, and $n_0 < 10^{31}$, so we only need find $< 3 \times 10^{13}$ primes, a quantity that is much more manageable. So the computational question becomes: How do we efficiently find a prime close to a given integer x?

Pseudoprime tests. We expect around 1 in every $\log x$ integers around x to be prime (as discussed in chapter 5), so we should not have to search for long if we simply test $x, x+1, \ldots$ for primality until we find a prime. Indeed Cramér's conjecture (see appendix 5C) implies that we should find a prime by the time we get to $x + 2(\log x)^2$. However finding a prime and proving that it is a prime are two different issues. As yet we have not seen a particularly efficient primality test for an arbitrary given integer n, though we could verify whether

(10.10.1) $$2^{n-1} \equiv 1 \pmod{n}.$$

If so, then n is either a prime or a base-2 pseudoprime; if not, it cannot be prime. Erdős proved that there are far fewer base-2 pseudoprimes than there are primes, so if a randomly selected integer n satisfies (10.10.1), then it is very likely to be prime. To be more precise, for any $A > 0$, there exists a constant $C > 0$ such that

$$\#\{n \leq x : n \text{ is a base-2 pseudoprime}\} \leq \frac{x}{(\log x)^A},$$

if x is sufficiently large. Let's take $A = 10$ so that a randomly chosen d digit integer (with d sufficiently large) is a base-2 pseudoprime with probability $< 1/d^{10}$, whereas we know that a randomly chosen d digit integer is prime with probability around $1/d$. For $d = 100$ the chance that n is not prime is miniscule.[16] However even if we find such an n we do not necessarily have an easy way to *prove* that it is prime, so we need some other tool to guarantee that an integer n that satisfies (10.10.1) really is prime.

We will construct a primality test that works efficiently for integers n of a certain form (though not for all n). There are enough n of this form, so that we can expect to be able to find such an n fairly rapidly in any given, sufficiently long, interval.

10.10. The Proth-Pocklington-Lehmer primality test

In section 10.4 we saw that one can prove that n is prime if one can find an integer g such that $g^{n-1} \equiv 1 \pmod{n}$ and $(g^{(n-1)/q} - 1, n) = 1$ for every prime q dividing n. This only works, in practice, if one can factor $n - 1$ in a reasonable time, something that is not guaranteed. So it is of interest to modify this method to try to be able to use less information to prove that n is prime. The key lemma is due to Pocklington (building on an idea of Proth):

Lemma 10.10.1. *Suppose that* $a^{n-1} \equiv 1 \pmod{n}$ *and* $(a^{\frac{n-1}{q}} - 1, n) = 1$ *where* $n \equiv 1 \pmod{q^e}$. *If prime p divides n, then* $p \equiv 1 \pmod{q^e}$.

Proof. Let m be the order of $a \pmod{p}$. Now $a^{n-1} \equiv 1 \pmod{p}$ but $a^{(n-1)/q} \not\equiv 1 \pmod{p}$ so that m divides $n-1$ but not $(n-1)/q$; that is, $q \nmid (n-1)/m$. Therefore if $q^e \| n-1$, then $q^e \| m$, and m divides $p - 1$. The result follows. □

In 1928, Lehmer realized that one can use this to test n for primality without needing to fully factor $n - 1$, but rather factor just "half of it".

Theorem 10.2. *Let* $n-1 = FR$ *where F is fully factored and* $F > \sqrt{n}$. *If* $g^{n-1} \equiv 1 \pmod{n}$ *and* $(g^{(n-1)/q} - 1, n) = 1$ *for every prime q dividing F, then n is prime.*

Proof. By Lemmma 10.10.1 we know that if p is a prime dividing n, then $p \equiv 1 \pmod{q^e}$ for each $q^e \| F$ and so $p \equiv 1 \pmod{F}$. Therefore $p > F > \sqrt{n}$, but n cannot have two such prime factors, so n must be prime. □

---

[16] Assuming that $d = 100$ qualifies as "sufficiently large".

Exercise 10.10.1 (Proth's Theorem). Suppose that $n = k \cdot 2^m + 1$ where $k < 2^m$. Show that n is prime if and only if there exists an integer a for which $a^{\frac{n-1}{2}} \equiv -1 \pmod{n}$.

Most large primes that have been found are of this form, because this is a relatively easy test to implement in practice, using fast exponentiation. Helfgott and Platt used Proth's Theorem in their calculations in resolving the ternary Goldbach problem.

Exercise 10.10.2. Suppose that $m > 1$.
  (a) Show that $n = 2^m + 1$ is prime if and only if $3^{2^{m-1}} \equiv -1 \pmod{n}$ if and only if $5^{2^{m-1}} \equiv 1 \pmod{n}$.
  (b) Let $u_0 = 3$ and then $u_{m+1} = u_m^2$ for all $n \geq 0$. Prove that $2^m + 1$ is prime if and only if $u_{m-1} \equiv -1 \pmod{2^m + 1}$. (This should be easy to implement algorithmically.)

What if one can factor a lot of $n - 1$ but not quite half of it? In 1975, Brillhart, Lehmer, and Selfridge showed that one can proceed if one can factor a third of it:

Theorem 10.3. *Let $n - 1 = FR$ where F is fully factored and $F > (2n)^{1/3}$. Let r be the least residue of $R \pmod F$ and $s = [R/F] = (R - r)/F$, and suppose that $r^2 - 4s$ is not a square. If $g^{n-1} \equiv 1 \pmod{n}$ and $(g^{(n-1)/q} - 1, n) = 1$ for every prime q dividing F, then n is prime.*

Proof. As in the proof of Theorem 10.2, we know that if p is a prime dividing n, then $p \equiv 1 \pmod{F}$. Therefore $p > n^{1/3}$, and so if n is not prime, then it must have exactly two prime factors; call them $1 + aF$ and $1 + bF$. Therefore $R = (n - 1)/F = (a + b) + abF$. Now $a, b \leq ab < n/F^2 < F/2$, so that $a + b < F$. Moreover $r \equiv R \equiv a + b \pmod{F}$, with $r < F$ and so $r = a + b$. We also have $s = (R - (a+b))/F = ab$, and so $r^2 - 4s = (a - b)^2$, contradicting the hypothesis. □

We have seen that we can verify whether n is prime if $n - 1$ is largely factored. If $n - 1$ is difficult to largely factor, then we can proceed when $n + 1$ is easy to mostly factor, by using second-order linear recurrence sequences.

Second-order linear recurrences

We can do something similar for second-order linear recurrences. Let $u_0 = 0$, $u_1 = 1$, and $u_{n+2} = au_{n+1} - u_n$ for all $n \geq 0$. Let $\Delta = a^2 - 4$, $\alpha = \frac{a+\sqrt{\Delta}}{2}$, and $\beta = \frac{a-\sqrt{\Delta}}{2}$ so that $\alpha\beta = 1$. Recall that $u_n = \frac{\alpha^n - \beta^n}{\alpha - \beta}$.

Theorem 10.4. *Suppose that $(n, 2\Delta) = 1$ and let $\delta = \left(\frac{\Delta}{n}\right)$. Let $n - \delta = 2FR$ where F is even, fully factored and $F > (\sqrt{n} + 1)/2$. If $u_{(n-\delta)/2} \equiv 0 \pmod{n}$ and $(u_{(n-\delta)/2q}, n) = 1$ for every prime q dividing F, then n is prime.*

Proof. Suppose that prime p divides n. Now as $(p, \alpha - \beta)$ divides $(n, \alpha - \beta)$ which divides $(n, \Delta) = 1$, we deduce that $u_r \equiv 0 \pmod{p}$ if and only if $\alpha^r \equiv \beta^r \pmod{p}$. (These congruences are between elements of $\mathbb{Z}[\sqrt{\Delta}]$; the congruence classes have representatives $a + b\sqrt{\Delta}$ with $0 \leq a, b \leq p - 1$.) As α is a unit, then $\alpha^r \equiv \beta^r \pmod{p}$ holds if and only if $\alpha^{2r} \equiv \alpha^r \beta^r \equiv 1 \pmod{p}$.

The hypothesis, together with the first paragraph, implies that $\alpha^{n-\delta} \equiv 1 \pmod{p}$ but $\alpha^{(n-\delta)/q} \not\equiv 1 \pmod{p}$. Therefore, if m is the order of $\alpha \pmod{p}$, then m divides $n - \delta$ but not $(n - \delta)/q$; that is, $q \nmid (n - \delta)/m$. Hence if $q^e \| n - \delta$, then $q^e \| m$, and so $2F$ divides m which divides $p - \left(\frac{\Delta}{p}\right)$ by Corollary 8.18.1. That is, $p \equiv \left(\frac{\Delta}{p}\right)$ (mod $2F$) and so $p \geq 2F - 1 > \sqrt{n}$. Therefore n cannot have two prime factors, so must be prime. □

There remains the question of finding such a recurrence sequence $(u_k)_{k \geq 0}$ when $n+1$ is partially factored. Morrison [2] showed that one can find second-order linear recurrence sequences that allow one to use a modification of Theorem 10.4 to prove the primality of n. (See theorem 4.2.4 in [**CP05**].)

Just as when $n - 1$ is factorable we can modify the ideas to test whether Fermat numbers are prime (as in exercise 10.10.2), we can use these new ideas to test whether Mersenne numbers are prime.

Corollary 10.10.1 (The Lucas-Lehmer primality test for Mersenne numbers). *Let $w_0 = 4$ and $w_{k+1} = w_k^2 - 2$ for all $k \geq 0$. Suppose that n is odd and ≥ 3, and let $M_n = 2^n - 1$ be a Mersenne number. If $w_{n-2} \equiv 0$ (mod M_n), then M_n is prime.*

It can also be shown that if M_n is prime, then $w_{n-2} \equiv 0$ (mod M_n).

Proof. We claim that an odd prime p divides at most one w_k, for if $w_k \equiv 0$ (mod p), then $w_{k+1} \equiv 0^2 - 2 \equiv -2$ (mod p) and $w_{k+2} \equiv (-2)^2 - 2 \equiv 2$ (mod p). From then on $w_{k+j} \equiv 2^2 - 2 \equiv 2$ (mod p) for all $j \geq 2$, and so $p \nmid w_n$ for all $n > k$.

Define the sequence $\{u_n\}_{n \geq 0}$ above with $a = 4$, so that $\Delta = 12$, $\alpha = 2 + \sqrt{3}$, and $\beta = 2 - \sqrt{3}$. Write $u_{2n} = u_n v_n$ so that $v_n = \alpha^n + \beta^n$. We claim that $w_k = v_{2^k}$ for all $k \geq 0$, since it is true for $k = 0$, and then $w_{k+1} = w_k^2 - 2 = v_{2^k}^2 - 2 = v_{2 \cdot 2^k}$ as $v_n^2 - 2 = \alpha^{2n} + 2(\alpha\beta)^n - 2 + \beta^{2n} = v_{2n}$ for all $n \geq 1$.

We deduce $u_{2^k} = w_{k-1} w_{k-2} \cdots w_0$ for all $k \geq 1$: It is obviously true for $k = 1$, and if it is true for $k - 1$, then $u_{2^k} = v_{2^{k-1}} u_{2^{k-1}} = w_{k-1} \cdot w_{k-2} \cdots w_0$.

Therefore $w_{k-1} \equiv 0$ (mod p) if and only if $u_{2^k} \equiv 0$ (mod p) and $(u_{2^{k-1}}, p) = 1$: By the previous paragraph, the right-hand side can be rewritten as $w_{k-1} \equiv 0$ (mod p) and $(w_j, p) = 1$ for all $j < k - 1$. By the first paragraph, this is the same as $w_{k-1} \equiv 0$ (mod p).

We apply Theorem 10.4 with n replaced by M_n. Then $\delta = \left(\frac{12}{2^n - 1}\right) = -\left(\frac{2^n - 1}{3}\right) = -\left(\frac{1}{3}\right) = -1$ so that $(2^n - 1) - \delta = 2^n = 2F$, where $F = 2^{n-1}$ is fully factored. Therefore if $w_{n-2} \equiv 0$ (mod M_n), then $u_{2^{n-1}} \equiv 0$ (mod M_n) and $(u_{2^{n-2}}, M_n) = 1$ by the previous paragraph, and so M_n is prime, by Theorem 10.4. □

References for this chapter

[1] John Brillhart, D. H. Lehmer, and J. L. Selfridge, *New primality criteria and factorizations of $2^m \pm 1$*, Math. Comp. 29 (1975), 620–647.

[2] M. A. Morrison, *A note on primality testing using Lucas sequences*, Math. Comp. 29 (1975), 181–182.

[3] Paulo Ribenboim, *The book of prime number records*, 2nd ed., Springer-Verlag, 1989.

Appendix 10D. Carmichael numbers

10.11. Constructing Carmichael numbers

We have discussed Carmichael numbers in sections 7.6, 10.4, and 10.8 of appendix 10A. In particular in exercise 7.10.19 we came up with a systematic way of finding infinitely many families of Carmichael numbers with three prime factors: For given pairwise coprime integers a, b, c we select the residue class $m_0 \pmod{abc}$ via the Chinese Remainder Theorem so that

$$m_0 \equiv \begin{cases} -\frac{1}{b} - \frac{1}{c} & \pmod{a}, \\ -\frac{1}{a} - \frac{1}{c} & \pmod{b}, \\ -\frac{1}{a} - \frac{1}{b} & \pmod{c}. \end{cases}$$

If $m \equiv m_0 \pmod{abc}$ and $am + 1$, $bm + 1$, and $cm + 1$ are all prime, then their product, $N = (am + 1)(bm + 1)(cm + 1)$, is a Carmichael number.

The prime k-tuplets conjecture implies (stated in the "Bonus Read" after appendix 5A) that there are infinitely many such prime triplets (creating infinite families of Carmichael numbers for any given pairwise coprime integers a, b, c) if the triplet is admissible. The triplet is admissible if for every prime p there exists a residue class $m \pmod{p}$ with $m \equiv m_0 \pmod{abc}$ for which p does not divide N: If prime $p \nmid abc$, let $m \equiv 0 \pmod{p}$ so that $N \equiv 1 \pmod{p}$. If $p | a$, then $bm + 1 \equiv b(-\frac{1}{b} - \frac{1}{c}) + 1 \equiv -b/c \pmod{p}$, and similarly $cm + 1 \equiv -c/b \pmod{p}$ so that $N \equiv 1 \cdot (-b/c) \cdot (-c/b) \equiv 1 \pmod{p}$. We get the same congruence, by the analogous argument, if $p | b$ or $p | c$.

Another way to construct Carmichael numbers is to begin with one, say $n = p_1 p_2 \ldots p_k$, and then prove that there are infinitely many Carmichael numbers $N = q_1 q_2 \ldots q_k$, where the q_j are each prime numbers with $q_j - 1 = m(p_j - 1)$ for $1 \leq j \leq k$, for some integer m. In exercise 7.18.4 we showed that composite r is a Carmichael number if and only if $\lambda(r)$ divides $r - 1$. Therefore

$$\lambda(n) = \text{lcm}[p_1 - 1, \ldots, p_k - 1] \text{ divides } n - 1.$$

Now if the q_j are indeed all primes, then

$$\lambda(N) = \text{lcm}[q_1 - 1, \ q_2 - 1, \ldots, \ q_k - 1] = m \ \text{lcm}[p_1 - 1, \ldots, \ p_k - 1] = m\lambda(n).$$

We select $m \equiv 1 \pmod{\lambda(n)}$ so that $(m, \lambda(n)) = 1$. Now

$$N = \prod_j (m(p_j - 1) + 1) \equiv 1 \pmod{m},$$

and

$$N = \prod_j (m(p_j - 1) + 1) \equiv \prod_j ((p_j - 1) + 1) = n \equiv 1 \pmod{\lambda(n)}.$$

Therefore $N \equiv 1 \pmod{m\lambda(n) = \lambda(N)}$. This implies that N is indeed a Carmichael number, by Lemma 7.6.1.

The prime k-tuplets conjecture implies that there are infinitely many such prime k-tuples, and hence infinitely many such Carmichael numbers, as the k-tuple is admissible: If prime $p \nmid \lambda(n)$, let $m \equiv 0 \pmod{p}$ so that $N \equiv 1 \pmod{p}$. If $p | \lambda(n)$, then $m \equiv 1 \pmod{p}$ and so $N \equiv n \equiv 1 \pmod{p}$.

10.12. Erdős's construction

Fix $\epsilon > 0$. It is believed that there are about $x^{\frac{1}{3}}$ Carmichael numbers up to x with three prime factors, about $x^{\frac{1}{4}}$ with four prime factors, etc. However we expect more than $x^{1-\epsilon}$ Carmichael numbers up to x in total if x is sufficiently large. How can this be? It seems unlikely that $x^{\frac{1}{3}} + x^{\frac{1}{4}} + x^{\frac{1}{5}} + \cdots$ could possibly equal $x^{1-\epsilon}$, so what is going on? The surprise is that the vast majority of Carmichael numbers have a "large" number of prime factors, in that the number of prime factors of a typical Carmichael number up to x goes to ∞ surprisingly rapidly as x grows.

In the construction of the previous section we fixed the number of prime factors of the Carmichael number N (as well as the ratios $q_i - 1 : q_j - 1$ of the distinct prime factors q_i, q_j of N), creating infinitely many families of Carmichael numbers, but atypical Carmichael numbers (as the number of prime factors is fixed). In our next construction the number of prime factors can vary:

Lemma 7.6.1 tells us that a squarefree, composite integer n is a Carmichael number if and only if $\lambda(n) = \text{lcm}[p - 1 : p|n]$ divides $n - 1$. Examples of Carmichael numbers n indicate that $\lambda(n)$ is typically much smaller than n, which is far from true for a typical integer n. Erdős reasoned if we are going to try to construct Carmichael numbers, then we should try to make sure that $\lambda(n)$ is surprisingly small compared to n. He approached this as follows:

- Select an integer L with lots of prime factors, for example the lcm of the integers $\leq y$.
- Find a large set $\mathcal{P} = \mathcal{P}(y)$ of primes $p > y$ for which $p - 1$ divides L.
- Find subsets $p_1, \ldots, p_r \in \mathcal{P}$ whose product $n = p_1 \cdots p_r$ is $\equiv 1 \pmod{L}$.

Then $\lambda(n)$ divides L, which divides $n - 1$, and so n is a Carmichael number.

10.12. Erdős's construction

For $L = 120$ we have $7, 11, 13, 31, 41, 61 \in \mathcal{P}$, and we easily identify that $41041 = 7 \times 11 \times 13 \times 41 \equiv 1 \pmod{120}$, $172081 = 7 \times 13 \times 31 \times 61 \equiv 1 \pmod{120}$, and $852841 = 11 \times 31 \times 41 \times 61 \equiv 1 \pmod{120}$ and so are all Carmichael numbers.

If we find k primes in $\mathcal{P}$, then there are $2^k - 1$ products to test in the last part of the algorithm. Assuming those products are randomly distributed mod L, we expect to construct about $2^k/L$ Carmichael numbers in this way. One problem though is that if L is large, then it requires a great deal of searching to find each such Carmichael number. Alford took a different approach.

- Find a subset $\mathcal{P}_0$ of $\mathcal{P}$ such that for every reduced residue $a \pmod{L}$ there exists $q_1, \ldots, q_k \in \mathcal{P}_0$ for which $q_1 \cdots q_k \equiv a \pmod{L}$.

Now if $p_1, \ldots, p_r$ is any non-trivial subset of $\mathcal{P} \setminus \mathcal{P}_0$, then let a be the inverse of $p_1 \cdots p_r \pmod{L}$. We have $q_1, \ldots, q_k \in \mathcal{P}_0$ for which $q_1 \cdots q_k \equiv a \pmod{L}$, and so

$$n = p_1 \cdots p_r \cdot q_1 \cdots q_k \equiv a^{-1} \cdot a \equiv 1 \pmod{L}.$$

Therefore n is a Carmichael number. So if there are ℓ primes in $\mathcal{P}_0$, and k primes in $\mathcal{P}$, then Alford's idea yields $2^{k-\ell} - 1$ Carmichael numbers.

Alford worked with the example $L = 2^6 \cdot 3^3 \cdot 5^2 \cdot 7^2 \cdot 11$ and found that there are 155 primes p for which $p - 1$ divides L (that is, in $\mathcal{P}$). By exhaustive calculations he showed that $\mathcal{P}_0$ can be taken to be the smallest 27 primes in $\mathcal{P}$, and therefore he deduced, on January 21, 1992, that there are at least $2^{128} - 1$ Carmichael numbers. This greatly increased the number known, from less than 2^{14} Carmichael numbers to many more Carmichael numbers than one could ever even hope to write down.[17] This inspired Granville and Pomerance to work with Alford to modify his argument suitably so as to establish that there are infinitely many Carmichael numbers: If x is sufficiently large, then there are more than $x^{2/7}$ Carmichael numbers up to x, which was subsequently improved by Harman to $x^{1/3}$. The ideas in the proof can be modified to show that for any integer B, there are infinitely many Carmichael numbers for which the least witness is $> B$.[18]

Carmichael numbers remain scarce all the way up to 10^{15}, which is surprising if Erdős's conjecture that there are more than $x^{1-\epsilon}$ Carmichael numbers up to x once x is sufficiently large is to be believed. Indeed Shanks [**Sha85**] challenged those who believe Erdős's conjecture to produce a value of x for which there are more than $x^{1/2}$ Carmichael numbers up to x.[19]

It is believed that there are about $\pi(x)/\phi(m)$ primes $\equiv 1 \pmod{m}$ up to x once $x \geq m^{1+\epsilon}$ (with m sufficiently large) and therefore the number should definitely be $> \pi(x)/2m$. In [**1**] it was shown that if this is true, then it implies Erdős's conjecture that there are more than $x^{1-\epsilon}$ Carmichael numbers up to x once x is sufficiently large. This shows that Erdős was almost certainly correct and Shanks wrong, but it still flies in the face of the computational evidence. In [**2**] the authors try to reconcile our theoretical understanding with the data.

---

[17] What one might call "infinity in practice".

[18] Certain well-known software packages did, at that time, assert that a given integer is prime if it is a strong pseudoprime for some given finite set of bases. This result shows that that software misidentified composites as primes.

[19] Up to $x = 10^{15}$, there are only a few more than $x^{1/3}$ Carmichael numbers.

The computational evidence

Richard Pinch has made extensive calculations of Carmichael numbers. For example the number of Carmichael numbers $C(x)$ up to x for $x = 10^j$ for $3 \leq j \leq 21$ is given in the following table:

n	10^3	10^4	10^5	10^6	10^7	10^8	10^9	10^{10}
$C(n)$	1	7	16	43	105	255	646	1547

n	10^{11}	10^{12}	10^{13}	10^{14}	10^{15}	10^{16}
$C(n)$	3605	8241	19279	44706	105212	246683

n	10^{17}	10^{18}	10^{19}	10^{20}	10^{21}
$C(n)$	585355	1401644	3381806	8220777	20138200

We expect $C(x) \geq x^{1-\epsilon}$, so it is troubling that $C(10^{21})$ is only a little bigger than $2x^{1/3}$. We do not know how large x needs to be to have $C(x) \geq \sqrt{x}$.

Another important issue is to understand how the number of prime factors of typical Carmichael numbers grow with x. It is believed that $C_k(x)$ is roughly $x^{1/k}$, for each *fixed* k once x is sufficiently large.

Carmichael numbers up to 10^{21} have at most 12 prime factors. In the next table we give (for typographical reasons) the values of $C_k(10^j)$ with $k \leq 9$:

x	$C_3(x)$	$C_4(x)$	$C_5(x)$	$C_6(x)$	$C_7(x)$	$C_8(x)$	$C_9(x)$
10^3	1	0	0	0	0	0	0
10^4	7	0	0	0	0	0	0
10^5	12	4	0	0	0	0	0
10^6	23	19	1	0	0	0	0
10^7	47	55	3	0	0	0	0
10^8	84	144	27	0	0	0	0
10^9	172	314	146	14	0	0	0
10^{10}	335	619	492	99	2	0	0
10^{11}	590	1179	1336	459	41	0	0
10^{12}	1000	2102	3156	1714	262	7	0
10^{13}	1858	3639	7082	5270	1340	89	1
10^{14}	3284	6042	14938	14401	5359	655	27
10^{15}	6083	9938	29282	36907	19210	3622	170
10^{16}	10816	16202	55012	86696	60150	16348	1436
10^{17}	19539	25758	100707	194306	172234	63635	8835
10^{18}	35586	40685	178063	414660	460553	223997	44993
10^{19}	65309	63343	306310	849564	1159167	720406	196391
10^{20}	120625	98253	514381	1681744	2774702	2148017	762963
10^{21}	224763	151566	846627	3230120	6363475	6015901	2714473

The median size of k, the number of prime factors, seems to grow quite rapidly. If each $C_k(x)$ is roughly $x^{1/k}$, we should have $C_3(x) > C_4(x) > \cdots > C_k(x)$ once x is sufficiently large, but we do not even have $C_3(x) > C_4(x)$ until $x > 10^{18}$.

References for this chapter

[1] W. R. Alford, Andrew Granville, and Carl Pomerance, *There are infinitely many Carmichael numbers*, Ann. of Math. 139 (1994), 703–722.

[2] Andrew Granville and Carl Pomerance, *Two contradictory conjectures concerning Carmichael numbers*, Math. Comp. 71 (2002), 883–908.

Appendix 10E. Cryptosystems based on discrete logarithms

Given a primitive root $g \bmod p$ and a reduced residue $a \pmod{p}$, we know that there exists an integer k for which $a \equiv g^k \pmod{p}$. We write $k := \operatorname{ind}_p(a)$, the *discrete log* of $a \bmod p$ in base g, which is well-defined mod $p-1$ (see section 7.16 of appendix 7B). The problem of efficiently determining k (given a, g, and p) is called the *discrete log problem* and seems to be difficult. It has therefore been used as the basis for various cryptographic protocols.

10.13. The Diffie-Hellman key exchange

Alice and Bob wish to create a secret number that they both know, without meeting. To do so, they must share information across on open channel with Oscar listening:

- They agree upon a large prime p and primitive root g.
- Alice picks her secret exponent a, and Bob picks his secret exponent b.
- Alice transmits the least positive residue of $g^a \pmod{p}$ to Bob, and Bob transmits the least positive residue of $g^b \pmod{p}$ to Alice.
- The secret key is the least residue of $g^{ab} \pmod{p}$. Alice computes it as $g^{ab} \equiv (g^b)^a \pmod{p}$, and Bob computes it as $g^{ab} \equiv (g^a)^b \pmod{p}$.

Oscar has access to p as well as to g, g^a, and $g^b \pmod{p}$, and he wishes to determine $g^{ab} \pmod{p}$. The only obvious way to proceed, with the information that he has, is to compute the discrete logarithms of g^a or g^b in base g, to recover a or b or both and hence to determine $(g^b)^a$ or $(g^a)^b$ or $g^{a \cdot b} \pmod{p}$, respectively. Notice that, in this exchange, Bob never knows Alice's secret exponent a, and Alice never knows Bob's secret exponent b.

There is a lot more that can be said, for example how to stop a "man-in-the-middle" attack. That is, Oscar can get in between Alice and Bob on their communication channel and hence pretend to be Bob when dealing with Alice and send her his own g^b and similarly pretend to be Alice when dealing with Bob and send him his own g^a. It is difficult to stop, or even to recognize, such fraudulent behavior, but there are well-established protocols for dealing with this, and other, difficult situations.

10.14. The El Gamal cryptosystem

This public key cryptosystem uses the secret key, g^{ab} (mod p), from above:

- Alice wishes to transmit a message M to Bob. She creates the ciphertext $x \equiv M/g^{ab}$ (mod p) and transmits that.
- Bob determines the original message M by computing xg^{ab} (mod p).

Oscar has access to p as well as g, g^a, and g^b (mod p), and the ciphertext x. Determining $M \equiv xg^{ab}$ (mod p) is therefore equivalent to determining g^{ab} (mod p). This is the same mathematical problem as in the Diffie-Hellman key exchange.

Why choose one cryptosystem over another? This is an important practical question, especially as we are unable to prove that any particular cryptosystem is truly secure (since, for all we know, there may be polynomial time algorithms for factoring or for solving the discrete log problem). Most people who are not directly involved in selling a particular product would guess that RSA is the safest, since factoring is a much better explored problem than discrete logs. However the El Gamal system has a distinct advantage over RSA, which is the quantity and difficulty of the calculations involved in implementing the algorithms: Let us compare the cryptosystems if Alice is regularly communicating with Bob. In RSA she must raise M to the power e each time she transmits a message, which requires around $\log e$ multiplications mod p. It would probably be best to choose e to be large, that is, of length comparable to the length of p, to ensure that RSA is most likely to be secure. On the other hand, in the El Gamal cryptosystem, Alice can compute g^a and g^{-ab} mod p, once and for all, so that when she transmits her ciphertext she simply multiplies M by g^{-ab} (mod p), one multiplication. This difference may not be so important if Alice works with a large computer, but many applications today use handheld devices, like a cellphone or a smartcard, which have limited computing capacity, so this time difference can be very significant.

To get the best of both worlds, one might choose to use a possibly not so secure but fast cryptosystem to exchange most messages, with regular changes of key, exchanged in a highly secure manner (because, traditionally, cryptographers have used similarities between many different ciphertexts created with the same key to expose flaws in cryptographic protocols). Therefore one might use the El Gamal cryptosystem on a day-to-day basis, while changing keys regularly using the Diffie-Hellman key exchange protocol.

Because of its high computational costs RSA is usually used only for highly secure messages, such as key exchanges. In practice, users take $e = 2^{16} + 1$, not too small but certainly not large, to simplify calculations (so we can efficiently compute the power M^e (mod n) as described in section 7.13 of appendix 7A).

For more about number theory algorithms, see the masterful book [**CP05**].

Appendix 10F. Running times of algorithms

10.15. P and NP

One should distinguish between a mathematical problem and the possible algorithms for resolving that problem. There may be many choices of algorithm and one wishes, of course, to find a fast one. We denote by P the class of problems that can be resolved by an algorithm that runs in polynomial time. There are very few mathematical problems which belong to P.

In section 10.4 we discussed problems that have been resolved, for which the answer can be quickly checked. For example one can exhibit factors of a given integer n to give a short proof that n is composite. We also saw Lucas's short proof that a number is prime based on the fact that only prime numbers n have primitive roots generating $n-1$ elements. By "short" we mean that the proof can be verified in polynomial time, and we say that such problems are in the class NP ("*non-deterministic polynomial time*"[20]). We are not suggesting that the proof can be found in polynomial time, only that the proof can be checked in polynomial time; indeed we have no idea whether it is possible to factor numbers in polynomial time, and this is now the outstanding number theory problem of this area.

By definition P⊆NP; and of course we believe that there are problems, for example the factoring problem, which are in NP, but not in P; however *this has not been proved*, and it is now perhaps the outstanding unresolved question of theoretical computer science. This is another of the Clay Mathematics Institute's million dollar problems, and perhaps the most likely to be resolved by someone with less formal training, since the experts seem to have few plausible ideas for attacking this question.

---

[20] Note that NP is **not** "non-polynomial time", a common source of confusion. In fact it is "non-deterministic polynomial time" because the method for discovering the proof is not necessarily determined.

It had better be the case that P≠NP, or else there is little chance that one can have safe public key cryptography (see, e.g., section 10.3) or that one could build a highly unpredictable pseudorandom number generator[21] or that we could have any one of several other necessary software tools for computers.

To have a good cryptosystem we want Bob to be able to decipher Alice's ciphertext quickly, and so the cryptosystem should be based on a number theory problem in the complexity class NP. However to stop Oscar being able to crack the cryptosystem, the number theory problem had better be difficult to solve, so should *not be* in the complexity class P. Therefore, to have safe cryptography, it appears that we need to have P≠NP. This question remains unresolved, and so no fast public key cryptographic protocol is, as yet, provably safe!

10.16. Difficult problems

There are only a finite number of possible commands for each line of a computer program, which therefore induce a finite number of possible states for the number and values of the variables.[22] It is therefore easy to show that most problems need exponential length programs to be solved:

We consider the set of problems where we input N bits and output one bit, that is, functions
$$f: \{0,1\}^N \to \{0,1\}.$$
Since there are 2^N possible inputs, and for each the function can have two possible outputs, hence the number of such functions is 2^{2^N}.

If a computer language allows M different possible statements on each line, then the number of programs containing k lines is M^k, and this is therefore a bound on the number of functions that can be calculated by a computer program that is k lines long. Therefore if $k < c_M \cdot 2^N$, where $c_M := \frac{\log 2}{2 \log M}$, then there are $\leq \sqrt{2^{2^N}}$ programs that are k lines long, and so we can compute no more than $\sqrt{2^{2^N}}$ functions with a k-line program in this language. Hence the vast majority of such problems require a program of length at least $c_M \cdot 2^N$. (Notice here that M, and thus c_M, is fixed by the computer language, and N is varying.)

Since almost all problems require such long programs, exponential in the length of the input, one would think that it would be easy to specify problems that need longish programs. However this is a wide open problem. Indeed even finding specific problems that cannot be resolved in polynomial time is open, or even problems that really require more than linear time! This is the pathetic state of our knowledge on lower bounds for running times, in practice. So if you ever hear claims that some secret code is provably difficult to break, that your secrets are perfectly safe, then either there has been a major scientific breakthrough or you are hearing a salesman's jibber-jabber, not mathematical proof.

---

[21] So-called "random number generators" cannot be random as they run on a computer where everything is designed to be determined! In reality they create a sequence of integers, determined in a totally predictable manner, but which appear to be random when subjected to "randomness tests" in which the tester does not know how the sequence was generated. See Oded Goldreich's, *Pseudorandomness*, Notices Amer. Math. Soc. **46** (1999), 1209–1216.

[22] Here we are talking about a classical computer. As yet impractical *quantum computers* face less restrictions and thus, perhaps, will allow more things to be computed rapidly.

Appendix 10G. The AKS test

Agrawal, Kayal, and Saxena's work starts with the following characterization of prime numbers.

Theorem 10.5. *n is prime if and only if $(x+1)^n \equiv x^n + 1 \pmod{n}$ in $\mathbb{Z}[x]$.*

Proof. Since $(x+1)^n - (x^n+1) = \sum_{1 \leq j \leq n-1} \binom{n}{j} x^j$, we have that $x^n + 1 \equiv (x+1)^n \pmod{n}$ if and only if n divides $\binom{n}{j}$ for all j in the range $1 \leq j \leq n-1$.

If $n = p$ is prime and $1 \leq j \leq p-1$, then p appears in the numerator of $\binom{p}{j}$ but is larger than, and so does not divide, any term in the denominator, and therefore p divides $\binom{p}{j}$.

If n is composite, let p be a prime dividing n. In the expansion

$$\binom{n}{p} = \frac{n(n-1)(n-2)\cdots(n-(p-1))}{p!}$$

the only terms divisible by p are the n in the numerator and the p in the denominator, and so if p^k is the largest power of p dividing n, then p^{k-1} is the largest power of p dividing $\binom{n}{p}$. Therefore n does not divide $\binom{n}{p}$, and so we deduce that $(x+1)^n \not\equiv x^n + 1 \pmod{n}$. □

This simple theorem is the basis of the AKS primality test. However we can't quickly calculate $(x+1)^n - (x^n+1) \pmod{n}$ and determine whether or not n divides each coefficient, since computing $(x+1)^n \pmod{n}$ is obviously slow since it will involve n coefficients. To reduce the number of coefficients involved, we might compute modulo some small degree polynomial as well as mod n, so that neither the coefficients nor the degree in the calculation gets large. The simplest polynomial of degree r is perhaps $x^r - 1$. So why not verify whether

$$(x+1)^n \equiv x^n + 1 \mod (n, x^r - 1)?$$

Here $f(x) \equiv g(x) \mod (n, h(x))$ for some $f(x), g(x), h(x) \in \mathbb{Z}[x]$ means that there exist polynomials $u(x), v(x) \in \mathbb{Z}[x]$ for which $f(x) - g(x) = nu(x) + h(x)v(x)$. In other words, $(n, h(x))$ is the ideal of $\mathbb{Z}[x]$ generated by n and $h(x)$.

This new congruence can be computed rapidly and it is true for any prime n (as a consequence of the theorem above), but it is unclear whether this fails to hold for all composite n and thus gives a true primality test. The main theorem of Agrawal, Kayal and Saxena provides a modification of this congruence, which can be shown to succeed for primes and fail for composites, thus providing a polynomial time primality test.

Exercise 10.17.1. Suppose that $(a,n) = 1$. Prove that n is prime if and only if $(x+a)^n \equiv x^n + a \pmod{n}$ in $\mathbb{Z}[x]$.

10.17. A computationally quicker characterization of the primes

Lemma 10.17.1. *For any given integer $n > 1$ there exists an integer r in the range $2 \leq r \leq R$ such that n has order $> \sqrt{R/\log n}$ modulo r.*

Proof. If n has order m mod r, then r divides $n^m - 1$. For each prime p let p^{e_p} be the largest power of p that is $\leq R$ and suppose that n has order $\leq M$ mod p^{e_p}. Therefore p^{e_p} divides $\prod_{m \leq M}(n^m - 1)$, and therefore, the product of the p^{e_p} also divides it. Therefore, by (5.5.6) we have

$$\frac{2^R}{R} \leq \operatorname{lcm}[m : m \leq R] = \prod_p p^{e_p} \leq \prod_{m \leq M} n^m \leq n^{(M^2+M)/2}.$$

The yields a contradiction if $R \geq M^2 \log n$ (as $\log 2 > \frac{1}{2}$). $\square$

Given an integer $n \geq 2$, let r be a positive integer for which n has order d modulo r, where $d > (3\log n)^2$. Lemma 10.17.1 implies that there is such an $r \leq 81(\log n)^5$. Since we wish to test whether n is prime we will assume that

- n is not a perfect power,
- n does not have any prime factor $\leq r$,
- $(x+a)^n \equiv x^n + a \mod (n, x^r - 1)$ for each integer a with $1 \leq a \leq A$ where $A := [3\sqrt{r} \log n]$.

All of these criteria hold for prime n (by exercise 10.17.1), and we wish to show that they do not hold for composite n, and therefore prove Theorem 10.1.

10.18. A set of extraordinary congruences

Let n be a composite integer as above, and let p be a prime dividing n so that n is not a power of p. Moreover

(10.18.1) $\qquad (x+a)^n \equiv x^n + a \mod (p, x^r - 1)$

for each integer a, $1 \leq a \leq A$. We can factor $x^r - 1$ into irreducibles in $\mathbb{Z}[x]$, as $\prod_{d|r} \Phi_d(x)$, where $\Phi_d(x)$ is the dth cyclotomic polynomial (as in appendix 4E), whose roots are the primitive dth roots of unity. Each $\Phi_r(x)$ is irreducible in $\mathbb{Z}[x]$

but may not be irreducible in $(\mathbb{Z}/p\mathbb{Z})[x]$; so let $h(x)$ be an irreducible factor of $\Phi_r(x) \pmod{p}$. Then (10.18.1) implies that

(10.18.2) $$(x+a)^n \equiv x^n + a \mod (p, h(x))$$

for each integer $a, 1 \le a \le A$, since the ideal $(p, h(x))$ divides the ideal $(p, x^r - 1)$.

The congruence classes mod $(p, h(x))$ can be viewed as the elements of a field $\mathbb{F}$ so that the congruences (10.18.2) are much easier to work with than (10.18.1), where the congruences do not correspond to a field. Note that x has order r in $\mathbb{F}$, as $h(x)$ divides $x^r - 1$ but not $x^d - 1$ for any $d < r$.

Let H be the elements mod $(p, x^r - 1)$ generated multiplicatively by x, $x+1$, $x+2, \ldots, x+A$. Let G be the (cyclic) subgroup of $\mathbb{F}$ generated multiplicatively by $x, x+1, x+2, \ldots, x+A$; in other words G is the reduction of H mod $(p, h(x))$. All of the elements of G are non-zero for if $x+a = 0$ in $\mathbb{F}$, then $x^n + a = (x+a)^n = 0$ in $\mathbb{F}$ by (10.18.2), so that $x^n = -a = x$ in $\mathbb{F}$, which would imply that $n \equiv 1 \pmod{r}$. However this would mean that n has order 1 mod r, contradicting our assumptions.

Our plan is to give upper and lower bounds on the size of G to establish a contradiction.

Upper bounds on $|G|$

In this subsection we will establish that

(10.18.3) $$|G| \le n^{2\sqrt{|R|}} - 1,$$

where R is the subgroup of the multiplicative group of reduced residues mod r, generated by n and p.

Define S to be the set of positive integers k for which
$$g(x^k) \equiv g(x)^k \mod (p, x^r - 1) \text{ for all } g \in H.$$
We can deduce that $g(x^k) = g(x)^k$ in $\mathbb{F}$ for each $k \in S$.

Lemma 10.18.1. *If $a, b \in S$, then $ab \in S$.*

Proof. If $g(x) \in H$, then $g(x^b) \equiv g(x)^b \mod (p, x^r - 1)$ since $b \in S$; and so, replacing x by x^a, we get $g((x^a)^b) \equiv g(x^a)^b \mod (p, (x^a)^r - 1)$, and therefore mod $(p, x^r - 1)$ since $x^r - 1$ divides $x^{ar} - 1$. Therefore, as $a \in S$ we have

$$g(x)^{ab} = (g(x)^a)^b \equiv g(x^a)^b \equiv g((x^a)^b) = g(x^{ab}) \mod (p, x^r - 1)$$

and so $ab \in S$ as desired. $\square$

Corollary 10.18.1. $R \subset S$.

Proof. Evidently $p \in S$. Moreover $n \in S$ since if $g(x) = \prod_{0 \le a \le A}(x+a)^{e_a} \in H$, then
$$g(x)^n = \prod_a ((x+a)^n)^{e_a} \equiv \prod_a (x^n + a)^{e_a} = g(x^n) \mod (p, x^r - 1)$$
by (10.18.1).

But then, $R \subset S$ by Lemma 10.18.1. $\square$

Lemma 10.18.2. *If $a, b \in S$ and $a \equiv b \mod r$, then $a \equiv b \mod |G|$.*

Proof. For any $g(x) \in \mathbb{Z}[x]$ we have that $u - v$ divides $g(u) - g(v)$. Therefore $x^r - 1$ divides $x^{a-b} - 1$, which divides $x^a - x^b$, which divides $g(x^a) - g(x^b)$; and so we deduce that if $g(x) \in H$, then $g(x)^a \equiv g(x^a) \equiv g(x^b) \equiv g(x)^b \mod (p, x^r - 1)$. Thus if $g(x) \in G$, then $g(x)^{a-b} = 1$ in $\mathbb{F}$.

As G is a cyclic group we can take g to be a generator of G. Then g has order $|G|$, and so $|G|$ divides $a - b$. □

Proof of (10.18.3). Since n is not a power of p, the integers $n^i p^j$ with $i, j \geq 0$ are distinct integers. There are $> |R|$ such integers with $0 \leq i, j \leq \sqrt{|R|}$, so two must be congruent (mod r), say

$$n^i p^j \equiv n^I p^J \pmod{r}.$$

By Corollary 10.18.1 these integers are both in S. By Lemma 10.18.2 their difference is divisible by $|G|$, and therefore

$$|G| \leq |n^i p^j - n^I p^J| \leq (np)^{\sqrt{|R|}} - 1 < n^{2\sqrt{|R|}} - 1,$$

as $p \leq n$, since p divides n. □

Lower bounds on $|G|$

Lemma 10.18.3. *Suppose that $f(x), g(x) \in \mathbb{Z}[x]$ both have degree $< |R|$, their reductions in $\mathbb{F}$ both belong to G, and $f(x) \equiv g(x) \mod (p, h(x))$. Then $f(x) \equiv g(x) \pmod{p}$.*

Proof. Consider the polynomial $\Delta(y) := f(y) - g(y)$ reduced in $\mathbb{F}$. If $k \in S$, then

$$\Delta(x^k) = f(x^k) - g(x^k) \equiv f(x)^k - g(x)^k \equiv 0 \mod (p, h(x)).$$

Therefore $\{x^k : k \in R\}$ are all distinct roots of the polynomial $\Delta(y)$ in $\mathbb{F}$, as x has order r in $\mathbb{F}$. Therefore $\Delta(y)$ has degree $< |R|$, with $\geq |R|$ distinct roots in $\mathbb{F}$, and so $\Delta(y) = 0$ in $\mathbb{F}$ by Lagrange's Theorem (Proposition 7.4.1). This implies that $\Delta(y) \equiv 0 \pmod{p}$ as its coefficients are independent of x. □

Now $1, n, n^2, \ldots, n^{d-1} \in R$, and so $|R| \geq d > (3 \log n)^2$. Therefore $A > B$ and $|R| > B$, where $B := [3\sqrt{|R|} \log n]$. The products $\prod_{a \in T}(x + a)$ lie in G for every subset T of $\{1, 2, \ldots, B\}$ and are distinct by Lemma 10.18.3. Therefore

$$|G| \geq 2^B > n^{2\sqrt{|R|}},$$

which contradicts (10.18.3). This completes the proof of Theorem 10.1. □

References for this chapter

[1] Andrew Granville, *It is easy to determine whether a given integer is prime*, Bulletin of the American Mathematical Society 42 (2005) 3–38.

Appendix 10H. Factoring algorithms for polynomials

10.19. Testing polynomials for irreducibility

There are few tools to prove that a polynomial $f(x) \in \mathbb{Z}[x]$ is irreducible (though we saw a particularly elegant criterion in Theorem 5.5). The most effective tools are based on the fact that if f is irreducible modulo some prime p, then f is irreducible (since if f is reducible, say $f = gh$, then $f \equiv gh \pmod{p}$ for every prime p). For example $x^2 + 1$ is irreducible mod 3, so it is irreducible in $\mathbb{Z}[x]$, as well as in $\mathbb{Q}[x]$ (by Lemma 3.22.2).

One can develop this idea further: f is irreducible if its possible factorizations into irreducibles, modulo two different primes, cannot possibly correspond to one another. For example $x^4 + x^2 + 2x - 1$ factors into irreducibles as $(x^2 + x + 1)^2$ (mod 2) and as $(x-1)(x^3 + x^2 - x + 1)$ (mod 3). So if $x^4 + x^2 + 2x - 1 = g(x)h(x)$, then $\deg g = \deg h = 2$ by the reduction mod 2, but this does not correspond to the factorization mod 3, and so $x^4 + x^2 + 2x - 1$ is irreducible.

These techniques are not guaranteed to always work; for example, one can factor the irreducible polynomial $x^4 + 1$ into the product of two degree-two polynomials modulo any prime p:

Exercise 10.19.1. (a) Factor $x^4 + 1 \pmod{2}$.
 (b) If prime $p \equiv 1 \pmod{4}$, show that we can factor $x^4 + 1$ as $(x^2 + b)(x^2 - b) \pmod{p}$ for some value of $b \pmod{p}$.
 (c) If prime $p \equiv 3 \pmod{4}$, show that we can factor $x^4 + 1$ as $(x^2 + bx + a)(x^2 - bx + a) \pmod{p}$, for some values of a and $b \pmod{p}$.

One of the more surprising irreducibility criteria was given by Eisenstein in the 1840s, which corresponds to reducing $f(x) \pmod{p^2}$ for an appropriate prime p.

Theorem 10.6 (Eisenstein's Irreducibility Criterion). *Suppose $f(x)$ is a polynomial with integer coefficients of degree $d \geq 1$, say $f(x) = \sum_{j=0}^{d} a_j x^j \in \mathbb{Z}[x]$. If p*

is a prime for which $f(x) \equiv a_d x^d \not\equiv 0 \pmod{p}$, that is, $a_0 \equiv a_1 \equiv \cdots \equiv a_{d-1} \equiv 0 \pmod{p}$, whereas $p^2 \nmid f(0) = a_0$, then $f(x)$ is irreducible in $\mathbb{Z}[x]$.

Proof. If $f = gh$ with $g, h \in \mathbb{Z}[x]$, then p divides $g(0)h(0) = f(0)$, but not p^2, and so p divides only one of $g(0)$ and $h(0)$, say $g(0)$. We write $g(x) = \sum_{j=0}^{m} b_j x^j$ and $h(x) = \sum_{j=0}^{n} c_j x^j$, where $m + n = d$ with $m, n \geq 1$ and $p \nmid c_0 = h(0)$. Select J to be the smallest integer for which p does not divide b_J. Now J exists as $g(x) \not\equiv 0 \pmod{p}$ (or else $f(x) = g(x)h(x) \equiv 0 \pmod{p}$ which is false by hypothesis), and $J \geq 1$ as $b_0 = g(0) \equiv 0 \pmod{p}$. Therefore $1 \leq J \leq m \leq d-1$ and we have $g(x) \equiv \sum_{j=J}^{m} b_j x^j \pmod{p}$ with $b_J \not\equiv 0 \pmod{p}$. This means that

$$f(x) = g(x)h(x) \equiv (b_m x^m + \cdots + b_J x^J)(c_n x^n + \cdots + c_0)$$
$$\equiv a_d x^d + \cdots + c_0 b_J x^J \pmod{p},$$

and so comparing the coefficients of x^J on either side yields that $a_J \equiv c_0 b_J \not\equiv 0 \pmod{p}$, contradicting the hypothesis. Therefore f is irreducible. □

Examples. Let $f(x) = x^5 - 6x + 3$. Then 3 divides all the coefficients of $f(x)$ except the first one, and 9 does not divide the coefficient of x^0. Hence $x^5 - 6x + 3$ is irreducible.

The pth cyclotomic polynomial (see appendix 4E) is given by

$$\phi_p(x) = x^{p-1} + x^{p-2} + \cdots + 1 = \frac{x^p - 1}{x - 1},$$

for p prime. The coefficients are all 1's so it does not seem that Eisenstein's Irreducibility Criterion is applicable. However a change of variable yields that $\phi_p(x+1) = \frac{(x+1)^p - 1}{x} = \sum_{j=1}^{p} \binom{p}{j} x^{j-1}$ by the binomial theorem. Now $p | \binom{p}{j}$ but $p^2 \nmid \binom{p}{j}$ for $1 \leq j \leq p-1$ by exercise 2.5.9(b), and so the conditions of Eisenstein's Irreducibility Criterion are satisfied. Therefore $\phi_p(x+1)$ is irreducible, and so $\phi_p(x)$ is irreducible. □

10.20. Testing whether a polynomial is squarefree

This is easy! In section 2.10 of appendix 2B we observed that $f(x) \in \mathbb{Z}[x]$ has repeated roots if and only if $f(x)$ and $f'(x)$ have a common factor, and in section 2.11 of appendix 2B that this can easily be determined by applying the Euclidean algorithm in $\mathbb{Z}[x]$ to $f(x)$ and $f'(x)$. The algorithm either terminates in an integer, the discriminant of f, or a polynomial which is the largest common polynomial factor of f and f'.

The analogous question for integers, to determine whether a given integer is squarefree, seems to be much more challenging. Indeed there is no known algorithm to test this that is any faster than fully factoring the integer and then seeing if any of the prime factors appear in the factorization to a power greater than 1.

10.21. Factoring a squarefree polynomial modulo p

We wish to factor a given polynomial $f(x)$ (mod p) of degree d, which we know has no repeated factors (as can be tested by determining the gcd of $f(x)$ and $f'(x)$ mod p). We now discuss Berlekamp's 1967 algorithm. Let

$$S(f) := \{g(x) \pmod{p} : \deg(g) \leq d-1 \text{ and } g(x)^p \equiv g(x) \pmod{(p, f(x))}\}.$$

We have $m \in S(f)$ for each constant m (mod p).

Lemma 10.21.1. *Suppose that $f(x) \equiv P_1 \cdots P_r$ (mod p) where the $P_j(x)$ are distinct and irreducible polynomials* mod p. *There is a bijection $\phi : S(f) \to (\mathbb{Z}/p\mathbb{Z})^r$ given by $\phi(g) = (n_1, \ldots, n_r)$ where $g(x) \equiv n_j \pmod{(p, P_j(x))}$ for $1 \leq j \leq r$.*

Proof. If $g \in S(f)$, then

$$P_1 \cdots P_r \equiv f(x) \text{ divides } g(x)^p - g(x) \equiv g(x)(g(x) - 1) \cdots (g(x) - (p-1)) \pmod{p}$$

and so each $P_j(x)$ divides some $g(x) - n_j$ mod p.

In the other direction, the congruences $g(x) \equiv n_j \pmod{(p, P_j)}$ for each j are satisfied by a unique congruence class modulo $(p, P_1 \cdots P_r) = (p, f)$, by the Chinese Remainder Theorem. Let $g(x)$ be the unique element of that congruence class of degree $< d$. For each j, we have

$$g(x)^p \equiv n_j^p \equiv n_j \equiv g(x) \pmod{(p, P_j)},$$

which implies that each P_j divides $g^p - g$ (mod p), and so f divides $g^p - g$ (mod p); that is, $g(x) \in S(f)$. $\square$

Hence, $S(f)$ is the set of constant functions mod p if and only if $r = 1$.

Lemma 10.21.2. *If $g \in S(f)$ but is not a constant, then there exists an integer n (mod p) such that*

$$\gcd(f(x), g(x) - n) \pmod{p}$$

is a proper factor of $f(x)$ (mod p).

Proof. In Lemma 10.21.1 we saw that each $P_j(x)$ divides $g(x) - n$ for some n (mod p). If so, then $\gcd(f(x), g(x) - n)$ is a proper factor of $f(x)$ (mod p) or else $f(x)$ divides $g(x) - n \not\equiv 0 \pmod{p}$ which is impossible as $\deg(f) < \deg(g - n)$. $\square$

We need to determine the elements of $S(f)$: For each k, $0 \leq k \leq p-1$, there exist $c_{k,j}$ (mod p) for which

$$x^{pk} \equiv \sum_{j=0}^{d-1} c_{k,j} x^j \pmod{(p, f(x))}.$$

Therefore if $g(x) = \sum_{j=0}^{d-1} g_j x^j$, then $g(x) \in S(f)$ if and only if

$$\sum_{j=0}^{d-1} g_j x^j = g(x) \equiv g(x)^p \equiv g(x^p) = \sum_{k=0}^{d-1} g_k x^{pk} \pmod{(p, f(x))}$$

$$\equiv \sum_{k=0}^{d-1} g_k \sum_{j=0}^{d-1} c_{k,j} x^j \equiv \sum_{j=0}^{d-1} \left(\sum_{k=0}^{d-1} g_k c_{k,j} \right) x^j \pmod{(p, f(x))}.$$

Comparing the coefficients of each side we deduce that
$$g \in S(f) \text{ if and only if } \mathbf{g}(C - I) \equiv 0 \pmod{p},$$
where $\mathbf{g} = (g_0, \ldots, g_{d-1}) \in (\mathbb{Z}/p\mathbb{Z})^d$ and $C = C(f)$ is the d-by-d matrix with i,jth entry $c_{i,j}$. Therefore $V(f) = \{\mathbf{g} : g(x) \in S(f)\}$ is the (right)-null space of $C(f) - I$ (mod p), something that is easy to calculate in practice using the tools of linear algebra. We deduce that $V(f)$ is a vector space of dimension r, and any basis which includes the vector $(1, 0, \ldots, 0)$ gives rise to a basis $1, g_2, \ldots, g_r$ for $S(f)$. In other words
$$S(f) = \{a_1 + a_2 g_2(x) + \cdots + a_r g_r(x) \pmod{p} : a_1, \ldots, a_r \pmod{p}\}.$$
(It is easy to show that $S(f)$ can be so generated for if $v, w \in S(f)$ and a, b are residues mod p, then
$$(av(x) + bw(x))^p \equiv a^p v(x)^p + b^p w(x)^p \equiv av(x) + bw(x) \pmod{p}$$
so that $av + bw \in S(f)$.)

We claim that for any $1 \le i < j \le r$ there exists an integer k, $1 \le k \le r$, such that $g_k(x) \equiv n_i \pmod{(P_i(x), p)}$ and $g_k(x) \equiv n_j \pmod{(P_j(x), p)}$ for some residues $n_i \not\equiv n_j \pmod{p}$. If not, then for each k there exists $n_k \pmod{p}$ such that $g_k(x) \equiv n_k \pmod{(P_i(x) P_j(x), p)}$. Any $g \in S(f)$ is a linear combination of $g_1, \ldots, g_r \pmod{p}$, and so there exists $n(g) \pmod{p}$ such that $g(x) \equiv n(g) \pmod{(P_i(x) P_j(x), p)}$. But this contradicts the fact that ϕ is a bijection, which we proved in Lemma 10.21.1.

We have therefore proved that there exists an integer k for which $P_i(x)$ divides $\gcd(f(x), g_k(x) - n_i) \pmod{p}$ and $P_j(x)$ divides $\gcd(f(x), g_k(x) - n_j) \pmod{p}$ with $n_i \not\equiv n_j \pmod{p}$. Therefore $P_i(x)$ and $P_j(x)$ appear in different gcds. So if we calculate the
$$\gcd(f(x), g_k(x) - n) \pmod{p} \text{ for } 2 \le k \le r \text{ and all } n \pmod{p},$$
and then the gcds of all these factors, we will obtain all of the P_j by exercise 10.21.1.

Exercise 10.21.1. (a) Suppose that $S_1, \ldots, S_m \subset \{1, \ldots, r\}$ with the property that for any $i \ne j$ there exists k such that $i \in S_k$ but $j \notin S_k$. Prove that for each h, $1 \le h \le r$, there is a subset $I_h \subset \{1, \ldots, m\}$ for which $\bigcap_{k \in I_h} S_k = \{h\}$.
(b) Let $P_1, \ldots, P_r$ be irreducible polynomials mod p. Suppose we are given a collection of polynomials $h_1(x), \ldots, h_m(x) \pmod{p}$ which are each products of some subset of the $P_i(x)$, with the property that for any $i \ne j$ there exists k such that P_i divides h_k but not P_j. Show that if we take all the possible gcds of the h_k, we will obtain each of the P_j.

We have established a technique to factor any given $f(x) \pmod{p}$ into irreducibles. In section 16.4 we will use this to provide an efficient algorithm to factor polynomials in $\mathbb{Z}[x]$.

References for this chapter

[1] Donald E. Knuth, *The art of computer programming*, 2nd ed., Addison-Wesley, **2** (1981), section 4.6.

[2] Susan Landau, *Factoring polynomials quickly*, Notices Amer. Math. Soc. **34** (1987), 3–8.

Chapter 11

Rational approximations to real numbers

How well can we approximate a real number by rational numbers? Obviously we can approximate π by $3, 3.1, 3.14$, etc., but there are even better approximations like $3, \frac{22}{7}, \frac{333}{106}, \frac{355}{113}, \ldots$ (see section 11.9 of appendix 11B for details). Are these the "best" approximations? And how do we measure how good an approximation is? We study these questions in detail in this chapter.

To start with we could ask how well we could approximate a rational number $\alpha = p/q$ with $(p,q) = 1$ and $q \geq 1$, by other, unequal, rational numbers. For any rational m/n with $n \geq 1$, which is $\neq p/q$, the difference is

$$(11.0.1) \qquad \left| \frac{p}{q} - \frac{m}{n} \right| = \frac{|pn - qm|}{qn} \geq \frac{1}{qn}$$

since $|pn - qm|$ is a non-negative integer that cannot be 0 as $p/q \neq m/n$, and so must be ≥ 1. We have therefore shown that the difference between rational α and an approximation m/n is at least some constant (in this case $1/q$) times $1/n$. We will see in the next section that one obtains much better approximations when α is real and irrational.

11.1. The pigeonhole principle

If real irrational α is very close to m/n, then $n\alpha$ must be close to m, so we are interested in how close the integer multiples of a given real number α can be to an integer. Dirichlet noted that one can get a surprisingly good answer to this question using the pigeonhole principle.

Theorem 11.1 (Dirichlet's Theorem). *Suppose that α is a given real number. For every integer $N \geq 1$ there exists a positive integer $n \leq N$ such that*
$$|n\alpha - m| < \frac{1}{N},$$
for some integer m. In other words,
$$\left|\alpha - \frac{m}{n}\right| < \frac{1}{nN}.$$

Proof. The $N+1$ numbers $\{0 \cdot \alpha\}, \{1 \cdot \alpha\}, \{2 \cdot \alpha\}, \ldots, \{N \cdot \alpha\}$ (where $\{t\}$ denotes the fractional part of t) all lie in the interval $[0, 1)$. The intervals
$$\left[0, \frac{1}{N}\right), \left[\frac{1}{N}, \frac{2}{N}\right), \ldots, \left[\frac{N-1}{N}, 1\right)$$
partition $[0, 1)$,[1] and so each of our $N + 1$ numbers lies in exactly one of the N intervals. Therefore some interval must contain at least two of our numbers by the pigeonhole principle, say $\{i\alpha\}$ and $\{j\alpha\}$ with $0 \leq i < j \leq N$, so that $|\{i\alpha\} - \{j\alpha\}| < \frac{1}{N}$. Therefore, if $n = j - i$, then $1 \leq n \leq N$, and if $m := [j\alpha] - [i\alpha] \in \mathbb{Z}$, then
$$n\alpha - m = (j\alpha - i\alpha) - ([j\alpha] - [i\alpha]) = \{j\alpha\} - \{i\alpha\},$$
and the first result follows by taking absolute values. The second result follows by dividing through by n. □

Exercise 11.1.1. Prove that for any irrational real number α there are arbitrarily small real numbers of the form $a + b\alpha$ with $a, b \in \mathbb{Z}$.

Corollary 11.1.1. *If α is a real irrational number, then there are infinitely many pairs m, n of coprime integers for which*
$$\left|\alpha - \frac{m}{n}\right| < \frac{1}{n^2}.$$

For large n this is a far better approximation of α than one can obtain for rational numbers, as we saw in (11.0.1).

Proof. Suppose that we are given a finite list, (m_j, n_j), $1 \leq j \leq k$, of solutions to this inequality. Since this is a finite list there is some solution with $|n_j\alpha - m_j|$ minimal, and $|n_j\alpha - m_j|$ must be > 0 as α is irrational. Therefore we can let N be the smallest integer $\geq 1/\min_{1 \leq j \leq k}\{|n_j\alpha - m_j|\}$. By Dirichlet's Theorem there exists $n \leq N$ such that
$$\left|\alpha - \frac{m}{n}\right| < \frac{1}{nN} \leq \frac{1}{n^2}.$$
Now
$$|n\alpha - m| < \frac{1}{N} \leq |n_j\alpha - m_j| \text{ for all } j,$$
and so (n, m) is another solution to the inequality, not included in the list. This implies that any finite list of solutions can be extended, and so there are infinitely many solutions. □

---

[1]That is, each point of $[0, 1)$ lies in exactly one of these intervals, and the union of these intervals exactly equals $[0, 1)$.

11.1. The pigeonhole principle

Dirichlet's Theorem is a very useful result as we will now exhibit by reproving two big results from earlier in the book:

Another proof of Corollary 3.5.2. [*If $(a, m) = 1$, then a has an inverse* mod m.] Take $m \geq 2$. Let $\alpha = \frac{a}{m}$ and $N = m - 1$ in Dirichlet's Theorem so that there exist integers r and s with $r \leq m - 1$ such that $|ra/m - s| < 1/(m-1)$; that is, $|ra - sm| < m/(m-1) \leq 2$. Hence $ra - sm = -1, 0$, or 1. It cannot equal 0 or else $m | sm = ar$ and $(m, a) = 1$ so that $m | r$ which is impossible as $r < m$. Hence $ra \equiv \pm 1 \pmod{m}$ and so $\pm r$ is the inverse of $a \pmod{m}$. $\square$

We saw an important use of the pigeonhole principle in number theory in the proof of Theorem 9.1, and this idea was generalized significantly by Minkowski and others. Now we reprove Theorem 9.1 using Dirichlet's Theorem:

Another proof of Theorem 9.1. [*If -1 is a square* mod n, *then n is the sum of two squares*.] Suppose that $r^2 \equiv -1 \pmod{n}$. By Dirichlet's Theorem there exists a positive integer $b < \sqrt{n}$ such that $|-\frac{r}{n} - \frac{c}{b}| < \frac{1}{b\sqrt{n}}$ for some integer c. Multiplying through by bn we deduce that $|a| < \sqrt{n}$ where $a = rb + cn$. Now $a \equiv rb \pmod{n}$ and so $a^2 + b^2 \equiv r^2 b^2 + b^2 = (r^2 + 1)b^2 \equiv 0 \pmod{n}$, and $0 < a^2 + b^2 < n + n = 2n$, and so we must have $a^2 + b^2 = n$. $\square$

For irrational α one might ask how the numbers $\{\alpha\}, \{2\alpha\}, \ldots, \{N\alpha\}$ are distributed in $[0, 1)$ as $N \to \infty$, for α irrational. In section 11.7 of appendix 11A we will show that the values are dense and even (roughly) equally distributed in $[0, 1)$. This ties in with the geometry of the torus and with exponential sum theory.

The next two exercises are multidimensional generalizations of Dirichlet's Theorem with not dissimilar proofs.

Exercise 11.1.2 (Simultaneous approximation). Suppose that $\alpha_1, \ldots, \alpha_k$ are given real numbers. Prove that for any positive integer N there exists a positive integer $n \leq N^k$ such that, for each j in the range $1 \leq j \leq k$, there exists an integer m_j for which

$$|n\alpha_j - m_j| < \frac{1}{N}.$$

Deduce that given $\alpha_1, \ldots, \alpha_k \in \mathbb{R}$ there exist integers q, $1 \leq q \leq Q$, and $p_1, \ldots, p_k$ such that

$$\left|\alpha_1 - \frac{p_1}{q}\right| \leq \frac{1}{q^{1+1/k}}, \quad \left|\alpha_2 - \frac{p_2}{q}\right| \leq \frac{1}{q^{1+1/k}}, \ldots, \left|\alpha_k - \frac{p_k}{q}\right| \leq \frac{1}{q^{1+1/k}}.$$

Exercise 11.1.3. Suppose that $\alpha_1, \ldots, \alpha_k$ are given real numbers. Prove that for any positive integer N there exist integers $n_1, n_2, \ldots, n_k$, not all zero, with each $|n_j| \leq N$, and an integer m for which

$$|n_1 \alpha_1 + n_2 \alpha_2 + \cdots + n_k \alpha_k - m| < \frac{1}{N^k}.$$

11.2. Pell's equation

Perhaps the most researched equation in the early history of number theory is the so-called *Pell equation*:[2] Are there non-trivial integer solutions x, y to

$$x^2 - dy^2 = 1?$$

(The "trivial solutions" are $x = \pm 1$ and $y = 0$.) The best-known ancient example comes from comparing the number of points in triangles of points, with the number of points in squares of points:

This triangle has $1 + 2 + 3 + 4 = 10$ points, whereas this square has $4 \times 4 = 16$. In general a triangle with m rows has $\frac{m(m+1)}{2}$ points, and a square with n rows has n^2 points. The numbers appearing in these two lists are mostly different, but there are exceptions, for example, 1, and then $36 = \frac{8 \cdot 9}{2} = 6^2$, and then $1225 = \frac{49 \cdot 50}{2} = 35^2$. So are there arbitrarily many "triangular numbers" that are also squares? More precisely, we are asking whether there are infinitely many pairs of integers m, n such that

$$\frac{m(m+1)}{2} = n^2.$$

It makes sense to clear denominators and to "complete the square" on the left side. Then we get

$$(2m+1)^2 = 4m^2 + 4m + 1 = 8 \cdot \frac{m(m+1)}{2} + 1 = 8n^2 + 1.$$

Taking $x = 2m+1$ and $y = 2n$ gives a solution to the Pell equation

$$x^2 - 2y^2 = 1.$$

On the other hand note that any solution to the Pell equation must have x odd, so is of the form $2m+1$, which implies that $2y^2 = x^2 - 1 \equiv 1 - 1 \equiv 0 \pmod{8}$ and so y is even and therefore must be of the form $2n$. (Our examples of triangular numbers above therefore correspond to the solutions $3^2 - 2 \cdot 2^2 = 1$, $17^2 - 2 \cdot 12^2 = 1$, and $99^2 - 2 \cdot 70^2 = 1$ to Pell's equation.) So we have proved that the set of triangular numbers that are also squares are in 1-to-1 correspondence with the positive integer solutions to this Pell equation.

We will show in Theorem 11.2 that there is a non-trivial solution to Pell's equation $x^2 - dy^2 = 1$ for every non-square integer $d > 1$. This was evidently known to Brahmagupta in India in 628 A.D., and one can guess that it was well

---

[2] In 1657 Fermat challenged Frénicle, Brouncker, Wallis, and "all mathematicians" to create a method for finding solutions to Pell's equation. Brouncker showed that he had done so by determining the smallest solution for $d = 313$, namely $x = 32188120829134849$, $y = 1819380158564160$. It seems that Euler attributed the equation to Pell because Rahn published an algebra book with Pell's help in 1658, which contained an example of this type of equation. The name stuck.

11.2. Pell's equation

understood by Archimedes far earlier, judging by his "Cattle Problem":

The Sun god's cattle, friend, apply thy care
to count their number, hast thy wisdom's share.
They grazed of old on the Thrinacian floor
of Sic'ly's island, herded into four,
colour by colour: one herd white as cream,
the next in coats glowing with ebon gleam,
brown-skinned the third, and stained with spots the last.
Each herd saw bulls in power unsurpassed,
in ratios these: count half the ebon-hued,
add one third more, then all the brown include;
thus, friend, canst thou the white bulls' number tell.
The ebon did the brown exceed as well,
now by a fourth and fifth part of the stained.
To know the spotted — all bulls that remained —
reckon again the brown bulls, and unite
these with a sixth and seventh of the white.
Among the cows, the tale of silver-haired
was, when with bulls and cows of black compared,
exactly one in three plus one in four.
The black cows counted one in four once more,
plus now a fifth, of the bespeckled breed
when, bulls withal, they wandered out to feed.
The speckled cows tallied a fifth and sixth

of all the brown-haired, males and females mixed.
Lastly, the brown cows numbered half a third
and one in seven of the silver herd.
Tell'st thou unfailingly how many head
the Sun possessed, o friend, both bulls well-fed
and cows of ev'ry colour – no-one will
deny that thou hast numbers' art and skill,
though not yet dost thou rank among the wise.
But come! also the foll'wing recognise.

Whene'er the Sun god's white bulls joined the black,
their multitude would gather in a pack
of equal length and breadth, and squarely throng
Thrinacia's territory broad and long.
But when the brown bulls mingled with the flecked,
in rows growing from one would they collect,
forming a perfect triangle, with ne'er
a diff'rent-coloured bull, and none to spare.
Friend, canst thou analyse this in thy mind,
and of these masses all the measures find,
go forth in glory! be assured all deem
thy wisdom in this discipline supreme!

— from an epigram written to ERATOSTHENES of Cyrene
by ARCHIMEDES (of Alexandria), 250 B.C.[3]

The first paragraph involves only linear equations. To resolve the second, one needs to find a non-trivial solution in integers u, v to

$$u^2 - 609 \cdot 7766 v^2 = 1.$$

The smallest solution is enormous, the smallest herd having about 7.76×10^{206544} cattle: It wasn't until 1965 that anyone was able to write down all 206545 decimal digits! How did Archimedes *know* that the solution would be ridiculously large? We don't know, though presumably he did not ask this question by chance.

The next result, the main result of this section, presumably known to many ancient mathematicians, is that there is always a solution to Pell's equation.

Theorem 11.2. *Let $d \geq 2$ be a given non-square integer. There exist integers x, y for which*

$$x^2 - dy^2 = 1,$$

with $y \neq 0$. If x_1, y_1 yields the smallest solution in positive integers,[4] then all other solutions are given by the recursion

$$x_{n+1} = x_1 x_n + d y_1 y_n \quad \text{and} \quad y_{n+1} = x_1 y_n + y_1 x_n \quad \text{for } n \geq 1.$$

We call the pair (x_1, y_1) the fundamental solution *to Pell's equation. Another way*

---

[3] *Archimedes, The Cattle Problem*, in English verse by S. J. P. Hillion & H. W. Lenstra Jr., Mercator, Santpoort, 1999.

[4] We measure the size of the solutions in positive integers x, y by the number $x + \sqrt{d}y$, though we would have the same ordering if we used either x or y.

to write the recursion is that
$$x_n + \sqrt{d}y_n = (x_1 + \sqrt{d}y_1)^n \text{ for every integer } n \geq 1,$$
where we match the coefficients of $\sqrt{d}$ on each side to determine y_n, and what remains, the coefficients of 1 on each side, to determine x_n.

Proof. We begin by showing that there always exists a solution to $x^2 - dy^2 = 1$ in integers with $y \neq 0$. By Corollary 11.1.1, there exist infinitely many pairs of integers (m, n) such that $|\sqrt{d} - \frac{m}{n}| < \frac{1}{n^2}$. For these pairs (m, n) we have

$$|m^2 - dn^2| = n^2 \left|\sqrt{d} - \frac{m}{n}\right| \cdot \left|\sqrt{d} + \frac{m}{n}\right| < \left|\sqrt{d} + \frac{m}{n}\right| \leq 2\sqrt{d} + \left|\sqrt{d} - \frac{m}{n}\right| < 2\sqrt{d} + 1.$$

This implies that $|m^2 - dn^2|$ must be an integer $< 2\sqrt{d} + 1$, so there must be some non-zero integer r, with $|r| < 2\sqrt{d} + 1$, for which there are infinitely many pairs of positive integers m, n such that $m^2 - dn^2 = r$. Pick the smallest such r. We can assume that each $(m, n) = 1$ or else if $(m, n) = g$ occurs infinitely often, then we have infinitely many solutions $(m/g)^2 - d(n/g)^2 = r/g^2$, contradicting the minimality of r.

Since there are only r^2 pairs of residue classes $(m \bmod r, n \bmod r)$ there must be some pair of residue classes a, b such that there are infinitely many pairs of integers m, n for which $m^2 - dn^2 = r$ with $m \equiv a \pmod{r}$ and $n \equiv b \pmod{r}$. Let m_1, n_1 be the smallest such pair, and m, n any other such pair, so that $m_1^2 - dn_1^2 = m^2 - dn^2 = r$ with $m_1 \equiv m \pmod{r}$ and $n_1 \equiv n \pmod{r}$. This implies that $r|(m_1 n - n_1 m)$ and

$$(m_1 m - dn_1 n)^2 - d(m_1 n - n_1 m)^2 = (m_1^2 - dn_1^2)(m^2 - dn^2) = r^2,$$

so that r^2 divides $r^2 + d(m_1 n - n_1 m)^2 = (m_1 m - dn_1 n)^2$, and thus $r|(m_1 m - dn_1 n)$. Therefore $x = |m_1 m - dn_1 n|/r$ and $y = |m_1 n - n_1 m|/r$ are integers for which $x^2 - dy^2 = 1$.

Exercise 11.2.1. Show that $y \neq 0$ using the fact that $(m, n) = 1$ for each such pair m, n.

We measure the size of solutions to Pell's equation, using the number $x + \sqrt{d}y$. If $x, y > 0$, then this is > 1. There are four solutions associated with each solution in positive integers u, v, and for these we have

$$u + \sqrt{d}v > 1 > u - \sqrt{d}v > 0 > -u + \sqrt{d}v > -1 > -u - \sqrt{d}v.$$

Therefore $x, y > 0$ if and only if $x + \sqrt{d}y > 1$.

Let x_1, y_1 be the solution to $x^2 - dy^2 = 1$ in positive integers with $x_1 + \sqrt{d}y_1$ minimal. We claim that all other solutions with $x, y > 0$ take the form $x + \sqrt{d}y = (x_1 + \sqrt{d}y_1)^n$. If not, let x, y be the counterexample with $x, y > 0$ for which $x + \sqrt{d}y$ is smallest. Now $x + \sqrt{d}y > x_1 + \sqrt{d}y_1$ since $x_1 + \sqrt{d}y_1$ is minimal.

If $X = x_1 x - dy_1 y$ and $Y = x_1 y - y_1 x$, then $X^2 - dY^2 = (x_1^2 - dy_1^2)(x^2 - dy^2) = 1$, and

$$X + \sqrt{d}Y = (x_1 - \sqrt{d}y_1)(x + \sqrt{d}y) = \frac{x + \sqrt{d}y}{x_1 + \sqrt{d}y_1},$$

which implies that

$$1 < X + \sqrt{d}Y < x + \sqrt{d}y.$$

11.2. Pell's equation

Hence $X, Y > 0$, and since x, y was the smallest counterexample, we deduce that
$$X + \sqrt{d}Y = (x_1 + \sqrt{d}y_1)^m \text{ for some integer } m \geq 1,$$
and therefore $x + \sqrt{d}y = (x_1 + \sqrt{d}y_1)(X + \sqrt{d}Y) = (x_1 + \sqrt{d}y_1)^{m+1}$, a contradiction.

If we define $x_n + \sqrt{d}y_n = (x_1 + \sqrt{d}y_1)^n$, then we obtain the recursion given in the theorem by an easy induction argument. We also deduce that the $x_n, y_n > 0$ and so $x_1 < x_2 < \cdots$ and $y_1 < y_2 < \cdots$ from the recursion formulas. □

Exercise 11.2.2. Prove that if $a + \sqrt{d}b = x + \sqrt{d}y$ where a, b, x, y, d are integers and d is not a square, then $a = x$ and $b = y$.

Exercise 11.2.3. Prove, by induction, that $x_{n+2} = 2x_1 x_{n+1} - x_n$ and $y_{n+2} = 2x_1 y_{n+1} - y_n$ for all $n \geq 0$.

Exercise 11.2.4. Show that all solutions to Pell's equation (not just the positive integer solutions) are given by the values $\pm(x_1 + \sqrt{d}y_1)^n$ (not just "+"), with $n \in \mathbb{Z}$ (not just $n \in \mathbb{N}$).

For technical reasons it is actually best to develop the analogous theory for the solutions to $x^2 - dy^2 = \pm 4$, as in appendix 11B, when we revisit Pell equations.

In the second half of the proof we saw how all of the solutions in positive integers can be generated from a fundamental solution. The proof is interesting in that it works by "descent": Given a solution we find a smaller one. This is a technique that we saw several times in chapter 6. We will see it play a central role in section 11.3, and later when we study elliptic curves in chapter 17.

The proof of Theorem 11.2 is not constructive, in that the proof does not indicate how to find a solution. In Lemma 11.11.2 of appendix 11B we will show how to find solutions using the continued fraction for $\sqrt{d}$ (as was known to all of the ancient mathematicians discussed here). How large is the smallest solution to Pell's equation? We saw that it can be surprisingly large, as in Archimedes's cattle problem. One can prove that the smallest solution is $\leq (8d)^{\sqrt{d}}$ (see section 13.7 of appendix 13B). However what is surprising is that the smallest solution seems to usually be this large. This is not something that has been proved; indeed understanding the distribution of sizes of the smallest solutions to Pell's equation is an outstanding open question in number theory.

In Theorem 11.2 we saw that if $d > 1$ is a non-square integer, then there are always solutions in integers $x, y > 0$ to Pell's equation $x^2 - dy^2 = 1$. This implies that
$$\sqrt{d}y(x - \sqrt{d}y) < (x + \sqrt{d}y)(x - \sqrt{d}y) = 1,$$
and so, dividing through by $\sqrt{d}y^2$, we exhibit rational approximations x/y to $\sqrt{d}$ that satisfy
$$\left|\sqrt{d} - \frac{x}{y}\right| < \frac{1}{\sqrt{d}y^2},$$
which are better approximations than those that are given by Corollary 11.1.1.

Another issue is whether there is a solution to $u^2 - dv^2 = -1$, the *negative Pell equation*. Notice, for example, that $2^2 - 5 \cdot 1^2 = -1$. Evidently if there is a solution, then -1 is a square mod d, so that d has no prime factors $\equiv -1 \pmod{4}$. Moreover d cannot be divisible by 4 or else $u^2 \equiv -1 \pmod 4$ which is impossible. We saw that $x^2 - dy^2 = 1$ has solutions for every non-square $d > 1$, and one might have guessed that there would be some simple criterion to decide whether there are solutions to $u^2 - dv^2 = -1$, but there does not appear to be. For example there are no solutions for $d = 34, 205$, or 221, yet in each case there is no congruence that easily explains why not. This is a subject of ongoing research. We will discuss the negative Pell equation in the next paragraph as well as in section 11.13 of appendix 11B.

The case $d = 5$ has many fascinating properties. For example
$$1^2 - 5 \cdot 1^2 = -4, \ 3^2 - 5 \cdot 1^2 = 4, \ 4^2 - 5 \cdot 2^2 = -4, \ 7^2 - 5 \cdot 3^2 = 4, \ldots.$$
All these solutions to $x^2 - 5y^2 = -4$ or 4 are given by $\frac{x+\sqrt{5}y}{2} = (\frac{1+\sqrt{5}}{2})^n$. If there are solutions to $x^2 - dy^2 = \pm 4$ with x, y both odd (as in this example), then $1 - d \equiv x^2 - dy^2 \equiv 0 \pmod 4$; that is, $d \equiv 1 \pmod 4$. If $d \equiv 1 \pmod 4$, then the proof of Theorem 11.2 can be used to prove there exist integers $u, v > 0$ such that:

All solutions to $x^2 - dy^2 = \pm 4$ with $x, y > 0$ are given by
$$\frac{x + \sqrt{d}y}{2} = \left(\frac{u + \sqrt{d}v}{2}\right)^n \text{ for some integer } n \geq 1.$$
To establish that there is at least one solution take $x = 2r, y = 2s$ from a solution to $r^2 - ds^2 = 1$ given by Theorem 11.2. Now select the solution to our equation with $\frac{u+\sqrt{d}v}{2} > 1$ but minimal. The proof of Theorem 11.2, suitably modified, then gives that all other solutions are given by a power of this first one.

We call $\frac{u+\sqrt{d}v}{2}$ the *fundamental solution* to Pell's equation and denote it by ϵ_d.

Exercise 11.2.5. The smallest solution to $x^2 - 2y^2 = 1$ is given by $(x, y) = (3, 2)$, which implies that 2^3 and 3^2 are consecutive *powerful numbers* (integer n is powerful if p^2 divides n whenever a prime p divides n). Use the theory of the solutions to $x^2 - 2y^2 = 1$ to prove that there are infinitely many pairs of consecutive powerful numbers.

11.3. Descent on solutions of $x^2 - dy^2 = n$, $d > 0$

Let x_1, y_1 be the fundamental solution to Pell's equation, and let $\epsilon_d = x_1 + y_1\sqrt{d}$ as in Theorem 11.2, so that $\epsilon_d > 1$.

Proposition 11.3.1. *Given integers $d, n > 0$, the integer solutions x, y to $x^2 - dy^2 = n$ are all given by $\pm\epsilon_d^k \beta$ for some integer k, where*
$$\beta \in B := \{u + \sqrt{d}v \in [\sqrt{n}, \sqrt{n}\epsilon_d) : \ u, v \geq 1 \text{ and } u^2 - dv^2 = n\}.$$

Proof. Given a solution to $x^2 - dy^2 = n$, let $\alpha = |x + y\sqrt{d}|$. As $\epsilon_d > 1$ the sequence of numbers $1, \epsilon_d, \epsilon_d^2, \ldots$ increases to infinity, and the sequence of numbers $1, \epsilon_d^{-1}, \epsilon_d^{-2}, \ldots$ decreases to 0. Therefore there exists a unique integer k such that
$$\epsilon_d^k \leq |\alpha|/\sqrt{n} < \epsilon_d^{k+1}.$$
Let $\beta := |\alpha|\epsilon_d^{-k}$, so that $\sqrt{n} \leq \beta < \sqrt{n}\epsilon_d$. Therefore α is of the form $\pm\beta\epsilon_d^k$, where $\beta \in [\sqrt{n}, \sqrt{n}\epsilon_d)$. Writing $\beta = u + \sqrt{d}v$ we obtain
$$u^2 - dv^2 = |(x + y\sqrt{d})(x - y\sqrt{d})| \cdot ((x_1 + y_1\sqrt{d})(x_1 - y_1\sqrt{d}))^{-k}$$
$$= (x^2 - dy^2)(x_1^2 - dy_1^2)^{-k} = n \cdot 1^{-k} = n.$$

11.4. Transcendental numbers

Moreover for a solution of $r^2 - ds^2 = n$ where $n > 0$, with $r, s \geq 0$, we have

$$\gamma := r + s\sqrt{d} > \sqrt{n} > n/\gamma = r - s\sqrt{d} > 0 > -r + s\sqrt{d} > -r - s\sqrt{d},$$

so of these four closely related solutions the unique one $> \sqrt{n}$ has both coordinates positive. In particular this implies that $u, v > 0$, so that $\beta \in B$. □

For $n = 1$ we have $B = \{1\}$. In some questions B can be empty; in others it can be large. For example, there are no solutions to $x^2 - dy^2 = n$ in integers if n is not a square mod d.

In the example $x^2 - 5y^2 = 209$, we have $\epsilon_5 = \left(\frac{1+\sqrt{5}}{2}\right)^6 = 9 + 4\sqrt{5}$ and, after a brief search we discover that $B = \{17 + 4\sqrt{5},\ 47 + 20\sqrt{5}\}$.

Exercise 11.3.1. Find all integer solutions x, y to (a) $x^2 - 5y^2 = -4$; (b) $x^2 - 5y^2 = 4$; (c) $x^2 - 5y^2 = -1$; (d) $x^2 - 5y^2 = 1$; (e) $x^2 - 5y^2 = -20$; (f) $x^2 - 5y^2 = -11$.

Exercise 11.3.2. Prove that for any non-square positive integer d and integer n there is either no solution or infinitely many solutions to $x^2 - dy^2 = n$.

11.4. Transcendental numbers

In section 3.4 we proved that $\sqrt{d}$ is irrational if d is an integer that is not the square of an integer. We can also prove that certain numbers are irrational simply by establishing how well they can be approximated by rationals:

Proposition 11.4.1. *Suppose that α is a given real number. Then α is irrational if and only if for every integer $q \geq 1$ there exist integers m, n such that*

$$0 < |n\alpha - m| < \frac{1}{q}.$$

Proof. If α is rational, then $\alpha = p/q$ for some coprime integers p, q with $q \geq 1$. For any integers m, n we then have $n\alpha - m = (np - mq)/q$. Now, the value of $np - mq$ is an integer $\equiv np \pmod{q}$. Hence $|np - mq| = 0$ or is an integer ≥ 1, and therefore $|n\alpha - m| = 0$ or is $\geq 1/q$.

If α is irrational, then Corollary 11.1.1 tells us that there are arbitrarily large coprime integers m, n for which $0 < |n\alpha - m| < \frac{1}{n}$. We select $n > q$ to prove the result claimed here. □

There are several other methods to prove that numbers are irrational, but it is more challenging to prove that a number is *transcendental*, that is, that the number is *not* the root of a polynomial with integer coefficients.[5] Next we show that algebraic numbers cannot be too well approximated by rationals. This suggests a method to identify a number as transcendental, generalizing how we identified irrationality in Proposition 11.4.1.

Theorem 11.3 (Liouville's Theorem). *Suppose that α is a root of an irreducible polynomial $f(x) \in \mathbb{Z}[x]$ of degree $d \geq 2$. There exists a constant $c_\alpha > 0$ (which*

---

[5] The root of a polynomial with integer coefficients is called *an algebraic number*.

depends only on α)[6] such that for any rational p/q with $(p,q) = 1$ and $q \geq 1$ we have

$$\left|\alpha - \frac{p}{q}\right| \geq \frac{c_\alpha}{q^d}.$$

Proof. Since $I := [\alpha - 1, \alpha + 1]$ is a closed interval, there exists a bound $B \geq 1$ for which $|f'(t)| \leq B$ for all $t \in I$. We will prove the result with $c_\alpha = 1/B$. If $p/q \notin I$, then $|\alpha - p/q| \geq 1 \geq c_\alpha \geq c_\alpha/q^d$ as desired. Henceforth we may assume that $p/q \in I$.

If $f(x) = \sum_{i=0}^d f_i x^i$ with each $f_i \in \mathbb{Z}$, then $q^d f(p/q) = \sum_{i=0}^d f_i p^i q^{d-i} \in \mathbb{Z}$. Now $f(p/q) \neq 0$ since f is irreducible of degree ≥ 2 and so $|q^d f(p/q)| \geq 1$.

The mean value theorem tells us that there exists t lying between α and p/q, and hence in I, such that

$$f'(t) = \frac{f(\alpha) - f(p/q)}{\alpha - p/q}.$$

Therefore, as $f(\alpha) = 0$,

$$\left|\alpha - \frac{p}{q}\right| = \frac{|q^d f(p/q)|}{q^d |f'(t)|} \geq \frac{1}{Bq^d} = \frac{c_\alpha}{q^d}. \qquad \square$$

Often students first learn to prove that there are transcendental numbers by showing that the set of real numbers is uncountable; in contrast, the set of algebraic numbers is countable, so the vast majority of real numbers are transcendental. This argument yields that most real numbers are transcendental, without actually constructing any! (See section 11.16 in appendix 11D.) The great advantage of Liouville's Theorem is that it can be used to actually construct transcendental numbers.

Corollary 11.4.1. *A* Liouville number *is an irrational real number α such that for every integer $n \geq 1$ there is a rational number p/q with $(p,q) = 1$ and $q > 1$ for which*

$$\left|\alpha - \frac{p}{q}\right| < \frac{1}{q^n}.$$

Every Liouville number is transcendental.

Proof. Let α be a Liouville number. Suppose that α is algebraic so that there exist d and c_α as in Liouville's Theorem. Select $n > d$ sufficiently large so that $2^{n-d} > 1/c_\alpha$. Then, selecting the approximation p/q with $q > 1$ as in the hypothesis we have

$$\frac{1}{q^n} > \left|\alpha - \frac{p}{q}\right| \geq \frac{c_\alpha}{q^d},$$

by Liouville's Theorem. Therefore $2^{n-d} > 1/c_\alpha > q^{n-d}$, contradicting that $q \geq 2$. Therefore α is not algebraic and so must be transcendental. $\qquad \square$

---

[6] In this chapter there are several constants like c_α which depend only on the variable given in the subscript. We do not attempt to be more precise about the constant because calculating a value for the constant will make things much more complicated, yet one will gain little from knowing its precise value.

11.4. Transcendental numbers

For example
$$\alpha = \frac{1}{10} + \frac{1}{10^{2!}} + \frac{1}{10^{3!}} + \cdots$$
is a Liouville number, since if p/q with $q = 10^{n!}$ is the sum of the first n terms, then $0 < \alpha - p/q < 2/q^{n+1} < 1/q^n$.

Liouville numbers are easily identifiable transcendental numbers, but there are many transcendental numbers which are not Liouville numbers, like π and e.

Liouville's Theorem has been improved to its, more or less, final form by Roth. To explain his result we have to introduce an ϵ and that sort of thing: For any fixed $\epsilon > 0$ (which should be thought of as being small), there exists a constant $\kappa_\epsilon > 0$, which depends on ϵ, and is chosen so it works in the proof.[7] In the notation in Roth's Theorem we have to go a little further than this since the constant also depends on the value of α we need to approximate, so our constant is $c_{\alpha,\epsilon}$, which depends on both α and ϵ, but nothing else. These dependencies do restrict our use of the inexplicit constants $c_{\alpha,\epsilon}$; for example, one cannot compare the constants that arise from different values of α.

Theorem 11.4 (Roth's Theorem, 1955). *Suppose that α is an irrational real algebraic number. For any fixed $\epsilon > 0$ there exists a constant $c_{\alpha,\epsilon} > 0$ such that for any rational p/q with $(p, q) = 1$ and $q \geq 1$ we have*
$$\left|\alpha - \frac{p}{q}\right| \geq \frac{c_{\alpha,\epsilon}}{q^{2+\epsilon}}.$$

The exponent "$2 + \epsilon$" in Roth's Theorem cannot be improved much since if α is irrational, then there are infinitely many p/q with $\left|\alpha - \frac{p}{q}\right| \leq \frac{1}{q^2}$, by Corollary 11.1.1. We will prove that approximations which are a little better than this must be convergents of the continued fraction of α (see Corollary 11.10.1 in section 11.10 of appendix 11B). The "worst approximable" irrational number is therefore $\frac{1+\sqrt{5}}{2}$, for which the best approximations are given by F_{n+1}/F_n where F_n is the nth Fibonacci number. One can show that the difference, $\left|\frac{1+\sqrt{5}}{2} - \frac{F_{n+1}}{F_n}\right|$, is roughly $1/(\sqrt{5}F_n^2)$ with an error $< 1/F_n^4$.

Exercise 11.4.1. Prove that if $\alpha \in \mathbb{C}\setminus\mathbb{R}$, then there exists a constant $\beta_\alpha > 0$ such that $|\alpha - p/q| \geq \beta_\alpha$ for all rational approximations p/q.

Exercise 11.4.2. Prove that if $f(t) = a_d \prod_{i=1}^d (t - \alpha_i)$, then $f'(\alpha_i) = a_d \prod_{j=1,\ j\neq i}^d (\alpha_i - \alpha_j)$.

There are many beautiful applications of Roth's Theorem to Diophantine equations. We highlight one:

Corollary 11.4.2 (Thue-Siegel Theorem). *Suppose that $f(t) = a_0 + a_1 t + \cdots + a_d t^d \in \mathbb{Z}[t]$ is an irreducible polynomial of degree $d \geq 3$. Then for any integer A there are only finitely many pairs of integers m, n for which*
$$n^d f(m/n) = a_0 n^d + a_1 n^{d-1} m + \cdots + a_d m^d = A.$$

---

[7] A proof that is far too involved for inclusion in this book.

Proof. If $A = 0$, the only solution is $m = n = 0$, as f is irreducible. So we may assume that $|A| \geq 1$ and write $f(t) = a_d \prod_{i=1}^{d}(t - \alpha_i)$; the α_i are distinct as $f(t)$ is irreducible. For any given pair of integers m, n select j so that $|\alpha_j - \frac{m}{n}|$ is minimized. If $i \neq j$, then

$$2\left|\alpha_i - \frac{m}{n}\right| \geq \left|\alpha_i - \frac{m}{n}\right| + \left|\alpha_j - \frac{m}{n}\right| \geq |\alpha_i - \alpha_j|,$$

so that, since $f'(\alpha_j) = a_d \prod_{1 \leq i \leq d, \, i \neq j}(\alpha_j - \alpha_i)$ (as in exercise 11.4.2),

$$\left|\alpha_j - \frac{m}{n}\right| \frac{|f'(\alpha_j)|}{2^{d-1}} = \left|\alpha_j - \frac{m}{n}\right| a_d \prod_{\substack{1 \leq i \leq d \\ i \neq j}} \frac{|\alpha_i - \alpha_j|}{2} \leq a_d \prod_{1 \leq i \leq d} \left|\alpha_i - \frac{m}{n}\right|$$

$$= |f(m/n)| = \frac{|a_0 n^d + a_1 n^{d-1} m + \cdots + a_d m^d|}{|n|^d} = \frac{|A|}{|n|^d}.$$

We now apply Roth's Theorem with $\alpha = \alpha_j$ and $\epsilon = \frac{1}{2}$, so that

$$\left|\alpha_j - \frac{m}{n}\right| \geq \frac{c_{\alpha_j, 1/2}}{|n|^{5/2}}.$$

Substituting this into the previous equation, then squaring both sides and multiplying through by denominators, we obtain either $|n| \leq 1$ or

$$|n|/2 \leq (|n|/2)^{2d-5} \leq B$$

where $B = 8 \max_j (A/c_{\alpha_j, 1/2} |f'(\alpha_j)|)^2$. Either way there are only finitely many possibilities for integer n, and for each such n there are at most d integers m which can be roots of the polynomial

$$a_d x^d + \cdots + a_1 n^{d-1} x + (a_0 n^d - A) = 0.$$

This proves the claimed result. $\square$

11.5. *The abc-conjecture*

In chapter 6 we discussed various Diophantine equations with three monomials like $x^2 + y^2 = z^2$, even $x^n + y^n = z^n$ for any integer $n \geq 2$, and there are others of interest like $x^p - y^q = 1$. So how do we determine which of these have infinitely many solutions in integers? This is not an easy question, and indeed the focus of a lot of research. One modern approach (motivated by deep considerations) is to study the prime powers dividing each term.

We begin by proving the following consequence of Roth's Theorem:

Corollary 11.5.1. *Let $F(x, y) \in \mathbb{Z}[x, y]$ be a homogenous polynomial of degree d, with no repeated linear factors. For each $\epsilon > 0$ there exists a constant $\kappa_{F, \epsilon} > 0$ such that for any coprime positive integers m, n:*

Either $F(m, n) = 0$ or $|F(m, n)| \geq \kappa_{F, \epsilon} |n|^{d-2-\epsilon}$.

In other words, either $F(m, n) = 0$ or $|F(m, n)|$ is large.

Proof. A *homogenous polynomial* in two variables takes the form

$$F(x, y) = \sum_{j=0}^{d} a_j x^j y^{d-j}.$$

11.5. The abc-conjecture

As there are no repeated factors, $F(x,y)$ can be divisible by y but not y^2. Then $f(t) = F(t,1)$ as a polynomial of degree $d-1$ or d (depending on whether $F(x,y)$ is divisible by y or not) and has no repeated roots (as F has no repeated linear factors).

Now if m and n are coprime integers, then either $F(m,n) = 0$ or, from the inequality in the proof of the Thue-Siegel Theorem,

$$\frac{|F(m,n)|}{|n|^d} = |F(m/n,1)| = |f(m/n)| \geq \frac{|f'(\alpha_j)|}{2^{d-1}} \cdot \left|\alpha_j - \frac{m}{n}\right| \geq \frac{\kappa_{F,\epsilon}}{|n|^{2+\epsilon}},$$

with $\kappa_{F,\epsilon} := \min_j c_{\alpha_j,\epsilon} |f'(\alpha_j)|/2^{d-1}$, where the last inequality follows from Roth's Theorem.[8] The result follows by multiplying each side through by $|n|^d$. □

Exercise 11.5.1. Let α be an algebraic number which is a root of $f(t) \in \mathbb{Z}[t]$, a polynomial of degree d. Let $F(x,y) = y^d f(x/y)$, and suppose that there exists a constant $\kappa > 0$ such that $|F(m,n)| \geq \kappa |n|^{d-2-\epsilon}$ for all integers m,n. Deduce that there exists a constant $c > 0$ such that $|\alpha - m/n| > c/n^{2+\epsilon}$ for all integers $m, n \neq 0$. (Thus Corollary 11.5.1 is "equivalent" to Roth's Theorem.)

We are going to move to what seems to be a rather different question but will eventually tie in closely with Corollary 11.5.1. We study pairwise coprime, positive integer solutions to the equation

$$a + b = c,$$

bounding the size of a, b, and c in terms of the product of the distinct primes that divide a, b, and c:

Conjecture 11.1 (The *abc*-conjecture). *Fix $\epsilon > 0$. There exists a constant $\kappa_\epsilon > 0$ such that if a and b are coprime positive integers with $c = a + b$, then*

$$\prod_{\substack{p \text{ prime} \\ p \text{ divides } abc}} p \geq \kappa_\epsilon c^{1-\epsilon}.$$

This is the *abc-conjecture*, one of the great open questions of modern mathematics.

For example, if we have a putative solution to Fermat's Last Theorem, like $x^n + y^n = z^n$ with $x,y,z > 0$, then we take $a = x^n$, $b = y^n$, and $c = z^n$. Now the product of the primes dividing $abc = (xyz)^n$ is the same as the product of the primes dividing xyz. Therefore the *abc*-conjecture with $\epsilon = 1/5$ implies for $n \geq 5$ that

$$\kappa(z^n)^{4/5} \leq \prod_{\substack{p \text{ prime} \\ p \text{ divides } x^n y^n z^n}} p = \prod_{\substack{p \text{ prime} \\ p \text{ divides } xyz}} p \leq xyz \leq z^3 \leq z^{3n/5},$$

where $\kappa = \kappa_{1/5}$, from which we deduce $z^n \leq 1/\kappa^5$. Since $x^n, y^n < z^n$ we deduce, from the *abc*-conjecture, that in every solution to $x^n + y^n = z^n$ with $n \geq 5$, the numbers x^n, y^n, and z^n are all bounded by some absolute constant, and therefore

---

[8] Yet again this seems like a lot of notation for a constant, especially an inexplicit constant, but the notation reflects what the constant depends on, and given the complicated derivation of this constant, it is certainly simpler not to try to be explicit about it.

there are only finitely many solutions. Therefore we have proved that the *abc*-conjecture implies that there are only finitely many solutions to $x^n + y^n = z^n$ with $(x, y) = 1$ and $n > 4$.

One can compare the *abc*-conjecture with the *abc*-theorem for polynomials (as in section 6.7 of appendix 6A). The size of the integers replaces the degrees of the polynomials; the prime divisors replace the irreducible polynomial factors. One cannot prove the *abc*-conjecture in the same way, since we relied heavily in our proof of the *abc*-theorem for polynomials on calculus, for which there is no analogy for numbers.

We now state a conjecture which implies both the *abc*-conjecture and Corollary 11.5.1 of Roth's Theorem:

Conjecture 11.2 (The *abc*-Roth conjecture). *Let $F(x, y) \in \mathbb{Z}[x, y]$ be a homogenous polynomial of degree d, with no repeated linear factors. For each $\epsilon > 0$ there exists a constant $\kappa_{F,\epsilon} > 0$ such that for any coprime positive integers m, n, either $F(m, n) = 0$ or*
$$\prod_{\substack{p \text{ prime} \\ p \text{ divides } F(m,n)}} p \geq \kappa_{F,\epsilon} |n|^{d-2-\epsilon}.$$

The *abc*-Roth conjecture implies both Corollary 11.5.1, since the product of the primes dividing non-zero $F(m, n)$ is $\leq |F(m, n)|$, and the *abc*-conjecture, taking $F(x, y) = xy(x+y)$ (since then $F(a, b) = abc$ when $a+b = c$). Quite remarkably Conjecture 11.2 follows from the *abc*-conjecture using some clever algebraic geometry. (See [2].)

Further reading for this chapter

[1] Edward B. Burger, *Diophantine olympics and world champions: Polynomials and primes down under*, Amer. Math. Monthly **107** (2000), 822–829.

[2] Andrew Granville and Thomas J. Tucker, *It's as easy as abc*, Notices Amer. Math. Soc **49** (2002), 1224–1231.

[3] Serge Lang, *Old and new conjectured Diophantine inequalities*, Bull. Amer. Math. Soc. **23** (1990), 37–75.

[4] H. W., Lenstra, Jr., *Solving the Pell equation*, Notices Amer. Math. Soc. **49** (2002), 182–192.

[5] Barry Mazur, *Questions about powers of numbers*, Notices Amer. Math. Soc. 47 (2000), no. 2, 195–202.

Additional exercises

Exercise 11.6.1. Suppose $(p, q) = 1$ and $q \geq 1$. Determine *all* rationals m/n for which $\left|\frac{p}{q} - \frac{m}{n}\right| = \frac{1}{qn}$.

Exercise 11.6.2. Reprove exercise 7.10.21(a) using (11.0.1).

Exercise 11.6.3.[†] Prove that there are infinitely many solutions to the Pell equation $u^2 - dv^2 = 1$ with $u \equiv 1 \pmod{d}$.

Exercise 11.6.4. Prove that if α is transcendental, then so is α^k for every non-zero integer k.

Exercise 11.6.5 (The "three gaps" theorem).‡ Given $\alpha \in \mathbb{R} \setminus \mathbb{Q}$, we put the fractional parts $\{\alpha\}, \{2\alpha\}, \ldots, \{N\alpha\} \in [0,1)$ in ascending order as $0 < \{a_1\alpha\} < \{a_2\alpha\} < \cdots < \{a_N\alpha\} < 1$ (so that $\{a_1, \ldots, a_N\}$ is a reordering of $\{1, \ldots, N\}$). We will prove that there are at most three distinct values in the set of consecutive differences, $D(A) := \{\{a_{j+1}\alpha\} - \{a_j\alpha\} : j = 1, \ldots, N-1\}$.
(a) Show that if $\{(a_{j+1} - 1)\alpha\} - \{(a_j - 1)\alpha\} \notin D(A)$, then either $a_j = 1$ or $a_{j+1} = 1$, or there exists k such that $\{(a_j - 1)\alpha\} < \{a_k\alpha\} < \{(a_{j+1} - 1)\alpha\}$.
(b) Show that if $\{(a_j - 1)\alpha\} < \{a_k\alpha\} < \{(a_{j+1} - 1)\alpha\}$, then $a_k = N$.
(c) Deduce from (a) and (b) that every element of $D(A)$ equals one of $\{a_1\alpha\}$, $1 - \{a_N\alpha\}$, or $\{a_1\alpha\} + 1 - \{a_N\alpha\}$.

Exercise 11.6.6. Suppose that a and b are given integers, with $3 \nmid a$.
(a) Show that we can select a congruence class $r \pmod{3}$ such that if integer $m \equiv r \pmod{3}$, then $x + y\sqrt{3} = (2 + \sqrt{3})^m (a + b\sqrt{3})$, then 3 divides y.
(b) Deduce that if integer N can be written in the form $a^2 - 3b^2$ where $3 \nmid N$, then there are infinitely many pairs of powerful numbers that differ by exactly N.

Exercise 11.6.7. Find an explicit value that can be used for c_α in Liouville's Theorem when $\alpha = \sqrt{D}$ where $D > 1$ is a squarefree positive integer.

Exercise 11.6.8. Fix $\epsilon > 0$, and integers $a_0, \ldots, a_d$. Deduce from Roth's Theorem that there are only finitely many pairs of coprime integers m, n for which $|a_0 n^d + a_1 n^{d-1} m + \cdots + a_d m^d| \leq \max\{|m|, |n|\}^{d-2-\epsilon}$.

Exercise 11.6.9. Assume the *abc*-conjecture to show that there are only finitely many sets of integers $x, y > 0$ and $p, q > 1$ for which $x^p - y^q = 1$.

Exercise 11.6.10. Suppose that $x^p + y^q = z^r$ with x, y, z pairwise coprime and $\frac{1}{p} + \frac{1}{q} + \frac{1}{r} < 1$.
(a) Prove that $\frac{1}{p} + \frac{1}{q} + \frac{1}{r} \leq \frac{41}{42}$.
(b) Assume the *abc*-conjecture. Prove that there exists a constant B for which $|x^p|, |y^q|, |z^r| < B$.

Exercise 11.6.11. The *abc*-conjecture is "best possible" in that one cannot take $\epsilon = 0$. To establish this, we need to find examples of solutions to $a + b = c$ in which $(1/c) \prod_{p | abc} p$ gets arbitrarily small.
(a) Prove that if $m^2 | b$, then $\prod_{p | b} p \leq b/m$.
(b) Prove that for any odd integer m there exists an integer n for which $2^n \equiv 1 \pmod{m^2}$.
(c)† Combine these two observations to show that for any $\epsilon > 0$ there exist coprime integers $a + b = c$ for which $\prod_{p | abc} p < \epsilon c$.

Appendix 11A. Uniform distribution

11.7. $n\alpha \bmod 1$

Dirichlet's Theorem, in section 11.1, implies that $n\alpha \bmod 1$ gets arbitrarily close to 0 as n runs through a sequence of integers n. One might also ask whether $n\alpha \bmod 1$ gets arbitrarily close to any given $\theta \in (0,1)$.

> **Theorem 11.5** (Kronecker's Theorem). *If α is a real irrational number, then the numbers $\{n\alpha\}$ are dense on $[0,1)$.*

Proof. Fix $\epsilon > 0$. By Dirichlet's Theorem there exists an integer n with $\|n\alpha\| < \epsilon$, where $\|t\|$ is the distance from t to the nearest integer. As α is irrational we also have that $\|n\alpha\| \neq 0$, and so $\{n\alpha\} \in (0,\epsilon)$ or $\{n\alpha\} \in (1-\epsilon, 1)$. We will assume that $\{n\alpha\} \in (0,\epsilon)$ (the case with $\{n\alpha\} \in (1-\epsilon, 1)$ being proved analogously).

Let $\delta = \{n\alpha\} \in (0,\epsilon)$. Select D to be the largest integer $< 1/\delta$ and so

$$\{n\alpha\},\ \{2n\alpha\},\ldots,\{Dn\alpha\} = \delta, 2\delta,\ldots,D\delta$$

is a set of points in $[0,1)$, consecutive points being spaced $\delta < \epsilon$ apart. Therefore if $\theta \in [0,1)$, then we let $k = [\theta/\delta]$ and so $\theta - k\delta \in [0,\delta)$, which implies that

$$\theta - \{kn\alpha\} = \theta - k\{n\alpha\} = \theta - k\delta \in [0,\delta) \subset [0,\epsilon).$$

That is, there are integer multiples of α in $\mathbb{R}/\mathbb{Z}$ that are arbitrarily close to θ. □

Exercise 11.7.1. Show that the conclusion of the theorem is not true if α is rational.

Exercise 11.7.2. Prove Kronecker's Theorem when $n\alpha \pmod 1 \in (1-\epsilon, 1)$.

Now we know that if α is irrational, then $n\alpha \bmod 1$ gets arbitrarily close to any given $\theta \in [0,1)$, we might ask how often $n\alpha \bmod 1$ gets close to each $\theta \in [0,1)$. Are the values of $n\alpha \bmod 1$ roughly equidistributed? To answer this question we must

11.7. $n\alpha$ mod 1

determine how often $\{n\alpha\} \in [\theta - \epsilon, \theta + \epsilon]$ for $\theta \in (0,1)$ and sufficiently small $\epsilon > 0$. If the numbers $\{n\alpha\}$ are equidistributed, then we might expect the frequency to be roughly proportional to the length of the interval. The analogous question can be asked for any sequence of numbers $x_1, x_2, \ldots \in [0, 1)$. We say that $\{x_n\}_{n \geq 1}$ is *uniformly distributed* mod 1 (or *equidistributed* mod 1) if for any $a < b \in [0, 1)$,

$$\lim_{N \to \infty} \frac{1}{N} \#\{n \leq N : a \leq x_n \leq b\} \text{ exists and equals } b - a.$$

The values of $x \pmod 1$ are in 1-to-1 correspondence with the values of $e(x)$ (where $e(t) := e^{2i\pi t}$) as its value depends on $x \pmod 1$ and not on x. Moreover the values $e(kx)$ for any given integer $k \neq 0$ remain consistent for x with any given value mod 1. That is, if $x = m + \delta$ with $0 \leq \delta < 1$, then $kx = km + k\delta$ so that $\{kx\} = \{k\delta\}$. This suggests that to study a sequence of values x_n mod 1, we might use Fourier analysis. This thinking leads to the famous theorem of Hermann Weyl (for more on this, including the proof, see [**GG**]):

Theorem 11.6 (Weyl's uniform distribution theorem). *The sequence $\{x_n\}_{n \geq 1}$ is uniformly distributed* mod 1 *if and only if for all non-zero integers k we have*

$$\lim_{N \to \infty} \frac{1}{N} \sum_{n=1}^{N} e(kx_n) \text{ exists and equals } 0.$$

Exercise 11.7.3. (a) Show that $\sum_{n=1}^{N} e(\{n\alpha\}) = \frac{e(N\alpha)-1}{1-e(-\alpha)}$ if $\alpha \notin \mathbb{Z}$, and then deduce that $|\sum_{n=1}^{N} e(\{n\alpha\})| \leq \frac{1}{|\sin \pi \alpha|}$.
(b) Use Weyl's uniform distribution theorem to deduce that if α is a real, irrational number, then $\{n\alpha\}_{n \geq 1}$ is uniformly distributed mod 1.

One can prove that $\{n\alpha\}$ is uniformly distributed mod 1 using fairly elementary ideas though it is not easy:

Exercise 11.7.4. Let $x_1, x_2, \ldots \in [0, 1)$ be a sequence of numbers. Suppose that there are arbitrarily large integers M for which

$$\lim_{N \to \infty} \frac{1}{N} \#\left\{n \leq N : \frac{m}{M} \leq x_n \leq \frac{m+1}{M}\right\} \text{ exists and equals } \frac{1}{M},$$

for $0 \leq m \leq M - 1$. Deduce that $\{x_n\}_{n \geq 1}$ is uniformly distributed mod 1.

Exercise 11.7.5.[‡] Let α be a real, irrational number. In this exercise we sketch a proof that $\{n\alpha\}_{n \geq 1}$ is uniformly distributed mod 1. Fix $\epsilon > 0$ arbitrarily small.
(a) Use Kronecker's Theorem to show that there exists an integer $N \geq 1$ such that $\{N\alpha\} = \delta \in (0, \epsilon)$.
(b) Prove that if $\{n\alpha\} < 1 - \delta$, then $\{(n + N)\alpha\} = \{n\alpha\} + \delta$. What if $\{n\alpha\} \geq 1 - \delta$?
(c) Suppose that $0 < t < 1 - 2\delta$. Show that $\{n\alpha\} \in [t, t+\delta]$ if and only if $\{(n+N)\alpha\} \in [t+\delta, t+2\delta]$, and so deduce that

$$|\#\{1 \leq n \leq x : t \leq \{n\alpha\} < t+\delta\} - \#\{1 \leq n \leq x : t+\delta \leq \{n\alpha\} < t+2\delta\}| \leq N.$$

Now let $\delta = 1/M$ for some large integer M.
(d)[†] Use (c) to show that if $0 \leq m \leq M - 1$, then

$$\left|\#\left\{1 \leq n \leq x : \frac{m}{M} \leq \{n\alpha\} < \frac{m+1}{M}\right\} - \frac{x}{M}\right| \leq MN.$$

(e) Deduce that $\{n\alpha\}_{n \geq 1}$ is uniformly distributed mod 1 using exercise 11.7.4.

Kronecker's Theorem in n dimensions. In exercise 11.1.2 we saw that Dirichlet's Theorem may be generalized to k dimensions; that is, given $\alpha_1, \ldots, \alpha_k \in \mathbb{R}$, for any $\epsilon > 0$ there exist infinitely many integers n such that each $\|n\alpha_j\| < \epsilon$. To generalize Kronecker's Theorem we would like that for $\theta_1, \ldots, \theta_k \in \mathbb{R}$ there are infinitely many n for which each $\|n\alpha_j - \theta_j\| < \epsilon$. However this is not true in all cases, even when $k = 1$: In the hypothesis of Theorem 11.5 we needed that α is irrational, and we showed that this is necessary in exercise 11.7.1. Another way to state that α is irrational is to insist that 1 and α are linearly independent over $\mathbb{Z}$.

In two dimensions we find another obstruction: Suppose that $\alpha_1 = \alpha$ and $\alpha_2 = 1 - \alpha$. If $\|n\alpha_j - \theta_j\| < \epsilon$ for each j, then

$$\|\theta_1 + \theta_2\| = \|n - \theta_1 - \theta_2\| \leq \|n\alpha_1 - \theta_1\| + \|n\alpha_2 - \theta_2\| < 2\epsilon.$$

But this should hold for any $\epsilon > 0$ which implies that $\theta_1 + \theta_2$ is an integer. Notice that in this example $1, \alpha_1, \alpha_2$ are not linearly independent over $\mathbb{Z}$.

Exercise 11.7.6. Let $\alpha_1, \ldots, \alpha_k, \theta_1, \ldots, \theta_k \in \mathbb{R}$ be given, and assume that there are integers $c_0, \ldots, c_k$ for which $c_0 + c_1\alpha_1 + \cdots + c_k\alpha_k = 0$. Suppose that for all $\epsilon > 0$ there are infinitely many n for which $\|n\alpha_j - \theta_j\| < \epsilon$ for $j = 1, 2, \ldots, k$. Prove that $c_1\theta_1 + \cdots + c_k\theta_k \in \mathbb{Z}$.

These are the only obstructions to the generalization:

Theorem 11.7 (Kronecker's Theorem in n dimensions). *Assume that the real numbers $1, \alpha_1, \ldots, \alpha_k$ are linearly independent over $\mathbb{Z}$. Then the points*

$$(n\alpha_1, \ldots, n\alpha_k)_{n \geq 1} \text{ are dense in } (\mathbb{R}/\mathbb{Z})^k.$$

In other words, for any given $\theta_1, \ldots, \theta_k \in \mathbb{R}$ and any $\epsilon > 0$ there are infinitely many integers n for which $\|n\alpha_j - \theta_j\| < \epsilon$ for all $j = 1, \ldots, k$.

This can be proved in several different ways that are accessible though tough. We refer the reader to sections 23.5–23.8 of [**HW08**].

11.8. Bouncing billiard balls

Billiards, snooker, and pool are all played on a rectangular table, hitting the ball along the surface. The sides of the table are cushioned so that the ball bounces off the side at the opposite angle to which it hits. That is, if it hits at angle $\alpha°$, then it bounces off at angle $(180 - \alpha)°$. Sometimes one miscues and the ball carries on around the table, coming to a stop without hitting another ball. Have you ever wondered what would happen if there were no friction, so that the ball never stops? Would your ball eventually hit the ball it is supposed to hit, no matter where that other ball is placed? Or could it go on bouncing forever without ever getting to the other ball? We could rephrase this question more mathematically by supposing that we play on a table in the complex plane, with two sides along the x- and y-axes. Say the table length is ℓ and width is w so that it is the rectangle with corners at $(0,0), (0, \ell), (w, 0), (w, \ell)$. Let us suppose that the ball is hit from the point (u, v) along a line with slope α (that is, at an angle α from the horizontal).

11.8. Bouncing billiard balls

As the line continues on indefinitely inside the box, does it get arbitrarily close to every point inside the box?

Exercise 11.8.1. Show that by rescaling with the map $x \to x/\ell$, $y \to y/w$ we can assume, without any loss of generality, that the billiards table is the unit square.

As a consequence of exercise 11.8.1, we may henceforth assume that $w = \ell = 1$.

The ball would run along the line $\mathcal{L} := \{(u+t, v+\alpha t), \ t \geq 0\}$ if it did not hit the sides of the table. Notice though that if after each time it hit a side, we reflected the true trajectory through the line that represents that side, then indeed the ball's trajectory would be $\mathcal{L}$.

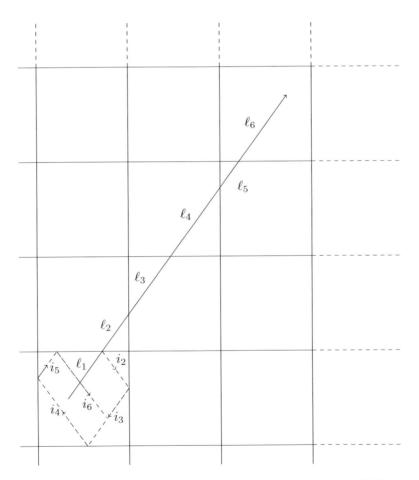

Figure 11.1. Billiards on the complex plane and on the unit square. Following a path inside the fundamental domain of a lattice: The path segment ℓ_j gets mapped to i_j for $j = 2, \ldots, 6$.

Develop this to prove:

Exercise 11.8.2. Show that the billiard ball is at (x, y) after time t, where x and y are given as follows:
Let $m = [u + t]$. If m is even, let $x = \{u + t\}$; if m is odd, let $x = 1 - \{u + t\}$.
Let $n = [v + \alpha t]$. If n is even, let $y = \{v + \alpha t\}$; if n is odd, let $y = 1 - \{v + \alpha t\}$.

Exercise 11.8.3. Show that if α is rational, then the ball eventually ends up exactly where it started from, and so it does not get arbitrarily close to every point on the table.

So how close does the trajectory get to the point (r, s), where $r, s \in [0, 1)$? Let us consider all of those values of t for which $x = r$, with m and n even to simplify matters (with m and n as in exercise 11.8.2), and see if we can determine whether y is ever close to s.

Exercise 11.8.4. Show that $[z]$ is even if and only if $\{z/2\} \in [0, 1/2)$. Deduce that $[z]$ is even and $\{z\} = r$ if and only if $\{z/2\} = r/2$.

Hence we want $(u+t)/2 = k+r/2$ for some integer k; that is, $t = 2k+(r-u)$, $k \in \mathbb{Z}$. In that case $v + \alpha t = 2\alpha k + \alpha(r - u) + v$ so we want $\{\alpha k + (\alpha(r - u) + v)/2\}$ close to $s/2$. That is, $k\alpha$ mod 1 should be close to $\theta := \{\frac{(s-v)+\alpha(u-r)}{2}\}$. Now, in Kronecker's Theorem (Theorem 11.5) we showed that the values $k\alpha$ mod 1 are dense in $[0, 1)$ when α is irrational, and so in particular there are values of k that allow $k\alpha$ mod 1 to be arbitrarily close to θ. Hence we have proved the difficult part of the following corollary:

Corollary 11.8.1. *If α is a real irrational number, then any ball moving at angle α (to the coordinate axes) will eventually get arbitrarily close to any point on a 1-by-1 billiards table.*

We finish with a challenge question to develop a similar theory of billiards played on a circular table!

Exercise 11.8.5. Imagine a trajectory inside the unit circle. A ball is hit and continues indefinitely. When it hits a side at angle θ (compared to the normal line at that point), it bounces off at angle $-\theta$.
  (a) Suppose that the first two points at which the ball hits the edge are at $e(\beta)$ and then at $e(\beta + \alpha)$. Show that the ball hits the edge at $e(\beta + n\alpha)$ for $n = 0, 1, 2, \ldots$.
  (b) Prove that the ball falls into a repeated trajectory if and only if α is rational.
  (c) Show that if α is irrational, then the points at which the ball hits the circle edge are dense (i.e., eventually the ball comes arbitrarily close to any point on the edge) but that it never hits the same edge point twice.
  (d) Prove that the ball's trajectory never comes inside the circle of radius $|\cos(\alpha/2)|$. Deduce that the trajectory of the ball is *never* dense inside the unit circle.
  (e) Prove that if α is irrational, then the trajectory of the ball is dense inside the ring between the circle of radius $|\cos(\alpha/2)|$ and the circle of radius 1. (The technical word for a ring is an *annulus*.)

Appendix 11B. Continued fractions

We introduced continued fractions in section 1.5. We noted that the partial quotients $1, 2, \frac{7}{4}, \frac{16}{9}, \frac{23}{13}$ of the continued fraction of $\frac{85}{48}$ "yield increasingly good approximations to $85/48$". Actually more is true: These are the best approximations to $\alpha := 85/48$ in the sense that if

$$\|q\alpha\| := \min_{p \in \mathbb{Z}} |q\alpha - p|,$$

then the smallest values so far in the sequence $\|\alpha\|, \|2\alpha\|, \ldots$ are always given by $\|q\alpha\|$ with $q = q_n$ for some n. When $\alpha = \frac{85}{48}$ our approximations yield the values

$$|\alpha - 2| = \frac{11}{48}, \quad |4\alpha - 7| = \frac{4}{48}, \quad |9\alpha - 16| = \frac{3}{48}, \quad |13\alpha - 23| = \frac{1}{48},$$

and these are the best approximations because $|q\alpha - p| > |q_n\alpha - p_n|$ for all $q < q_n$.

Knowing how to identify the best approximations to a given real number is very useful. For example if we wish to find a non-trivial solution to Pell's equation $p^2 - dq^2 = 1$, then we note that $p^2 \approx dq^2$ and so taking square roots we have $p \approx \sqrt{d}q$; that is, p/q must be a good approximation to $\sqrt{d}$. In fact it is such a good approximation that we find every solution to Pell's equation in positive integers among the partial quotients of the continued fraction for $\sqrt{d}$. This yields a remarkably efficient algorithm for finding solutions to Pell's equation (much like the Euclidean algorithm is such an efficient algorithm for finding gcds; the connection of the Euclidean algorithm to continued fractions was discussed in detail in section 1.5 and in section 1.8 of appendix 1A).

We will develop these remarks in detail in this appendix.

11.9. Continued fractions for real numbers

We have the following algorithm for constructing the continued fraction $[a_0, a_1, \ldots]$ of a given real number α, where each a_i is an integer, and $a_m \geq 1$ for all $m \geq 1$: Suppose that we have already determined integers $a_0, \ldots, a_{m-1}$ and real number α_m such that $\alpha = [a_0, a_1, \ldots, a_{m-1}, \alpha_m]$, starting with $\alpha_0 = \alpha$.

- Let $a_m := [\alpha_m]$. We will show that $\alpha_m \geq 1$ and so $a_m \geq 1$ for each $m \geq 1$.
- If $\alpha_m = a_m$, then α_m is an integer. Stop.
- If $\alpha_m > a_m$, then let $\alpha_{m+1} = 1/(\alpha_m - a_m)$.

Repeat this 3-step algorithm with m replaced by $m+1$.

Now $a_m + 1 > \alpha_m \geq a_m$ and so if $\alpha_m > a_m$, then $1 > \alpha_m - a_m > 0$ and therefore α_{m+1} is a real number > 1. Finally note that $\alpha_m = a_m + 1/\alpha_{m+1}$ and so
$$\alpha = [a_0, a_1, \ldots, a_{m-1}, \alpha_m] = [a_0, a_1, \ldots, a_{m-1}, a_m, \alpha_{m+1}].$$
This yields a unique continued fraction for each real number α.

Exercise 11.9.1. Explain why if α has a finite length continued fraction, then the last term is an integer ≥ 2.

The *convergents* $p_n/q_n := [a_0, a_1, \ldots, a_n]$ for each $n \geq 0$ are defined as in section 1.5 and given by

$$(11.9.1) \qquad \begin{pmatrix} p_n & p_{n-1} \\ q_n & q_{n-1} \end{pmatrix} = \begin{pmatrix} a_0 & 1 \\ 1 & 0 \end{pmatrix} \begin{pmatrix} a_1 & 1 \\ 1 & 0 \end{pmatrix} \cdots \begin{pmatrix} a_n & 1 \\ 1 & 0 \end{pmatrix}.$$

Note that

$$(11.9.2) \qquad p_n = a_n p_{n-1} + p_{n-2} \quad \text{and} \quad q_n = a_n q_{n-1} + q_{n-2} \quad \text{for all } n \geq 2,$$

so that the sequences $p_1, p_2, \ldots$ and $q_1, q_2, \ldots$ are increasing. Taking determinants yields that

$$(11.9.3) \qquad \frac{p_n}{q_n} - \frac{p_{n-1}}{q_{n-1}} = \frac{(-1)^{n+1}}{q_{n-1} q_n}$$

for each $n \geq 1$.

Next we show that $[a_0, a_1, a_2, \ldots]$ really does converge to α. Now $\alpha = [a_0, a_1, a_2, \ldots, a_n, \alpha_{n+1}]$, so that $\alpha = R/S$ where

$$\begin{pmatrix} R & p_n \\ S & q_n \end{pmatrix} = \begin{pmatrix} p_n & p_{n-1} \\ q_n & q_{n-1} \end{pmatrix} \begin{pmatrix} \alpha_{n+1} & 1 \\ 1 & 0 \end{pmatrix},$$

and therefore

$$(11.9.4) \qquad \alpha = \frac{R}{S} = \frac{\alpha_{n+1} p_n + p_{n-1}}{\alpha_{n+1} q_n + q_{n-1}},$$

which lies between $\frac{p_{n-1}}{q_{n-1}}$ and $\frac{p_n}{q_n}$ for each $n \geq 1$, by exercise 11.9.2:

Exercise 11.9.2. Show that if a, b, A, B, u, v are positive reals, then $\frac{au+Av}{bu+Bv}$ lies between $\frac{a}{b}$ and $\frac{A}{B}$.

11.9. Continued fractions for real numbers

By (11.9.4) and then (11.9.3) we have

$$\alpha - \frac{p_n}{q_n} = \frac{\alpha_{n+1}p_n + p_{n-1}}{\alpha_{n+1}q_n + q_{n-1}} - \frac{p_n}{q_n} = \frac{(-1)^n}{q_n(\alpha_{n+1}q_n + q_{n-1})}.$$

This certainly $\to 0$ as $n \to \infty$ implying that $p_n/q_n \to \alpha$ as $n \to \infty$; that is, $[a_0, a_1, a_2, \ldots]$ converges to α. Moreover this implies that $\alpha < p_n/q_n$ if n is odd, and $\alpha > p_n/q_n$ if n is even.

Exercise 11.9.3. Deduce that

$$\frac{p_0}{q_0} < \frac{p_2}{q_2} < \cdots < \frac{p_{2j}}{q_{2j}} < \cdots < \frac{p_{2j+1}}{q_{2j+1}} < \cdots < \frac{p_3}{q_3} < \frac{p_1}{q_1}$$

and that p_n/q_n tends to a limit as $n \to \infty$.

The last displayed estimate gives us very precise information about how fast the quotients converge to α: As $a_{n+1} \leq \alpha_{n+1} < a_{n+1} + 1$ we obtain

$$q_{n+1} \leq \alpha_{n+1}q_n + q_{n-1} < (a_{n+1}+1)q_n + q_{n-1} = q_{n+1} + q_n < 2q_{n+1},$$

yielding

(11.9.5) $$\frac{1}{2q_n q_{n+1}} < \left|\alpha - \frac{p_n}{q_n}\right| \leq \frac{1}{q_n q_{n+1}}.$$

The most famous example $\pi = [3, 7, 15, 1, 292, 1, \ldots]$, leads to the convergents

$$3 < \frac{333}{106} < \cdots < \pi < \cdots < \frac{355}{113} < \frac{22}{7}.$$

Having such a large partial quotient as $a_4 = 292$ means that the next convergent, $\frac{103993}{33102}$, has a far larger denominator, and therefore

$$\left|\pi - \frac{355}{113}\right| < \frac{1}{113 \cdot 33102} < 3 \cdot 10^{-7},$$

a fantastic approximation. This was known to Archimedes in the third century B.C.[9] The continued fraction for e displays an interesting pattern:

$$e = [2, 1, 2, 1, 1, 4, 1, 1, 6, 1, 1, 8, \ldots].$$

One can generalize the notion of continued fractions to obtain

$$\frac{\pi}{4} = \cfrac{1}{1+\cfrac{1^2}{2+\cfrac{3^2}{2+\cfrac{5^2}{2+\cdots}}}} \quad \text{or} \quad \pi = \cfrac{4}{1+\cfrac{1^2}{3+\cfrac{2^2}{5+\cfrac{3^2}{7+\cdots}}}}.$$

We will not pursue such representations here, as fun as they are.

Exercise 11.9.4. Show how to use the continued fraction to determine, for any irrational real number α, arbitrarily small real numbers of the form $a + b\alpha$ with $a, b \in \mathbb{Z}$.

---

[9] Around 1650 B.C., ancient Egyptians approximated π by regular octagons obtaining 256/81, a method developed further by Archimedes in Greece, and then Liu Hui in China in the third century A.D. In 1168 B.C. the Talmudic scholar Maimonides claimed that π can only be known approximately, that is, that π is irrational. In the ninth century B.C. the Indian astronomer Yajnavalkya arguably gave the approximation 333/106 in *Shatapatha Brahmana*; in the 14th century A.D., Madhava of the Kerala school in India showed how to get arbitrarily good approximations to π.

11.10. How good are these approximations?

Equation (11.9.5) shows that the convergents of the continued fraction give excellent approximations to α. Lagrange showed that one cannot really do better in that $\min_{q \leq Q} \|q\alpha\|$ (where $\|t\|$ denotes the distance from t to the nearest integer) is always attained by $q = q_n$, and no other q, which follows from the next result.

Theorem 11.8. *If $1 \leq q < q_{n+1}$, then $|q_n\alpha - p_n| \leq |q\alpha - p|$, with equality only if $q = q_n$ and $p = p_n$.*

Proof. Let $x = (-1)^n(qp_{n+1} - pq_{n+1})$ and $y = (-1)^n(qp_n - pq_n)$, so that

$$p_n x - p_{n+1} y = p \quad \text{and} \quad q_n x - q_{n+1} y = q$$

as $q_n p_{n+1} - p_n q_{n+1} = (-1)^n$. We observe that $x \neq 0$ or else q_{n+1} divides $q_{n+1} y = -q$ so that $q_{n+1} \leq q$, contradicting the hypothesis. We may assume that $y \neq 0$ or else $p = xp_n, q = xq_n$ and the result follows.

Now $q_n x = q_{n+1} y + q$ where $q < q_{n+1} \leq |q_{n+1} y|$, and so $q_n x$ and $q_{n+1} y$ have the same sign, and therefore x and y have the same sign. We saw earlier that $q_n\alpha - p_n$ and $q_{n+1}\alpha - p_{n+1}$ have opposite signs, and so $x(q_n\alpha - p_n)$ and $y(q_{n+1}\alpha - p_{n+1})$ have opposite signs. Now $q\alpha - p = x(q_n\alpha - p_n) - y(q_{n+1}\alpha - p_{n+1})$ and so

$$|q\alpha - p| = |x(q_n\alpha - p_n)| + |y(q_{n+1}\alpha - p_{n+1})| > |q_n\alpha - p_n|.$$

The result follows. □

Exercise 11.10.1. Deduce that if $1 \leq q < q_n$, then $\left|\alpha - \frac{p_n}{q_n}\right| < \left|\alpha - \frac{p}{q}\right|$.

We next show that if p/q is a good enough approximation to α, then it must be a convergent.

Corollary 11.10.1. *If $\left|\alpha - \frac{p}{q}\right| < \frac{1}{2q^2}$, then $\frac{p}{q}$ is a convergent for α.*

Proof. If $q_n \leq q < q_{n+1}$, then $|q_n\alpha - p_n| < 1/2q$ by Theorem 11.8. Hence $p/q = p_n/q_n$ or else

$$\frac{1}{qq_n} \leq \left|\frac{p}{q} - \frac{p_n}{q_n}\right| \leq \left|\alpha - \frac{p}{q}\right| + \left|\alpha - \frac{p_n}{q_n}\right| < \frac{1}{2q^2} + \frac{1}{2qq_n} \leq \frac{1}{qq_n},$$

a contradiction. □

Exercise 11.10.2. Show that if $-\sqrt{d} + \frac{1}{2} < p^2 - dq^2 \leq \sqrt{d}$ with $p, q \geq 1$, then p/q is a convergent in the continued fraction for $\sqrt{d}$.

Exercise 11.10.3. Show that if $d \equiv 1 \pmod{4}$ and $\frac{-\sqrt{d}+1}{2} < p^2 - pq + \left(\frac{1-d}{4}\right)q^2 \leq \frac{\sqrt{d}}{2}$ with $p, q \geq 1$, then p/q is a convergent in the continued fraction for $\frac{1+\sqrt{d}}{2}$.

In 1904 Fatou showed that if $\left|\alpha - \frac{p}{q}\right| < \frac{1}{q^2}$, then we cannot quite deduce that p/q is a convergent but we can come close, deducing that there exists an integer n for which

$$\frac{p}{q} = \frac{p_n}{q_n} \quad \text{or} \quad \frac{p_n - p_{n-1}}{q_n - q_{n-1}} \quad \text{or} \quad \frac{p_n + p_{n-1}}{q_n + q_{n-1}}.$$

At least every other convergent yields as good an approximation as required in Corollary 11.10.1:

Lemma 11.10.1. *The inequality* $\left|\alpha - \frac{p}{q}\right| \leq \frac{1}{2q^2}$ *is satisfied for at least one of* $\frac{p}{q} = \frac{p_n}{q_n}$ *or* $\frac{p_{n+1}}{q_{n+1}}$ *for each* $n \geq 0$.

Proof. If not, then, since $\alpha - \frac{p_n}{q_n}$ and $\alpha - \frac{p_{n+1}}{q_{n+1}}$ have opposite signs,

$$\frac{1}{q_n q_{n+1}} = \left|\frac{p_n}{q_n} - \frac{p_{n+1}}{q_{n+1}}\right| = \left|\alpha - \frac{p_n}{q_n}\right| + \left|\alpha - \frac{p_{n+1}}{q_{n+1}}\right| > \frac{1}{2q_n^2} + \frac{1}{2q_{n+1}^2},$$

which is false for any distinct reals q_n, q_{n+1}. □

Hurwitz showed that for at least one of every three convergents one can improve this to $\leq 1/(\sqrt{5}q^2)$. However this is the best possible in general since the convergents for $\frac{1+\sqrt{5}}{2}$ are $\frac{F_{n+1}}{F_n}$ and, as we will prove in the next exercise, $F_n^2 \left|\frac{1+\sqrt{5}}{2} - \frac{F_{n+1}}{F_n}\right| \to \frac{1}{\sqrt{5}}$ as $n \to \infty$.

Exercise 11.10.4. Show that $\frac{1+\sqrt{5}}{2} = [1, 1, 1, 1, \ldots]$ and so the convergents are F_{n+1}/F_n where F_n is the nth Fibonacci numbers. By using the general formula for Fibonacci numbers, determine how good these approximations are; i.e., prove a strong version of the formula in section 11.4:

$$\left|\frac{1+\sqrt{5}}{2} - \frac{F_{n+1}}{F_n} + \frac{(-1)^n}{\sqrt{5}F_n^2}\right| \leq \frac{1}{5F_n^4}.$$

11.11. Periodic continued fractions and Pell's equation

Given a solution to Pell's equation, $p^2 - dq^2 = \pm 4$ with $p, q > 0$, we have $\sqrt{d} + p/q > \sqrt{d}$ so that

$$\left|\sqrt{d} - \frac{p}{q}\right| = \frac{|p^2 - dq^2|}{q^2(\sqrt{d} + p/q)} < \frac{4}{\sqrt{d}q^2}.$$

If $d \geq 64$, then this $< 1/2q^2$ and so p/q is a convergent for $\sqrt{d}$ by Corollary 11.10.1.

What about in the other direction? If p/q is a convergent in the continued fraction $\sqrt{d}$, then $|\sqrt{d} - p/q| < 1/q^2$ and therefore $|\sqrt{d} + p/q| \leq |\sqrt{d} - p/q| + 2\sqrt{d} \leq 2\sqrt{d} + 1$. Hence

$$|p^2 - dq^2| = q^2 \left|\sqrt{d} - \frac{p}{q}\right| \cdot \left|\sqrt{d} + \frac{p}{q}\right| < 2\sqrt{d} + 1.$$

Not quite a solution to Pell's equation but it at least guarantees that $p^2 - dq^2$ is a small integer in absolute value.

In this section we study the continued fractions of quadratic irrationals. We will prove that, like the continued fraction for $\frac{1+\sqrt{5}}{2} = [1, 1, 1 \ldots]$, they are all eventually periodic: We say that $[a_0, a_1, \ldots]$ is *periodic* with period $a_0, \ldots, a_{m-1}$ where $m \geq 1$ if $a_{n+m} = a_n$ for all $n \geq 0$, and write

$$[\overline{a_0, a_1, \ldots, a_{m-1}}].$$

Lemma 11.11.1. *If α has a periodic continued fraction, then α is a real quadratic irrational.*

For example if $\alpha = [\overline{a}]$, then $\alpha = a + \frac{1}{\alpha}$, which implies $\alpha^2 - a\alpha - 1 = 0$ so that $\alpha = (a \pm \sqrt{d})/2$, where $d = a^2 + 4$. Therefore $a^2 - d \cdot 1^2 = -4$, a solution to the Pell equation for d.

Exercise 11.11.1. What numbers have continued fraction $[a, b, a, b, a, \ldots]$ where a and b are integers ≥ 1? Can you use this to find solutions to a family of Pell equations?

Proof. As α has a periodic continued fraction, with period length m, say, then $\alpha_m = \alpha$ and $\alpha = [a_0, \ldots, a_{m-1}, \alpha]$ which, as above, implies that

$$(11.11.1) \qquad \alpha = \frac{\alpha p_{m-1} + p_{m-2}}{\alpha q_{m-1} + q_{m-2}}.$$

Multiplying through by the denominator we find that α satisfies the quadratic equation

$$(11.11.2) \qquad q_{m-1}\alpha^2 + (q_{m-2} - p_{m-1})\alpha - p_{m-2} = 0.$$

Therefore α is quadratic irrational, for it cannot be rational or else the continued fraction would have finite length (as in section 1.5). $\square$

Lemma 11.11.2. *If $\alpha = u + v\sqrt{d}$, with d squarefree and $u, v > 0$, has a periodic continued fraction of period length m, then we have positive integer solutions u_n, v_n to Pell's equation*

$$u_n^2 - dv_n^2 = 4(-1)^n$$

whenever m divides n, taking

$$u_n = q_{n-2} + p_{n-1} \quad \text{and} \quad v_n = 2q_{n-1}v.$$

If we write $\epsilon_n = \frac{1}{2}(u_n + \sqrt{d}v_n)$, then $\epsilon_n = \epsilon_m^k$ where $n = mk$ for all integers $k \geq 1$.

In the next section we will prove that any α of the form $\alpha = u + v\sqrt{d}$ has an "eventually" periodic continued fraction.

Proof. Now (11.11.1) is satisfied, with n replaced by m, whenever n is a multiple of m. Solving (11.11.2) and noting that α is the positive root, we find that

$$(11.11.3) \qquad u + v\sqrt{d} = \alpha = \frac{p_{n-1} - q_{n-2} + \sqrt{(q_{n-2} - p_{n-1})^2 + 4p_{n-2}q_{n-1}}}{2q_{n-1}}.$$

Multiplying through by the denominator and comparing the square of the square root term on each side, we obtain

$$(q_{n-2} + p_{n-1})^2 - 4(-1)^n = (q_{n-2} - p_{n-1})^2 + 4p_{n-2}q_{n-1} = d(2q_{n-1}v)^2.$$

Since the left side is an integer, so is the right side, and the first part of the result follows.

The matrix $\begin{pmatrix} p_{n-1} & p_{n-2} \\ q_{n-1} & q_{n-2} \end{pmatrix}$ has eigenvalues ϵ_n and $\epsilon_n' = \frac{1}{2}(u_n - \sqrt{d}v_n)$. By (11.9.1) we see that if $n = mk$, then

$$\begin{pmatrix} p_{n-1} & p_{n-2} \\ q_{n-1} & q_{n-2} \end{pmatrix} = \begin{pmatrix} p_{m-1} & p_{m-2} \\ q_{m-1} & q_{m-2} \end{pmatrix}^k,$$

and therefore, comparing eigenvalues and noting that $u_n, v_n > 0$, we deduce that $\epsilon_n = \epsilon_m^k$ for all $k \geq 1$. $\square$

11.12. Quadratic irrationals and periodic continued fractions

One can show that these generate all the solutions to Pell's equation. Moreover we see that there is a solution to the negative Pell equation only if m is odd.

11.12. Quadratic irrationals and periodic continued fractions

In the previous section we worked with periodic continued fractions to find solutions to Pell's equation, but is there really a periodic continued fraction for each $\sqrt{d}$? In this subsection we develop a theory that allows us to easily determine whether a given number of the form $u + v\sqrt{d}$, with u, v rational, has a periodic continued fraction.

We say that $\alpha = [a_0, a_1, \ldots]$ is *eventually periodic* if $a_{n+m} = a_n$ for all $n \geq r$, for some integer $r \geq 0$, and we write $\alpha = [a_0, \ldots, a_{r-1}, \overline{a_r, \ldots, a_{r+m-1}}]$. This α is a real quadratic irrational, since $\gamma := [\overline{a_r, \ldots, a_{r+m-1}}]$ is real quadratic irrational by Lemma 11.11.1, and

$$(11.12.1) \qquad \alpha = \frac{\gamma p_{r-1} + p_{r-2}}{\gamma q_{r-1} + q_{r-2}}.$$

Theorem 11.9. *The continued fraction of a quadratic irrational real number is eventually periodic.*

Proof. Suppose that α has minimal polynomial $ax^2 + bx + c = a(x - \alpha)(x - \beta)$, and define

$$f(x, y) := ax^2 + bxy + cy^2 = \begin{pmatrix} x & y \end{pmatrix} \begin{pmatrix} a & b/2 \\ b/2 & c \end{pmatrix} \begin{pmatrix} x \\ y \end{pmatrix}.$$

By (11.11.1), $\begin{pmatrix} \alpha \\ 1 \end{pmatrix} = \kappa \begin{pmatrix} p_n & p_{n-1} \\ q_n & q_{n-1} \end{pmatrix} \begin{pmatrix} \alpha_{n+1} \\ 1 \end{pmatrix}$ for some $\kappa \neq 0$, and so if we define

$$\begin{pmatrix} A_n & B_n/2 \\ B_n/2 & C_n \end{pmatrix} := \begin{pmatrix} p_n & q_n \\ p_{n-1} & q_{n-1} \end{pmatrix} \begin{pmatrix} a & b/2 \\ b/2 & c \end{pmatrix} \begin{pmatrix} p_n & p_{n-1} \\ q_n & q_{n-1} \end{pmatrix}$$

(so that $d := b^2 - 4ac = B_n^2 - 4A_n C_n$ by taking determinants of both sides), then

$$A_n \alpha_{n+1}^2 + B_n \alpha_{n+1} + C_n = \begin{pmatrix} \alpha_{n+1} & 1 \end{pmatrix} \begin{pmatrix} A_n & B_n/2 \\ B_n/2 & C_n \end{pmatrix} \begin{pmatrix} \alpha_{n+1} \\ 1 \end{pmatrix}$$

$$= \begin{pmatrix} \alpha_{n+1} & 1 \end{pmatrix} \begin{pmatrix} p_n & q_n \\ p_{n-1} & q_{n-1} \end{pmatrix} \begin{pmatrix} a & b/2 \\ b/2 & c \end{pmatrix} \begin{pmatrix} p_n & p_{n-1} \\ q_n & q_{n-1} \end{pmatrix} \begin{pmatrix} \alpha_{n+1} \\ 1 \end{pmatrix}$$

$$= \kappa^2 \begin{pmatrix} \alpha & 1 \end{pmatrix} \begin{pmatrix} a & b/2 \\ b/2 & c \end{pmatrix} \begin{pmatrix} \alpha \\ 1 \end{pmatrix} = \kappa^2 f(\alpha, 1) = 0.$$

Therefore $f_n(x) := A_n x^2 + B_n x + C_n$ has root α_{n+1}. Now $A_n = f(p_n, q_n)$ and $C_n = A_{n-1}$. By (11.9.5), $\left| \alpha - \frac{p_n}{q_n} \right| < \frac{1}{q_n^2} \leq 1$, and

$$\left| \beta - \frac{p_n}{q_n} \right| \leq |\beta - \alpha| + \left| \alpha - \frac{p_n}{q_n} \right| \leq |\beta - \alpha| + 1,$$

so that, for all $n \geq 1$, we have

$$|A_n| = |f(p_n, q_n)| = aq_n^2 \left| \alpha - \frac{p_n}{q_n} \right| \left| \beta - \frac{p_n}{q_n} \right| \leq a(|\beta - \alpha| + 1) = a + \sqrt{d}.$$

Since the A_n are all integers, there are only finitely many possibilities for the values of A_n and C_n once $n \geq 2$. Moreover, given these values there are only two possibilities for B_n, as $B_n^2 = d + 4A_nC_n$. Hence there are only finitely many possible polynomials $f_n(x)$ and each corresponds to at most two roots, so one such root must repeat infinitely often. That is, there exists $m < n$ for which $\alpha_m = \alpha_n$. The continued fraction for α_m is $[a_m, a_{m+1}, \ldots, a_{n-1}, \alpha_n]$, and so α_m is periodic with period $n - m$.

Exercise 11.12.1. Deduce that the continued fraction for α is eventually periodic.

Moreover if $n \geq 2$, then $B_n^2 = d + 4A_nC_n \leq d + 4(a + \sqrt{d})^2$ so that $|B_n| < 2a + 3\sqrt{d}$. Now $a_n = [\alpha_n]$ and $\alpha_{n+1} = \frac{-B_n + \sqrt{d}}{2A_n}$. Therefore, as each $|A_n| \geq 1$, we have

(11.12.2) $$a_n \leq \frac{|B_{n-1}| + \sqrt{d}}{2} \leq a + 2\sqrt{d} \quad \text{for all } n \geq 3. \qquad \square$$

Exercise 11.12.2. Prove that if $a = 1$, then $a_n \leq \sqrt{d} + 1$ for all $n \geq 3$ in (11.12.2).

We have shown that α has an eventually periodic continued fraction if and only if α is a quadratic irrational real number. However to apply Lemma 11.11.2 we need α to have a periodic continued fraction. In the next result we give an easily verified criterion to determine whether a given α has a periodic continued fraction. We define the *conjugate* of $u + v\sqrt{d}$ to be $u - v\sqrt{d}$.

Proposition 11.12.1. *Suppose α is a real quadratic irrational number with conjugate $-\beta$. Then α has a periodic continued fraction if and only if $\alpha > 1 > \beta > 0$.*

Let $d > 1$ be an integer that is not a square. Let $\alpha = [\sqrt{d}] + \sqrt{d}$, which is obviously > 1. Moreover α has conjugate $-(\sqrt{d} - [\sqrt{d}]) = -\{\sqrt{d}\}$, and $\{\sqrt{d}\}$ lies in $(0, 1)$. Therefore α has a periodic continued fraction, which means we can use Lemma 11.11.2 in order to find the solutions to Pell's equation for d.

Proof. By Theorem 11.9, the continued fraction of α is eventually periodic.

Assume that $\alpha > 1 > \beta > 0$. This implies that $\alpha_n > a_n \geq 1$ for all $n \geq 0$, and we now show that $0 < \beta_n < 1$ for all $n \geq 0$ by induction. It is true for $n = 0$ as $\beta_0 = \beta$ and so now take $n \geq 1$. Since $\alpha_{n-1} = a_{n-1} + 1/\alpha_n$ by definition, we have $-\beta_{n-1} = a_{n-1} - 1/\beta_n$ by taking conjugates. This means that $a_{n-1} = 1/\beta_n - \beta_{n-1}$ is an integer in $(1/\beta_n - 1, 1/\beta_n)$ by the induction hypothesis, and so $a_{n-1} = [1/\beta_n]$ and hence $1/\beta_n > 1$ implying that $0 < \beta_n < 1$ as required. Since the continued fraction for α is eventually periodic, there exists $0 \leq m < n$ with $\alpha_m = \alpha_n$; select m to be the minimal integer ≥ 0 for which such an n exists. Then $m = 0$ or else taking conjugates gives $\beta_m = \beta_n$, so that $a_{m-1} = [1/\beta_m] = [1/\beta_n] = a_{n-1}$ and hence $\alpha_{m-1} = a_{m-1} + 1/\alpha_m = a_{n-1} + 1/\alpha_n = \alpha_{n-1}$, contradicting the minimality of m. We have therefore proved that α has a periodic continued fraction.

On the other hand if the continued fraction of α is purely periodic of period length n, then $\alpha > a_0 = a_n \geq 1$. Let $f(x) := q_{n-1}x^2 + (q_{n-2} - p_{n-1})x - p_{n-2} = 0$ which has roots $x = \alpha$ and $-\beta$, as noted above. Now $f(0) = -p_{n-2} < 0$ and $f(-1) = (q_{n-1} - q_{n-2}) + (p_{n-1} - p_{n-2}) > 0$, and so f has a root in $(-1, 0)$ by

11.13. Solutions to Pell's equation from a well-selected continued fraction

the intermediate value theorem. This root cannot be α which is > 1 so it must be $-\beta$. □

Exercise 11.12.3. Prove that $\sqrt{d} + [\sqrt{d}]$ has a periodic continued fraction for any squarefree integer $d > 1$.

Lemma 11.12.1. *If $\alpha = [\overline{a_0, \ldots, a_{m-1}}]$, then the conjugate of α is $-\beta$ where $\beta = [0, \overline{a_{m-1}, a_{m-2}, \ldots, a_0}]$.*

If we write $\alpha = u + v\sqrt{d}$, then $u, v > 0$ with $v\sqrt{d} + u > 1 > v\sqrt{d} - u > 0$, by Proposition 11.12.1.

Proof. Let $\delta = 1/\gamma$ where $\gamma = [a_{m-1}, a_{m-2}, \ldots, a_0, \gamma]$, and

$$\begin{pmatrix} a_{m-1} & 1 \\ 1 & 0 \end{pmatrix} \cdots \begin{pmatrix} a_0 & 1 \\ 1 & 0 \end{pmatrix} = \left(\begin{pmatrix} a_0 & 1 \\ 1 & 0 \end{pmatrix} \cdots \begin{pmatrix} a_{m-1} & 1 \\ 1 & 0 \end{pmatrix} \right)^T = \begin{pmatrix} p_{m-1} & q_{m-1} \\ p_{m-2} & q_{m-2} \end{pmatrix},$$

so that

$$\gamma = \frac{\gamma p_{m-1} + q_{m-1}}{\gamma p_{m-2} + q_{m-2}}.$$

Multiplying through by the denominator this implies that $-\delta$ satisfies the same quadratic equation (11.11.2) as α, and they must be distinct roots as $\alpha > 0 > -\delta$, and so $\delta = \beta$. □

11.13. Solutions to Pell's equation from a well-selected continued fraction

The above results combine to show that the continued fraction of $\sqrt{d}$ has some very special structure:

Corollary 11.13.1. *If $d > 1$ is an integer but not a square then one can write*

$$\sqrt{d} = [b_0, \overline{b_1, \ldots, b_m}] \text{ where } b_j = b_{m-j} \text{ for } j = 1, 2, \ldots, m-1,$$

and $b_0 = [\sqrt{d}]$ with $b_m = 2b_0$. Let m be the smallest such integer, and suppose that the nth convergent of the continued fraction is P_n/Q_n. The fundamental solution to the Pell equation $x^2 - dy^2 = \pm 1$ is given by $(x, y) = (P_{m-1}, Q_{m-1})$, with

$$P_{m-1}^2 - dQ_{m-1}^2 = (-1)^m.$$

All other solutions to Pell's equation with $x, y > 0$ are given by

$$P_{n-1}^2 - dQ_{n-1}^2 = (-1)^n$$

whenever m divides n, so that $n = mk$ for some integer $k \geq 1$, and

$$P_{n-1} + \sqrt{d}Q_{n-1} = (P_{m-1} + \sqrt{d}Q_{m-1})^k.$$

Proof. Let $\alpha = b_0 + \sqrt{d}$, which has conjugate $-\beta$ where $\beta = \sqrt{d} - b_0 = \alpha - 2b_0$. As $a_0 = 2b_0$, Lemma 11.12.1 then implies that $b_j = a_j = a_{m-j} = b_{m-j}$ for all $j = 1, 2, \ldots, m-1$. Moreover $b_m = a_m = 2b_0$.

By definition the nth convergent in the continued fraction for α is $p_n/q_n = b_0 + P_n/Q_n$, and so $Q_n = q_n$ and $P_n = p_n - b_0 q_n$ for all $n \geq 1$. We apply Lemma 11.11.2. Now (11.11.3) implies that $v = 1$ so that $v_n = 2q_{n-1} = 2Q_{n-1}$, and $2q_{n-1}b_0 = p_{n-1} - q_{n-2}$, so that $P_{n-1} = b_0 Q_{n-1} + Q_{n-2}$, and therefore $u_n = p_{n-1} + q_{n-2} = P_{n-1} + b_0 Q_{n-1} + Q_{n-2} = 2P_{n-1}$. Therefore $P_{n-1}^2 - dQ_{n-1}^2 = (-1)^n$ whenever m divides n, by Lemma 11.11.2.

Exercise 11.13.1. Prove that if $b_{n+j} = b_j$ for all $j \geq 1$, then m divides n.

Now suppose that $x^2 - dy^2 = \pm 1$ with x, y positive integers. If $d > 2$, then, by exercise 11.10.2, we see that x/y is a convergent in the continued fraction for $\sqrt{d}$, say p_{r-1}/q_{r-1} for some $r \geq 2$. If we write $\sqrt{d} = [b_0, \ldots, b_{r-1}, \gamma]$, then

$$\sqrt{d} = \frac{p_{r-1}\gamma + p_{r-2}}{q_{r-1}\gamma + q_{r-2}}$$

so that $(p_{r-1} - \sqrt{d}q_{r-1})\gamma = -(p_{r-2} - \sqrt{d}q_{r-2})$. Multiplying this through by $p_{r-1} + \sqrt{d}q_{r-1}$, we obtain

$$(-1)^r \gamma = (p_{r-1}^2 - dq_{r-1}^2)\gamma = -(p_{r-2} - \sqrt{d}q_{r-2})(p_{r-1} + \sqrt{d}q_{r-1})$$
$$= (p_{r-1}q_{r-2} - p_{r-2}q_{r-1})\sqrt{d} + (dp_{r-1}q_{r-2} - p_{r-1}p_{r-2}),$$

so that $\gamma = \sqrt{d} + c$ for some integer c. By definition γ has a periodic continued fraction and so by exercise 11.12.3 we know that $c = b_0$. This implies that $b_r = 2b_0$ and $b_{r+j} = b_j$ for all $j \geq 1$. Therefore m divides r by exercise 11.13.1. The result follows (other than this last part when $d = 2$ which follows from finding the fundamental unit $1^2 - 2 \cdot 1^2 = -1$ from the first convergent of the continued fraction $\sqrt{2} = [1, \overline{2}]$). □

We deduce from Corollary 11.13.1 that there is a solution to the negative Pell equation $x^2 - dy^2 = -1$ if and only if m, the minimum period length of the continued fraction, is odd.

Here are some examples of the continued fraction for $\sqrt{d}$:

$$\sqrt{2} = [1, \overline{2}], \quad \sqrt{3} = [1, \overline{1, 2}], \quad \sqrt{5} = [2, \overline{4}], \quad \sqrt{6} = [2, \overline{2, 4}], \quad \sqrt{7} = [2, \overline{1, 1, 1, 4}],$$
$$\sqrt{8} = [2, \overline{1, 4}], \quad \sqrt{10} = [3, \overline{6}], \quad \sqrt{11} = [3, \overline{3, 6}], \quad \sqrt{12} = [3, \overline{2, 6}],$$
$$\sqrt{13} = [3, \overline{1, 1, 1, 1, 6}], \quad \sqrt{14} = [3, \overline{1, 2, 1, 6}], \ldots$$

Next we give only the longest continued fractions and the largest fundamental solutions for discriminants up to 135:

$$\sqrt{2} = [1, \overline{2}], \quad 1^2 - 2 \cdot 1^2 = -1,$$
$$\sqrt{3} = [1, \overline{1, 2}], \quad 2^2 - 3 \cdot 1^2 = 1,$$
$$\sqrt{6} = [2, \overline{2, 4}], \quad 5^2 - 6 \cdot 2^2 = 1,$$
$$\sqrt{7} = [2, \overline{1, 1, 1, 4}], \quad 8^2 - 7 \cdot 3^2 = 1,$$
$$\sqrt{13} = [3, \overline{1, 1, 1, 1, 6}], \quad 18^2 - 13 \cdot 5^2 = -1,$$
$$\sqrt{19} = [4, \overline{2, 1, 3, 1, 2, 8}], \quad 170^2 - 19 \cdot 39^2 = 1,$$
$$\sqrt{22} = [4, \overline{1, 2, 4, 2, 1, 8}], \quad 197^2 - 22 \cdot 42^2 = 1,$$
$$\sqrt{31} = [5, \overline{1, 1, 3, 5, 3, 1, 1, 10}], \quad 1520^2 - 31 \cdot 273^2 = 1,$$
$$\sqrt{43} = [6, \overline{1, 1, 3, 1, 5, 1, 3, 1, 1, 12}], \quad 3482^2 - 43 \cdot 531^2 = 1,$$

11.13. Solutions to Pell's equation from a well-selected continued fraction

$$\sqrt{46} = [6, \overline{1,3,1,1,2,6,2,1,1,3,1,12}], \quad 24335^2 - 46 \cdot 3588^2 = 1,$$
$$\sqrt{76} = [8, \overline{1,2,1,1,5,4,5,1,1,2,1,16}], \quad 57799^2 - 76 \cdot 6630^2 = 1,$$
$$\sqrt{94} = [9, \overline{1,2,3,1,1,5,1,8,1,5,1,1,3,2,1,18}], \quad 2143295^2 - 94 \cdot 221064^2 = 1,$$
$$\sqrt{124} = [11, \overline{7,2,1,1,1,3,1,4,1,3,1,1,1,2,7,22}], \quad 4620799^2 - 124 \cdot 414960^2 = 1,$$
$$\sqrt{133} = [11, \overline{1,1,7,5,1,1,1,2,1,1,1,5,7,1,1,22}], \quad 2588599^2 - 133 \cdot 224460^2 = 1.$$

Now we give the largest fundamental solutions, and their continued fraction lengths, for discriminants up to 750:

Length = 18: $\quad 77563250^2 - 139 \cdot 6578829^2 = 1,$

Length = 20: $\quad 1728148040^2 - 151 \cdot 140634693^2 = 1,$

Length = 22: $\quad 1700902565^2 - 166 \cdot 132015642^2 = 1,$

Length = 26: $\quad 278354373650^2 - 211 \cdot 19162705353^2 = 1,$

Length = 26: $\quad 695359189925^2 - 214 \cdot 47533775646^2 = 1,$

Length = 26: $\quad 5883392537695^2 - 301 \cdot 339113108232^2 = 1,$

Length = 34: $\quad 2785589801443970^2 - 331 \cdot 153109862634573^2 = 1,$

Length = 37: $\quad 44042445696821418^2 - 421 \cdot 2146497463530785^2 = -1,$

Length = 40: $\quad 84056091546952933775^2 - 526 \cdot 3665019757324295532^2 = 1,$

Length = 42: $\quad 181124355061630786130^2 - 571 \cdot 7579818350628982587^2 = 1,$

Length = 44: $\quad 5972991296311683199^2 - 604 \cdot 243037569063951720^2 = 1,$

Length = 48: $\quad 48961575312998650035560^2 - 631 \cdot 1949129537575151036427^2 = 1.$

The period length of the continued fractions of the $\sqrt{d}$ seems to mostly be around size $2\sqrt{d}$, and the size of the fundamental solutions mostly seems to be around $10^{\sqrt{d}}$. We believe that something like these estimates are true in general, extrapolating from this computational evidence, though these claims are not yet proven. There are some "thin" classes of discriminants for which the period length of the continued fractions and the fundamental solutions are much smaller. (For example, if $d = m^2 + 1$, then $\mathbb{Q}(\sqrt{d})$ has the fundamental solution $m + \sqrt{d}$ to Pell's equation, and $\sqrt{d} = [m, \overline{2m}]$ has periodic length one, for all integers $m \geq 1$.)

Corollary 11.13.2 (Relation between the continued fraction length and the size of the fundamental solution). *Suppose that $\sqrt{d}$ has a continued fraction that is eventually periodic of period m. Let $\epsilon_d = u + v\sqrt{d}$ where u, v give the first solution to $u^2 - dv^2 = 1$ as in Corollary 11.13.1. Then $(\frac{1+\sqrt{5}}{2})^m < \epsilon_d < 2(\sqrt{d}+2)^m$.*

Proof. Let $\tau_r := P_r + \sqrt{d}Q_r$ for all $r \geq 0$, so that $\tau_0 = b_0 + \sqrt{d}$ and $\tau_{n+1} = b_n\tau_n + \tau_{n-1}$ for all $n \geq 1$, starting with $\tau_{-1} = 1$. Since the minimum polynomial for $\sqrt{d}$ is $x^2 - d$, exercise 11.12.2 implies that each $b_n \leq \sqrt{d} + 1$ (for $n = 1, 2$ we

deduce this as $b_n = b_{n+m}$). The τ_n are obviously increasing, and so
$$\tau_n < (b_{n-1}+1)\tau_{n-1} < \cdots < (\sqrt{d}+2)^n \tau_0 < 2(\sqrt{d}+2)^{n+1}.$$
The upper bound follows as $\epsilon_d = \tau_{m-1}$ by Corollary 11.13.1. The lower bound follows from noting that $\tau_n \geq F_{n+3}$ for $n = -1, 0$, and then by induction for all n, and so $\tau_{m-1} \geq F_{m+2} > (\frac{1+\sqrt{5}}{2})^m$. □

In section 11.2 we saw that the fundamental units can be of the form $\frac{u+v\sqrt{d}}{2}$ with u and v odd when $d \equiv 1 \pmod{4}$. To find these we prove the analogy to Corollary 11.13.1 for the continued fraction of $\frac{1+\sqrt{d}}{2}$ (in place of the continued fraction of $\sqrt{d}$).

Corollary 11.13.3. *If $d > 1$ is $\equiv 1 \pmod{4}$ but is not a square, then*
$$\frac{1+\sqrt{d}}{2} = [b_0, \overline{b_1, \ldots, b_m}] \text{ where } b_j = b_{m-j} \text{ for } j = 1, 2, \ldots, m-1,$$
and $b_m = 2b_0 - 1$ is the largest odd integer $< \sqrt{d}$. Let m be the smallest such integer, and suppose that the nth convergent of the continued fraction is P_n/Q_n. The fundamental solution to the Pell equation $x^2 - dy^2 = \pm 4$ is given by $(x, y) = (2P_{m-1} - Q_{m-1}, Q_{m-1})$, with
$$(2P_{m-1} - Q_{m-1})^2 - dQ_{m-1}^2 = 4(-1)^m,$$
which can be rewritten as
$$P_{m-1}^2 - P_{m-1}Q_{m-1} + \left(\frac{1-d}{4}\right)Q_{m-1}^2 = (-1)^m.$$
All other solutions to Pell's equation with $x, y > 0$ are given by
$$(2P_{n-1} - Q_{n-1})^2 - dQ_{n-1}^2 = 4(-1)^n$$
whenever m divides n, so that $n = mk$ for some integer $k \geq 1$, and
$$P_{n-1} - \left(\frac{1+\sqrt{d}}{2}\right)Q_{n-1} = \left(P_{m-1} - \left(\frac{1+\sqrt{d}}{2}\right)Q_{m-1}\right)^k.$$

Proof. Let a_0 be the largest odd integer $< \sqrt{d}$ and $\alpha = \frac{a_0+\sqrt{d}}{2}$, so that $[\alpha] = a_0$. Now α has conjugate $-\beta$ where $\beta = \alpha - a_0 \in (0,1)$, and so $\alpha = [\overline{a_0, \ldots, a_{m-1}}]$ with $\beta = [0, \overline{a_{m-1}, a_{m-2}, \ldots, a_0}]$ by Lemma 11.12.1. Equating $\beta = \frac{\sqrt{d}-a_0}{2} = \alpha - a_0 = [0, \overline{a_1, \ldots, a_m}]$, we see that $a_j = a_{m-j}$ for $j = 0, 1, \ldots, m$. Now $\frac{1+\sqrt{d}}{2} = \alpha - \frac{a_0-1}{2}$, and so $b_0 = \frac{a_0+1}{2}$ and $b_j = a_j$ for all $j \geq 1$. In particular $b_m = a_m = a_0 = 2b_0 - 1$.

By definition $p_n/q_n = (b_0-1) + P_n/Q_n$, and so $Q_n = q_n$ and $P_n = p_n - (b_0-1)q_n$ for all $n \geq 1$. We apply Lemma 11.11.2, so (11.11.3) implies that $2v = 1$ with $v_n = q_{n-1} = Q_{n-1}$ and $q_{n-1}(2b_0-1) = q_{n-1}a_0 = p_{n-1} - q_{n-2}$, so that $P_{n-1} = b_0 Q_{n-1} + Q_{n-2}$. Therefore
$$u_n = p_{n-1} + q_{n-2} = P_{n-1} + (b_0-1)Q_{n-1} + Q_{n-2} = 2P_{n-1} - Q_{n-1}.$$

Now suppose that $x^2 - dy^2 = \pm 4$ with $x, y > 0$, so that $x \equiv y \pmod 2$. If $d > 9$, then take $p = \frac{x+y}{2}$ and $q = y$ in exercise 11.10.3; we see that p/q is a

11.13. Solutions to Pell's equation from a well-selected continued fraction

convergent in the continued fraction for $\frac{1+\sqrt{d}}{2}$, say p_{r-1}/q_{r-1} for some $r \geq 2$. If we write $\frac{1+\sqrt{d}}{2} = [b_0, \ldots, b_{r-1}, \gamma]$, then

$$\frac{1+\sqrt{d}}{2} = \frac{p_{r-1}\gamma + p_{r-2}}{q_{r-1}\gamma + q_{r-2}}.$$

Proceeding analogously to the proof of Corollary 11.13.1, we deduce that $\gamma = \frac{1+\sqrt{d}}{2} + c$ for some integer c. By definition γ has a periodic continued fraction and so we must have $c = b_0 - 1$. This implies that $b_r = 2b_0 - 1$ and $b_{r+j} = b_j$ for all $j \geq 1$. Therefore m divides r by exercise 11.13.1. The result follows (other than this last part when $d = 5$ which follows from finding the fundamental unit $1^2 - 5 \cdot 1^2 = -4$ from the first convergent of the continued fraction $\frac{1+\sqrt{5}}{2} = [\overline{1}]$). □

The first few examples of the continued fraction for $\frac{1+\sqrt{d}}{2}$ are

$$\frac{1+\sqrt{5}}{2} = [\overline{1}], \quad \frac{1+\sqrt{13}}{2} = [2, \overline{3}], \quad \frac{1+\sqrt{17}}{2} = [2, \overline{1, 1, 3}], \quad \frac{1+\sqrt{21}}{2} = [2, \overline{1, 3}],$$

$$\frac{1+\sqrt{29}}{2} = [3, \overline{5}], \quad \frac{1+\sqrt{33}}{2} = [3, \overline{2, 1, 2, 5}], \quad \frac{1+\sqrt{37}}{2} = [3, \overline{1, 1, 5}],$$

$$\frac{1+\sqrt{41}}{2} = [3, \overline{1, 2, 2, 1, 5}], \quad \frac{1+\sqrt{45}}{2} = [3, \overline{1, 5}], \ldots.$$

Now we give only the longest continued fractions:

$$\frac{1+\sqrt{57}}{2} = [4, \overline{3, 1, 1, 1, 3, 7}], \quad \frac{1+\sqrt{73}}{2} = [4, \overline{1, 3, 2, 1, 1, 2, 3, 1, 7}],$$

$$\frac{1+\sqrt{97}}{2} = [5, \overline{2, 2, 1, 4, 4, 1, 2, 2, 9}], \quad \frac{1+\sqrt{129}}{2} = [6, \overline{5, 1, 1, 2, 3, 2, 1, 1, 5, 11}],$$

$$\frac{1+\sqrt{161}}{2} = [6, \overline{1, 5, 2, 2, 1, 2, 2, 5, 1, 11}],$$

$$\frac{1+\sqrt{177}}{2} = [7, \overline{6, 1, 1, 2, 1, 3, 1, 2, 1, 1, 6, 13}],$$

$$\frac{1+\sqrt{193}}{2} = [7, \overline{2, 4, 6, 1, 2, 1, 1, 1, 1, 2, 1, 6, 4, 2, 13}],$$

$$\frac{1+\sqrt{217}}{2} = [7, \overline{1, 6, 2, 3, 4, 1, 1, 1, 1, 4, 3, 2, 6, 1, 13}],$$

$$\frac{1+\sqrt{241}}{2} = [8, \overline{3, 1, 4, 2, 2, 1, 1, 1, 7, 7, 1, 1, 1, 2, 2, 4, 1, 3, 15}],$$

$$\frac{1+\sqrt{313}}{2} = [9, \overline{2, 1, 8, 5, 1, 3, 1, 1, 2, 2, 1, 1, 3, 1, 5, 8, 1, 2, 17}],$$

$$\frac{1+\sqrt{337}}{2} = [9, \overline{1, 2, 8, 1, 5, 4, 2, 2, 1, 1, 1, 1, 2, 2, 4, 5, 1, 8, 2, 1, 17}],$$

$$\frac{1+\sqrt{409}}{2} = [10, \overline{1, 1, 1, 1, 2, 1, 3, 3, 9, 1, 4, 6, 1, 1, 6, 4, 1, 9, 3, 3, 1, 2, 1, 1, 1, 1, 19}].$$

The period in this last continued fraction has length 27, and so the fundamental solution to Pell's equation is

$$111921796968^2 - 409 \cdot 5534176685^2 = -1.$$

11.14. Sums of two squares from continued fractions

Corollary 11.14.1. *The continued fraction $\frac{b+\sqrt{d}}{a} = [\overline{a_0, \ldots, a_{m-1}}]$ is symmetric (that is, $a_j = a_{m-1-j}$ for $j = 0, \ldots, m-1$) if and only if $a^2 + b^2 = d$.*

Proof. If $\alpha = \frac{b+\sqrt{d}}{a} = [\overline{a_0, \ldots, a_{m-1}}]$ is symmetric, with conjugate β, then $\alpha\beta = 1$ by Lemma 11.12.1. Therefore

$$\frac{b^2 - d}{a^2} = \frac{b+\sqrt{d}}{a} \cdot \frac{b-\sqrt{d}}{a} = \alpha \cdot (-\beta) = -1;$$

that is, $a^2 + b^2 = d$.

On the other hand if $a^2 + b^2 = d$ with $a, b > 0$, then let $\alpha = \frac{b+\sqrt{d}}{a}$. Now $a < \sqrt{d}$ so that $\alpha > 1$. Moreover $-\beta$ is the conjugate of α where $\beta = \frac{\sqrt{d}-b}{a}$. As $b < \sqrt{d} < a + b$ we deduce that $0 < \beta < 1$. Therefore α has a periodic continued fraction by Proposition 11.12.1. Now $\alpha = 1/\beta$ and so the continued fraction is symmetric by Lemma 11.12.1. $\square$

Proposition 11.14.1. *Suppose that $A \neq 0$ and B are integers for which A divides $B^2 - d$ and $\frac{B+\sqrt{d}}{A} = [b_0, \overline{b_1, \ldots, b_{2k+1}}]$ where $b_j = b_{2k+1-j}$ for $j = 1, 2, \ldots, 2k$. Then*

$$[\overline{b_{k+1}, b_{k+2}, \ldots, b_{2k}, b_{2k+1}, b_1, \ldots, b_{k-1}, b_k}] = \frac{b+\sqrt{d}}{a}$$

for some integers $a \neq 0$ and b for which $a^2 + b^2 = d$.

Proof. If $\frac{r+\sqrt{d}}{s} = [b_0, b_1, \ldots]$ where r and s are integers for which s divides $r^2 - d$, then, for $R = sb_0 - r$ and $S = (d - R^2)/s$, we have

$$[b_1, b_2, \ldots] = 1 \bigg/ \left(\frac{\sqrt{d}-R}{s}\right) = \frac{R+\sqrt{d}}{S}.$$

We note that S is an integer since $d - R^2 \equiv d - r^2 \equiv 0 \pmod{s}$, and that S divides $R^2 - d$ as $d - R^2 = sS$. Iterating this we see that $[b_{k+1}, b_{k+2}, \ldots]$ takes the form $\frac{b+\sqrt{d}}{a}$, as claimed. Now this continued fraction is also purely periodic and symmetric, which implies that $a^2 + b^2 = d$ by Corollary 11.14.1. $\square$

For examples, $\frac{1+\sqrt{13}}{2} = [2, \overline{3}]$ and so $[\overline{3}] = \frac{3+\sqrt{13}}{2}$, and $\sqrt{13} = [3, \overline{1,1,1,1,6}]$ and $[\overline{1,1,6,1,1}] = \frac{2+\sqrt{13}}{3}$. Either way we find that $13 = 2^2 + 3^2$.

Typically we apply Proposition 11.14.1 with the continued fractions for $\sqrt{d}$ or $\frac{1+\sqrt{d}}{2}$, which works provided there is a solution to the negative Pell equation (as the period in Proposition 11.14.1 must be odd).

Additional exercises.

Exercise 11.14.1. The number $\phi = \frac{1+\sqrt{5}}{2}$ satisfies the equation $\phi = 1 + 1/\phi$.
(a) Iterate this to obtain $\phi = 1 + 1/(1 + 1/\phi)$, and then reprove the first part of exercise 11.10.4.
(b) Show that if α is the positive root of $x^2 - ax - b = 0$ where a and b are positive integers, then
$$\alpha = a + \cfrac{b}{a + \cfrac{b}{a + \cfrac{b}{a + \cfrac{b}{\cdots}}}}.$$

Exercise 11.14.2. Let $\phi = \frac{1+\sqrt{5}}{2}$.
(a) Show that $\phi = \sqrt{1 + \phi}$.
(b) Iterate this to obtain $\phi = \sqrt{1 + \sqrt{1 + \phi}}$, and then prove that
$$\frac{1 + \sqrt{5}}{2} = \sqrt{1 + \sqrt{1 + \sqrt{1 + \sqrt{1 + \cdots}}}}.$$
(c) Show that if α is the positive root of $x^2 - ax - b = 0$ where a and b are positive integers, then
$$\alpha = \sqrt{b + a\sqrt{b + a\sqrt{b + a\sqrt{b + \cdots}}}}.$$

Exercise 11.14.3 (Hermite, Serret 1848).[‡] We give an efficient way to determine, for a given prime $p \equiv 1 \pmod 4$, integers a and b for which $p = a^2 + b^2$: Let $r^2 \equiv -1 \pmod p$ with $0 < r < p/2$, and write $p/r = [a_0, \ldots, a_n]$. We will show that $n = 2m + 1$ is odd and a and b can be determined from $a/c = [a_0, a_1, \ldots, a_m]$ and $b/d = [a_0, a_1, \ldots, a_{m-1}]$.

(a) Use that $a_n \geq 2$ to deduce that $0 < s < p/2$ where $s/\ell = [a_0, a_1, \ldots, a_{n-1}]$.
(b) Prove that $s = r$ and $n = 2m + 1$ for some integer m.
(c) Show that $a_j = a_{n-j}$ for $j = 0, 1, \ldots, n$.
(d) Show that $p = a^2 + b^2$.

Modern uses of continued fractions

[1] W. Duke, *Continued fractions and modular functions*, Bull. Amer. Math. Soc. **42** (2005), 137–162.

Appendix 11C. Two-variable quadratic equations

11.15. Integer solutions to 2-variable quadratics

Theorem 11.10. *If there is one solution in integers to*
$$ax^2 + bxy + cy^2 + dx + ey + f = 0,$$
where $a, b, c, d, e, f \in \mathbb{Z}$, then there are infinitely many, except perhaps when $a = b = c = 0$ or when $b^2 - 4ac = 0$.

Proof. If $a \neq 0$, we multiply through by $4a$ to complete the square so that if $z = 2ax + by + d$, then
$$z^2 = py^2 + 2qy + r$$
where $p = b^2 - 4ac \neq 0$, $\quad q = bd - 2ae$, $\quad$ and $\quad r = d^2 - 4af$.

We multiply through by p to complete the square so that if $w = py + q$, then
$$w^2 - pz^2 = q^2 - pr.$$

Taking any solution to Pell's equation $u^2 - pv^2 = 1$ with $u \equiv 1 \pmod{p}$ from exercise 11.6.3, we create a new solution to our equation by taking

$$W = wu + pvz \quad \text{and} \quad Z = uz + wv, \quad \text{since } W + \sqrt{p}Z = (u + \sqrt{p}v)(w + \sqrt{p}z),$$

which yields a new integer solution X, Y to the original equation given by

$$X = (u - bv)x - 2cvy - ev + \frac{(u-1)}{p}(be - 2cd),$$

$$Y = (u + bv)y + 2avx + dv + \frac{(u-1)}{p}(bd - 2ae).$$

If $a = 0$ but $c \neq 0$, we swap the roles of the variables x and y. If $a = c = 0$, then $b \neq 0$ and we replace y by $x + y$. □

Exercise 11.15.1. What happens in the exceptional cases $a = b = c = 0$ and $b^2 - 4ac = 0$?

438

Appendix 11D.
Transcendental numbers

11.16. Diagonalization

In section 11.4 we exhibited a transcendental number and proved that it is transcendental using Liouville's Theorem. In general it is difficult to prove that a given number is transcendental (which is why the Liouville number exhibited in section 11.4 looks so obscure) though one expects that any number that is not rather obviously algebraic will be transcendental. We now give Cantor's "diagonalization" proof that there are far more transcendental numbers than algebraic (after appreciating how one type of infinity can be larger than another):

A set S is said to be *countable* if one can assign an order to the elements of S, say $s_1, s_2, \ldots, s_n, \ldots$, using the positive integers as indices, and this sequence includes all of the elements of S. It is evident that the integers are countable (one can order them as $0, 1, -1, 2, -2, \ldots$), and also the rationals (one can order them by listing all p/q with $q > 0$ and $(p, q) = 1$ such that $|p| + |q| = 1$, then such that $|p| + |q| = 2, 3$, etc.). This gives some idea of how to show the algebraic numbers are countable:

Exercise 11.16.1. Any given algebraic number α has a minimum polynomial $\sum_{i=0}^{d} a_i x^i \in \mathbb{Z}[x]$ with $d \geq 1$. Assign a weight $w(\alpha) := d \sum_{i=0}^{d} |a_i|$.
(a) Prove that there are finitely many α of any given weight.
(b) Deduce that the algebraic numbers are countable.

Exercise 11.16.2. Reprove the last result using the bijection $\mathbb{Z}[x] \leftrightarrow \mathbb{Q}$ defined by

$$a_0 + a_1 x + \cdots + a_d x^d \leftrightarrow 2^{a_0} 3^{a_1} \cdots p_{d+1}^{a_d},$$

where $p_1 = 2 < p_2 < \cdots$ is the sequence of primes. Can you construct a bijection $\mathbb{Z}[x] \leftrightarrow \mathbb{Z}$?

We now show that the real numbers are *uncountable* (that is, not countable), and hence there are more of them than just the algebraic numbers; in other words there are many transcendental real numbers. So, let us assume that the real numbers are countable and therefore can all be put in a list as $r_1, r_2, \ldots$. We now construct a real number $\beta \in (0, 1)$, that is not on the list, by selecting digits for β written in decimal, one at a time. The trick is that the jth digit of β should be different from the jth digit of r_j.[10] But then $\beta \neq r_j$ for any j (since they differ in the jth digit), which contradicts the assumption that the r_j's were a list of all the real numbers (as the list does not include β). Therefore the real numbers are not countable, and so uncountable.

Philosophically this is quite an extraordinary result as it tells us that there are (at least) two very different types of infinity, the countable infinity like $\mathbb{Z}$, and the uncountable infinity like $\mathbb{R}$. Are there other infinities? This is an important question but we start to tread too far away from number theory.

11.17. The hunt for transcendental numbers

Proving that any specific number is transcendental is tough. The two numbers that we most wanted to understand were, for a long time, π and e. Even proving that they are irrational is not so easy:

Exercise 11.17.1. Suppose that e is rational, say $e = p/q$, and let $r = q!(1 + 1/1! + \cdots + 1/q!)$.
(a) Show that r is an integer and therefore $q! \cdot \frac{p}{q} - r$ is an integer.
(b) Prove that $q!e - r \in (0, 1)$.
(c) Use these remarks to establish a contradiction.

Exercise 11.17.2. Given $f(x) \in \mathbb{C}[x]$, define $F(x) := f(x) - f^{(2)}(x) + f^{(4)}(x) - f^{(6)}(x) + \cdots$ where $f^{(k)}(x) = \frac{d^k f}{dx^k}$.
(a) Prove that $\frac{d}{dx}(F'(x)\sin x - F(x)\cos x) = f(x)\sin x$.
(b) Deduce that $\int_0^\pi f(x)\sin x \, dx = F(\pi) + F(0)$.
(c) Show that if $f(x) = f(\pi - x)$, then $F(\pi) = F(0)$.
(d) Now, suppose that π is rational, say $\pi = p/q$, and let $f(x) = x^n(p - qx)^n/n!$ for some integer $n \geq 1$. By establishing that $f^{(k)}(0)$ is an integer for all $k \geq 0$, prove that $F(0)$ and $F(\pi)$ are integers.
(e) Show that if $0 < x < \pi$, then $0 < f(x) < (\frac{\pi^2}{4}q)^n/n!$.
   Assume that $n > \frac{\pi^2}{4} \cdot eq$.
(f) In exercise 4.14.2 we proved that $n! > 2n^n/e^n$. Deduce that if $0 < x < \pi$, then $0 < f(x) < \frac{1}{2}$.
(g) Prove that $\int_0^\pi |\sin x| dx = 2$, and deduce that $0 < \int_0^\pi f(x)\sin x \, dx < 1$.
(h) Combine the above to deduce that π is irrational.

In 1873 Hermite succeeded in proving that e is transcendental, and then, nine years later, Lindemann succeeded in proving that π is transcendental. (The known proofs are probably a little too difficult for this book, but look at the discussion in section 6.6 of [**Bak84**].) Lindemann's Theorem implied that it is impossible to "square the circle" using a compass and ruler (see section 0.18 of appendix 0G). Lindemann actually proved the extraordinary result that for any distinct algebraic numbers $\alpha_1, \ldots, \alpha_n$ there do not exist non-zero algebraic numbers $\beta_1, \ldots, \beta_n$ for which

$$\beta_1 e^{\alpha_1} + \beta_2 e^{\alpha_2} + \cdots + \beta_n e^{\alpha_n} = 0.$$

---

[10]Any suitable protocol will do. For example, we can let the jth digit of β be 1, unless the jth digit of r_j is 1, in which case we let the jth digit of β be 7.

11.18. Normal numbers

One deduces that π is transcendental, since $e^{i\pi} + e^0 = 0$ and so $i\pi$ cannot be algebraic. One also can deduce that e is transcendental and that $\log \alpha$ is transcendental for any algebraic $\alpha \neq 0, 1$.

Exercise 11.17.3. Let $\alpha := \sqrt{2}^{\sqrt{2}}$. We wish to show that there exist irrational numbers x, y such that x^y is rational. Use either α or $\alpha^{\sqrt{2}}$ to prove this.

Gelfond and Schnieder showed, in 1934, that if α is algebraic and $\neq 0, 1$, and β is algebraic and irrational, then α^β is transcendental. In particular this implies that both $\sqrt{2}^{\sqrt{2}}$ and e^π are transcendental, which had been famous questions of Hilbert.

Finally, in 1966 Baker established that any non-vanishing linear form
$$\beta_1 \log \alpha_1 + \cdots + \beta_n \log \alpha_n$$
is transcendental, where the α_i's and β_i's denote non-zero algebraic numbers, with $\alpha_i \neq 1$. Examples include $\log(p/q)$ for any rational number $p/q \neq 0, 1$.

There are many numbers that have resisted attack, in that we do not know whether they are irrational, for example γ (the Euler-Mascheroni constant from section 4.14 of appendix 4D) and $\zeta(2m+1)$ for all $m \geq 2$.[11] (We do know that $\zeta(2m) = (-1)^{m+1}(2\pi)^{2m}B_{2m}/2(2m)!$ for each $m \geq 2$, where B_n is the nth Bernoulli number. This is a rational multiple of π^{2m} and therefore transcendental.) Moreover, even though we know that e and π are both transcendental, we know nothing about $e + \pi$; for all we know it could be rational, though that seems very unlikely.

We expect more or less any class of interesting numbers to be transcendental and for their sums to be transcendental. Of great interest is the set of "non-trivial" zeros of the Riemann zeta function (see sections 5.16 and 5.17 of appendix 5D). These can be written as $\frac{1}{2} \pm i\gamma_1, \frac{1}{2} \pm i\gamma_2, \frac{1}{2} \pm i\gamma_3, \ldots$ with $0 < \gamma_1 \leq \gamma_2 \leq \ldots$, assuming the Riemann Hypothesis. We believe that each of the γ_j's is transcendental and that they are linearly independent over the rationals (the *Linear Independence Hypothesis*). However no part of these beliefs has been proved, and there is very little evidence either way for the algebraic nature of the γ_j's.

Exercise 11.17.4. Prove that if β is transcendental, then $a_0 + a_1\beta + \cdots + a_n\beta^n$ is also transcendental whenever the $a_j \in \overline{\mathbb{Q}}$ (the set of algebraic numbers, which is the *algebraic closure* of $\mathbb{Q}$) with $n \geq 1$ and $a_n \neq 0$.

11.18. Normal numbers

What do the digits of π look like? We know π is a transcendental number and believe that it has no particular algebraic properties, so we might guess that there is nothing special about its digits. For example, we might ask whether there are more 3's than 8's when we write out the digits of π, or whether the five consecutive digits pattern 31415 occurs more often than any another pattern. It seems not, and it seems unlikely that the digits of π will favor any particular digit pattern, but we really

---

[11]There have been one or two successes on this question: In 1981 Apéry showed that $\zeta(3)$ is irrational, though we do not know whether it is transcendental. In 2001, Ball and Rivoal showed that infinitely many of the $\zeta(2m+1)$ are irrational, and in 2004 Zudilin showed that at least one of $\zeta(5), \zeta(7), \zeta(9), \zeta(11)$ is irrational.

do not know. Moreover, we asked these questions in the base 10 representation of π. What about the base 2 representation? Or the base 37 representation?[12]

We define n to be a *normal number in base b* as follows: Write the fractional part of n as $\{n\} = n_1/b + n_2/b^2 + \cdots$ in base b, where the n_i's are the base b digits (and so are integers between 0 and $b-1$). For each integer $d \geq 1$ and for every d-digit integer in base b (that is, the integers r in the range $0 \leq r \leq b^d - 1$), we should have that

$$\lim_{M \to \infty} \frac{1}{M} \#\{m \leq M : n_m b^{d-1} + n_{m+1} b^{d-2} + \cdots + n_{m+d-1} b^0 = r\} = \frac{1}{b^d}.$$

That is, for every $d \geq 1$, every base b digit pattern of length d eventually occurs roughly equally often. A number is *absolutely normal* if it is normal in every base $b > 1$. It is not difficult to show that "almost all" real numbers are normal in base b; that is, if one picks a real number at random, then it is normal in base b.

There are some fun examples of normal numbers, like Champernowne's number

$$0.12345678910111213141516 17\ldots,$$

obtained by concatenating the decimal expansions of the positive integers, which is normal in base 10; and similarly

$$0.23571113171923293137 4143\ldots,$$

obtained by concatenating the prime numbers, is normal in base 10, and the same construction works in every base $b > 1$.

The definition of a normal number is complicated and can be reworked as follows: The condition $n_m b^{d-1} + n_{m+1} b^{d-2} + \cdots + n_{m+d-1} b^0 = r$ will now be rephrased, by dividing through by b^d, as $n_m/b + n_{m+1}/b^2 + \cdots + n_{m+d-1}/b^d = r/b^d$, which is the same as

$$(11.18.1) \quad \{nb^{m-1}\} = \frac{n_m}{b} + \frac{n_{m+1}}{b^2} + \cdots + \frac{n_{m+d-1}}{b^d} + \frac{n_{m+d}}{b^{d+1}} + \cdots \in \left[\frac{r}{b^d}, \frac{r+1}{b^d}\right).$$

Therefore n is a normal number in base b if and only if the sequence $\{nb^{m-1}\}_{m \geq 1}$ is uniformly distributed mod 1. By Weyl's criterion (section 11.7 of appendix 11A) this holds if and only if for every integer $k \geq 1$ we have

$$\lim_{M \to \infty} \frac{1}{M} \sum_{1 \leq k \leq M} e(nb^k) \text{ exists and equals } 0$$

(where $e(t) := e^{2i\pi t}$ as usual).

Exercise 11.18.1. Complete the details of the proof of the equivalence of normality in base b, and the sequence $\{nb^{m-1}\}_{m \geq 1}$ being uniformly distributed mod 1.

Normal digits of numbers

[1] Greg Martin, *Absolutely abnormal numbers*, Amer. Math. Monthly **108** (2001), 746–754.

---

[12] In the novel *Contact* by Carl Sagan, a long string of 0's and 1's is found far out in the base 11 expansion of π. This is then used to create a picture, which suggests (in the context of the novel) that π was created at the convenience of some supernatural being. It's a fun read.

Chapter 12

Binary quadratic forms

Let a, b, and c be given integers. We saw in Corollary 1.3.1 that the integers that can be represented by the binary linear form $ax + by$ are those integers divisible by $\gcd(a, b)$. We are now interested in what integers can be represented by the *binary quadratic form*,[1]

$$f(x, y) := ax^2 + bxy + cy^2.$$

As in the linear case, we can immediately reduce our considerations to the case that $\gcd(a, b, c) = 1$.

The first important result of this type was given by Fermat for the particular example $f(x, y) = x^2 + y^2$, as discussed in section 9.1. The two main results were that an odd prime p can be represented by $f(x, y)$ if and only if $p \equiv 1 \pmod{4}$, and that the product of two integers that can be written as the sum of two squares can also be written as the sum of two squares, a consequence of the identity (9.1.1). One can combine these two facts to classify exactly which integers are represented by the binary quadratic form $x^2 + y^2$.

At first sight it looks like it might be difficult to work with the example $f(x, y) = x^2 + 20xy + 101y^2$. However, this can be rewritten as $(x + 10y)^2 + y^2$ and so represents exactly the same integers as $g(x, y) = x^2 + y^2$. In other words

$$n = f(u, v) \text{ if and only if } n = g(r, s), \text{ where } \begin{pmatrix} r \\ s \end{pmatrix} = \begin{pmatrix} 1 & 10 \\ 0 & 1 \end{pmatrix} \begin{pmatrix} u \\ v \end{pmatrix}.$$

This 2-by-2 matrix is invertible over the integers, so we can express u and v as integer linear combinations of r and s. Thus every representation of n by f corresponds to one by g, and vice versa, a 1-*to*-1 *correspondence*, obtained using the invertible *linear transformation* $u, v \to u + 10v, v$. Such a pair of quadratic forms, f and g, are said to be *equivalent*; and we have just seen how equivalent binary quadratic forms represent exactly the same integers. The *discriminant* of

---

[1] "Binary" as in the **two** variables x and y, and "quadratic" as in degree **two**. The monomials ax^2, bxy, cy^2 each have degree two, since the degree of a term is given by the degree in x plus the degree in y.

443

$ax^2+bxy+cy^2$ is b^2-4ac. We will show that equivalent binary quadratic forms have the same discriminant, so that it is an *invariant* of the equivalence class of binary quadratic forms. All of this will be discussed in this chapter and, in appendix 12A, we will study generalizations of the identity (9.1.1).

12.1. Representation of integers by binary quadratic forms

An integer N is *represented* by f if there exist integers m, n for which $N = f(m, n)$, and N is *properly represented* if $(m, n) = 1$ (see exercise 3.9.13 for the same question for linear forms).

Exercise 12.1.1. Prove that if N is squarefree, then all representations of N are proper.

What integers can be properly represented by $ax^2 + bxy + cy^2$? That is, for what integers N do there exist coprime integers m, n such that

(12.1.1) $$N = am^2 + bmn + cn^2?$$

We may reduce to the case that $\gcd(a, b, c) = 1$ by dividing though by $\gcd(a, b, c)$. (If $\gcd(a, b, c) = 1$, then f is a *primitive* binary quadratic form.) One idea is to complete the square to obtain

(12.1.2) $$4aN = (2am + bn)^2 - dn^2$$

where the *discriminant* $d := b^2 - 4ac$. This implies that the discriminant always satisfies

$$d \equiv 0 \text{ or } 1 \pmod{4}.$$

There is always at least one binary quadratic form of discriminant d, for such d, which we call the *principal form*:

$$\begin{cases} x^2 - (d/4)y^2 & \text{when } d \equiv 0 \pmod{4}, \\ x^2 + xy + \frac{(1-d)}{4}y^2 & \text{when } d \equiv 1 \pmod{4}. \end{cases}$$

We call d a *fundamental discriminant* if $d = D \equiv 1 \pmod 4$, or $d = 4D$ with $D \equiv 2$ or $3 \pmod 4$, and if $D = d/(d, 4)$ is squarefree. These are precisely the discriminants for which every binary quadratic form is primitive (see exercise 12.1.3). We met this notion already in exercise 8.16.4 of appendix 8D, when classifying the genuinely different Jacobi symbols.

When $d < 0$ the right side of (12.1.2) can only take positive values, which makes our discussion easier than when $d > 0$. For this reason we will restrict ourselves to the case $d < 0$ here and revisit the case $d > 0$ in appendix 12C. If $d < 0$ and $a < 0$, we replace a, b, c by $-a, -b, -c$, so as to ensure that $am^2 + bmn + cn^2$ is always ≥ 0; in this case, we call $ax^2 + bxy + cy^2$ a *positive definite* binary quadratic form.

At the start of this chapter we worked through one example of equivalence of binary quadratic forms, and here is another: The binary quadratic form $x^2 + y^2$ represents the same integers as $X^2 + 2XY + 2Y^2$, for if $N = m^2 + n^2$, then $N = (m-n)^2 + 2(m-n)n + 2n^2$, and similarly if $N = u^2 + 2uv + 2v^2$, then $N = (u+v)^2 + v^2$. The reason is that the substitution

$$\begin{pmatrix} x \\ y \end{pmatrix} = M \begin{pmatrix} X \\ Y \end{pmatrix} \text{ where } M = \begin{pmatrix} 1 & 1 \\ 0 & 1 \end{pmatrix}$$

transforms x^2+y^2 into $X^2+2XY+2Y^2$, and the transformation is invertible, since $\det M = 1$. We therefore say that $x^2 + y^2$ and $X^2 + 2XY + 2Y^2$ are *equivalent* which we denote by
$$x^2 + y^2 \sim X^2 + 2XY + 2Y^2.$$

Much more generally define
$$\mathrm{SL}(2,\mathbb{Z}) = \left\{ \begin{pmatrix} \alpha & \beta \\ \gamma & \delta \end{pmatrix} : \alpha, \beta, \gamma, \delta \in \mathbb{Z} \text{ and } \alpha\delta - \beta\gamma = 1 \right\}.$$

We can represent the binary quadratic form as
$$ax^2 + bxy + cy^2 = \begin{pmatrix} x & y \end{pmatrix} \begin{pmatrix} a & b/2 \\ b/2 & c \end{pmatrix} \begin{pmatrix} x \\ y \end{pmatrix}.$$

Its discriminant is -4 times the determinant of $\begin{pmatrix} a & b/2 \\ b/2 & c \end{pmatrix}$. We deduce that if
$$\begin{pmatrix} x \\ y \end{pmatrix} = M \begin{pmatrix} X \\ Y \end{pmatrix} \text{ where } M = \begin{pmatrix} \alpha & \beta \\ \gamma & \delta \end{pmatrix} \in \mathrm{SL}(2,\mathbb{Z}),$$
then
$$AX^2 + BXY + CY^2 = \begin{pmatrix} X & Y \end{pmatrix} M^{\mathrm{T}} \begin{pmatrix} a & b/2 \\ b/2 & c \end{pmatrix} M \begin{pmatrix} X \\ Y \end{pmatrix},$$
so that
$$(12.1.3) \qquad \begin{pmatrix} A & B/2 \\ B/2 & C \end{pmatrix} = M^{\mathrm{T}} \begin{pmatrix} a & b/2 \\ b/2 & c \end{pmatrix} M,$$
which yields the somewhat painful looking explicit formulas
$$(12.1.4) \qquad \begin{cases} A &= f(\alpha, \gamma) = a\alpha^2 + b\alpha\gamma + c\gamma^2, \\ B &= 2\alpha\beta a + (\alpha\delta + \beta\gamma)b + 2\gamma\delta c, \\ C &= f(\beta, \delta) = a\beta^2 + b\beta\delta + c\delta^2. \end{cases}$$

When working with binary quadratic forms it is convenient to represent $ax^2 + bxy + cy^2$ by the notation $[a, b, c]$. We have just proven the following.

Proposition 12.1.1. *If $f = [a,b,c] \sim F = [A,B,C]$, then there exist integers $\alpha, \beta, \gamma, \delta$ with $\alpha\delta - \beta\gamma = 1$ for which $A = f(\alpha, \gamma)$ and $C = f(\beta, \delta)$. Moreover f and F represent the same integers, and there is a 1-to-1 correspondence between their representations and proper representations of a given integer.*

Exercise 12.1.2. (a) Suppose that d is a fundamental discriminant. Prove that the character $(d/\cdot)$ has conductor dividing d.
(b) Prove that for any non-zero integer d, the character $(d/\cdot)$ has conductor that divides $4d$.

The *conductor* of $f(\cdot)$ is the minimum $p > 0$ such that $f(n+p) = f(n)$ for all integers n.

Exercise 12.1.3. Suppose that $d \equiv 0$ or $1 \pmod 4$. Show that every binary quadratic form of discriminant d is primitive if and only if d is a fundamental discriminant.

Exercise 12.1.4. (a) Show that if $d < 0$, then $am^2 + bmn + cn^2$ has the same sign as a, no matter what the choices of integers m and n.
(b) Show that if $ax^2 + bxy + cy^2$ is positive definite, then $a, c > 0$.
(c) Show that if $d > 0$, then $am^2 + bmn + cn^2$ can take both positive and negative values, by making explicit choices of integers m, n.

Exercise 12.1.5. Use (12.1.3) to show that two equivalent binary quadratic forms have the same discriminant.

Exercise 12.1.6. Show that the principal form is equivalent to every binary quadratic form $x^2 + bxy + cy^2$ with leading coefficient 1, up to equivalence.

Exercise 12.1.7. In each part, determine whether the two binary quadratic forms are equivalent. If so, make the equivalence explicit; if not, explain why not.
(a) $y^2 + xy + 4x^2$ and $x^2 - 5xy + 10y^2$.
(b) $x^2 + 3xy + 5y^2$ and $3x^2 - 4xy + 11y^2$.

12.2. Equivalence classes of binary quadratic forms

In this section we will develop an algorithm that will allow us to show, for example, that $29X^2 + 82XY + 58Y^2$ is equivalent to $x^2 + y^2$. We do this as it is surely more intuitive to work with the latter form rather than the former. Gauss observed that every equivalence class of binary quadratic forms (with $d < 0$) contains a unique smallest representative, called the *reduced* representative, which we now prove: The quadratic form $ax^2 + bxy + cy^2$ with discriminant $d < 0$ is *reduced* if

$$-a < b \leq a \leq c \text{ and } b \geq 0 \text{ whenever } a = c.$$

Theorem 12.1. *Every positive definite binary quadratic form is equivalent to a reduced form.*

Proof. We will define a sequence of properly equivalent forms; the algorithm terminates when we reach one that is reduced. Given a form $[a, b, c]$, we use one of three transformations, described in terms of matrices from $\mathrm{SL}(2, \mathbb{Z})$:

(i) If $c < a$, the transformation

$$\begin{pmatrix} x \\ y \end{pmatrix} = \begin{pmatrix} 0 & -1 \\ 1 & 0 \end{pmatrix} \begin{pmatrix} X \\ Y \end{pmatrix}$$

yields the form $[c, -b, a]$ which is properly equivalent to $[a, b, c]$ (as $ax^2 + bxy + cy^2 = a(-Y)^2 + b(-Y)(X) + c(X)^2 = cX^2 - bXY + aY^2$). Hence $A = c < a = C$.

(ii) If $b > a$ or $b \leq -a$, then select B to be the absolutely least residue of b (mod $2a$), so that $-a < B \leq a$, say $B = b - 2ka$. The transformation matrix will be

$$\begin{pmatrix} x \\ y \end{pmatrix} = \begin{pmatrix} 1 & -k \\ 0 & 1 \end{pmatrix} \begin{pmatrix} X \\ Y \end{pmatrix}.$$

The resulting form $[A, B, C]$ with $A = a$ is properly equivalent to $[a, b, c]$, where $-A < B \leq A$.

(iii) If $c = a$ and $-a < b < 0$, then the transformation

$$\begin{pmatrix} x \\ y \end{pmatrix} = \begin{pmatrix} 0 & -1 \\ 1 & 0 \end{pmatrix} \begin{pmatrix} X \\ Y \end{pmatrix}$$

yields the form $[A, B, A]$ with $A = a$ and $B = -b$, so that $0 < B < A$.

If the resulting form is not reduced, then repeat the algorithm. If none of these hypotheses holds, then one can easily verify that the form is reduced. To prove that the algorithm terminates in finitely many steps we follow the leading coefficient

12.3. Congruence restrictions on the values of a binary quadratic form

a: a starts as a positive integer. Each transformation of type (i) reduces the size of a. It stays the same after transformations of type (ii) or (iii), but after a type (iii) transformation the algorithm terminates, and after a type (ii) transformation we either have another type (i) transformation or else the algorithm stops after at most one more transformation. Hence the algorithm finishes in no more than $2a+1$ steps. □

Examples. Applying the reduction algorithm to the form $[76, 217, 155]$ of discriminant -31, one finds the sequence of forms

$$[76, 65, 14], [14, -65, 76], [14, -9, 2], [2, 9, 14], [2, 1, 4],$$

the sought-after reduced form. Similarly the form $[11, 49, 55]$ of discriminant -19 gives the sequence of forms $[11, 5, 1], [1, -5, 11], [1, 1, 5]$.

This proof of Theorem 12.1 can be rephrased to prove Theorem 1.2 of section 1.10 (of appendix 1A), that every matrix in $\mathrm{SL}(2,\mathbb{Z})$ can be represented as the product of powers of the matrices $S = \begin{pmatrix} 1 & 1 \\ 0 & 1 \end{pmatrix}$ and $T = \begin{pmatrix} 0 & 1 \\ -1 & 0 \end{pmatrix}$. The matrices used in the transformations in the proof of Theorem 12.1 are $\begin{pmatrix} 0 & -1 \\ 1 & 0 \end{pmatrix} = T^{-1}$ and $\begin{pmatrix} 1 & -k \\ 0 & 1 \end{pmatrix} = S^{-k}$.

The very precise conditions in the definition of "reduced" were chosen so that every positive definite binary quadratic form is properly equivalent to a *unique* reduced form. The key to proving uniqueness is exercise 12.6.1; the (messy) details are completed in exercise 12.6.2.

12.3. Congruence restrictions on the values of a binary quadratic form

What restrictions are there on the values that can be taken by a binary quadratic form (in analogy to Theorem 9.2)?

Proposition 12.3.1. *Let $d = b^2 - 4ac$ where $(a, b, c) = 1$.*

(i) *If integer N is properly represented by $ax^2 + bxy + cy^2$, then d is a square mod $4N$.*

(ii) *If d is a square mod $4N$, then there exists a binary quadratic form of discriminant d that properly represents N.*

Proof. (ii) If $d \equiv b^2 \pmod{4N}$, then $d = b^2 - 4Nc$ for some integer c, and so $Nx^2 + bxy + cy^2$ is a quadratic form of discriminant d which represents $N = N \cdot 1^2 + b \cdot 1 \cdot 0 + c \cdot 0^2$.

(i) Suppose that $N = am^2 + bmn + cn^2$ with $(m, n) = 1$. Then $(2am + bn)^2 - dn^2 = 4aN$ so that $dn^2 \equiv (2am + bn)^2 \pmod{4N}$; that is, dn^2 is a square mod $4N$ and, analogously, dm^2 is a square mod $4N$. Now if p is a prime such that $p^k \| 4N$, then p does not divide at least one of m and n, as $(m, n) = 1$. We deduce that d is a square mod p^k from the fact that dn^2 is a square mod p^k if p does not divide n, and from the fact that dm^2 is a square mod p^k if p does not divide m. The result, that d is a square mod $4N$ now follows from the Chinese Remainder Theorem. □

For a given odd prime p, Proposition 12.3.1 tells us that p is represented by some binary quadratic form of discriminant d if and only if $(d/p) = 1$ or 0. However it does not tell us which binary quadratic form. In section 9.6 we could not immediately determine which of the two reduced binary quadratic forms of discriminant -20, namely $x^2 + 5y^2$ and $2x^2 + 2xy + 3y^2$, represents which primes p with $(-20/p) = 1$. There we found we could distinguish which prime was represented by which form by also studying the values of (p/d). We now see how this works out in general.

We can appeal to Corollary 9.4.1 to restrict the possibilities for the binary quadratic forms of discriminant d that represent N. Given a primitive binary quadratic form f of discriminant d we define, for each odd prime p dividing d,

$$\sigma_f(p) = \left(\frac{a}{p}\right) \text{ if } p \text{ does not divide } a, \text{ and } \sigma_f(p) = \left(\frac{c}{p}\right) \text{ if } p \text{ does divide } a.$$

If p divides a, then p divides $d + 4ac = b^2$ and so divides b, and therefore cannot divide c as f is primitive. Therefore $\sigma_f(p)$ equals 1 or -1 for each such p.

Exercise 12.3.1.[†] Prove that if $f \sim g$, then $\sigma_f(p) = \sigma_g(p)$ for all odd primes p dividing d.

Corollary 12.3.1. *Suppose that d is a fundamental discriminant and that N is a squarefree integer for which $(N, d) = 1$. If d is a square mod $4N$, then there exists a binary quadratic form f of discriminant d that properly represents N such that $\sigma_f(p) = \left(\frac{N}{p}\right)$ for every odd prime p dividing d.*

Proof. There exists a binary quadratic form f of discriminant d that properly represents N, by Proposition 12.3.1(ii). Therefore N is represented by inserting rationals into f and this happens, by Corollary 9.4.1, if and only if $\left(\frac{N}{p}\right) = \sigma_f(p)$ for every odd prime p dividing d. □

When $d = -20$ we have $\sigma_f(5) = 1$ for $f = x^2 + 5y^2$ and $\sigma_f(5) = (2/5) = -1$ for $f = 2x^2 + 2xy + 3y^2$. This can certainly settle such issues in several cases.

There are three reduced quadratic forms $[1, 1, 6], [2, \pm 1, 3]$ with $d = -23$. However $\sigma_f(23) = 1$ for each of these, so this does not help us to distinguish between the integers represented by these quadratic forms. This case is **much** more complicated and beyond the scope of this book.

We develop these ideas further in section 12.11 of appendix 12B.

Exercise 12.3.2. Prove that if $p_1, \ldots, p_k$ are distinct primes that are each represented by some form of discriminant d, then $p_1 \cdots p_k$ is also represented by some form of discriminant d.

12.4. Class numbers

Theorem 12.2. *If $d < 0$, then there are only finitely many reduced binary quadratic forms of discriminant d.*

Proof. For a reduced binary quadratic form, $|d| = 4ac - (|b|)^2 \geq 4a \cdot a - a^2 = 3a^2$ and so a is a positive integer for which

$$a \leq \sqrt{|d|/3}.$$

Therefore for a given $d < 0$ there are only finitely many a, and so b (as $|b| \leq a$), but then $c = (b^2 - d)/4a$ is determined, and so there are only finitely many reduced binary quadratic forms of discriminant d. □

Let $h(d)$ denote the *class number*, the number of equivalence classes of binary quadratic forms of discriminant d. We have just shown $h(d)$ is finite, and the proof of Theorem 12.2 even describes an algorithm to easily find all the reduced binary quadratic forms of a given discriminant $d < 0$. In fact $h(d) \geq 1$ since we always have the principal form. If $h(d) = 1$, then all binary quadratic forms are equivalent to the principal form.

Example. If $d = -163$, then $|b| \leq a \leq \sqrt{163/3} < 8$. But b is odd, since $b \equiv b^2 = d + 4ac \equiv d \pmod 2$, so $|b| = 1, 3, 5$, or 7. Therefore $ac = (b^2 + 163)/4 = 41, 43, 47$, or 53, a prime, with $0 < a < c$ and hence $a = 1$. Since b is odd and $-a < b \leq a$, we deduce that $b = 1$ and so $c = 41$. Hence $x^2 + xy + 41y^2$ is the only reduced binary quadratic form of discriminant -163, and therefore $h(-163) = 1$.

Exercise 12.4.1. Determine all of the reduced binary quadratic forms of discriminant d for $-20 \leq d \leq -1$ as well as for $d = -28, -43, -67, -167$, and -171.

Exercise 12.4.2. Determine all of the reduced binary quadratic forms of discriminant d for $d = -3, -15, -23, -39, -47, -87, -71$, and -95.

Exercise 12.4.3. Determine all of the reduced binary quadratic forms of discriminant d for $d = -4, -20, -56$, and -104.

Exercise 12.4.4. Prove that if $ax^2 + bxy + cy^2$ is a reduced binary quadratic of discriminant $d < 0$, then $|c| \geq \sqrt{|d|}/2$.

12.5. Class number one

Corollary 12.5.1. *Suppose that $h(d) = 1$. Then N is properly represented by the form of discriminant d if and only if d is a square* mod $4N$.

Proof. This follows immediately from Proposition 12.3.1, since there is just one equivalence class of quadratic forms of discriminant d, and forms in the same equivalence class represent the same integers by Proposition 12.1.1. □

We have $h(-4) = 1$ and so Corollary 12.5.1 implies that integer N is properly represented by $x^2 + y^2$ if and only if -4 is a square mod $4N$. This is more or less Theorem 9.2 (and can be deduced from its proof).

In the example in section 12.4 we showed that $x^2 + xy + 41y^2$ is the only binary quadratic form of discriminant -163. This implies, by Corollary 12.5.1, that if prime $p \neq 2$ or 163, then it can be represented by the binary quadratic form $x^2 + xy + 41y^2$ if and only if $(-163/p) = 1$.

In exercise 12.4.1 we exhibited nine fundamental discriminants $d < 0$ with $h(d) = 1$, namely $d = -3, -4, -7, -8, -11, -19, -43, -67$, as well as -163. It turns out these are the only ones with class number one.[2] Therefore, as in the example above, if $p \nmid 2d$, then

p is represented by $x^2 + y^2$ if and only if $(-1/p) = 1$;
p is represented by $x^2 + 2y^2$ if and only if $(-2/p) = 1$;
p is represented by $x^2 + xy + y^2$ if and only if $(-3/p) = 1$;
p is represented by $x^2 + xy + 2y^2$ if and only if $(-7/p) = 1$;
p is represented by $x^2 + xy + 3y^2$ if and only if $(-11/p) = 1$;
p is represented by $x^2 + xy + 5y^2$ if and only if $(-19/p) = 1$;
p is represented by $x^2 + xy + 11y^2$ if and only if $(-43/p) = 1$;
p is represented by $x^2 + xy + 17y^2$ if and only if $(-67/p) = 1$;
p is represented by $x^2 + xy + 41y^2$ if and only if $(-163/p) = 1$.

Euler noticed that the polynomial $x^2 + x + 41$ is prime for $x = 0, 1, 2, \ldots, 39$, and similarly the other polynomials above. Rabinowiscz proved that this is an "if and only if" condition:

Theorem 12.3 (Rabinowiscz's criterion). *We have $h(1 - 4A) = 1$ for $A \geq 2$ if and only if $x^2 + x + A$ is prime for $n = 0, 1, 2, \ldots, A - 2$.*

At $n = A - 1$ the polynomial takes value $(A-1)^2 + (A-1) + A = A^2$ which is composite. We will prove Rabinowiscz's criterion below.

What about when the class number is not one? In the example with $d = -20$ we have $h(-20) = 2$; the two reduced forms are $x^2 + 5y^2$ and $2x^2 + 2xy + 3y^2$. By Proposition 12.3.1, p is represented by at least one of these two forms if and only if $(-5/p) = 0$ or 1, that is, if $p \equiv 1, 3, 7,$ or $9 \pmod{20}$ or $p = 2$ or 5. Can we decide which of these primes are represented by which of the two forms? Note that if $p = x^2 + 5y^2$, then $(p/5) = 0$ or 1 and so $p = 5$ or $p \equiv \pm 1 \pmod 5$, and thus $p \equiv 1$ or $9 \pmod{20}$. If $p = 2x^2 + 2xy + 3y^2$, then $2p = (2x+y)^2 + 5y^2$ and so $p = 2$ or $(2p/5) = 1$; that is, $(p/5) = -1$, and hence $p \equiv 3$ or $7 \pmod{20}$. Hence we have proved

p is represented by $x^2 + 5y^2$ if and only if $p = 5$, or $p \equiv 1$ or $9 \pmod{20}$;

p is represented by $2x^2 + 2xy + 3y^2$ if and only if $p = 2$, or $p \equiv 3$ or $7 \pmod{20}$.

That is, we can distinguish which primes can be represented by which binary quadratic form of discriminant -20, through congruence conditions, despite the fact that the class number is not one. However we cannot always do this; that is, we cannot always distinguish which primes are represented by which binary quadratic form of discriminant d. It is understood how to recognize those discriminants d for which we can determine which binary quadratic forms of discriminant d represent

---

[2] The proof that the above list gives all of the $d < 0$, for which $h(d) = 1$, has an interesting history. By 1934 it was known that there is no more than one further such d, but that putative d could not be ruled out by the method. In 1952, Kurt Heegner, a German school teacher proposed an extraordinary proof that there are no further d. At the time his paper was ignored since it was based on a result from an old book (of Weber) whose proof was known to be incomplete. In 1966 Alan Baker gave a very different (and more obviously correct) proof that this was the complete list of discriminants with class number one, and this was widely acknowledged to be correct. However, soon afterwards Stark realized that the proofs in Weber are easily corrected, so that Heegner's work had been fundamentally correct. Heegner was subsequently given credit for solving this famous problem, but sadly only after he had died. Heegner's paper contains a most extraordinary construction, widely regarded to be one of the most creative and influential in modern number theory.

which integers simply through congruence conditions (see section 12.11 of appendix 12B). These *idoneal numbers* were recognized by Euler. He found 65 of them, and no more are known—it is an open conjecture as to whether Euler's list is complete. It is known that there can be at most one further undiscovered idoneal number, but it seems unlikely whether the techniques used can rule out this putative example.[3]

Exercise 12.5.1. (a) Determine the two reduced binary quadratic forms of discriminant -15.
(b) Determine which reduced residue classes can be represented by some form of discriminant -15?
(c) Distinguish which primes are represented by which form (with proof).

Proof of Rabinowiscz's criterion. We begin by showing that $f(n) := n^2+n+A$ is composite for some integer n in the range $0 \leq n \leq A-2$, if and only if $d = 1-4A$ is a square mod $4p$ for some prime $p < A$. For if $n^2 + n + A$ is composite, let p be its smallest prime factor so that $p \leq f(n)^{1/2} < f(A-1)^{1/2} = A$. Then $(2n+1)^2 - d = 4(n^2+n+A) \equiv 0 \pmod{4p}$ so that d is a square mod $4p$. On the other hand if d is a square mod $4p$ where p is a prime $\leq A-1$, select m to be the smallest positive integer such that $d \equiv m^2 \pmod{4p}$. Then $m < 2p$ (or else replace m by $4p - m$) and m is odd (as d is odd), so write $m = 2n+1$ and then $0 \leq n \leq p-1 \leq A-2$ with $d \equiv (2n+1)^2 \bmod 4p$. Therefore p divides n^2+n+A with $p < A = f(0) < f(n)$ so that n^2+n+A is composite.

Now we show that $h(d) > 1$ if and only if $d = 1-4A$ is a square mod $4p$ for some prime $p < A$. If $h(d) > 1$, then there exists a reduced binary quadratic $ax^2+bxy+cy^2$ of discriminant d with $1 < a \leq \sqrt{|d|/3} < A$ by the proof of Theorem 12.2. If p is a prime factor of a, then $p \leq a < A$ and $d = b^2 - 4ac$ is a square mod $4p$. On the other hand if d is a square mod $4p$ for some prime $p < A$, and $h(d) = 1$, then p is represented by $x^2+xy+Ay^2$ by Proposition 12.3.1(ii). Now $y \neq 0$ as p is not a square. Therefore $4p = (2x+y)^2 + |d|y^2 \geq 0^2 + |d| \cdot 1^2 = |d|$; that is, $p \geq A$, a contradiction. (We will extend this proof to obtain more on the small values taken by any binary quadratic form of negative discriminant, in exercise 12.6.1(a).) Hence $h(d) > 1$.

Putting these two results together, we deduce that $h(d) > 1$ if and only if $f(n) := n^2+n+A$ is composite for some integer n in the range $0 \leq n \leq A-2$, which implies Rabinowiscz's criterion. □

Exercise 12.5.2.[†] Prove that if n^2+n+A is prime for all integers n in the range $0 \leq n \leq B$, where $1 \leq B < (A-1)/2$, then $\left(\frac{1-4A}{p}\right) = -1$ for all primes $p \leq 2B+1$.

The class number one problem for even negative fundamental discriminants is not difficult:

Theorem 12.4. *If $h(d) = 1$ with $d = -4n$ for $n \in \mathbb{N}$, then $n = 1, 2, 3, 4$, or 7.*

Proof. Suppose that $h(-4n) = 1$. Then n must be a prime power or else there exist coprime integers $1 < a \leq c$ for which $ac = n$ and so $[a, 0, c]$ is a non-principal reduced form of discriminant $-4n$. Moreover $n+1$ must be an odd prime or a power of 2 or else there exist integers $1 < a \leq c$ with $\gcd(a, 2, c) = 1$ for which $ac = n+1$ and so $[a, 2, c]$ is a non-principal reduced form of discriminant $-4n$.

---
[3] We therefore find ourselves in much the same situation as for class number one before Heegner's work, as discussed in the last footnote.

One of n and $n+1$ is even and hence must be a power of 2 (from the previous paragraph). If $n = 2^k$ with $k \geq 4$, then we have the non-principal reduced form $[4, 4, 2^{k-2}+1]$, and if $n+1 = 2^k$ with $k \geq 6$, then we have the non-principal reduced form $[8, 6, 2^{k-3} + 1]$.

Therefore if $h(-4n) = 1$, then $n = 1, 2, 4,$ or 8 or $n + 1 = 2, 4, 8, 16,$ or 32. We can rule out $n = 15$ (as 15 is composite) and $n = 8$ (as 9 is not an odd prime) and $n = 31$ (as $[5, 4, 7]$ is a non-principal reduced form of discriminant -124). We know that $h(-4n) = 1$ for $n = 1, 2, 3, 4,$ and 7 by exercise 12.4.1. □

These discriminants have a beautiful property.

Corollary 12.5.2. *Let $n = 1, 2, 3, 4,$ or 7. If p is a prime that does not divide $4n$, then p can be written as $u^2 + nv^2$ if and only if $\left(\frac{-n}{p}\right) = 1$.*

Proof. As we just discussed $h(-4n) = 1$, and so all binary quadratic forms of discriminant $-4n$ are equivalent to $x^2 + ny^2$. By Proposition 12.3.1, p can be represented by some form of discriminant $-4n$ if and only if $-4n$ is a square mod p, and the result follows. □

We had already discussed representations of p by $x^2 + y^2$, $x^2 + 2y^2$, $x^2 + 3y^2$ in sections 9.1 and 9.2, and $x^2 + 4y^2 = x^2 + (2y)^2$ follows easily from $x^2 + y^2$. This leaves only the most interesting of the cases of Corollary 12.5.2:

$$p = x^2 + 7y^2 \text{ if and only if } p \equiv 1, 9, 11, 15, 23, \text{ or } 25 \pmod{28}.$$

Exercise 12.5.3. Let q be a prime $\equiv -1 \pmod 4$. Prove that $\left(\frac{p}{q}\right) = -1$ for all primes $p < \frac{q+1}{4}$ if and only if $h(-q) = 1$. This result suggests that finding a small prime p with $\left(\frac{p}{q}\right) = 1$ can be a deep problem (see appendix 8B for a discussion of small quadratic residues).

For much more on the values taken by binary quadratic forms, particularly the prime values, we recommend David Cox's wonderful book [**1**].

References for this chapter

[1] David A. Cox, *Primes of the form $x^2 + ny^2$*, Wiley, 1989.
[2] Dorian Goldfeld, *Gauss's class number problem for imaginary quadratic fields*, Bull. Amer. Math. Soc. **13** (1985), 23–37.

Additional exercises

These last questions get considerably more involved but may be of interest to students interested in further pursuing number theory.

Exercise 12.6.1. Suppose that $f(x, y) = ax^2 + bxy + cy^2$ is a reduced binary quadratic form.
(a) Show that if $am^2 + bmn + cn^2 \leq a - |b| + c$ with $(m, n) = 1$, then $|m|, |n| \leq 1$.
(b) Prove that the least values properly represented by f are $a \leq c \leq a - |b| + c$, the first two properly represented twice, the last twice unless $b = 0$, in which case it is properly represented four times.

Exercise 12.6.2. We now use the results of exercise 12.6.1 to understand equivalences between primitive reduced binary quadratic forms. The idea is to recognize a reduced binary quadratic form by the smallest values it properly represents.

(a) Prove that:
- If $0 < |b| < a < c$, then $[a, b, c]$ properly represents a, c, and $a - |b| + c$ in exactly 2, 2, and 2 different ways, respectively.
- If $0 < |b| = a < c$, then $[a, b, c]$ properly represents a, and $c = a - |b| + c$ in exactly 2, and 4 different ways, respectively.
- If $0 < |b| < a = c$, then $[a, b, c]$ properly represents $a = c$, and $a - |b| + c$ in exactly 4, and 2 different ways, respectively.
- If $0 = |b| < a < c$, then $[a, b, c]$ properly represents a, c, and $a - |b| + c$ in exactly 2, 2, and 4 different ways, respectively.
- $[1, 1, 1]$ properly represents 1 in exactly six different ways.
- $[1, 0, 1]$ properly represents both 1 and 2 in exactly four different ways.

(b) Deduce that if $[a, b, c]$, and $[A, B, C]$ are equivalent primitive reduced binary quadratic forms, then $A = a$, $C = c$, and $B = b$ or $-b$.

(c) Use exercise 12.6.1(a) to show that the entries of a matrix representing such an equivalence must each be -1, 0, or 1.

(d) Prove that distinct primitive reduced binary quadratic forms are all inequivalent. Together with Theorem 12.1 this implies that every positive definite binary quadratic form is properly equivalent to a unique reduced form.

(e) Suppose that $M \in \mathrm{SL}(2, \mathbb{Z})$ transforms a primitive reduced binary quadratic form to itself (this is an *automorphism*). Show that $M = \pm I$, except in the following two cases:
- $[1, 1, 1]$ has automorphisms given by $\pm I$, $\pm \begin{pmatrix} 0 & -1 \\ 1 & 1 \end{pmatrix}$, and $\pm \begin{pmatrix} 1 & 1 \\ -1 & 0 \end{pmatrix}$.
- $[1, 0, 1]$ has automorphisms given by $\pm I$ and $\pm \begin{pmatrix} 0 & 1 \\ -1 & 0 \end{pmatrix}$.

Exercise 12.6.3. (a) Show that if $[A, B, C] \sim [a, b, c]$, then $[A, -B, C] \sim [a, -b, c]$.

(b) Use exercise 12.6.2(d) to show that if $[a, b, c]$ is reduced, then $[a, b, c] \sim [a, -b, c]$ if and only if $b = 0$, $b = a$, or $a = c$.

(c) Deduce that $[A, B, C] \sim [A, -B, C]$ if and only if they are equivalent to a quadratic form $[a, 0, c]$, $[a, a, c]$, or $[a, b, a]$.

(d) Prove that $[a, a, c] \sim [c, 2c - a, c]$.

(e) If $d < 0$ is odd, then show that the primitive reduced forms are given by taking each factorization $-d = rs$ with $0 < r \leq s$ and $(r, s) = 1$,

$$\begin{cases} [a, a, c] & \text{if } s \geq 3r \text{ where } a = r \text{ and } c = (r + s)/4, \\ [a, b, a] & \text{if } s < 3r \text{ where } a = (r + s)/4 \text{ and } b = (s - r)/2. \end{cases}$$

(f) If $d < 0$ is even, then show that the primitive reduced forms are given by taking each factorization $-d/4 = rs$ with $0 < r \leq s$ and $(r, s) = 1$,

$$\begin{cases} [a, 0, c] & \text{with } a = r \text{ and } c = s, \\ [a, a, c] & \text{if } s > 3r \text{ where } a = 2r \text{ and } c = (r + s)/2, \\ [a, b, a] & \text{if } s < 3r \text{ where } a = (r + s)/2 \text{ and } b = s - r. \end{cases}$$

Note that the last two cases hold only if $d/4$ is odd.

(g) Show that each binary quadratic form either represents both r and s, or both $2r$ and $2s$. (In (d), take $f(1, -2) = s$ in the first case; $f(1, 1) = s$, $f(1, -1) = r$ in the second case.)

(h) Deduce that if $d < 0$ is a fundamental discriminant, then there are exactly 2^{t-1} reduced binary quadratic forms for which $[a, b, c] \sim [a, -b, c]$, where t is the number of odd prime divisors of $|d|$, unless $4 \| d$ in which case there are 2^t.

Exercise 12.6.4.[†] (a) Prove that $x^2 + 6y^2$ and $2x^2 + 3y^2$ are the only binary quadratic forms, up to equivalence, of discriminant -24.

(b) Prove that prime p can be written in the form $a^2 + 6b^2$ if and only if $p \equiv 1$ or $7 \pmod{24}$.

(c) Prove that prime p can be written in the form $2u^2 + 3v^2$ if and only if $p = 2$ or 3, or $p \equiv 5$ or $11 \pmod{24}$.

We can refine this further:
(d) Prove that prime p can be written in the form $a^2 + 24B^2$ if and only if $p \equiv 1 \pmod{24}$.
(e) Prove that prime p can be written in the form $8U^2 + 3v^2$ if and only if $p = 3$, or $p \equiv 11 \pmod{24}$.

Automorphisms of binary quadratic forms.

Exercise 12.6.5. Suppose that $f \sim g$ via the transformation M and that G is the group of automorphisms of f.
(a) Prove that $M^{-1}GM$ is the group of automorphisms of g.
(b) Prove that MG is the set of transformations yielding g from f.
(c) Deduce that there are $\omega(d)$ automorphisms of every primitive quadratic form of discriminant d, where $\omega(-3) = 6, \omega(-4) = 4$, and $\omega(d) = 2$ for all other discriminants $d < 0$.

Exercise 12.6.6. (a) If $N = f(a, b)$, then $N = f(-a, -b)$. If $N = a^2 + b^2$, then $N = b^2 + (-a)^2 = (-a)^2 + (-b)^2 = (-b)^2 + a^2$. If $N = a^2 + ab + b^2$, then find five other representations of N by the quadratic form $x^2 + xy + y^2$.
(b) Explain how these representations correspond to the automorphisms of the quadratic form.
(c) Why did we not include $N = (-a)^2 + b^2$ in the representations in part (a)?

Exercise 12.6.7. (a) Let $\alpha, \beta, \gamma, \delta$ be given integers for which $\alpha\delta - \beta\gamma = 1$. Prove that β', δ' are integers for which $\alpha\delta' - \beta'\gamma = 1$ if and only if there exists an integer k such that
$$\begin{pmatrix} \alpha & \beta' \\ \gamma & \delta' \end{pmatrix} = \begin{pmatrix} \alpha & \beta \\ \gamma & \delta \end{pmatrix} \begin{pmatrix} 1 & k \\ 0 & 1 \end{pmatrix}.$$
(b) If $A = f(\alpha, \gamma)$ with $(\alpha, \gamma) = 1$, then prove that there exists a unique pair of integers β, δ such that $f \sim [A, B, C]$ using the matrix $M = \begin{pmatrix} \alpha & \beta \\ \gamma & \delta \end{pmatrix} \in \mathrm{SL}(2, \mathbb{Z})$ for some integer B in the range $-A < B \leq A$.
(c) Deduce that the proper representations of the integer A by reduced binary quadratic forms of discriminant d are in $\omega(d)$-to-1 correspondence with the solutions to $B^2 \equiv d \pmod{4A}$ with $-A < B \leq A$.

Exercise 12.6.8. Let $f_1, \ldots, f_h$ be the $h = h(d)$ distinct reduced binary quadratic forms of discriminant d, where $d \equiv 0$ or $1 \pmod 4$. Let $r_j(A)$ denote the number of proper representations of A by f_j. Prove that
$$r_1(A) + \cdots + r_h(A) = \frac{1}{2}\omega(d) \cdot \#\{B \pmod{4A} : B^2 \equiv d \pmod{4A}\}$$
and that this equals $\omega(d) \cdot \prod_{p|A}\left(1 + \left(\frac{d}{p}\right)\right)$ unless perhaps $4|(A, d)$.

Exercise 12.6.9. Suppose that p is an odd prime for which $(d/p) = 1$. Prove that p is properly represented either by only the principal form of discriminant d, or by only two non-principal, reduced, binary quadratic forms of discriminant d, one, say, $ax^2 + bxy + cy^2$, the other $ax^2 - bxy + cy^2$.

Transformations of the upper half-plane. Let $\mathcal{H} := \{z \in \mathbb{C} : \mathrm{Im}(z) > 0\}$ be the *upper half-plane*. We consider transformations with $M = \mathrm{SL}(2, \mathbb{Z})$ acting on $z \in \mathbb{C}$ by taking $M\begin{pmatrix} z \\ 1 \end{pmatrix} = \begin{pmatrix} u \\ v \end{pmatrix}$ and considering this to be the map $z \to u/v$. In Theorem 1.2 we saw that that every matrix in $\mathrm{SL}(2, \mathbb{Z})$ can be represented as a product of the two fundamental matrices $S = \begin{pmatrix} 1 & 1 \\ 0 & 1 \end{pmatrix}$ and $T = \begin{pmatrix} 0 & 1 \\ -1 & 0 \end{pmatrix}$.

Exercise 12.6.10. Prove that S represents the transformation $z \to z + 1$ and that T represents the transformation $z \to -1/z$.

We define
$$\mathcal{F} = \left\{ z \in \mathbb{C} : |z| > 1 \text{ and } -\frac{1}{2} \leq \operatorname{Re}(z) < \frac{1}{2} \right\} \cup \left\{ z \in \mathbb{C} : |z| = 1 \text{ and } -\frac{1}{2} \leq \operatorname{Re}(z) \leq 0 \right\}.$$

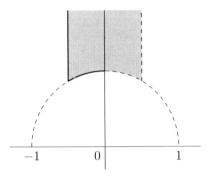

Figure 12.1. The shaded region is the *fundamental domain* $\mathcal{F} \subset \mathcal{H}$.

Exercise 12.6.11.[†] Prove that the binary quadratic form $ax^2 + bxy + cy^2$ with discriminant $d < 0$ is reduced if and only if $\frac{-b+\sqrt{d}}{2a} \in \mathcal{F}$.

Exercise 12.6.12.[†] Prove that for every $z \in \mathbb{C}$ there exists $M \in \operatorname{SL}(2, \mathbb{Z})$ such that $Mz \in \mathcal{F}$. Prove that M is unique.

Exercise 12.6.13.[‡] Show that $\{M\mathcal{F} : M \in \operatorname{SL}(2, \mathbb{Z})\}$ is a partition of $\mathcal{H}$ into disjoint sets.

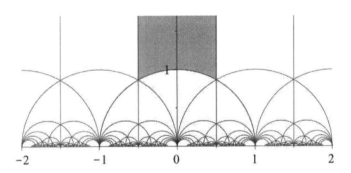

The shaded region is $\mathcal{F}$. Each enclosed region is a domain $M\mathcal{F}$ for some $M \in \operatorname{SL}(2, \mathbb{Z})$.

Appendix 12A. Composition rules: Gauss, Dirichlet, and Bhargava

We study generalizations of the identity (9.1.1), which leads to a notion of "multiplying" binary quadratic forms together, and hence to the group structure discovered by Gauss. We go on to study the reformulations of Dirichlet and Bhargava.

12.7. Composition and Gauss

In (9.1.1) we see that the product of any two integers represented by the binary quadratic form $x^2 + y^2$ is also an integer represented by that binary quadratic form. We now look for further such identities. One easy generalization is given by

(12.7.1) $(u^2 + Dv^2)(r^2 + Ds^2) = x^2 + Dy^2$ where $x = ur + Dvs$ and $y = us - vr$.

Therefore the product of any two integers represented by the binary quadratic form $x^2 + Dy^2$ is also an integer represented by that binary quadratic form. For general *diagonal* binary quadratic forms (that is, having no "cross term" bxy) we have

(12.7.2) $(au^2 + cv^2)(ar^2 + cs^2) = x^2 + acy^2$ where $x = aur + cvs$ and $y = us - vr$.

Notice here that the quadratic form on the right-hand side is different from those on the left; that is, the product of any two integers represented by the binary quadratic form $ax^2 + cy^2$ is an integer represented by the binary quadratic form $x^2 + acy^2$.

One can come up with a similar identity no matter what the quadratic form, though one proceeds slightly differently depending on whether the coefficient b is odd or even. The discriminant $d = b^2 - 4ac$ has the same parity as b. If d is even, then

(12.7.3) $$(au^2 + buv + cv^2)(ar^2 + brs + cs^2) = x^2 - \frac{d}{4}y^2,$$

where $x = aur + \frac{b}{2}(vr + us) + cvs$ and $y = rv - su$.

If d is odd, then

(12.7.4) $$(au^2 + buv + cv^2)(ar^2 + brs + cs^2) = x^2 + xy - \frac{d-1}{4}y^2,$$

where $x = aur + \frac{b-1}{2}vr + \frac{b+1}{2}us + cvs$ and $y = rv - su$.

That is, the product of two integers represented by the same binary quadratic form can be represented by the principal binary quadratic form of the same discriminant.

Exercise 12.7.1. (a) Prove that if n is represented by $ax^2 + bxy + cy^2$, then an is represented by the principal form of the same discriminant.
(b) Suppose that $d < 0$. Deduce that if d is a square mod $4n$, then there is a multiple an of n which is represented by the principal form of discriminant d, with $1 \leq a \leq \sqrt{|d|/3}$.
(c) We obtained the bound $1 \leq a \leq \sqrt{|d|}$ when d is even in section 9.6. Use that method to find a bound in the case that d is odd.

What about the product of the values of two different binary quadratic forms? If d is even, we have

(12.7.5) $$(au^2 + buv + cv^2)(r^2 - \frac{d}{4}s^2) = ax^2 + bxy + cy^2,$$

where $x = ur + \frac{b}{2}su + cvs$ and $y = vr - asu - \frac{b}{2}vs$.

If d is odd, then

(12.7.6) $$(au^2 + buv + cv^2)(r^2 + rs - \frac{d-1}{4}s^2) = ax^2 + bxy + cy^2,$$

where $x = ur + \frac{b+1}{2}su + cvs$ and $y = vr - asu - \frac{b-1}{2}vs$.

That is, the product of an integer that can be represented by a binary quadratic form f and an integer that can be represented by the principal binary quadratic form of the same discriminant can be represented by f.

Exercise 12.7.2. Suppose that a is a prime and $d = b^2 - 4ac$ is even. Let $D = -d/4$.
(a) Show that if a divides $r^2 + Ds^2$, then a divides either $r + (b/2)s$ or $r - (b/2)s$.
(b) Prove that if $r^2 + Ds^2 = an$, then there exist integers X, Y for which $n = aX^2 + bXY + cY^2$.

If n is prime, then this result is true whether or not a is prime, but we will not prove that here. Assume though that is so.
(c) Suppose that $(d/p) = 1$ and that ap is the smallest multiple of p that is represented by the principal form. Prove that a here must take the same value as in exercise 12.6.9.
(d) Prove that $1 \leq a \leq \sqrt{|d|/3}$ and then use exercises 12.4.4 and 12.6.1(b) to prove that if $p < \sqrt{|d|}/2$, then $a = p$.

What about two different binary quadratic forms with no particular structure? For example,

$$(4u^2 + 3uv + 5v^2)(3r^2 + rs + 6s^2) = 2x^2 + xy + 9y^2$$

by taking $x = ur - 3us - 2vr - 3vs$ and $y = ur + us + vr - vs$. These are three inequivalent binary quadratic forms of discriminant -71. Gauss called this *composition*, that is, finding, for given binary quadratic forms f and g of the same discriminant, a third binary quadratic form h of the same discriminant for which

$$f(u,v)g(r,s) = h(x,y),$$

where x and y are quadratic polynomials in $u, v, r,$ and s.

These constructions suggest many questions. For example, are the identities that we found for two given quadratic forms the only possibility? Could the product of two sums of two squares always equal the value of some entirely different quadratic form? When we are given two quadratic forms of the same discriminant, is it true that there is always some third quadratic form of the same discriminant such that the product of the values of the first two always equals a value of the third? That is, is there always a composition of two given binary quadratic forms of the same discriminant? If so, can we determine the third quadratic form quickly?

Gauss proved that one *can always* find the composition of two binary quadratic forms of the same discriminant. The formulas above can mislead one into guessing that this is simply a question of finding the right generalization, but that is far from the truth. All of the examples, (12.7.1) through to (12.7.6), are so explicit only because they are very special cases in the theory. In Gauss's proof he had to prove that various other equations could be solved in integers in order to find h and the quadratic polynomials x and y (which are polynomials in $u, v, r,$ and s). This was so complicated that some of the intermediate formulas took two pages to write down and are very difficult to make sense of.[4] We will prove Gauss's theorem though we will approach it in a somewhat different way.

Exercise 12.7.3. Given non-zero integers a, b, c, d prove that there exist integers m, n such that the set of integers that can be represented by $(ar + bs)(cu + dv)$ as r, s, u, v run over the integers is the same as the set of integers that can be represented by $mx + ny$ as x, y run over the integers.

We finish this section by presenting a fairly general composition.

Proposition 12.7.1. *Suppose that $a_i x^2 + b_i xy + c_i y^2$ for $i = 1, 2$ are binary quadratic forms of discriminant d such that $q = (a_1, a_2)$ divides $\frac{b_1+b_2}{2}$. Then*

$$(12.7.7) \quad (a_1 x_1^2 + b_1 x_1 y_1 + c_1 y_1^2)(a_2 x_2^2 + b_2 x_2 y_2 + c_2 y_2^2) = a_3 x_3^2 + b_3 x_3 y_3 + c_3 y_3^2$$

where $a_3 = a_1 a_2/q^2$ and b_3 is any integer simultaneously satisfying the following (solvable) set of congruences:

$$b_3^2 \equiv d \pmod{4a_1 a_2/q^2},$$
$$b_3 \equiv b_1 \pmod{2a_1/q}, \qquad b_3 \equiv b_2 \pmod{2a_2/q},$$
$$b_3(b_1 + b_2) \equiv b_1 b_2 + d \pmod{4a_1 a_2/q},$$

and c_3 is chosen so that the discriminant of $a_3 x_3^2 + b_3 x_3 y_3 + c_3 y_3^2$ is d.

Exercise 12.7.4. Show that the above congruences for b_3 can be solved.

Proposition 12.7.1 implies that we can always compose two binary quadratic forms f and g of the same discriminant, whose leading coefficients are coprime.

---

[4]See article 234 and beyond in Gauss's book *Disquisitiones Arithmeticae* (1804).

12.8. Dirichlet composition

Proof sketch. Computer software verifies that (12.7.7) holds taking $a_3 = a_1 a_2/q^2$, for any integer q dividing (a_1, a_2), with

$$x_3 = qx_1x_2 + \frac{b_2 - b_3}{2a_2/q} \cdot x_1 y_2 + \frac{b_1 - b_3}{2a_1/q} \cdot x_2 y_1 + \frac{b_1 b_2 + d - b_3 b_1 - b_3 b_2}{4a_1 a_2/q} \cdot y_1 y_2,$$

$$\text{and} \quad y_3 = \frac{a_1}{q} \cdot x_1 y_2 + \frac{a_2}{q} \cdot x_2 y_1 + \frac{b_1 + b_2}{2q} \cdot y_1 y_2.$$

To ensure that we are always working with integers, the coefficients of x_3 and y_3 must be integers. So this formula works if we can find integers q and b_3 for which q divides a_1, a_2, and $\frac{b_1+b_2}{2}$, and the above four congruences hold simultaneously for integer b_3. It is difficult to determine whether there is such a b_3 for an arbitrary q, but not so challenging if $q = (a_1, a_2)$ divides $\frac{b_1+b_2}{2}$. □

Corollary 12.7.1. *For any given integers a, b, c, h, k we have*

$$(ab, hk, ch) \cdot (ac, hk, bh) \sim (ah, hk, bc).$$

Proof. We multiply (ab, hk, ch) and $(ac, hk, bh) \sim (bh, -hk, ac)$ using the proof of Proposition 12.7.1. We take $q = b$ so that $a_3 = ah$ and $2q|b_1 + b_2 = 0$. Selecting $b_3 = hk$ we find that the congruences of Proposition 12.7.1 reduce to $d \equiv (hk)^2$ (mod $4abh$), which follows from $d = (hk)^2 - 4abch$. Hence we have that $(ab, hk, ch) \cdot (ac, hk, bh) \sim (ab, hk, ch) \cdot (bh, -hk, ac) \sim (ah, hk, bc)$.

To get more symmetry in the statement of the result we note that $(ah, hk, bc) \cdot (bc, hk, ah) = 1$, and so

$$(ab, hk, ch) \cdot (ac, hk, bh) \cdot (bc, hk, ah) \sim 1. \qquad \square$$

12.8. Dirichlet composition

Dirichlet claimed that when he was a student, working with Gauss, he slept with a copy of *Disquisitiones* under his pillow every night for three years. It worked, as Dirichlet found a way to better understand Gauss's proof of composition, which amounts to a straightforward algorithm to determine the composition of two given binary quadratic forms f and g of the same discriminant.

Exercise 12.8.1. Given any primitive binary quadratic form $f(x, y) \in \mathbb{Z}[x, y]$ and non-zero integer A, prove that there exist integers r and s such that $f(r, s)$ is coprime to A. Deduce that there exists a binary quadratic form g, for which $f \sim g$, with $(g(1, 0), A) = 1$.

Exercise 12.8.2. Suppose that $f(x, y), F(X, Y)$ are two binary quadratic forms, with disc$(f) \equiv$ disc(F) (mod 2), for which $f(1, 0) = a$ is coprime to $F(1, 0) = A$. Prove that there exist quadratic forms $g = ax^2 + bxy + cy^2$ and $G = AX^2 + bXY + CY^2$ with the same middle coefficient, such that $f \sim g$ and $F \sim G$.

Now suppose we begin with two quadratic forms of the same discriminant. Let A be the leading coefficient of one of them. Then the other is equivalent to a quadratic form with leading coefficient a, for some integer a coprime to A, by exercise 12.8.1. Then these are equivalent to quadratic forms $g = ax^2 + bxy + cy^2$ and $G = AX^2 + bXY + CY^2$, respectively, by exercise 12.8.2. Since these have the

same discriminant we deduce that $ac = AC$ and so there exists an integer h for which
$$g(x,y) = ax^2 + bxy + Ahy^2 \quad \text{and} \quad G(x,y) = Ax^2 + bxy + ahy^2.$$
Then
$$H(m,n) = g(u,v)G(r,s) \text{ with } H(x,y) = aAx^2 + bxy + hy^2,$$
where $m = ur - hvs$ and $n = aus + Avr + bvs$.

Dirichlet went on to interpret this in terms of what we would today call *ideals*; and this in turn led to the birth of modern algebra by Dedekind. In this theory one is typically not so much interested in the identity, writing H as a product of g and G (which is typically very complicated and none too enlightening), but rather in how to determine H from g and G. Dirichlet's proof goes as follows:

The ideal $I\left(\frac{-b+\sqrt{d}}{2}, a\right)$ is associated to a given binary quadratic form $ax^2 + bxy + cy^2$ (see section 12.10 of appendix 12B). Therefore when we multiply together g and G, we multiply together their associated ideals to obtain
$$J := I\left(\frac{-b+\sqrt{d}}{2}, a\right) \cdot I\left(\frac{-b+\sqrt{d}}{2}, A\right),$$
which contains aA as well as both $a \cdot \frac{-b+\sqrt{d}}{2}$ and $A \cdot \frac{-b+\sqrt{d}}{2}$. Since $(a, A) = 1$ there exist integers r, s for which $ar + As = 1$ and so our new ideal contains
$$r \cdot a \cdot \frac{-b+\sqrt{d}}{2} + s \cdot A \cdot \frac{-b+\sqrt{d}}{2} = \frac{-b+\sqrt{d}}{2}.$$
Therefore
$$J = I\left(\frac{-b+\sqrt{d}}{2}, aA\right)$$
which is the ideal associated with the binary quadratic form H.

Defining the class group. We now know that we can multiply together the values of any two quadratic forms of the same discriminant and get another. Since there are only finitely many equivalence classes of binary quadratic forms of a given discriminant this might seem to lead to a group structure, under multiplication. To prove this we will need to know that the usual group properties hold (most importantly, associativity), and also that the values of a binary quadratic form classifies the form. Unfortunately this is not quite true. In exercise 12.6.2 we saw that the only issue in distinguishing between the values taken by forms is perhaps the values taken by $ax^2 + bxy + cy^2$ and $au^2 - buv + cv^2$. However there is an automorphism $u = x, v = -y$ between their sets of values so they cannot be distinguished in this way. On the other hand, the ideals
$$I\left(\frac{-b+\sqrt{d}}{2}, a\right) \quad \text{and} \quad I\left(\frac{b+\sqrt{d}}{2}, a\right)$$
are quite distinct, and so multiplying ideals (and therefore forms) using Dirichlet's technique leads one immediately to being able to determine a group structure. This is called the *class group*, since the group acts on equivalence classes of ideals (and

so of forms). In this approach, associativity follows easily, as multiplication of the numbers in the ideals multiply associatively, and it is similarly evident that the class group is commutative. Therefore the class group is a commutative group, acting on the ideal classes of a given discriminant, with identity element given by the class of principal ideals (which correspond to the principal form).

We will now give a useful criterion to determine how to take square roots inside the class group.

Proposition 12.8.1. *If f is a binary quadratic form of fundamental discriminant d which represents the square of an odd integer, then there exists a binary quadratic form g of discriminant d for which $g \cdot g \sim f$.*

Proof. We begin by squaring the primitive form $ax^2 + bxy + acy^2$. Then

$$J := I\left(\frac{-b+\sqrt{d}}{2}, a\right)^2$$

contains $a^2, a \cdot \frac{-b+\sqrt{d}}{2}$, and $(\frac{-b+\sqrt{d}}{2})^2 = -a^2c - b(\frac{-b+\sqrt{d}}{2})$. Therefore J contains $a \cdot \frac{-b+\sqrt{d}}{2}$ and $b \cdot \frac{-b+\sqrt{d}}{2}$. Now $(a,b) = 1$ or else our original form was not primitive, and so J contains $\frac{-b+\sqrt{d}}{2}$. Therefore

$$J = I\left(\frac{-b+\sqrt{d}}{2}, a^2\right)$$

and the corresponding binary quadratic form is $a^2x^2 + bxy + cy^2$.

One can justify this by finding a suitable multiplication of forms, namely,

$$(ar^2 + brs + acs^2)(au^2 + buv + acv^2) = a^2x^2 + bxy + cy^2,$$

where $x = ru - csv$ and $y = asu + arv + bsv$.

Now if f represents a^2 with $(a,d) = 1$, then there exist integers b, c such that the quadratic form $F := a^2x^2 + bxy + cy^2$ is equivalent to f. Note that $(a,b)^2$ divides $d = b^2 - 4a^2c$, which is a fundamental discriminant and so squarefree except perhaps a power of 2. However a is odd and so $(a,b) = 1$. Therefore we let $g = ax^2 + bxy + acy^2$ so that, as in the previous paragraph $g \cdot g \sim F \sim f$. □

12.9. Bhargava composition[5]

Let us begin with one further explicit composition, a tiny variant on (12.7.3) (letting $s \to -s$ there):

$$(au^2 + 2Buv + cv^2)(ar^2 - 2Brs + cs^2) = x^2 + (ac - B^2)y^2$$

where $x = aur + B(vr - us) - cvs$ and $y = us + vr$.

Combining this with the results of the previous section suggests that if the discriminant d is divisible by 4 (which is equivalent to b being even), then

(12.9.1) $$F(u,v)G(r,s)H(m,-n) = P(x,y)$$

---

[5] Although there is no Nobel Prize in mathematics, there is the *Fields Medal*, awarded every four years, only to people 40 years of age or younger. In 2014, in Korea, one of the laureates was Manjul Bhargava for a body of work that begins with his version of composition, as discussed here, and allows us to much better understand many classes of equations, especially cubic.

where $P(x,y) = x^2 - \frac{d}{4}y^2$ is the principal form and x and y are cubic polynomials in m, n, r, s, u, v. Analogous remarks can be made if the discriminant is odd.

In 2004 Bhargava came up with an entirely new way to find all of the triples F, G, H of binary quadratic forms of the same discriminant for which (12.9.1) holds: We begin with a 2-by-2-by-2 cube, the corners of which are labeled with the integers a, b, c, d, e, f, g, h.

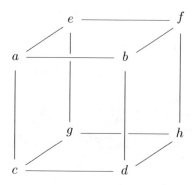

Figure 12.2. Bhargava's Rubik-type cube.

There are six faces of a cube, and these can be split into three parallel pairs. To each such parallel pair consider the pair of 2-by-2 matrices given by taking the entries in each face, those entries corresponding to opposite corners of the cube, always starting with a. Hence we get the pairs

$$M_1(x,y) := \begin{pmatrix} a & b \\ c & d \end{pmatrix} x + \begin{pmatrix} e & f \\ g & h \end{pmatrix} y = \begin{pmatrix} ax+ey & bx+fy \\ cx+gy & dx+hy \end{pmatrix},$$

$$M_2(x,y) := \begin{pmatrix} a & c \\ e & g \end{pmatrix} x + \begin{pmatrix} b & d \\ f & h \end{pmatrix} y = \begin{pmatrix} ax+by & cx+dy \\ ex+fy & gx+hy \end{pmatrix},$$

$$M_3(x,y) := \begin{pmatrix} a & b \\ e & f \end{pmatrix} x + \begin{pmatrix} c & d \\ g & h \end{pmatrix} y = \begin{pmatrix} ax+cy & bx+dy \\ ex+gy & fx+hy \end{pmatrix},$$

where we have, in each, appended the variables, x, y, to create matrix functions of x and y. The determinant, $-Q_j(x,y)$, of each $M_j(x,y)$ is a quadratic form in x and y. Incredibly Q_1, Q_2, and Q_3 all have the same discriminant and their composition equals P, the principal form, just as in (12.9.1). We present two proofs. First, by substitution, one can exhibit that

$$Q_1(x,-y) = Q_2(x_2, y_2) Q_3(x_3, y_3)$$

where

$$y = \begin{pmatrix} x_3 & y_3 \end{pmatrix} \begin{pmatrix} a & b \\ c & d \end{pmatrix} \begin{pmatrix} x_2 \\ y_2 \end{pmatrix} \quad \text{and} \quad x = \begin{pmatrix} x_3 & y_3 \end{pmatrix} \begin{pmatrix} e & f \\ g & h \end{pmatrix} \begin{pmatrix} x_2 \\ y_2 \end{pmatrix}.$$

Let's work though an example: Plot the cube in three dimensions, take the Cartesian coordinates of every corner (each 0 or 1), and then label the corner

12.9. Bhargava composition

(x, y, z), with $2^2 x + 2y + z$, squared. Hence
$$a, b, c, d, e, f, g, h = 2^2, 6^2, 0^2, 4^2, 3^2, 7^2, 1^2, 5^2,$$
yielding the cube in Figure 12.3.

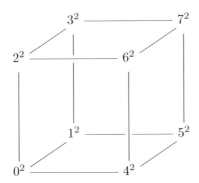

Figure 12.3. The construction of three binary quadratic forms using Bhargava's cube.

This cube leads to three binary quadratic forms of discriminant $-7 \cdot 4^4$:
$Q_1 = -4^2(4x^2+13xy+11y^2)$, $Q_2 = -2^2(x^2-2xy+29y^2)$, and $Q_3 = 4^2(8x^2+5xy+y^2)$.
After some work one can verify that
$$Q_1(m,n)Q_2(r,s)Q_3(u,v) = 4(x^2 + 4^3 \cdot 7y^2),$$
where x and y are the following cubic polynomials in m, n, r, s, u, v:
$$x = 8(-11mru - 3mrv + 25msu + 17msv - 17nru - 4nrv + 59nsu + 32nsv)$$
and $y = mru + mrv + 21msu + 5msv + 3nru + 2nrv + 31nsu + 6nsv$.

Bhargava proves his theorem, inspired by a 2-by-2-by-2 Rubik's cube. His idea is to apply one invertible linear transformation at a time, simultaneously to a pair of opposite sides, and to slowly "reduce" the numbers involved, while retaining the equivalence classes of Q_1, Q_2, and Q_3, until one reduces to a cube and a triple of binary quadratic forms with coefficients having a convenient structure.

Lemma 12.9.1. *If one applies an invertible linear transformation to a pair of opposite sides, then the associated binary quadratic form is transformed in the usual way, whereas the other two quadratic forms remain the same.*

Therefore we can act on our cube by such $\mathrm{SL}(2, \mathbb{Z})$-transformations, in each direction, and the three binary quadratic forms each remain in the same equivalence class.

Proof. If $\begin{pmatrix} \alpha & \beta \\ \gamma & \delta \end{pmatrix} \in \mathrm{SL}(2, \mathbb{Z})$, then we replace the face

$\begin{pmatrix} a & b \\ c & d \end{pmatrix}$ by $\begin{pmatrix} a & b \\ c & d \end{pmatrix}\alpha + \begin{pmatrix} e & f \\ g & h \end{pmatrix}\beta$; and $\begin{pmatrix} e & f \\ g & h \end{pmatrix}$ by $\begin{pmatrix} a & b \\ c & d \end{pmatrix}\gamma + \begin{pmatrix} e & f \\ g & h \end{pmatrix}\delta$.

Then $M_1(x,y)$ gets mapped to

$$\left\{\begin{pmatrix} a & b \\ c & d \end{pmatrix}\alpha + \begin{pmatrix} e & f \\ g & h \end{pmatrix}\beta\right\}x + \left\{\begin{pmatrix} a & b \\ c & d \end{pmatrix}\gamma + \begin{pmatrix} e & f \\ g & h \end{pmatrix}\delta\right\}y,$$

that is, $M_1(\alpha x + \gamma y, \beta x + \delta y)$. Therefore the quadratic form $Q_1(x,y)$ gets mapped to $Q_1(\alpha x + \gamma y, \beta x + \delta y)$ which is equivalent to $Q_1(x,y)$. Now $M_2(x,y)$ gets mapped to

$$\begin{pmatrix} a\alpha + e\beta & c\alpha + g\beta \\ a\gamma + e\delta & c\gamma + g\delta \end{pmatrix} x + \begin{pmatrix} b\alpha + f\beta & d\alpha + h\beta \\ b\gamma + f\delta & d\gamma + h\delta \end{pmatrix} y = \begin{pmatrix} \alpha & \beta \\ \gamma & \delta \end{pmatrix} M_2(x,y);$$

hence the determinant, $-Q_2(x,y)$, is unchanged. An analogous calculation reveals that $M_3(x,y)$ gets mapped to $\begin{pmatrix} \alpha & \beta \\ \gamma & \delta \end{pmatrix} M_3(x,y)$ and the determinant, $-Q_3(x,y)$, is also unchanged. □

The previous lemma allows one to proceed in "reducing" the three binary quadratic forms to equivalent forms that are easy to work with (rather as in Dirichlet's proof).

Proof of the Bhargava composition. We will simplify the entries in the cube by the following reduction algorithm:

- We select the corner that is to be a so that $a \neq 0$.
- We will transform the cube to ensure that a divides b, c, and e. If not, say a does not divide e, then select integers α, β so that $a\alpha + e\beta = (a,e)$, and then let $\gamma = -e/(a,e)$, $\delta = a/(a,e)$. In the transformed matrix we have $a' = (a,e)$, $e' = 0$, and $1 \leq a' \leq a - 1$. It may well now be that a' does not divide b' or c', so we repeat the process. Each time we do this we reduce the value of a by at least 1; and since it remains positive this can only happen a finite number of times. At the end of the process a divides b, c, and e.
- We will transform the cube to ensure that $b = c = e = 0$. We already have that $a|b,c,e$. Now select $\alpha = 1$, $\beta = 0, \gamma = -e/a$, $\delta = 1$, so that $e' = 0, b' = b, c' = c$. We repeat this in each of the three directions to ensure that $b = c = e = 0$.

Replacing a by $-a$, we have that the three matrices are

$$M_1(x,y) := \begin{pmatrix} -a & 0 \\ 0 & d \end{pmatrix} x + \begin{pmatrix} 0 & f \\ g & h \end{pmatrix} y, \text{ so that } Q_1(x,y) = adx^2 + ahxy + fgy^2,$$

$$M_2(x,y) := \begin{pmatrix} -a & 0 \\ 0 & g \end{pmatrix} x + \begin{pmatrix} 0 & d \\ f & h \end{pmatrix} y, \text{ so that } Q_2(x,y) = agx^2 + ahxy + dfy^2,$$

$$M_3(x,y) := \begin{pmatrix} -a & 0 \\ 0 & f \end{pmatrix} x + \begin{pmatrix} 0 & d \\ g & h \end{pmatrix} y, \text{ so that } Q_3(x,y) = afx^2 + ahxy + dgy^2.$$

All three Q_j have discriminant $(ah)^2 - 4adfg$, and we observe that

$$Q_1(fy_2x_3 + gx_2y_3 + hy_2y_3, ax_2x_3 - dy_2y_3) = Q_2(x_2,y_2)Q_3(x_3,y_3)$$

where $x_1 = fy_2x_3 + gx_2y_3 + hy_2y_3$ and $y_1 = ax_2x_3 - dy_2y_3$. □

This brings to mind the twists of the Rubik's cube, though in that case one has only finitely many possible transformations, whereas here there are infinitely many possibilities, as there are infinitely many invertible linear transformations over $\mathbb{Z}$.

Appendix 12B. The class group

12.10. A dictionary between binary quadratic forms and ideals

In section 12.8 of appendix 12A we were reminded that we associate the ideal

$$I_{\mathbb{Z}}(2a, -b+\sqrt{d}) = \{2ax + (-b+\sqrt{d})y : x, y \in \mathbb{Z}\}$$

with the binary quadratic form $ax^2 + bxy + cy^2$, denoted $[a, b, c]$. To see why, note that if we multiply any element of the ideal with its conjugate, we obtain

$$(2ax + (-b+\sqrt{d})y) \cdot (2ax + (-b-\sqrt{d})y) = 4a(ax^2 + bxy + cy^2).$$

We now investigate how the key operations of equivalence and composition translate into the ideal setting:

- All equivalences of binary quadratic forms can be broken down into the two linear transformations $x \to x + ky$, $y \to y$, $k \in \mathbb{Z}$, and $x \to -y$, $y \to x$:
  - The linear transformation $x \to x + ky$, $y \to y$, $k \in \mathbb{Z}$ converts the linear form to
    $$2a(x+ky) + (-b+\sqrt{d})y = 2ax + (-B+\sqrt{d})y;$$
    that is, $(2a, -b+\sqrt{d}) = (2a, -B+\sqrt{d})$, where $B = b-2ak$. This new ideal corresponds to the binary quadratic form $[a, B, c]$, which is equivalent to $[a, b, c]$, via the inverse of this linear transformation.
  - The linear transformation $x \to -y$, $y \to x$ converts the linear form to $2a(-y) + (-b+\sqrt{d})x$. If we multiply this through by $b+\sqrt{d}$, we obtain
    $$(b+\sqrt{d}) \cdot (2a(-y) + (-b+\sqrt{d})x) = -2a(b+\sqrt{d})y + (d-b^2)x$$
    $$= -4acx - 2a(b+\sqrt{d})y$$
    $$= -2a(2cx + (b+\sqrt{d})y),$$

465

so that $(b + \sqrt{d}) \cdot (2a, -b + \sqrt{d}) = 2a \cdot (2c, b + \sqrt{d})$. This new ideal, $(2c, b + \sqrt{d})$, corresponds to the binary quadratic form $[c, -b, a]$ which is equivalent to $[a, b, c]$, via the inverse of this linear transformation.

It therefore makes sense to define:

Ideals I and J in $\mathbb{Q}(\sqrt{d})$ are *equivalent* if there exist $\alpha, \beta \in \mathbb{Q}(\sqrt{d})$ such that $\alpha I = \beta J$.

We have shown that if $f \sim g$, then the ideals corresponding to f and g are equivalent.

- In section 12.8 of appendix 12A we sketched Dirichlet's proof that the composition of two binary quadratic forms of the same discriminant corresponds to the product of their ideals. Although composition is a rather mysterious operation, multiplication is not! We know that it is commutative and associative. We have seen that composition has a 1 (the principal form corresponding to the principal ideals) and has inverses, so that

$$(2a, -b + \sqrt{d}) \cdot (2a, b + \sqrt{d}) \sim (1).$$

In Theorem 12.2 we showed that the number of ideal classes (that is, the number of classes of ideals, modulo this equivalence relation) is finite, and therefore the ideal classes form a finite abelian group under multiplication, called the *class group*.

Exercise 12.10.1. (a) Show that if a and b are represented by f of discriminant d, then ab is represented by the principal quadratic form of discriminant d.
(b) Show that if a is represented by f of discriminant d, and b is represented by the principal quadratic form of discriminant d, then ab is represented by f.
(c) Prove that if $f(x, y)$ represents integers a, b, and c, then it represents abc.

In section 12.5 we discussed Heegner's Theorem exhibiting the nine fundamental negative discriminants $d < 0$ of class number one. By the dictionary in this section this shows that there are exactly nine fields of the form $\mathbb{Q}(\sqrt{d})$ with $d < 0$ in which there are only principal ideals in the ring of integers, $R = \mathbb{Z}[\sqrt{d}]$. In other words these are the only d for which R is a principal ideal domain; and therefore these are the only d for which $\mathbb{Z}[\sqrt{d}]$ has unique factorization (as discussed in section 3.19 of appendix 3D).

12.11. Elements of order two in the class group

We saw the benefit of studying the subgroup of squares, and their cosets, when working with the group of reduced residues mod n. Similarly when G is the class group, we let $G^2 := \{g^2 : g \in G\}$ and H be the set of elements in G of order one or two; then G^2 and H form subgroups of G with $G/G^2 \cong H$.

As $[a, -b, c]$ is the inverse of $[a, b, c]$ we deduce that $[a, b, c]$ has order one or two if and only if $[a, b, c] \sim [a, -b, c]$ where $[a, b, c]$ is reduced, without loss of generality. We proved that reduced forms are inequivalent so this can only happen if $b = 0$ (so that the forms are identical) or if $[a, -b, c]$ is not reduced. Now if $d < 0$ and $[a, b, c]$ is reduced, but not $[a, -b, c]$, then either $b = a$, or $b > 0$ and $a = c$. One can verify that $[a, b, c] \sim [a, -b, c]$ in each of these cases via the transformations $x \to x - y$, $y \to y$, and $x \to -y$, $y \to x$, respectively. Therefore we have a set of

12.11. Elements of order two in the class group

representatives for H:

$$\{[a,b,c] \text{ reduced of discriminant } d : b = 0 \text{ or } b = a \text{ or } a = c\}.$$

When G is the group of reduced residues mod p, the Legendre symbol allows us to determine which coset of G^2 a given reduced residue belongs to. We now develop the same idea modified to this setting. Corollary 12.3.1 tells us that if N is coprime to d and is properly represented by a binary quadratic form f of discriminant d, then $\sigma_f(p) = \left(\frac{N}{p}\right)$ for every odd prime p dividing d. Now suppose that f and g are two binary quadratic forms of discriminant d, and let $h = fg$ (that is, h is the composition of f and g, their product inside the class group). If m is represented by f and n is represented by g, then mn is represented by $fg = h$, and so

$$\sigma_h(p) = \left(\frac{mn}{p}\right) = \left(\frac{m}{p}\right)\left(\frac{n}{p}\right) = \sigma_f(p)\sigma_g(p).$$

This implies that $\sigma : G \to \{-1, 1\}^r$, where r is the number of odd prime factors of d, is actually a group homomorphism. In particular if $f = g$ so that $h = f^2$, then $\sigma_h(p) = 1$ for all p, and so h is in the kernel of the map; that is, $G^2 \subset \ker \sigma$. In fact one can prove that G^2 is the kernel of the map, and therefore

$$\{\sigma_f(p) : p|d\} \text{ is isomorphic to } H.$$

If d is even, then the value of $\sigma_f(2)$ can be determined by the $\sigma_f(p)$ with p odd, as the following exercise shows us.

Exercise 12.11.1. Use Corollary 9.4.1 and quadratic reciprocity to prove that $\prod_{p|d} \sigma_f(p) = 1$.

Exercise 12.11.2. Prove that if $f \sim g$, then $\sigma_f(2) = \sigma_g(2)$ when d is even.

Exercise 12.11.1 implies that $\text{Im}(\sigma) \subseteq \{(\delta_p)_{p|d} \in \{-1, 1\} : \prod_{p|d} \delta_p = 1\}$, and one can prove that this is the image.

If d is a discriminant for which G^2 contains only the identity element, then all elements of G have order one or two. Therefore there is a unique equivalence class of binary quadratic forms f of discriminant d for which $\sigma_f(p) = \delta_p$ for all primes p, for any given $(\delta_p)_{p|d} \in \text{Im}(\sigma)$. We can therefore determine exactly which equivalence class of binary quadratic forms represents each given prime p for which $(d/p) = 1$, using only Legendre symbols, by Corollary 12.3.1. These discriminants d are Euler's idoneal numbers discussed in section 12.5.

Finally we prove a result in the converse direction to exercise 12.3.1.

Exercise 12.11.3. Suppose that $f = [a, b, c]$ and $F = [A, B, C]$ are reduced, primitive binary quadratic forms of fundamental discriminant $d < 0$ for which $\sigma_f(p) = \sigma_F(p)$ for all primes p dividing d. Prove that f and F are equivalent over the rationals. (In other words, there exist $\alpha, \beta, \gamma, \delta \in \mathbb{Q}$ with $\alpha\delta - \beta\gamma = 1$ for which $F = f(\alpha x + \beta y, \gamma x + \delta y)$.)

References for this chapter

[1] J. W. S. Cassels, *Rational quadratic forms*, Dover, 1978.

[2] Harvey Cohn *Advanced number theory*, Dover, 1962.

[3] David A. Cox, *Primes of the form $x^2 + ny^2$*, Wiley, 1989.

[4] J.-P. Serre, *A course in arithmetic*, Graduate Texts in Mathematics **7**, Springer-Verlag, 1973.

Appendix 12C. Binary quadratic forms of positive discriminant

12.12. Binary quadratic forms with positive discriminant, and continued fractions

When $d > 0$, Gauss defined $ax^2 + bxy + cy^2$ to be *reduced* when

(12.12.1) $$0 < \sqrt{d} - b < 2|a| < \sqrt{d} + b.$$

This implies that $0 < b < \sqrt{d}$ so that $|a| < \sqrt{d}$ and therefore there are only finitely many reduced forms of positive discriminant d. Since $d - b^2 = 4ac$ we have $ac = (b^2 - d)/4 < 0$, and

$$0 < \sqrt{d} - b < 2|c| < \sqrt{d} + b,$$

and so $ax^2 + bxy + cy^2$ is reduced if and only if $cx^2 + bxy + ay^2$ is.

Forms $f(x,y) = ax^2 + bxy + cy^2$ and $g(x,y) = cx^2 + Bxy + Cy^2$ are *neighbors* (and equivalent) if they have the same discriminant and $b + B \equiv 0 \pmod{2c}$. They are equivalent under the transformation $\begin{pmatrix} x \\ y \end{pmatrix} \to \begin{pmatrix} 0 & -1 \\ 1 & k \end{pmatrix} \begin{pmatrix} x \\ y \end{pmatrix}$ where $b + B = 2ck$.

Gauss's reduction algorithm goes as follows: Given $ax^2 + bxy + cy^2$ we select a neighbor as follows: Let b' be the absolutely least residue of $-b \pmod{2c}$ so that $|b'| \le c$.

- If $|b'| > \sqrt{d}$, then let $B = b'$.
  Note that $0 < B^2 - d \le c^2 - d$. Therefore $|C| = (B^2 - d)/4|c| < |c|/4$.
- If $|b'| < \sqrt{d}$, then select $B \equiv b' \equiv -b \pmod{2c}$ with B as large as possible with $B < \sqrt{d}$. Therefore $B = -b + 2k|c|$ where $k = [\alpha_f]$ and $\alpha_f := \frac{b + \sqrt{d}}{2|c|}$.
  Note that $-d \le B^2 - d = 4cC < 0$.

12.12. Positive discriminant quadratic forms and continued fractions

- If $2|c| > \sqrt{d}$, then $|C| \leq |d/4c| < |c|$.
- Otherwise $\sqrt{d} \geq 2|c|$ and $\sqrt{d}-2|c| < B < \sqrt{d}$, and therefore $0 < \sqrt{d}-B < 2|c| < \sqrt{d} + B$, so that the neighbor is reduced.

We repeat this algorithm until we obtain a reduced form. This is guaranteed to terminate since if $f \sim g$, then $|c_g| = |C| < |c| = |c_f|$; that is, the absolute value of the final coefficient of the binary quadratic form, which is an integer, is reduced at each step until the binary quadratic form is itself reduced.

For example, the quadratic form $[38, 101, 67]$ has discriminant 17. Here $b' = 33$ and so $|b'| > \sqrt{d}$ and therefore we have $[38, 101, 67] \sim [67, 33, 4]$. For this next form $b' = -1$ and so $|b'| < \sqrt{d}$. However $2|c| > \sqrt{d}$, so we are in the second case of the algorithm and $[67, 33, 4] \sim [4, -1, -1]$. For this form, $b' = 1$ so that $|b'| < \sqrt{d}$, and $B = 3$. Now as $2|c| < \sqrt{d}$ we are in the third case of the algorithm and deduce that $[4, -1, -1] \sim [-1, 3, 2]$, which is reduced. What happens when we apply the algorithm to $[-1, 3, 2]$? Surprisingly we find that $[-1, 3, 2]$ is equivalent to a different reduced form, $[2, 1, -2]$, of discriminant 17. This is quite different from the $d < 0$ case when there was a unique reduced form in each equivalence class. We can repeat the algorithm obtaining $[2, 1, -2] \sim [-2, 3, 1]$, and then $[-2, 3, 1] \sim [1, 3, -2]$, followed by $[1, 3, -2] \sim [-2, 1, 2]$, $[-2, 1, 2] \sim [2, 3, -1]$, and finally $[2, 3, -1] \sim [-1, 3, 2]$, where we get back to our original reduced form. This is known as a *cycle of reduced forms* and can be written as

$$[-1, 3, 2] \sim [2, 1, -2] \sim [-2, 3, 1] \sim [1, 3, -2] \sim [-2, 1, 2] \sim [2, 3, -1] \sim [-1, 3, 2],$$

or more succinctly as

(12.12.2) $\qquad -1 \ ^3 \ 2 \ ^1 \ -2 \ ^3 \ 1 \ ^3 \ -2 \ ^1 \ 2 \ ^3 \ -1 \ ^3 \ldots .$

We can guarantee that there are only finitely many forms in a cycle because there are only finitely many reduced forms of a given discriminant.

In general if $[a_0, b_0, a_1] \sim [a_1, b_1, a_2] \sim [a_2, b_2, a_3] \sim \cdots$, then we can write

$$a_0 \ ^{b_0} \ a_1 \ ^{b_1} \ a_2 \ ^{b_2} \ a_3 \ \ldots ,$$

always selecting each successive b_m by Gauss's reduction algorithm and then letting $a_{m+1} = (b_m^2 - d)/4a_m$. For another example, when $d = 816$,

$$5 \ ^{26} \ -7 \ ^{16} \ 20 \ ^{24} \ -3 \ ^{24} \ 20 \ ^{16} \ -7 \ ^{26} \ 5 \ ^{24} \ -12 \ ^{24} \ 5 \ ^{26} \ -7 \ \ldots$$

which is a cycle of period 8.

Exercise 12.12.1. Prove that every reduced form of positive discriminant has a unique reduced neighboring form to the left and a unique reduced neighboring form to the right.

We can explain this surprising feature of forms of positive discriminant by understanding a remarkable connection between Gauss's reduction algorithm and the continued fraction for $\alpha_f = \frac{b+\sqrt{d}}{2|c|}$. We showed that $k = [\alpha_f]$ and $\alpha_f - k = \frac{-B+\sqrt{d}}{2|c|}$, and so

$$(\alpha_f - k)\alpha_g = \frac{-B+\sqrt{d}}{2|c|} \cdot \frac{B+\sqrt{d}}{2|C|} = \frac{d-B^2}{4|ac|} = 1$$

as $B < \sqrt{d}$, which implies that $\alpha_f = [k, \alpha_g]$. Thus we can trace Gauss's reduction algorithm by following the continued fraction algorithm. For instance, in our first example above we look at the continued fraction of $\frac{33+\sqrt{17}}{8} = [4, 1, \overline{1, 1, 3}]$. Then, in the notation of appendix 11B, $\alpha = \alpha_0$ and $\alpha_5 = \alpha_2$ with

$$\alpha_0 = \frac{33+\sqrt{17}}{8}, \ \alpha_1 = \frac{-1+\sqrt{17}}{2}, \ \alpha_2 = \frac{3+\sqrt{17}}{4}, \ \alpha_3 = \frac{1+\sqrt{17}}{4}, \ \alpha_4 = \frac{3+\sqrt{17}}{2}.$$

The cycle length here is 3, not 6, since $\alpha_2 = \alpha_5$ corresponds to the quadratic form $[-1, 3, 2]$ as well as $[1, 3, -2]$. The linear transformations that gives rise to each successive equivalence, in the cycle (12.12.2), are

(12.12.3) $\begin{pmatrix} 0 & -1 \\ 1 & 1 \end{pmatrix}, \begin{pmatrix} 0 & -1 \\ 1 & -1 \end{pmatrix}, \begin{pmatrix} 0 & -1 \\ 1 & 3 \end{pmatrix}, \begin{pmatrix} 0 & -1 \\ 1 & -1 \end{pmatrix}, \begin{pmatrix} 0 & -1 \\ 1 & 1 \end{pmatrix}, \begin{pmatrix} 0 & -1 \\ 1 & -3 \end{pmatrix}.$

The entry in the bottom right-hand corner of each matrix comes from the periodic part of the continued fraction for α.

12.13. The set of automorphisms

If $M = \begin{pmatrix} \alpha & \beta \\ \gamma & \delta \end{pmatrix}$ is an automorphism of $ax^2 + bxy + cy^2$, and if z is a root of $ax^2 + bx + c$, then $Mz = z$. Therefore $\gamma z^2 + (\delta - \alpha)z - \beta = 0$ which implies that there exists an integer v such that

$$\gamma = av, \ \delta - \alpha = bv, \ -\beta = cv.$$

Now if $u = \delta + \alpha$, then

$$u^2 - dv^2 = u^2 - (bv)^2 + 4(av)(cv) = (\delta+\alpha)^2 - (\delta-\alpha)^2 - 4\gamma\beta = 4(\alpha\delta - \beta\gamma) = 4$$

and the automorphism takes the form

(12.13.1) $\begin{pmatrix} \frac{u-bv}{2} & -cv \\ av & \frac{u+bv}{2} \end{pmatrix}.$

We found all solutions to Pell's equation $u^2 - dv^2 = 4$ in appendix 11B so we can determine all of the automorphisms of $[a, b, c]$ from (12.13.1). In fact the solutions with $u, v > 0$ are all given by $\frac{u_n+v_n\sqrt{d}}{2} = (\frac{u+v\sqrt{d}}{2})^n$ for $n \geq 1$ where u, v gives the smallest solution to Pell's equation with $\frac{u+v\sqrt{d}}{2} > 1$. One can verify that

$$\begin{pmatrix} \frac{u_n-bv_n}{2} & -cv_n \\ av_n & \frac{u_n+bv_n}{2} \end{pmatrix} = \pm \begin{pmatrix} \frac{u-bv}{2} & -cv \\ av & \frac{u+bv}{2} \end{pmatrix}^n.$$

Going back to our first example, the product of the six matrices in (12.12.3) is $\begin{pmatrix} -9 & 32 \\ 16 & -57 \end{pmatrix}$, mapping $[-1, 3, 2]$ to itself, which can be presented as minus (12.13.1), with $a = -1, b = 3, c = 2$ and $u = 66, v = 16$ (and $u^2 - 17v^2 = 4$).

On the other hand the morphism of $[a, b, c]$ to $[-a, b, -c]$ has transformation given by the product of the first three matrices in (12.12.3) which has transformation matrix $\begin{pmatrix} \frac{u-bv}{2} & cv \\ av & \frac{-u-bv}{2} \end{pmatrix}$ with $a = -1, b = 3, c = 2$ and $u = 8, v = 2$. Notice here that $u^2 - 17v^2 = -4$, a solution to the negative Pell equation, and $33 + 8\sqrt{17} = (-4 - \sqrt{17})^2$.

Appendix 12D. Sums of three squares

We have seen which integers are representable as the sum of two squares. What integers are representable as the sum of three squares?

12.14. Connection between sums of 3 squares and $h(d)$

We may restrict our attention to integers that are not divisible by 4: Since the only squares mod 4 are 0 and 1, if n is divisible by 4 and is the sum of three squares, then all three squares must be even. Hence if $n = 4m$, then to obtain every representation of n as the sum of three squares, we just take every representation of m as the sum of three squares, and double the numbers that are being squared.

The only squares mod 8 are 0, 1, and 4. Therefore no integer $\equiv 7 \pmod 8$ can be written as the sum of three squares (of integers). By the previous remark no integer of the form $4^k(8m+7)$ can be written as the sum of three squares. Legendre proved that these are the only exceptions (we will not prove this here as all known proofs are too complicated for a first course):

Theorem 12.5 (Legendre's Theorem, 1798). *A positive integer n can be written as the sum of three squares of integers if and only if n is not of the form $4^k(8m+7)$.*

In section 11.2, we introduced Pell's equation by asking for those triangular numbers that are also a square. The young Gauss was intrigued by which integers can be written as the sum of three triangular numbers; that is,

$$n = \frac{a(a+1)}{2} + \frac{b(b+1)}{2} + \frac{c(c+1)}{2}.$$

We can multiply through by 8 and add 3 to get the equivalent equation

$$8n + 3 = (2a+1)^2 + (2b+1)^2 + (2c+1)^2.$$

On the other hand, if an integer $\equiv 3 \pmod 8$ is to be written as the sum of three squares, then those three squares must each be odd, and therefore we have this equation. In other words we have proved: n is the sum of three triangular numbers if and only if $8n + 3$ is the sum of three squares. Legendre's Theorem tells us that every integer $\equiv 3 \pmod 8$ is the sum of three squares, and so every positive integer is the sum of three triangular numbers. When Gauss discovered this in 1796 he wrote in his diary:

$$\text{EYPHKA! num} = \triangle + \triangle + \triangle.$$

One might ask how many ways one can write an integer as the sum of three squares. Gauss proved the following remarkable theorem (for which there is still no easy proof): If n is squarefree, the number of representations of n as the sum of three squares is

$$\begin{cases} 24h(-n) & \text{if } n \equiv 3 \pmod 8, \\ 12h(-4n) & \text{if } n \equiv 1 \text{ or } 2 \pmod 8, \ n > 1. \end{cases}$$

Here $h(-d)$ is the class number of $\mathbb{Q}(\sqrt{-d})$.

Confronted by such a surprising formula I typically verify the claim by computing. In so doing I not only verified the above claim but found the following. I give it as an exercise but I do not know how to prove it using only elementary methods!

Exercise 12.14.1.‡ If p is a prime and $p = 8n \pm 1$ or ± 5, then there are exactly n solutions to $p^2 = a^2 + b^2 + c^2$ with $1 \leq a \leq b \leq c$.

12.15. Dirichlet's class number formula

In 1832 Jacobi was interested in the sum

$$J_p := \sum_{k=1}^{p-1} k \left(\frac{k}{p} \right),$$

for odd primes p. This is the sum of the quadratic residues mod p up to p, minus the sum of the quadratic non-residues mod p up to p. Calculations yield that

$$J_3 = -1, \ J_5 = 0, \ J_7 = -7, \ J_{11} = -11, \ J_{13} = 0, \ J_{17} = 0, \ J_{19} = -19, \ J_{23} = -69,$$

and patterns begin to emerge. The two most obvious are that $J_p = 0$ if $p \equiv 1 \pmod 4$, and that J_p seems to be divisible by p if $p \equiv 3 \pmod 4$ and $p > 3$.

Exercise 12.15.1. (a) Prove that $J_p = 0$ if $p \equiv 1 \pmod 4$.
(b) Prove that J_p is divisible by p if p is a prime > 3.

For $p \equiv 3 \pmod 4$ and $p > 3$ we find that the values of $j_p := J_p/p$ are

$$j_7 = -1, \ j_{11} = -1, \ j_{19} = -1, \ j_{23} = -3, \ j_{31} = -3, \ j_{43} = -1, \ j_{47} = -5, \ j_{59} = -3, \ldots.$$

Can you see any patterns and perhaps guess at the value? Write a program to get more data. At this point Jacobi did something unexpected: He computed the class number $h(-p)$ for primes $p \equiv 3 \pmod 4$ (see, e.g., section 12.4) and compared. Do you see a pattern now?

12.15. Dirichlet's class number formula

In 1832 Jacobi conjectured that the class number $h(-p)$, when $p \equiv 3 \pmod 4$ and $p > 3$, is given by $-J_p/p$; that is,

$$(12.15.1) \qquad h(-p) = -\frac{1}{p} \sum_{n=1}^{p-1} \left(\frac{n}{p}\right) n.$$

In 1838 Dirichlet gave a proof of Jacobi's conjecture and much more. His miraculous class number formula links algebra and analysis in an unforeseen way that was a foretaste of many of the most important subsequent works in number theory, including Wiles's proof of Fermat's Last Theorem. We will simply state the formulas here, which apply whenever d is a fundamental discriminant. If $d > 0$, then

$$h(d) \, \log \epsilon_d = \sqrt{d} \sum_{n \geq 1} \frac{1}{n} \left(\frac{d}{n}\right).$$

If $-d < -4$, then

$$h(-d) = \frac{\sqrt{d}}{\pi} \sum_{n \geq 1} \frac{1}{n} \left(\frac{-d}{n}\right).$$

It is not obvious that this is equivalent to (12.15.1) where $d = -p \equiv 1 \pmod 4$, but this will be confirmed in section 14.8 of appendix 14A.

We encountered the infinite sum on the right-hand side of these formulas in section 8.17 in connection with Dirichlet's work on primes in arithmetic progressions. It is not obvious that this sum converges, but we will give a proof that it converges in section 13.7. Now $h(d) \geq 1$ for all d (positive or negative) since $h(d)$ counts the binary quadratic forms of discriminant d, and one always has the principal form. Hence the formulas imply that

$$\sum_{n \geq 1} \frac{1}{n} \left(\frac{d}{n}\right) > 0,$$

which was a crucial step in the proof that there are infinitely many primes in arithmetic progressions (see section 8.17). By taking $h(d) \geq 1$ in these formulas we obtain the lower bounds

$$\sum_{n \geq 1} \frac{1}{n} \left(\frac{-d}{n}\right) \geq \frac{\pi}{\sqrt{d}}$$

for $-d < -4$ and, when $d > 0$ then

$$\sum_{n \geq 1} \frac{1}{n} \left(\frac{d}{n}\right) \geq \frac{\log \epsilon_d}{\sqrt{d}} \geq \frac{\log d}{2\sqrt{d}}.$$

In section 12.5 we saw that $h(-d) = 1$ for only a few values of d, the largest of which is 163. Dirichlet's class number formula provides an explanation: We expect a lot of cancellation in our infinite sum, $\sum_{n \geq 1} \frac{1}{n}(\frac{-d}{n})$, so its value should be neither too large nor too small. Therefore we expect $h(-d)$ to typically be close to $\sqrt{d}$ (that is, within, say, a factor of 10), and so $h(-d)$ should be much larger than 1, once d is sufficiently large.

How about $h(d)$ when $d > 0$? We again expect a lot of cancellation in our infinite sum, $\sum_{n \geq 1} \frac{1}{n}(\frac{d}{n})$, so that we expect $h(d) \log \epsilon_d$ to be around $\sqrt{d}$. One of

$h(d)$ and $\log \epsilon_d$ could be small and the other large. As we mentioned in section 11.13 of appendix 11B, calculations suggest that $\log \epsilon_d$ is usually around $\sqrt{d}$ so that we expect $h(d)$ to typically be "small", perhaps even bounded by a constant. However there are values of d for which ϵ_d is small. For example, if $d = m^2 + 1$ is squarefree, then $\epsilon_d = m + \sqrt{d}$, so for these discriminants we expect that $h(d)$ grows like $\sqrt{d}/\log d$ with d.

How often is $h(d) = 1$ when $d > 0$? Here is a list of the fundamental discriminants $d, 1 \leq d \leq 100$, for which $h(d) = 1$:[6]

$$5, 8, 12, 13, 17, 21, 24, 28, 29, 33, 37, 41, 44,$$
$$53, 56, 57, 61, 69, 73, 76, 77, 88, 89, 92, 93, 97,$$

that is, 26 of the 30 fundamental discriminants up to 100, a surprisingly high proportion. We believe that $h(p) = 1$ for roughly 3 out of every 4 primes $p \equiv 1 \pmod{4}$, but this is an open conjecture, and we do not even know how to prove that there are infinitely many fundamental discriminants d, never mind primes, for which $h(d) = 1$.

Exercise 12.15.2. Let $S := \sum_{n=1}^{(p-1)/2} \left(\frac{n}{p}\right)$ and $T := \sum_{n=1}^{(p-1)/2} \left(\frac{n}{p}\right) n$.
  (a) Show that $S = 0$ when $p \equiv 1 \pmod{4}$. Henceforth assume that $p \equiv 3 \pmod{4}$ with $p > 3$.
  (b) Note that $\left(\frac{p-n}{p}\right)(p-n) = \left(\frac{n}{p}\right)(n-p)$. Use this to evaluate the sum $\sum_{n=1}^{p-1} \left(\frac{n}{p}\right) n$ in terms of S and T by pairing up the nth and $(p-n)$th term, for $n = 1, 2, \ldots, \frac{p-1}{2}$.
  (c) Do this taking $n = 2m$, $m = 1, 2, \ldots, \frac{p-1}{2}$ to deduce from (12.15.1) that

$$h(-p) = \frac{1}{2 - \left(\frac{2}{p}\right)} \sum_{n=1}^{(p-1)/2} \left(\frac{n}{p}\right).$$

Exercise 12.15.3. Let p be an odd prime and $x \equiv \frac{p-1}{2}! \pmod{p}$. In exercise 7.4.3 we showed that $x^2 \equiv -\left(\frac{-1}{p}\right) \pmod{p}$.
  (a) Prove that $x^{\frac{p-1}{2}} \equiv (-1)^N \pmod{p}$ where $N \equiv \frac{1}{2} \sum_{n=1}^{(p-1)/2} \left(1 - \left(\frac{n}{p}\right)\right) \pmod{2}$.
  (b) Show that if $p \equiv 3 \pmod{4}$, then $h(-p)$ is odd and $N \equiv \frac{1}{2}(h(-p)+1) \pmod{2}$ using exercise 12.15.2(c).
  (c) Deduce that if $p \equiv 3 \pmod 4$, then $\frac{p-1}{2}! \equiv (-1)^{\frac{h(-p)+1}{2}} \pmod p$.

---

[6] Here we are looking at the class number of $\mathbb{Q}(\sqrt{d})$ as discussed in appendix 12B. For a fundamental discriminant D we take $d = D/(D, 4)$, so if $d \equiv 2$ or $3 \pmod 4$ in our list, then this means that there is just one equivalence class of binary quadratic forms of discriminant $4d$, namely the principal class.

Appendix 12E. Sums of four squares

12.16. Sums of four squares

Next we discuss a beautiful and surprising theorem that Euler tried, and failed, to prove for many years.

Theorem 12.6 (Lagrange's Theorem). *Every positive integer is the sum of four squares.*

Proof. We start from the identity

(12.16.1) $$(a^2 + b^2 + c^2 + d^2)(u^2 + v^2 + w^2 + x^2) = A^2 + B^2 + C^2 + D^2,$$

where $A := au + bv + cw + dx$, $B := -av + bu - cx + dw$,
$C := -aw + bx + cu - dv$, $D := -ax - bw + cv + du$.

Much as for the sum of two squares, if we can show that every prime is the sum of four squares, then every integer, written as the product of its prime factors, is the sum of four squares by repeated applications of (12.16.1).

Now $2 = 1^2 + 1^2 + 0^2 + 0^2$ so we focus on odd primes p. There exist integers a, b, c, d, not all zero, such that $a^2 + b^2 + c^2 + d^2 \equiv 0 \pmod{p}$ by exercise 8.9.9(b) (and taking $d = 0$). We select a, b, c, d so that $mp = a^2 + b^2 + c^2 + d^2$ where m is the minimal such integer ≥ 1. Our goal is to show that $m = 1$.

Exercise 12.16.1. Prove that we may take $|a|, |b|, |c|, |d| < p/2$, so that $m < p$.

Exercise 12.16.2. Show that if m is even, then we can reorder a, b, c, d so that $a - b$ and $c - d$ are both even. Using the identity

$$\left(\frac{a-b}{2}\right)^2 + \left(\frac{a+b}{2}\right)^2 + \left(\frac{c-d}{2}\right)^2 + \left(\frac{c+d}{2}\right)^2 = \frac{1}{2}(a^2 + b^2 + c^2 + d^2),$$

prove that m must be odd.

475

Let u, v, w, x be the absolutely least residues of a, b, c, d (mod m), respectively. Therefore $u^2 + v^2 + w^2 + x^2 \equiv a^2 + b^2 + c^2 + d^2 \equiv 0$ (mod m). Moreover $|u|, |v|, |w|, |x| < m/2$ (since m is odd), and so $u^2 + v^2 + w^2 + x^2 < 4(m/2)^2 = m^2$. Hence we can write $u^2 + v^2 + w^2 + x^2 = mn$ for some integer $n < m$.

Exercise 12.16.3. Prove that $A \equiv B \equiv C \equiv D \equiv 0$ (mod m).

Since $A/m, B/m, C/m, D/m$ are integers by the last exercise, we obtain
$$(A/m)^2 + (B/m)^2 + (C/m)^2 + (D/m)^2 = \frac{(a^2+b^2+c^2+d^2)}{m} \frac{(u^2+v^2+w^2+x^2)}{m}$$
$$= np.$$
This contradicts the minimality of m unless $n = 0$ in which case $u = v = w = x = 0$ so that $a \equiv b \equiv c \equiv d \equiv 0$ (mod m) and then $m^2 | a^2 + b^2 + c^2 + d^2 = mp$. Therefore $m = 1$ as $m < p$, which is what we wished to prove. □

Exercise 12.16.4. Prove that no integer of the form 2^{2k+1} is the sum of three or four positive squares, and the only such representation of 2^{2k} is $(2^{k-1})^2 + (2^{k-1})^2 + (2^{k-1})^2 + (2^{k-1})^2$.

12.17. Quaternions

The identity $(a^2 + b^2)(u^2 + v^2) = (x^2 + y^2)$ where $x = au - bv$ and $y = av + bu$, which was used in the result about representations by the sum of two squares, is perhaps most naturally obtained by taking norms in the product $(a + ib)(u + iv) = x + iy$ where $i^2 = -1$. The identity (12.16.1) can be obtained by taking norms in the *quaternions*. This is a non-commutative ring. There are three square roots of -1 which do not commute with one another, specifically: We let i, j, k be "imaginary numbers" such that
$$i^2 = j^2 = k^2 = -1,$$
$$ij = k = -ji, \quad jk = i = -kj, \quad ki = j = -ik.$$
The *quaternions* are the ring $\{a + bi + cj + dk : a, b, c, d \in \mathbb{Z}\}$. Now
$\text{Norm}(a + bi + cj + dk) := (a + bi + cj + dk)(a - bi - cj - dk) = a^2 + b^2 + c^2 + d^2$
since $ij + ji = jk + kj = ki + ik = 0$. Moreover
$$(a + bi + cj + dk)(u - vi - wj - xk) = A + Bi + Cj + Dk$$
with A, B, C, D defined as in (12.16.1). Our proof that primes can be written as the sum of four squares can now be translated into the language of quaternions.

12.18. The number of representations

Let $r_4(n)$ be the number of representations of n as a sum of four squares. In 1834 Jacobi showed that $r_4(n) = 8\sigma(n)$ if n is odd, and $r_4(n) = 24\sigma(m)$ if n is even, writing $n = 2^k m$, $k \geq 1$, where m is odd and $\sigma(n) = \sum_{d|n} d$.

Exercise 12.18.1. Prove that this can rewritten as follows:
$$\sum_{N \geq 0} r_4(N) x^N := \left(\sum_{n \in \mathbb{Z}} x^{n^2} \right)^4 = 8 \sum_{\substack{d \geq 1 \\ 4 \nmid d}} \frac{d x^d}{1 - x^d} = 1 + 8 \left(\sum_{n \geq 1} \frac{x^n}{(1-x^n)^2} - \sum_{m \geq 1} \frac{4 x^{4m}}{(1-x^{4m})^2} \right).$$

12.18. The number of representations

Exercise 12.18.2. (a) Show that if $8|N$ and $N = a^2 + b^2 + c^2 + d^2$, then a, b, c, d are all even; and so deduce that $r_4(2^k m) = r_4(2^{k-2} m)$ if $k \geq 3$.
(b)† Prove that if m is odd, then $r_4(2^k m) = 3 r_4(m)$ for all $k \geq 1$.

We now prove that $r_4(n) = 8\sigma(n)$ when n is odd. The result for n even follows from exercise 12.18.2.

We saw that $r_2(0) = 1$ and

$$r_2(n) = 4 \sum_{d|n} \left(\frac{-1}{d}\right) = 4 \left(\sum_{\substack{ab=n \\ a \equiv 1 \pmod 4}} 1 - \sum_{\substack{ab=n \\ a \equiv -1 \pmod 4}} 1 \right)$$

for all $n \geq 1$ (writing $a = d$ and $b = n/d$), so that

$$r_4(n) - 2r_2(n) = \sum_{\substack{m+M=n \\ m,M \geq 1}} r_2(m) r_2(M) = 16 \left(\sum_{\substack{ab+AB=n \\ a \equiv A \pmod 4 \\ a \text{ odd}}} 1 - \sum_{\substack{ab+AB=n \\ a \equiv -A \pmod 4 \\ a \text{ odd}}} 1 \right)$$

$$= 16 \left(\sum_{\substack{ab+AB=n \\ a \equiv A \pmod 4}} 1 - \sum_{\substack{ab+AB=n \\ a \equiv -A \pmod 4}} 1 \right).$$

Here we write $m = ab$ and $M = AB$ and apply the above formula for $r_2(m)$ and $r_2(M)$. We have added into both sums the terms for which a is even, since then the condition $a \equiv A \pmod 4$ is the same as the condition $a \equiv -A \pmod 4$.

We split the first sum into three parts: If $a > A$, write $a = A + 4c$ with $c \geq 1$, so we have $n = ab + AB = (A + 4c)b + AB = Ad + 4cb$ with $d = b + B > b$ and Ad odd. Therefore

$$\sum_{\substack{ab+AB=n \\ a \equiv A \pmod 4 \\ a > A}} 1 = \sum_{\substack{ad+4bc=n \\ d > b}} 1.$$

We take the same approach and get the same quantity if $a < A$. If $a = A$, then we have $n = ab + AB = a(b + B)$. This is then a sum over divisors d of n, writing $a = n/d$, so there are $d - 1$ choices of pairs $b + B = d$. This yields, in total, $\sum_{d|n} d - 1 = \sigma(n) - \tau(n)$ possibilities. Therefore, changing variables, we have

$$\sum_{\substack{ab+AB=n \\ a \equiv A \pmod 4}} 1 = 2 \sum_{\substack{ab+4AB=n \\ a > B}} 1 + \sigma(n) - \tau(n).$$

We also split the second sum into three parts: If $b > B$, write $b = B + C$ so we have
$$n = ab + AB = a(B + C) + AB = (a + A)B + aC = 4dB + aC$$
as $a + A \equiv 0 \pmod 4$, with $4d > a$, analogously if $b < B$. If $b = B$, then we have $n = ab + AB = (a + A)B$. This though is impossible as 4 divides $a + A$, which divides the odd n. Therefore, changing variables, we have

$$\sum_{\substack{ab+AB=n \\ a \equiv -A \pmod 4}} 1 = 2 \sum_{\substack{ab+4AB=n \\ b < 4B}} 1.$$

Now if $ab + 4AB = n$ with $a < A$, write $A = a + c$ to have
$$n = ab + 4(a+c)B = a(b+4B) + 4Bc = ad + 4Bc$$
where $d = b + 4B > 4B$, and so
$$\sum_{\substack{ab+4AB=n \\ a<A}} 1 = \sum_{\substack{ab+4AB=n \\ b>4B}} 1.$$

Subtracting these quantities from the total number of solutions to $ab + 4AB = n$ in positive integers, we get
$$\sum_{\substack{ab+4AB=n \\ a \geq A}} 1 = \sum_{\substack{ab+4AB=n \\ b \leq 4B}} 1.$$

We cannot have $b = 4B$ or else 4 divides n, so combining the above equations gives that
$$r_4(n) - 2r_2(n) = 16\left(\sigma(n) - \tau(n) - 2\sum_{\substack{ab+4AB=n \\ a=A}} 1\right).$$

If $ab + 4AB = n$ with $a = A$, then $n = a(b + 4B)$, and so if $d|n$ with $a = n/d$, then there are
$$\left[\frac{d-1}{4}\right] = \frac{d-2}{4} + \frac{1}{4}\left(\frac{-1}{d}\right)$$
possibilities for $b + 4B = d$, as d is odd. Therefore
$$\sum_{\substack{ab+4AB=n \\ a=A}} 1 = \sum_{d|n} \frac{d-2}{4} + \frac{1}{4}\sum_{d|n}\left(\frac{-1}{d}\right) = \frac{1}{4}\sigma(n) - \frac{1}{2}\tau(n) + \frac{r_2(n)}{16}.$$

Substituting this in above we obtain
$$r_4(n) = 8\sigma(n),$$
as desired. □

Exercise 12.18.3. (a) Verify the identity
$4(a^2 + b^2 + c^2 + d^2) = (a+b+c+d)^2 + (a+b-c-d)^2 + (a-b+c-d)^2 + (a-b-c+d)^2$.

(b)† Suppose that $m \equiv n \pmod{2}$. Prove that there exist integers a, b, c, d for which $n = a^2 + b^2 + c^2 + d^2$ with $m = a + b + c + d$ if and only if $4n - m^2$ can be written as the sum of three squares.

Henceforth let n be odd.

(c) Deduce that for every odd integer m with $|m| < 2\sqrt{n}$ there exist integers a, b, c, d for which $n = a^2 + b^2 + c^2 + d^2$ and $m = a + b + c + d$.

(d) Show that there is a 1-to-1 correspondence between solutions to $4n - m^2 = u^2 + v^2 + w^2$ with $u \equiv v \equiv w \equiv m \pmod 4$ and solutions to $n = a^2 + b^2 + c^2 + d^2$, $m = a+b+c+d$ with $a \not\equiv b \equiv c \equiv d \pmod 2$.

(e) Deduce that there is a 2-to-1 correspondence between solutions to $4n - m^2 = u^2 + v^2 + w^2$ and solutions to $n = a^2 + b^2 + c^2 + d^2$, $m = a+b+c+d$.

(f) Using Gauss's result mentioned at the end of section 12.14 of appendix 12D, deduce that if n is odd, then there are $12h(m^2 - 4n)$ solutions to $n = a^2 + b^2 + c^2 + d^2$, $m = a+b+c+d$.

Appendix 12F. Universality

12.19. Universality of quadratic forms

A *universal* quadratic form represents all positive integers. We have seen that $x^2 + y^2 + z^2 + w^2$ is universal, while $x^2 + y^2 + z^2$ is not. One can show that no binary quadratic form is universal as a consequence of Proposition 12.3.1(i). Using a suitable generalization of the local-global principle (in the form of the Hilbert symbol—see appendix 9B) one can show that no positive definite ternary quadratic form is universal, though it is easy to show that $x^2 - y^2 + 2z^2$ is universal.

Lemma 12.19.1. *No positive definite diagonal ternary quadratic form $ax^2 + by^2 + cz^2$ is universal.*

Proof. Assume $1 \leq a \leq b \leq c$ without loss of generality. To represent 1 we must have $a = 1$. To represent 2 we therefore need that $b = 1$ or 2. The quadratic form $x^2 + y^2 + cz^2$ does not represent 3 if $c > 3$, and it does not represent 7, 14, or 6 if $c = 1, 2$, or 3, respectively. The quadratic form $x^2 + 2y^2 + cz^2$ does not represent 5 if $c > 5$, and it does not represent 7, 10, 14, or 10 if $c = 2, 3, 4$, or 5, respectively. □

Let $[a, b, c, d]$ denote the quadratic form $ax^2 + by^2 + cz^2 + dw^2$ with $a, b, c, d > 0$. In 1916 Ramanujan[7] asserted that the following quaternary quadratic forms are

---

[7]Srinivasa Ramanujan was a self-taught Indian mathematician who sent letters with his discoveries but no proofs to prominent mathematicians, most of whom largely ignored him. In 1913 he wrote to G. H. Hardy, a professor at Trinity College, Cambridge, who was impressed by some of these formulas, including

$$1 - 5\left(\frac{1}{2^2}\binom{2}{1}\right)^3 + 9\left(\frac{1}{2^4}\binom{4}{2}\right)^3 - 13\left(\frac{1}{2^6}\binom{6}{3}\right)^3 + \cdots = \frac{2}{\pi}.$$

Although Hardy could not prove this, nor several of Ramanujan's other results, he figured "they must be true, because ... no one would have the imagination to invent them." Hardy brought Ramanujan to Cambridge, England, in 1914. Ramanujan spent the next five years proving and publishing great theorems, in a wide variety of subjects, under Hardy's guidance (several of which will appear in the next few chapters). Sadly, Ramanujan died in 1920 of tuberculosis at the age of 32, having suffered several health problems in the damp English climate. Most of the results in his notebooks, stated there without proof, were only proved and published many years later. We are still not sure of the intuition that led him to many of his results; researchers continue to try recreate and appreciate his thinking up to the present day. The world's leading prize for a young number theorist is the Sastra-Ramanujan

universal: $[1,1,1,k]$, $[1,2,2,k]$ for $1 \leq k \leq 7$; $[1,1,2,k]$, $[1,2,4,k]$ for $1 \leq k \leq 14$; $[1,1,3,k]$ for $1 \leq k \leq 6$; $[1,2,3,k], [1,2,5,k]$ for $1 \leq k \leq 10$; though this is not quite true for $[1,2,5,5]$, that is, $x^2 + 2y^2 + 5z^2 + 5w^2$, since it represents every positive integer except 15. This is a lovely theorem but difficult to state or remember. In 1938 Halmos observed that a much simpler and more memorable way to recall Ramanujan's result is to note that $ax^2 + by^2 + cz^2 + dw^2$ is universal if and only if it represents every positive integer ≤ 15.

It is of interest to classify all universal positive definite quadratic forms in no matter how many variables. In 1993 Conway observed that Halmos's observation extends to this question.

Theorem 12.7 (The Fifteen Criterion, I). *Suppose that f is a positive definite diagonal quadratic form. Then f represents all positive integers if and only if f represents all positive integers ≤ 15.*

Proof assuming Ramanujan's Theorem. Suppose that $f = a_1 x_1^2 + a_2 x_2^2 + \cdots + a_d x_d^2$ with $1 \leq a_1 \leq a_2 \leq \cdots \leq a_d$ represents all positive integers. Since f represents 1 we must have $a_1 = 1$. Since f represents 2 we must have $a_2 = 1$ or 2. If $a_1 = a_2 = 1$, then, since f represents 3 we must have $a_3 = 1, 2,$ or 3. If $a_1 = 1$, $a_2 = 2$, then, since f represents 5 we must have $a_3 = 2, 3, 4,$ or 5. Now

$x_1^2 + x_2^2 + x_3^2$ represents m, $1 \leq m \leq 6$, but not 7, and so $1 \leq a_4 \leq 7$;

$x_1^2 + x_2^2 + 2x_3^2$ represents m, $1 \leq m \leq 13$, but not 14, and so $1 \leq a_4 \leq 14$;

$x_1^2 + x_2^2 + 3x_3^2$ represents m, $1 \leq m \leq 5$, but not 6, and so $1 \leq a_4 \leq 6$;

$x_1^2 + 2x_2^2 + 2x_3^2$ represents m, $1 \leq m \leq 6$, but not 7, and so $1 \leq a_4 \leq 7$;

$x_1^2 + 2x_2^2 + 3x_3^2$ represents m, $1 \leq m \leq 9$, but not 10, and so $1 \leq a_4 \leq 10$;

$x_1^2 + 2x_2^2 + 4x_3^2$ represents m, $1 \leq m \leq 13$, but not 14, and so $1 \leq a_4 \leq 14$;

$x_1^2 + 2x_2^2 + 5x_3^2$ represents m, $1 \leq m \leq 9$, but not 10, and so $1 \leq a_4 \leq 10$.

Ramanujan's result implies that, for these possibilities for a_1, a_2, a_3, a_4, the quadratic form $a_1 x_1^2 + a_2 x_2^2 + a_3 x_3^2 + a_4 x_4^2$ represents every positive integer except perhaps 15, and the result follows. □

Actually one only needs to verify that

$$1, 2, 3, 5, 6, 7, 10, 14, \text{ and } 15$$

are represented; that is, f represents all positive integers if and only if f represents each of $1, 2, 3, 5, 6, 7, 10, 14,$ and 15. Conway and Schneeberger observed that this result generalizes rather nicely in that one can decide whether *any* positive definite quadratic form with even cross coefficients is universal in much the same way:

Theorem 12.8 (The Fifteen Criterion, II). *Suppose that f is a positive definite quadratic form, which is diagonal mod 2. Then f represents all positive integers if and only if f represents $1, 2, 3, 5, 6, 7, 10, 14,$ and 15.*

---

prize, presented only to researchers who are no older than Ramanujan was, at the time of his death. Ramanujan's story is well told in the 2015 movie *The man who knew infinity*.

This is sharp in the sense that for each integer m in the list there is such a quadratic form that represents every integer except m. For example, we already saw that $x^2 + 2y^2 + 5z^2 + 5w^2$ represents every positive integer other than 15, and $2x^2 + 3y^2 + 4z^2 + 5w^2$ represents every positive integer other than 1, etc.

This was extended to all quadratic forms by Bhargava and Hanke:

Theorem 12.9 (The 290 Criterion). *Suppose that f is a positive definite quadratic form. Then f represents all positive integers if and only if f represents all positive integers ≤ 290.*

This is sharp since $x^2 + xy + 2y^2 + xz + 4z^2 + 29(a^2 + ab + b^2)$ represents every positive integer other than 290. In fact f represents all positive integers if and only if it represents the 29 integers

$$1, 2, 3, 5, 6, 7, 10, 13, 14, 15, 17, 19, 21, 22, 23, 26, 29,$$
$$30, 31, 34, 35, 37, 42, 58, 93, 110, 145, 203, \text{ and } 290.$$

This set is also minimal in that for each integer m in the list there is a positive definite quadratic form that represents all positive integers except m.

The Fifteen Criterion has been generalized to representations of any set S of positive integers by Bhargava: There exists a finite subset T of S such that a positive definite quadratic form f, which is diagonal mod 2, represents every integer in S if and only if it represents every integer in T. Hence Bhargava has reduced any such classification problem to a finite problem. For example, such an f represents all primes if and only if f represents the 15 primes up to 47 as well as 67 and 73.

Here are a few nice examples that have been worked out explicitly:

- (Bhargava) A positive definite quadratic form, which is diagonal mod 2, represents every odd positive integer if and only if it represents 1, 3, 5, 7, 11, 15, and 33.

- (Rouse, under certain mild assumptions) A positive definite quadratic form represents every odd positive integer if and only if it represents every odd integer ≤ 451.

- (Crystel Bujold) A positive definite ternary quadratic form, which is diagonal mod 2, represents every odd positive integer if and only if it represents 1, 3, 5, 7, 11, and 15.

In 2015, Kenneth Williams wrote a delightful article in which he observed some further such results: If $a, b, c > 0$, then $ax^2 + by^2 + cz^2$ represents all odd positive integers if and only if it represents 1, 3, 5, and 15. Moreover $ax^2 + by^2 + cz^2$ represents all positive integers $\equiv 2 \pmod{4}$ if and only if it represents 2, 6, 10, 14, and 30.

References for this appendix

[1] Manjul Bhargava and Jonathan Hanke, *Universal quadratic forms and the 290-Theorem* (preprint).
[2] John H. Conway, *The sensual (quadratic) form*, Carus Mathematical Monographs (26), 1997.
[3] Kenneth S. Williams, *A "four integers" theorem and a "five integers" theorem*, Amer. Math. Monthly 122 (2015), 528–536.

Appendix 12G. Integers represented in Apollonian circle packings

Sarnak studied the set of curvatures of circles surrounding a fixed circle in any given integer Apollonian circle packing (see section 9.15 of appendix 9D). We will suppose we are given integers a, b, c, d satisfying $2(a^2+b^2+c^2+d^2) = (a+b+c+d)^2$, so that $a+b+c+d$ is even. We fix the circle of curvature a and make the change of variables

$$x = a+b, \quad y = a+c \quad \text{and} \quad z = \frac{a+b+c-d}{2}$$

so our equation becomes

$$a^2 + z^2 - xy = 0.$$

The transformations of (a, b, c, d) given by $b \to 2(a+c+d)-b$, $c \to 2(a+b+d)-c$, and $d \to 2(a+b+c)-d$ become the transformations of (x, z, y) given by

$$\begin{pmatrix} x \\ z \\ y \end{pmatrix} \to M \begin{pmatrix} x \\ z \\ y \end{pmatrix}$$

for $M = M_b, M_c, M_d$, respectively, with

$$M_b := \begin{pmatrix} 1 & -4 & 4 \\ 0 & -1 & 2 \\ 0 & 0 & 1 \end{pmatrix}, \quad M_c := \begin{pmatrix} 1 & 0 & 0 \\ 2 & -1 & 0 \\ 4 & -4 & 1 \end{pmatrix}, \quad \text{and} \quad M_d := \begin{pmatrix} 1 & 0 & 0 \\ 0 & -1 & 0 \\ 0 & 0 & 1 \end{pmatrix}.$$

12.20. Combining these linear transformations

We wish to determine which matrices are given by the products of the M_b, M_c, M_d taken arbitrarily often in arbitrary order, or at least to find a large and interesting subset of the set of products. Sarnak found a way to fit this question into known work on matrices by making an extraordinary observation (which appears in rather different form in Gauss's work): There is a map from 2-by-2 matrices to certain 3-by-3 matrices given by

$$\rho : \begin{pmatrix} \alpha & \beta \\ \gamma & \delta \end{pmatrix} \longrightarrow \frac{1}{\alpha\delta - \beta\gamma} \begin{pmatrix} \alpha^2 & 2\alpha\gamma & \gamma^2 \\ \alpha\beta & \alpha\delta + \beta\gamma & \gamma\delta \\ \beta^2 & 2\beta\delta & \delta^2 \end{pmatrix}$$

which preserves multiplication though reverses the order, $\rho(AB) = \rho(B)\rho(A)$. The images of

$$\begin{pmatrix} 1 & 2 \\ 0 & 1 \end{pmatrix} \text{ and } \begin{pmatrix} 1 & 0 \\ -2 & 1 \end{pmatrix}$$

under the map ρ yield the matrices $M_d M_b$ and $M_d M_c$, respectively.

These two matrices, together with $-I$ (and note that $\rho(-I) = I$) generate the *congruence subgroup*,

$$G := \left\{ \begin{pmatrix} \alpha & \beta \\ \gamma & \delta \end{pmatrix} : \alpha, \beta, \gamma, \delta \in \mathbb{Z}, \ \alpha\delta - \beta\gamma = 1, \text{ and } \begin{pmatrix} \alpha & \beta \\ \gamma & \delta \end{pmatrix} \equiv I \pmod{2} \right\}.$$

We prove this by something akin to the Euclidean algorithm: We begin with a matrix $\begin{pmatrix} a & b \\ c & d \end{pmatrix} \in G$ and we will post multiply this matrix by other elements of G so as to reduce the pair of integers a, b. If $|a| > |b| > 0$, let A be the absolutely least residue of $a \pmod{2|b|}$ so that $A = a - 2kb$ for some integer k. Then

$$\begin{pmatrix} a & b \\ c & d \end{pmatrix} \begin{pmatrix} 1 & 0 \\ -2 & 1 \end{pmatrix}^k = \begin{pmatrix} a & b \\ c & d \end{pmatrix} \begin{pmatrix} 1 & 0 \\ -2k & 1 \end{pmatrix} = \begin{pmatrix} A & b \\ C & d \end{pmatrix}$$

where $C = c - 2kd$, which also belongs to G (as G is a group and so closed under multiplication). What is important is that $|A| \leq |b|$.

Similarly if $|b| > |a| > 0$, then we write $B = b - 2ka$ and post-multiply by $\begin{pmatrix} 1 & -2 \\ 0 & 1 \end{pmatrix}^k$, and so we replace b by B in the matrix, where $|B| \leq |a|$. This algorithm terminates in a finite number of steps, when either $|a| = |b|$ or one of a and b is 0. But a is odd and b is even, and so $b = 0$ and $ad = 1$ since the determinant is 1, implying that either $a = d = 1$ or $a = d = -1$. If $a = d = -1$, we multiply through by $-I$ and so we have $a = d = 1$. Since c is even we post-multiply by $\begin{pmatrix} 1 & 0 \\ -2 & 1 \end{pmatrix}^{c/2}$ to obtain the identity. The result is proved. $\square$

We have therefore proved that the matrices generated by $M_d M_b$ and $M_d M_c$ all take the form

$$\begin{pmatrix} \alpha^2 & 2\alpha\gamma & \gamma^2 \\ \alpha\beta & \alpha\delta + \beta\gamma & \gamma\delta \\ \beta^2 & 2\beta\delta & \delta^2 \end{pmatrix} \text{ where } \begin{pmatrix} \alpha & \beta \\ \gamma & \delta \end{pmatrix} \in G.$$

We wish to extend this to the group generated by M_b, M_c, M_d, which means we have to take into account the effect of M_d only, as $M_d^2 = I$. Now if we pre-multiply one of our matrices by M_d, this has the same effect as changing the signs of α and γ; and if we post-multiply one of our matrices by M_d, this has the same effect as changing the signs of α and β. Both cases are accounted for simply by replacing G

by the group
$$H := \left\{ \begin{pmatrix} \alpha & \beta \\ \gamma & \delta \end{pmatrix} : \alpha, \beta, \gamma, \delta \in \mathbb{Z}, \ \alpha\delta - \beta\gamma = 1 \text{ or } -1, \text{ and } \begin{pmatrix} \alpha & \beta \\ \gamma & \delta \end{pmatrix} \equiv I \bmod 2 \right\}.$$

Tracing back through the argument above, we find that a new triple of circles mutually tangent with each other and the one of curvature a have curvatures B, C, and D given by
$$\begin{pmatrix} B \\ D \\ C \end{pmatrix} = \begin{pmatrix} 1 & 0 & 0 \\ 1 & -2 & 1 \\ 0 & 0 & 1 \end{pmatrix} \begin{pmatrix} \alpha^2 & 2\alpha\gamma & \gamma^2 \\ \alpha\beta & \alpha\delta + \beta\gamma & \gamma\delta \\ \beta^2 & 2\beta\delta & \delta^2 \end{pmatrix} \begin{pmatrix} a+b \\ (a+b+c-d)/2 \\ a+c \end{pmatrix} - \begin{pmatrix} a \\ a \\ a \end{pmatrix}$$
for any $\begin{pmatrix} \alpha & \beta \\ \gamma & \delta \end{pmatrix} \in H$. We deduce that B equals
$$(\alpha^2 + \alpha\gamma + \gamma^2 - 1)a + (\alpha^2 + \alpha\gamma)b + (\alpha\gamma + \gamma^2)c - \alpha\gamma d,$$
C the same with α, γ replaced by β, δ, and D the same with α, γ replaced by $\alpha - \beta, \gamma - \delta$, respectively. Modulo 2, these are all different pairs.

Now consider the quadratic polynomial
$$f(m,n) = (m^2 + mn + n^2 - 1)a + (m^2 + mn)b + (mn + n^2)c - mnd,$$
when m and n are coprime integers.

(i) If m is odd and n is even, let $\alpha = m$ and $\gamma = n$ so there exist integers η, δ for which $\alpha\delta - \eta \cdot 2\gamma = 1$, and let $\beta = 2\eta$. Thus we have an element of H and $f(m,n) = B$.

(ii) If m is even and n is odd, let $\beta = m$ and $\delta = n$ so there exist integers η, α for which $\alpha\delta - \eta \cdot 2\beta = 1$, and let $\gamma = 2\eta$. Thus we have an element of H and $f(m,n) = C$.

(iii) If m and n are both odd, then there exist integers r, s for which $mr - ns = 1$. If r is odd, then s is even and we let $\delta = r, \beta = s$; if r is even, then we let $\delta = r + n, \beta = s + m$. Either way, δ is odd and β is even. We let $\alpha = \beta + m, \gamma = \delta + n$. Thus we have an element of H and $f(m,n) = B$.

Therefore the set of curvatures of the circles generated in this way is precisely the set of values of $f(m,n)$ as m and n vary over coprime integers. We can write $f(m,n) = g(m,n) - a$ where
$$g(m,n) = (a+b)m^2 + (a+b+c-d)mn + (a+c)n^2,$$
a binary quadratic form of discriminant $-4a^2$. Note that $(a+b)g(m,n) = M^2 + (an)^2$ where $M = (a+b)m + \frac{(a+b+c-d)}{2}n$.

In our 21, 24, 28 example this leads to the values of $45m^2 - 84mn + 49n^2 - 21$. Letting $m \to m - n$ gives $45m^2 + 6mn + 10n^2 - 21$.

In 1974 Iwaniec proved that any quadratic polynomial in two variables m and n, subject to certain obvious necessary conditions, takes on infinitely many prime values.[8] In our case this means that $f(m,n)$ takes on infinitely many prime values with $(m,n) = 1$ provided $(a,b,c,d) = 1$ and $a \neq 0$. We therefore deduce that for any circle of curvature a in our packing, there are infinitely many circles in the packing

---

[8]Iwaniec does not insist that $(m,n) = 1$ but this restriction can be incorporated into his proof.

that are tangent to a and have prime curvature. Take any one of those circles of prime curvature and make that the new value for a. Then the same calculation reveals that there are infinitely many circles of prime curvature in the packing that are tangent to our original circle of prime curvature. We have therefore proved the *Apollonian twin primes conjecture*.

Exercise 12.20.1. Suppose that we are given four mutually tangent circles with integer radii, and create an Apollonian circle packing from them. Prove that there is an infinite chain of distinct circles in the packing with *prime* curvatues $p_1 < p_2 < \cdots$ where the circles of curvatures p_m and p_{m+1} are mutually tangent, for every $m \geq 1$.

Further reading on Apollonian packings

[1] Elena Fuchs, *Counting problems in Apollonian packings*, Bull. Amer. Math. Soc. **50** (2013), 229–266.

[2] Alex Kontorovich, *From Apollonius to Zaremba: Local-global phenomena in thin orbits*, Bull. Amer. Math. Soc. **50** (2013), 187–228.

[3] Peter Sarnak, *Integral Apollonian packings*, Amer. Math. Monthly **118** (2011), 291–306.

Chapter 13

The anatomy of integers

One studies an organism's anatomy by breaking it up into its smallest possible meaningful indecomposable components. In biology that is the DNA, and then a subject is easily identified as each has unique DNA. For integers we study their prime factors: Every integer may be broken up in a unique way into primes, and every given set of primes (including repetitions) gives rise to a unique integer. Similar remarks might be made about polynomials (the components being the irreducible polynomials), and even permutations, which can be partitioned up into cycles.

In this chapter we study what the prime factors of a typical integer n look like: their quantity, their size, how they combine to make divisors of n, etc.; and we discuss several natural consequences of these musings.

13.1. Rough estimates for the number of integers with a fixed number of prime factors

The prime number theorem (as discussed in section 5.4) tells us that there are about $x/\log x$ integers up to x with exactly one prime factor. In other words, a proportion of 1 in $\log x$ of the integers up to x is prime, which is a vanishing proportion as $x \to \infty$. So primes are rare. One can use this approximation to estimate the number of integers up to x that are the product of two primes: If $pq \leq x$, where $p \leq q$ are prime, then $p \leq \sqrt{x}$ and $p \leq q \leq x/p$, and so the number of such integers is

$$\sum_{\substack{p \leq \sqrt{x} \\ p \text{ prime}}} \sum_{\substack{p \leq q \leq x/p \\ q \text{ prime}}} 1 = \sum_{\substack{p \leq \sqrt{x} \\ p \text{ prime}}} (\pi(x/p) - \pi(p) + 1).$$

Using the prime number theorem on the first term in this last sum, this is roughly

$$\sum_{\substack{p \leq \sqrt{x} \\ p \text{ prime}}} \frac{x/p}{\log x/p}, \text{ which is } \approx \frac{x}{\log x} \sum_{\substack{p \leq \sqrt{x} \\ p \text{ prime}}} \frac{1}{p},$$

as $\frac{1}{2}\log x \le \log x/p < \log x$ for primes $p \le \sqrt{x}$. This last sum is about $\log\log x$, by (5.12.4). The final two terms in the above sum add up to

$$\le \sum_{\substack{p \le \sqrt{x} \\ p \text{ prime}}} \pi(p) \le \pi(\sqrt{x})^2 \le \frac{cx}{(\log x)^2}$$

for some constant c, by (5.5.1). This is much smaller than the other term. Combining these estimates (and being more precise, though at the cost of quite a few details) one can prove that there are

$$\sim \frac{x}{\log x} \log\log x$$

integers up to x that are the product of two primes.

One can further develop this argument to prove by induction that, for any fixed integer $k \ge 1$, $N(x,k)$, the number of integers up to x with exactly k prime factors is

$$\sim \frac{x}{\log x} \cdot \frac{(\log\log x)^{k-1}}{(k-1)!}.$$

In other words, the probability that a randomly chosen positive integer $\le x$ has exactly $k = m+1$ prime factors, $\frac{1}{x}N(x, m+1) \sim e^{-\lambda}\lambda^m/m!$ where $\lambda = \log\log x$. This looks much like the *Poisson distribution*, from probability theory, with mean $\log\log x$. If this formula holds for all m (which can be allowed to grow as x grows), then we can deduce that a typical integer has around $\log\log x$ prime factors. It is hard to prove that this formula (or something like this formula) holds for all relevant m, but we can establish that a typical integer has around $\log\log x$ prime factors by different, simpler methods:

Exercise 13.1.1. Use Stirling's formula (exercise 4.15.1) to show that if m is the nearest integer to λ, then $e^{-\lambda}\lambda^m/m!$ is roughly $1/\sqrt{2\pi\lambda}$. This suggests the fact that if m is the closest integer to $\log\log x$, then there are roughly $x/\sqrt{2\pi \log\log x}$ integers up to x with exactly m prime factors.

13.2. The number of prime factors of a typical integer

The number $12 = 2^2 \times 3$ has two or three prime factors depending on whether one counts the 2^2 as one or two primes. So define

$$\omega(n) = \sum_{\substack{p \text{ prime} \\ p|n}} 1 \quad \text{and} \quad \Omega(n) = \sum_{\substack{p \text{ prime}, a \ge 1 \\ p^a|n}} 1:$$

$\omega(n)$ counts the number of distinct prime factors of n, while $\Omega(n)$ counts the number of distinct prime powers that divide n, so that $\omega(12) = 2$ and $\Omega(12) = 3$. On average

13.2. The number of prime factors of a typical integer

the difference between these two is

$$\frac{1}{x} \sum_{n \leq x} \sum_{\substack{p \text{ prime}, a \geq 2 \\ p^a | n}} 1 = \frac{1}{x} \sum_{\substack{p \text{ prime} \\ a \geq 2}} \sum_{\substack{n \leq x \\ p^a | n}} 1 = \frac{1}{x} \sum_{\substack{p \text{ prime} \\ a \geq 2}} \left[\frac{x}{p^a}\right]$$

$$\leq \sum_{p \text{ prime}, a \geq 2} \frac{1}{p^a} = \sum_{p \text{ prime}} \frac{1}{p(p-1)} \leq \sum_{n \geq 2} \frac{1}{n(n-1)} = 1$$

so we can work with either, as is convenient. It will make little difference.

The average number of distinct prime factors of an integer $\leq x$ can be obtained from calculating

$$\sum_{n \leq x} \omega(n) = \sum_{n \leq x} \sum_{\substack{p \text{ prime} \\ p|n}} 1 = \sum_{p \text{ prime}} \sum_{\substack{n \leq x \\ p|n}} 1 = \sum_{p \text{ prime}} \left[\frac{x}{p}\right].$$

Hence the average is approximately

$$\frac{1}{x} \sum_{\substack{p \text{ prime} \\ p \leq x}} \frac{x}{p} = \sum_{\substack{p \text{ prime} \\ p \leq x}} \frac{1}{p} \approx \log \log x$$

by (5.12.4).

Exercise 13.2.1. (a) Prove that

$$\sum_{\substack{p \text{ prime} \\ p \leq x}} \left(\frac{x}{p} - \left[\frac{x}{p}\right]\right) \leq \pi(x).$$

(b) Deduce that

$$\lim_{x \to \infty} \left| \frac{1}{x} \sum_{n \leq x} \omega(n) - \sum_{\substack{p \text{ prime} \\ p \leq x}} \frac{1}{p} \right| = 0.$$

Calculations suggest that most integers have very few prime factors, but sometimes one gets limited information from calculations, especially for a function that we expect to grow like $\log \log x$. As Dan Shanks wrote: "*Numbers at infinity are quite different from those that we see down here: the average number of their prime divisors increases like $\log \log x$ and, while that increases very slowly, it increases without bound.*"

We are going to go one step further and ask how much $\omega(n)$ varies from its mean; that is, we are going to compute the statistical quantity, the *variance*, using the following standard identity:

Exercise 13.2.2. Show that if $a_1, \ldots, a_N$ have mean m, then

$$\frac{1}{N} \sum_{n \leq N} (a_n - m)^2 = \frac{1}{N} \sum_{n \leq N} a_n^2 - m^2.$$

Therefore the variance of $\omega(n)$ from its mean, over the integers $n \leq x$, is

$$(13.2.1) \quad \frac{1}{x}\sum_{n\leq x}\left(\omega(n) - \frac{1}{x}\sum_{m\leq x}\omega(m)\right)^2 = \frac{1}{x}\sum_{n\leq x}\omega(n)^2 - \left(\frac{1}{x}\sum_{m\leq x}\omega(m)\right)^2.$$

The first term here is

$$\frac{1}{x}\sum_{n\leq x}\omega(n)^2 = \frac{1}{x}\sum_{n\leq x}\sum_{\substack{p\text{ prime}\\p|n}}\sum_{\substack{q\text{ prime}\\q|n}} 1 = \frac{1}{x}\sum_{p\text{ prime}}\left(\sum_{\substack{q\text{ prime}\\q=p}}\left[\frac{x}{p}\right] + \sum_{\substack{q\text{ prime}\\q\neq p}}\left[\frac{x}{pq}\right]\right)$$

$$\leq \sum_{\substack{p\text{ prime}\\p\leq x}}\frac{1}{p} + \sum_{p\text{ prime}}\sum_{\substack{q\text{ prime},\,q\neq p\\pq\leq x}}\frac{1}{pq} \leq \sum_{\substack{p\text{ prime}\\p\leq x}}\frac{1}{p} + \left(\sum_{\substack{p\text{ prime}\\p\leq x}}\frac{1}{p}\right)^2.$$

The second term is roughly $\left(\frac{1}{x}\sum_{m\leq x}\omega(m)\right)^2$, and so the variance is bounded by the first term. Making this argument more precise one can show that

$$(13.2.2) \quad \frac{1}{x}\sum_{n\leq x}(\omega(n) - \log\log x)^2 \leq 2\log\log x.$$

Exercise 13.2.3. ("Almost all" integers n have about $\log\log n$ prime factors.) Show that (13.2.2) implies that there are $< 2x/(\log\log x)^{1/3}$ integers $n \leq x$ for which $|\omega(n) - \log\log x| \geq (\log\log x)^{2/3}$. In other words, we have $|\omega(n) - \log\log x| < (\log\log x)^{2/3}$ for all but at most $< 2x/(\log\log x)^{1/3}$ integers $n \leq x$. This is a famous result of Hardy and Ramanujan (we will develop their proof in section 13.4).

In section 4.15 we saw that although the value of $\tau(n)$ bounces around a lot, the mean value looks like $\log n + 2\gamma - 1$, with a very small error term. By exercise 4.3.9 we have

$$2^{\omega(n)} \leq \tau(n) \leq 2^{\Omega(n)},$$

and we have just shown that $\omega(n)$ and $\Omega(n)$ are close to $\log\log n$ for almost all integers n in exercise 13.2.3. This implies that

$$\tau(n) \text{ is roughly } 2^{\log\log n} = (\log n)^{\log 2} \text{ for almost all integers } n.$$

Since $\log 2 = 0.693\ldots$, this seems wildly inconsistent compared to what we know about the mean value of $\tau(n)$.

Exercise 13.2.4. Explain, by creating a simpler but analogous example, how it is possible that $\tau(n)$ can usually take values around $(\log n)^{\log 2}$, but averages about $\log n$. (You might think of 100 students taking an exam in which most do poorly, but one does well.)

13.3. The multiplication table problem

When you were young, you probably had to learn the multiplication table by heart, with perhaps all the values of $a \times b$ for $1 \leq a, b \leq 12$ written in a grid like

×	1	2	3	4	5	6	7	8	9	10	11	12
1	1	2	3	4	5	6	7	8	9	10	11	12
2	2	4	6	8	10	12	14	16	18	20	22	24
3	3	6	9	12	15	18	21	24	27	30	33	36
4	4	8	12	16	20	24	28	32	36	40	44	48
5	5	10	15	20	25	30	35	40	45	50	55	60
6	6	12	18	24	30	36	42	48	54	60	66	72
7	7	14	21	28	35	42	49	56	63	70	77	84
8	8	16	24	32	40	48	56	64	72	80	88	96
9	9	18	27	36	45	54	63	72	81	90	99	108
10	10	20	30	40	50	60	70	80	90	100	110	120
11	11	22	33	44	55	66	77	88	99	110	121	132
12	12	24	36	48	60	72	84	96	108	120	132	144

If you were Paul Erdős, you might have quickly become bored waiting for the others to learn it and asked yourself other questions. For example, how many different integers are there in the table? There is the obvious symmetry down the diagonal (as $a \times b = b \times a$), meaning that one only need look at the upper triangle for distinct entries. One spots other coincidences, like $3 \times 4 = 2 \times 6$ and $4 \times 5 = 2 \times 10$, and wonders how many there are. In other words, what percentage of the integers up to N^2 appear in the N-by-N multiplication table? More precisely define

$$p(N) := \frac{1}{N^2} \#\{n \leq N^2 : \text{there exist } a, b \leq N \text{ for which } ab = n\}.$$

For $N = 6$ there are 18 distinct entries; that is, $p(6) = 1/2$. For $N = 10$ there are 42 distinct entries, so that $p(10) = .42$. For $N = 12$ we have 59, so $p(12) \approx .41$. Then $p(25) = .36$, $p(50) = .32$, $p(75) \approx .306$, $p(100) \approx .291$, $p(250) \approx .270$, $p(500) \approx .259$, $p(1000) \approx .248$. Do these proportions tend to a limit and, if so, can one guess what the limit is? Erdős proved that the limit exists and equals 0, and his proof is beautiful:[1]

The idea is simply that almost all integers up to N have about $\log \log N$ prime factors and so the product of two such integers has about $2 \log \log N$ prime factors. However almost all integers up to N^2 have about $\log \log N^2 = \log \log N + \log 2$ prime factors and so cannot be the product of two typical integers $\leq N$.

Exercise 13.3.1. Give a more formal version of Erdős's proof.

---

[1] True mathematicians are motivated by elegant proofs, none more so than the great Paul Erdős. He used to say that "the supreme being" keeps a book which contains all of the most beautiful proofs of each theorem. We mortals are only occasionally allowed to glimpse this book, when we discover an extraordinary proof. Erdős's proof of the multiplication table theorem is truly from the book (and see [**AZ18**] for more examples).

13.4. Hardy and Ramanujan's inequality

Hardy and Ramanujan proved the general upper bound: For all $k \geq 1$ we have

$$(13.4.1) \qquad \pi(x,k) := \#\{n \leq x : \omega(n) = k\} \leq c_0 \frac{x}{\log x} \frac{(\log \log x + c_1)^{k-1}}{(k-1)!}$$

for certain constants c_0 and c_1. This is not difficult to prove by induction on $k \geq 1$.

For $k = 1$ this follows from Chebyshev's upper bound on $\pi(x)$ given in (5.5.1).

For larger k, suppose that $n = p_1^{e_1} \cdots p_k^{e_k} \leq x$ with $p_1^{e_1} < \cdots < p_k^{e_k}$. If $j \leq k-1$, then $(p_j^{e_j})^2 < p_j^{e_j} p_k^{e_k} \leq n \leq x$ and so $p_j^{e_j} \leq \sqrt{x}$. Moreover $m_j = n/p_j^{e_j}$ satisfies $m_j \leq x/p_j^{e_j}$ and $\omega(m_j) = k-1$. Therefore

$$(k-1)\pi(x,k) \leq \sum_{p^e \leq \sqrt{x}} \pi(x/p^e, k-1) \leq \sum_{p^e \leq \sqrt{x}} c_0 \frac{x/p^e}{\log x/p^e} \frac{(\log \log(x/p^e) + c_1)^{k-2}}{(k-2)!},$$

by the induction hypothesis. We now use the bound $\log \log(x/p^e) \leq \log \log x$, and

$$(13.4.2) \qquad \sum_{p^e \leq \sqrt{x}} \frac{1}{p^e \log(x/p^e)} \leq \frac{\log \log x + c_1}{\log x},$$

which follows (after some work) from (5.12.4) with c_1 sufficiently large, to obtain the desired bound (13.4.1). □

Exercise 13.4.1.[†] Give another proof that "almost all" integers n have about $\log \log n$ prime factors using (13.4.1).

We noted earlier that $\omega(n)$ and $\Omega(n)$ are typically close together, so we might expect that $\pi(x,k)$ and $N(x,k) := \#\{n \leq x : \Omega(n) = k\}$ are close. This is true when $k \leq (1+\epsilon) \log \log x$. For larger k we now give a lower bound on $N(x,k)$ that is far larger than the upper bound (13.4.1) on $\pi(x,k)$:

Consider the set of integers $n \leq x$ of the form $2^{k-\ell} m$ where $\Omega(m) = \ell$ and $m \leq x/2^{k-\ell}$. Therefore $\Omega(n) = k-\ell + \Omega(m) = k$ and so

$$N(x,k) \geq N(x/2^{k-\ell}, \ell).$$

We noted earlier (see exercise 13.1.1) that the largest values of $\pi(X, \ell)$ and $N(X, \ell)$ occur when ℓ is close to $\log \log X$ in which case there exists a constant $c_3 > 0$ such that $\pi(X, \ell), N(X, \ell) \geq c_3 X/\sqrt{\log X}$ if $X = x/2^{k-\ell}$ is sufficiently large. Now suppose that $\sqrt{\log x} > k > \ell$, so that if $\epsilon > 0$, then $x > X = x/2^{k-\ell} > x^{1-\epsilon}$ once x is sufficiently large. Therefore, we deduce that

$$(13.4.3) \qquad N(x,k) \geq \frac{c_3 x}{2^{k-\ell} \sqrt{\log \log x}} \geq c_4 \frac{x}{2^k} (\log x)^{\log 2},$$

for some constants $c_3, c_4 > 0$, when $\ell > \log \log x + \log \log \log x$.

Exercise 13.4.2.[‡] Let $k = [A \log \log x]$. Use (13.4.1) together with exercise 4.15.1(d) to give an upper bound on $\pi(x,k)$, and then use (13.4.3) to give a lower bound on $N(x,k)$. Deduce that once A satisfies $1 + A(\log A - 1) > (A-1)\log 2$, then $N(x,k) > 2\pi(x,k)$ for x sufficiently large.

Appendix 13A. Other anatomies

There are features of the anatomies of certain other mathematical objects, when broken up into their indecomposable components, that are very similar to the integers. We explore that briefly here; for more information, presented in a rather different format, see the graphic novel [**2**].

13.5. The anatomy of polynomials in finite fields

Monic polynomials (over $\mathbb{C}$ or in $\mathbb{F}_p$) can be factored in a unique way (up to order) into monic irreducible polynomials. There are p^n monic polynomials of degree n in $\mathbb{F}_p$, and in (4.12.3) of appendix 4C, we showed the number of monic irreducible polynomials of degree n in $\mathbb{F}_p$ is $\frac{1}{n}\sum_{d|n}\mu(d)p^{n/d}$. This is close to p^n/n; that is, roughly 1 out of every n polynomials of degree n is irreducible. We "calibrate" this with the proportion $1/\log x$ of integers up to x that are prime to compare the anatomies of polynomials in finite fields with those of integers.

In appendix 4D we showed that, on average, integers $\leq x$ have about $\log x$ divisors. For the analogous result, note that if a given monic polynomial $f(x)$ of degree m divides a monic polynomial of degree n, then it can be written as $f(x)g(x)$ for some monic polynomial $g(x)$ of degree $n-m$. Therefore, the average number of monic polynomials dividing a monic polynomial of degree n is

$$\frac{1}{p^n} \sum_{\substack{h(x) \text{ monic} \\ \text{of degree } n}} \sum_{m=0}^{n} \sum_{\substack{f(x) \text{ monic} \\ \text{of degree } m \\ f(x) \text{ divides } h(x)}} 1 = \frac{1}{p^n} \sum_{m=0}^{n} \sum_{\substack{f(x) \text{ monic} \\ \text{of degree } m}} \sum_{\substack{g(x) \text{ monic} \\ \text{of degree } n-m}} 1$$

$$= \frac{1}{p^n} \sum_{m=0}^{n} p^m \cdot p^{n-m} = n+1,$$

which was calibrated with $\log x + 1$.

The number of monic polynomials of degree n with exactly two irreducible monic polynomial factors is

$$\frac{1}{2}\sum_{d=1}^{n-1} \sum_{\substack{f(x) \text{ monic irreducible} \\ \text{of degree } d}} 1 \cdot \sum_{\substack{g(x) \text{ monic irreducible} \\ \text{of degree } n-d}} 1$$

less the cases where $f = g$. Now the formula for the number of monic irreducibles is complicated so let's just work with main terms, so we see that the above is roughly

$$\frac{1}{2}\sum_{d=1}^{n-1} \frac{p^d}{d} \cdot \frac{p^{n-d}}{n-d} - \frac{p^{n/2}}{n/2} \approx \frac{p^n}{2}\sum_{d=1}^{n-1} \frac{1}{d(n-d)} = \frac{p^n}{n}\sum_{d=1}^{n-1} \frac{1}{d} \approx \frac{p^n}{n}\log n.$$

This can be compared with the number of integers $\leq x$ with exactly two prime factors, $\approx \frac{x}{\log x} \log\log x$, which gives the same proportion of the total number, replacing n by $\log x$. A similar argument yields that the number of monic polynomials of degree n in $\mathbb{F}_p$ with exactly k irreducible monic polynomial factors is roughly $\frac{p^n}{n} \cdot \frac{(\log n)^{k-1}}{(k-1)!}$, at least if k is not too large (in terms of n).

Let $\omega(F)$ denote the number of distinct monic irreducible factors of F. The mean value of $\omega(F)$, over monic F of degree n, is

$$= \frac{1}{p^n} \sum_{\substack{F(x) \text{ monic} \\ \text{of degree } n}} \sum_{m=1}^{n} \sum_{\substack{g(x) \text{ monic irreducible} \\ \text{of degree } m \\ g \text{ divides } F}} 1 = \frac{1}{p^n} \sum_{m=1}^{n} \sum_{\substack{g(x) \text{ monic irreducible} \\ \text{of degree } m}} \sum_{\substack{F(x) \text{ monic} \\ \text{of degree } n \\ g \text{ divides } F}} 1$$

$$= \frac{1}{p^n}\sum_{m=1}^{n} \sum_{\substack{g(x) \text{ monic irreducible} \\ \text{of degree } m}} p^{n-m} = \sum_{m=1}^{n} \frac{1}{p^m} \cdot \frac{1}{m}\sum_{d|m} \mu(d)p^{m/d} \approx \sum_{m=1}^{n} \frac{1}{m} \approx \log n,$$

taking only the $d = 1$ terms.

One can then prove, in one of several ways (analogous to how we approached the prime factors of integers), that the variance is also about $\log n$, and so almost all polynomials of degree n in $\mathbb{F}_p$ have about $\log n$ distinct monic irreducible factors.

Exercise 13.5.1. Sketch a proof that almost all polynomials in $\mathbb{F}_p$ of degree $2d$ are *not* the product of two polynomials of degree d, as d gets large.

13.6. The anatomy of permutations

Permutations can be represented as a product of cycles in a unique way, and a given set of cycles defines a permutation. A cycle is an irreducible permutation. Let S_N be the set of permutations on N letters. The number of permutations on N letters is $|S_N| = N!$. There are $(N-1)!$ cycles on N letters, since the first letter can be sent to any of the other $N-1$ letters, that letter to any of the $N-2$ remaining letters, etc. The cycles form a proportion $(N-1)!/|S_N| = (N-1)!/N! = 1/N$ of all the permutations in S_N. We "calibrate" this with the proportion $1/\log x$ of integers up to x that are prime to compare the anatomies of permutations with those of integers.

If $n = p_1 \cdots p_k$, the factorization of squarefree n into primes, then each divisor can be written as $p_{j(1)} \cdots p_{j(\ell)}$ for some $\{j(1), \ldots, j(\ell)\}$ of $\{1, \ldots, k\}$ (and each such

13.6. The anatomy of permutations

product gives a divisor of n). In this language, the analogy for permutations would therefore be: If $\sigma = C_1 \cdots C_k$, the factorization of σ into cycles, then each divisor can be written as $C_{j(1)} \cdots C_{j(\ell)}$ for some subset $\{j(1), \ldots, j(\ell)\}$ of $\{1, \ldots, k\}$. This set of cycles acts on some subset S of the N letters, permuting the elements of S (and of the complementary set, T). Thus the *divisors* of $\sigma \in S_N$ are the subsets S of the N letters that are fixed by σ. If σ is a cycle, then the only subsets it fixes are $\emptyset$ and itself, very much in analogy with how we define primes. The average number of divisors of a permutation of a set Λ of N letters is

$$\frac{1}{N!} \sum_{\sigma \in S_N} \sum_{\substack{S \cup T = \Lambda \\ \sigma \text{ fixes } S \text{ and } T}} 1.$$

If $|S| = k$, then there are $k! \cdot (N-k)!$ permutations of S and T, and so this is the number of $\sigma \in \Lambda$ which fix S and T. Therefore the above equals

$$\frac{1}{N!} \sum_{k=0}^{N} \sum_{\substack{S \subset \Lambda \\ |S|=k}} k! \cdot (N-k)! = \frac{1}{N!} \sum_{k=0}^{N} \binom{N}{k} \cdot k! \cdot (N-k)! = \sum_{k=0}^{N} 1 = N+1.$$

The number of permutations with exactly two cycles is

$$= \frac{1}{2} \sum_{k=1}^{N-1} (k-1)!(N-k-1)! \sum_{\substack{S \cup T = \Lambda \\ |S|=k}} 1 = \frac{1}{2} \sum_{k=1}^{N-1} (k-1)!(N-k-1)! \binom{N}{k}$$

$$= \frac{N!}{2} \sum_{k=1}^{N-1} \frac{1}{k(N-k)} = \frac{N!}{2N} \sum_{k=1}^{N-1} \left(\frac{1}{k} + \frac{1}{N-k} \right) = \frac{N!}{N} \sum_{k=1}^{N-1} \frac{1}{k} \approx \frac{N!}{N} \cdot \log N.$$

A similar argument yields that the number of permutations with exactly k cycles is

$$\frac{N!}{N} \frac{1}{(k-1)!} \sum_{\substack{a_1, \ldots, a_{k-1} \geq 1 \\ a_1 + \cdots + a_{k-1} \leq N-1}} \frac{1}{a_1 \cdots a_{k-1}} \approx \frac{N!}{N} \cdot \frac{(\log N)^{k-1}}{(k-1)!},$$

at least if k is not too large (in terms of N), as we prove in the following exercise:

Exercise 13.6.1. (a) Prove that

$$\left(\sum_{a \leq A/m} \frac{1}{a} \right)^m \leq \sum_{\substack{a_1, \ldots, a_m \geq 1 \\ a_1 + \cdots + a_m \leq A}} \frac{1}{a_1 \cdots a_{k-1}} \leq \left(\sum_{a \leq A} \frac{1}{a} \right)^m.$$

(b)‡ Prove that if $m \leq \frac{\log A}{(\log \log A)^2}$, then the two terms at either end of the inequalities in (a) differ by a multiplicative factor which gets arbitrarily close to 1 as A grows.

We will now determine the average number of cycles in a permutation. First note that the number of permutations containing a given cycle C of length k is $(N-k)!$, since one determines all the ways that σ can act on the letters not acted on by C. The number of cycles of length k is $\binom{N}{k}(k-1)! = \frac{N!}{(N-k)!\, k}$. Therefore the average number of cycles per permutation of S_N is

$$\frac{1}{N!}\sum_{\sigma\in S_N}\sum_{\substack{C\text{ a cycle}\\C\in\sigma}}1 = \frac{1}{N!}\sum_{\sigma\in S_N}\sum_{k=1}^{N}\sum_{\substack{C\text{ a cycle}\\|C|=k\\C\in\sigma}}1 = \frac{1}{N!}\sum_{k=1}^{N}\sum_{|C|=k}\sum_{\substack{\sigma\in S_N\\C\in\sigma}}1$$

$$= \frac{1}{N!}\sum_{k=1}^{N}\frac{N!}{(N-k)!\,k}\cdot(N-k)! = \sum_{k=1}^{N}\frac{1}{k}\approx\log N.$$

To determine the variance, we calculate

$$\frac{1}{N!}\sum_{\sigma\in S_N}\left(\sum_{\substack{C\text{ a cycle}\\C\in\sigma}}1\right)^2 = \frac{1}{N!}\sum_{\sigma\in S_N}\sum_{\substack{C\text{ a cycle}\\C\in\sigma}}1 + \frac{1}{N!}\sum_{\sigma\in S_N}\sum_{\substack{C\cup D\text{ disjoint cycles}\\C\cup D\in\sigma}}1.$$

We just calculated the first term. For the second we note that given $C\cup D$ with $|C|=k$, $|D|=\ell$, the number of $\sigma\in S_N$ with $C\cup D\in\sigma$ in $(N-(k+\ell))!$. The number of pairs of disjoint cycles $C\cup D$ with $|C|=k$, $|D|=\ell$ is $\frac{N!}{(N-k)!\,k}\cdot\frac{(N-k)!}{(N-k-\ell)!\,\ell} = \frac{N!}{(N-k-\ell)!\,k\ell}$. Therefore the second term in the last displayed equation equals

$$\frac{1}{N!}\sum_{\substack{k,\ell\geq 1\\k+\ell\leq N}}\frac{N!}{(N-k-\ell)!\,k\ell}\cdot(N-(k+\ell))! = \sum_{\substack{k,\ell\geq 1\\k+\ell\leq N}}\frac{1}{k\ell}.$$

Therefore the variance equals

$$\sum_{k=1}^{N}\frac{1}{k}+\sum_{\substack{k,\ell\geq 1\\k+\ell\leq N}}\frac{1}{k\ell}-\left(\sum_{k=1}^{N}\frac{1}{k}\right)^2 < \sum_{k=1}^{N}\frac{1}{k}\approx\log N.$$

We deduce that almost all permutations on N letters have about $\log N$ cycles.

Exercise 13.6.2. Prove, by taking $m=k+\ell$, that

$$0<\left(\sum_{k=1}^{N}\frac{1}{k}\right)^2-\sum_{\substack{k,\ell\geq 1\\k+\ell\leq N}}\frac{1}{k\ell} = \sum_{\substack{1\leq k,\ell\leq N\\k+\ell>N}}\frac{1}{k\ell} = 2\sum_{m=N+1}^{2N}\frac{1}{m}\sum_{k=m-N}^{N}\frac{1}{k} = 2\sum_{k=1}^{N}\frac{1}{k}\sum_{m=N+1}^{N+k}\frac{1}{m} < 2.$$

There are many other aspects of the anatomies of polynomials in finite fields, and of permutations, that mirror the anatomy of integers.

More on mathematical anatomies

[1] Richard Arratia, A. D. Barbour, and Simon Tavaré, *Random combinatorial structures and prime factorizations*, Notices Amer. Math. Soc. **44** (1997), 903–910.

[2] Andrew Granville, Jennifer Granville, and Robert J. Lewis, *Prime suspects: The anatomy of integers and permutations*, Princeton University Press, 2019.

[3] Anatoly M. Vershik, *Asymptotic combinatorics and algebraic analysis*, Proc ICM Zurich (1994), 1384–1394.

Appendix 13B. Dirichlet L-functions

In section 8.17 of appendix 8D we noted that the infinite series

$$L(s,\chi) := \sum_{n\geq 1} \frac{\chi(n)}{n^s}$$

for the Dirichlet L-function $L(s,\chi)$ is absolutely convergent whenever $\mathrm{Re}(s) > 1$. However, in the subsequent sketch of Dirichlet's proof that there are infinitely many primes in arithmetic progressions $a \pmod{q}$ with $(a,q) = 1$, it was necessary to establish that every $L(s,\chi)$ with χ non-principal can be analytically continued into a region that contains $s = 1$. We will prove this here.

13.7. Dirichlet series

To determine the value of $L(1,\chi)$, we sum its series in order, taking q terms at a time:

$$\sum_{n=kq+1}^{kq+q} \frac{\chi(n)}{n} = \sum_{n=kq+1}^{kq+q} \left(\frac{\chi(n)}{kq+1} - \chi(n)\left(\frac{1}{kq+1} - \frac{1}{n}\right) \right).$$

Summing up the first term here gives $\frac{1}{kq+1} \sum_{n=kq+1}^{kq+q} \chi(n) = 0$ as we saw in (8.16.2). We bound the second term by noting that $|-\chi(n)| \leq 1$, so that

$$\left| \sum_{n=kq+1}^{kq+q} \frac{\chi(n)}{n} \right| \leq \sum_{n=kq+1}^{kq+q} \left(\frac{1}{kq+1} - \frac{1}{n} \right) < q\left(\frac{1}{kq+1} - \frac{1}{kq+q+1} \right).$$

Therefore, applying this for $k = 1, 2, 3, \ldots$ we obtain

$$\left| \sum_{n \geq 1} \frac{\chi(n)}{n} \right| \leq \sum_{n=1}^{q} \frac{1}{n} + \sum_{k \geq 1} \left| \sum_{n=kq+1}^{kq+q} \frac{\chi(n)}{n} \right|$$

$$< (\log q + 1) + q \sum_{k \geq 1} \left(\frac{1}{kq+1} - \frac{1}{kq+q+1} \right) < \log q + 2$$

since the last sum is telescoping. We have proved that if one sums the series q terms at a time, then the new sum is absolutely convergent.

We deduce from Dirichlet's class number formula (section 12.15 of appendix 12D) that if $d > 0$, then

$$\log \epsilon_d \leq h(d) \log \epsilon_d = \sqrt{d}\, L(1, (d/\cdot)) \leq \sqrt{d}(\log d + 2),$$

as $h(d) \geq 1$, and so $\epsilon_d \leq (8d)^{\sqrt{d}}$, which is of a similar order of magnitude to what we observed from the data in section 11.13 in appendix 11B.

More general s. Taking q terms at a time we can also give a valid definition of $L(s, \chi)$ for any $s \in \mathbb{C}$ with $\text{Re}(s) > 0$ and χ non-principal, and we can justify this by much the same argument. Therefore we define

(13.7.1) $$L(s, \chi) = \sum_{k \geq 0} A_k(s, \chi), \text{ where } A_k(s, \chi) := \sum_{j=1}^{q} \frac{\chi(kq+j)}{(kq+j)^s}.$$

Since $\chi(kq + j) = \chi(j)$ for each j, we have

$$A_k(s, \chi) = \sum_{j=1}^{q} \left(\frac{\chi(j)}{(kq+1)^s} + \chi(j) \left(\frac{1}{(kq+j)^s} - \frac{1}{(kq+1)^s} \right) \right).$$

The first terms sum to 0. For the second term,

$$\frac{1}{(kq+j)^s} - \frac{1}{(kq+1)^s} = -s \int_{1}^{j} \frac{dt}{(kq+t)^{s+1}}.$$

Taking absolute values, where $\sigma = \text{Re}(s) > 0$, this is

$$\leq |s| \int_{1}^{j} \frac{dt}{(kq+t)^{\sigma+1}} = \frac{|s|}{\sigma} \left(\frac{1}{(kq+1)^\sigma} - \frac{1}{(kq+j)^\sigma} \right)$$

since $|n^s| = n^\sigma$. Substituting this bound in above and noting that each $|\chi(j)| \leq 1$, we have

$$|A_k(s, \chi)| \leq \sum_{j=1}^{q} \frac{|s|}{\sigma} \left(\frac{1}{(kq+1)^\sigma} - \frac{1}{(kq+j)^\sigma} \right) < \frac{q|s|}{\sigma} \left(\frac{1}{(kq+1)^\sigma} - \frac{1}{(kq+q+1)^\sigma} \right).$$

Summing this over each $k \geq 0$ we deduce that

$$\sum_{k \geq 0} |A_k(s, \chi)| < \frac{q|s|}{\sigma} \sum_{k \geq 0} \left(\frac{1}{(kq+1)^\sigma} - \frac{1}{(kq+q+1)^\sigma} \right) = \frac{q|s|}{\sigma}.$$

Therefore the new sum in (13.7.1) defining $L(s, \chi)$ is absolutely convergent for any s with $\text{Re}(s) > 0$, meaning that we can unambiguously decide on its value throughout

13.7. Dirichlet series

this domain.[2] We have *analytically continued* these L-functions; to do so for all $s \in \mathbb{C}$ is beyond the scope of this book (but see [**Graa**]).

Exercise 13.7.1. Let $\sigma = \mathrm{Re}(s) > 0$. Prove that

$$|L(s,\chi)| \leq \begin{cases} \frac{\sigma}{\sigma-1} & \text{if } \sigma > 1, \\ \log q + |s| + 1 & \text{if } \sigma = 1, \\ \frac{\sigma}{\sigma-1} + q^{1-\sigma}\left(\frac{1}{1-\sigma} + \frac{|s|}{\sigma}\right) & \text{if } 0 < \sigma < 1. \end{cases}$$

In section 8.17 of appendix 8D we sketched Dirichlet's proof that there are infinitely many primes in arithmetic progressions. One begins with the (absolutely convergent) Euler product representation

$$L(s,\chi) = \prod_{p \text{ prime}} \left(1 - \frac{\chi(p)}{p^s}\right)^{-1},$$

for $\mathrm{Re}(s) > 1$ (see section 4.9 of appendix 4B), and takes the logarithm to obtain

$$\log L(s,\chi) = \sum_{\substack{p \text{ prime} \\ p \leq x}} \sum_{k \geq 1} \frac{\chi(p^k)}{kp^{ks}}.$$

We have seen that $|L(1,\chi)|$ is bounded. If, also, $L(1,\chi) \neq 0$, then we know that $\log L(1,\chi)$ takes a bounded value, and so we can let $s \to 1^+$ above. The terms with $k \geq 2$ contribute $\leq \sum_{n \geq 2} \sum_{k \geq 1} \frac{1}{n^k} = \sum_{n \geq 2} \frac{1}{n(n-1)} = 1$. Therefore, we deduce that

$$\sum_{\substack{p \text{ prime} \\ p \leq x}} \frac{\chi(p)}{p} \quad \text{converges as} \quad x \to \infty$$

(as in (8.17.2)), and from there Dirichlet's proof is given in section 8.17 of appendix 8D. It remains to prove that $L(1,\chi) \neq 0$ for every non-principal Dirchlet character $\chi \pmod{q}$.

Exercise 13.7.2.[†] (a) Prove that if $\sigma > 1$, then

$$\frac{1}{\phi(q)} \sum_{\chi \pmod{q}} \log L(\sigma, \chi) = \sum_{p^k \equiv 1 \pmod q} \frac{1}{kp^{k\sigma}}.$$

(b) Deduce that $\prod_{\chi \pmod q} L(s,\chi)$ is non-zero at $s=1$.
(c) Prove that if $L(1,\chi) = 0$, then $L(1,\overline{\chi}) = 0$.
(d)[‡] Deduce that if $L(1,\chi) = 0$, then χ is real.

The last exercise leaves us only needing to prove that $L(1,(d/\cdot)) \neq 0$. This task presented an enormous challenge to Dirichlet but culminated in his brilliant class number formulas (see section 12.15 of appendix 12D), which established that $L(1,(d/\cdot)) \neq 0$ as each class number $h(d) \geq 1$ (since one always has the principal form of fundamental discriminant d).

---

[2] What is remarkable is that no matter how we rewrite $L(s,\chi)$, the new expression takes the same value in the wider domain as any other, provided the new definition converges absolutely in that domain.

Further exercises

Exercise 13.7.3.[†] (a) Show that for any integer $a, 1 \leq a \leq 9$, there are $\sim x/9$ integers $\leq x$ whose leading digit is a, where $x = 10^n$ and integer $n \to \infty$.
(b) Show that there are $\sim 5x/9$ integers $\leq x$ whose leading digit is 1, where $x = 2 \cdot 10^n$ and integer $n \to \infty$.
(c) What can we say about the density of integers whose leading digit is 1?
(d) The *logarithmic density* of a set S of positive integers up to x is given by $\frac{1}{\log x} \sum_{n \leq x,\ n \in S} \frac{1}{n}$. For any given integer a, $1 \leq a \leq 9$, let S_a be the set of integers with leading coefficient a. Prove that the logarithmic density of S_a, namely

$$\lim_{x \to \infty} \frac{1}{\log x} \sum_{n \leq x,\ n \in S_a} \frac{1}{n} \text{ exists and equals } \frac{\log(1 + 1/a)}{\log 10}.$$

Exercise 13.7.4. Show that $\prod_{n=2}^{N} \frac{n^3 - 1}{n^3 + 1} = \frac{2}{3}\left(1 + \frac{1}{N(N+1)}\right)$.

Exercise 13.7.5.[†] (a) Use Theorem 5.3 to establish that there exist constants $0 < c_1 < c_2$ such that if $x \geq 2$, then

$$c_1 < \sum_{x < p \leq 3x} \frac{\log p}{p} < c_2.$$

(b) Deduce that there exist constants $0 < c_3 < c_4$ such that if $x \geq 6$, then

$$c_3 \log x < \sum_{p \leq x} \frac{\log p}{p} < c_4 \log x.$$

(c) In section 13.1 we claimed that $\sum_{\substack{p \leq \sqrt{x} \\ p \text{ prime}}} \frac{x/p}{\log x/p}$ is well-approximated by $\frac{x}{\log x} \sum_{\substack{p \leq \sqrt{x} \\ p \text{ prime}}} \frac{1}{p}$. Show that there exists a constant $c_5 > 0$ such that the difference between these two expressions is $\leq c_5 x$.
(d) Prove (13.4.2).

Exercise 13.7.6. (a) Show that every integer n can be written as mr where m is powerful, r is squarefree, and $(m, r) = 1$; and deduce that $\Omega(n) - \omega(n) = \Omega(m) - \omega(m)$.
(b) Prove that there are $\leq x/m$ integers $n \leq x$ of the form mr as in (a).
(c) Prove that if $\Omega(m) - \omega(m) \geq k$, then $m \geq 2^{k+1}$. Deduce that

$$\frac{1}{x} \#\{n \leq x : \Omega(n) - \omega(n) \geq k\} \leq \sum_{\substack{m \text{ powerful} \\ m \geq 2^{k+1}}} \frac{1}{m}.$$

(d) Prove that every powerful number m can be written as $a^2 b^3$ for some integers a and b.
(e) Deduce that if $a^2 b^3 \geq 2^{k+1}$, then $a \geq 2^{k/4}$ or $b \geq 2^{k/6}$, and therefore that

$$\sum_{\substack{a,b \geq 1 \\ a^2 b^3 \geq 2^{k+1}}} \frac{1}{a^2 b^3} \leq \sum_{a \geq 2^{k/4}} \frac{1}{a^2} \sum_{b \geq 1} \frac{1}{b^3} + \sum_{a \geq 1} \frac{1}{a^2} \sum_{b \geq 2^{k/6}} \frac{1}{b^3} < \frac{5}{2^{k/4}}.$$

(f) Deduce that there are $< x/2^\ell$ integers $n \leq x$ for which $\Omega(n) - \omega(n) \geq 4\ell + 3$.

Exercise 13.7.7. (a) Use (13.4.1) to show that

$$\#\{a, b \leq x : \omega(a) + \omega(b) = k\} \leq c_0^2 \frac{x^2}{(\log x)^2} \frac{(2(\log\log x + c_1))^{k-2}}{(k-2)!}.$$

(b) Write $K := \left[\frac{\log \log x}{\log 2}\right]$ and let $\delta = 1 - \frac{1 + \log \log 2}{\log 2} = .086071\ldots$. Use Stirling's formula to prove that there exists a constant c_2 such that

$$\#\{a, b \leq x : \omega(a) + \omega(b) \leq K\} + \#\{n \leq x^2 : \omega(n) > K\} \leq c_2 \frac{x^2}{(\log x)^\delta}.$$

(c) Use this result to more or less justify the claim that there are $\leq c_2 N^2/(\log N)^\delta$ distinct integers in the N-by-N multiplication table.

Chapter 14

Counting integral and rational points on curves, modulo p

A *planar curve* is defined to be an equation $f(x,y) = 0$ in two variables, x and y. In this chapter we are interested in identifying the number of solutions $u, v \pmod p$ to $f(u,v) \equiv 0 \pmod p$.

For linear equations, this is easy. If p does not not divide a, say, then

$$\#\{(x,y) \pmod p : ax + by \equiv c \pmod p\} = p,$$

since we have a unique value of $x \pmod p$, namely $(c - by)/a \pmod p$, for any given $y \pmod p$.

14.1. Diagonal quadratics

We can take the same approach to the congruence $ax^2 + by^2 \equiv c \pmod p$: The number of $x \pmod p$ that satisfy this equation for a given value of $y \pmod p$ is $1 + \left(\frac{(c-by^2)/a}{p}\right)$ by Corollary 8.1.1, and if we can sum these up, over all y, then we have our answer. This seems like a daunting task so let's take a different approach.

We saw in exercise 6.1.1 that the rational points (u,v) on the curve $u^2 + v^2 = 1$ are in 1-to-2 correspondence with the integer solutions to the Pythagorean equation $x^2 + y^2 = z^2$ with $(x,y) = 1$ (taking $u = x/z$ and $v = y/z$), so it seems natural that there should be some relationship between the number of solutions to either equation mod p.[1]

Theorem 14.1. *If p is a prime that does not divide integers a, b, c, then*

$$N := \#\{(x,y,z) \pmod p : ax^2 + by^2 + cz^2 \equiv 0 \pmod p\} = p^2.$$

---

[1] However a note of caution: If we take the same approach to identify rational points on the curve $u^2 - v^2 = 1$ with coprime integral solutions to $x^2 - y^2 = z^2$, then we have to be more careful when $z = 0$ (which did not occur in the first example).

Proof. We look at the solutions to $u+v+w \equiv 0 \pmod{p}$ and then count the number of $(x,y,z) \pmod{p}$ with $u \equiv ax^2$, $v \equiv by^2$, and $w \equiv cz^2 \pmod{p}$. The number of x with $x^2 \equiv u/a \pmod{p}$ is $1 + \left(\frac{u/a}{p}\right) = 1 + \left(\frac{a}{p}\right)\left(\frac{u}{p}\right)$. Therefore

$$N = \sum_{\substack{u,v,w \pmod{p} \\ u+v+w \equiv 0 \pmod{p}}} \left(1 + \left(\frac{a}{p}\right)\left(\frac{u}{p}\right)\right)\left(1 + \left(\frac{b}{p}\right)\left(\frac{v}{p}\right)\right)\left(1 + \left(\frac{c}{p}\right)\left(\frac{w}{p}\right)\right).$$

We now multiply out the brackets and end up with eight terms. We claim that all but the first and last obviously equal 0: To see this suppose that our term includes $\left(\frac{u}{p}\right)$, but not for $\left(\frac{w}{p}\right)$. Then the condition "$u+v+w \equiv 0 \pmod{p}$" is redundant, since we take $w \equiv -u-v \pmod{p}$ and its value does not affect the sum, and so we sum over all possible u and $v \pmod{p}$. In particular $\sum_{u \pmod{p}} \left(\frac{u}{p}\right) = 0$.

There are p^2 terms in the sum for N (since we can pick u and v at will, and then $w \equiv -u-v \pmod{p}$), and so

$$N = p^2 + \left(\frac{abc}{p}\right) \sum_{\substack{u,v,w \pmod{p} \\ u+v+w \equiv 0 \pmod{p}}} \left(\frac{uvw}{p}\right).$$

We will quantify the solutions in terms of u. If $u=0$, then the summand is 0, so we may assume that $u \not\equiv 0 \pmod{p}$, and then we can write $r \equiv v/u$ and $s \equiv w/u \pmod{p}$. Therefore $\left(\frac{uvw}{p}\right) = \left(\frac{u \cdot ur \cdot us}{p}\right) = \left(\frac{u}{p}\right)\left(\frac{rs}{p}\right)$ and $u+v+w \equiv u(1+r+s) \pmod{p}$. Therefore

$$\sum_{\substack{u,v,w \pmod{p} \\ u+v+w \equiv 0 \pmod{p}}} \left(\frac{uvw}{p}\right) = \sum_{u \pmod{p}} \left(\frac{u}{p}\right) \cdot \sum_{\substack{rs \pmod{p} \\ 1+r+s \equiv 0 \pmod{p}}} \left(\frac{rs}{p}\right) = 0,$$

as the first sum is 0. The result follows. $\square$

Corollary 14.1.1. *If p is a prime that does not divide integers a, b, c, then*

$$A(c) := \#\{(u,v) \pmod{p}: au^2 + bv^2 + c \equiv 0 \pmod{p}\} = p - \left(\frac{-ab}{p}\right).$$

Proof. We partition the solutions counted by N in the proof of Theorem 14.1 into the solutions in which $z \equiv 0 \pmod{p}$, and those with $z \not\equiv 0 \pmod{p}$. There are exactly $A(0)$ with $z \equiv 0 \pmod{p}$ (taking $x = u$ and $y = v$). Otherwise we can divide through by $z \pmod{p}$ to obtain a solution $(x/z, y/z) \pmod{p}$ in $A(c)$; and if we have a solution $(u,v) \pmod{p}$ in $A(c)$, we obtain $p-1$ solutions (uz, vz, z) counted in N. Therefore

$$N = A(0) + (p-1)A(c).$$

Now for the case $c = 0$. We always have the solution $u \equiv v \equiv 0 \pmod{p}$. Otherwise, given any non-zero solution $(u,v) \pmod{p}$, we may divide through by v to get a solution to $w^2 \equiv -b/a \pmod{p}$ where $w \equiv u/v \pmod{p}$. From any such solution $w \pmod{p}$ we can recover all non-zero solutions $(u,v) \equiv (vw, v) \pmod{p}$

to $au^2 + bv^2 \equiv 0 \pmod{p}$, as v runs through the reduced residues mod p. Therefore
$$A(0) = 1 + (p-1)\#\{w \pmod{p}: w^2 \equiv -b/a \pmod{p}\} = 1 + (p-1)\left(1 + \left(\frac{-ab}{p}\right)\right),$$
and the result follows. $\square$

One pertinent example is that
$$\#\{(x,y) \pmod{p}: x^2 - dy^2 \equiv 1 \pmod{p}\} = p - \left(\frac{d}{p}\right).$$

Exercise 14.1.1. Prove that if odd prime p does not divide n, then
$$\sum_{x \pmod{p}} \left(\frac{x^2 - n}{p}\right) = \sum_{y \pmod{p}} \left(\frac{y(y+n)}{p}\right) = -1.$$

Exercise 14.1.2. Suppose that odd prime p does not divide n.
(a) Show that there are $\frac{1}{4}(p - 3 - (\frac{n}{p})(1 + (\frac{-1}{p})))$ residues $m \pmod{p}$ for which m and $m+n$ are both quadratic residues mod p.
(b) Show that there are $\frac{1}{4}(p - 3 + (\frac{n}{p})(1 + (\frac{-1}{p})))$ residues $\ell \pmod{p}$ for which ℓ and $\ell + n$ are both quadratic non-residues mod p.

14.2. Counting solutions to a quadratic equation and another proof of quadratic reciprocity

We now present a proof due to Wouter Castryck, based on Lebesgue's 1838 proof.

Given an odd prime q, we define N_n for any odd integer n to be the number of solutions $(x_1, \ldots, x_n) \pmod{q}$ to the congruence
$$x_1^2 - x_2^2 + x_3^2 - \cdots + x_n^2 \equiv 1 \pmod{q}.$$
We first prove, by induction, that
$$N_n = q^{n-1} + q^{\frac{n-1}{2}} \text{ for all odd integers } n \geq 1.$$
For $n = 1$ this follows as there are two square roots of $1 \pmod{q}$. Otherwise we make the invertible change of variables $X_1 = x_1 - x_2$ and $X_j = x_j$ for all $j \geq 2$, to obtain the congruence
$$2X_1 X_2 + X_1^2 + X_3^2 - \cdots + X_n^2 \equiv 1 \pmod{q},$$
with the same number of solutions. Now if we select any $X_3, \ldots, X_n \pmod{q}$ and any $X_1 \not\equiv 0 \pmod{q}$, then there is a unique choice of $X_2 \pmod{q}$, a total of $(q-1)q^{n-2}$ solutions. Otherwise if $X_1 \equiv 0 \pmod{q}$, then we can choose any $X_2 \pmod{q}$, and $X_3, \ldots, X_n \pmod{q}$ satisfy $X_3^2 - \cdots + X_n^2 \equiv 1 \pmod{q}$; this has qN_{n-2} solutions. Therefore we have proved that
$$N_n = (q-1)q^{n-2} + qN_{n-2}$$
and then the claimed result follows by the induction hypothesis.

We now determine N_p, for odd prime p, in a rather different way. If we let $x_1^2 \equiv t_1 \pmod{q}, x_2^2 \equiv -t_2 \pmod{q}, x_3^2 \equiv t_3 \pmod{q}, \ldots$, then we have
$$N_p = \sum_{t_1 + t_2 + \cdots + t_p \equiv 1 \pmod{q}} \left(1 + \left(\frac{t_1}{q}\right)\right)\left(1 + \left(\frac{-t_2}{q}\right)\right) \cdots \left(1 + \left(\frac{t_p}{q}\right)\right).$$

If we multiply this out and sum each of the 2^p terms, we find that all but the first and last terms trivially sum to 0 (just as in the proof of Theorem 14.1). Therefore

$$N_p = q^{p-1} + \left(\frac{-1}{q}\right)^{\frac{p-1}{2}} \sum_{t_1+\cdots+t_p\equiv 1 \pmod{q}} \left(\frac{t_1 t_2 \cdots t_p}{q}\right).$$

Comparing our two evaluations of N_p we obtain

$$\sum_{t_1+\cdots+t_p\equiv 1 \pmod{q}} \left(\frac{t_1 t_2 \cdots t_p}{q}\right) = \left(\left(\frac{-1}{q}\right) q\right)^{\frac{p-1}{2}}.$$

We will now approach this sum, mod p, using the same idea as in the necklace proof of Fermat's Little Theorem (see section 7.12 of appendix 7A). The idea is to group together the solutions $(t_1, t_2, \ldots, t_p), (t_2, t_3, \ldots, t_p, t_1), \ldots$ which each contribute the same amount to our sum. These are p distinct solutions unless each $t_j = t_1 = t$, say, and so

$$\sum_{t_1+\cdots+t_p\equiv 1 \pmod{q}} \left(\frac{t_1 t_2 \cdots t_p}{q}\right) \equiv \sum_{pt\equiv 1 \pmod{q}} \left(\frac{t^p}{q}\right) = \left(\frac{p}{q}\right)^{-p} = \left(\frac{p}{q}\right) \pmod{p}.$$

Comparing the two expressions for the sum and recalling that $\left(\frac{-1}{q}\right) = (-1)^{\frac{q-1}{2}}$, and that $q^{\frac{p-1}{2}} \equiv \left(\frac{q}{p}\right) \pmod{p}$ by Euler's criterion, we obtain the law of quadratic reciprocity.

14.3. Cubic equations modulo p

Given our results for linear and quadratic equations, we might guess that if $p \nmid 6abc$, then $N(a, b, c) := \#\{(x, y, z) \pmod{p} : ax^3 + by^3 + cz^3 \equiv 0 \pmod{p}\} = p^2$. We can test this guess with $p = 7$. A calculation reveals that $N(a, b, c)$ equals one of 19, 55, or 73 for every a, b, c, so our guess is wrong, and there is something to be understood.

Exercise 14.3.1. Let p be a prime $\equiv 2 \pmod 3$.
  (a) Prove that for every $a \pmod p$ there is exactly one $b \pmod p$ for which $b^3 \equiv a \pmod p$.
  (b) Deduce that if $p \nmid (a, b, c)$, then $N(a, b, c) = p^2$.

Henceforth we will assume that $p \equiv 1 \pmod 3$. We can proceed as in the proof of Theorem 14.1 which leads us to Jacobi sums (which we will explore in sections 14.9 and 14.10 of appendix 14A) though here we will take a different route.

Exercise 14.3.2. Let p be a prime $\equiv 1 \pmod 3$ with a primitive root g, and suppose $p \nmid abc$.
  (a) Show that if $p \nmid rst$, then $N(ar^3, bs^3, ct^3) = N(a, b, c)$.
  (b) Show that $N(1, 1, 1) + N(1, 1, g) + N(1, 1, g^2) = p^2$.
  (c) Show that $N(1, g, 1) + N(1, g, g) + N(1, g, g^2) = p^2$.
  (d) Deduce that $N(1, g, g^2) = N(1, 1, 1)$.
  (e) Show that $N(a, b, c) = N(1, 1, abc)$.

Therefore we can reduce our question to counting the solutions (x, y, z) to

$$x^3 + y^3 + dz^3 \equiv 0 \pmod{p}.$$

14.4. The equation $E_b : y^2 = x^3 + b$

We will transform this problem to counting solutions on a curve: If $x + y \not\equiv 0$ (mod p), then let $u = (x+y)/72d \not\equiv 0$ (mod p), and $v = 36d - y/u$, $w = -z/6u$, to obtain
$$v^2 \equiv w^3 - 432d^2 \pmod{p}.$$
Given a solution (v, w) we can reverse this process and create $p-1$ solutions to the original equation. If $x + y \equiv 0$ (mod p), then $z \equiv 0$ (mod p), and so there are p such solutions. Therefore we have proved that

(14.3.1) $\quad N(1, 1, d) = p + (p-1) \# \{x, y \pmod{p} : y^2 \equiv x^3 - 3(12d)^2 \pmod{p}\}.$

The problem of determining $N(a, b, c)$ has now been reduced to special cases of determining the number of mod p-points satisfying the following equation:

14.4. The equation $E_b : y^2 = x^3 + b$

Let p be a prime $\equiv 1$ (mod 3). By the theory of binary quadratic forms (see chapters 9 and 12), we know there are unique integers a and b, up to sign, for which $p = a^2 + 3b^2$. We select the sign of a so that $a \equiv 2$ (mod 3).[2]

Theorem 14.2. *Let p be a prime $\equiv 1$ (mod 3) with a primitive root g, and define a and b as above. If p does not divide ℓ, then*
$$N_p(\ell) := \#\{(m, n) \pmod{p} : m^2 \equiv n^3 + \ell \pmod{p}\}$$
$$= p + \left(\frac{\ell}{p}\right) u \text{ where } u := \begin{cases} 2a & \text{if } \ell \text{ is a cube mod } p, \\ -a + 3b & \text{if } \ell/g \text{ is a cube mod } p, \\ -a - 3b & \text{if } \ell/g^2 \text{ is a cube mod } p. \end{cases}$$

There are three possible ways to write $4p$ in the form $u^2 + 3v^2$ with $u \equiv 1$ (mod 3):
$$4p = (2a)^2 + 3(2a)^2 = (-a+3b)^2 + 3(a+b)^2 = (-a-3b)^2 + 3(a-b)^2.$$

In each case we have $|u| \leq 2\sqrt{p}$ and so $|N_p(k) - p| \leq 2\sqrt{p}$ for every non-zero k.

Proof. We have
$$N_p(\ell) = \sum_{n \pmod p} \left\{ 1 + \left(\frac{n^3+\ell}{p} \right) \right\} = p + S_\ell \text{ where } S_\ell := \sum_{n=1}^{p} \left(\frac{n^3+\ell}{p} \right).$$

Exercise 14.4.1. Let p be a prime $\equiv 1$ (mod 3) with a primitive root g and suppose $(\ell, p) = 1$.
  (a) Prove that if $L \equiv \ell r^3$ (mod p), then $S_L = \left(\frac{r}{p}\right) S_\ell$.
  Define $T_\ell := \left(\frac{\ell}{p}\right) S_\ell$, so that $T_L = T_\ell$.
  (b) Prove that if $\ell \equiv g^k$ (mod p) and i is the least residue of k (mod 3), then $T_\ell = T_{g^i}$.
  (c) Prove that T_1 is even, whereas T_g and T_{g^2} are odd.
  (d) Prove that each $T_\ell \equiv 1$ (mod 3).
Therefore there exist integers A, B, C such that $T_1 = 2A, T_g = -A + 3B$, and $T_{g^2} = -A - 3C$.

---

[2] It is difficult to determine the sign of b in Theorem 14.2 by the elementary methods employed here.

We will sum up the T_ℓ in two different ways: By exercise 14.4.1(b) we have

$$\sum_{\ell \pmod p} T_\ell = \frac{p-1}{3}(T_1 + T_g + T_{g^2}) = (p-1)(B-C).$$

By definition and then exercise 14.1.1 we have

$$\sum_{\ell \pmod p} T_\ell = \sum_{1 \le n \le p} \sum_{\ell \pmod p} \left(\frac{\ell(\ell+n^3)}{p}\right) = (p-1) + \sum_{1 \le n \le p-1}(-1) = 0.$$

Together these equations imply that $C = B$.

We will sum up the T_ℓ^2 in two different ways: First,

$$\sum_{\ell \pmod p} T_\ell^2 \equiv \frac{p-1}{3}((2A)^2 + (-A+3B)^2 + (-A-3B)^2) = 2(p-1)(A^2 + 3B^2).$$

Then, by definition and exercise 14.1.1, we have

$$\sum_{\ell \pmod p} T_\ell^2 = \sum_{m,n \pmod p} \sum_{\substack{\ell \pmod p \\ \ell \not\equiv 0 \pmod p}} \left(\frac{(\ell+m^3)(\ell+n^3)}{p}\right)$$

$$= p \sum_{\substack{m,n \pmod p \\ m^3 \equiv n^3 \pmod p}} 1 - \sum_{m,n \pmod p} 1 - \sum_{m,n \pmod p}\left(\frac{m^3 n^3}{p}\right)$$

$$= p(1 + 3(p-1)) - p^2 - 0 = 2p(p-1).$$

Together these equations imply that $p = A^2 + 3B^2$.

Therefore $A = \pm a$ and $B = \pm b$ (as these are unique up to sign), and by exercise 14.4.1(d) we have $2A \equiv 1 \pmod 3$; that is, $A \equiv 2 \pmod 3$ so that $A = a$. □

By inserting Theorem 14.2 into (14.3.1), and using exercise 14.3.2(e) we deduce that

$$N(a,b,c) = p^2 + (p-1)u \text{ where } u := \begin{cases} 2a & \text{if } d/2 \text{ is a cube mod } p, \\ -a+3b & \text{if } dg/2 \text{ is a cube mod } p, \\ -a-3b & \text{if } dg^2/2 \text{ is a cube mod } p. \end{cases}$$

In our example above with $p = 7 = 2^2 + 3 \cdot 1^2$, the three possible values for u are 4, -5, or 1, so that $N(a,b,c)$ equals one of 73, 19, or 55, confirming our calculations.

Exercise 14.4.2.[†] (a) Prove that if p does not divide abc, then

$$\#\{x,y \pmod p : ax^3 + by^3 \equiv c \pmod p\} = p - \chi(a/b) - \chi(b/a) + u,$$

where the character $\chi \pmod p$ has order 3.
(b) Deduce that $\#\{w,x,y,z \pmod p : x^3 + y^3 \equiv w^3 + z^3 \pmod p\} = p^3 + 6p(p-1)$.

In general, it is of great interest to determine

$$\#\{x,y \pmod p : y^2 \equiv x^3 + ax + b \pmod p\}.$$

All such curves have the order 2 automorphism $x \to x, y \to -y$ We just succeeded in counting the solutions when $a = 0$. This is known to be a relatively simple case[3] because the curve has the "extra" order 3 automorphism, $x \to \omega x, y \to y$. Similarly when $b = 0$ the curve has the extra order 4 automorphism, $x \to -x, y \to iy$, and we can also attack our problem by elementary means:

14.5. The equation $y^2 = x^3 + ax$

Let $N_p(a)$ be the number of pairs (x, y) (mod p) such that $y^2 \equiv x^3 + ax$ (mod p). We know that $N_p(a) = p + S_a$ where

$$S_a := \sum_{1 \leq n \leq p} \left(\frac{n^3 + an}{p} \right).$$

Exercise 14.5.1. Let p be a prime $\equiv 3$ (mod 4). Observing that $(-n)^3 + a(-n) = -(n^3 + an)$ deduce that $S_a = 0$, so that $N_p(a) = p$.

Henceforth assume that $p \equiv 1$ (mod 4). Calculations with $a = -1$ (that is, the curve $y^2 = x^3 - x$ (mod p)) reveal that, for $N_p = N_p(-1)$, we have

$$N_5 = 7, \ N_{13} = 7, \ N_{17} = 15, \ N_{29} = 39, \ N_{37} = 39,$$

$$N_{41} = 31, \ N_{53} = 39, \ N_{61} = 71, \ N_{73} = 79, \ldots.$$

These are all odd and close to p, so we compute $E_p = E_p(-1) := (N_p - p)/2 = S_{-1}/2$:

$$E_5 = 1, \ E_{13} = -3, \ E_{17} = -1, \ E_{29} = 5, \ E_{37} = 1,$$

$$E_{41} = -5, \ E_{53} = -7, \ E_{61} = 5, \ E_{73} = 3, \ldots.$$

These are all odd numbers ... any guesses as to what they are? Bearing in mind that we are only dealing with primes $\equiv 1$ (mod 4), we might expect these odd numbers to reflect a property that primes $\equiv 1$ (mod 4) have, but not primes $\equiv 3$ (mod 4). If you can predict $|E_p|$, then try to predict the sign of E_p. The reader will learn more from playing with the data for a while than peeking immediately at the following result:

Let p be a prime $\equiv 1$ (mod 4). By the theory of binary quadratic forms (see chapters 9 and 12), we know there are unique integers a and b, up to sign, for which $p = a^2 + b^2$, where a is odd and b is even. We select the sign of a so that $a \equiv 3$ (mod 4). Since $b^2 \equiv p - 1$ (mod 8), we deduce that $b/2 \equiv \frac{p-1}{4}$ (mod 2).[4]

Theorem 14.3. *Let p be a prime $\equiv 1$ (mod 4) with primitive root g, and let a and b be as above. If $\left(\frac{\ell}{p}\right) = 1$, then select r (mod p) for which $r^2 \equiv \ell$ (mod p). Then*

$$N_p(\ell) = p + 2a \left(\frac{r}{p} \right).$$

---

[3]Though this extra automorphism doesn't obviously help our elementary approach other than to render trivial the $p \equiv 2 \mod 3$ case.

[4]It is difficult to determine the sign of b in Theorem 14.3 by the elementary methods employed here.

If $\left(\frac{\ell}{p}\right) = -1$, then select $r \pmod{p}$ for which $r^2 \equiv \ell/g \pmod{p}$. Then

$$N_p(\ell) = p + 2b\left(\frac{r}{p}\right).$$

In each case we have $|N_p(\ell) - p| \leq 2\sqrt{p}$ for every non-zero ℓ.

Proof. We will sum the S_ℓ^2 in two different ways, using the following:

Exercise 14.5.2. Let p be a prime $\equiv 1 \pmod 4$ with a primitive root g and suppose $(\ell, p) = 1$.
(a) Prove that S_ℓ is even, so there exist integers A, B such that $S_1 = 2A$ and $S_g = 2B$.
(b) Using that $\left(\frac{(-n)^3 + \ell(-n)}{p}\right) = \left(\frac{n^3 + \ell n}{p}\right)$ establish that $S_\ell \equiv 3 - \left(\frac{\ell}{p}\right) \pmod 4$.
(c) Deduce that A is odd and B is even.
(d) Prove that if $c \equiv \ell r^2 \pmod p$, then $S_c = \left(\frac{r}{p}\right) S_\ell$.
(e) Deduce that $S_{-1} = (-1)^{\frac{p-1}{4}} \cdot 2A$.

By exercise 14.5.2(d) if $\left(\frac{\ell}{p}\right) = 1$, then $S_\ell^2 = S_1^2 = (2A)^2$, and if $\left(\frac{\ell}{p}\right) = -1$, then $S_\ell^2 = S_g^2 = (2B)^2$. Therefore

$$\sum_{\ell \pmod p} S_\ell^2 = \frac{p-1}{2}((2A)^2 + (2B)^2) = 2(p-1)(A^2 + B^2).$$

On the other hand, by definition and exercise 14.1.1, we have

$$\sum_{\ell \pmod p} S_\ell^2 = \sum_{\ell \pmod p} \sum_{m,n \pmod p} \left(\frac{m^3 + \ell m}{p}\right)\left(\frac{n^3 + \ell n}{p}\right)$$

$$= \sum_{m,n \pmod p} \left(\frac{mn}{p}\right) \sum_{\ell \pmod p} \left(\frac{(\ell + m^2)(\ell + n^2)}{p}\right)$$

$$= p \sum_{\substack{m,n \pmod p \\ m^2 \equiv n^2 \pmod p}} \left(\frac{mn}{p}\right) - \sum_{m,n \pmod p} \left(\frac{mn}{p}\right).$$

The second sum is 0. In the first sum we get 0 when $n = 0$. Otherwise $m \equiv \pm n \pmod p$ so $\left(\frac{mn}{p}\right) = \left(\frac{\pm n^2}{p}\right) = 1$, and so the sum equals $(p-1) \cdot 2$.

Together these two equations imply that $A^2 + B^2 = p$. To complete the proof we need to determine the value of $A \pmod 4$ (and therefore its sign). This is perhaps easiest via another transformation: If $y^2 \equiv n^3 - n \pmod p$ with $n \not\equiv 0 \pmod p$, then $n^2 - 1 \equiv nz^2 \pmod p$ with $y \equiv nz \pmod p$. This map is invertible and, since $y \equiv 0 \pmod p$ when $n \equiv 0 \pmod p$, we deduce that

$$N_p(-1) = 1 + \#\{(n,z) \pmod p : n^2 - z^2 n - 1 \equiv 0 \pmod p\}.$$

For each given $z \pmod p$ this has $1 + \left(\frac{z^4 + 4}{p}\right)$ solutions $n \pmod p$. This equals 2 if $z = 0$. If $r \not\equiv 0 \pmod p$, then there are either 0 or 4 solutions to $z^4 \equiv r \pmod p$

(as $p \equiv 1 \pmod 4$), and so

$$N_p(-1) = 3 + \sum_{r=1}^{p-1}\left(1+\left(\frac{r+4}{p}\right)\right)\#\{z:\ z^4 \equiv r \pmod p\}$$
$$\equiv 3 + \#\{z:\ z^4 \equiv -4 \pmod p\} \pmod 8.$$

Now $x^4 + 4 \equiv (x-1+i)(x-1-i)(x+1+i)(x+1-i) \pmod p$ where $i^2 \equiv -1 \pmod p$, and so $N_p(-1) \equiv 3+4 \equiv -1 \pmod 8$. We therefore deduce from exercise 14.5.2(e) that $N_p(-1) \equiv p + 2A(-1)^{\frac{p-1}{4}} \equiv 7 \pmod 8$, so that $A \equiv 3 \pmod 4$, and therefore $A = a$. □

14.6. A more general viewpoint on counting solutions modulo p

These results (Theorems 14.2 and 14.3) generalize nicely: Suppose that p does not divide the discriminant, $4a^3 + 27b^2$, of the polynomial $x^3 + ax + b$. Let

$$N_p := \#\{(m,n) \pmod p :\ m^2 \equiv n^3 + an + b \pmod p\}.$$

The Hasse-Weil Theorem states that there exists an algebraic integer $\frac{1}{2}(u + v\sqrt{-d})$ such that $u^2 + dv^2 = 4p$, for which $N_p = p + u$. This implies that $|N_p - p| < 2\sqrt{p}$. This will be proved in [**Grab**].

In general, for the curve $f(x,y) = 0$ one might guess that for arbitrary m and $n \pmod p$, the value of $f(m,n) \pmod p$ is equally likely to be in any given residue class; in particular, the probability that it is $0 \pmod p$ is $1/p$, so we might *expect* that $N_p(f) := \#\{(m,n) \pmod p :\ f(m,n) \equiv 0 \pmod p\}$ is close to p. We have proved this for linear and quadratic f and some examples of cubic f, with an error term no bigger than $2\sqrt{p}$. This is the sort of error term one would obtain from a suitable probabilistic model, and so we might conjecture that a similar error term holds more generally. Indeed it is known that if a curve $f(x,y) = 0$ has *genus* $g \geq 1$ (the cubic curves have genus 1), then $|N_p(f) - p| < 2g\sqrt{p}$. The notion of genus is beyond the scope of this book, but when $h(x) \pmod p$ is a polynomial of degree $d \geq 2$ without repeated roots and $f(x,y) = y^2 - h(x)$, then we can deduce that

(14.6.1) $$|N_p(f) - p| < (d-1)\sqrt{p}.$$

We conclude this section by proving a surprisingly general result.

Let $f(x_1, x_2, \ldots, x_n) \in \mathbb{Z}[x_1, x_2, \ldots, x_n]$ be of degree d.[5] The number of solutions to $f \equiv 0 \pmod p$ is congruent to

(14.6.2) $$\sum_{m_1, \ldots, m_n \pmod p} \left(1 - f(m_1, \ldots, m_n)^{p-1}\right) \pmod p,$$

since

$$1 - f(m)^{p-1} \equiv \begin{cases} 1 \pmod p & \text{if } f(m) \equiv 0 \pmod p, \\ 0 \pmod p & \text{if } f(m) \not\equiv 0 \pmod p, \end{cases}$$

by Fermat's Little Theorem. The first term, all the 1's, evidently sums to $p^n \equiv 0 \pmod p$. When we expand the second term, we get a sum of terms of the form $m_1^{k_1} \cdots m_n^{k_n}$, each of total degree $k_1 + \cdots + k_n \leq d(p-1)$. If such a term, summed

---

[5] The degree of the monomial $cx_1^{e_1} x_2^{e_2} \ldots x_n^{e_n}$ is $e_1 + e_2 + \cdots + e_n$. The degree of a polynomial in several variables is the largest of the degrees of the monomials from which the polynomial is constructed.

over m_i (mod p), is non-zero mod p, then $k_i \geq p-1$ by Corollary 7.5.2. Hence if the sum over each m_i is non-zero, then $d(p-1) \geq k_1 + \cdots + k_n \geq n(p-1)$, and so $d \geq n$. We can therefore deduce:

Theorem 14.4 (Chevalley-Warning Theorem). *If $f(x_1, \ldots, x_n) \in \mathbb{Z}[x_1, \ldots, x_n]$ has degree $< n$, then*

$$\#\{m_1, \ldots, m_n \pmod{p} : f(m_1, \ldots, m_n) \equiv 0 \pmod{p}\} \equiv 0 \pmod{p}.$$

Therefore if $f(0, 0, \ldots, 0) = 0$, that is, the constant term of f is 0, then there are at least $p-1$ distinct non-zero solutions to $f(m_1, \ldots, m_n) \equiv 0 \pmod{p}$.

One example is the equation $ax^2 + by^2 + cz^2 \equiv 0 \pmod{p}$ since here $n = 3 > d = 2$ and $f(0,0,0) = 0$.

Exercise 14.6.1. Use (14.6.1) to show that if $h(x)$ (mod p) is a polynomial of degree $d \geq 2$ without repeated roots, then

$$\left| \sum_{n \pmod{p}} \left(\frac{h(n)}{p} \right) \right| < (d-1)\sqrt{p}.$$

Exercise 14.6.2. Prove that the number of n (mod p) for which $\left(\frac{n^2+1}{p}\right) = \left(\frac{n^2+2}{p}\right) = 1$ equals $p/4$ plus or minus an error of at most $4\sqrt{p}$.

Exercise 14.6.3.† Let p be an odd prime which can be written as $p = a^2 + b^2$ with $a \equiv 3$ (mod 4). Prove that the number of n (mod p) for which $\left(\frac{n-1}{p}\right) = \left(\frac{n}{p}\right) = \left(\frac{n+1}{p}\right) = 1$ equals $\frac{p+1+2a}{8} - 2$ if $p \equiv 1 \pmod 8$, and equals $\frac{p+1-2a}{8} - 1$ if $p \equiv 5 \pmod 8$.

Exercise 14.6.4. Prove that the number of n (mod p) for which $\left(\frac{n+1}{p}\right) = \cdots = \left(\frac{n+k}{p}\right) = 1$ with $k \leq \log p$, equals $p/2^k$ plus or minus an error of at most $k\sqrt{p}$.

Exercise 14.6.5. Let $\delta_1, \ldots, \delta_k$ be an arbitrary sequence of 1's and -1's, with $k \leq \log p$. Prove that the number of n (mod p) for which $\left(\frac{n+j}{p}\right) = \delta_j$ for $j = 1, 2, \ldots, k$, equals $p/2^k$ plus or minus an error of at most $k\sqrt{p}$.

This is an example of an important family of questions: Let $f(n)$ be a multiplicative function that only takes values -1 and 1. Let $\delta_1, \ldots, \delta_k$ be an arbitrary sequence of 1's and -1's. Do there exist infinitely many integers n for which $f(n+j) = \delta_j$ for $j = 1, 2, \ldots, k$? And, if so, how often does this occur? Does each *sign pattern* occur (more or less) equally often? We have resolved this problem when $f(\cdot)$ is a Legendre symbol, in the last two problems.

The outstanding open question of this type is when $f(\cdot)$ is the Liouville function. There has been some progress recently, indeed there are many interesting applications, and it is the subject of much on-going research.

Appendix 14A. Gauss sums

For a given Dirichlet character $\chi \pmod{q}$ we define the *Gauss sum*, $g(\chi)$, by

$$g(\chi) := \sum_{a=1}^{q} \chi(a) e^{2i\pi a/q}.$$

Note that the summand $\chi(a)e^{2i\pi a/q}$ depends only on the value of $a \pmod{q}$. Gauss sums play an important role in number theory and have some beautiful properties.

14.7. Identities for Gauss sums

By making the change of variable $a \equiv nb \pmod{q}$ for some integer n with $(n,q) = 1$, the variable b runs through a complete system of residues mod q as a does. Therefore we obtain the surprising identity

$$(14.7.1) \qquad \sum_{b=1}^{q} \chi(b) e^{2i\pi nb/q} = \overline{\chi(n)} \sum_{b=1}^{q} \chi(nb) e^{2i\pi nb/q} = \overline{\chi}(n) g(\chi).$$

Therefore if q is prime and χ is non-principal, then

$$\phi(q)|g(\chi)|^2 = \sum_{\substack{1 \leq n \leq q \\ (n,q)=1}} |\overline{\chi}(n) g(\chi)|^2 = \sum_{n=0}^{q-1} \left| \sum_{b=1}^{q} \chi(b) e^{2i\pi nb/q} \right|^2,$$

since the $n = 0$ sum equals $\sum_{b=1}^{q} \chi(b) = 0$. Expanding the square we obtain

$$\sum_{b=1}^{q} \chi(b) \sum_{c=1}^{q} \overline{\chi}(c) \sum_{n=0}^{q-1} e^{2i\pi n(b-c)/q} = q \sum_{b=1}^{q} |\chi(b)|^2 = q\phi(q),$$

since $\sum_{n=0}^{q-1} e^{2i\pi na/q} = 0$ unless q divides a. Therefore we have proved that

$$|g(\chi)|^2 = q.$$

To better use this we have

$$\overline{g(\chi)} = \sum_{a=1}^{q} \overline{\chi}(a) e^{-2i\pi a/q} = \chi(-1) g(\overline{\chi})$$

by (14.7.1), so that $g(\chi)g(\overline{\chi}) = \chi(-1)|g(\chi)|^2 = \chi(-1)q$. In particular if $\chi = (\cdot/q)$, so that $\chi = \overline{\chi}$, then

$$g((\cdot/q))^2 = (-1/q)q.$$

Taking the square root, it remains to determine which sign gives the value of $g((\cdot/q))$. It took Gauss four years to figure this out, so we will simply state his result:

(14.7.2) $$g((\cdot/q)) = \begin{cases} \sqrt{q} & \text{if } q \equiv 1 \pmod{4}, \\ i\sqrt{q} & \text{if } q \equiv 3 \pmod{4}. \end{cases}$$

Another proof of the law of quadratic reciprocity. Let $q^* = (-1/q)q$ and $g = g((\cdot/q))$ so that $g^2 = g^*$. Now

$$g^p = \left(\sum_{a=1}^{q} \left(\frac{a}{q} \right) e^{2i\pi a/q} \right)^p \equiv \sum_{a=1}^{q} \left(\frac{a}{q} \right)^p e^{2i\pi ap/q} \pmod{p},$$

as $(x_1 + \cdots + x_q)^p \equiv x_1^p + \cdots + x_q^p \pmod{p}$. Then, by (14.7.1), we have

$$g^p \equiv \sum_{a=1}^{q} \left(\frac{a}{q} \right) e^{2i\pi ap/q} = \left(\frac{p}{q} \right) g \pmod{p}.$$

We may divide through by g as $(g^2, p) = (q, p) = 1$, so that

$$\left(\frac{p}{q} \right) \equiv g^{p-1} = (g^2)^{(p-1)/2} = (q^*)^{(p-1)/2}$$

$$= (-1)^{\frac{p-1}{2} \cdot \frac{q-1}{2}} q^{(p-1)/2} \equiv (-1)^{\frac{p-1}{2} \cdot \frac{q-1}{2}} \left(\frac{q}{p} \right) \pmod{p},$$

by Euler's criterion. Both sides are integers equal to 1 or -1 and differ by a multiple of p, which is ≥ 3, and so they must be equal. That is, we obtain Theorem 8.5, the law of quadratic reciprocity.

14.8. Dirichlet L-functions at $s = 1$

We now use (14.7.1) to try to find a simple expression for $L(1, \chi)$. We again let q be prime so that $\sum_{b=1}^{q} \chi(b) = 0$, and therefore the identity (14.7.1) holds for all integers n (not just those n coprime to q). Assuming that there are no convergence issues in swapping the orders of summation, we have

$$g(\chi) L(1, \overline{\chi}) = \sum_{n \geq 1} \frac{g(\chi) \overline{\chi}(n)}{n} = \sum_{n \geq 1} \frac{\sum_{b=1}^{q-1} \chi(b) e^{2i\pi nb/q}}{n} = \sum_{b=1}^{q-1} \chi(b) \sum_{n \geq 1} \frac{e^{2i\pi nb/q}}{n}.$$

The sum over n is the Taylor series for $-\log(1-t)$ with $t = e^{2i\pi nb/q}$ (since each $t \neq 1$). Therefore

$$g(\chi) L(1, \overline{\chi}) = -\sum_{b=1}^{q-1} \chi(b) \log(1 - e^{2i\pi nb/q}).$$

Exercise 14.8.1. (a) Prove that $\arg(1 - e^{i\theta}) \in (-\frac{\pi}{2}, \frac{\pi}{2})$.
(b) Deduce that if $0 < \theta < 2\pi$, then $\log(1 - e^{i\theta}) - \log(1 - e^{-i\theta}) = i(\theta - \pi) \in (-\pi, \pi)$.

Now assume that $\chi(-1) = -1$ and add the b and $q - b$ terms in the sum above, so that by the last exercise we have

$$2g(\chi)L(1,\overline{\chi}) = -\sum_{b=1}^{q-1} \chi(b)(\log(1-e^{2i\pi b/q}) - \log(1-e^{-2i\pi b/q})) = -i\sum_{b=1}^{q-1} \chi(b)(2\pi b/q - \pi).$$

The second sum on the right-hand side is 0, and so multiplying through by $g(\overline{\chi})$, we obtain

$$qL(1,\overline{\chi}) = \frac{i\pi g(\overline{\chi})}{q} \sum_{b=1}^{q-1} \chi(b)b$$

as $-g(\chi)g(\overline{\chi}) = q$.

Now let $\chi = (\cdot/q)$ with prime $q \equiv 3 \pmod 4$ where $q > 3$, so that $\overline{\chi} = \chi$. Dirichlet's class number formula (given in section 12.15 of appendix 12D with $d = -q$) reads $\pi h(-q) = \sqrt{q} L(1,\chi)$ and therefore the last displayed formula becomes

$$h(-q) = -\frac{1}{q} \sum_{b=1}^{q-1} \chi(b)b$$

since $g((\cdot/q)) = i\sqrt{q}$ by (14.7.2). This is Jacobi's conjecture, stated as (12.15.1).

14.9. Jacobi sums

Let χ and ψ be characters mod q and define the *Jacobi sum*

$$j(\chi,\psi) := \sum_{\substack{r,s \ (\bmod\ q) \\ r+s \equiv 1 \ (\bmod\ q)}} \chi(r)\psi(s).$$

To evaluate this sum we state the condition "$r + s \equiv 1 \pmod q$" in term of a sum, so that

$$j(\chi,\psi) = \sum_{r,s \ (\bmod\ q)} \chi(r)\psi(s) \cdot \frac{1}{q}\sum_{k=0}^{q-1} e^{2i\pi \frac{k}{q}(r+s-1)}$$

$$= \frac{1}{q}\sum_{k=0}^{q-1} e^{-2i\pi \frac{k}{q}} \left(\sum_{r \ (\bmod\ q)} \chi(r)e^{2i\pi \frac{kr}{q}}\right)\left(\sum_{s \ (\bmod\ q)} \psi(s)e^{2i\pi \frac{ks}{q}}\right)$$

$$= \frac{1}{q}\sum_{k=0}^{q-1} e^{-2i\pi \frac{k}{q}} (\overline{\chi}(k)g(\chi))(\overline{\psi}(k)g(\psi))$$

$$= \frac{\overline{\chi\psi}(-1)}{q} g(\overline{\chi\psi})g(\chi)g(\psi).$$

If q is prime and each of χ, ψ, and $\overline{\chi\psi}$ is non-principal, then we know that $|g(\overline{\chi\psi})| = |g(\chi)| = |g(\psi)| = \sqrt{q}$, so that $|j(\chi,\psi)| = \sqrt{q}$. By its definition $j(\chi,\psi)$ is an algebraic integer and belongs to the field defined by the values of χ and ψ.

14.10. The diagonal cubic, revisited

Let p be a prime $\equiv 1 \pmod 3$. Since the group of characters mod p is isomorphic to the multiplicative group of reduced residues mod p, we know that there are two characters $\chi, \chi^2 \pmod p$ of order 3. We can establish the analogy to Corollary 8.1.1 for cubic residues:

Exercise 14.10.1. Prove that if $p \nmid a$, then
$$\#\{x \pmod p : ax^3 \equiv u \pmod p\} = 1 + \chi(a^{-1}u) + \chi^2(a^{-1}u).$$

By exercise 14.10.1, $N(a,b,c)$ equals the sum over triples $u, v, w \pmod p$ for which $u + v + w \equiv 0 \pmod p$, of
$$\left(1 + \chi(a^{-1}u) + \chi^2(a^{-1}u)\right)\left(1 + \chi(b^{-1}v) + \chi^2(b^{-1}v)\right)\left(1 + \chi(c^{-1}w) + \chi^2(c^{-1}w)\right).$$
We again multiply the triples out. The first product, $1 \cdot 1 \cdot 1$, sums to p^2. Any other product that contains a 1 sums to 0, since the remaining variables can be summed independently (and each independent sum is of the shape $\sum_t \chi(t) = 0$). Therefore
$$N(a,b,c) = p^2 + \sum_{\substack{1 \leq i,j,k \leq 2}} \sum_{\substack{u,v,w \pmod p \\ u+v+w \equiv 0 \pmod p}} \chi^i(a^{-1}u)\chi^j(b^{-1}v)\chi^k(c^{-1}w).$$

We may assume $u \not\equiv 0 \pmod p$ since those summands equal 0. Therefore we can write $v = -ur, w = -us$ and separate out the sum $\sum_{u \pmod p} \chi^{i+j+k}(u)$. This equals 0 when 3 does not divide $i+j+k$. This therefore leaves us with only the terms where $i = j = k$, in which case the sum over u equals $p-1$. For $i = j = k = 1$ we have
$$\sum_{\substack{u,v,w \pmod p \\ u+v+w \equiv 0 \pmod p}} \chi(uvw) = (p-1) \sum_{\substack{r,s \pmod p \\ r+s \equiv 1 \pmod p}} \chi(rs) = (p-1)j(\chi,\chi),$$
and likewise for χ^2. Therefore
$$N(a,b,c) = p^2 + (p-1)(\overline{\chi}(d)j(\chi,\chi) + \chi(d)j(\overline{\chi},\overline{\chi})),$$
where $d = abc$. In section 14.9 we proved that $j(\chi,\chi)$ is an algebraic integer in $\mathbb{Q}(\frac{1+\sqrt{-3}}{2})$ of norm p, so we can write $\overline{\chi}(d)j(\chi,\chi) = \frac{u+v\sqrt{-3}}{2}$ with $u \equiv v \pmod 2$, and $u^2 + 3v^2 = 4p$. We therefore recover the result,
$$N(a,b,c) = p^2 + (p-1)u,$$
that we established in section 14.4. Moreover by calculating $j(\chi,\chi)$ we can determine the sign of b in Theorem 14.2.

Chapter 15

Combinatorial number theory

Combinatorial number theory, one of the oldest topics in mathematical research, is still an area of active interest today. In this chapter we sample several topics, both old and new from the more combinatorial side of number theory.

15.1. Partitions

Let $p(n)$ denote the number of ways of partitioning n into positive integers. For example $p(7) = 15$ since

$$7 = 6+1 = 5+2 = 5+1+1 = 4+3 = 4+2+1 = 4+1+1+1$$
$$= 3+3+1 = 3+2+2 = 3+2+1+1 = 3+1+1+1+1 = 2+2+2+1$$
$$= 2+2+1+1+1 = 2+1+1+1+1+1 = 1+1+1+1+1+1+1.$$

Euler observed that there is a beautiful generating function for $p(n)$: In the generating function $p(n)$ is the coefficient of t^n, and for each partition $n = a_1 + \cdots + a_k$ we can think of t^n as $t^{a_1} \cdots t^{a_k}$. Splitting this product up into the values of the a_i, but taking $(t^a)^j$ if j of the a_i's equal a, we see that

$$\sum_{n \geq 0} p(n) t^n = \prod_{a \geq 1} \left(\sum_{j \geq 0} (t^a)^j \right) = \frac{1}{(1-t)(1-t^2)(1-t^3) \cdots}.$$

Similarly the generating function for the number of partitions into odd parts is

$$\frac{1}{(1-t)(1-t^3)(1-t^5) \cdots},$$

and the generating function for the number of partitions with no repeated parts is

$$(1+t)(1+t^2)(1+t^3) \cdots.$$

Exercise 15.1.1. Deduce that the number of partitions of n into odd parts is equal to the number of partitions of n with no repeated parts.

515

Partitions can be represented by rows and columns of dots in a *Ferrers diagram*; for example $27 = 11 + 7 + 3 + 3 + 2 + 1$ is represented by

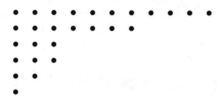

the first row having 11 dots, the second 7, etc. Now, reading the diagram in the other direction yields the partition $27 = 6 + 5 + 4 + 2 + 2 + 2 + 2 + 1 + 1 + 1 + 1$. This bijection between partitions is at the heart of many beautiful theorems about partitions. For example if a partition has m parts, then its "conjugate" has largest part m. Using generating functions, we therefore find that the number of partitions with $\leq m$ parts equals the number of partitions with all parts $\leq m$, which has generating function

$$\frac{1}{(1-t)(1-t^2)(1-t^3)\cdots(1-t^m)}.$$

Looking at Ferrers diagrams, partitions come in pairs, each partition with its conjugate, other than those that are *self-conjugate* (that is, those partitions in which the conjugate partition is the same as the original partition). Self-conjugate Ferrers diagrams have a symmetry about the diagonal axis of the diagram. Hence a self-conjugate Ferrers diagram looks like:

```
• • • • • •        1 1 1 1 1 1 1
• • • • •          1 2 2 2 2
• • •              1 2 3
• •                1 2
• •                1 2
•                  1
```

A self-conjugate partition in 1-to-1 correspondence with another type of partition.

This example yields $19 = 6+5+3+2+2+1$. We have constructed another partition of 19, using the same Ferrers diagram. The top row and first column contain 11 entries, marked by "1"'s in the diagram, what's left of the second row and second column have 7 remaining entries, marked by "2"'s, and finally one element is left (marked by a "3"), yielding the partition $19 = 11 + 7 + 1$.

Exercise 15.1.2. Prove that there is a bijection between self-conjugate partitions and partitions where all the entries are odd and distinct. Give an elegant form for the generating function for the number of self-conjugate partitions.

The sequence $p(n)$ begins $p(1) = 1, 2, 3, 5, 7, 11, 15, 22, 30$; $p(10) = 42, 56, 77, 101, 135, 176, 231, 297, 385, 490$; $p(20) = 627,\ldots$; with $p(100) = 190569292$ and $p(1000) \approx 2.4 \times 10^{31}$. Ramanujan was intrigued by these numbers, both by their growth (which seems quite fast) and by their congruence conditions. For $n = 1000$ we see that there are roughly $10^{\sqrt{n}}$ partitions, which is an unusual function in mathematics. Hardy and Ramanujan proved the extraordinary asymptotic

(15.1.1) $$p(n) \sim \frac{1}{4n\sqrt{3}} e^{\pi\sqrt{2n/3}},$$

and Rademacher developed their idea to give an exact formula.[1]

Ramanujan also noted several congruences for the $p(n)$:

$$p(5k+4) \equiv 0 \pmod 5, \quad p(7k+5) \equiv 0 \pmod 7, \quad p(11k+6) \equiv 0 \pmod{11},$$

for all k. Notice that these are all of the form $p(n) \equiv 0 \pmod q$ whenever q divides $24n - 1$ and, for a long time, these seemed to be the only such congruences. However, Ken Ono recently found many more such congruences, only a little more complicated: For any prime $q \geq 5$ there exist primes ℓ such that $p(n) \equiv 0 \pmod q$ whenever $q\ell^3$ divides $24n - 1$.

15.2. Jacobi's triple product identity

There are many beautiful identities involving the power series from partitions. One of the most extraordinary is Jacobi's powerful *triple product identity*: If $|x| < 1$, then

$$\boxed{\prod_{n \geq 1}(1-x^{2n})(1+x^{2n-1}z)(1+x^{2n-1}z^{-1}) = \sum_{m \in \mathbb{Z}} x^{m^2} z^m.}$$

We will prove this in an exercise at the end of this section. For now, we shall determine some useful consequences: If we let $x = t$ and $z = 1$, we obtain

$$\prod_{n \geq 1}(1-t^{2n})(1+t^{2n-1})^2 = \sum_{m \in \mathbb{Z}} t^{m^2};$$

the sum on the right-hand side appeared when we were studying the number of representations of integers as the sums of four squares in section 12.18 of appendix 12E. Now, if we let $x = t$ and $z = -1$, then we obtain

(15.2.1) $$\prod_{n \geq 1}(1-t^{2n})(1-t^{2n-1})^2 = \sum_{m \in \mathbb{Z}}(-1)^m t^{m^2}.$$

Exercise 15.2.1. Prove that

$$\prod_{\ell \geq 1}(1-t^{\ell})(1-t^{2\ell-1}) = \prod_{n \geq 1}(1-t^{2n})(1-t^{2n-1})^2 = \sum_{m \in \mathbb{Z}}(-1)^m t^{m^2}.$$

Much more generally we can take $x = t^a$, $z = t^b$, and $n = k+1$ in Jacobi's triple product identity to obtain

$$\prod_{k \geq 0}(1-t^{2ak+2a})(1+t^{2ak+a+b})(1+t^{2ak+a-b}) = \sum_{m \in \mathbb{Z}} t^{am^2+bm}.$$

The special case $a = b = \frac{1}{2}$ yields

$$\prod_{k \geq 0}(1+t^k)(1-t^{2k+2}) = \sum_{m \in \mathbb{Z}} t^{\frac{m^2+m}{2}}.$$

---

[1] This proof gave birth to the *circle method*, still one of the most important techniques in number theory. Proving this result is too difficult for this book, but see [**Dav05**] for many uses of the circle method.

Exercise 15.2.2. By writing $1 + t^k$ as $(1-t^{2k})/(1-t^k)$ or otherwise, deduce that
$$\sum_{m \geq 0} t^{\frac{m^2+m}{2}} = \frac{(1-t^2)(1-t^4)(1-t^6)\cdots}{(1-t)(1-t^3)(1-t^5)\cdots}.$$

Letting $x = t^a$, $z = -t^b$, and $n = k+1$ in Jacobi's triple product identity we obtain
$$\prod_{k \geq 0}(1-t^{2ak+2a})(1-t^{2ak+a+b})(1-t^{2ak+a-b}) = \sum_{m \in \mathbb{Z}}(-1)^m t^{am^2+bm}.$$

The special case $a = \frac{3}{2}$, $b = \frac{1}{2}$ yields Euler's identity,
$$\prod_{n \geq 1}(1-t^n) = \sum_{m \in \mathbb{Z}}(-1)^m t^{\frac{3m^2+m}{2}}.$$

Exercise 15.2.3. Interpret this combinatorially, in terms of the number of partitions of m into unequal parts.

Exercise 15.2.4. Show that if $(12/\cdot)$ is the Jacobi symbol, then
$$t\prod_{n \geq 1}(1-t^{24n}) = \sum_{m \geq 1}\left(\frac{12}{m}\right)t^{m^2}.$$

Now let $z = -ux$ in Jacobi's triple product identity to obtain
$$\prod_{n \geq 1}(1-x^{2n})(1-ux^{2n})(1-x^{2n-2}u^{-1}) = \sum_{m \in \mathbb{Z}} x^{m^2+m}(-u)^m.$$

The third factor in the product on the left-hand side is $1 - u^{-1}$ at $n = 1$. The sum of the m and $-m-1$ terms on the right-hand side is
$$(-1)^m x^{m^2+m}(1-u^{-1})(u^m + u^{m-1} + \cdots + u^{-m}).$$

Dividing through by $1 - u^{-1}$ and then letting $u = 1$, we obtain
$$\prod_{n \geq 1}(1-x^{2n})^3 = \sum_{m \geq 0}(-1)^m(2m+1)x^{m^2+m}.$$

Writing $x = t^4$, with $a = 2m+1$, we have

(15.2.2)
$$t\prod_{n \geq 1}(1-t^{8n})^3 = \sum_{\substack{a \geq 1 \\ a \text{ odd}}}(-1)^{\frac{a-1}{2}} a t^{a^2}.$$

We now let $x = t^4$ and $b = 2m$ in exercise 15.2.1, and multiply this by (15.2.2) to obtain

(15.2.3)
$$t\prod_{n \geq 1}(1-t^{4n})^2(1-t^{8n})^2 = \sum_{\substack{a \geq 1, \, a \text{ odd} \\ b \in \mathbb{Z}, \, b \text{ even}}}(-1)^{\frac{a+b-1}{2}} a\, t^{a^2+b^2}.$$

What are the coefficients of t^p in this power series, when p is a prime? We know that $p = a^2 + b^2$ with a odd and b even, if and only if $p \equiv 1 \pmod 4$. In that case, a and b are unique up to sign, with $b \equiv \frac{p-1}{2} \pmod 4$, and if $A = (-1)^{\frac{a-1}{2}}a$, then $A \equiv 1 \pmod 4$. We deduce from Theorem 14.3 and exercise 14.5.2(e) that the coefficient of t^p equals (accounting for the two possible signs of b)
$$p - \#\{(x,y) \pmod p : y^2 \equiv x^3 - x \pmod p\}.$$

This is also true (but less interesting) for $p = 2$ and for the primes $p \equiv 3 \pmod 4$, hence for all primes p. This might seem like a not terribly interesting coincidence but surprisingly, this is not a coincidence, and it generalizes to an extraordinary extent: For any cubic polynomial $f(x)$ without repeated roots mod p, the quantity

$$p - \#\{(x,y) \pmod p : y^2 \equiv f(x) \pmod p\}$$

is the coefficient of t^p in a certain power series with very special properties. Proving a precise version of this is the key to Andrew Wiles's proof of Fermat's Last Theorem. There is no way for the reader to see the connection from what I have written here; that would take a whole other book: We will understand much more about this strange link between seemingly very different types of questions in [**Grab**].

Exercise 15.2.5. Write the power series on the right-hand side of (15.2.3) as $\sum_{n\geq 1} f(n)t^n$.
(a)† Prove that $f(pq) = f(p)f(q)$ for any distinct primes p, q.
(b)‡ Prove that $f(\cdot)$ is a multiplicative function.

Exercise 15.2.6 (Proof of Jacobi's triple product identity). Define

$$P_N(x,z) := \prod_{k=1}^{N} (1-x^{2k})(1+x^{2k-1}z)(1+x^{2k-1}z^{-1}) = \sum_{n\in\mathbb{Z}} c_{N,n}(x)\, z^n.$$

(a) Prove that $c_{N,n}(x) \in \mathbb{Z}[x]$; $c_{N,n}(x) = 0$ for $n > N$, and $c_{N,-n}(x) = c_{N,n}(x)$.
(b) Prove that $P_{N+1}(x,z) = (1-x^{2N+2})(1+x^{2N+1}z)(1+x^{2N+1}z^{-1})P_N(x,z)$.
(c)† Deduce that
$$c_{N+1,n}(x) = (1-x^{2N+2})(x^{2N+1}c_{N,n-1}(x) + (1+x^{4N+2})c_{N,n}(x) + x^{2N+1}c_{N,n+1}(x)).$$
(d) Prove that $c_{N,n}(x) = x^{n^2}\prod_{m=N-n+1}^{N}(1-x^{2m}) \cdot \prod_{m=N+n+1}^{2N}(1-x^{2m})$ for $0 \leq n \leq N$, for all $N \geq 1$.
(e) Show that if $|x| < 1$, then $\lim_{N\to\infty} c_{N,n}(x) = x^{n^2}$ for every integer n.
(f) Deduce Jacobi's triple product identity.

15.3. The Freiman-Ruzsa Theorem

We pass now to a much more modern theme in combinatorial number theory, a new subject called *additive combinatorics*, which combines older themes with new perspectives. The main question is to determine when sets of integers contain subsets with special structure. For example, any subset S of the integers up to N contains two consecutive integers if $|S| \geq \frac{N}{2} + 1$, and contains three consecutive integers if $|S| \geq \frac{2N}{3} + 2$, and these results are best possible. But in this latter case, three consecutive integers are a special example of three integers in an arithmetic progression, so we might ask how big S needs to be to guarantee that it contains a non-trivial three-term arithmetic progression, or even a k-term arithmetic progression. Another key theme is to start with a sparse set of integers, like, say, the squares, which have no four-term arithmetic progression (as we will establish in Theorem 17.5), and then ask whether we will obtain longer arithmetic progressions among the set of integers that are the sum of two squares, that is, by adding together elements of our original set.

For any finite sets of integers A and B, we define their *sumset*

$$A + B := \{a+b : a\in A, b\in B\}.$$

We write $2A$ for $A+A$. It is easy to see that $|2A| \leq |A|(|A|+1)/2$, since the distinct elements of $A + A$ are a subset of $\{a_i + a_j : 1 \leq i \leq j \leq |A|\}$.

Exercise 15.3.1. Give an example of a set A of n integers for which $|2A| = n(n+1)/2$.

Typically $|A + A|$ is large and is only small in very special circumstances:

Lemma 15.3.1. *If A and B are finite subsets of $\mathbb{Z}$, then $|A + B| \geq |A| + |B| - 1$. Equality holds if and only if A and B are each complete finite segments of an arithmetic progression to the same modulus (that is, $A = \{a, a+d, \ldots, a+(r-1)d\}$ and $B = \{b, b+d, \ldots, b+(s-1)d\}$ for some $a, b, r, s,$ and $d \geq 1$).*

Proof. Write the elements of A as $a_1 < a_2 < \cdots < a_r$, and those of B as $b_1 < b_2 < \cdots < b_s$. Then $A + B$ contains the $r + s - 1$ distinct integers

$$a_1 + b_1 < a_1 + b_2 < a_1 + b_3 < \cdots < a_1 + b_s < a_2 + b_s < a_3 + b_s < \cdots < a_r + b_s.$$

If $|A + B| = r + s - 1$, then these must be the same integers, in the same order, as

$$a_1 + b_1 < a_2 + b_1 < a_2 + b_2 < a_2 + b_3 < \cdots < a_2 + b_s < a_3 + b_s < \cdots < a_r + b_s.$$

Comparing terms, we have $a_1 + b_{i+1} = a_2 + b_i$ for $1 \leq i \leq s - 1$; that is, $b_j = b_1 + (j-1)d$ where $d = a_2 - a_1$. A similar argument with the roles of a and b swapped reveals our result. □

If $A + B$ is small, but not quite as small as $|A| + |B| - 1$, then we might expect some similar structure. However the combinatorics of comparing different sums quickly becomes very complicated.

Looking for other examples in which $A + B$ is small, one soon finds the possibility that A and B are each large subsets of complete finite segments of an arithmetic progression to the same modulus. For example, if A contains $2m$ integers from $\{1, 2, 3, \ldots, 3m\}$, then $A + A$ is a subset of $\{2, 3, 4, \ldots, 6m - 1, 6m\}$, and so $|A + A| < 3|A|$. One can find a criterion similar to Lemma 15.3.1: If $|A| \geq |B|$ and $|A + B| \leq |A| + 2|B| - 4$, then A and B are each subsets of arithmetic progressions with the same common difference, of lengths $\leq |A + B| - |B| + 1$ and $\leq |A + B| - |A| + 1$, respectively.

A further interesting example is given by

$$A = B = \{1, 2, \ldots, 10, 101, 102, \ldots, 110, 201, 202, \ldots, 210\}$$

or its large subsets. This can be written as $1 + \{0, 1, 2, \ldots, 9\} + \{0, 100, 200\}$, a translate of the sum of complete finite segments of two arithmetic progressions. More generally, define a *generalized arithmetic progression* $C = C(a_0, \ldots, a_k; N_1, \ldots, N_k)$ as

$$C := \{a_0 + a_1 n_1 + a_2 n_2 + \cdots + a_k n_k : 0 \leq n_j \leq N_j - 1 \text{ for } 1 \leq j \leq k\}$$

where $a_0, a_1, \ldots, a_k$ are given integers and $N_1, \ldots, N_k$ are given positive integers. Note that $C(a_0, a_1, \ldots, a_k; N_1, N_2, \ldots, N_k) = a_0 + \sum_{i=1}^{k} a_i \cdot \{0, 1, \ldots, N_i - 1\}$. This generalized arithmetic progression is said to have *dimension* k and *volume* $N_1 \cdots N_k$. Notice that

$$2C(a_0, a_1, \ldots, a_k; N_1, N_2, \ldots, N_k) = C(2a_0, a_1, \ldots, a_k; 2N_1 - 1, 2N_2 - 1, \ldots, 2N_k - 1),$$

so that $|2C| < 2^k |C|$ (as long as the elements of C are distinct). We think of C as an image of the k-dimensional lattice segment

$$S := \{0, 1, \ldots, N_1 - 1\} \times \{0, 1, \ldots, N_2 - 1\} \times \cdots \times \{0, 1, \ldots, N_k - 1\} \subset \mathbb{Z}^k.$$

15.3. The Freiman-Ruzsa Theorem

Here $|2S| < 2^k|S|$ and S is the set of lattice points inside a k-dimensional rectangle. We can replace the rectangle by the lattice points inside a "never too thin" convex, compact region of $\mathbb{R}^k$ and get the same bound $|2S| < 2^k|S|$.[2]

We can combine these two constructions so that if A is a large subset of a generalized arithmetic progression, then $|A+A| < K|A|$ for some smallish constant K.

Are there any other examples of sets A and B for which $A+B$ is small? Freiman showed that the answer is "no", having the extraordinary insight to suggest and prove that $A + A$ can be "small" if and only if it is a "large" subset of a "low"-dimensional generalized arithmetic progression of "not too big" volume.[3]

Theorem 15.1 (The Freiman-Ruzsa Theorem[4]). *For any constant $K \geq 2$ there exist constants $d(K)$ and $V(K)$ such that if A is a set of integers for which $|A+A| < K|A|$, then A is a subset of d-dimensional generalized arithmetic progression of volume $V \cdot |A|$, where $d \leq d(K)$ and $V \leq V(K)$.*

Define the *product set* $A \cdot B := \{ab : a \in A, b \in B\}$.

Exercise 15.3.2. Explain the bijection between $A \cdot B$ and $\log A + \log B$.

We have seen that if $A + A$ is small, then A has a lot of additive structure; that is, it is a low-dimensional subset of a low-volume generalized arithmetic progression, by the Freiman-Ruzsa Theorem. If $A \cdot A$ is small, then $\log A$ has a lot of additive structure by the Freiman-Ruzsa Theorem; that is, A has a lot of multiplicative structure. Can a set have both types of structure at once?

When A is the set of integers $\{1, 2, \ldots, N\}$, we have $|A + A| \leq 2N$, so it is small. We saw that $|A \cdot A| < \epsilon N^2$ (during our discussion of Erdős's multiplication table problem, in section 13.3). By taking products of pairs of primes $\leq N$, we see that $|A \cdot A| \geq \pi(N)^2/2 > N^2/3(\log N)^2$, so that $A \cdot A$ is not much smaller than N^2. One might guess that $A \cdot A$ is large whenever $A + A$ is small.

Erdős and Szemerédi conjectured that a set cannot have this kind of additive structure and multiplicative structure simultaneously, by predicting the *sum-product inequality*,

$$\max\{|A + A|, |A \cdot A|\} \geq c_\epsilon |A|^{2-\epsilon}$$

for some constant $c_\epsilon > 0$ for any $\epsilon > 0$. More daringly one might guess, from the same reasoning, that either $|A + B| \geq c_\epsilon(|A||B|)^{1-\epsilon}$ or $|A \cdot C| \geq c_\epsilon(|A||C|)^{1-\epsilon}$ for any finite sets of integers A, B, C. In 2009, Solymosi showed that if A and B are two finite sets of real numbers with $|A| \geq |B| > 1$, then

$$|A \cdot B||A + A||B + B| \geq c\frac{(|A||B|)^2}{\log |B|},$$

---

[2] We need to be cautious with thin regions. For example, the rectangle $R := \{(x, y) : \frac{1}{4} \leq x \leq \frac{3}{4}, 0 \leq y \leq N\}$ has volume $N/2$ and contains 0 lattice points, whereas $2R$ has volume $2N$ and contains $2N + 1$ lattice points.

[3] Freiman's 1962 proof is both long and difficult to understand. Ruzsa's 1994 proof of Freiman's result is elegant and introduced new techniques, heralding an explosion of ideas in this area.

[4] Who should get credit for a theorem? Only the person who had the insight and power to prove it? Or perhaps someone who gave a later proof that inspired others to be interested? There is no rule. Here we have chosen to give Ruzsa credit, as well as Freiman who had the original insight. In chapter 17 we discuss Mordell's Theorem, which is usually called the "Mordell-Weil Theorem", thanks to several brilliant insights that Weil brought to the subject, as well as an extraordinary generalization.

for some constant $c > 0$. We can deduce from this, for example, that if $|A + A| \leq K|A|$ and $|B + B| \leq K|B|$, then $|A \cdot B| \geq c' \frac{|A||B|}{\log |B|}$ where $c' = c/K^2$.

Exercise 15.3.3. Deduce the sum-product inequality for any $\epsilon > 2/3$.

15.4. Expansion and the Plünnecke-Ruzsa inequality

Suppose that A is an additive set with $|2A| \leq K|A|$. By the Freiman-Ruzsa Theorem, we know that $A \subset P$, where P is a generalized arithmetic progression of small dimension and A is a large portion of P. Adding several copies of the inclusion $A \subset P$ together, it follows that $kA \subset kP$ for any integer $k \geq 2$. The size of $|kP|$ grows in a fairly controlled manner as a function of k, which implies that $|kA|$ cannot grow too rapidly. Such a statement can be proven directly:[5]

Theorem 15.2 (Plünnecke-Ruzsa Theorem). *If A and B are finite sets of integers for which $|A + B| \leq K|A|$, then $|mB - nB| \leq K^{m+n}|A|$ for all integers $m, n \geq 0$.*

In particular, taking $B = A$ we see that if $|A + A| \leq K|A|$, then $|nA| \leq K^n|A|$ for all $n \geq 1$. Also, taking $B = A$ or $B = -A$ yields that if $|A \pm A| \leq K|A|$, then $|mA - nA| \leq K^{m+n}|A|$.

A first result in adding sets is the *Ruzsa triangle inequality* which states that for any given finite sets of integers U, V, W one has

(15.4.1) $$|V - W||U| \leq |V - U||U - W|.$$

Proof. We will define a map $\phi : (V - W) \times U \to (V - U) \times (U - W)$ and prove that it is an injection, which implies the result. Given $d \in V - W$ select a pair $v_d \in V, w_d \in W$ for which $d = v_d - w_d$ (there may be more than one such pair, but for each d we make a definite choice). Then define

$$\phi(d, u) = (v_d - u, u - w_d)$$

for each $d \in V - W$ and $u \in U$. To prove that ϕ is an injection, suppose that $(x, y) \in \text{Im}(\phi) \subseteq (V - U) \times (U - W)$. If $\phi(d, u) = (x, y)$, then $x + y = (v_d - u) + (u - w_d) = v_d - w_d = d$, and therefore we can determine d and hence v_d and w_d from (x, y). And we also determine u as $u = -x + v_d (= y - w_d)$. □

Our proof of the Plünnecke-Ruzsa Theorem (due to George Petridis) rests on the following "expansion" result.

Proposition 15.4.1. *Suppose A and B are sets of integers for which $|A + B| \leq K|A|$. Let X be that subset of A for which the ratio $|X + B|/|X|$ is minimal. Then*

$$|S + X + B| \leq K|S + X|,$$

for every finite set of integers S.

A fairly short proof of this result may be found in [**GG**]. An easy consequence is that

(15.4.2) $$|X + kB| \leq K^k|X|.$$

---

[5]Rather than as a consequence of the Freiman-Ruzsa Theorem. Indeed it is an important ingredient in Ruzsa's proof of the Freiman-Ruzsa Theorem.

We prove this by induction: It is trivial for $k = 0$. If it is true for $k - 1$, then, by Proposition 15.4.1 with $S = (k-1)B$, we have

$$|X + kB| = |X + S + B| \leq K|X + S| \leq K \cdot K^{k-1}|X| = K^k|X|.$$

Proof of the Plünnecke-Ruzsa Theorem. By Ruzsa's triangle inequality with U, V, W replaced by $X, -mB, -nB$, respectively, and then (15.4.2), we have

$$|mB - nB| \, |X| \leq |X + mB| \cdot |X + nB| \leq K^{m+n}|X|^2.$$

Therefore $|mB - nB| \leq K^{m+n}|X| \leq K^{m+n}|A|$ as $X \subseteq A$. $\square$

Another use of Proposition 15.4.1 comes in noting that if $x \in X$, then $B + S + x \subset B + S + X$, so that $|B + S| = |B + S + x| \leq |B + S + X| \leq K|X + S| \leq K|A + S|$; that is, $|B + S||A| \leq |A + B||A + S|$. Changing notation gives

(15.4.3) $$|V + W||U| \leq |V + U||U + W|,$$

the *Petridis triangle inequality*, which nicely complements the Ruzsa triangle inequality.

15.5. Schnirel′man's Theorem

We call a set of non-negative integers A an *asymptotic basis of order h* if every sufficiently large integer is the sum of at most h elements of A. The most famous example is Goldbach's conjecture which asserts that every even integer ≥ 4 is the sum of two primes. This implies that every integer ≥ 2 is the sum of at most 3 primes (by taking $n - 3$ as the sum of two primes for every odd integer $n \geq 7$). Therefore the Goldbach conjecture implies that the primes form an asymptotic basis of order 3.

Suppose that A contains 0. Then

$$A \subseteq 2A \subseteq 3A \subseteq \cdots \subseteq hA.$$

Therefore A is an asymptotic basis of order h if and only if every sufficiently large integer is the sum of exactly h elements of A (that is, $\mathbb{N} \setminus hA$ is finite). A pervasive phenomenon in additive number theory is the enrichment of structure that one sees when moving from A to $2A$, to $3A$, and so on. In this section we explore ideas of Schnirel′man which allow us to quantify the intuition that the sets jA should get bigger as j increase.

We can quantify an infinite set of integers, A, by various notions of *density*. The most useful for studying the addition of sets is *Schnirel′man density*. The Schnirel′man density of a set $A \subset \mathbb{Z}$ is denoted by $\sigma(A)$, and it is defined by

$$\sigma(A) := \inf_{n \geq 1} \frac{A[n]}{n},$$

where $A[n] := \#\{a \in A : 1 \leq a \leq n\}$. One obvious consequence of the definition of Schnirel′man density is that

$$A[n] \geq n\sigma(A) \text{ for all } n \geq 1.$$

Taking $n = 1$, we immediately see that the Schnirel′man density has the slightly strange property that $\sigma(A) = 0$ unless $1 \in A$.

Using the pigeonhole principle we prove the following simple, but useful, fact about Schnirel′man density.

Lemma 15.5.1. *Let A and B be sets of integers, both containing 0. If $\sigma(A) + \sigma(B) \geq 1$, then $A + B$ contains all nonnegative integers.*

Proof. We have $0 = 0 + 0 \in A + B$. Suppose that $n \geq 1$ and $n \notin A + B$. Then $n \notin A$, or else $n = 0 + n \in A + B$. Similarly $n \notin B$. Moreover we cannot have $a \in A$ and $n - a \in B$, for any a, $0 \leq a \leq n$, or else $n = a + (n - a) \in A + B$. This implies that $1_A(a) + 1_B(n - a) \leq 1$ for all such a (where 1_A is the characteristic function for A). These two observations together imply that

$$A[n] + B[n] = A[n-1] + B[n-1] = \sum_{a=1}^{n-1} \bigl(1_A(a) + 1_B(n-a)\bigr) \leq n - 1.$$

On the other hand

$$A[n] + B[n] \geq \sigma(A)n + \sigma(B)n \geq n.$$

These two inequalities are contradictory, and so the result follows. □

Next we come to a more serious fact, which quantifies the intuition that $A + B$ is measurably bigger than A and B.

Theorem 15.3 (Schnirel′man's Theorem). *Let A and B be sets of integers, and suppose that $1 \in A$ and $0 \in B$. Then $\sigma(A + B) \geq \sigma(A) + \sigma(B) - \sigma(A)\sigma(B)$.*

This can be reformulated as $1 - \sigma(A + B) \leq (1 - \sigma(A))(1 - \sigma(B))$.

Proof. We denote the elements of A up to n as $1 = a_1 < \cdots < a_k \leq n$, so that $k \geq \sigma(A)n$. We write $a_{k+1} := n + 1$ for convenience in this proof (even though this element might not lie in A). We obtain a lower bound on $(A + B)[n]$ by counting only those elements of $A + B$ that lie in $[a_i, a_{i+1})$ which take the form $a_i + b$ with $b \in B$, so that $0 \leq b \leq a_{i+1} - a_i - 1$. Therefore

$$|(A+B) \cap [a_i, a_{i+1})| \geq |B \cap \{0, \ldots, a_{i+1} - a_i - 1\}| = 1 + B[a_{i+1} - a_i - 1]$$
$$\geq 1 + \sigma(B)(a_{i+1} - a_i - 1) = \sigma(B)(a_{i+1} - a_i) + 1 - \sigma(B).$$

Since the intervals $[a_i, a_{i+1})$, $1 \leq i \leq k$, partition the integers in $[1, n]$, we deduce that

$$(A+B)[n] = \sum_{i=1}^{k} |(A+B) \cap [a_i, a_{i+1})|$$
$$\geq \sum_{i=1}^{k} (\sigma(B)(a_{i+1} - a_i) + 1 - \sigma(B))$$
$$= \sigma(B)n + k(1 - \sigma(B)),$$

and the result follows as $k \geq \sigma(A)n$. □

An immediate consequence of the above two results is that any set containing 0, with positive Schnirel′man density, is an asymptotic basis of some finite order h depending only on $\sigma(A)$. In fact, better than this, hA contains *all* nonnegative integers.

Corollary 15.5.1. *Let A be a set of integers with $0 \in A$ and $\sigma(A) > 0$. Then $hA = \mathbb{N}$ whenever $h \geq 2\lceil (\log 2)/(-\log(1-\sigma(A)))\rceil$. (Here $\lceil t \rceil$ denotes the smallest integer $\geq t$.)*

Proof. As $\sigma(A) > 0$ we know that $1 \in A$, and hence (since $0 \in A$) $1 \in kA$ for all $k \geq 1$. We deduce, by induction from Schnirel'man's Theorem, that $1 - \sigma(kA) \leq (1-\sigma(A))^k$ for all $k \geq 1$. If $k \geq \lceil (\log 2)/(-\log(1-\sigma(A)))\rceil$, then $(1-\sigma(A))^k \leq \frac{1}{2}$, and therefore $\sigma(kA) \geq 1/2$. It follows from Lemma 15.5.1 that, for any such k, $2kA = kA + kA$ contains all nonnegative integers. $\square$

In 1942 Mann improved Schnirel'man's Theorem to
$$\sigma(A+B) \geq \min\{\sigma(A)+\sigma(B), 1\}$$
which is an optimal result of this type.

Schnirel'man's Theorem leads to the first and easiest proof that there exists an integer k such that every integer ≥ 2 is the sum of at most k primes; that is, $\mathcal{P}$, the set of primes, is an asymptotic basis for the positive integers. To prove this one first establishes that $\sigma(2\mathcal{P}-4) > 0$ (a result which is proved using sieve theory; see [**Grab**]). The result then follows from Corollary 15.5.1. In 1937 Vinogradov proved that every sufficiently large odd integer is the sum of three primes (which is proved in [**Graa**]), and so $\mathcal{P}$ is an asymptotic basis of order 4 for the positive integers.

15.6. Classical additive number theory

Given a largish subset of the integers up to N one can ask whether it contains certain simple structures, because of its large size. Or, rather than quantify "largish", one might instead partition the integers into two (or more) subsets and ask whether any of these subsets have the given structure. This is a familiar theme from combinatorics, and ideas from that subject will allow us to give a first answer to these questions.

We begin with a well-known result from graph theory:

Lemma 15.6.1. *For all integers $r \geq 1$ there exists a constant $N(r)$ such that if the edges of the complete graph with N vertices are colored with r colors, where $N \geq N(r)$, then there is a monochromatic triangle (that is, the edges of some triangle in the graph all have the same color).*

Proof. By induction on $r \geq 1$. Evidently $N(1) = 3$. For larger r consider the $N-1$ edges attached to any one vertex v. If $N-1 \geq rN(r-1)$, then, by the pigeonhole principle, there must be some color c for which there are $\geq N(r-1)$ edges adjacent to v of color c. Let H be those vertices that share an edge of color c with v. If there are any two vertices in H that are attached by an edge of color c, then these two vertices along with v form a monochromatic triangle. Otherwise the edges of H are colored by just $r-1$ colors and the result follows by induction. $\square$

Exercise 15.6.1. Justify that if $N \geq r(N(r-1)-1)+2$, then there must be some color c for which there are $\geq N(r-1)$ edges adjacent to v of color c. Show by induction that we may take $N(r) \leq 3\,r!$.

This is a typical Ramsey theory proof, involving a greedy-type algorithm, and leads to bounds that are probably far larger than the best bounds possible. However there are questions in the subject in which the bounds must grow extraordinarily fast and cannot be given in terms of *primitive recursive functions*.

There is a quite beautiful corollary:

Theorem 15.4 (Schur's Theorem). *For all integers $r \geq 1$ there exists a constant $N(r)$ such that if the integers up to N are colored with r colors, where $N \geq N(r)$, then there is a monochromatic solution to $x+y = z$ in positive integers $x, y, z \leq N$.*

Proof. We construct the complete graph on N vertices, labeled $1, 2, \ldots, N$, and joining vertices i and j with an edge that has the color of $|j - i|$. Lemma 15.6.1 tells us that there is a monochromatic triangle joining, say, the vertices with labels $i < j < k$. Therefore if $x = j - i$, $y = k - j$, and $z = k - i$, we know that these positive integers all have the same color and indeed satisfy $x + y = z$. □

In 1927 van der Waerden answered a conjecture of Schur, by showing that if the positive integers are partitioned into two sets, then one set must contain arbitrarily long arithmetic progressions.

Theorem 15.5 (van der Waerden's Theorem, 1927). *For any given integers $r \geq 2$ and $k \geq 3$ there exists a constant $W(r, k)$ such that however we color the positive integers $\leq W(r, k)$ with r colors, there is a monochromatic k-term arithmetic progression.*

Exercise 15.6.2. Prove that $W(2, 3) = 9$.

Exercise 15.6.3. Deduce that if the positive integers are partitioned into r sets, then at least one of the sets must contain arbitrarily long arithmetic progressions.

Exercise 15.6.4. Prove that if we color any arithmetic progression of integers of length $W(r, k)$ with r colors, then it will contain a monochromatic k-term arithmetic progression.

Proof for all $r > 2$ given that the $W(2, k)$ exist. (This simple proof explains why Schur only asked for two colors.) For each fixed k we prove our result by induction on $r \geq 2$, assuming the $r = 2$ case.

For $r > 2$, let $W := W(r - 1, k)$. Suppose that the positive integers $\leq N := W(2, W)$ are colored with the colors $1, 2, \ldots, r$. We will combine colors $1, 2, \ldots, r-1$ into a new color which we will call color 0, so the integers up to N are colored by the two colors 0 and r. By the definition of N, there is either a W-term arithmetic progression of color r, which is far more than we were seeking (as $W = W(r-1, k) > k$), or there is a W-term arithmetic progression of color 0. Reverting back to our original coloring, this means that there is an arithmetic progression of integers of length $W(r - 1, k)$, amongst the positive integers $\leq N$, which is colored with $r - 1$ colors. The result then follows from exercise 15.6.4. □

Exercise 15.6.5. Partition the integers into two sets neither of which has an infinitely long arithmetic progression.

A set of integers $A \subset \mathbb{N}$ has *positive lower density* if there exists $\delta > 0$ such that $A[n] > \delta n$ for all sufficiently large n. The most important result in this area

15.6. Classical additive number theory

is the following result of Szemerédi. It tells us that any set of integers of positive lower density contains arithmetic progressions of *arbitrary* (finite) length.

Theorem 15.6 (Szemerédi's Theorem, 1975). *For any $\delta > 0$ and integer $k \geq 3$, there exists an integer $N_{k,\delta}$ such that if $N \geq N_{k,\delta}$ and $A \subset \{1, 2, \ldots, N\}$ with $|A| \geq \delta N$, then A contains an arithmetic progression of length k.*

Exercise 15.6.6. Show that if $N \geq N_{k,\delta}$ and A is a subset of an arithmetic progression of length N, with $|A| \geq \delta N$, then A contains an arithmetic progression of length k.

Exercise 15.6.7. Deduce van der Waerden's Theorem from Szemerédi's Theorem.

The $k = 3$ case was first proved in a cunning proof by Roth in 1952 using Fourier analysis. In 1969 Szemerédi proved the $k = 4$ case by combinatorial methods and extended this to all k in 1975. In 1977 Furstenberg reproved Szemerédi's Theorem in a very surprising manner, using ergodic theory. It was not until 1998 that Tim Gowers finally gave an analytic proof of Szemerédi's Theorem, the proof based on the overall plan of Roth, but involving a new kind of higher-dimensional analysis (partly based on the Freiman-Ruzsa Theorem). Gowers's proof was the starting point for the proof of the following theorem:

Theorem 15.7 (Green and Tao, 2008). *For any integer N one can find (infinitely many different) pairs of integers $a, d \geq 1$, such that $a, a + d, \ldots, a + (N - 1)d$ are all primes.*

We discussed this great result, and its corollaries, in various appendices of chapter 5.

How much further can one develop Szemerédi's Theorem? Erdős conjectured that any set A of positive integers for which

$$\sum_{a \in A} \frac{1}{a} = \infty$$

must contain arbitrarily long arithmetic progressions, a question that is still very open today. Erdős stated this conjecture as a means to prove that there are arbitrarily long arithmetic progressions of primes (but this is not how Green and Tao proceeded). Can Erdős's conjecture even be proved in the $k = 3$ case?

How large is the largest subset $S(N)$ of $\{1, 2, \ldots, N\}$ that has no three-term arithmetic progression? If one could show that $|S(N)| < N/\log N$, then one would know that there are infinitely many three-term arithmetic progressions of primes. Recently Thomas Bloom, improving on work of Tom Sanders, came agonizingly close to this goal by showing that $|S(N)| < c(\log \log N)^4 \cdot N/\log N$ for some constant $c > 0$. Is this close to the true size of $S(N)$? The best we know is the far smaller lower bound $S(N) > Ne^{-c\sqrt{\log N}}$ for some $c > 0$ given by a beautiful construction of Behrend:

Exercise 15.6.8. Show that a, b, c are in arithmetic progression if and only if $a + c = 2b$.

Exercise 15.6.9. Write $a = \sum_{i=1}^{k} a_i (2m)^{i-1} \in C := C(0, 1, 2m, \ldots, (2m)^{k-1}; m, \ldots, m)$.
 (a) Show that a is an integer in the range $0 \leq a \leq (2m)^k - 1$.
 (b) Show that $a, b, c \in C$ are in arithmetic progression if and only if the vectors $\mathbf{a} = (a_1, \ldots, a_k)$, $\mathbf{b}$, and $\mathbf{c}$ are collinear.
 (c) Show that $|\mathbf{a}|^2$ is an integer in the range $0 \leq |\mathbf{a}|^2 \leq km^2$.

(d) Let $C_r = \{\mathbf{a} \in C : |\mathbf{a}| = r\}$. Show no three distinct elements of C_r are collinear.
(e) Prove that there exists an r for which C_r contains $\geq m^k/(1+km^2)$ elements.
(f)[†] By selecting $k = [\sqrt{\log N}]$ and $m = [N^{1/k}/2]$ prove that if N is sufficiently large, then $S(N) > Ne^{-c\sqrt{\log N}}$ for some constant $c > 0$.

In this last exercise we found a large subset of $\{0, 1, \ldots, 2m-1\}^k \subset \mathbb{Z}^k$ with no three-term arithmetic progression and used this to construct a large subset of $\{1, 2, \ldots, N\}$ that has no three-term arithmetic progression. This suggests extending the problem in several directions, most fruitfully to $\mathbb{F}_3^k$, where $\mathbb{F}_3$ is the field of 3 elements (see section 7.25 in appendix 7E for details). This differs from $\{0,1,2\}^k$ because, for example, $(0,0), (1,2), (2,1)$ is a three-term arithmetic progression in $\mathbb{F}_3^2$ but not in $\{0,1,2\}^2$. It is easier to do arithmetic in a field than in the integers (which is only a ring), and so researchers have long developed techniques for bounding the largest subset of $\{1, 2, \ldots, N\}$ that has no three-term arithmetic progression, by bounding the largest subset of $\mathbb{F}_3^k$ that has no three-term arithmetic progression, where 3^k is about size N. However this approach disintegrated in 2016 when Croot, Lev, and Pach showed that much smaller sets in $\mathbb{F}_q^k$ must contain three-term arithmetic progressions. Indeed it is now known [5] that if A is a subset of $\mathbb{F}_3^k$ with $\geq 2.756^k$ elements, then A contains a three-term arithmetic progression. (Since $\mathbb{F}_3^k$ contains $N := 3^k$ elements, then the bound here is $\geq N^{.923}$, much smaller than $\frac{N}{\log N}$ which is the much sought-after bound in the integer case.)

15.7. Challenging problems

Exercise 15.7.1.[†] Prove that for every set of integers $a_1, \ldots, a_n$ there exists a non-empty subset which sums to an integer divisible by n.

Exercise 15.7.2.[†] Suppose that $1 \leq a_1 < \cdots < a_n$ are positive integers, and let the integers $0 = b_1 < b_2 \leq \cdots \leq b_{2^n}$ be all the sums of subsets of the a_i, that is, the numbers $\sum_{i \in I} a_i$ for each $I \subset \{1, 2, \ldots, n\}$.
(a) Write down the generating function, $\sum_{j=1}^{2^n} x^{b_j}$, in terms of polynomials involving the a_i. Henceforth we will assume that the b_j are all distinct.
(b) Prove that $b_j \geq j - 1$ for each j.
(c) Deduce that if $0 \leq x \leq 1$, then $\sum_{j=1}^{2^n} x^{b_j} \leq \frac{1-x^{2^n}}{1-x}$.
(d) Deduce further that $\prod_{i=1}^n (1 + x^{a_i}) \leq \prod_{k=1}^n (1 + x^{2^{k-1}})$.
(e) Prove that if $a > 0$, then $\int_0^1 \frac{\log(1+x^a)}{x} dx = \frac{1}{a} \int_0^1 \frac{\log(1+y)}{y} dy$.
(f) Deduce that $\sum_{i=1}^n \frac{1}{a_i} \leq 2 - \frac{1}{2^n}$ with equality only when each $a_i = 2^{i-1}$.

Exercise 15.7.3.[†] Let $a_1, \ldots, a_N$ be a sequence of distinct real numbers. Prove that if m is the length of the longest decreasing subsequence, and n is the length of the longest increasing subsequence, then $mn \geq N$.

We now prove the following related result using techniques from section 14.6.

Theorem 15.8 (Erdős-Ginzburg-Ziv). *Let p be prime. Any sequence of $2p-1$ integers contains a subsequence of size p that sums to an integer divisible by p. In fact we show that the number of such subsequences is $\equiv 1 \pmod{p}$.*[6]

---

[6]Erdős-Ginzburg-Ziv actually showed that for any integer n, any sequence of $2n-1$ integers contains a subsequence of size n that sums to an integer divisible by n.

Proof. The number of subsets of $\{a_1, \ldots, a_{2p-1}\}$ of size p that sum to 0 (mod p) is

$$\equiv \sum_{\substack{I \subset \{1,\ldots,2p-1\} \\ |I|=p}} \left(1 - \left(\sum_{i \in I} a_i\right)^{p-1}\right) = \binom{2p-1}{p} - \sum_e c_e a_{i_1}^{e_1} \cdots a_{i_k}^{e_k} \pmod{p}.$$

Here the $e_i \geq 1$ and sum to $p-1$ (so that $1 \leq k \leq p-1$). The coefficient c_e consists of $\frac{(p-1)!}{e_1! \cdots e_k!}$ from expanding the power, multiplied by how often this coefficient occurs. This is $\binom{2p-1-k}{p-k}$, the number of subsets $I \subset \{1, \ldots, 2p-1\}$ of size p which contain $\{i_1, \ldots, i_k\}$. But $\binom{2p-1-k}{p-k} \equiv 0 \pmod{p}$ for $1 \leq k \leq p-1$ as p divides the numerator but not the denominator, and $\binom{2p-1}{p} \equiv 1 \pmod{p}$. $\square$

Further reading for chapter 15

[1] Scott Ahlgren and Ken Ono, *Addition and counting: The arithmetic of partitions*, Notices Amer. Math. Soc. **48** (2001), 978–984.

[2] Henry L. Alder, *Partition identities—from Euler to the present*, Amer. Math. Monthly **76** (1969), 733–746.

[3] George E. Andrews, *Euler's "De Partitio numerorum"*, Bull. Amer. Math. Soc. **44** (2007), 561–573.

[4] George E. Andrews and F. G. Garvan, *Dyson's crank of a partition*, Bull. Amer. Math. Soc. **18** (1988), 167–171.

[5] Jordan S. Ellenberg and Dion Gijswijt, *On large subsets of $\mathbb{F}_q^n$ with no three-term arithmetic progression*, Annals Math. **185** (2017), 339–343.

[6] Izabella Łaba, *From harmonic analysis to arithmetic combinatorics*, Bull. Amer. Math. Soc. **45** (2008), 77–115.

Appendix 15A. Summing sets modulo p

We have seen in Lemma 15.3.1 that if we add two finite sets of integers A and B, then $|A+B| \geq |A|+|B|-1$. Moreover much of the rest of chapter 15 has focussed on the expansion properties of adding the same set to itself several times. Here we focus on summing sets (mod n).

15.8. The Cauchy-Davenport Theorem

Theorem 15.9 (Cauchy-Davenport)**.** *Let p be a prime, and suppose that A and B are non-empty subsets of $\mathbb{Z}/p\mathbb{Z}$. Then*

$$|A+B| \geq \min(|A|+|B|-1, p).$$

Proof. The theorem is unaffected by translating A and B, that is, replacing A by $A+x$ and B by $B+y$. Sometimes it will be convenient to perform such translations.

Another useful transformation is to replace the pair of sets A, B by their *union* $U := A \cup B$ and their *intersection* $I := A \cap B$. Note that $|U|+|I| = |A|+|B|$ and $U + I \subseteq A + B$, so that $|A+B| \geq |U+I|$. Therefore if we can prove the theorem for the pair I, U, then it follows for the original pair A, B.

Suppose without loss of generality that $|A| \geq |B|$. By translating A and B appropriately we may assume that $0 \in A, B$. We will prove our result by induction on $|B|$. If $|B| = 1$, then the result is trivial. If $|B| > 1$ and there exist $a \in A, b \in B$ for which $B - b \not\subseteq A - a$, then the result follows by applying the induction hypothesis to the pair $I = (A-a) \cap (B-b)$, $U = (A-a) \cup (B-b)$, since in this case $|I| < |B|$. Hence we may assume that $B - b \subseteq A - a$, and therefore that $B - b + a \subseteq A$ for all $a \in A$ and $b \in B$. Taking the union over all $a \in A, b \in B$ we deduce that $B - B + A \subseteq A$.

From this we see that $2B - 2B + A = B - B + (B - B + A) \subseteq B - B + A \subseteq A$, and then, by induction, that $kB - kB + A \subseteq A$ for all $k \geq 1$. Since $0 \in B$ we deduce

15.8. The Cauchy-Davenport Theorem

that if H is the subgroup of $\mathbb{Z}/p\mathbb{Z}$ generated additively by B, then $A + H = A$. But there are only two subgroups of $\mathbb{Z}/p\mathbb{Z}$: either $H = \{0\}$ or $H = \mathbb{Z}/p\mathbb{Z}$. In the first case we have $B = \{0\}$ since $B \subseteq H$. The theorem is trivial in this case. In the second case we have $A = \mathbb{Z}/p\mathbb{Z}$, in which case the result is also trivial. □

Exercise 15.8.1. Suppose that A is a subset of $\mathbb{Z}/p\mathbb{Z}$ with at least two elements. Show that if $n \geq \frac{p-1}{|A|-1}$, then $nA = \mathbb{Z}/p\mathbb{Z}$.

Exercise 15.8.2. Use Theorem 15.9 to show that if A and B are finite sets of integers, then $|A + B| \geq |A| + |B| - 1$.

We can try to modify this argument to work in $\mathbb{Z}/n\mathbb{Z}$ or other additive groups G. Some care is needed here for, if H is a subgroup of G, then $H + H = H$, so there is no expansion. Moreover if $A = A_0 + H$ and $B = B_0 + H$ are unions of cosets of H (with A_0 and B_0 minimal), then $A + B = (A_0 + B_0) + H$, and therefore

$$|A + B| - |A| - |B| = |H|(|A_0 + B_0| - |A_0| - |B_0|);$$

in the particular case that $|A_0| = |B_0| = 1$, we obtain $|A + B| = |A| + |B| - |H|$.

To state the best possible result in general finite additive groups G, we need to define the *stabilizer* of A, a subset of G, to be

$$\text{Stab}(A) := \{g \in G : g + A = A\}.$$

One can see that $H := \text{Stab}(A)$ is a subgroup of G, with the property that $A + H = A$, and so A is a union of cosets of $\text{Stab}(A)$.

Theorem 15.10 (Kneser's Theorem). *If A and B are finite subsets of an additive group G and $H = \text{Stab}(A + B)$, then*

$$|A + B| \geq |A + H| + |B + H| - |H|.$$

Exercise 15.8.3. Suppose that A is a subset of $\mathbb{Z}/N\mathbb{Z}$ and that A additively generates all of $\mathbb{Z}/N\mathbb{Z}$; that is, there exists r for which $rA = \mathbb{Z}/N\mathbb{Z}$. Prove that $NA = \mathbb{Z}/N\mathbb{Z}$.

Exercise 15.8.4. We give another proof of Theorem 15.8.
 (a) Show that any sequence of $2m$ (not necessarily distinct) residues mod p either has $m + 1$ identical residues, or can be partitioned into m sets of two distinct residues.
 (b) Prove that if $A_1, \ldots, A_{p-1}$ are subsets of $\mathbb{Z}/p\mathbb{Z}$ which each contain two distinct residues, then $A_1 + \cdots + A_{p-1} = \mathbb{Z}/p\mathbb{Z}$.
 (c) Deduce Theorem 15.8.

Appendix 15B. Summing sets of integers

15.9. The Frobenius postage stamp problem, III

If we are allowed to use no more than N postage stamps, which have face values a and b where $(a, b) = 1$, then what amounts of postage can we cover? In other words, can we understand the set

$$\{am + bn : m, n \in \mathbb{Z}, \ m, n \geq 0, \ m + n \leq N\}?$$

This can be rephrased in the language of additive combinatorics: Let $A = \{0, a, b\}$. We wish to determine NA. Note that $A \subset 2A \subset 3A \subset \cdots$. Recall that in Theorem 3.12, we proved that every positive integer outside the set $\mathcal{E}(a,b)$ can be written as $am + bn$ with $m, n \geq 0$, where $\mathcal{E}(a, b)$ is a subset of the integers $\leq mn - m - n$.

Theorem 15.11. *Suppose that $1 \leq a < b$ are coprime integers. If $N \geq b$, then*

$$NA = \{n \in \mathbb{Z} : \ 0 \leq n \leq bN\} \setminus \mathcal{E}(a, b) \setminus (bN - \mathcal{E}(b - a, b)).$$

Here $bN - \mathcal{E}(c, b) = \{bN - r : \ r \in \mathcal{E}(c, b)\}$.

Proof. If $r \in NA$, then $0 = a \cdot 0 + b \cdot 0 \leq am + bn = r \leq a \cdot 0 + b \cdot N = bN$. We now prove that if r is any integer in the range

$$ab - a - b < r < bN - ((b-a)b - (b-a) - b),$$

then $r \in NA$. Select m to be the least non-negative residue of $r/a \pmod{b}$, so that $0 \leq m \leq b - 1$ and $r \equiv am \pmod{b}$. This implies that there exists n for which $r = am + bn$. Now $bn = r - am > a(b - 1 - m) - b \geq -b$ as $r > ab - a - b$, and so $n \geq 0$. On the other hand,

$$bn = r - am < bN - (b-a)(b-1) + b - am \leq bN - (b-a)m + b - am = b(N+1-m)$$

and so $m + n \leq N$. Therefore $r \in NA$.

Now suppose that $0 \leq r \leq ab - a - b$ and $r \notin \mathcal{E}(a,b)$. As $r \notin \mathcal{E}(a,b)$ we can write $r = am + bn$ for some integers $m, n \geq 0$. Therefore $a(m+n) \leq r \leq ab - a - b < a(b-2)$ and so $m + n \leq b - 3$. Hence there exists $\ell \leq b - 3$ for which $r \in \ell A \subset bA \subset NA$.

Now suppose that $bN - ((b-a)b - (b-a) - b) \leq R \leq bN$ and $R \notin bN - \mathcal{E}(b-a, b)$. Let $r = bN - R$ so that $0 \leq r \leq (b-a)b - (b-a) - b$ with $r \notin \mathcal{E}(b-a, b)$. By the previous paragraph (with a replaced by $b - a$), we see that $r = u(b-a) + vb$ with $u + v \leq b - 3$. Therefore
$$R = bN - r = ua + (N - u - v)b = ma + nb$$
with $0 \leq m = u$, $n = N - u - v$, and $m + n = u + (N - u - v) = N - v \leq N$ so that $R \in NA$. $\square$

Exercise 15.9.1. Let $A = \{a, b, c\}$ where $a < b < c$ are integers for which $(b-a, c-b) = 1$. Prove that if N is sufficiently large, then
$$NA = \{n \in \mathbb{Z} \colon aN \leq n \leq cN\} \setminus (aN + \mathcal{E}(b-a, c-a)) \setminus (cN - \mathcal{E}(c-b, c-a)).$$

By exercise 3.25.5(b) we know that if $a_1 < \cdots < a_k$ are positive integers for which $(a_1, \ldots, a_k) = 1$, then there exists a finite set $\mathcal{E}(a_1, \ldots, a_k)$ for which
$$\mathcal{P}(a_1, \ldots, a_k) := \{a_1 n_1 + \cdots + a_k n_k \colon n_1, \ldots, n_k \in \mathbb{Z}_{\geq 0}\} = \{m \geq 0\} - \mathcal{E}(a_1, \ldots, a_k).$$

Exercise 15.9.2. Suppose that $a_1 = 0$.
  (a) Use exercise 15.8.3 to show that if $m \geq a_k^2$, then $m \in \mathcal{P}(a_1, \ldots, a_k)$.
  (b) Let $N \geq 2a_k$. Prove that if $a_k^2 \leq m \leq a_k(N - a_k)$, then $m \in NA$.
  (c) Let $N \geq a_k^2$. Prove that if $m \leq a_k^2$ and $m \in \mathcal{P}(a_1, \ldots, a_k)$, then $m \in NA$.
  (d) Let $N \geq a_k^2$. Deduce that if $m \geq a_k N - a_k^2$ and $m \in a_k N - \mathcal{P}(a_k - a_1, \ldots, a_k - a_k)$, then $m \in NA$.
  (e) Prove that if N is sufficiently large, then
$$NA = \{n \in \mathbb{Z} \colon 0 \leq n \leq a_k N\} \setminus \mathcal{E}(a_1, \ldots, a_k) \setminus (a_k N - \mathcal{E}(a_k - a_1, \ldots, a_k - a_k)).$$
  (f) State the general result when a_1 is not necessarily 0.

Chapter 16

The p-adic numbers

16.1. The p-adic norm

For a given prime number p and integer n we define $v_p(n)$ to be the maximum power of p that divides n; that is, $p^{v_p(n)}\|n$. For example $v_2(60) = 2$ whereas $v_3(60) = 1$, $v_5(60) = 1$, and $v_7(60) = 0$, since $60 = 2^2 \cdot 3^1 \cdot 5^1 \cdot 7^0$. We can extend this definition to all rational numbers r, for if we write $r = a/b$ where a and b are integers, then let $v_p(r) = v_p(a) - v_p(b)$. For example $v_2(49/60) = -2$, $v_3(49/60) = -1$, $v_5(49/60) = -1$, and $v_7(49/60) = 2$. Notice that, in this definition, it does not matter if a and b have a common factor. Moreover we can always write $r = p^{v_p(r)} m/n$ where m and n are integers, neither of which are divisible by p. We define $\mathbb{Z}_{(p)} = \{r \in \mathbb{Q} : v_p(r) \geq 0\}$, which is a subring of $\mathbb{Q}$.

The size, $|r|$, of a given number r, measures how far that number is from 0. We now define a notion of size in terms of the power of p dividing the given rational number r, measuring how far away r is from 0 in terms of p-divisibility. Since 0 is divisible by any arbitrarily large power of p, the higher the power of p that divides r, the closer it is to 0, and thus the smaller it is in the p-adic norm. This justifies defining the *p-adic norm*, $|.|_p$, by

$$|r|_p = p^{-v_p(r)}.$$

For example, $|49/60|_2 = 2^2$, $|49/60|_3 = 3$, $|49/60|_5 = 5$, $|49/60|_2 = 7^{-2}$, and $|49/60|_p = 1$ for all primes $p > 7$. The inverses of all of these prime powers appear in the factorization $49/60 = 2^{-2} 3^{-1} 5^{-1} 7^2$, which leads us to the *product formula*:

(16.1.1) $$|r| \cdot \prod_{p \text{ prime}} |r|_p = 1.$$

In this context, it is convenient to write $|r|_\infty$ in place of $|r|$. (16.1.1) is reminiscent of Hilbert's product formula, which was stated in Theorem 9.6 of section 9.10 in appendix 9B.

Exercise 16.1.1. Prove the product formula for all non-zero rational numbers r.

Any norm introduces a notion of distance. The key to the Euclidean norm is the triangle inequality

(16.1.2) $$|a - c| \leq |a - b| + |b - c|$$

so that the distance from a to c is no more than the distance from a to b plus the distance from b to c.[1] For the p-adic norm, suppose that $a - b = p^k r$ and $b - c = p^\ell s$, where $v_p(r) = v_p(s) = 0$. Swapping a and c if necessary, we may suppose that $k \leq \ell$. Therefore

$$a - c = (a - b) + (b - c) = p^k r + p^\ell s = p^k(r + p^{\ell-k} s).$$

The denominators of r and s are not divisible by p, and so neither is the denominator of $p^{\ell-k} s$ (since $\ell - k \geq 0$) nor the denominator of $r + p^{\ell-k} s$. Therefore $v_p(a - c) \geq k$, which implies that

(16.1.3) $$|a - c|_p \leq p^{-k} = \max\{|a - b|_p, |b - c|_p\}.$$

This is quite a different triangle inequality from (16.1.2), sometimes called the *ultrametric triangle inequality*; and a norm satisfying (16.1.3) is called *non-Archimedean*, whereas those satisfying (16.1.2) are *Archimedean*. Non-Archimedean norms arise naturally in this context, but the inequality (16.1.3) has a lot of unintuitive consequences. For example, suppose that the p-adic distance $|a - c|_p$, between a and c, is the largest of the three edges of the triangle with vertices a, b, c. If $|a - b|_p$ is the second largest, then, by (16.1.3), we deduce that $|a - c|_p \leq |a - b|_p$, and therefore these two longest edges must be equally long. This implies that *every triangle is isosceles* in the p-adic norm.

Exercise 16.1.2.[†] A circle in $\mathbb{C}$ takes the form $B(a; r) := \{x : |x - a| \leq r\}$ where a is the center of the circle and r is its radius. We define a p-adic circle to be $B_p(a; r) := \{x : |x - a|_p \leq r\}$.
  (a) Prove that if $b \in B_p(a; r)$, then $B_p(a; r) = B_p(b; r)$. (In other words, every point inside a p-adic circle can be taken to be its center.)
  (b) Prove that if two circles $B_p(a; r)$ and $B_p(A; R)$, with $r \leq R$, have a point b in common, then $B_p(a; r) \subset B_p(A; R)$. (That is, two p-adic circles are either disjoint or one is contained in the other.)

16.2. p-adic expansions

Rational numbers are easily defined from the axioms of integer arithmetic. Each real number is then defined as the limit of a convergent sequence of rational numbers; indeed one of the key uses of decimal notation is to make this concept intuitively obvious.[2] Thus for a real number r with decimal expansion

$$r = a_m/10^m + a_{m+1}/10^{m+1} + \cdots \text{ with each } a_j \in \{0, 1, \ldots, 9\},$$

for some integer m, we define $r_k := a_m/10^m + \cdots + a_k/10^k \in \mathbb{Q}$ for each integer $k \geq m$. The sequence (r_k) converges since it is a Cauchy sequence, as $|r_k - r_\ell| < 10^{-k}$ whenever $\ell \geq k$, and it converges to the limit r, as $|r - r_k| < 10^{-k}$ for all $k \geq m$. Here the distance is measured by the Euclidean norm $|.| = |.|_\infty$; we call this process *completing the rationals with respect to the Euclidean norm* $|.|_\infty$. We

---

[1] That is, the shortest distance between two points is a straight line.
[2] By understanding what goes into that intuition we construct a technique that is useful in other contexts, as we see in this section.

now try to complete the rationals with respect to the p-adic norm $|.|_p$, beginning by constructing an analogous expansion for $3/7$ in the 5-adics:

The fraction $3/7 \equiv 4 \pmod 5$, and then $3/7 \equiv 4 \pmod{25}$. For the next power of 5 we have $3/7 \equiv 54 \pmod{125}$. Since 25 divides 125 we must have $54 \equiv 3/7 \equiv 4 \pmod{25}$, and indeed $54 = 4 + 2 \cdot 5^2$. Next we have $3/7 \equiv 179 \pmod{5^4}$ and $179 = 4 + 2 \cdot 5^2 + 5^3$. As we keep on going, through increasingly higher powers of 5, we build a 5-*adic expansion* for $3/7$:

$$3/7 = 4 + 2 \cdot 5^2 + 5^3 + 4 \cdot 5^4 + 2 \cdot 5^5 + 3 \cdot 5^6 + 2 \cdot 5^8 + 5^9 + 4 \cdot 5^{10} + 2 \cdot 5^{11} + 3 \cdot 5^{12} + \cdots.$$

We can write this "in base 5" (without writing the powers of 5 each time) as

$$(3/7)_5 = 40214230214230214230214230214230214230214230214230\ldots.$$

Notice that this is eventually periodic (with period 021423), as are the p-adic expansions of all rational numbers:

Exercise 16.2.1. (a) Prove that the p-adic expansion of any non-zero rational number a/n with $(a, n) = 1$ is eventually periodic.
(b) Show that if $p \nmid n$, then the length of the period divides $\text{ord}_n(p)$.

For any expansion of the form

$$\alpha := a_m p^m + a_{m+1} p^{m+1} + \cdots$$

with each $a_j \in \{0, 1, 2, \ldots, p-1\}$, let $\alpha_k = a_m p^m + \cdots + a_k p^k \in \mathbb{Q}$ for each $k \geq m$. Then $|\alpha - \alpha_k|_p \leq p^{-k-1}$ and so α is the limit of the rational numbers α_k measuring proximity using the p-adic norm. These expansions are the p-*adic numbers*, denoted by $\mathbb{Q}_p$. If $m \geq 0$, then these are the p-*adic integers*, denoted by $\mathbb{Z}_p$. Its elements can all be written in the form

$$a_0 + a_1 p + a_2 p^2 + \cdots \text{ with each } 0 \leq a_i \leq p-1.$$

The rules of arithmetic for p-adic numbers and integers are straightforward enough.

Exercise 16.2.2. Suppose that $(r_k)_{k \geq 1}$ is any sequence of rationals such that for any $\epsilon > 0$, if k and ℓ are sufficiently large, then $|r_k - r_\ell|_p < \epsilon$. Prove that the r_k tend to a limit with a p-adic expansion.

16.3. p-adic roots of polynomials

Suppose that prime p does not divide a or n. Starting with the root $b_1 \pmod p$ to $x^n \equiv a \pmod p$, Proposition 7.20.1 gives us a unique root b_k to $x^n \equiv a \pmod{p^k}$ for every $k \geq 1$, where $b_k \equiv b_j \pmod{p^i}$ for all $j, k \geq i$. This implies that $|b_k - b_j|_p \leq p^{-i}$ whenever $j, k \geq i$, so that $\lim_{k \to \infty} b_k$ exists in the p-adic integers, by exercise 16.2.2. This p-adic root, call it β, satisfies $\beta^n = a$ in the p-adics. Moreover, this implies that the roots β of $x^n = a$ in the p-adics are in 1-to-1 correspondence with the solutions $b_1 \pmod p$ to $x^n \equiv a \pmod p$. We can find p-adic roots of most equations:

Theorem 16.1 (Hensel's Lifting Lemma). *Suppose that $f(x) \in \mathbb{Z}[x]$ and that p is an odd prime. If a is an integer for which $f(a) \equiv 0 \pmod p$ and $f'(a) \not\equiv 0 \pmod p$, then there is a unique p-adic root α to $f(\alpha) = 0$ with $\alpha \equiv a \pmod p$. On the other hand if α is a p-adic root of $f(\alpha) = 0$ with $|f'(\alpha)|_p = 1$, then $f(a) \equiv 0 \pmod p$ where $a \equiv \alpha \pmod p$ with $f'(a) \not\equiv 0 \pmod p$.*

This follows immediately from the following proposition, which generalizes the statement and proof of Proposition 7.20.1.

Proposition 16.3.1. *Suppose that $f(x) \in \mathbb{Z}[x]$ and that p is an odd prime. If $f(a) \equiv 0 \pmod{p}$ and $f'(a) \not\equiv 0 \pmod{p}$, then for each integer k there exists a unique residue class $a_k \pmod{p^k}$ with $a_k \equiv a \pmod{p}$ for which $f(a_k) \equiv 0 \pmod{p^k}$.*

In Hensel's Lifting Lemma we take $\alpha = \lim_{k \to \infty} a_k$ in the p-adics.

Proof. The Taylor expansion of the polynomial $f(x)$ at $x = a$ is simply the expansion of f as a polynomial in $x - a$. If f has degree d, then

$$f(x) = f(a) + f'(a)(x-a) + f^{(2)}(a)\frac{(x-a)^2}{2!} + \cdots + f^{(d)}(a)\frac{(x-a)^d}{d!}.$$

We now proceed by induction on $k \geq 2$. Suppose that $f(A) \equiv 0 \pmod{p^k}$ with $A \equiv a \pmod{p}$. Then $f(A) \equiv 0 \pmod{p^{k-1}}$ with $A \equiv a \pmod{p}$ and so $A \equiv a_{k-1} \pmod{p^{k-1}}$ by the induction hypothesis. Writing $A = a_{k-1} + rp^{k-1}$ for some integer r, we use the Taylor expansion to deduce that

$$0 \equiv f(A) \equiv f(a_{k-1} + rp^{k-1}) \equiv f(a_{k-1}) + f'(a_{k-1})rp^{k-1} \pmod{p^k},$$

as p is odd. Hence r is determined by

$$r \equiv -f(a_{k-1})/(f'(a_{k-1})p^{k-1}) \equiv -(f(a_{k-1})/p^{k-1})/f'(a) \pmod{p},$$

as p^{k-1} divides $f(a_{k-1})$ by the induction hypothesis, and $f'(a_{k-1}) \equiv f'(a) \not\equiv 0 \pmod{p}$ as $a_{k-1} \equiv a \pmod{p}$, so that $f'(a)$ is invertible mod p. Hence r belongs to a unique residue class mod p, and therefore A is given uniquely $\pmod{p^k}$. $\square$

Exercise 16.3.1. Show that if $f(x) \in \mathbb{Z}[x]$ has no repeated roots, then there are only finitely many primes p for which there exists an integer a_p with $f(a_p) \equiv f'(a_p) \equiv 0 \pmod{p}$.

Exercise 16.3.2. Prove that if odd prime p does not divide a, then there are exactly $1 + \left(\frac{a}{p}\right)$ square roots of $a \pmod{p}$.

Exercise 16.3.3. (a) If prime $p \nmid a$, show that the sequence $a_n = a^{p^n}$ converges in the p-adics.
(b) Show that $\alpha := \lim_{n \to \infty} a_n$ is a $(p-1)$st root of unity and that all solutions to $x^{p-1} - 1$ in $\mathbb{Z}_p$ can be obtained in this way.
(c) Conclude that $i := \lim_{n \to \infty} 2^{5^n}$ is a square root of -1 in $\mathbb{Q}_5$.

What are the square roots of 7 in the 3-adics? Let $f(x) = x^2 - 7$ and $p = 3$. For $a = 1$ and -1 we have $f(a) \equiv 0 \pmod{3}$ but $f'(a) \not\equiv 0 \pmod{3}$, so we can apply Hensel's Lemma. We can lift the root which is 1 $\pmod 3$ as follows:

We write $a = 1 + 3k$ for some integer k. Then $7 \equiv a^2 = (1+3k)^2 \equiv 1 + 6k \pmod{9}$ and so $k \equiv 1 \pmod{3}$, and $a \equiv 4 \pmod{9}$.

We write $a = 4 + 9\ell$ for some integer ℓ. Then $7 \equiv a^2 = (4+9\ell)^2 \equiv 16 - 9\ell \pmod{27}$ and so $\ell \equiv 1 \pmod{3}$, and $a \equiv 13 \pmod{27}$.

We write $a = 13 + 27m$ for some integer m. Then $7 \equiv a^2 = (13+27m)^2 \equiv 7 - 27m \pmod{81}$ and so $m \equiv 0 \pmod{3}$, and $a \equiv 13 \pmod{81}$.

In summary we have established that if $\alpha^2 = 7$ in the 3-adics with $\alpha \equiv 1$ (mod 3), then $\alpha \equiv 13 = 1 \cdot 3^0 + 1 \cdot 3^1 + 1 \cdot 3^2 + 0 \cdot 3^3$ (mod 3^4). We can continue in this way to calculate the digits of α, though we now introduce a more efficient root-finding technique:

The *Newton-Raphson method* is an iterative technique to find a root of a polynomial $f(x) \in \mathbb{R}[x]$. The idea is to start with a first guess a_0 at a root of $f(x) = 0$ and then to construct a sequence of increasingly good approximations $a_1, a_2, \ldots$ by taking
$$a_{n+1} = a_n - \frac{f(a_n)}{f'(a_n)},$$
which is the correction suggested by the slope of f at a_n. We can adapt this algorithm to lifting solutions modulo powers of p.

Proposition 16.3.2. *Suppose that $f(x) \in \mathbb{Z}[x]$ and that p is an odd prime. If a is an integer for which $f(a) \equiv 0$ (mod p^k) with $k \geq 1$ and $f'(a) \not\equiv 0$ (mod p), then $f(A) \equiv 0$ (mod p^{2k}) with $A \equiv a$ (mod p^k) where $A = a - f(a)/f'(a)$.*

Proof. Let $\Delta = -f(a)/f'(a)$ so that $A = a + \Delta$, and $v_p(\Delta) \geq k$ so that $A \equiv a$ (mod p^k). If $j \geq 2$, then $v_p(\Delta^j/j!) \geq 2k$, and the Taylor expansion gives
$$f(A) = f(a + \Delta) = f(a) + \Delta f'(a) + \frac{\Delta^2}{2} f^{(2)}(a) + \cdots$$
$$\equiv f(a) + \Delta f'(a) = 0 \pmod{p^{2k}}. \qquad \square$$

The convergence here is much more rapid than in our previous algorithm, for if $|a_0|_p \leq p^{-1}$, then one can prove by induction that $|a_n| \leq p^{-2^n}$ for all $n \geq 1$. In our previous example, finding a square root of 7 in the 3-adics starting with $a_0 = 1$, we obtain, as $f(x) = x^2 - 7$, the transformation
$$a \to a - \frac{a^2 - 7}{2a} = \frac{1}{2}\left(a + \frac{7}{a}\right).$$
Therefore $a_1 = 4$; then $a_2 = 23/8 \equiv 13$ (mod 81), and $a_3 = 88/13$ (mod 6561). We have $a_3^2 \equiv 7$ (mod 6561) as $88^2 - 7 \cdot 13^2 = 6561$. Therefore if $\alpha^2 = 7$ in the 3-adics with $\alpha \equiv 1$ (mod 3), then we have determined that $\alpha \equiv 88/13$ (mod 3^8), via an algorithm that converges far faster than the method we used above.

16.4. p-adic factors of a polynomial

Theorem 16.2. *Suppose that $f(x) \in \mathbb{Z}_{(p)}[x]$ and can be factored as $g_1(x) \cdots g_m(x)$ (mod p) where $\deg(f - g_1 \cdots g_m) < \deg(f)$, and $g_i(x)$ and $g_j(x)$ have no common polynomial factor mod p, whenever $i \neq j$. Then, there exist unique polynomials $G_1(x), \ldots, G_m(x) \in \mathbb{Z}_{(p)}[x]$ with $G_j(x) \equiv g_j(x)$ (mod p) and $\deg(G_j - g_j) < \deg(g_j)$ for each j, for which $f(x) = G_1(x) \cdots G_m(x)$ in the p-adics.*

Proof. We will prove this with $m = 2$ and then the result for larger m follows by induction. So we write $f(x) \equiv g(x)h(x)$ (mod p) where $g(x)$ and $h(x)$ have no common polynomial factor mod p, and therefore there exist $u(x), v(x) \in \mathbb{Z}[x]$ for which $g(x)u(x) + h(x)v(x) \equiv 1$ (mod p) with $\deg(u) < \deg(h)$ and $\deg(v) < \deg(g)$.

We will prove, by induction, that for each $k \geq 1$, there exist unique polynomials $g_k(x), h_k(x) \pmod{p^k}$, with $g_k(x) \equiv g(x) \pmod{p}$, $h_k(x) \equiv h(x) \pmod{p}$, and $f(x) \equiv g_k(x)h_k(x) \pmod{p^k}$ where $\deg(g - g_k) < \deg(g)$ and $\deg(h - h_k) < \deg(h)$. The result then follows with $f(x) = G(x)H(x)$ where $G(x)$ and $H(x)$ are the p-adic limits, as $k \to \infty$, for $g_k(x)$ and $h_k(x)$, respectively.

This is true by definition for $k = 1$, so now let $k \geq 1$. If $f(x) \equiv A(x)B(x) \pmod{p^{k+1}}$ with $A(x) \equiv g(x) \pmod{p}$, $B(x) \equiv h(x) \pmod{p}$ and $\deg(A - g) < \deg(g), \deg(B - h) < \deg(h)$, then $f(x) \equiv A(x)B(x) \pmod{p^k}$ and so $A(x) \equiv g_k(x) \pmod{p^k}$ and $B(x) \equiv h_k(x) \pmod{p^k}$ by the uniqueness of g_k and h_k. Therefore we can write $A(x) = g_k(x) + p^k a(x)$ and $B(x) = h_k(x) + p^k b(x)$ with $\deg(a) < \deg(g)$ and $\deg(b) < \deg(h)$. This implies that

$$f(x) \equiv A(x)B(x) = (g_k(x) + p^k a(x))(h_k(x) + p^k b(x))$$
$$\equiv g_k(x)h_k(x) + p^k(a(x)h_k(x) + b(x)g_k(x))$$
$$\equiv g_k(x)h_k(x) + p^k(a(x)h(x) + b(x)g(x)) \pmod{p^{k+1}}.$$

Let $\Delta_k(x) = (f(x) - g_k(x)h_k(x))/p^k \in \mathbb{Z}_{(p)}[x]$. We select $a_k(x)$ to be the polynomial of minimal degree with

$$a_k(x) \equiv v(x)\Delta_k(x) \pmod{(p, g(x))}$$

so that $\deg(a_k) < \deg(g)$. Writing $a_k \equiv v\Delta_k + \ell(x)g(x) \pmod{p}$ for some $\ell(x) \in \mathbb{Z}[x]$, we let $b_k(x) = u(x)\Delta_k(x) - \ell(x)h(x)$, so that

$$a_k(x)h(x) + b_k(x)g(x) \equiv (v\Delta_k + \ell g)h + (u\Delta_k - \ell h)g = (ug + vh)\Delta_k \equiv \Delta_k \pmod{p}.$$

Therefore there must exist a solution to $f(x) \equiv A(x)B(x) \pmod{p^{k+1}}$ satisfying all the desired hypotheses. It remains to show that this solution is unique: If we have any other solution, then $ah + bg \equiv a_k h + b_k g \pmod{p}$ so that $(a - a_k)h = (b_k - b)g \pmod{p}$. Since g and h have no common factor we deduce that g divides $a - a_k$ and h divides $b - b_k$, and so $\deg(a) \geq \deg(g)$ and $\deg(b) \geq \deg(h)$, which are impossible. The result therefore follows with $g_{k+1} = g_k + p^k a_k$ and $h_{k+1} = h_k + p^k b_k$. □

Factoring polynomials in $\mathbb{Z}[x]$ efficiently

We wish to factor a given polynomial $f(x) \in \mathbb{Z}[x]$. It is not difficult to find any factors that appear to a power > 1 by calculating (f, f'), so we may assume that $f(x)$ has no repeated factors. We will need a bound on the size of the coefficients of any potential factor $g(x)$ of $f(x)$. To do this we need to measure the "size" of a given polynomial

$$f(x) = \sum_{i=0}^{d} f_i x^i = a \prod_{j=1}^{d}(x - \alpha_j) \in \mathbb{C}[x].$$

Two ways to do this are given by the 2-*norm* $\|f\|$, a non-negative real number which is defined by

$$\|f\|^2 := \sum_{i=0}^{d} |f_i|^2, \text{ and the } Mahler\ measure\ M(f) = |a| \prod_{j=1}^{d} \max\{1, |\alpha_j|\}.$$

We need to compare these two:

Exercise 16.4.1. Assume $g(x) \in \mathbb{C}[x]$ and $\alpha \in \mathbb{C}$ with $\alpha \neq 0$.
(a) Prove that $\|(x-\alpha)g(x)\| = \|(\overline{\alpha}x - 1)g(x)\|$ and $M((x-\alpha)g(x)) = M((\overline{\alpha}x-1)g(x))$.
(b) If $|\alpha| \leq 1$ whenever $f(\alpha) = 0$, prove that $M(f) \leq \|f\| \leq M(f)\left(\sum_{i=0}^{n}\binom{n}{i}^2\right)^{1/2}$, where f has degree n. (We note that $\sum_{i=0}^{n}\binom{n}{i}^2 = \binom{2n}{n} \leq 2^{2n}$.)
(c) Deduce that if f has degree n, then $M(f) \leq \|f\| \leq 2^n M(f)$.
(d) Suppose that $g(x) \in \mathbb{Z}[x]$ divides $f(x) \in \mathbb{Z}[x]$. Prove that $M(g) \leq M(f)$.
(e) Deduce that if g has degree d, then $\|g\| \leq 2^d \|f\|$.

Here is the Berlekamp-Zassenhaus algorithm for factoring a given squarefree polynomial $f(x) \in \mathbb{Z}[x]$ of degree n:

- Calculate $\Delta(f)$, the discriminant of $f(x)$, by using the Euclidean algorithm over $\mathbb{Z}[x]$. This is a non-zero integer as $f(x)$ has no repeated factors (as proved in sections 2.10 and 2.11 of appendix 2B).
- Select a prime p which does not divide $\Delta(f)$. Therefore $f(x) \pmod{p}$ is squarefree, by Corollary 3.23.1.
- Use the algorithm of section 10.21 (of appendix 10H) to efficiently factor $f(x) \pmod{p}$ into irreducibles.
- Select k to be the smallest integer for which $p^k > 2^n \|f\|$.
- By Theorem 16.2 and its proof, we can efficiently lift the factors of $f(x) \pmod{p}$, uniquely, to a factorization of $f(x) \equiv g_1(x)\cdots g_m(x) \pmod{p^k}$.
- For each $S \subset \{1,\ldots,m\}$ let $g_S(x) \equiv \prod_{i \in S} g_i(x) \pmod{p^k}$, a polynomial with integer coefficients for which each coefficient lies in $(-\frac{p^k}{2}, \frac{p^k}{2}]$.
- Test whether each $g_S(x)$ is a factor of $f(x)$ in $\mathbb{Z}[x]$.

This gives all the factors of $f(x)$. For if $g(x)$ properly divides $f(x)$, then exercise 16.4.1(e) implies that each coefficient g_j of g satisfies $|g_j| \leq 2^{n-1}\|f\| < \frac{p^k}{2}$ as $d \leq n-1$. Moreover if we reduce the equality $f(x) = g(x)h(x)$, modulo p^k, then we see that there must exist some proper subset S of $\{1,\ldots,m\}$ for which $g(x) \equiv g_S(x) \pmod{p^k}$. The coefficients of their difference is $< p^k$ and so $g(x) = g_S(x)$.

Further reading on factoring polynomials

[1] Susan Landau, *Factoring polynomials quickly*, Notices Amer. Math. Soc. **34** (1987), 3–8.

16.5. Possible norms on the rationals

A *norm* on $\mathbb{Q}$ is a map $|.| : \mathbb{Q} \to \mathbb{R}$ for which $|x| \geq 0$ for all $x \in \mathbb{R}$, with $|x| = 0$ if and only if $x = 0$; $|xy| = |x| \cdot |y|$; and $|x+y| \leq |x| + |y|$.

Exercise 16.5.1. Prove that $|1| = |-1| = 1$.

We have defined the Euclidean norm $|.|_\infty$ and the p-adic norm $|.|_p$. We now prove that, up to taking powers, these are the only possibilities:

Theorem 16.3 (Ostrowski's Theorem). *A norm on the rational numbers is either $|.|_\infty^\kappa$ for some constant κ, $0 \leq \kappa \leq 1$, or $|.|_p^\kappa$ for some prime p, for some constant $\kappa > 0$.*

Proof. Suppose that $|m| > 1$ for some integer $m > 1$. Fix any other integer $b > 1$, and then define $B := \max\{|c| : 0 \leq c \leq b-1\}$. Any integer N can be written in base b as $m_0 + m_1 b + \cdots + m_d b^d$ where each $m_j \in \{0, 1, 2, \ldots, b-1\}$ and $m_d \geq 1$. Therefore

$$|N| \leq \sum_{i=0}^{d} |m_i b^i| = \sum_{i=0}^{d} |m_i||b|^i \leq (d+1) \, B \, \max\{1, |b|^d\},$$

as there are $d+1$ terms in the sum, each $|m_i| \leq B$, and each $|b|^i \leq 1$ if $|b| \leq 1$, with each $|b|^i \leq |b|^d$ if $|b| > 1$.

We let $N = m^n$ for any integer $n \geq 1$, so that $|N| = |m|^n$. Now $b^d \leq N$ and so $d \leq \frac{\log N}{\log b} = n \frac{\log m}{\log b}$. Substituting this into the inequality above gives

$$|m|^n \leq \left(1 + n \frac{\log m}{\log b}\right) B \, \max\{1, |b|^{n \frac{\log m}{\log b}}\}.$$

We will take nth roots of both sides and let $n \to \infty$. We notice that $(u+nv)^{1/n} \to 1$ as $n \to \infty$ for any $v > 0$ and so deduce that

$$|m| \leq \max\{1, |b|^{\frac{\log m}{\log b}}\}.$$

Since $|m| > 1$ this implies that $|b| > 1$ and therefore, taking logarithms, we have

$$\frac{\log |m|}{\log m} \leq \frac{\log |b|}{\log b}.$$

Since $|b| > 1$ we may run the same argument with the roles of b and m reversed and obtain the opposite inequality, and combining these we get equality. But this holds for all integers $b > 1$. Hence there exists a number κ for which $|b| = b^\kappa = |b|_\infty^\kappa$ for all integers $b > 1$.

As $|m| > 1$ we deduce that $\kappa > 0$. Since $2^\kappa = |2|_\infty^\kappa \leq |1|_\infty^\kappa + |1|_\infty^\kappa = 2$, therefore $\kappa \leq 1$. One can show that the triangle inequality holds whenever $0 < \kappa \leq 1$.

If $|n| = 1$ for all $n > 1$, then $|n| = |n|_\infty^0$.

We may now assume that $|n| \leq 1$ for all integers $n > 1$, and that $|m| < 1$ for some $m > 1$. By multiplicativity we know that $|p| < 1$ for some prime p dividing m. There can be no other prime q with $|q| < 1$, or else we select a large power e, so large that $|q|^e < 1 - |p|$. Since $(p, q) = 1$ there exist integers a, b for which $ap + bq^e = 1$ and therefore

$$1 = |1| = |ap + bq^e| \leq |ap| + |bq^e| = |a||p| + |b||q|^e \leq |p| + |q|^e < 1,$$

a contradiction. Therefore, if $|p| = p^{-\kappa} = |p|_p^\kappa$, then $|.| = |.|_p^\kappa$. Taking powers of (16.1.3) we see that any norm $|.|_p^\kappa$ satisfies the ultrametric triangle inequality and therefore satisfies the Euclidean triangle inequality. $\square$

16.6. Power series convergence and the p-adic logarithm

Theorem 16.4. *Let $(a_n)_{n \geq 0}$ be an infinite sequence of p-adic numbers. The infinite sum $\sum_{n \geq 0} a_n$ converges to some number L (in the p-adics) if and only if $a_n \to 0$ as $n \to \infty$.*

16.6. Power series convergence and the p-adic logarithm

Proof. Suppose that $a_n \to 0$ as $n \to \infty$. Fix $\epsilon > 0$ and M_0 so that $|a_n| < \epsilon$ whenever $n \geq M_0$. Using the ultrametric inequality we have that if $N > M \geq M_0$, then

$$\left| \sum_{n \leq N} a_n - \sum_{n \leq M} a_n \right|_p \leq \max_{M < n \leq N} |a_n|_p < \epsilon.$$

Therefore the partial sum of the a_n form a Cauchy sequence, and therefore $\sum_{n \geq 0} a_n$ converges to some number L.

On the other hand if $\sum_{n \geq 0} a_n$ converges to some number L, then for any $\epsilon > 0$ there exists M_1 such that $|\sum_{n \leq M} a_n - L| < \epsilon$ whenever $M \geq M_1$. Therefore if $N \geq M_1 + 1$, then we have $a_N = (\sum_{n \leq N} a_n - L) - (\sum_{n \leq N-1} a_n - L)$ so that

$$|a_N|_p \leq \max \left\{ \left| \sum_{n \leq N} a_n - L \right|_p, \left| \sum_{n \leq N-1} a_n - L \right|_p \right\} < \epsilon.$$

We deduce that $a_n \to 0$ as $n \to \infty$. $\square$

Exercise 16.6.1. Let p be a given prime.
(a) Prove that $\sum_{n \geq 0} z^n/a_n$ converges when $|z|_p < p^{-\tau}$, where $\tau = \limsup_{n \to \infty} v_p(a_n)/n$.
(b) Deduce that $\sum_{n \geq 1} z^n/n$ converges if $|z|_p < 1$. (In $\mathbb{C}$ this also converges inside $|z| < 1$.)
Exercise 3.10.1(b) states that $v_p(n!) = \frac{n - s_p(n)}{p-1}$ where $s_p(n)$ is the sum of the digits of n when written in base p.
(c) Deduce that $\sum_{n \geq 1} z^n/n!$ converges if $|z|_p < p^{-1/(p-1)}$.

We define the *p-adic logarithm* to be

$$\log_p(z) := -\sum_{n \geq 1} \frac{(1-z)^n}{n}$$

whenever $|1-z|_p < 1$ with $z \in \mathbb{Z}_p$ (this sum converges by exercise 16.6.1(b)). Similarly we define the *p-adic exponential* to be

$$\exp_p(z) := \sum_{n \geq 1} \frac{z^n}{n!}$$

wherever $|z|_p < p^{-1/(p-1)}$ (this sum converges by exercise 16.6.1(c)).

Exercise 16.6.2. Suppose that $|1-a|_p, |1-b|_p < p^{-1/(p-1)}$.
(a) Prove that $|1-ab|_p < p^{-1/(p-1)}$.
(b) Deduce that $\exp_p(ab) = \exp_p(a) \exp_p(b)$.

In exercise 2.5.9(b) we saw that $\frac{1}{p}\binom{p}{j} \equiv (-1)^{j-1}/j \pmod{p}$ for $1 \leq j \leq p-1$, so that if $z \in \mathbb{Z}_{(p)}$, then

$$\sum_{n=1}^{p-1} \frac{z^n}{n} \equiv -\frac{1}{p} \sum_{n=1}^{p-1} \binom{p}{n}(-z)^n = -\frac{1}{p}((1-z)^p - 1 + z^p) \pmod{p}.$$

This suggests that there might be a convenient expression of this type for $\log_p(z)$.

Exercise 16.6.3. Suppose that $v_p(x) > 0$.
(a) Prove that if $p^k \leq m < p^{k+1}$, then $v_p(x^m/m) \geq p^k v_p(x) - k$, for each integer $k \geq 0$.
(b) Suppose that $v_p(x) \geq 2r/p^r$ for some integer $r \geq 1$. Deduce that $v_p(x^m/m) \geq k$ for all $m \geq p^k$, whenever $k \geq r$.

Lemma 16.6.1. For any $z \in \mathbb{Z}_{(p)}$ for which $|1 - z|_p < 1$, we have
$$\log_p(z) = \lim_{k \to \infty} \frac{z^{p^k} - 1}{p^k}.$$

Proof. Let $x = 1 - z$ so that $v_p(x) > 0$. We can select an integer $r \geq 1$ for which $v_p(x) \geq 2r/p^r$, as $i/p^i \to 0$ as $i \to \infty$. Let k be any integer $\geq 3r$ and let ℓ be the largest integer $\leq k/3$, so that $\ell \geq r$. Therefore if $m \geq p^{\ell+1}$, then $v_p(x^m/m) \geq \ell + 1 > k/3$ by exercise 16.6.3(b).

For $1 \leq m \leq p^k$ we have
$$(-1)^{m-1} \frac{1}{p^k} \binom{p^k}{m} = (-1)^{m-1} \prod_{j=1}^{m-1} \frac{p^k - j}{j} \cdot \frac{1}{m} = \frac{c_m}{m} \text{ where } c_m := \prod_{j=1}^{m-1} \left(1 - \frac{p^k}{j}\right).$$

We let $c_m = 0$ for all $m > p^k$, so that $-\log_p(z) = -\log_p(1-x) = \sum_{m \geq 1} \frac{x^m}{m}$, and
$$\frac{1 - z^{p^k}}{p^k} = \frac{1 - (1-x)^{p^k}}{p^k} = \sum_{m=1}^{p^k} (-1)^{m-1} \frac{1}{p^k} \binom{p^k}{m} x^m = \sum_{m=1}^{p^k} c_m \frac{x^m}{m}.$$

Therefore
$$-\log_p(z) - \frac{1 - z^{p^k}}{p^k} = \sum_{m \geq 1} (1 - c_m) \frac{x^m}{m}.$$

Now if $1 \leq j < m \leq p^k$, then $1 - \frac{p^k}{j} \in \mathbb{Z}_{(p)}$, and so $1 - c_m \in \mathbb{Z}_{(p)}$ for all $m \geq 1$; that is, $|1 - c_m|_p \leq 1$. Therefore if $m \geq p^{\ell+1}$, then
$$\left|(1 - c_m) \frac{x^m}{m}\right|_p \leq \left|\frac{x^m}{m}\right|_p < p^{-k/3},$$
by the first paragraph.

If $1 \leq j < m < p^{\ell+1}$, then $1 - \frac{p^k}{j} \equiv 1 \pmod{p^{k-\ell}}$, and so $|1 - c_m| \leq p^{\ell-k}$. Now $|m|_p \geq p^{-\ell}$ and so
$$\left|(1 - c_m) \frac{x^m}{m}\right|_p \leq p^{2\ell - k} \leq p^{-k/3}.$$

Combining these last two estimates, we deduce that
$$\left|\log_p(z) + \frac{1 - z^{p^k}}{p^k}\right|_p = \left|\sum_{m \geq 1} (1 - c_m) \frac{x^m}{m}\right|_p \leq \max_{m \geq 1} \left|(1 - c_m) \frac{x^m}{m}\right|_p \leq p^{-k/3},$$

and the result follows, letting $k \to \infty$. $\square$

Exercise 16.6.4. Prove that $\sum_{m \geq 1} 2^m/m = 0$ in the 2-adics.

Exercise 16.6.5. Assume that $|a - 1|_p, |b - 1|_p < 1$.
  (a) Prove that $\lim_{k \to \infty} a^{p^k} = 1$.
  (b) Deduce that $\log_p(ab) = \log_p(a) + \log_p(b)$.
  (c) Deduce that if $a = b^n$, then $\log_p(a) = n \log_p(b)$.
  (d)† Suggest an algorithm for the discrete log problem in the p-adics.

At the moment the function $\log_p(z)$ is defined only when $|z-1|_p < 1$. For any $\beta \in \mathbb{Z}_p$ with $|\beta|_p = 1$, there exists an integer b with $b \equiv \beta \pmod{p}$ and $b \not\equiv 0 \pmod{p}$. Therefore, by Fermat's Little Theorem, $\beta^{p-1} \equiv b^{p-1} \equiv 1 \pmod{p}$, and so $\log_p(\beta^{p-1})$ is well-defined. Taking our lead from exercise 16.6.5(c) we therefore define

$$\log_p(\beta) := \frac{\log_p(\beta^{p-1})}{p-1} = \lim_{k \to \infty} \frac{\beta^{p^k(p-1)} - 1}{p^k(p-1)}.$$

Exercise 16.6.6. Assume that $\alpha, \beta \in \mathbb{Z}_p$.
(a) Prove that $\log_p(-\alpha) = \log_p(\alpha)$.
(b) Prove that $\log_p(\alpha\beta) = \log_p(\alpha) + \log_p(\beta)$.

Any $\gamma \in \mathbb{Z}_p$ can be written in the form $\gamma = p^e \beta$ where $|\beta|_p = 1$, so we define[3]

$$\log_p(\gamma) = e \log_p(p) + \log_p(\beta).$$

16.7. The p-adic dilogarithm

For each $k \geq 1$, define

$$\mathcal{L}_k(x) := \sum_{m \geq 1} \frac{x^m}{m^k}.$$

The case $k = 2$ is the *dilogarithm function*.

Exercise 16.7.1. (a) Prove that the sum defining $\mathcal{L}_k(x)$ converges for all $x \in \mathbb{C}$ with $|x|_\infty \leq 1$ for all $k \geq 2$, and for $|x|_p < 1$ in the p-adics.
(b) Establish that $\mathcal{L}_k(x) + \mathcal{L}_k(-x) = 2^{1-k} \mathcal{L}_k(x^2)$ when $|x|_p < 1$.

Theorem 16.5. *If $|1-z|_p < 1$, then*

$$(16.7.1) \qquad \mathcal{L}_2(1-z) + \mathcal{L}_2(1-z^{-1}) = -\frac{1}{2}(\log_p z)^2.$$

In particular $\mathcal{L}_2(2) = 0$ in the 2-adics.

Proof. For $|x|_p < 1$, we have

$$\frac{d\mathcal{L}_2(x)}{dx} = \frac{1}{x} \sum_{m \geq 1} \frac{x^m}{m} = -\frac{\log_p(1-x)}{x},$$

and so, by the chain rule, we have

$$\frac{d}{dz}(\mathcal{L}_2(1-z) + \mathcal{L}_2(1-z^{-1})) = -\mathcal{L}_2'(1-z) + z^{-2}\mathcal{L}_2'(1-z^{-1})$$

$$= \frac{\log_p(z)}{1-z} - z^{-2}\frac{\log_p(z^{-1})}{1-z^{-1}} = -\frac{\log_p z}{z}.$$

Integrating yields $\mathcal{L}_2(1-z) + \mathcal{L}_2(1-z^{-1}) = -\frac{1}{2}(\log_p z)^2 + C$ for some constant C. Taking $z = 1$ we see that $C = 0$, yielding (16.7.1).

Replacing z by z^2, we obtain

$$\mathcal{L}_2(1-z^2) + \mathcal{L}_2(1-z^{-2}) = -2(\log_p z)^2 = 4(\mathcal{L}_2(1-z) + \mathcal{L}_2(1-z^{-1})).$$

When $p = 2$ we may take $z = -1$ in this equation and so $8\mathcal{L}_2(2) = 2\mathcal{L}_2(0) = 0$. □

---

[3] We can select any value for $\log_p(p)$ as is convenient in context; we do not have to let it be 1.

Exercise 16.7.2. Let $p = 2$ and $|z - 1|_2 < 1$.
 (a) Prove that $\mathcal{L}_2(1-z) + \mathcal{L}_2(1+z) = \frac{1}{2}\mathcal{L}_2(1-z^2) + C$ for some constant C.
 (b) Prove that $C = 0$ using (16.7.1).
 (c) Deduce (again) that $\mathcal{L}_2(2) = 0$.

We have now seen that
$$\sum_{n \geq 1} \frac{2^n}{n} = \sum_{n \geq 1} \frac{2^n}{n^2} = 0$$
in the 2-adics. It is interesting to see how rapidly this convergence happens. If $n \geq N \geq 2^k$, then $v_2(2^n/n) \geq 2^k - k$ so that
$$\sum_{n < N} \frac{2^n}{n} = -\sum_{n \geq N} \frac{2^n}{n} \equiv 0 \pmod{2^{2^k - k}}$$
and similarly
$$\sum_{n < N} \frac{2^n}{n^2} \equiv 0 \pmod{2^{2^k - 2k}}.$$

It looks like there might be a pattern here. How about $\sum_{n \geq 1} 2^n/n^3$? Unfortunately the $n = 4$ term gives the unique maximum, 2^2, of $|2^n/n^3|_2$, and so $|\sum_{n \geq 1} 2^n/n^3|_2 = 4$, not 0.

Exercise 16.7.3. Prove that if $|x|_p, |y|_p < 1$, then
$$\mathcal{L}_2(x) + \mathcal{L}_2(y) - \mathcal{L}_2(xy) - \mathcal{L}_2\left(\frac{x(1-y)}{1-xy}\right) - \mathcal{L}_2\left(\frac{y(1-x)}{1-xy}\right) = \log_p\left(\frac{1-x}{1-xy}\right)\log_p\left(\frac{1-y}{1-xy}\right).$$

Further reading on p-adics

[1] Richard M. Hill, *Introduction to number theory*, chapter 4, World Scientific, Singapore, 2018.

Chapter 17

Rational points on elliptic curves

Any equation of degree 3 in two variables with a rational solution can be transformed, via an invertible transformation with rational coefficients, to an equation of the form
$$E: y^2 = x^3 + ax + b.$$
We call the curve E, given by such an equation, an *elliptic curve*. When one draws the real points (x, y) on this curve, denoted by $E(\mathbb{R})$, there are two possible shapes, depending on the number of real zeros of $x^3 + ax + b$:

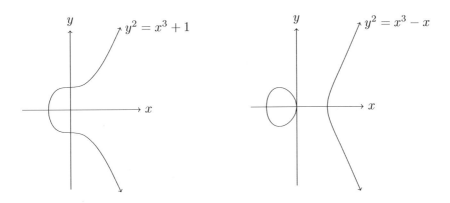

Figure 17.1. Elliptic curves with one and three real zeros, respectively.

There can be no real points with an x-coordinate for which $x^3 + ax + b < 0$, in particular, to the left of the graph.

Exercise 17.0.1. Let $\Delta = 4a^3 + 27b^2$. Show that if $a > 0$ or if $\Delta > 0$, then $x^3 + ax + b = 0$ has just one real root. Show that if $a, \Delta < 0$, then $x^3 + ax + b = 0$ has three real roots. Sketch the shape of the curve $y^2 = x^3 + ax + b$ in the two cases.

547

17.1. The group of rational points on an elliptic curve

The rational solutions of the original equation of degree 3 in two variables are in 1-to-1 correspondence with the rational solutions of $y^2 = x^3 + ax + b$, as long as we include the point(s) at ∞.[1] If the original equation had rational coefficients, then a and b are integers. We are interested in understanding the rational points on E, that is, those (x, y) on E with $x, y \in \mathbb{Q}$, which we denote by $E(\mathbb{Q})$.

Suppose that we are given two distinct points $P_1 = (x_1, y_1)$ and $P_2 = (x_2, y_2) \in E(\mathbb{Q})$. The line between them is either vertical (when $x_1 = x_2$) or of the form $y = mx + \nu$ with $m, \nu \in \mathbb{Q}$.[2] These two points are both intersections of the line with the elliptic curve $y^2 = x^3 + ax + b$. If the line is $y = mx + \nu$, then x_1, x_2 satisfy

$$(mx + \nu)^2 = y^2 = x^3 + ax + b;$$

in other words x_1 and x_2 are two of the three roots of the cubic polynomial

$$x^3 - m^2 x^2 + (a - 2m\nu)x + (b - \nu^2) = 0.$$

If the third root is x_3, then $x_3 = m^2 - x_1 - x_2 \in \mathbb{Q}$, and if $y_3 = mx_3 + \nu$, then $P_3 = (x_3, y_3) \in E(\mathbb{Q})$ is the third intersection of the line with E. This method for generating a third rational point on E from two given ones goes back to Fermat.[3]

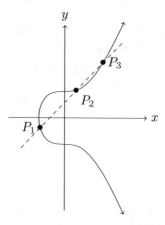

Figure 17.2. Obtaining a new rational point from two old ones: Draw the line through P_1 and P_2. The third point of intersection is P_3. If P_1 and P_2 have rational coordinates, then so does P_3.

If $x_1 = x_2$, then $y_2 = -y_1$, and the vertical line $(x = x_1)$ between P_1 and P_2 appears to meet the curve at only these two points. Is there a third point

---

[1] This same issue occurs in linear transformations of all equations, even of degree 1. For example, the rational points (x, y) on $x + 2y = 3$ are in 1-to-1 correspondence with the rational points (u, v) on $u + 2 = 3v$, via the invertible transformation $u = x/y, v = 1/y$. However there is a problem with the point $(x, y) = (3, 0)$ as this transforms to $(u, v) = (\infty, \infty)$; and similarly the point $(u, v) = (-2, 0)$ transforms back to $(x, y) = (\infty, \infty)$. To take these properly into account we can work with integer solutions to the equation $a + 2b = 3c$, with a, b, c not all 0. These are called the *projective coordinates*, and solutions for which the ratios $a : b : c$ are the same are considered to be equivalent. (In the original two *affine* equations, the solutions are given by $x = a/c, y = b/c$, and $u = a/b, v = c/b$.)

[2] In section 6.1 we saw that two rational points on the unit circle give rise to a line with rational coefficients and vice versa; this allowed us to find all the rational points on the unit circle.

[3] And compare this with the method for parameterizing rational points on a quadratic curve given, for example, by exercise 6.1.2.

17.1. The group of rational points on an elliptic curve

of intersection? There is no other point of intersection on the graph (that is, on the real plane), but the line stretches to infinity, and indeed the third point is, rather surprisingly, the point at infinity, which we denote by $\mathcal{O}$. The best way to see this is to rewrite the curve in *projective coordinates*: That is, we change $(x, y) \to (x/z, y/z)$ and multiply the equation of the curve through by z^3 to obtain $y^2 z = x^3 + axz^2 + bz^3$. The point at infinity, $\mathcal{O}$, is $(0, 1, 0)$, the only point with $z = 0$, up to scalar multiplication of all three coordinates. We see that it lies on the (projective) line $x = x_1 z$.

One can do even better and generate a second point from a given one: If $P_1 = (x_1, y_1) \in E(\mathbb{Q})$, let $y = mx + \nu$ be the equation of the tangent line to $y^2 = x^3 + ax + b$ at (x_1, y_1). To calculate this, we simply differentiate to obtain $2y_1 m = 3x_1^2 + a$ and then $\nu = y_1 - mx_1$. In this case, our cubic polynomial has a double root at $x = x_1$ and we again compute a third point by taking $x_3 = m^2 - 2x_1$, $y_3 = mx_3 + \nu$ so that $P_3 = (x_3, y_3) \in E(\mathbb{Q})$.

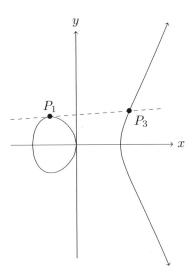

Figure 17.3. Obtaining a new rational point from an old one: Draw the tangent line through P_1. The other point of intersection is P_3. If P_1 has rational coordinates, then so does P_3.

Exercise 17.1.1. Prove that there cannot be four points of $E(\mathbb{Q})$ on the same line.

Poincaré made an extraordinary observation: If we take any three points P, Q, R of E on the same line, then we can define a group by taking

$$P + Q + R = 0.$$

This implies that in the first example above, $(x_1, y_1) + (x_2, y_2) + (x_3, y_3) = 0$. The line at infinity tells us that $\mathcal{O} + \mathcal{O} + \mathcal{O} = \mathcal{O}$, and therefore the point $\mathcal{O}$ is the zero of this group. The group equation becomes

$$P + Q + R = \mathcal{O}.$$

Moreover we have seen that $(x,y) + (x,-y) + \mathcal{O} = \mathcal{O}$ which implies that $(x,y) + (x,-y) = \mathcal{O}$, and therefore if $P = (x,y)$, then $-P = (x,-y)$.
Returning to the first example above, this implies that
$$(x_1, y_1) + (x_2, y_2) = (x_3, -y_3).$$

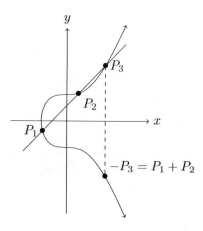

Figure 17.4. The group law: Adding two points P_1 and P_2 to obtain, geometrically, $P_1 + P_2$.

The addition operation is evidently closed; that is, adding points in $E(\mathbb{R})$, including $\mathcal{O}$, yields another point in $E(\mathbb{R})$. Similarly, addition is closed within the subgroup $E(\mathbb{Q})$. It is complicated to establish that addition is associative, which we must do to prove that we indeed have a group (see [**Grab**] for details). It is also obvious that the addition law is commutative.

Exercise 17.1.2. In Figure 17.3 the polynomial $x^3 + ax + b$ has three real roots, $r_1 < r_2 < r_3$, and so the points of $E(\mathbb{R})$ come in two continuous components, the *egg* and the *infinite part* $E^+(\mathbb{R}) = \{(x,y) \in E(\mathbb{R}) : x > r_3\} \cup \{\mathcal{O}\}$.
(a) Prove that a straight line intersects the egg in exactly 0 or 2 points (counted with multiplicity).
(b) Prove that the tangent at any point $P \in E(\mathbb{R})$ hits E again in $E^+(\mathbb{R})$.
(c) Deduce that $E^+(\mathbb{R})$ is a subgroup of $E(\mathbb{R})$.
(d) Deduce that the egg is the coset $(r_2, 0) + E^+(\mathbb{R})$ (that is, a coset of $E^+(\mathbb{R})$ in $E(\mathbb{R})$).

Suppose that we have a point $P \in E(\mathbb{R})$. Take the tangent, find the third point of intersection of the tangent line with E to obtain $-2P$, and then reflect in the x-axis to obtain $2P$. Fermat suggested that if we repeat this process over and over again, then we are unlikely to come back again to the same point. If we never return to the same point, then we say that P has *infinite order*; otherwise P has finite order, the order being the minimum positive integer n for which $nP = 0$ (points of finite order are known as *torsion points*).

Exercise 17.1.3. Prove that if $2P = \mathcal{O}$, then either $P = \mathcal{O}$ or $P = (x, 0)$ where $x^3 + ax + b = 0$. Deduce that the number of rational points of order 1 or 2 is one plus the number of integer roots of $x^3 + ax + b$ and therefore equals 1, 2, or 4.

Exercise 17.1.4. Prove that the torsion points form a subgroup of $E(\mathbb{C})$.

Exercise 17.1.5. Suppose that $x^3 + ax + b$ has three real roots.
(a) Prove that if $P \in E(\mathbb{R})$ is a torsion point of odd order, then $P \in E^+(\mathbb{R})$.
(b) Prove that if a torsion point P of E lies on the egg, then it is one of the points of order 2 at either end of the egg.

So far this discussion has largely been geometric, focusing on the real points of E. Number theorists wish to identify the structure of the group of rational points, $E(\mathbb{Q})$. Is $E(\mathbb{Q})$ finite or infinite? *Mordell's Theorem* states that $E(\mathbb{Q})$ is finitely generated, and so it can be written in the form $T \oplus \mathbb{Z}^r$ for some integer $r \geq 0$, where T is the set of points of finite order (the torsion points) and is finite. In other words, there exist a set of points $P_1, \ldots, P_r$ of infinite order, such that

$$E(\mathbb{Q}) = \{t + a_1 P_1 + \cdots + a_r P_r : t \in T \text{ and each } a_j \in \mathbb{Z}_j\}.$$

Each of these points are distinct. We say that $E(\mathbb{Q})$ has *rank* r, though strictly speaking we are only referring to the rank of the quotient group $E(\mathbb{Q})/T$.

Exercise 17.1.6. Deduce from Mordell's Theorem that $E(\mathbb{Q})$ is finite if and only if its rank $r = 0$.

We have already seen this group structure, with $r = 1$, when we were considering solutions to Pell's equation in section 11.2. There, all solutions take the form $\pm \epsilon_d^a$ with $a \in \mathbb{Z}$, so that the group of units of $\mathbb{Q}(\sqrt{d})$, when $d > 0$, is generated by -1 and ϵ_d and has structure $\mathbb{Z}/2\mathbb{Z} \times \mathbb{Z}$, the ± 1 being torsion. There can be more torsion in the group of units of $\mathbb{Q}(\sqrt{d})$ than just ± 1, when $d > 0$: For example, in $\mathbb{Q}[i]$ we also have the units $\pm i$ so the unit group structure is $\mathbb{Z}/4\mathbb{Z}$, generated by i.

Lutz and Nagell showed that if (x,y) is a torsion point in $E(\mathbb{Q})$, then x and y are integers, and y^2 divides $4a^3 + 27b^2$, the discriminant of $x^3 + ax + b$. Therefore there are only finitely many rational torsion points. Mazur improved this by showing that the torsion subgroup of $E(\mathbb{Q})$ contains at most 16 points: It either has the structure $\mathbb{Z}/N\mathbb{Z}$ for some $1 \leq N \leq 10$ or $N = 12$, or the structure $\mathbb{Z}/2\mathbb{Z} \times \mathbb{Z}/2N\mathbb{Z}$ for some $1 \leq N \leq 4$.

The proof of Mordell's Theorem is the main focus of the second part of our book [**Grab**] but is a bit too complicated to develop here. Instead, to get some idea of how we prove Mordell's Theorem, we focus on a very special class of elliptic curves:

17.2. Congruent number curves

A *Pythagorean* triangle is a right-angled, rational-sided triangle. Is there a Pythagorean triangle with area equal to a given rational number A? In section 6.1 we saw that Pythagorean triangles can be parametrized by edge lengths $g(t^2 - 1), 2gt$, $g(t^2 + 1)$ where $g, t \in \mathbb{Q}$. (We also need that $t > 1$ and $g > 0$ for these lengths to be positive.) This triangle has area $\frac{1}{2} \cdot g(t^2 - 1) \cdot 2gt = g^2(t^3 - t)$, which equals A if $(At, A^2/g)$ is a rational point on the elliptic curve:

$$E_A : y^2 = x^3 - A^2 x.$$

Moreover any rational point (x, y) on this curve, with $x > A$ (so that $(x, y) \in E_A^+(\mathbb{Q})$) and $y > 0$, generates a Pythagorean triangle of area A by taking $t = x/A, g = A^2/y$.

Exercise 17.2.1. Prove that if $A = ar^2$ for some rational r, then $E_a(\mathbb{Q})$ is isomorphic to $E_A(\mathbb{Q})$. (We may therefore restrict our attention to E_A where A is a squarefree positive integer.)

The elliptic curves E_A are called *congruent number curves*. They each have three points, $(-A, 0), (0, 0), (A, 0)$, of order two.

Exercise 17.2.2. Suppose that $P \in E_A(\mathbb{Q})$ has order > 2. Prove that exactly one of points P and $P + (0,0)$ belongs to $E_A^+(\mathbb{Q})$. Therefore this point yields a Pythagorean triangle of area A, with parameters $t = x/A, g = A^2/|y|$).

Exercise 17.2.2 implies that the set of Pythagorean triangles of area A has a structure that can be best understood via the rational points of E_A.

Theorem 17.1. *Let A be a squarefree, positive integer. If there is one Pythagorean triangle of area A, then there are infinitely many. Starting with one of area A with parameter $t > 1$, we obtain infinitely many by iterating the map*

$$t \to T = \frac{(t^2 + 1)^2}{4(t^3 - t)}.$$

Proposition 17.2.1. *Let A be a squarefree, positive integer. If $(x, y) \in E_A(\mathbb{Q})$, then we may write x and y as $x = m/n^2$ and $y = \ell/n^3$ where $\ell^2 = m(m^2 - A^2 n^4)$ with $(\ell m, n) = 1$ and $n > 0$. If $P = (x, y) \in E_A(\mathbb{Q})$ with $y \neq 0$, then $2P = (X, Y)$ where*

$$X = \left(\frac{x^2 + A^2}{2y}\right)^2 = \left(\frac{m^2 + A^2 n^4}{2\ell n}\right)^2.$$

This also gives

$$X - A = \left(\frac{x^2 - 2Ax - A^2}{2y}\right)^2 \quad \text{and} \quad X + A = \left(\frac{x^2 + 2Ax - A^2}{2y}\right)^2,$$

so that $2P \in E_A^+(\mathbb{Q})$. We may write $X = M/N^2$ where $(M, AN) = 1$ and $N \geq n$.

Given a Pythagorean triangle with parameter $t = m/An^2$, this proposition yields another Pythagorean triangle of area A, this one with parameter $T = \frac{(t^2+1)^2}{4(t^3-t)}$.

Proof. If $(x, y) \in E_A(\mathbb{Q})$, then we may write x and y as $x = m/u$ and $y = \ell/v$ where $(m, u) = (\ell, v) = 1$ and $u, v > 0$. Then

$$u^3 \ell^2 = u^3(vy)^2 = v^2 \cdot u^3 y^2 = v^2 \cdot u^3(x^3 + ax + b) = v^2(m^3 + amu^2 + bu^3).$$

Now $(u, m^3 + amu^2 + bu^3) = (u, m^3) = 1$ as $(u, m) = 1$, so that, since u^3 divides $v^2(m^3 + amu^2 + bu^3)$ we deduce that u^3 divides v^2 by Euclid's lemma. An analogous argument reveals that v^2 divides u^3, and so $v^2 = u^3$, since they are both positive. This implies that there exists a positive integer n for which $v = n^3$ and $u = n^2$, and the first claim follows.

The second claim, that $X, X - A$, and $X + A$ are all squares, is simply a calculation. We can write $X = M/N^2$ with $(M, N) = 1$ just as we did with x. We need to prove that $(M, A) = 1$. If not, then there exists a prime p dividing M and A, so p divides $M = XN^2$ and $M - AN^2 = (X - A)N^2$, which are each squares. Therefore p^2 divides them both and so their difference AN^2. However p cannot divide N as $(p, N) = 1$ and so p^2 divides A, which contradicts the fact that A is squarefree.

From our formula for X we have $N = 2\ell n/g$ where $g = (2\ell n, m^2 + A^2 n^4)$. Now $(n, m^2 + A^2 n^4) = (n, m^2) = 1$, and so g divides 2ℓ. Therefore $N = (2\ell/g)n \geq n$. □

17.4. The group of rational points of $y^2 = x^3 - x$

Exercise 17.2.3. Let $x_P = m/n^2 > A$ and $x_{2P} = M/N^2$ with $(m,n) = (M,N) = 1$.
(a) Prove that $M < 4m^4$.
(b) Prove that if $(m,A) = 1$, then $(m^2 + A^2n^4, 2\ell n) = 1$ or 2, and N is even.
(c) Deduce that if $(m,A) = 1$, then $M > m^4/4$, and also $M > m^4$ if n is even.
(d) Deduce that, in general, $(m^2 + A^2n^4, 2\ell n)$ divides $2(m,A)^2$.
(e)† Prove that $(m^2 + A^2n^4, 2\ell n) = (m,A)^2$ or $2(m,A)^2$.

Proof of Theorem 17.1. A Pythagorean triangle of area A leads to a rational point on $P = (x_P, y_P) \in E_A(\mathbb{Q})$ with $x_P > A$. We will write $x_{2^k P} = m_k/(n_k)^2$ for all $k \geq 0$. By Proposition 17.2.1 we can write $x_P = At$ where $t := m_0/An_0^2 > 1$. Moreover $x_{2P} = AT$ where $T := (t^2+1)^2/4(t^3-t) = m_1/An_1^2 > 1$ with $(m_1, An_1) = 1$ and $n_1 \geq n_0$.

We deduce from Proposition 17.2.1, by induction on each $k \geq 1$, that $x_{2^k P} > A$ with $(m_k, An_k) = 1$, and so $m_{k+1} > m_k^4/4$ by exercise 17.2.3. Therefore the m_k are increasing, and so the $2^k P$ are distinct. We can translate each such point, via exercise 17.2.2, to obtain an infinite sequence of distinct Pythagorean triangles of area A. □

Exercise 17.2.4. Prove that the torsion subgroup of $E_A(\mathbb{Q})$ is isomorphic to $\mathbb{Z}/2\mathbb{Z} \times \mathbb{Z}/2\mathbb{Z}$.

Our goal now is to understand the points of infinite order on congruent number curves. We begin though with an easier example.

17.3. No non-trivial rational points by descent

We can show that there are some elliptic curves which have no rational points by an easy descent: Suppose that $x, y \in \mathbb{Q}$ such that $y^2 = x^3 + x$, so that $x = m/n^2, y = \ell/n^3$ with $(\ell m, n) = 1$ and
$$\ell^2 = m(m^2 + n^4).$$
Now $(m, m^2 + n^4) = (m, n^4) = 1$ so that both m and $m^2 + n^4$ are squares, say, $m = u^2$ and $m^2 + n^4 = w^2$. Therefore
$$n^4 + u^4 = m^2 + n^4 = w^2,$$
which has no non-trivial solutions, by Theorem 6.1. The trivial solutions have either $n = 0$ (corresponding to the point $\mathcal{O}$ at ∞) or $u = 0$ (corresponding to the point $(0,0)$ of order two). Hence $E(\mathbb{Q}) \cong \mathbb{Z}/2\mathbb{Z}$.

17.4. The group of rational points of $y^2 = x^3 - x$

We apply ideas from elementary number theory to rational points on the elliptic curve
$$E: y^2 = x^3 - x.$$
There are several obvious rational points: the point at infinity, as well as the three points of order two, $(-1,0), (0,0), (1,0)$. Are there any others? We can write $x = m/n^2$, $y = \ell/n^3$ with $(\ell m, n) = 1$ to obtain
$$(m - n^2)m(m + n^2) = \ell^2.$$
Here, the product of three integers equals a square, so we can write each of them as a squarefree integer times a square, where the product of the three squarefree integers must also equal a square. Note that the three squarefree integers cannot

have a common factor, or else that factor cubed is a square and so the integers were not squarefree. This means that we can write
$$m - n^2 = pqu^2, \quad m = prv^2, \quad m + n^2 = qrw^2,$$
for some squarefree integers p, q, r which are pairwise coprime. Moreover since the product of the integers is positive, and as $m-n^2 < m < m+n^2$, we see that $m+n^2 > 0$ and $m - n^2$ and m have the same sign. Hence we may assume that p has the same sign as m, and q and r are positive. Now note that $(m \pm n^2, m) = (n^2, m) = 1$ and so $|p| = r = 1$. Moreover $(m - n^2, m + n^2) = (m - n^2, 2n^2) = (m - n^2, 2)$ since $(m - n^2, n^2) = (m, n^2) = 1$. To summarize, we have proved that there are four possibilities for the value of (p, q, r), since $p = -1$ or 1, $q = 1$ or 2, and $r = 1$. This leads to four possible sets of equations:

$m - n^2$	$=$	u^2	$-u^2$	$2u^2$	$-2u^2$
m	$=$	v^2	$-v^2$	v^2	$-v^2$
$m + n^2$	$=$	w^2	w^2	$2w^2$	$2w^2$

Table 17.1. One of these cases must hold, for some integers u, v, and w.

The solutions with $(m, n) = (1, 0), (0, 1), (1, 1), (-1, 1)$, obtained from the four rational points $\mathcal{O}, (0, 0), (1, 0), (-1, 0)$, correspond to these four cases, respectively.

Suppose we have another rational point on the curve, say, $P = (m/n^2, \ell/n^3)$. Let $Q = P + \mathcal{O}, P + (0, 0), P + (1, 0),$ or $P + (-1, 0)$, according to which of the four ways that $(m - n^2)m(m + n^2)$ factors. Writing $Q = (M/N^2, L/N^3)$ we find that Q always belongs to the first class; that is, each of $M - N^2$, M, and $M + N^2$ are squares. For example, the line between P and $(0, 0)$ is $mny = \ell x$. If $P + (0, 0) = (u, v)$, then $u + m/n^2 + 0 = (\ell/mn)^2$, so that $u = (\ell^2 - m^3)/m^2 n^2 = -n^2/m$ since $\ell^2 = m^3 - mn^4$. Hence we see that m and n^2 in the second equation must be replaced by $M = n^2$ and $N^2 = -m$, respectively, so that $M - N^2 = n^2 + m = w^2$, $M = n^2$, $M + N^2 = n^2 - m = u^2$, respectively, as claimed.

Therefore any rational point leads to a solution in the first case, and from there to an integer solution to the equation
$$X^4 - Y^4 = Z^2 \quad \text{with } (X, Y) = 1 \text{ and } X + Y \text{ odd},$$
taking $(X, Y, Z) = (v, n, uw)$. But this was shown to have no non-trivial solutions in Theorem 6.2 by constructing a smaller solution from any given solution. Therefore
$$E(\mathbb{Q}) = \{\mathcal{O}, (0, 0), (1, 0), (-1, 0)\}.$$

This approach generalizes to all elliptic curves, though we will focus on the congruent number curves.

17.5. Mordell's Theorem: $E_A(\mathbb{Q})$ is finitely generated

We now indicate how to generalize the above proof to rational points $P \in E_A(\mathbb{Q})$, for arbitrary squarefree, positive integers A. If $P \neq \mathcal{O}$, then we can write $P = (x_P, y_P)$ with $x_P = m/n^2$ where $(m, n) = 1$, as we saw in Proposition 17.2.1, and such that
$$(m - An^2)m(m + An^2) = \ell^2$$

17.5. Mordell's Theorem: $E_A(\mathbb{Q})$ is finitely generated

for some integer ℓ. As before we find that $(m, m - An^2) = (m, m + An^2) = (m, A)$ divides A, and $(m - An^2, m + An^2)$ divides $2A$. Therefore we can write

(17.5.1) $\qquad m - An^2 = pqu^2, \quad m = prv^2, \quad m + An^2 = qrw^2,$

for some squarefree integers p, q, r which are pairwise coprime, with $q, r > 0$. Since pqr divides $2A$, there are only finitely many possible such triples.

There were two key steps of the argument in the previous section that will need to be generalized:

- For the "smallest" $P_0 \in E_A(\mathbb{Q})$ with the same p, q, r-values as P, show that if $P + P_0 = Q$, then x_Q, $x_Q - A$, and $x_Q + A$ are all squares.
- If $Q \in E_A(\mathbb{Q})$ with $p = q = r = 1$, then construct a smaller point of $E_A(\mathbb{Q})$ by a descent argument.

The technical setup for the first step follows from the explicit formulas for adding points:

Proposition 17.5.1. *If $P, Q, R = P + Q \in E_A(\mathbb{Q})$, with $P, Q, R \ne \mathcal{O}$, for which*

$$x_P x_Q, \quad (x_P - A)(x_Q - A), \quad \text{and} \quad (x_P + A)(x_Q + A)$$

are each squares, then x_R, $x_R - A$, and $x_R + A$ are all squares.

Proof. This holds when $P = Q$ by Proposition 17.2.1. Now assume that $P \ne Q$. We have

$$x_R = \left(\frac{y_Q - y_P}{x_Q - x_P}\right)^2 - (x_P + x_Q) = \frac{1}{x_P x_Q} \cdot \left(\frac{x_P y_Q - x_Q y_P}{x_Q - x_P}\right)^2.$$

Therefore if $x_P x_Q$ is a square, then so is x_R. Similarly

$$x_R - A = \frac{1}{(x_P - A)(x_Q - A)} \cdot \left(\frac{(x_P - A)y_Q - (x_Q - A)y_P}{x_Q - x_P}\right)^2,$$

and the analogous expression with A replaced by $-A$. The result follows immediately from these formulas provided that $y_P y_Q \ne 0$ (so that each of $x_P x_Q$, $(x_P - A)(x_Q - A)$, $(x_P + A)(x_Q + A)$ are non-zero).

Now assume that $y_P = 0$. Then $y_Q \ne 0$, or else either $Q = P$ so that $P + Q = \mathcal{O}$, or Q is another point of order 2, and the hypothesis is not satisfied. We also have $y_R \ne 0$ or else either $Q = \mathcal{O}$ or $y_Q = 0$. Therefore $x_R, x_R - A, x_R + A$ are all non-zero.

Two of the three integers $x_P x_Q$, $(x_P - A)(x_Q - A)$, $(x_P + A)(x_Q + A)$ are non-zero squares, which implies that two of the three integers $x_R, x_R - A, x_R + A$ are non-zero squares. Moreover their product is the non-zero square y_R^2, and so the third of these integers is also a non-zero square. $\square$

Exercise 17.5.1. Suppose that $P \in E_A^+(\mathbb{Q})$ and that it corresponds to a Pythagorean triangle of area A, with parameter t. Prove that $Q := P + (A, 0) \in E_A^+(\mathbb{Q})$ and that it corresponds to *the same* Pythagorean triangle of area A, but with parameter T, where $T = \frac{t+1}{t-1}$.

An appropriate construction for the second step of our plan is obtained from the following converse theorem, which allows us to construct a smaller solution from a larger:

Theorem 17.2 (The converse theorem). *If $Q = (X,Y) \in E_A(\mathbb{Q})$ with X, $X - A$, and $X + A$ each non-zero squares, then there exists $P \in E_A^+(\mathbb{Q})$ for which $Q = 2P$.*

Proof. Write $X - A = u^2$, $X = v^2$, and $X + A = w^2$ so that $u^2 + w^2 = 2v^2$ and $w^2 - u^2 = 2A$. We now let

$$x = A \cdot \frac{w - u}{2v - u - w} \quad \text{and} \quad y = \frac{2A^2}{2v - u - w}.$$

Substituting we obtain

$$\frac{x^3 - A^2 x}{y^2} = \frac{(u - w)(v - w)(v - u)}{A(2v - u - w)}.$$

Now $2(v - u)(v - w) = 2v^2 - 2v(u + w) + 2uw = u^2 + w^2 - 2v(u + w) + 2uw = (u + w)(u + w - 2v)$, and so the last line becomes $(w^2 - u^2)/2A = 1$. Therefore $P_0 = (x, y) \in E_A(\mathbb{Q})$, and we can verify $Q = 2P_0$ using the equations of Proposition 17.2.1. We select $P = \pm P_0$ or $\pm P_0 + (0, 0)$, as one of these is in $E_A^+(\mathbb{Q})$ by exercise 17.2.2, and we select the sign "$\pm$" so as to get the sign of Y correct. □

<u>A plan for the descent</u>, based on these last two results:

Let $\mathcal{S}_A$ be the set of triples $\tau = (p, q, r)$ for which there exists some $P \in E_A^+(\mathbb{Q})$ such that (17.5.1) holds. Select $P_\tau \in E_A^+(\mathbb{Q})$ to be that $P \in E_A^+(\mathbb{Q})$ satisfying (17.5.1) for which m is minimal. Now, given point $P \in E_A^+(\mathbb{Q})$ satisfying (17.5.1) for some given τ, let $Q = P + P_\tau$, so that $\tau_Q = (1, 1, 1)$ by Proposition 17.5.1. Then we can determine $R \in E_A^+(\mathbb{Q})$ for which $Q = 2R$, by Theorem 17.2. Therefore $2R = Q = P + P_\tau$, so we expect that m_R is smaller than m_P. If so, then we can prove Mordell's Theorem by induction on m_P: If $m_R, m_{P_\tau} < m_P$, then we know that R and P_τ can be expressed in terms of our finite basis, and therefore $P = 2R - P_\tau$ also can. To make this plan work we need to compare the m_R with m_P.

Height bounds. Expanding the square in the second formula for x_R in the proof of Proposition 17.5.1 we have

$$\frac{m_Q}{n_Q^2} = \frac{m_P n_P^2(m_{P_\tau}^2 - A^2 n_{P_\tau}^4) + m_{P_\tau} n_{P_\tau}^2(m_P^2 - A^2 n_P^4) - 2n_P n_{P_\tau} \ell_P \ell_{P_\tau}}{(m_{P_\tau} n_P^2 - m_P n_{P_\tau}^2)^2},$$

so that m_Q divides the numerator of the expression on the right. Now $m_P > An_P^2$ so that $A(n_P \ell_P)^2 = (An_P^2) m_P (m_P^2 - A^2 n_P^4) < m_P^4$, and therefore

$$A m_Q < 4 m_{P_\tau}^2 m_P^2.$$

The proof of Theorem 17.1 gives that $m_Q = ((m_R^2 + A^2 n_R^4)/g)^2$, where $g := (2\ell_R n_R, m_R^2 + A^2 n_R^4)$, and $g \leq 2(m_R, A)^2 \leq 2A^2$ by exercise 17.2.3(d). We deduce that

(17.5.2) $\qquad m_Q = ((m_R^2 + A^2 n_R^4)/g)^2 \geq (m_R^2/2A^2)^2.$

Combining our two estimates, we find that
$$(m_R^2/2A^2)^2 \leq m_Q < 4m_{P_\tau}^2 m_P^2/A,$$
and so $m_R < m_P$ provided $m_P \geq 4A^2 m_{P_\tau}$. We have proved the following result:

Theorem 17.3. *For each $\tau \in \mathcal{S}_A$ select $P_\tau \in E_A^+(\mathbb{Q})$ to be that $P \in E_A^+(\mathbb{Q})$ satisfying (17.5.1) for which m_P is minimal. Let $C_A = 4A^2 \max_{\tau \in \mathcal{S}_A} m_{P_\tau}$. Let $\mathcal{G}_A := \{P \in E_A^+(\mathbb{Q}) : m_P < C_A\}$. Then $E_A^+(\mathbb{Q})$ is generated, additively, by the points in $\mathcal{G}_A$.*

Since $\mathcal{G}_A$ is finite we deduce that $E_A(\mathbb{Q})$ is finitely generated, moreover that $E_A(\mathbb{Q})$ can be written in the form $T \oplus \mathbb{Z}^r$ where T is the set of torsion points, for some integer $r \geq 0$, the rank of $E_A(\mathbb{Q})$, which is the number of independent generators of infinite order.[4] This is Mordell's Theorem for congruent number curves.

One can prove Mordell's Theorem for general elliptic curves by suitably modifying this proof, as we will discuss in appendix 17A.

Completely characterizing which integers A are the area of some rational right-angled triangle remains an open question of current interest. The beautiful book [3] highlights the following surprising result:

Theorem 17.4 (Tunnell's Theorem). *Assume certain widely believed conjectures about elliptic curves.[5] Then squarefree integer A is the area of a Pythagorean triangle if and only if*
$$\#\{(a,b,c) \in \mathbb{Z}^3 : A = a^2 + 2b^2 + 8c^2 \text{ with } c \text{ odd}\}$$
$$= \#\{(a,b,c) \in \mathbb{Z}^3 : A = a^2 + 2b^2 + 8c^2 \text{ with } c \text{ even}\}.$$

The connection between Pythagorean triangles of area A and the number of representations of A by certain quadratic forms is quite mysterious and goes through some amazing links between elliptic curves and modular forms.

Much of the discussion in this chapter is developed from:

[1] Stephanie Chan, *Rational right triangles of a given area*, Amer. Math. Monthly **125** (2018), 689–703.

Exercise 17.5.2 (Areas of rational-sided triangles, I). Let T be a triangle with rational side lengths.
   (a) Prove that rational-sided isosceles triangles of area $2A$ are in 1-to-1 correspondence with rational-sided right-angled triangles of area A.
   (b) Show that if T has rational area A, then it has rational height, no matter which side is the base.
   (c) Draw a perpendicular line from the base of the triangle to the top triangle vertex. This splits the triangle into two rational-sided right-angled triangles. Prove that we can parameterize

---

[4] It is worth remarking that there is no simple structure theorem for infinite abelian groups, analogous to the result discussed in section 3.16 of appendix 3C for finite abelian groups. For example, the group $(\mathbb{Q}, +)$ looks very different in that $2\mathbb{Q} = \mathbb{Q}$ yet it has infinite rank.

[5] This needs to be vague as there is considerable work in defining and explaining the terminology in the statements of these conjectures.

a rational scalar multiple of T as follows, where a and b are rational numbers > 1:

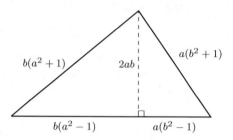

(d) Prove that rational number A is the area of a triangle with rational side lengths if and only if there exist rational numbers a, b, c for which $A = abc^2(a+b)(ab-1)$.

(e) Verify that for given A and b there is a 1-to-1 correspondence between such triangles and rational points on the elliptic curve $E_{A,b} : y^2 = x(x+Ab)(x-A/b)$ (taking $x = Aa, y = A^2/bc$) with $x > A/b$.

(f) Show that if we are given a point $(Aa, A^2/bc) \in E_{A,b}(\mathbb{Q})$, then we can determine another point $(X, Y) \in E_{A,b}(\mathbb{Q})$ with $X = (\frac{(a^2+1)bc}{2})^2$.

Exercise 17.5.3 (Areas of rational-sided triangles, II). Fix a squarefree positive integer A. Let $t = A^2 u^4$ for some $u \in \mathbb{Q}$ and then

$$a = \frac{2(t+1)^2}{9Au^2} \quad \text{and} \quad b = \frac{6A(t+1)u^2}{(2t-1)^2}.$$

Use these parametrizations to prove that there are infinitely many distinct rational-sided triangles of area A. Exhibit several of area 1.

17.6. Some nice examples

Diagonal cubic surface. G. H. Hardy, visiting Ramanujan as he lay ill from pneumonia in an English hospital, remarked that the number 1729 of the taxicab he had ridden from the train station to the hospital was extremely dull. Ramanujan contradicted him by noting that it is the smallest number which is the sum of two cubes in two different ways:

$$1^3 + 12^3 = 9^3 + 10^3 = 1729.$$

(Ramanujan might also have mentioned that it is the third smallest Carmichael number!).[6]

For a given integer $m \geq 1$, we are interested in integers a and b for which $a^3 + b^3 = m$. The rational solutions of $a^3 + b^3 = m$ ($\neq 0$) are in 1-to-1 correspondence with the rational points (x, y) on the elliptic curve

$$E'_m : y^2 = x^3 - 3(12m)^2,$$

via the transformation $u = a+b$, $v = a-b$ and then $y = 36mv/u$, $x = 12m/u$.[7] So questions about the representation of integers as the sums of two cubes are also question about special points on elliptic curves. However, it seems unlikely that this will help much with understanding the integer solutions to $a^3 + b^3 = m$.

---

[6] Littlewood, speaking of Ramanujan's encyclopaedic knowledge of the properties of many different numbers, claimed, "Every positive integer was one of his personal friends."

[7] A transformation that we highlighted in section 14.3.

Exercise 17.6.1. Prove that if a and b are integers for which $a^3 + b^3 = m$, then $|a|, |b| \leq (4m/3)^{1/2}$.

We have seen that 1729 is the smallest integer that can be represented in two ways. Are there integers that can be represented in three ways or four ways or...? This is not difficult to answer using the doubling process on the cubic curve $x^3 + y^3 = 1729$. One can take a rational point on this curve, transform this to a rational point on the elliptic curve E'_{1729}, double, and then transform back, but it is easier to proceed directly. In general we begin with the identity

$$t(t-2)^3 + (1-t)(1+t)^3 = (1-2t)^3.$$

Given $ax^3 + by^3 = cz^3$ let $t = ax^3/cz^3$ and multiply through by $(cz^3)^4$ to obtain

(17.6.1) $\quad a(x(by^3 + cz^3))^3 + b(-y(ax^3 + cz^3))^3 = c(z(ax^3 - by^3))^3,$

a new rational point on $au^3 + bv^3 = c$. In particular if $u^3 + v^3 = 1729$, then $A^3 + B^3 = 1729$ where

(17.6.2) $\quad A = u \cdot \dfrac{u^3 - 3458}{1729 - 2u^3} \quad$ and $\quad B = v \cdot \dfrac{u^3 + 1729}{1729 - 2u^3}.$

So, starting from the solution $(12, 1)$, we get further solutions

$$\left(\frac{20760}{1727}, \frac{-3457}{1727} \right), \left(\frac{184026330892850640}{15522982448334911}, \frac{61717391872243199}{15522982448334911} \right),$$

and it is pointless to write down the next solution since each ordinate has seventy digits! The main point is that there are infinitely many different solutions, and we write them as $(p_j/q_j, r_j/q_j)$, $j = 1, 2, \ldots$. From (17.6.2) one can deduce that q_j divides q_{j+1} for all $j \geq 1$ and so we have N solutions to $a^3 + b^3 = 1729 q_N^3$ taking $a_i = p_i(q_N/q_i)$ and $b_i = r_i(q_N/q_i)$ for $i = 1, \ldots, N$.

Scaling up rational points seems like a bit of a cheat, so let's ask whether there exists an integer m that can be written in N ways as the sum of two cubes of coprime integers. People have found examples for $N = 3$ and 4 but not beyond, and this remains an open question.

Magic squares and elliptic curves. We discussed magic squares in appendix 1A. In Figure 1.2 of section 1.13 we parametrized magic squares in terms of 5 variables, x_1, x_2, x_3, a, b. The entries of the magic square are integers if our variables are all integers. Every entry is the square of an integer if and only if each $x_j, x_j + a, x_j + b$ is a square, for $j = 1, 2, 3$. Therefore, magic squares in which every entry is the square of an integer are in 1-to-1 correspondence with

$$(E_{a,b}, P_1, P_2, P_3)$$

where $E_{a,b}$ denotes the elliptic curve $E_{a,b} : y^2 = x(x+a)(x+b)$, each $P_i \in E(\mathbb{Z})$ and $2P_i = (x_i, y_i)$.

Further reading on the basics of elliptic curves

[1] Edray H. Goins, *The ubiquity of elliptic curves*, Notices Amer. Math. Soc **66** (2019), 169–174.

[2] Fernando Q. Gouvea, *A marvelous proof*, Amer. Math. Monthly **101** (1994), 203–222.

[3] Neil Koblitz, *Introduction to Elliptic Curves and Modular Forms*, Graduate Texts in Mathematics 97, Springer-Verlag, New York, 1993.

[4] Barry Mazur, *Number theory as gadfly*, Amer. Math. Monthly, **98** (1991), 593–610.

[5] Joseph H. Silverman, *Taxicabs and sums of two cubes*, Amer. Math. Monthly **100** (1993), 331–340.

Appendix 17A. General Mordell's Theorem

We need to be a bit more formal about the possible values of p, q, r in (17.5.1). Let $H := \mathbb{Q}^*/(\mathbb{Q}^*)^2$, which means that two non-zero rational numbers whose ratio is a square are considered to be equal in H. We define a map

$$\phi : E_A(\mathbb{Q}) \to \{(a,b,c) \in H : abc = 1 \text{ in } H\} \text{ by } \phi(x,y) = (x - A, x, x + A).$$

if $x - A, x, x + A$ are all non-zero. (So if $x = m/n^2$ then $x - A = m - An^2$ in H, and this $= pq$ in H in (17.5.1).)

If $x + A = 0$, then $x - A, x$ are non-zero, and we let $\phi(P) = (x - A, x, x(x - A))$, and we define $\phi(P)$ analogously if $x = 0$ or $x - A = 0$. The identities in Proposition 17.2.1, and the proof of Proposition 17.5.1, imply that ϕ is a homomorphism. The converse theorem (Theorem 17.2) implies that $\ker \phi = 2E_A(\mathbb{Q})$. The first isomorphism theorem then implies that the image of ϕ

$$\phi(E_A(\mathbb{Q})) \cong E_A(\mathbb{Q})/2E_A(\mathbb{Q}).$$

We have been working with $E_A^+(\mathbb{Q})$. The condition $x > A$ implies that if $P \in E_A^+(\mathbb{Q})$, each coordinate of $\phi(P)$ is positive, and so $\phi(E_A^+(\mathbb{Q}))$ is a subgroup of $\phi(E_A(\mathbb{Q}))$, and the quotient group, $\frac{\phi(E_A(\mathbb{Q}))}{\phi(E_A^+(\mathbb{Q}))}$, is isomorphic to $\{(1,1,1), (-1,-1,1)\}$.

Mordell's Theorem implies that $E(\mathbb{Q})/2E(\mathbb{Q}) = T/2T \oplus (\mathbb{Z}/2\mathbb{Z})^r$. We saw above how to restrict the elements of $\phi(E(\mathbb{Q}))$ to a finite set where each entry is a divisor of $2A$. It was Weil who first fully developed the role of the map ϕ. In honor of their work the group of points $E(\mathbb{Q})$ is known as the *Mordell-Weil group*.

17.7. The growth of points

Another important issue is the growth of the size of the coordinates after successive doubling. We define the *height* of a rational number a/b with $(a,b) = 1$ to be $H(a/b) := \max\{|a|, |b|\}$. We extend this to a point $P \in E(\mathbb{Q})$ by letting

$H(P) := H(x(P))$. If $P \in E_A^+(\mathbb{Q})$, then $x(P) = m/n^2 > A$ and so $H(P) = m_P$. By Proposition 17.2.1 and exercise 17.2.3(a),(c) we have that if $Q = 2^k P$ with $k \geq 2$, then $\frac{1}{4}H(Q)^4 < H(2Q) < 4H(Q)^4$.

Exercise 17.7.1.‡ Let $P \in E_A^+(\mathbb{Q})$ and $Q = 4P$.
(a) Prove that $(4^{-1/3}H(Q))^{4^r} < H(2^r Q) < (4^{1/3}H(Q))^{4^r}$ for all $r \geq 1$.
(b) Prove that $\lim_{k \to \infty} H(2^k P)^{1/4^k}$ exists, which we denote by $\hat{H}(P)$, the *Néron-Tate height*.
(c) Prove that $\hat{H}(2P) = \hat{H}(P)^4$.
(d) Prove that $4^{-1/3}H(Q) \leq \hat{H}(Q) \leq 4^{1/3}H(Q)$.

One can similarly define the Néron-Tate height for points on an arbitrary elliptic curve. However it is much more challenging to obtain suitable upper and lower bounds on $H(2P)$ in terms of $H(P)$, for the general elliptic curve.

Four squares in an arithmetic progression. If $a - d$, a, $a + d$, and $a + 2d$ are all squares, say, u_{-1}^2, u_0^2, u_1^2, u_2^2, then $(-2d/a - 1, 2u_{-1}u_0 u_1 u_2/a^2)$ is a point on the elliptic curve
$$E : y^2 = (x-1)x(x+3).$$

In this case, Image(ϕ) is a subgroup of $\langle (-1,-1,1), (2,1,2), (1,3,3) \rangle$, which is itself a subgroup of $\{(a,b,c) \in H : abc = 1 \text{ in } H\}$. We have the elements $\phi((-1,2)) = (-2,-1,2)$ and $\phi((-3,0)) = (-1,-3,3)$ of Image(ϕ), and we now prove that $(-1,-1,1)$ is not in Image(ϕ). If it were, there would exist integers m, n, u, v, w with $(m,n) = 1$ for which $m - n^2 = -u^2$, $m = -v^2$, and $m + 3n^2 = w^2$. But then $-v^2 + 3n^2 = w^2$, and so $-v^2 \equiv w^2 \pmod{3}$, but as $\left(\frac{-1}{3}\right) = -1$ this implies that 3 divides v and w, and so n. Therefore 3 divides $(m,n) = 1$, a contradiction.

We deduce that Image(ϕ) $\cong (\mathbb{Z}/2\mathbb{Z})^2$ and so $E(\mathbb{Q})$ is all torsion, since we have four rational points of order dividing 2. One can show that
$$E(\mathbb{Q}) = \{\mathcal{O}, (1,0), (0,0), (-3,0), (3, \pm 6), (-1, \pm 2)\} \cong \mathbb{Z}/2\mathbb{Z} \oplus \mathbb{Z}/4\mathbb{Z}.$$
Translating this back to the original question yields:

Theorem 17.5 (Fermat's Theorem). *There are no four squares in an arithmetic progression $a - d$, a, $a + d, a + 2d$.*

Exercise 17.7.2. Use Szemerédi's Theorem (Theorem 15.6 from section 15.6) and Fermat's Theorem to deduce the following: For any $\delta > 0$ there exists a constant M_δ such that if $N \geq M_\delta$, then any arithmetic progression of length N contains $< \delta N$ squares. (It is conjectured that the N-term arithmetic progression with the most squares is $1, 1 + 24, 1 + 24 \cdot 2, \ldots, 1 + 24(N-1)$, which contains about $\sqrt{8N/3}$ squares; the best bound proved to date is at most a little more than $N^{3/5}$ squares.)

Exercise 17.7.3 (Another proof that there are infinitely many primes). Suppose not and that $p_1, \ldots, p_k$ is the complete set of primes. We will color the positive integers as follows: By the Fundamental Theorem of Arithmetic one can write every positive integer n in the form $p_1^{e_1} \cdots p_k^{e_k}$ where the e_j are integers ≥ 0. We will color n as $c(n) = p_1^{r_1} \cdots p_k^{r_k}$, where r_j is the least non-negative residue of $e_j \pmod 2$ for $j = 1, \ldots, k$.
(a) Establish that $c(n)$ provides a coloring of the positive integers with 2^k colors.
(b) Use van der Waerden's Theorem (Theorem 15.5 from section 15.6) to establish that there is a four-term arithmetic progression of integers $A, A + D, A + 2D, A + 3D$ for which $c(A) = c(A + D) = c(A + 2D) = c(A + 3D)$.
(c) Let $a = A/c(A)$ and $d = D/c(A)$. Prove that each of $a, a + d, a + 2d, a + 3d$ is a square.
(d) Establish a contradiction using Fermat's Theorem.

Appendix 17B. Pythagorean triangles of area 6

We begin with the $(3,4,5)$-triangle, which yields an infinite sequence of Pythagorean triangles of area 6 by Theorem 17.1. Are there any other Pythagorean triangles of area 6? To answer this question we need to determine the Mordell-Weil group $E_6^+(\mathbb{Q})$. By considering the divisors of 12, we know that $\phi(E_6^+(\mathbb{Q}))$ is a subgroup of the group Γ generated (multiplicatively) by $(2,2,1), (2,1,2), (3,3,1), (3,1,3)$, which is itself a subgroup of $\{(a,b,c) \in H : abc = 1 \text{ in } H\}$.

We already know several points in $E_6^+(\mathbb{Q})$, and therefore several elements of $\phi(E_6^+(\mathbb{Q}))$: We always have $\phi(\mathcal{O}) = (1,1,1)$. The torsion point $(6,0)$ yields $\phi((6,0)) = (2,6,3)$. The $(3,4,5)$-triangle yields $P = (12,36) \in E_6^+(\mathbb{Q})$ and $\phi(P) = (6,3,2)$, and so $\phi(P + (6,0)) = \phi(P)\phi((6,0)) = (3,2,6)$. We claim that

$$\phi(E_6^+(\mathbb{Q})) = \{(1,1,1), (2,6,3), (6,3,2), (3,2,6)\}.$$

To prove this we need to show that the remaining elements of $\Gamma \cong (\mathbb{Z}/2\mathbb{Z})^4$ do not belong to $\text{Image}(\phi)$. Since these are both groups, we need only show that two independent generators of $\Gamma/\langle (2,6,3), (6,3,2) \rangle$ do not belong to $\text{Image}(\phi)$:

If $(2,2,1) \in \text{Image}(\phi)$, then there exist integers m,n,u,v,w with $(m,n) = 1$ for which

$$m - 6n^2 = 2u^2, \quad m = 2v^2, \quad m + 6n^2 = w^2.$$

This leads to $2v^2 + 6n^2 = w^2$ and so $w^2 \equiv 2v^2 \pmod{3}$. However $\left(\frac{2}{3}\right) = -1$ and so 3 divides v and w, and hence n, implying that 3 divides $(m,n) = 1$, a contradiction.

If $(2,1,2) \in \text{Image}(\phi)$, then there exist integers m,n,u,v,w with $(m,n) = 1$ for which

$$m - 6n^2 = 2u^2, \quad m = v^2, \quad m + 6n^2 = 2w^2.$$

This leads to $v^2 + 6n^2 = 2w^2$ and so $v^2 \equiv 2w^2 \pmod{3}$. However $\left(\frac{2}{3}\right) = -1$ and so 3 divides v and w, and hence n, implying that 3 divides $(m,n) = 1$, a contradiction.

This establishes the claim, and so since $\phi(E_6^+(\mathbb{Q}))$ includes the image of the torsion point $(6,0)$ we deduce that $E_6^+(\mathbb{Q})$, and so $E_6(\mathbb{Q})$, has rank 1, the infinite part generated by P. Finally $E_6(\mathbb{Q}) \cong (\mathbb{Z}/2\mathbb{Z})^2 \oplus \mathbb{Z}$.

Therefore all of the Pythagorean triangles of area 6 are generated by $nP = (x_n, y_n)$, $n \geq 1$, taking $t = x_n/6$ and $g = 36/|y_n|$.

Integer points

In appendix 6C we need all of the $(x,y) \in E_6(\mathbb{Z})$ with x divisible by 6. The values $t = 2, 3$, and 49 given there correspond to the points $(12, \pm 36)$, $(18, \pm 72)$, and $(294, \pm 5040) \in E_6(\mathbb{Z})$. These are the only such integer points on $E_6(\mathbb{Z})$, but this is difficult to prove. Siegel's Theorem tells us that $E(\mathbb{Z})$ is always finite but finding all of its elements can be quite a challenge. Bennett [1] determines $E_A(\mathbb{Z})$ whenever $A = p$ or $2p$ for some odd prime p:

There are several "families" of integral points that only occur in very special circumstances. For example, if p can be written as the sum of two fourth powers, say, $p = r^4 + s^4$, then $(-(2rs)^2, \pm 4rs(r^4 - s^4)) \in E_{2p}(\mathbb{Z})$. Or if $p^2 = 2m^2 - 1$ for some integer m, then $(m^2, \pm(m^3 - m)) \in E_p(\mathbb{Z})$. None of these types of special circumstances occur when $p \equiv \pm 3 \pmod 8$. In that case, if $(x,y) \in E_A(\mathbb{Z})$, then either $y = 0$ or we have one of the points

$$(-3, \pm 9), (-2, \pm 8), (6 \cdot 2, \pm 6^2), (6 \cdot 3, \pm 6^2 \cdot 2), (6 \cdot 49, \pm 6^2 \cdot 140) \in E_6(\mathbb{Z}),$$
$$(-4, \pm 6), (5 \cdot 9, \pm 5^2 \cdot 12) \in E_5(\mathbb{Z}), \ (22 \cdot 99, \pm 22^2 \cdot 210) \in E_{22}(\mathbb{Z}),$$

or $(29 \cdot 9801, \pm 29^2 \cdot 180180) \in E_{29}(\mathbb{Z})$.

We observe that A divides x for most of these points $(x,y) \in E_A(\mathbb{Z})$. For such points we can provide a good bound on x, assuming the *abc*-Roth conjecture from section 11.5.

Theorem 17.6. *Assume the abc-Roth conjecture. Fix $\delta > 0$. There exists a constant $c_\delta > 0$ such that if $(x,y) \in E_A(\mathbb{Z})$ with x divisible by squarefree A, then $|x| \leq c_\delta A^{2+\delta}$.*

Proof. Select ϵ such that $(1-2\epsilon)(1+\delta) = 1$. Write $x = AX$ so that $y^2 = x^3 - A^2 x = A^3(X^3 - X)$. Therefore A^2 divides y as A is squarefree, and so $X^3 - X = AY^2$ where $y = A^2 Y$. We now apply the *abc*-Roth conjecture with $F(u,v) = uv(u^2 - v^2)$, so that $F(-1, X) = AY^2$, which yields

$$\kappa_{F,\epsilon} |X|^{2-\epsilon} \leq \prod_{p | AY} p \leq A|Y| = (A \cdot AY^2)^{1/2} < (AX^3)^{1/2}.$$

We deduce that $|X| \leq c_\delta A^{1+\delta}$ where $c_\delta = \kappa_{F,\epsilon}^{-2-2\delta}$, and the result follows. □

There are many techniques to limit integer points in:

[1] Michael Bennett, *Integral points on congruent number curves*, Int. J. Number theory **9** (2013), 1619–1640.

Appendix 17C. 2-parts of abelian groups

17.8. 2-parts of abelian, arithmetic groups

Let G be a *finite* abelian group, written additively, so that, by the Fundamental Theorem of Abelian Groups (as discussed in section 3.16 of appendix 3C)

$$G \cong H \oplus \mathbb{Z}/2^{e_1}\mathbb{Z} \oplus \cdots \oplus \mathbb{Z}/2^{e_\ell}\mathbb{Z}$$

where H is an abelian group of odd order h, say, for some integers $e_1, e_2, \ldots, e_\ell \geq 1$. If $g = (a, b_1, \ldots, b_\ell) \in G$ with $2g = 0$, then $2a = 0$ in H, so that $a = \frac{h+1}{2} \cdot 2a = 0$ in H; and each $2b_j \equiv 0 \pmod{2^{e_j}}$, so that $b_j \equiv 0$ or $2^{e_j - 1} \pmod{2^{e_j}}$. Therefore

$$G/2G \cong (\mathbb{Z}/2\mathbb{Z})^\ell.$$

We call ℓ the *2-rank of G*.

More generally if G is a *finitely generated* abelian group, then we can write

$$G \cong T \oplus \langle g_1, \ldots, g_r \rangle \cong T \oplus \mathbb{Z}^r,$$

where T contains the elements of G of finite order, and $g_1, \ldots, g_r$ are linearly independent elements of G of infinite order. Now T must be a finite abelian group and therefore has the structure above. We call r the *rank of the infinite part of G*. The rank of G is r plus the rank of T. Also

$$G/2G \cong T/2T \oplus (\mathbb{Z}/2\mathbb{Z})^r,$$

a finite group, so that the 2-rank of G equals $r + \ell$, where ℓ is the 2-rank of T (and, as above, $T/2T \oplus (\mathbb{Z}/2\mathbb{Z})^\ell$). In the theory of elliptic curves we wish to determine the rank of the infinite part of the Mordell-Weil group $E(\mathbb{Q})$, which equals the 2-rank of $E(\mathbb{Q})$, minus 0, 1, or 2, depending on whether there are 0, 1, or 3 elements of order 2 in $E(\mathbb{Q})$.

Appendix 17D. Waring's problem

17.9. Waring's problem

In Theorem 12.6 we proved that every positive integer is the sum of four squares.

How about sums of cubes? If $-2 \leq a \leq 3$ and integer $n \equiv a \equiv a^3 \pmod{6}$, then

$$n = (x+1)^3 + (x-1)^3 + (-x)^3 + (-x)^3 + a^3 = 6x + a^3,$$

a sum of five cubes. We can ask about sums of non-negative cubes instead. More generally, for each integer $k \geq 2$, Hilbert showed in 1909 that there exists an integer $g(k)$ such that every integer is the sum of $g(k)$ kth powers of non-negative integers, resolving a 1770 problem of Waring. We have seen that $g(2) = 4$, and it is known that $g(3) = 9$, $g(4) = 19$, $g(5) = 37, g(6) = 73, \ldots$. Actually $g(k)$ grows fast, because one can need a large number of kth powers to represent small integers:

Exercise 17.9.1. Let $s_k(n)$ be the smallest number of positive integers $a_1, \ldots, a_s$ for which $n = a_1^k + \cdots + a_s^k$.
(a) Prove that $s_k(2^k - 1) = 2^k - 1$.
(b) Prove that if $n = 2^k m - 1$ where $m = [(3/2)^k]$, then $s_k(n) = 2^k + [(3/2)^k] - 2$.
Euler's son conjectured that $g(k) = 2^k + [(3/2)^k] - 2$.
(c)‡ Prove that if $2^k \{(3/2)^k\} + [(3/2)^k] \leq 2^k$, then Euler Jr's conjecture is true.
This inequality follows if $\{(3/2)^k\} < 1 - (3/4)^k$ which is probably true for all integers $k \geq 2$ and is known to hold for $2 \leq k < 10^8$.

Let $G(k)$ be the smallest integer such that every sufficiently large integer is the sum of $G(k)$ kth powers of non-negative integers. If it is true that $s_k(n)$ is only large for small k, then we might expect $G(k)$ to be significantly smaller than $g(k)$. By Theorem 12.5 we know that $s_2(7 \cdot 4^j) \geq 4$ for all $j \geq 0$, and so $G(2) = 4 = g(2)$, no improvement. But for $k = 3$ we have $4 \leq G(3) \leq 7$ smaller than $g(3) = 9$. In general $G(k) \leq k \log k + k \log \log k + Ck$ for some constant C, much smaller than $g(k)$, which is $\geq 2^k$.

17.9. Waring's problem

Stop the press. As we finish proofreading this book, in September 2019, there have been extraordinary developments on the question of which integers can be represented as the sum of three cubes. Since every cube is $\equiv -1, 0$, or $1 \pmod 9$, a sum of three cubes cannot be $\equiv 4$ or $5 \pmod 9$ (see exercise 6.5.10). Heath-Brown has conjectured that every integer $\not\equiv 4$ or $5 \pmod 9$ can be written in infinitely many different ways as the sum of three cubes of integers (positive or negative). Until this month we had only known the very small solutions $1^3 + 1^3 + 1^3 = 4^3 + 4^3 - 5^3$ for 3, and feared there might be no more, but a widely distributed computation (on the "Charity Engine") found a new solution where the integers being cubed have up to twenty digits each. Moreover the first solutions were found for the only remaining cases up to < 100 without a known solution, namely

$$33 = 8866128975287528^3 - 8778405442862239^3 - 2736111468807040^3$$

and $42 = 80435758145817515^3 + 12602123297335631^3 - 80538738812075974^3.$

This lends evidence to Heath-Brown's conjecture.

We finish this section by giving Liouville's proof that $g(4) \leq 53$:

Theorem 17.7 (Liouville). *Every positive integer can be written as the sum of* 53 *fourth powers.*

Proof. Lagrange's Theorem (Theorem 12.6) states that any positive integer m can be written as the sum of four squares, say, $m = n_1^2 + n_2^2 + n_3^2 + n_4^2$. Therefore

$$6m^2 = 6(n_1^2 + n_2^2 + n_3^2 + n_4^2)^2 = \sum_{1 \leq i < j \leq 4} (n_i + n_j)^4 + (n_i - n_j)^4,$$

the sum of 12 fourth powers. Now any positive integer q can be written as $m_1^2 + m_2^2 + m_3^2 + m_4^2$, and so $6q$ can be written as the sum of $4 \times 12 = 48$ fourth powers. Finally any positive integer n can be written as $6q + r$ for some r, $0 \leq r \leq 5$, where $6q$ can be written as the sum of 48 fourth powers, and r as the sum of at most 5 fourth powers, and the result follows. □

Exercise 17.9.2. By proving that each of 0, 1, 2, 81, 16, and 17 can be written as the sum of at most 2 fourth powers, deduce that $g(4) \leq 50$.

Further reading on Waring's problem

[1] H. M. Davenport, *Analytic methods for Diophantine equations and Diophantine inequalities*, Cambridge University Press, 2010.

[2] W. J. Ellison, *Waring's problem*, Amer. Math. Monthly **78** (1971), 10–36.

[3] R. C. Vaughan, *The Hardy-Littlewood method*, Cambridge Tracts in Mathematics, 125 (1997), Cambridge University Press.

Hints for exercises

EXERCISES IN CHAPTER 0

Exercise 0.1.1(b). The key observation is that if $\alpha = \frac{1+\sqrt{5}}{2}$ or $\frac{1-\sqrt{5}}{2}$, then $\alpha^2 = \alpha + 1$ and so, multiplying through by α^{n-2}, we have $\alpha^n = \alpha^{n-1} + \alpha^{n-2}$ for all $n \geq 2$.

Exercise 0.1.3(b). Multiplying through by ϕ we have $\phi^{n+1} = F_n \phi^2 + F_{n-1}\phi$. Now use (a).

Exercise 0.1.5(b). Determine a and b in terms of α and then c and d in terms of $\alpha, x_0,$ and x_1.

Exercise 0.2.1(a). Note that $N^2 + (2N+1) = (N+1)^2$.

Exercise 0.3.1. In both parts use induction on n.

Exercise 0.4.2. Use (0.1.1) to establish that $|F_n - \phi^n/\sqrt{5}| < \frac{1}{2}$ for all $n \geq 0$.

Exercise 0.4.7. If the first character in a string in A_n is a 0, what must the subsequent string look like? What if the string begins with a 1?

Exercise 0.4.8. Use Gauss's trick to show that $\sum_{a<n\leq b} n = \binom{b+1}{2} - \binom{a+1}{2} = \frac{(b-a)(b+a+1)}{2}$, a product of two integers of opposite parity, both > 1. Show that if N is not a power of 2 (so that it has an odd divisor $m > 1$), then it is a product of two integers of opposite parity, both > 1. Determine a and b in terms of N and m.

Exercise 0.4.10(a). Verify this for $k = 1$ and 2, and then for larger k by induction.
 (b) Select k and m as functions of n.

Exercise 0.4.16. By (0.1.1), $\sqrt{5}F_n = \phi^n - \overline{\phi}^n$, and so $(\sqrt{5}F_n)^k = \sum_{j=0}^{k} \binom{k}{j}(-1)^j \rho_j^n$ where $\rho_j := \overline{\phi}^j \phi^{k-j}$. Let $x^{k+1} - \sum_{i=0}^{k} c_i x^i = \prod_{j=0}^{k}(x - \rho_j)$. Therefore

$$\sum_{i=0}^{k} c_i(\sqrt{5}F_{n+i})^k = \sum_{j=0}^{k}\binom{k}{j}(-1)^j \rho_j^n \cdot \sum_{i=0}^{k} c_i \rho_j^i = \sum_{j=0}^{k}\binom{k}{j}(-1)^j \rho_j^n \cdot \rho_j^{k+1} = (\sqrt{5}F_{n+k})^k.$$

The result follows after dividing through by $(\sqrt{5})^k$.

Exercise 0.6.1(a). Prove this for $k = 0$, and then by induction on k, using differential calculus.

Exercise 0.18.3. Substitute the value of y given by the line, into the equation of the circle.

Exercise 0.18.4. Subtract the equations for the two circles, and use exercise 0.18.3.

EXERCISES IN CHAPTER 1

Exercise 1.1.1(a). Write $a = db$ for some integer d. Show that if $d \ne 0$, then $|d| \ge 1$. (b) Prove that if u and v are integers for which $uv = 1$, then either $u = v = 1$ or $u = v = -1$. (c) Write $b = ma$ and $c = na$ and show that $bx + cy = max + nay$ is divisible by a.

Exercise 1.1.2. Use Lemma 1.1.1 and induction on a for fixed b.

Exercise 1.2.1(a). By exercise 1.1.1(c) we know that d divides $au + bv$ for any integers u and v. Now use Theorem 1.1. (d) First note that a divides b if and only if $-a$ divides b. If $|a| = \gcd(a, b)$, then $|a|$ divides both a and b, and so a divides b. On the other hand if a divides b, then $|a| \le \gcd(a, b) \le |a|$ by (c).

Exercise 1.2.4(b). Let $g = \gcd(a, b)$ and write $a = gA, b = gB$ for some integers A and B. What is the value of $Au + Bv$? Now apply (a).

Exercise 1.2.5(a). Use Theorem 1.1.

Exercise 1.4.2. Use Lemma 1.4.1.

Exercise 1.7.5(e). Write $r = m + \delta$ where $0 < \delta < 1$, so that $[r] = m$ and $a - r = a - m - \delta$ so that $[a - r] = ?$.

Exercise 1.7.10. Given any solution, determine u using Lemma 1.1.1.

Exercise 1.7.11. One might apply Corollary 1.2.2.

Exercise 1.7.14(d). Use exercise 1.7.10.

Exercise 1.7.22. For each given $m \ge 1$, prove that $x_m | x_{mr}$ for all $r \ge 1$, by induction on r, using exercise 0.4.10(a) with $k = rm$.

Exercise 1.7.23(a). Prove that $\gcd(x_n, b) = \gcd(ax_{n-1}, b)$ for all $n \ge 2$, and then use induction on $n \ge 1$, together with Corollary 1.2.2. (b) Prove that $\gcd(x_n, x_{n-1}) = \gcd(bx_{n-2}, x_{n-1})$ for all $n \ge 2$, and then use induction on $n \ge 1$, together with Corollary 1.2.2. (c) Use exercise 0.4.10(a) with $k = n - m$ and then (b). (d) Follow the steps of the Euclidean algorithm using (c).

Exercise 1.9.1. Use the matrix transformation for $(u_j, u_{j+1}) \to (u_{j+1}, u_{j+2})$.

Exercise 1.14.1(c). If n is odd, take $a = b = c = 1, d = -1$. Show that if n is even, then a, b, c, d are odd so that $ad - bc$ is even.

Exercise 1.17.1. Divide the representation of $2/n$ above by an appropriate power of 2. Be careful when b is a power of 2.

EXERCISES IN CHAPTER 2

Exercise 2.1.4(b). Write the integers in the congruence class $a \pmod{d}$ as $a + nd$ as n varies over the integers, and partition the integers n into the congruences classes mod k.

Exercise 2.1.5. Write the congruence in terms of integers and then use exercise 1.1.1(c).

Exercise 2.1.6. Write the congruence in terms of integers and then use exercise 1.1.1(e).

Exercise 2.4.1(c). Factor 1001.

Exercise 2.5.4(a). Split the integers into k blocks of m consecutive integers, and use the main idea from the first proof of Theorem 2.1. (b) Write $N = km + r$ with $0 \le r \le m - 1$. Use (a) to get k such integers in the first km consecutive integers, and at most one in the remaining r. Compare k or $k + 1$ to the result required.

Exercise 2.5.6(b). Use the results for $m = 4$ from (a). (d) Use the same idea as in (c). (e) Study squares mod 8.

Exercise 2.5.9(b). Use that $\frac{1}{j}\binom{p-1}{j-1} = \frac{1}{p}\binom{p}{j}$.

Exercise 2.5.10(a). Treat the cases $a \geq b$ and $a < b$ separately. (b) Treat the cases $c \geq d$ and $c < d$ separately.

Exercise 2.5.13. Proceed by induction on $k \geq 1$.

Exercise 2.5.15(b). Use induction.

Exercise 2.5.16(a). Try a proof by contradiction. Start by assuming that the kth pigeonhole contains a_k letters for each k, and determine a bound on the total number of letters if each $a_k \leq 1$. (b) Use the pigeonhole principle. (c) Use induction.

Exercise 2.5.17(a). Use the pigeonhole principle on pairs $(x_r \pmod{d}, x_{r+1} \pmod{d})$. (d) Use exercise 1.7.24.

Exercise 2.10.2. Use induction.

Exercises in chapter 3

Exercise 3.0.1. The only divisors of p are 1 and p. Therefore $\gcd(p, a) = 1$ or p, and so $\gcd(p, a) = p$ if and only if p divides a. This implies that $\gcd(p, a) = 1$ if and only if p does not divide a.

Exercise 3.1.1. Use induction and the fact that every integer > 1 has a prime divisor, as proved in the "prerequisites" section. (The proof will appear as part of the proof of Theorem 3.2.)

Exercise 3.1.2(a). Apply Theorem 3.1 with $a = a_1 \cdots a_{k-1}$ and $b = a_k$, and if p divides a, then proceed by induction.

(b) p divides some q_j by (a), and as q_j only has divisors 1 and q_j, and as $p > 1$, we deduce that $p = q_j$.

Exercise 3.1.3(b). Write $n = 2^k m$ with m odd. Then n has an odd prime factor if and only if $m > 1$. Therefore if n has no odd prime factor, then $n = 2^k$.

Exercise 3.2.1. We have $[a, b] = ab$ by Corollary 3.2.2. The result follows from Lemma 1.4.1.

Exercise 3.3.1. Look at this first in the case that m and n are both powers of p, say, $m = p^a$ and $n = p^b$. If d divides m and n, then $d = p^c$, say, with $c \leq a$ and $c \leq b$. The maximum c that satisfies both of these inequalities is $\min\{a, b\}$. Similarly if m and n divide $L = p^e$, then $a \leq e$ and $b \leq e$ and so the minimum e that satisfies both of these inequalities is $\max\{a, b\}$. Now use this idea when m and n are arbitrary integers.

Exercise 3.3.5. Use exercise 3.3.3(d).

Exercise 3.3.7(c). Use exercise 3.3.3(c).

Exercise 3.5.1(a). Show that the $aj + b$ are distinct mod m.

Exercise 3.5.2. Prove that the $r_j \pmod{m}$ are all reduced residues, and then that they are distinct.

Exercise 3.5.3. If $ar \equiv c \pmod{b}$, then b divides $ar - c$. Therefore $\gcd(a, b)$ divides $ar - c$ and so c. In the other direction, we write $g = \gcd(a, b)$ and so $a = gA, b = gB, c = gC$, and we are looking for solutions to $Ar \equiv C \pmod{B}$. Then use exercise 3.5.1(b).

Exercise 3.5.5. Use the second proof of Corollary 3.5.2.

Exercise 3.6.4. If $am + bn = c$, then $am + bn \equiv c \pmod{b}$ (or indeed mod any integer $r \geq 1$). On the other hand if $au + bv \equiv c \pmod{b}$ and m is any integer $\equiv u \pmod{b}$, then $am \equiv au + bv \equiv c \pmod{b}$ and so there exists an integer n for which $am + bn = c$.

Exercise 3.7.2(a) We proceed by induction on the number of moduli using exercise 3.2.1.

(b) Replace m in (a) by $m - n$.

Exercise 3.7.8(a). Work with the prime power divisors of m and use the Chinese Remainder Theorem.

Exercise 3.8.3. Calculate the product mod p^e, for every prime power $p^e \| m$.

Exercise 3.9.1. Use exercise 1.7.20(a).

Exercise 3.9.3(a). If $2k+1 = n/m$, take $u = \alpha^m$ and $v = \beta^m$ in

$$\frac{u^{2k+1} + v^{2k+1}}{u+v} = (-uv)^k + \sum_{j=1}^{k}(-uv)^{k-j}(u^{2j} + v^{2j}),$$

so that y_n/y_m is a linear polynomial in the y_{2jm} with coefficients that are $\pm$ powers of b.

Exercise 3.9.6. Use exercise 3.3.7(c), and factor $gA^2 - gB^2$.

Exercise 3.9.7(a). Write $\frac{z^p - y^p}{z-y}$ as a polynomial in y and z.

Exercise 3.9.10(a). $\sqrt{2} + \sqrt{3}$ is a root of $x^4 - 10x^2 + 1$. Use Theorem 3.4.

(b) $\sqrt{a} + \sqrt{b}$ is a root of $x^4 - 2(a+b)x^2 + (a-b)^2$. Therefore the rational root $m = \sqrt{a} + \sqrt{b}$ must be an integer, and then m divides $a - b$. Writing $a = b + mk$ we have $k = \sqrt{a} - \sqrt{b}$ so that $b = (\frac{m-k}{2})^2$ and $a = (\frac{m+k}{2})^2$.

Exercise 3.9.11(b). Prove that $(\sqrt{d} + m)(\sqrt{d} - m)$ is an integer

Exercise 3.9.15(b). Use Corollary 2.3.1.

Exercise 3.9.17(b). Write $m = gM$ and $n = gN$ where $g = \gcd(m,n)$ so that $(M,N) = 1$, and then use exercise 3.7.7 (or exercise 3.9.16(b), for a less complete solution).

Exercise 3.10.2. Write the trinomial coefficient as the product of binomial coefficients.

Exercise 3.11.1. Prove this by induction on $n \geq 1$, using the observation in the paragraph immediately above.

Exercise 3.15.2. Let $G = \mathbb{Z}/4\mathbb{Z}$ and H be the subgroup of order two. Determine the maximal order of an element of $H \oplus G/H$, as well as of G.

Exercise 3.19.1. Use exercise 0.14.6.

Exercise 3.19.2. R is a Euclidean domain if there exists $w : R \to \mathbb{Z}_{\geq 0}$ such that for any $a, b \in R$ with $b \neq 0$, there exists $q \in R$ such that if $r = a - qb$, then $w(r) < w(b)$. Given any ideal I of R, let $b \in I$, $b \neq 0$, with $w(b)$ minimal. For any $a \in I$ let $r = a - qb \in I$ so that $w(r) < w(b)$ which contradicts the minimality of $w(b)$.

Exercise 3.22.1. Use Proposition 2.10.1 and adapt the proof of Euclid's Lemma.

Exercise 3.24.1(c). $f(x) = 3x + 1$ has the rational root $-\frac{1}{3}$ yet $f(n) \equiv 1 \pmod{3}$, for all integers n.

Exercises in chapter 4

Exercise 4.0.1. One can proceed by induction on the number of distinct prime factors of n, using the definition of multiplicative.

Exercise 4.1.3. Pair m with $n - m$, and then m with n/m.

Exercise 4.1.5. If the prime factors of n are $p_1 < p_2 < \cdots < p_k$, then $p_j \geq k + j$ and so $\frac{\phi(n)}{n} = \prod_{j=1}^{k} \frac{p_j - 1}{p_j} \geq \prod_{j=1}^{k} \frac{k+j-1}{k+j} = \frac{k}{2k} = \frac{1}{2}$.

Exercise 4.2.2. Let $\ell = (d, a)$ so that $\ell | a$ and therefore $d/\ell | (a/\ell)b$ with $(d/\ell, a/\ell) = 1$ and therefore $m = d/\ell | b$.

Exercise 4.2.3(b). What is the power of 2 in $\sigma(n)$?

Exercise 4.2.4. Give a general lower bound on $\sigma(n)$.

Exercise 4.2.5(a). If $p^e \| n$, then $1 + \frac{1}{p} \leq \sigma(p^e)/p^e < 1 + \frac{1}{p} + \frac{1}{p^2} + \cdots = \frac{p}{p-1}$.

Hints for exercises 573

(b) If n is a perfect number, then $\sigma(n)/n = 2$, and if it is odd with ≤ 2 prime factors, then $\prod_{p|n} \frac{p}{p-1} \leq \frac{3}{2} \cdot \frac{5}{4}$ which is < 2, contradicting (a).

Exercise 4.3.7(a). Use exercise 3.9.15(a).

Exercise 4.3.11(a). Prove this when a and b are both powers of a fixed prime and then use multiplicativity.

Exercise 4.3.12. In both parts write, for each $d|n$, the integers $m = an/d$ with $(a,d) = 1$. Use exercise 4.1.3.

Exercise 4.3.13(a). You could use the second part of exercise 4.1.3.

Exercise 4.3.15(b). Use multiplicativity. (e) Use exercise 4.2.5.

Exercise 4.5.1(a). Use the binomial theorem. (b) Let $m = \prod_{p|n} p$ and $x = -1$ in (a).

Exercise 4.5.2. Expand the right-hand side.

Exercise 4.6.2. Let $r = (a, m)$ and then $s = a/r$ and $t = m/r$ which therefore must be coprime. Now $a = rs$ divides $mn = rtn$, so that s divides tn and therefore s divides n as $(s,t) = 1$. Let $u = n/s$ and we finally deduce $b = mn/a = tu$.

Exercise 4.8.2. Use the expansion $\phi(n) = \sum_{d|n} \mu(n/d) d$ from the proof of Theorem 4.1 in section 4.4, and a similar expression for σ.

Exercise 4.9.1(b). How large is the set $\{n \geq 1 : |f(n)| > n^\sigma\}$?

Exercise 4.11.1. For $n = 1, 2, \ldots$ determine the coefficient of t^n in $a(t)b(t) = 1$ as a polynomial in the a_i and b_j, and then find where a given b_m first appears.

Exercise 4.14.1(c). Use that $\{t\} = t - [t]$, and cut the integral up into intervals $[n, n+1)$, taking the integral up to integer N and then letting $N \to \infty$.

Exercise 4.15.1(a). Write $t = n + u$, subtract $\log n$, and then integrate the first few terms of the resulting Taylor series.

Exercise 4.16.1(b). Use the Fundamental Theorem of Algebra.

Exercise 4.16.4. Use the Möbius inversion formula from section 4.6.

Exercises in Chapter 5

Exercise 5.1.4. Show that if $2^{2^{n-1}} < x \leq 2^{2^n}$, then there are $\geq n$ primes up to x. Then give a lower bound for n as a function of x.

Exercise 5.3.2. Show that if every prime factor of n is $\equiv 0$ or $1 \pmod 3$, then $n \equiv 0$ or $1 \pmod 3$.

Exercise 5.3.4. Consider splitting arithmetic progressions mod 3 into several arithmetic progressions mod 6.

Exercise 5.3.5. One might use exercise 3.1.4(b) in this proof.

Exercise 5.4.1(b). We wish to show that $\pi(x+\epsilon x) > \pi(x)$. By (5.4.2) (and footnote 14) we know that for any fixed $\delta > 0$ we have $(1-\delta)\frac{x}{\log x} < \pi(x) < (1+\delta)\frac{x}{\log x}$ if x is sufficiently large. The result will then follow if the middle inequality holds in

$$\pi(x) < (1+\delta)\frac{x}{\log x} < (1-\delta)\frac{x+\epsilon x}{\log(x+\epsilon x)} < \pi(x+\epsilon x).$$

Now $\frac{\log(x+\epsilon x)}{\log x} < 1 + \frac{\epsilon}{\log x}$ as $\log(1+\epsilon) < \epsilon$, and so the middle inequality follows if $1 + \frac{\epsilon}{\log x} < (1-\delta)(1+\epsilon)/(1+\delta)$. Selecting, say, $\delta = \epsilon/3$ this holds if x is sufficiently large.

Exercise 5.8.11. Use l'Hôpital's rule.

Exercise 5.8.12. First prove that $\left(\text{Li}(x) - \frac{x}{\log x}\right) / \frac{x}{(\log x)^2} \to 1$ as $x \to \infty$.

Exercise 5.8.14(a). Use Corollary 2.3.1.

Exercise 5.9.1. Either use Kummer's Theorem (Theorem 3.7) or consider directly how often p divides the numerator and denominator of $\binom{2n}{n}$.

Exercise 5.9.3. Use induction to show that, for each $n \geq 6$, every integer in $[7, 2N+6]$ is the sum of distinct primes in $\{2, 3, \ldots, 2N\}$, by induction on $N \geq 1$.

Exercise 5.9.6. Let p be a prime in $[2n, 4n]$. Now construct all the pairs you can that sum to p. Proceed.

Exercise 5.10.1. Maximize the log of the ratio using calculus.

Exercise 5.10.2. Use Proposition 5.10.1.

Exercise 5.10.3(a). If $r \leq s/2$, then by Bertrand's postulate there is a prime $p \in (s/2, s] \subset (r, s]$. Otherwise $k = s - r \leq r$. In either case, by Bertrand's postulate or the Sylvester-Schur Theorem, one term has a prime factor $p > k$, and so this is the only term that can be divisible by p.

Exercise 5.11.8(b). Use the Fundamental Theorem of Algebra mod p (see Lagrange's Theorem, Proposition 7.4.1).

Exercise 5.11.9(a). Can be proved by induction on k. For $k = 0$ this is trivial. For larger k, let $T \subset \{1, 2, \ldots, m-1\}$ and we pair together the terms for $S = T$ and $S = T \cup \{m\}$ in our sum. The sum therefore becomes

$$\sum_{T \subset \{1,2,\ldots,m-1\}} (-1)^{|T|} \left(\left(x_m + x_0 + \sum_{j \in T} x_j \right)^k - \left(x_0 + \sum_{j \in T} x_j \right)^k \right)$$

$$= \sum_{i=0}^{k-1} \binom{k}{i} x_m^{k-i} \sum_{T \subset \{1,2,\ldots,m-1\}} (-1)^{|T|} \left(x_0 + \sum_{j \in T} x_j \right)^i$$

and the result follows by induction, as $m - 1 > k - 1 \geq i$.

(b) Let $x_0 = \log n$ and if n has prime factors $p_1, \ldots, p_m$, then let $x_j = -\log p_j$ for each $j \geq 1$.

(c) We get $k! x_1 \ldots x_k$ in (a) and so $(-1)^k k! \prod_{p|n} \log p$ in (b). We prove this by induction using the proof in (a), since in the induction step only the $i = k - 1$ term remains, which is the result from the previous step multiplied by $k x_k$.

Exercise 5.16.1. If $\operatorname{Re}(s) > 1$, then the Euler product for $\zeta(s)$ is absolutely convergent (as we proved) and so $\zeta(s) = 0$ if and only if some $1 - p^{-s} = 0$.

Exercise 5.28.3. We will show how to find all the possible orbits when the period has length 1:

(a) Suppose we have the orbit $0 \to b \to a \to a \to \cdots$ with $0, a, b$ distinct integers. By Corollary 2.3.1, we have that $b = b - 0$ divides $f(b) - f(0) = a - b$ and so b divides a. Moreover, $a = a - 0$ divides $f(a) - f(0) = a - b$ and so a divides b. Therefore $|b| = |a|$, and so $b = -a$ as a and b are distinct. To find the polynomials $f(.)$ for which $f(0) = -a, f(a) = f(-a) = a$ we extrapolate. The second two conditions give that $f(x) = a + (x - a)(x + a)h(x)$ for some $h(x) \in \mathbb{Z}[x]$. Substituting in $x = 0$ gives $-a = f(0) = a - a^2 h(0)$, so that a^2 divides $2a$; that is, $a = -2, -1, 1,$ or 2 (as $a \neq 0$).

(b) Can an orbit have the shape $0 \to u \to v \to w \to w \to \cdots$ with $0, u, v, w$ distinct integers? Let $g(x) = f(x + u) - u$ so that this orbit becomes $-u \to 0 \to b \to a \to a \to \cdots$ where $b = v - u$ and $a = w - u$. Then $b = -a = -2, -1, 1,$ or 2 by (a). By Corollary 2.3.1, we have $u = 0 - (-u)$ divides $b - 0 = -a$ and so $u = \pm 1$ and $a = \pm 2$. Next $b + u = u - a$ divides $a - 0 = a$ and so $a = 2u$, which is impossible or else $a - (-u) = 3u$ divides $a - 0 = 2u$.

Hints for exercises

Exercise 5.28.4(c). If x_0 is periodic, then $x_{n+2} = x_n$ for all $n \geq 0$ as the period length is either one or two. Moreover, if x_0 is strictly preperiodic, then 0 is strictly preperiodic for the map $x \to f(x + x_0) - x_0$, with orbit $0 \to b_1 \to b_2 \to \cdots$ where $b_n = x_n - x_0$. In all four cases of our classification $b_2 = b_4$, and so $x_2 = x_4$.

EXERCISES IN CHAPTER 6

Exercise 6.1.2. Study where lines of rational slope, going through the point $(2,1)$, hit the curve again.

Exercise 6.1.5. Write down an equation that identifies when three given squares are in arithmetic progression.

Exercise 6.3.1(a). By (6.1.1) the area is $g^2 rs(r^2 - s^2)$ where $r > s \geq 1$ and $(r, s) = 1$. If this is a square, then each of r, s, and $r^2 - s^2$ must be squares; call them x^2, y^2, and z^2, respectively, so that $x^4 - y^4 = z^2$, which contradicts Theorem 6.2.

(c) Consider a right-angled triangle with sides $x^2, 2y^2, z$.

Exercise 6.5.3. Here b is the hypotenuse, and c is the area. Further hint: We need $b^2 - 4c$ and $b^2 + 4c$ to be integer squares, say, u^2 and v^2, so that $4c = b^2 - u^2 = v^2 - b^2$. Therefore $2b^2 = u^2 + v^2$, so u, v have the same parity and therefore $(\frac{u+v}{2})^2 + (\frac{u-v}{2})^2 = b^2$. This is our Pythagorean triangle, which has area $\frac{1}{2} \cdot \frac{u+v}{2} \cdot \frac{v-u}{2} = \frac{v^2 - b^2 + b^2 - u^2}{8} = c$.

Exercise 6.5.6. Let $\alpha = p/q$ with $(p, q) = 1$ so that $\alpha = (a\alpha + b)/\alpha = (ap + bq)/p$. Now $(p, q) = 1$ so comparing denominators we must have $q = 1$, and p divides $ap + bq$, so that p divides bq, and therefore b.

Exercise 6.5.7. By (6.1.1) the perimeter of such a triangle has length $2grs + g(r^2 - s^2) + g(r^2 + s^2) = 2gr(r + s)$ where $r > s > 0$. Therefore n has divisors r and $r + s$, where $r < r + s < 2r$. On the other hand if n has divisors d_1, d_2 for which $d_1 < d_2 < 2d_1$, then we may assume they are coprime, by dividing through by any common factor. Therefore $d_1 d_2$ divides n and so we can let $r = d_1$, $s = d_2 - d_1$, and $g = n/d_1 d_2$.

Exercise 6.5.9. Prove that if $n \geq 13$, then $(n+1)^2 + 128 < 2n^2$. Then proceed by induction on n for $m \in [n^2 + 129, 2n^2)$.

Exercise 6.5.10. What values can cubes take mod 9?

Exercise 6.8.1. By simple geometry things must look like the following diagram.

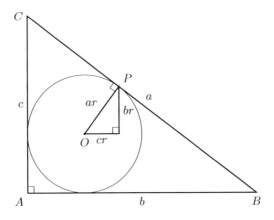

Figure. A circle inscribed inside a right-angled triangle.

If $A = (0,0)$, then BC is the line $by + cx = bc$. The point $P = (ar + cr, br + ar)$ lies on this line, and so $bc = r(b(b+a) + c(a+c)) = r(b^2 + ab + c^2 + ac) = ar(a+b+c)$. Therefore the radius of the circle is

$$ar = \frac{bc}{a+b+c} = \frac{bc(b+c-a)}{(b+c+a)(b+c-a)} = \frac{bc(b+c-a)}{b^2+c^2-a^2+2bc} = \frac{b+c-a}{2}.$$

Exercise 6.10.1. First prove that $(n+i)(n+j) \ne (n+I)(n+J)$ for $k \ge I, J, i, j \ge 1$ unless $\{I, J\} = \{i, j\}$: Suppose that $(n+i)(n+j) = (n+I)(n+J)$ so that $(i+j-I-J)n = IJ-ij$. If $I = J = k$ or $I = J = 1$, then evidently $i = j = I = J$. Therefore $k(k-1) \ge IJ \ge 2$, and similarly ij, so that $IJ - ij \equiv 0 \pmod{n}$ and $n > k^2 - k > |IJ - ij|$. Therefore $IJ = ij$ and so $I + J = i + j$, which implies that $\{I, J\} = \{i, j\}$. Now if $a_I a_J = a_i a_j$ with $\{I, J\} \ne \{i, j\}$, we may suppose that $(n+I)(n+J) > (n+i)(n+j)$. Therefore

$$(n+I)(n+J) - (n+i)(n+j) = a_i a_j ((m_I m_J)^\ell - (m_i m_j)^\ell)$$
$$\ge a_i a_j ((m_i m_j + 1)^\ell - (m_i m_j)^\ell) > \ell a_i m_i^{\ell-1} a_j m_j^{\ell-1}$$
$$> \ell((n+i)(n+j))^{1-1/\ell} > 3n^{4/3} > 3n(k-1)$$

as $n > k^\ell - k > (k-1)^3$. Therefore

$$3n(k-1) < (n+I)(n+J) - (n+i)(n+j) = (I+J-i-j)n + (IJ-ij) \le (k-1)(2n+k+1),$$

which is a contradiction.

Exercises in chapter 7

Exercise 7.1.2(b). Use the technique in the proof of Lemma 7.1.1

Exercise 7.2.2. Let $k := \text{ord}_m(a)$ and $A = \{1, a, a^2, \ldots, a^{k-1} \pmod{m}\}$. Show that if b and b' are any two reduced residues mod m, then either bA and $b'A$ are disjoint or are equal. Therefore the sets of the form bA, where b is a reduced residue mod m, which are each of size k, partition the $\phi(m)$ reduced residues mod m. This implies that k divides $\phi(m)$ as desired.

Exercise 7.3.1. Let $k := \text{ord}_q(2)$. We have $2^p \equiv 1 \pmod{q}$ and so k divides p by Lemma 7.1.2. Therefore $k = 1$ or p, but $k \ne 1$ as $2^1 \not\equiv 1 \pmod{q}$.

Exercise 7.4.1(a) If n is not of the form p or p^2, write $n = ab$ with $1 < a < b$. If $n = p^2$, then n divides $p \cdot 2p$.

Exercise 7.4.3(a). If $Q = \frac{p-1}{2}$, then

$$(p-1)!/Q! = (p-1)(p-2) \cdots (p-Q) \equiv (-1)(-2) \cdots (-Q) = (-1)^Q Q! \pmod{p}.$$

Exercise 7.5.2(b). As $(g^{\frac{p-1}{2}})^2 = g^{p-1} \equiv 1 \pmod{p}$, so $g^{\frac{p-1}{2}}$ is a square root of $1 \pmod{p}$; that is, $g^{\frac{p-1}{2}} \equiv 1$ or $-1 \pmod{p}$. But g has order $p-1$ and so $g^{\frac{p-1}{2}} \not\equiv 1 \pmod{p}$.

Exercise 7.10.2. Use Proposition 7.4.1.

Exercise 7.10.4. In every solution $n, n-1, n-2$ have prime factors $2, 3, p$ for some $p > 3$. At most one of these integers is divisible by p. Show that the other two lead to a solution to $2^n - 3^m = \pm 1$ and use exercise 7.10.3.

Exercise 7.10.5(b). Use Theorem 7.1.

(d) Make sure a is chosen so that $(q, a-1) = 1$.

Exercise 7.10.6(a). The trick is to write $z^p = ((z-y) + y)^p$ and then use the binomial theorem. One can also write $x_n = \frac{z^n - y^n}{z - y}$ and use exercise 2.5.20(a).

Exercise 7.10.12. Take the j and $p - j$ terms together.

Exercise 7.10.13. Let $M = a_0 + 1$ so that $a_n = 2^n M - 1$ for all $n \ge 0$. Let p be an odd prime dividing a_1. Then p divides a_p.

Exercise 7.10.16(b). Since n is not a Carmichael number, the subgroup in (a) is proper and so contains at most half the reduced residues. (c) Let $q = 2p-1$. Now $n-1 \equiv p-1$ (mod $2p-2$), so that if $(a,n) = 1$, then $a^{n-1} \equiv a^{p-1} \equiv 1$ (mod p) and $a^{n-1} \equiv a^{p-1} \equiv a^{\frac{q-1}{2}} \equiv \pm 1$ (mod q).

Exercise 7.10.17(a). $M_p - 1 = 2^p - 2$ is divisible by p.

Exercise 7.12.1(b). Let $f(x_1, \ldots, x_p) = (x_2, \ldots, x_p, x_1)$ in part (a).

Exercise 7.17.4(c). Consider $\gcd(q^\ell - 1, (q^n - 1)/(q^\ell - 1))$. (d) Use Lemma 7.17.1.

Exercise 7.18.1. Let $g = \gcd(\lambda(p^e), \lambda(q^f))$ where p^e and q^f are powers of the two different primes dividing m (where $f \geq 2$ if $q = 2$), so that 2 divides g. Now

$$\lambda(m) = \operatorname{lcm}[\lambda(p^e) : p^e \| m] \leq \frac{1}{g} \prod_{p^e \| m} \lambda(p^e) \leq \frac{1}{g} \prod_{p^e \| m} \phi(p^e) = \frac{\phi(m)}{g} < \phi(m).$$

Exercise 7.18.5(a). Recall exercise 4.3.7.

Exercise 7.19.1. In one direction let $y = x^{n/g}$. In the other, write $g = an + b\lambda(m)$ so that if $x = y^a$, then $x^n \equiv (y^a)^n (y^{\lambda(m)})^b \equiv y^g$ (mod m).

Exercise 7.25.1(b). You might show that if $p \cdot 1 = q \cdot 1 = 0$ where p and q are distinct primes, then $1 = 0$.

Exercise 7.25.3(a). You might use the ideas in the proof of Theorem 7.6.

Exercise 7.28.6. Consider divisibility by F_r where 2^r is the highest power of 2 dividing $k - \ell$. Then we must have $p = F_r$ and so $2^{2^n} - 1 = 2^{2^r} + 1 + 2^{\ell+j2^r} + 2^\ell$. Consider this equation mod 2^4 and we severely limit the possibilities.

Exercise 7.30.1(b). Multiply through by $1 - x^{m_k}$ and then substitute in x to be a root of $\phi_{m_k}(x)$.

Exercise 7.33.1(b). Show that $|\phi_n(a)| > n$ if $\phi(n) > 2$, and for $\phi(n) = 2$ with $|a| > 2$. Analyze carefully the remaining few cases.

Exercise 7.33.2(b). Prove that $\Delta = (\alpha - \beta)^2$. (c) Use exercise 2.5.20, and be careful when p divides a.

Exercises in Chapter 8

Exercise 8.1.2(b). Use Lemma 8.1.1.

Exercise 8.1.3(a). Use that $\left(\frac{b^2}{p}\right) = \left(\frac{b}{p}\right)^2 = (\pm 1)^2 = 1$.

Exercise 8.1.6(a). The residues $1, g^2, g^4, \ldots, g^{p-3}$ (mod p) are evidently distinct and non-zero squares. As there are $\frac{p-1}{2}$ of them, they are all of the quadratic residues by Lemma 8.1.1.

(b) We see above that $g = g^1$ is not one of the quadratic residues.

Exercise 8.2.1. There are two solutions to $r^2 \equiv a$ (mod p), say, r and $-r$ (mod p), whose product is $r \cdot (-r) \equiv -a$ (mod p). Note also that $|S| = \frac{p-3}{2}$.

Exercise 8.4.1. r is the largest integer with $2r - 1 \leq \frac{q-1}{2}$; that is, $r \leq \frac{q+1}{4}$.

Exercise 8.4.5. Look at $(2/p)$.

Exercise 8.7.2(a). Use the Chinese Remainder Theorem and exercise 8.1.2(b).

Exercise 8.7.5. If a is odd, then $a = 1 + 2 \cdot \frac{a-1}{2}$, and so

$$1 + 2 \cdot \frac{ab-1}{2} = ab = \left(1 + 2 \cdot \frac{a-1}{2}\right)\left(1 + 2 \cdot \frac{b-1}{2}\right) \equiv 1 + 2 \cdot \left(\frac{a-1}{2} + \frac{b-1}{2}\right) \pmod{4}.$$

Exercise 8.7.6. Select $a^2 \equiv -2 \pmod{p}$ with a odd and minimal, so that $1 \le a \le p-1$. Write $a^2 + 2 = pr$. Evidently $pr \equiv a^2 + 2 \equiv 3 \pmod 8$ and so $r \equiv 3p \equiv 5$ or $7 \pmod 8$. But then $a^2 \equiv -2 \pmod r$ and so $\left(\frac{-2}{r}\right) = 1$ with $r = \frac{a^2+2}{p} < p$. This contradicts the induction hypothesis, and so $\left(\frac{-2}{p}\right) = -1$.

Exercise 8.8.1. Suppose that $k > \ell \ge 1$. If r is a quadratic residue mod p^k, then r is a quadratic residue mod p^ℓ, trivially. On the other hand if r is a quadratic residue mod p^ℓ, then it is a quadratic residue mod $p^{\ell+1}$ by Proposition 8.8.1, then mod $p^{\ell+2}$ by Proposition 8.8.1, etc., up to mod p^k. We take $\ell = 1$ if p is odd, and $\ell = 3$ if $p = 2$ and note that if r is a quadratic residue mod 8, then $r \equiv 1 \pmod 8$.

Exercise 8.9.5(a). Write $n = 3^a m$ where $3 \nmid m$.

Exercise 8.9.9(a). Consider the size of the set of residues $\{a^2 \pmod p\}$ and of the set of residues $\{m - b^2 \pmod p\}$, as a and b vary.

(b) Take $m = -1$.

(c) Prove there is a solution u, v to $au^2 + bv^2 \equiv -c \pmod p$ and then multiply through by any $z \pmod p$.

Exercise 8.9.10(e). Apply Gauss's trick as in the proof of Corollary 7.5.2.

Exercise 8.9.12. For each solution to $y^2 \equiv b \pmod p$, consider whether there are solutions to $x^2 \equiv y \pmod p$.

Exercise 8.9.14. Let $b^2 \equiv -1 \pmod p$ and study $(1+b)^2 \pmod p$.

Exercise 8.9.15. Show that if a has order $m \pmod p$, then $\sigma_{a,p}$ consists of $\frac{p-1}{m}$ cycles of length m.

Exercise 8.9.16(a). Use exercise 1.7.20(c). (b) Use exercise 1.7.20(b).

Exercise 8.9.17. Select integer m with $(m/n) = -1$. Consider the prime divisors of integers of the form $kn + m$ for well-chosen values of k.

Exercise 8.9.18(a). Modify the ideas in Euclid's proof that there are infinitely many primes. (b) $n = -3$. (c) Look at $4m^2 + 3$ with m odd. (d) $n = 3$. Note $(m^2 - 3)/2 \equiv 2 \pmod 3$. (e) $n = -4$. Note $m^2 + 4 \equiv 5 \pmod 8$. (f) $n = 2$. Note $m^2 - 2 \equiv 7 \pmod 8$. (g) $n = -2$. Note $m^2 + 2 \equiv 3 \pmod 8$. (h) $n = -4$ with $(m, 6) = 1$.

Exercise 8.9.24. Therefore $\left(\frac{2}{n}\right) = \left(\frac{2}{n-2}\right)$ if $n \equiv 1 \pmod 4$, and $\left(\frac{2}{n}\right) = -\left(\frac{2}{n-2}\right)$ if $n \equiv 3 \pmod 4$, and so the result follows by the induction hypothesis.

Exercise 8.10.2. If $N = pq + m$ where $0 \le m \le p - 1$, then $N - p[N/p] = N - pq = m$. If $r \ge 0$, then $m = r$; and if $r < 0$, then $m = p + r$.

Exercise 8.16.3(f). Prove this first when n is a prime power; and then note that if $m \not\equiv 1 \pmod n$, then $m \not\equiv 1 \pmod{p^k}$ for some prime power $p^k \| n$.

Exercise 8.17.2(a). Use the law of quadratic reciprocity. (c) Look back at exercise 8.9.5.

Exercise 8.18.1. Calculate $F_{p-1} \pmod p$ using F_p and F_{p+1}, and then proceed by induction.

Exercise 8.18.2. Prove that $F_{n+p+1} \equiv -F_n \pmod p$ by induction.

Exercise 8.18.3. Proceed analogously to as in the two previous exercises.

Exercise 8.18.6(a). Let $m = kr$ in exercise 0.4.10, and use the congruence in exercise 2.5.19(c) with $r = 1$ and 2. (b) When does $F_{kr}, F_{kr+1} \equiv 0, 1 \pmod{p^2}$?

Exercises in Chapter 9

Exercise 9.1.2. If p does not divide a, then $(b/a)^2 \equiv -1 \pmod{p}$. Therefore $p = 2$ or $p \equiv 1 \pmod 4$. We get the same conclusion if p does not divide b and, otherwise, p divides (a, b).

Exercise 9.1.4. By induction on $k \geq 1$: It is trivial for $k = 1$ and otherwise let $n_k = a^2 + b^2$ and $n_1 \cdots n_{k-1} = c^2 + d^2$ (by the induction hypothesis), and then the result follows from (9.1.1).

Exercise 9.1.7(d). Use (a) to prove that $|ac - bd|, |ad - bc| < p$.

Exercise 9.3.1. Proceed as in the geometric proof of (6.1.1), or as in the proof of Proposition 9.1.2.

Exercise 9.7.2(b). Replace a and b by their absolutely least residues mod p.

Exercise 9.7.3(b). Select any b with $\left(\frac{b}{p}\right) = -1$ in (a), and let $m = r$ or s.

Exercise 9.7.7. We know that n is the length of the hypotenuse of a primitive Pythagorean triple iff there exist coprime integers r, s of different parity with $n = r^2 + s^2$. Hence all of n's prime factors are $\equiv 1 \pmod 4$, and we know we get at least two representations of n if it has at least two distinct prime factors.

Exercise 9.7.9. Since $m^2 \pm 2$ are odd they must be $\equiv 3 \pmod 4$, and so must be divisible by a prime $\equiv 3 \pmod 4$.

Exercise 9.7.10(a). In what domains do each of the ranges of ϕ lie? (b) We must be in the middle case (as $y, z \neq 0$) so that $x = y$ in which case $x(x + 4z) = p$. Since p can only be factored in one way into positive integers, we have $x = 1, z = \frac{p-1}{4}$; that is, $v = (1, 1, \frac{p-1}{4})$. (c) Pair up the elements of S using ϕ.

Exercise 9.9.2. Try $a = b = n = 1$.

Exercise 9.11.1(d). Use exercise 8.9.9(c).

Exercise 9.12.2. Use (9.12.1).

Exercise 9.13.2. Use the characteristic polynomial for A, which is the polynomial $x^2 - tx + d$ satisfied by A, where t is the trace of A and d is the determinant.

Exercises in Chapter 10

Exercise 10.3.2. Hopefully $n = pq$ and $\phi(n) = de - 1 = 29 \times 197 - 1 = 5712$; if so, then $p + q = n + 1 - \phi(n) = 180$. Therefore $(x - p)(x - q) = x^2 - 180x + 5891$ which we factor to obtain p and q.

Exercise 10.4.2(b). Use Corollary 7.5.3.

Exercise 10.7.5. Since n is a Carmichael number we know that it is squarefree and has prime divisors p and q, by Lemma 7.6.1. If $a^{(n-1)/2} \equiv -1 \pmod n$, then let $b \equiv 1 \pmod p$ and $b \equiv a \pmod q$, and determine the value of $b^{(n-1)/2} \pmod{pq}$.

Exercise 10.8.6(a). Factor $4x^4 + 1$ and substitute in $x = 2^n$.

Exercise 10.19.1(c). Use the quadratic reciprocity law for 2 and -2.

Exercises in Chapter 11

Exercise 11.2.1. If $y = 0$, then $m_1 n = n_1 m$. Now $(m, n) = (m_1, n_1) = 1$ and so $m_1 = m$ and $n_1 = n$ contradicting our construction of the pair m, n.

Exercise 11.2.5. Consecutive powerful numbers of the form $2^3 a^2$ followed by b^2, for some integers a and b.

Exercise 11.4.2. Use the product rule to compute the derivative.

Exercise 11.6.3. Given a smallest solution to $x^2 - dy^2 = 1$ expand $(x + \sqrt{d} y)^{\phi(d)} \pmod d$.

Exercise 11.6.11(c). Consider the example $1 + (2^n - 1) = 2^n$ with $m \geq 2/\epsilon$.

Exercise 11.14.3(a). Write $\begin{pmatrix} p & s \\ r & \ell \end{pmatrix} = \begin{pmatrix} a_0 & 1 \\ 1 & 0 \end{pmatrix} \cdots \begin{pmatrix} a_n & 1 \\ 1 & 0 \end{pmatrix}$ and use that $a_n \geq 2$.
(b) Take determinants of the matrix equation so that $rs \equiv (-1)^n \pmod{p}$, and therefore $s = r$ or $p - r$. (c) Take the transpose of $\begin{pmatrix} p & r \\ r & \ell \end{pmatrix}$. (d) Write $\begin{pmatrix} a & b \\ c & d \end{pmatrix} = \begin{pmatrix} a_0 & 1 \\ 1 & 0 \end{pmatrix}$
$\cdots \begin{pmatrix} a_m & 1 \\ 1 & 0 \end{pmatrix}$ so that $\begin{pmatrix} p & r \\ r & \ell \end{pmatrix} = \begin{pmatrix} a & b \\ c & d \end{pmatrix} \begin{pmatrix} a & c \\ b & d \end{pmatrix}$.

Exercise 11.16.2. One thought is to take 2^{a_0} if $a_0 \geq 0$ and 3^{-a_0} if $a_0 < 0$, and then use the primes 5 and 7 for a_1, etc.

Exercises in chapter 12

Exercise 12.1.3. Suppose that d is a fundamental discriminant and $[a, b, c]$ is an imprimitive form of discriminant d. If $h|(a, b, c)$, then $h^2|d$, so that $h = 2$. But then $D = d/h^2 \equiv 0$ or $1 \pmod 4$, a contradiction. Now suppose that d is not a fundamental discriminant. Then there exists a prime p such that $d = p^2 D$, where $D \equiv 0$ or $1 \pmod 4$. There is always a form g of discriminant D and so pg is an imprimitive form of discriminant d.

Exercise 12.1.4(c). Study the right-hand side of (12.1.2).

Exercise 12.1.5. Take determinants of both sides.

Exercise 12.1.6. First note that $b \equiv d \mod 2$, and that if $b = 2k+\delta$ with δ the least residue of $d \pmod 2$, then the change of variable $x \to x - ky$ shows that $[1, b, c] \sim [1, \delta, A]$, the principal form. The value of A must be $(\delta - d)/4$, so that the discriminant is $d = b^2 - 4c$.

Exercise 12.4.1. One example is $d = -171$. We begin by noting that $|b| \leq a \leq \sqrt{171/3} = \sqrt{57} < 8$ and b is odd. If $b = \pm 1$, then $ac = (1 + 171)/4 = 43$ with $a \leq c$ so that $a = 1$. If $b = \pm 3$, then $ac = (9 + 171)/4 = 45$ with $a \leq c$ so that $a = 1, 3, 5$ and $1 < |b|$. If $b = \pm 5$, then $ac = (25 + 171)/4 = 49$ with $a \leq c$ so that $a = 1, 7$ and $1 < |b|$. If $b = \pm 7$, then $ac = (49 + 171)/4 = 55$ with $a \leq c$ so that $a = 1, 5$ which are both $< |b|$, so we are left with $[1, 1, 43]$, $[3, 3, 15]$, $[5, 3, 9]$, $[5, -3, 9]$, $[7, 5, 7]$, and $[3, 3, 15]$ which is imprimitive.

Exercise 12.4.2. These are the smallest negative fundamental discriminants of class numbers 1 to 8:
For $d = -3$ we have $[1, 1, 1]$. For $d = -15$ we have $[1, 1, 4]$, $[2, 1, 2]$.
For $d = -23$ we have $[1, 1, 6]$, $[2, \pm 1, 3]$.
For $d = -39$ we have $[1, 1, 10]$, $[2, \pm 1, 5]$, $[3, 3, 4]$.
For $d = -47$ we have $[1, 1, 12]$, $[2, \pm 1, 6]$, $[3, \pm 1, 4]$.
For $d = -87$ we have $[1, 1, 22]$, $[2, \pm 1, 11]$, $[3, 3, 8]$, $[4, \pm 3, 6]$.
For $d = -71$ we have $[1, 1, 18]$, $[2, \pm 1, 9]$, $[3, \pm 1, 6]$, $[4, \pm 3, 5]$.
For $d = -95$ we have $[1, 1, 24]$, $[2, \pm 1, 12]$, $[3, \pm 1, 8]$, $[4, \pm 1, 6]$, $[5, 5, 6]$.

Exercise 12.4.3. These are the smallest even negative fundamental discriminants of class numbers 1 to 6: For $d = -4$ we have $[1, 0, 1]$; for $d = -20$ we have $[1, 0, 5]$, $[2, 2, 3]$; for $d = -56$ we have $[1, 0, 14]$, $[2, 0, 7]$, $[3, \pm 2, 5]$; for $d = -104$ we have $[1, 0, 26]$, $[2, 0, 13]$, $[3, \pm 2, 9]$, $[5, \pm 4, 6]$.

Exercise 12.5.3. Use Rabinowicz's criterion, and quadratic reciprocity.

Exercise 12.6.1. Prove and use the inequality $am^2 + bmn + cn^2 \geq am^2 - |b|\max\{|m|, |n|\}^2 + cn^2$.

Exercise 12.6.2(b). Use the smallest values properly represented by each form.

Exercise 12.6.5(c). Use exercise 12.6.2(e).

Hints for exercises 581

Exercise 12.6.7(c). Given a solution B, let $C = (B^2 - d)/4A$ and then $[A, B, C]$ represents A properly (by $(1,0)$). Find reduced $f \sim [A, B, C]$ and use the transformation matrix to find the representation as in (b).

Exercise 12.8.1. Prove this one prime factor of A at a time and then use the Chinese Remainder Theorem. For each prime p, try $f(1,0)$, $f(0,1)$, and then $f(1,1)$.

Exercise 12.8.2 If $f = [a, r, u]$, then the transformation $x \to x + ky, y \to y$ yields that $f \sim [a, b, c]$ where $b = r + 2ka$; that is, we can take b to be any value $\equiv r \pmod{2a}$. Similarly if $F = [A, s, v]$, then we can take b to be any value $\equiv s \pmod{2A}$. Such a b exists by the Chinese Remainder Theorem provided $r \equiv s \pmod{2}$, and r and s have the same parity as the discriminants of f and F.

Exercise 12.11.3. Now $d = b^2 - 4ac = B^2 - 4AC$, and so if $p|4aA$, then $(d/p) = 0$ or 1. We will now prove that there are rational points on the curve $aAu^2 = v^2 - dw^2$, by using Legendre's version of the local-global principle. There are obviously real solutions with $u = 0$. If odd prime p divides aA but not d, then we have seen that $(d/p) = 1$. If odd prime p divides d but not aA, then $(aA/p) = (a/p)(A/p) = \sigma_f(p)\sigma_F(p) = 1$. Finally we have the case in which p divides a and d. Hence p divides b, and $p^1 \| d$ as d is fundamental, and so $p \nmid (a/p)c$. So writing $a = pa', b = pb', d = pD$ we have $D = p(b')^2 - 4a'c$ which implies that $(-a'cD/p) = 1$. We also have $(Ac/p) = \sigma_f(p)\sigma_F(p) = 1$, and so $(-a'AD/p) = (-a'cD/p)(Ac/p) = 1$ as needed. Dividing through by u we have $aA = t^2 - d\gamma^2$ for some rationals t, γ; letting $t = 2a\alpha + b\gamma$ we deduce that $A = f(\alpha, \gamma)$ for some $\alpha, \gamma \in \mathbb{Q}$. We can select any $\beta, \delta \in \mathbb{Q}$ for which $\alpha\delta - \beta\gamma = 1$ to obtain a transformation for f to a form $Ax^2 + b'xy + c'y^2$. We now let $x = X + kY, y = Y$ where k is chosen so that $2AK + b' = B$ to obtain a form $Ax^2 + Bxy + C'y^2$. Since both transformations have determinant 1, we see that $B^2 - AC' = d$ and so $C' = C$. Hence f and F are equivalent over the rationals.

Exercise 12.15.1(a). Use Euler's criterion and Corollary 7.5.2.

Exercise 12.15.3(c). Use exercise 12.15.2(c).

Exercise 12.18.2(a). If even $N = a^2 + b^2 + c^2 + d^2$ with $a \equiv b \pmod{2}$ and $c \equiv d \pmod{2}$, then $N/2 = (\frac{a+b}{2})^2 + (\frac{a-b}{2})^2 + (\frac{c+d}{2})^2 + (\frac{c-d}{2})^2$. If $N \equiv 1 \pmod{4}$ with $N = a^2 + b^2 + c^2 + d^2$, then we may let a be odd, the rest even. To obtain representations of $2N$ we have the first two squares as $(a+b)^2 + (a-b)^2$, the other two even. This yields back a and the choice of b and so it is a 1-to-3 map. We have a similar construction if $N \equiv 3 \pmod{4}$.

Exercise 12.18.3(c). Use Legendre's Theorem (Theorem 12.5). (d) Let $u = a+b-c-d, v = a-b+c-d, w = a-b-c+d$, etc. (e) Be careful with the cases where $u = v$ etc.

Exercises in Chapter 13

Exercise 13.7.3(d). If $x = 10^K - 1$, then the integers in S_a that are $\leq x$ are the union of the sets $\{a \cdot 10^k \leq n < (a+1) \cdot 10^k\}$ for $k = 0, 1, 2, \ldots, K-1$.

Exercise 13.7.4 One way is to factor the numerator and denominator and note that $(m+1)^2 - (m+1) + 1 = m^2 + m + 1$.

Exercises in Chapter 15

Exercise 15.2.6(d). Proceed by induction, using the recursion in (c).

Exercise 15.7.1. Consider the sequence $0, a_1, a_1 + a_2, \ldots, a_1 + \cdots + a_n \pmod{n}$, and use the pigeonhole principle.

Exercise 15.7.3. Remove the first greedily constructed increasing subsequence starting with a_1, so the second term is a_r, where r is the minimal integer for which $a_r > a_1$. Suppose this has length ℓ. If the remaining subsequence has parameters m', n' show that $n \geq n' + 1$ and $m \geq \max\{m', \ell\}$. Then complete the proof by an induction hypothesis.

Exercises in Chapter 16

Exercise 16.2.1. Assume $(ab, p) = 1$; let d be the order of p (mod b). Let $A = a \cdot (p^d - 1)/b$, and then $C = p^d - 1 - A$. Write C (mod p^d) in base p as $c_0 + c_1 p + \cdots + c_{d-1} p^{d-1}$. Then $a/b = 1 + \sum_{j \geq 0} c_j p^j$ where $c_j = c_k$ when k is the least non-negative residue of j (mod d).

Exercise 16.6.5(b). Use the identity $AB - 1 = (A - 1) + A(B - 1)$.

Exercise 16.7.2(a). Differentiate this expression.

Recommended further reading

[AZ18]   Martin Aigner and Günter M. Ziegler, *Proofs from The Book*, sixth ed., Springer, Berlin, 2018.

[Bak84] Alan Baker, *A concise introduction to the theory of numbers*, Cambridge University Press, Cambridge, 1984. MR781734

[BB09]   Arthur T. Benjamin and Ezra Brown (eds.), *Biscuits of number theory*, The Dolciani Mathematical Expositions, vol. 34, Mathematical Association of America, Washington, DC, 2009. MR2516529

[Cas78] J. W. S. Cassels, *Rational quadratic forms*, London Mathematical Society Monographs, vol. 13, Academic Press, Inc. [Harcourt Brace Jovanovich, Publishers], London-New York, 1978. MR522835

[CG96]   John H. Conway and Richard K. Guy, *The book of numbers*, Copernicus, New York, 1996. MR1411676

[Cox13] David A. Cox, *Primes of the form $x^2 + ny^2$*, second ed., Pure and Applied Mathematics (Hoboken), John Wiley & Sons, Inc., Hoboken, NJ, 2013. MR3236783

[CP05]   Richard Crandall and Carl Pomerance, *Prime numbers: A computational perspective*, second ed., Springer, New York, 2005. MR2156291

[Dav80] Harold M. Davenport, *Multiplicative number theory*, Springer-Verlag, New York, 1980.

[Dav05] H. Davenport, *Analytic methods for Diophantine equations and Diophantine inequalities*, second ed., Cambridge Mathematical Library, Cambridge University Press, Cambridge, 2005.

[DF04]   David S. Dummit and Richard M. Foote, *Abstract algebra*, third ed., John Wiley & Sons, Inc., Hoboken, NJ, 2004. MR2286236

[Edw01] H. M. Edwards, *Riemann's zeta function*, Dover Publications, Inc., Mineola, NY, 2001. MR1854455

[GG]   Andrew Granville and Ben Green, *Additive combinatorics*, American Mathematical Society (to appear).

[Graa]   Andrew Granville, *The distribution of primes: Analytic number theory revealed*, American Mathematical Society (to appear).

[Grab]   Andrew Granville, *Rational points on curves: Arithmetic geometry revealed*, American Mathematical Society (to appear).

[GS]   Andrew Granville and K. Soundararajan, *The pretentious approach to analytic number theory*, Cambridge University Press (to appear).

[Guy04] Richard K. Guy, *Unsolved problems in number theory*, third ed., Problem Books in Mathematics, Springer-Verlag, New York, 2004. MR2076335

[HW08]   G. H. Hardy and E. M. Wright, *An introduction to the theory of numbers*, sixth ed., revised by D. R. Heath-Brown and J. H. Silverman, Oxford University Press, Oxford, 2008. MR2445243

[IR90]   Kenneth Ireland and Michael Rosen, *A classical introduction to modern number theory*, second ed., Graduate Texts in Mathematics, vol. 84, Springer-Verlag, New York, 1990. MR1070716

[Knu98]   Donald E. Knuth, *The art of computer programming. Vol. 2*, Seminumerical algorithms, third edition [of MR0286318], Addison-Wesley, Reading, MA, 1998. MR3077153

[LeV96]   William J. LeVeque, *Fundamentals of number theory*, reprint of the 1977 original, Dover Publications, Inc., Mineola, NY, 1996. MR1382656

[NZM91]   Ivan Niven, Herbert S. Zuckerman, and Hugh L. Montgomery, *An introduction to the theory of numbers*, fifth ed., John Wiley & Sons, Inc., New York, 1991. MR1083765

[Rib91]   Paulo Ribenboim, *The little book of big primes*, Springer-Verlag, New York, 1991. MR1118843

[Sha85]   Daniel Shanks, *Solved and unsolved problems in number theory*, third ed., Chelsea Publishing Co., New York, 1985. MR798284

[ST15]   Joseph H. Silverman and John T. Tate, *Rational points on elliptic curves*, second ed., Undergraduate Texts in Mathematics, Springer, Cham, 2015. MR3363545

[Ste09]   William Stein, *Elementary number theory: Primes, congruences, and secrets. A computational approach*, Undergraduate Texts in Mathematics, Springer, New York, 2009. MR2464052

[Tig16]   Jean-Pierre Tignol, *Galois' theory of algebraic equations*, second ed., World Scientific Publishing Co. Pte. Ltd., Hackensack, NJ, 2016. MR3444922

[TMF00]   Gérald Tenenbaum and Michel Mendès France, *The prime numbers and their distribution*, translated from the 1997 French original by Philip G. Spain, Student Mathematical Library, vol. 6, American Mathematical Society, Providence, RI, 2000. MR1756233

[VE87]   Charles Vanden Eynden, *Elementary number theory*, The Random House/Birkhäuser Mathematics Series, Random House, Inc., New York, 1987. MR943119

[Wat14]   John J. Watkins, *Number theory: A historical approach*, Princeton University Press, Princeton, NJ, 2014. MR3237512

[Zei17]   Paul Zeitz, *The art and craft of problem solving*, third edition [of MR1674658], John Wiley & Sons, Inc., Hoboken, NJ, 2017. MR3617426

Index

p-adics, 535–537, 539, 542–545
abc-conjecture, 223, 226, 414, 416

Algebraic numbers and integers, 23, 29
Algebraic units, 23, 29

Bernoulli numbers and polynomials, 11, 193, 286, 289
Binary quadratic forms, 340, 344, 345, 353, 380, 426, 443, 444, 446–452, 456, 458, 459, 465, 466, 468–470, 484, 506, 508
Binomial coefficients, 4, 9, 68, 99, 172, 233, 241, 285, 320, 321

Carmichael λ-function, 261, 388
Carmichael numbers, 245, 250, 378, 387, 388
Catalan equation, 223, 417
Chinese Remainder Theorem, 92, 97, 104, 106, 132, 309, 366
Class group and composition, 456–461, 463–466
Class number, 448–451, 471–474, 513
Computation and running times, 52, 255, 256, 322, 366, 370, 372, 378, 382, 389, 390, 393, 394
Congruent number problem, 221, 229, 551–553, 557
Constructibility and pre-Galois theory, 17, 25, 30, 103
Continued fractions, 39, 47, 381, 423, 424, 426–429, 431, 434, 436, 437, 470

Convolutions, 136, 141, 143
Covering systems, 280, 281
Cryptography, 366, 368, 370, 391, 392
Cyclotomic polynomials, 153, 248, 290

Descent, 219, 223, 229, 360, 361, 410, 553, 554, 556, 561, 562
Diophantine problems, 215, 218, 220, 233, 249, 263, 309, 339, 360, 361, 406, 410, 413, 414, 444, 452, 501, 504, 508, 509, 514, 547, 558
Dirichlet L-functions, 140, 192, 330, 473, 497, 499, 512
Dirichlet characters, 321, 326, 327, 330, 511
Discrete logs, 260
Discriminants, 15, 28, 77, 121
Divisibility tests, 66, 246
Divisors (incl. gcds), 34, 85, 129, 147, 150, 490, 493
Dynamics, 48, 208, 360, 361, 364, 420, 482

Egyptian fractions, 59, 124
Elliptic curves, 220, 504–506, 508, 514, 519, 547, 548, 550, 551, 553, 554, 556–559, 561–563
Euclidean algorithm, 33, 40, 41, 45, 47, 51, 75, 90, 118
Euclidean domains, 78
Euler's ϕ-function, 128
Euler's criterion, 298, 305

Factoring methods, 373, 380, 382, 399

Fermat numbers, 8, 96, 103, 156, 240, 249, 280, 303, 313, 372, 379
Fermat quotients, 284, 289, 293, 543–545
Fermat's Last Theorem, 96, 218, 221, 225, 249, 279
Fermat's Little Theorem, 238, 254, 270, 288, 297
Fermat-Catalan conjecture, 223, 228, 417
Fibonacci numbers, 1, 14, 42, 47, 69, 96, 281, 333, 335, 427, 437
Finite fields, 144, 273, 493, 501, 503–506, 508, 509, 514
Frobenius postage stamp problem, 57, 123, 532
Fundamental discriminants, 329, 444, 445, 448, 450, 451, 466, 467, 473, 474
Fundamental Theorem of Algebra, 117
Fundamental Theorem of Arithmetic, 81, 82, 112, 117

Generating functions, 11, 14, 515, 518, 528, 542
Groups, 19, 71, 74, 109, 269, 271, 326, 328, 361, 397, 398, 408, 454, 494, 550, 565

Heuristics, xxvi, 187, 289, 336, 383

Ideals, 37, 43, 47, 113, 115, 460, 465
Irrational numbers, 24, 87, 97, 223, 404, 418, 429

Jacobi symbol, 307, 311, 321, 327, 329

Lattice points, 317, 338, 347–349, 351, 353, 356, 358, 404, 420, 455
Legendre symbol, 296, 321, 326
Lifting solutions mod p^k, 266, 310, 538–540
Linear algebra, 35, 37, 49, 56, 90, 96, 104, 123, 226, 276, 360, 374, 443, 445
Local-global principle, 92, 97, 342, 343, 345, 348, 350, 351, 353, 355, 447
Lucas sequence, 2

Magic squares, 54, 201, 559
Mahler measure, 540
Matrices and matrix groups, 20, 28, 45, 47, 49, 105, 180, 276, 361, 364, 424, 428, 445, 447, 454, 462, 470, 482–484
Mersenne numbers, 8, 44, 69, 96, 130, 157, 240, 303, 336, 386
Möbius function, 135, 177, 249
Modularity, 519
Multiplication table problem, 491, 494

Orders (of elements), 73, 236, 240, 242, 246, 252, 259, 261, 263, 265, 269

Pascal's triangle, 5, 99
Pell's equation, 406, 408, 427, 428, 431, 434, 470
Pell's equation; negative, 410, 428, 431, 432
Perfect numbers, 129, 131
Polynomial properties, 10, 13, 23, 66, 75, 87, 96, 117, 119, 120, 144, 153, 168, 200, 223, 225, 226, 240, 275, 288, 299, 311, 373, 381, 396, 399–401, 493
Power residues, 235, 298
Primality testing, 157, 371, 372, 375, 383, 395, 396
Prime k-tuplets conjecture, 178, 190, 199, 485
Prime factors: number of, 487, 488, 490, 492
Primes in arithmetic progressions, 159, 163, 169, 179, 198, 248, 312, 331, 527
Primes: infinitely many, 155, 156, 160, 182, 202, 203, 208, 562
Primes: number of, 157, 160, 163, 171, 184, 185, 187, 194
Primitive roots, 242, 243, 246, 258, 260, 263, 313, 371
Pseudoprimes, 245, 250, 375–379
Pythagorean triangle, 215, 221, 223, 229, 347, 551, 552, 557, 563

Quadratic fields, 115, 357, 460, 461, 465
Quadratic forms, 343, 355, 363, 479–482, 501, 503
Quadratic reciprocity (Law of), 300, 301, 303, 305, 309, 313, 315, 323, 333, 341, 355, 503, 512
Quadratic residues / non-residues, 296, 299
Quadratic residues and non-residues; least, 297, 310, 319

Index

Residues (mod n), 61, 71, 89, 236, 238, 262, 295, 305, 315, 328, 405
Resultants and discriminants, 77, 120
Riemann zeta-function, 13, 141, 142, 182, 184, 186, 192
Rings and fields, 22, 73, 109
Roots of polynomials, 15, 117, 122, 265, 267, 275, 537, 540

Second-order linear recurrence sequences, 2, 7, 14, 44, 69, 96, 281, 290, 334, 385
Square roots (mod n), 94, 267, 299, 309–311, 365, 376
Sums of (more than two) squares, 471, 475, 476, 478, 479, 567
Sums of powers of integers, 3, 9, 286, 287, 292
Sums of two squares, 337, 339, 347, 356, 405, 436, 437, 456
Sumsets, 519, 521–524, 530, 532

Tiling, 41
Transcendental numbers, 24, 411, 412, 416, 439, 440

Unique factorization, 112, 113, 117, 120, 272

Waring's problem, 224, 566
Wilson's Theorem, 241, 270, 287